Practical Skills in Biomolecular Sciences

Practical Skills in Biomolecular Sciences

Fifth edition

Rob Reed

David Holmes

Jonathan Weyers

Allan Jones

PEARSON

Harlow, England • London • New York • Boston • San Francisco • Toronto • Sydney • Auckland • Singapore • Hong Kong
Tokyo • Seoul • Taipei • New Delhi • Cape Town • São Paulo • Mexico City • Madrid • Amsterdam • Munich • Paris • Milan

Pearson Education Limited

Edinburgh Gate
Harlow CM20 2JE
United Kingdom
Tel: +44 (0)1279 623623
Web: www.pearson.com/uk

First published 1998 by Addison Wesley Longman Limited (print)
Second edition 2003 (print)
Third edition 2007 (print)
Fourth edition 2013 (print and electronic)
Fifth edition published 2016 (print and electronic)

ISBN: 978-1-292-10073-9 (print)
         978-1-292-09529-5 (PDF)
         978-1-292-16121-1 (ePub)

British Library Cataloguing-in-Publication Data
A catalogue record for the print edition is available from the British Library

Library of Congress Cataloging-in-Publication Data
Names: Reed, Robert (Robert H.), author. | Holmes, David, 1946- , author. |
    Weyers, Jonathan D. B., author. | Jones, A. M. (Allan M.), 1945- , author.
Title: Practical skills in biomolecular sciences / Rob Reed, David Holmes,
    Jonathan Weyers, Allan Jones.
Description: Fifth edition. | Harlow, England ; New York : Pearson, 2016. |
    Preceded by: Practical skills in biomolecular sciences / Rob Reed ... [et al.].
    4th ed. 2013. | Includes bibliographical references and index.
Identifiers: LCCN 2016003540| ISBN 9781292100739 (print) | ISBN 9781292095295
    (PDF) | ISBN 9781292095264 (eText)
Subjects: | MESH: Clinical Laboratory Techniques | Biochemical Phenomena
Classification: LCC QH506 | NLM QY 25 | DDC 572.8—dc23 LC record available at http://lccn.loc.gov/2016003540

10 9 8 7 6 5
20 19 18

Cover image © Shutterstock
Print edition typeset in 10/12pt Times NR MT Pro by Lumina Datamatics
Print edition printed in Great Britain by CPI Group (UK) Ltd, Croydon CR0 4YY

Contents

Contents

Companion Website

ON THE WEBSITE

For open-access **student resources** specifically written to complement this textbook and support your learning, please visit www.pearsoned.co.uk/practicalskills

Lecturer Resources

For password-protected online resources tailored to support the use of this textbook in teaching, please visit www.pearsoned.co.uk/practicalskills

List of boxes

Preface

'...there is seen to be a need to re-emphasise the practical nature of the biosciences, through laboratory and fieldwork; and the need for significant levels of numeracy for a subject that is both complex and analytical. ...there is an explicit understanding that the biosciences are practical subjects, and cannot be effectively delivered without significant and extensive learning, teaching and experience in a field and/or laboratory environment.'

Foreword, QAAHE Subject Benchmark Statement for Biosciences (QAAHE, 2007)

Practical work forms the cornerstone of scientific knowledge and understanding. Consequently, practical work is an important component of training in the biosciences and successful students must develop a number of skills, ranging from those required to observe, measure and record accurately to those associated with operating up-to-date analytical equipment, alongside broader skills involved in teamwork and effective study. In creating this edition, we have maintained the approach of the earlier versions, aiming to support students (and lecturers) in courses where cellular and molecular biosciences form a major component of the syllabus, e.g. biochemistry, biomedical sciences, biotechnology, genetics, microbiology and molecular biology. As before, this support is provided in a concise but user-friendly manner, with key points and definitions, illustrations, worked examples, tips and hints, 'how to' boxes and checklists.

This new edition consolidates the changes made for the fourth edition, which included additional material covering bioethics, stable isotopes, photosynthesis and respiration, and aspects of microbiology, and also includes updated material on molecular genetics, with additional detail on DNA profiling, next-generation sequencing and other aspects. We have completely revised and updated the text references and sources for further study, and incorporated over 70 new tips, figures, tables and boxes. Throughout the book we have changed the guidance regarding Microsoft *Office* software to apply in a generic sense rather than to any specific version. This may mean that readers may need to adjust commands if these are not appropriate.

Guidance on specific commands and their syntax can usually be found using the software's help facility. Boxes giving details of approaches based on *Office 2003* and *Office 2007* that appeared in previous editions will be made available via the book's online resource at www.pearsoned.co.uk/practicalskills. This website continues to host the answers to the study exercises as well as text references and sources for further study – with live web links where applicable.

We would like to take this opportunity to thank our wives and families for their continued support, and to recognise the following colleagues and friends who have provided assistance, comment and food for thought at various points during the production of all editions: James Abbott, Margaret Adamson, Chris Baldwin, Gary Black, Geoff Bosson, Eldridge Buultjens, Richard Campbell, Bob Cherry, Mirela Cuculescu, Steve Cummings, John Dean, Jackie Eager, Brian Eddy, Charmain Elder, Neil Fleming, Alan Grant, Howard Griffiths, Rod Herbert, Steve Hitchin, Helen Hooper, Jane Illes, Andy Johnston, Alan Jones, Ian Kill, Rhonda Knox, Lisa Lee-Jones, Phil Manning, Pete Maskrey, Fiona McKie-Bell, Steve Millam, Kirsty Millar, Stephen Moore, Rachel Morris, Lorna Moxham, Bob Newby, Fiona O'Donnell, John Raven, Steve Reed, Pete Rowell, David Sillars, Liz Smith, Peter Sprent, Bill Tomlinson, Ruth Valentine, Lorraine Walsh, Dave Wealleans, Mark White, Will Whitfield, Ian Winship, Bob Young and Hilary-Kay Young. We would also like to thank the staff of Pearson Education for their friendly support over the years, and would wish to acknowledge Richelle Zakrzewski, Rufus Curnow, Pat Bond, Owen Knight, Simon Lake, Alex Seabrook and Pauline Gillett for their encouragement and commitment to the *Practical Skills* series. Our thanks are also extended to Louise Attwood, Gary Hall, Julie Jackson and Mary Lince for their excellent work during the preparation of the new edition. As with the previous editions, we would be grateful to hear of any errors you might notice, so that these can be put right at the earliest opportunity.

ROB REED (r.reed@cqu.edu.au)
DAVID HOLMES (david.holmes@northumbria.ac.uk)
JONATHAN WEYERS (j.d.b.weyers@dundee.ac.uk)
ALLAN JONES (allan.jones9@btinternet.com)

List of abbreviations

A	absorbance (e.g. A_{260} = absorbance at 260 nm)
AC	affinity chromatography
ACDP	Advisory Committee on Dangerous Pathogens
ADP	adenosine diphosphate
ANOVA	analysis of variance
ATP	adenosine triphosphate
BSA	bovine serum albumin
CCCP	carbonylcyanide m-chlorophenylhydrazone
CE	capillary electrophoresis
CFU	colony-forming unit
CGE	capillary gel electrophoresis
COSHH	Control of Substances Hazardous to Health
CTP	cytosine triphosphate
CZE	capillary zone electrophoresis
ddNTP	dideoxyribonucleotide triphosphate
DMSO	dimethyl sulfoxide
DNA	deoxyribonucleic acid
dNTP	deoxyribonucleoside triphosphate
d.p.m.	disintegrations per minute
dsDNA	double-stranded DNA
ECD	electron capture detector
EDTA	ethylenediaminetetraacetic acid
EI	electron impact ionisation
EIA	enzyme immunoassay
ELISA	enzyme-linked immunosorbent assay
EMR	electromagnetic radiation
EOF	electro-osmotic flow
ESR	electron spin resonance
F	Faraday constant
FIA	fluorescence immunoassay
FID	flame ionisation detector
FPLC	fast protein liquid chromatography
FT	Fourier transformation
g	acceleration due to gravity
GC	gas chromatography
GPC	gel permeation chromatography
HEPES	N-[2-hydroxyethyl]piperazine-N'-[ethanesulfonic acid]
HIC	hydrophobic interaction chromatography
HPLC	high-performance liquid chromatography
IEC	ion-exchange chromatography
IEF	isoelectric focusing
Ig	immunoglobulin
IMAC	immobilised metal affinity chromatography
IR	infrared (radiation)
IRGA	infrared gas analyser
IRMA	immunoradiometric assay
IRMS	isotope ratio mass spectroscopy
ISE	ion-selective electrode
K_m	Michaelis constant
K_w	ionisation constant of water
LDH	lactate dehydrogenase
LSD	least significant difference
MEKC	micellar electrokinetic chromatography
MPN	most probable number
M_r	relative molecular mass
MRI	magnetic resonance imaging
MS	mass spectrometry
NAD^+	nicotinamide adenine dinucleotide (oxidised form)
NADH	nicotinamide adenine dinucleotide (reduced form)
$NADP^+$	nicotinamide adenine dinucleotide phosphate (oxidised form)
NADPH	nicotinamide adenine dinucleotide phosphate (reduced form)
NH	null hypothesis
NMR	nuclear magnetic resonance
PAGE	polyacrylamide gel electrophoresis
PAR	photosynthetically active radiation
PCR	polymerase chain reaction
PDP	personal development planning
PEG	polyethylene glycol
PFD	photon flux density
PFGE	pulsed field gel electrophoresis
PFU	plaque-forming unit
pH	$-\log_{10}$ proton concentration (activity), in mol l^{-1}
PI	photosynthetic irradiance
PPFD	photosynthetic photon flux density
PPi	pyrophosphate (inorganic)
PVA	polyvinyl alcohol
PY-MS	pyrolysis-mass spectrometry
R	universal gas constant
RCF	relative centrifugal field
R_F	relative frontal mobility
RIA	radioimmunoassay
RID	radioimmunodiffusion
RNA	ribonucleic acid
RP-HPLC	reverse phase high-performance liquid chromatography
r.p.m.	revolutions per minute
RT	reverse transcriptase

SDS	sodium dodecyl sulfate		TLC	thin-layer chromatography
SE	standard error (of the sample mean)		TRIS	tris(hydroxymethyl)aminomethane
SEM	scanning electron microscopy		TTP	thymidine triphosphate
SI	Systeme International d'Unités			
ssRNA	single-stranded RNA		UNG	uracil-N-glycosylase
STP	standard temperature and pressure		URL	uniform resource locator
			UV	ultraviolet (radiation)
TCA	trichloroacetic acid			
TCD	thermal conductivity detector		V_{max}	maximum velocity
TEM	transmission electron microscopy			
TEMED	N,N,N',N'-tetramethylethylenediamine		z	net charge on an ion

Acknowledgements

We are grateful to the following for permission to reproduce copyright material:

Figures

Figures 46.1 and 46.3 from *Methods for Physical and Chemical Analysis of Fresh Waters; International Biological Programme, no 8*, 2 ed., Blackwell Scientific (Golterman, H.L., Clymo, R.S. And Ohnstad, M.A.M. 1978) John Wiley & Sons (UK), reproduced with permission of Blackwell Publishing Ltd.

Tables

Table 46.2 from *'Light' in The Biology of Seaweeds*, Blackwell Scientific, Oxford (Luning, K.J., Lobban and Wynne, N.J. (eds) 1981) pp.326–55, John Wiley & Sons (UK), reproduced with Permission of Blackwell Publishing Ltd; Table 54.1 adapted from *Tables of Standard Electrode Potentials*, Wiley, London (Milazzo, G., Caroli, S. and Sharma, V.K. 1978) John Wiley and Sons (UK), reproduced with permission of John Wiley & Sons Ltd.

Text

Box 5.1 adapted from A Guide to Learning Preferences, vark-learn.com, © Copyright Version 7.1 (2011) held by Neil D. Fleming, Christchurch, New Zealand.

Photographs

American Physiological Society: 302; Image courtesy of S. Charnock: 409; GE Healthcare: 353, 361; KNAUER, Germany. www.knauer.net: 167; Supplied by Microscopy, an operating division of KeyMed (Medical & Industrial Equipment) Ltd.: 181, 186; Courtesy of PP Systems, Amesbury, MA, USA (http://wwww.ppsystems.com): 430; Rob Reed: 224, 234; Skye Instruments Ltd: 308; Courtesy of Heinz Walz GmbH, Effeltrich, Germany (http://www.walz.com): 432; Jonathan Weyers: 177, 181b, 183, 185, 213.

For the student

This book aims to provide guidance and support over the broad range of your undergraduate course, including laboratory classes, project work, lectures, tutorials, seminars and examinations, as outlined below.

Chapters 1–8 cover general skills

These chapters include a number of transferable skills that you will develop during your course, for example: self-evaluation; time management; teamwork; preparing for exams; creating a CV. They also provide guidance on how to study effectively and how to approach examinations and other assessments.

Chapters 9–20 deal with IT, library resources and communication

These chapters will help you get the most out of the resources and information available in your library, and on the web, as well as providing helpful guidance on the use of software packages for data analysis, preparing assignments, essays and laboratory reports, alongside support in relation to oral, visual and written forms of communication. The ability to evaluate information is an increasingly important skill in contemporary society, and practical guidance is provided here, as well as more specific advice, e.g. on bioinformatics resources available via the Internet.

Chapters 21–68 cover a wide range of specific practical skills required in biomolecular sciences

These chapters are based on the authors' experience of the questions students often ask in practical classes, and the support that is needed in order to get the most out of particular exercises. The text includes tips, hints, definitions, worked examples and 'how to' boxes that set out the key procedures in a step- by-step manner, with appropriate comments on safe working practice. The material ranges from basic laboratory procedures, such as preparing solutions, through specimen collection, identification and manipulation to the more advanced practical procedures that you might use during a final-year project, e.g. radioisotope work and more advanced analytical methods.

Chapters 69–74 explain data analysis and presentation

This will be an important component of your course and you will find that these chapters guide you through the skills and techniques required, ranging from the presentation of results as graphs or tables through to the application of statistical tests. Worked examples are used to reinforce the numerical aspects wherever possible.

Study exercises

We added these following comments from students and staff at UK universities, who felt that they would provide a useful opportunity to practise some of the skills covered in the book and a check on the understanding of the material. We hope that the exercises will be useful both to learners and to their tutors: some of the exercises are based on material contained within the corresponding chapter, while others provide opportunities to develop understanding in a particular topic area beyond the basic materials. In general, the more straightforward exercises have been placed first, with more advanced problems at the end of each section.

Most of the exercises and problems assume that students are working on their own, using the information supplied; however, tutors might wish to provide alternative starting material (e.g. a set of data from a practical class). We have also assumed that students will have access to a scientific calculator and, sometimes, to a networked PC with typical 'office' programs (especially word processor and spreadsheet), plus Internet access via a modem and browser. Where a library is mentioned, this is assumed to include access to standard reference works and a selection of scientific journals.

We recommend that students work together for some exercises - this is a valuable means of learning and, where there is no single correct answer to a problem, teamwork provides a mechanism for checking and discussing different approaches. Answers are provided on the book's website at www.pearsoned.co.uk/practicalskills. For numerical problems, the working out is shown with the final answer, while, for non-numerical exercises, 'answers' are provided in the form of tips, general guidance or illustrative examples, etc.

We hope that you will find this book and its companion website a helpful guide throughout your course, and beyond.

Study and examination skills

1 The importance of transferable skills

Skills terminology – different phrases may be used to describe transferable skills and associated personal qualities, depending on place or context. These include: 'graduate attributes', 'personal transferable skills' (PTS), 'key skills', 'core skills' and 'competences'.

This chapter outlines the range of transferable skills and their significance to biomolecular scientists. It also indicates where practical skills fit into this scheme. Having a good understanding of this topic will help you place your university studies in a wider context. You will also gain an insight into the qualities that employers expect you to have developed by the time you graduate. Awareness of these matters will help when carrying out personal development planning (PDP) as part of your studies.

The range of transferable skills

Table 1.1 provides a comprehensive listing of university-level transferable skills under six skill categories. There are many possible classifications – and a different one may be used in your institution or field of study. Note particularly that 'study skills', while important, and rightly emphasised at the start of many courses, constitute only one area of skills acquired by most university students.

The phrase *'Practical Skills'* in the title of this book indicates that there is a special subset of transferable skills related to work in the laboratory. However, although this text deals primarily with skills and techniques required for laboratory practicals and associated studies, a broader range of material is included. This is because the skills concerned are important, not only in the biosciences but also in the wider world. Examples include time management, evaluating information and communicating effectively.

Using course materials – study your course handbook and the schedules for each practical session to find out what skills you are expected to develop at each point in the curriculum. Usually the learning outcomes/objectives (p. 30) will outline the skills involved.

> **KEY POINT** Biomolecular sciences are essentially practical subjects, and therefore involve highly developed laboratory skills. The importance that your lecturers place on practical skills will probably be evident from the large proportion of curriculum time you will spend on practical work in your course.

The word 'skill' implies much more than the robotic learning of, for example, a laboratory routine. Of course, some of the tasks you will be asked to carry out in practical classes *will* be repetitive. Certain techniques require manual dexterity and attention to detail if accuracy and precision are to be attained, and the necessary competence often requires practice to make perfect. However, a deeper understanding of the context of a technique is important if the skill is to be appreciated fully and then transferred to a new situation. That is why this text is not simply a 'recipe book' of methods and protocols and why it includes background information, tips and worked examples, as well as study exercises to test your understanding.

Example The skills involved in teamwork cannot be developed fully without a deeper understanding of the interrelationships involved in successful groups. The context will be different for every group and a flexible approach will always be required, according to the individuals involved and the nature of the task.

Transferability of skills

'Transferability' implies that someone with knowledge, understanding or ability gained in one situation can adapt or extend this for application in a different context. In some cases, the transfer of a skill is immediately obvious. Take, for example, the ability to use a spreadsheet to summarise biological data and create a graph to illustrate results. Once the key concepts and commands are learned (Chapter 13), they can be applied to many instances outside the biosciences where this type of output is used. This is not only true for similar data sets, but also in unrelated situations, such as making up a financial balance sheet and creating a pie chart to

Table 1.1 Transferable skills identified as important in the biosciences. The list is based on several sources, including the most recent UK Quality Assurance Agency for Higher Education *Subject Benchmark Statement for the Biosciences and for Biomedical Sciences* (QAA, 2007). Particularly relevant chapters are shown for the skills covered by this book (numbers in **bold coloured** text indicate a deeper, or more extensive, treatment).

Skill category	Examples of skills and competences	Relevant chapters in this textbook
Generic skills for bioscientists	Having an appreciation of the complexity and diversity of life and life processes	12, 30, 35–40, 56–59
	Reading and evaluating biological literature with a full and critical understanding	4, 9, **10**
	Capacity to communicate a clear and accurate account of a biological topic, both verbally and in writing	**15**, **16**, **17**, **18–20**
	Applying critical and analytical skills to evaluate evidence regarding theories and hypotheses	10, **32**
	Using a variety of methods for studying the biosciences	35–68
	Having the ability to think independently, set personal tasks and solve problems	32, 34, 72
Intellectual skills	Recognising and applying biological theories, concepts and principles	10, 32
	Analysing, synthesising and summarising information critically	10, 20, 70–74
	Obtaining evidence to formulate and test hypotheses; applying knowledge to address familiar and unfamiliar problems	30–34, 74
	Recognising and explaining moral, ethical and legal issues in biological research	**22**, 23, 35, 36, 40
Experimental (practical) and observational skills	Carrying out basic laboratory techniques and understanding the principles that underlie them	**21**, 22–31, 42–47, 55, 64
	Working in the laboratory safely, responsibly and legally, with due attention to ethical aspects	**21**, 23, 34–41
	Designing, planning, conducting and reporting on biological investigations and data arising from them	15, 16, 19, **32**, 34
	Obtaining, recording, collating and analysing biological data	30–34, 42–54, 69–74
	Carrying out basic techniques relevant to core subjects in biomedical science (biochemistry, molecular genetics, immunology, microbiology)	21–29, 30–41, 42–54, 64–68
Numeracy, communication and IT skills	Understanding and using data in several forms (e.g. numerical, textual, verbal and graphical)	4, 10, 70–74
	Communicating in written, verbal, graphical and visual forms	**15**, **16**, **17**, 18–20, **70**, **71**, 72
	Citing and referencing the work of others in an appropriate manner	**9**, 10, 20
	Obtaining data, including the concepts behind sampling and sampling errors, calibration and types of error	29, 30–34, **42**, 72–74
	Processing, interpreting and presenting data, and applying appropriate statistical methods for summarising and analysing data	12, 70–72, **73**, **74**
	Solving problems with calculators and computers, including the use of tools such as spreadsheets	11, **12**, 13, 21, **72**
	Using computer technology to communicate and as a source of biological information	**11**, **12**, 13, **14**
Interpersonal and teamwork skills	Working individually or in teams as appropriate; identifying individual and group goals and acting responsibly and appropriately to achieve them	**3**
	Recognising and respecting the views and opinions of others	3
	Evaluating your own performance and that of others	3, 8
	Appreciating the interdisciplinary nature of contemporary biosciences	1, 20
Self-management and professional development skills	Working independently, managing time and organising activities	2, 32, 34
	Identifying and working towards targets for personal, academic and career development	1, **8**
	Developing an adaptable and effective approach to study and work (including revision and exam technique)	2, 4, 5, 6, **7**

show sources of expenditure. Similarly, knowing the requirements for good graph drawing and tabulation (Chapters 70 and 71), perhaps practised by hand in earlier work, might help you use spreadsheet commands to make the output suit your needs.

Other cases may be less clear but equally valid. For example, towards the end of your undergraduate studies you may be involved in designing experiments as part of your project work. This task will draw on several skills gained at earlier stages in your course, such as preparing solutions (Chapters 24–27), deciding about numbers of replicates and experimental layout (Chapters 32 and 34) and perhaps carrying out some particular method of observation, measurement or analysis (Chapters 42–68). How and when might you transfer this complex set of skills? In the workplace, it is unlikely that you would be asked to repeat the same process, but in critically evaluating a problem or in planning a complex project for a new employer, you will need to use many of the time management, organisational and analytical skills developed when designing and carrying out experiments. The same applies to information retrieval and evaluation and writing essays and dissertations, when transferred to the task of analysing or writing a business report.

Personal development planning

Many universities have schemes for personal development planning (PDP), which may go under slightly different names such as progress file or professional development plan. You will usually be expected to create a portfolio of evidence on your progress, then reflect on this, and subsequently set yourself plans for the future, including targets and action points. Analysis of your transferable skills profile will probably form part of your PDP (Box 1.1). Other aspects commonly included are:

- **your aspirations, goals, interests and motivations;**
- **your learning style or preference** (see p. 25);
- **your assessment transcript or academic profile information** (e.g. record of grades in your modules);
- **your developing CV** (see p. 45).

Taking part in PDP can help you to focus your thoughts about your university studies and future career. This is important, as many biosciences degrees do not lead only to a single, specific occupation. The PDP process will introduce you to some new terms and will help you to describe your personality and abilities. This will be useful when constructing your CV and when applying for jobs.

Graduate attributes and employability

The skills emphasised in biology courses (Table 1.1) are sometimes considered alongside a university-wide framework of graduate attributes that are intended to summarise the qualities and skills that an employer might expect in those with qualifications from your institution. The associated notion of 'graduateness' summarises the effect of degree-level experience and learning on an individual. This in turn is connected with the concept of 'employability', which encompasses those skills and qualities required to gain and maintain employment. An understanding of these notions is important for every student, as this not only leads to a better understanding of the value of certain activities and assessments, but also

---

Opportunities to develop and practise skills in your private or social life – you could, for example, practise spreadsheet skills by organising personal or club finances using Microsoft *Excel*, or teamwork skills within any university clubs or societies you may join (see Chapter 7).

Types of PDP portfolio and their benefits – some PDP schemes are centred on academic and learning skills, while others are more focused on career planning. Some are carried out independently and others in tandem with a personal tutor or advisory system. Some PDP schemes involve creating an online portfolio, while others are primarily paper-based. Each method has specific goals and advantages, but whichever way your scheme operates, maximum benefit will be gained from being fully involved with the process.

Definition

Employability – the 'combination of in-depth subject knowledge, work awareness, subject-specific, generic and career management skills, and personal attributes and attitudes that enable a student to secure suitable employment and perform excellently throughout a career spanning a range of employers and occupations.' (Anon, 2015: *Higher Education Academy Centre for Bioscience* definition of employability for bioscientists)

Box 1.1   How to carry out a personal skills audit

1. **Create a list of appropriate skills.** As noted on p. 3, there are many systems for categorising skills. If your university publishes a specific skill set, e.g. as part of its framework for personal development planning (PDP) or graduate attributes, then you should use that. If not, you could adapt the listing in Table 1.2 or consult a text such as McMillan and Weyers (2012). Your list should relate to you personally, your intended career and any specific skills associated with your intended qualification.

2. **Lay out your list in table format.** You will need to create a table using a word processor or spreadsheet program. Your table should have four columns, as shown in Table 1.2.

3. **Rate your skills.** This may be challenging for many students as it is difficult to be objective and tough to gauge employer expectations. A confident student may rate a certain skill strongly, whereas a self-critical person may consider the same level of skill to be deficient. However, this does not matter too much as you will effectively be comparing yourself at different stages in your learning, rather than judging yourself against an outside standard. The suggested method is to use a scale of 1 to 10, with low values indicating that the skill 'needs lots of development' and high values indicating that, *for the time being*, you feel that your competence is 'well above average'.

4. **Note actions.** This especially applies to skills with low scores in the previous column – and you may wish to prioritise certain ones. You will need to think about ways in which you could improve, and this may require some research on your part. Is there a book you could read? Is there a training workshop you could attend? Could an extracurricular activity help you to develop? Should you sign up to speak to a skills advisor? It is important that you recognise that the solution to any deficiencies you perceive lies in your own hands. At university, no one will do the work for you.

5. **Add comments and progress notes.** Here is where you can add any comments to amplify or assist with the action points. The addition of progress notes implies that you will revisit the list from time to time. If your university PDP system allows you to add the list to a portfolio, then do this.

Inevitably, your skills audit will become out of date after a period. It will still be useful, however, to look back at it so that you can see how you have progressed. This will give a sense of achievement and self-awareness that could be valuable when speaking to academic tutors, careers advisors and potential employers. You may wish to set up a new list at intervals, perhaps at the start of each academic year.

Table 1.2 One possible way of creating a personal skills audit. The second row provides guidance about the content of each column. The third row provides an example of possible content.

Skill	Rating at [date] with notes	Proposed actions	Comments and notes on progress
You should be quite specific. It may be a good idea to sub-divide complex skills such as 'communication'	Provide a realistic evaluation of your competence in the skill at a specific point in time	This column will note what you intend to do to try to improve the skill. You might tick these off as completed	This column will summarise your progress. You may wish to add a revised rating
Giving spoken presentations	4/10 [3 March 2011] wasn't satisfied with presentation to tutorial group – nervous, a little disorganised and ppt too 'wordy'	1. Read Ch 14 in Practical Skills in Biology ✔ 2. Learn how to use advanced features of PowerPoint ✔ 3. Ask more questions in tutorials ✔	Gave second presentation to tutorial group; went well, although quite nervous at start. Slides much better. Make sure not to rush the introduction next time. 7/10

provides a specialised vocabulary and gives insights about personal and career development.

At the end of your course, which may seem some time away, you will aim to get a job and start on your chosen career path. You will need to sell yourself to your future employer, firstly in your application form and curriculum vitae (Chapter 8), and perhaps later at interview. Companies rarely employ bioscience graduates simply because they know how to carry out a particular lab routine or because they can recall specific facts about their chosen degree subject. Instead, they will be looking for a range of graduate-level skills and attributes. Typically, for example, they will seek employees who can demonstrate the ability to work in a team, to speak effectively and write clearly about their work. All of these skills and attributes can be developed at different stages during your university studies.

> **KEY POINT** Factual knowledge is important in degrees with a strong vocational element, but understanding how to find and evaluate information is usually rated more highly by employers than the ability to memorise facts.

Most likely, your future employer(s) will seek someone with an organised yet flexible mind, capable of demonstrating a logical approach to problems – someone who has a range of skills and who can transfer these skills to new situations. Many competing applicants will probably have similar qualifications. If you want the job, you will have to show that your additional skills and personal attributes place you above the other candidates.

Text references

Anon. (2015) *Employability in the Context of Teaching Bioscience.* Available: http://www.bioscience.heacademy.ac.uk/ftp/events/empforum/definition.pdf Last accessed 01/09/15. [Part of the HE Academy Centre for Bioscience website.]

McMillan, K.M. and Weyers, J.D.B. (2012) *The Study Skills Book,* 3rd edn. Pearson Education, Harlow.

Quality Assurance Agency (2007) *Subject Benchmark Statement – Biosciences 2007.* Available: http://www.qaa.ac.uk/academicinfrastructure/benchmark/ Last accessed 01/09/15. [Part of the Quality Assurance Agency Academic Infrastructure.]

Sources for further study

Drew, S. and Bingham, R. (2010) *The Guide to Learning and Study Skills.* Gower Publishing Ltd, Aldershot.

Race, P. (2007) *How to Get a Good Degree: Making the Most of Your Time at University,* 2nd edn. Open University Press, Buckingham.

STUDY EXERCISES

1.1 Evaluate your skills. Examine the list of skill topics shown in Table 1.1 (p. 4). Now create a new table with two columns, like the one shown opposite. The first half of this table should indicate *five* skills you feel confident about and show where you demonstrated the skill (for example, 'working in a team' and 'in a first-year group project in molecular biology'). The second half of the table should show *five* skills you do not feel confident about, or that you recognise need development (e.g. 'communicating in verbal form'). List these and then list ways in which you think the course material and activities in your current modules will provide opportunities to develop these skills, or what activities you might take to improve them (e.g. 'forming a study group with colleagues').

Skills I feel confident about	Where demonstrated
1.	
2.	
3.	
4.	
5.	

Skills that I could develop	Opportunities for development
6.	
7.	
8.	
9.	
10.	

1.2 Find skills resources. For at least one of the skills in the second half of Table 1.1, check your university's library database to see if there are any texts on that subject. Alternatively, carry out a search for relevant websites (there are many); decide which are useful and bookmark them for future use (Chapter 11).

1.3 Analyse your goals and aspirations. Spend a little time thinking about what you hope to gain from university. See if your friends have the same aspirations. Think about and/or discuss how these goals can be achieved, while keeping the necessary balance between university work, paid employment and your social life.

Answers to these study exercises are available at www.pearsoned.co.uk/practicalskills.

2   Managing your time

One of the most important activities that you can do is to organise your personal and working time effectively. There is a lot to do at university and a common complaint is that there isn't enough time to accomplish everything. In fact, research shows that most people use up a lot of their time without realising it through ineffective study or activities such as extended coffee breaks. Developing your time management skills will help you achieve more in work, rest and play, but it is important to remember that putting time management techniques into practice is an individual matter, requiring a level of self-discipline not unlike that required for dieting. A new system won't always work perfectly straight away, but through time you can develop a system that is effective for you. An inability to organise your time effectively, of course, results in feelings of failure, frustration, guilt and being out of control in your life.

Setting your goals

The first step is to identify clearly what you want to achieve, both in work and in your personal life. We all have a general idea of what we are aiming for, but to be effective, your goals must be clearly identified and priorities allocated. Clear, concise objectives can provide you with a framework in which to make these choices. Try using the 'SMART' approach, in which objectives should be:

- **Specific** – clear and unambiguous, including what, when, where, how and why.
- **Measurable** – having quantified targets and benefits to provide an understanding of progress.
- **Achievable** – being attainable within your resources.
- **Realistic** – being within your abilities and expectations.
- **Timed** – stating the time period for completion.

Having identified your goals, you can now move on to answer four very important questions:

1. **Where does your time go?**
2. **Where should your time go?**
3. **What are your time-wasting activities?**
4. **What strategies can help you?**

Analysing your current activities

The key to successful development of time management is a realistic knowledge of how you currently spend your time. Start by keeping a detailed time log for a typical week (Fig. 2.1), but you will need to be truthful in this process. Once you have completed the log, consider the following questions:

- **How many hours do I work in total and how many hours do I use for relaxation?**
- **What range of activities do I do?**

Definition

Time management – a system for controlling and using time as efficiently and as effectively as possible.

Example   The objective 'to spend an extra hour each week on directed study in microbiology next term' fulfils the SMART criteria, in contrast to a general intention 'to study more'.

Advantages of time management – these include:

- a feeling of much greater control over your activities;
- avoidance of stress;
- improved productivity – achieve more in a shorter period;
- improved performance – work to higher standards because you are in charge;
- increase in time available for nonwork matters – work hard, but play hard too.

	Activity									Notes
Time slots										
7.00–7.15										
7.15–7.30										
7.30–7.45										
7.45–8.00										
8.00–8.15										
8.15–8.30										
8.30–8.45										
8.45–9.00										
9.00–9.15										

Fig. 2.1 Example of how to lay out a time log. Write activities along the top of the page, and divide the day into 15-min segments as shown. Think beforehand how you will categorise the different things you do, from the mundane (laundry, having a shower, drinking coffee, etc.) to the well-timetabled (tutorial meeting, sports club meeting) and add supplementary notes if required. At the end of each day, place a dot in the relevant column for each activity and sum the dots to give a total at the bottom of the page. You will need to keep a diary like this for at least a week before you see patterns emerging.

- **How long do I spend on each activity?**
- **What do I spend most of my time doing?**
- **What do I spend the least amount of my time doing?**
- **Are my allocations of time in proportion to the importance of my activities?**
- **How much of my time is ineffectively used, e.g. for unplanned socialising or interruptions?**

If you wish, you could use a spreadsheet (Chapter 13) to produce graphical summaries of time allocations in different categories as an aid to analysis and management. Divide your time into:

- **Committed time** – timetabled activities involving your main objectives/goals.
- **Maintenance time** – time spent supporting your general life activities (shopping, cleaning, laundry, etc.).
- **Discretionary time** – time for you to use as you wish, e.g. recreation, sport, hobbies, socialising.

Avoiding time-wasting activities

Look carefully at those tasks that could be identified as time-wasting activities. They include gossiping, over-long breaks, uninvited interruptions and even ineffective study periods. Try to reduce these to a minimum, but do not count on eliminating them entirely. Remember also that some relaxation *should* be programmed into your daily schedule.

Organising your tasks

Having analysed your time usage, you can now use this information, together with your objectives and prioritised goals, to organise your activities, both on a short-term and a long-term basis. Consider using a diary-based system (such as those produced by Filofax, TMI and Day-Timer) that will help you plan ahead and analyse your progress.

Quality in time management – avoid spending a lot of time doing unproductive studying, e.g. reading a textbook without specific objectives for that reading. Make sure you test your recall of the material, if you are working towards an examination (p. 30).

Being assertive – if friends and colleagues continually interrupt you, find a way of controlling them before they control you. Indicate clearly on your door that you do not wish to be disturbed and explain why. Otherwise, try to work away from disturbance.

WEEKLY DIARY Week beginning:

	Sunday	Monday	Tuesday	Wednesday	Thursday	Friday	Saturday
DATE							

	Sunday	Monday	Tuesday	Wednesday	Thursday	Friday	Saturday
7–8 am		Breakfast	Breakfast	Breakfast	Breakfast	Breakfast	
8–9		Preparation	Preparation	Preparation	Preparation	Preparation	Breakfast
9–10	Breakfast	PE112(L)	PE112(L)	PE112(L)	PE112(L)	BIOL(P)	Travel
10–11	FREE	CHEM(L)	CHEM(L)	CHEM(L)	CHEM(L)	BIOL(P)	WORK
11–12	STUDY	STUDY	STUDY	STUDY	STUDY	BIOL(P)	WORK
12–1 pm	STUDY	BIOL(L)	BIOL(L)	BIOL(L)	BIOL(L)	TUTORIAL	WORK
1–2	Lunch	Lunch	Lunch	Lunch	Lunch	Lunch	Lunch
2–3	(VOLLEY-	CHEM(P)	STUDY	SPORT	PE112(P)	STUDY	WORK
3–4	BALL	CHEM(P)	STUDY	(VOLLEY-	PE112(P)	STUDY	WORK
4–5	MATCH)	CHEM(P)	STUDY	BALL	PE112(P)	SHOPPING	WORK
5–6	FREE	STUDY	STUDY	CLUB)	STUDY	TEA ROTA	WORK
6–7	Tea	Tea	Tea	Tea	Tea	Tea	Tea
7–8	FREE*	STUDY	STUDY	FREE*	STUDY	FREE*	FREE
8–9	FREE*	STUDY	STUDY	FREE*	STUDY	FREE*	FREE
9–10	FREE*	FREE*	STUDY	FREE*	STUDY	FREE*	FREE

	Sunday	Monday	Tuesday	Wednesday	Thursday	Friday	Saturday
Study (h)	2	10	11	4	11	6	0
Other (h)	13	5	4	11	4	9	15

Total study time = 44 h

Fig 2.2 A weekly diary with examples of entries for a first-year science student with a Saturday job and active membership of a volleyball club. Note that 'free time' can change to 'study time' in sessions marked with an asterisk when assessed work is to be produced or during revision for exams. Study time (including attendance at lectures, practicals and tutorials) thus represents between 42 and 50% of the total time.

Divide your tasks into several categories, such as:

- **Urgent** – must be done as a top priority and at short notice (e.g. doctor's appointment).
- **Routine** – predictable and regular and therefore easily scheduled (e.g. attending lectures or playing sport).
- **One-off activities** – usually with rather shorter deadlines and which may be of high priority (e.g. a tutorial assignment or seeking advice on a specific issue).
- **Long-term tasks** – sometimes referred to as 'elephant tasks' that are too large to 'eat' in one go (e.g. learning a language). These are best managed by scheduling frequent small 'bites' to achieve the task over a longer timescale.

You should make a weekly plan (Fig. 2.2) for the routine activities, with gaps for less predictable tasks. This should be supplemented by individual daily checklists, preferably written at the end of the previous working day. Such plans and checklists should be flexible, forming the basis for most of your activities except when exceptional circumstances intervene. The planning must be kept brief, however, and should be scheduled into your activities. Box 2.1 provides tips for effective time management during your studies.

Use checklists as often as possible – post your lists in places where they are easily and frequently visible, such as in front of your desk. Ticking things off as they are completed gives you a feeling of accomplishment and progress, increasing motivation.

Matching your work to your body's rhythm – everyone has times of day when they feel more alert and able to work. Decide when these times are for you and programme your work accordingly. Plan relaxation events for periods when you tend to be less alert.

KEY POINT Review each day's plan at the end of the previous day, making such modifications as are required by circumstances, e.g. adding an uncompleted task from the previous day or a new and urgent task.

Box 2.1 Tips for effective planning and working

- Set guidelines and review expectations regularly.
- Don't procrastinate: don't keep putting off doing things you know are important – they will not go away but they will increase to crisis point.
- Don't be a perfectionist – perfection is paralysing.
- Learn from past experience – review your management system regularly.
- Don't set yourself unrealistic goals and objectives – this will lead to procrastination and feelings of failure.
- Avoid recurring crises – they are telling you that something is not working properly and needs to be changed.
- Learn to concentrate effectively – do not let yourself be distracted by casual interruptions.

- Learn to say 'no' firmly but graciously when appropriate.
- Know your own body rhythms: e.g. are you a morning person or an evening person?
- Learn to recognise the benefits of rest and relaxation at appropriate times.
- Take short but complete breaks from your tasks – come back feeling refreshed in mind and body.
- Work in suitable study areas – keep your own workspace organised.
- Avoid clutter (physical and mental).
- Learn to access and use information effectively (Chapter 10).
- Learn to read and write accurately and quickly (Chapters 4 and 17).

Sources for further study

Anon. *Day-Timer.* Available: http://www.daytimer.co.uk
Last accessed 01/09/15.
[Website for products of Day-Timer Europe Ltd, Fairfax Road, Newton Abbot, TQ12 6UD, UK.]

Anon. *Filofax.* Available: http://www.filofax.co.uk
Last accessed 01/09/15.
[Website for products of Filofax UK, Broadgate Tower, Primrose Street, London, EC2A 2EW, UK.]

Anon. *Study Guides and Strategies.* Available:
http://www.studygs.net/timman.htm
Last accessed 01/09/15.
[Website guides to time management and other study skills for students.]

Evans, C. (2008) *Time Management for Dummies,* UK edn. Wiley, Chichester.

STUDY EXERCISES

2.1 Evaluate your time usage. Compile a spreadsheet to keep a record of your daily activities in 15-min segments for a week. Analyse this graphically and identify areas for improvement.

2.2 List your short-, medium- and long-term tasks and allocate priorities. Produce several lists, one for each of the three timescales, and prioritise each item. Use this list to plan your time management, by scheduling

high priority tasks and leave low priority activities to 'fill in' the spare time that you may identify. This task should be done on a regular (monthly) basis to allow for changing situations.

2.3 Plan an 'elephant' task. Spend some time planning how to carry out a large or difficult task (learning a language or learning to use a complex computer program) by breaking it down into achievable segments ('bites').

Answers to these study exercises are available at www.pearsoned.co.uk/practicalskills.

Definitions

Team – a team is not simply a group of people with job titles, but a congregation of individuals, each of whom has a role within a shared task which is understood by other members. Members of a team may seek out certain roles and they may perform most effectively in the ones that are most natural to them.

Team role – a tendency to behave, contribute and interrelate with others in a particular way.

(both after Belbin, 2010)

Peer assessment – this term applies to marking schemes in which all or a proportion of the marks for a teamwork exercise are allocated by the team members themselves. Read the instructions carefully before embarking on the exercise, so you know which aspects of your work your fellow team members will be assessing. When deciding what marks to allocate yourself, try to be as fair as possible with your marking.

Gaining confidence through experience – the more you take part in teamwork, the more you know how teams operate and how to make teamwork effective for you.

It is highly likely that you will be expected to work with fellow students during practicals, tutorials and group study exercises. This might take the form of sharing tasks or casual collaboration through discussion, or it might be formally directed teamwork such as problem-based learning (Box 7.1) or preparing a poster (Chapter 15). Interacting with others in this way can be extremely rewarding and realistically represents the professional world, where teamworking is common. The advantages of such collaborative work with others include:

- **Synergistic interactions** – teamwork often results in better ideas, produced by the interchange of views, and better output, due to the complementary skills of team members.
- **Support provided to individuals by other members of the team.**
- **Enhanced levels of personal commitment.**
- **Sharing of responsibilities for complex or difficult tasks.**

However, some people approach teamwork with negative feelings. Reasons for this include:

- **Initial reservations about working with strangers** – not knowing whether you will be able to form a friendly and productive relationship.
- **Lack of personal commitment to the specified task** – perhaps due to unfamiliarity or perceived dislike concerning the topic.
- **Apprehension that others may not do their share of the work** – especially if you hope to do well in the course.
- **Lack of previous experience** – worries about the kinds of personal interactions likely to occur and the team role likely to suit you best.
- **Concerns about the outcomes of peer assessment** – in particular, whether others will give you a fair mark for your efforts.

In most cases, these feelings can be transformed through positive participation and a common desire to perform well, so that the team as a whole achieves its target.

Teamwork skills

Some of the key skills you will need to develop to maximise the success of your teamworking activities include:

- **Interpersonal skills.** These include the way in which you react to new people, how you listen to and communicate with others, and how you deal with conflicts and disagreements.
- **Delegation and sharing of tasks.** This implies that you are willing to share effort and responsibility, are willing to trust your fellow team members, and have mechanisms to deal with unexpected outcomes or even failure to contribute.
- **Effective listening.** This involves giving others the opportunity and time to express their views and to take these seriously, however expressed.
- **Speaking clearly and concisely.** Effective communication is a vital part of teamwork, not only between team members, but also when presenting team outcomes to others (see Chapter 16).

- **Providing constructive criticism**. It is all too easy to be negative but only constructive criticism will have a positive effect on interactions with others.

> Web-based resources and support for brainstorming – websites such as http://www.belbin.com give further information and practical advice for teamworking.

The dynamics of teamworking

It is important that team activities are properly structured so that each member knows what is expected of them. Allocation of responsibilities usually requires the clear identification of a leader. Several studies of groups have identified different team roles that derive from differences in personality. You should be aware of such categorisations, both in terms of your own predispositions and those of your fellow team members, as it will help the group to interact more productively. Belbin (2010) identified eight such roles, recently extended to nine, as shown in Table 3.1. Several of the categories shown in Table 3.1 are suitable for a leader, including 'coordinator' and 'shaper'.

In formal team situations, your course organiser may deal with these issues; even if they do not, it is important that you are aware of these roles and their potential impact on the success or failure of teamwork. You should try to identify your own 'natural' role.

If asked to form a team, bear these different roles in mind during your selection of colleagues and your interactions with them. The ideal team

Table 3.1 A summary of the team roles described by Belbin (2010). A good team requires members who are able to undertake appropriate roles at different times. Each role provides important strengths to a team, and its compensatory weaknesses should be accepted within the group framework.

Team role	Personality characteristics	Typical function in a team	Strengths	Allowable weaknesses
Coordinator	Self-confident, calm and controlled	Leading: causing others to work towards shared goals	Good at spotting others' talents and delegating activities	Often less creative or intellectual than others in the group
Shaper	Strong need for achievement; outgoing; dynamic; highly strung	Leading: generating action within team; imposing shape and pattern to work	Providing drive and realism to group activities	Can be headstrong, emotional and less patient than others
Innovator	Individualistic, serious-minded; often unorthodox	Generating action; imposing shape and pattern to work activities	Creative, innovative and knowledgeable	Tendency to work in isolation; ideas may not always be practical
Monitor–evaluator	Sober, unemotional and prudent	Analysing problems and evaluating ideas	Shrewd judgement	May work slowly; not usually a good motivator
Implementer	Well-organised and self-disciplined, with practical common sense	Doing what needs to be done	Organising abilities and common sense	Lack of flexibility and tendency to resist new ideas
Teamworker	Sociable, mild and sensitive	Being supportive, perceptive and diplomatic; keeping the team going	Good listener; reliable and flexible; promotes team spirit	Not comfortable when leading; may be indecisive
Resource investigator	Extrovert, enthusiastic, curious and communicative	Exploiting opportunities; finding resources; external relations	Quick thinking; good at developing others' ideas	May lose interest rapidly
Completer–finisher	Introvert and anxious; painstaking, orderly and conscientious	Ensuring completion of activity to high standard	Good focus on fulfilling objectives and goals	Obsessive about details; may wish to do all the work to control quality
Specialist	Professional, self-motivated and dedicated	Providing essential skills	Commitment and technical knowledge	Contribute on a narrow aspect of project; tend to be single-minded

should contain members capable of adopting most of these roles. However, you should also note the following points:

- **People will probably best fit one of these roles** naturally as a function of their personality and skills.
- **Group members may be suited to more than one role.**
- **In some circumstances, team members may be required to adapt** and take a different role from the one that they feel suits them.
- **No one role is 'better' than any other.** For good teamwork, the group should have a balance of personality types present.
- **People may have to adopt multiple roles,** especially if the team size is small.

> **KEY POINT** In formal teamwork situations, be clear as to how individual contributions are to be identified and recognised. This might require discussion with the course organiser. Make sure that recognition, including assessment, is truly reflective of effort. Failure to ensure that this is the case can lead to disputes and feelings of unfairness.

Your lab partner(s)

Many laboratory sessions in the life sciences involve working in pairs or small groups. In some cases, you may work with the same partner(s) for a series of practicals or for a complete module. The relationship you develop as a team is important to your progress, and can enhance your understanding of the material and the grades you obtain. Tips for building a constructive partnership include:

- **Introduce yourselves at the first session** and take a continuing interest in each other's interests and progress at university.
- **At appropriate points, discuss the practical** (both theory and tasks) and your understanding of what is expected of you.
- **Work jointly to complete the practical effectively,** avoiding the situation where either partner dominates the activities and gains most from the practical experience.
- **Share tasks according to your strengths,** but do this in such a way that one partner can learn new skills and knowledge from the other.
- **Make sure you ask questions of each other** and communicate any doubts about what you have to do.
- **Discuss other aspects of your course,** e.g. by comparing notes from lectures or ideas about in-course assessments.
- **Consider meeting up outside the practical sessions** to study, revise and discuss exams.

Tutorial groups

Tutorials are a group learning environment. Discussions in tutorials can help you to develop your ideas by observing others' approaches, views and ideas and assimilating the facts and concepts that they introduce (see p. 27). This depends on all present participating fully and enthusiastically, while at the same time respecting the right of others to make a contribution and express their opinions.

Recording group discussions – make sure you structure meetings (including writing agendas) and note their outcomes (taking minutes and noting action points).

Making the most of lab partnerships – make sure that you share the different practical tasks across all members of your lab group, otherwise some will miss out on the opportunities afforded by practical classes. This is particularly important in subjects involving a practical exam at the end of term.

Making the most of tutorials – most tutorials require preparation in advance of each session: this can include pre-tutorial reading(s), literature-based research and/ or the completion of specific tutorial 'questions' (see also p. 27).

If you feel you need to team up with someone for study, but lack a suitable 'study buddy', then remember that your fellow tutees might be feeling the same. Tutorial meetings are therefore chances to set up further informal meetings for study or exam revision, as discussed below.

Collaboration for learning

Much collaboration is informal and consists of pairs or groups of individuals getting together to exchange materials and ideas while studying. It may consist of a 'brainstorming' session for a topic or piece of work, or sharing efforts to research a topic. This has much to commend it and is generally encouraged. However, it is vital that this collaborative learning is distinguished from the collaborative writing of individually assessed documents: the latter is not usually acceptable and, in its most extreme form, is plagiarism, usually with a heavy potential punishment in university assessment systems. Make sure you know what plagiarism is, what unacceptable collaboration is (p. 59), and how they are treated within your institution.

Studying with others – teaming up with someone else on your course for revision (a 'study buddy') is a potentially valuable activity and may especially suit some types of learners (p. 25). It can help keep your morale high when things get tough. You might consider:

- sharing notes, textbooks and other information;
- going through past papers together, dissecting the questions and planning answers;
- talking to each other about a topic (good for aural learners; Box 5.1);
- giving tutorials to each other about parts of the course that have not been fully grasped.

> **KEY POINT** Collaboration is inappropriate during the final phase of an assessed piece of work unless you have been directed to produce a group report. Collaboration is encouraged during research and learning activities but the final write-up must normally be your own work. The extreme of producing copycat write-ups is regarded as plagiarism (p. 59) and will be punished accordingly.

Text reference

Belbin, R.M. (2010) *Team Roles at Work,* 2nd edn. Butterworth–Heinemann, Oxford.

Source for further study

Belbin, R.M. *Belbin® Team Roles.* Available: http://www.belbin.com/ Last accessed 01/09/15.

STUDY EXERCISES

3.1 Evaluate your 'natural' team role(s). Using Table 3.1 as a source, decide which team role best fits your personality.

3.2 Keep a journal during a group activity. Record your feelings and observations about experiences of working with other students. After the event, review the journal, then draw up a strategy for developing aspects where you feel you might have done better.

3.3 Reflect upon your teamwork abilities. Draw up a list of your reactions to previous efforts at collaboration or teamwork and analyse your strengths and weaknesses. How could these interactions have been improved or supported more effectively?

Answers to these study exercises are available at www.pearsoned.co.uk/practicalskills.

4 Taking notes from lectures and texts

Note-taking is an essential skill that you will require in many different situations, such as:

- **listening to staff in lectures, seminars and practical classes;**
- **attending meetings and tutorials;**
- **reading texts and research papers;**
- **finding information on the web.**

> **KEY POINT**   Good performance in assessments and exams is built on effective learning and revision (Chapters 5, 6 and 7). However, both ultimately depend on the quality of your notes.

Taking notes from lectures

Taking legible and meaningful lecture notes is essential if you are to make sense of them later, but many students find it difficult when starting their university studies. Begin by noting the date, course, topic and lecturer on the first page of each day's notes. Number every page in case they get mixed up later. The most popular way of taking notes is to write in a linear sequence down the page, emphasising the underlying structure using headings, as in Fig. 17.3. However, the 'pattern' and 'Mind Map' methods (Figs 4.1 and 4.2) have their advocates: experiment, to see which you prefer.

Whatever technique you use, don't try to take down all of the lecturer's words, except when an important definition or example is being given, or when the lecturer has made it clear that he/she is dictating. Listen first, then write. Your goal should be to take down the structure and reasoning behind the lecturer's approach in as few words and phrases as possible. At this stage, follow the lecturer's sequence of delivery. Use headings and leave plenty of space, but don't worry too much about being tidy – it is more important that you get down the appropriate information in a readable form. Use abbreviations to save time. Recognise that you may need to alter your note-taking technique to suit different lecturers' styles.

Choose note-taking methods appropriately – the method you choose to take notes might depend on the subject; the lecturer and their style of delivery; and your own preference.

Compare lecture notes with a colleague – looking at different sets of notes for the same lecture may reveal interesting differences in approach, depth and detail.

Adjusting to the styles of your lecturers – recognise that different approaches to lecture delivery demand different approaches to note-taking. For example, if a lecturer seems to tell lots of anecdotes or spend much of the time on examples during a lecture, do not switch off – you still need to be listening carefully to recognise the important take-home messages. Similarly, if a lecture includes a section consisting mainly of images, you should still try to take notes – names of organisms, locations, key features, even quick sketches. These will help prompt your memory when revising. Do not be deterred by lecturers' idiosyncrasies; in every case you still need to focus and take useful notes (see Box 5.2).

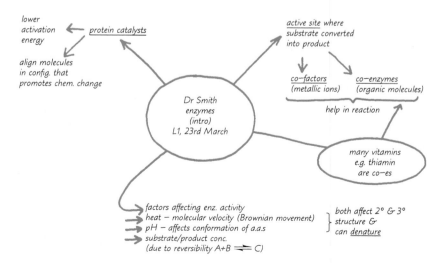

Fig. 4.1 An example of 'pattern' notes, an alternative to the more commonly used 'linear' format. Note the similarity to the 'spider diagram' method of brainstorming ideas (Fig. 17.2).

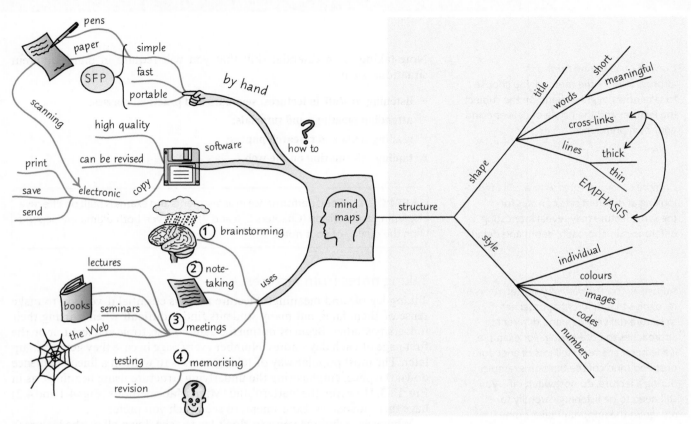

Fig. 4.2 Example of the 'Mind Map' approach to note-taking and 'brainstorming'. Start at the centre with the overall topic title, adding branches and sub-branches for themes and subsidiary topics. 'Basic' maps consist of a branched hierarchy overwritten with key words (e.g. shaded portion, above). Connections should be indicated with arrows; numbering and abbreviations are encouraged. To aid recall and creativity, Buzan and Buzan (2009) recommend use of colour, different fonts, 3-dimensional doodles and other forms of emphasis (e.g. non-shaded portion, above).

Example Commonly used abbreviations include:	
$\exists$	there are, there exist(s)
$\therefore$	therefore
$\because$	because
$\propto$	is proportional to
$\rightarrow$	leads to, into
$\leftarrow$	comes from, from
$\rightarrowtail$	involves several processes in a sequence
$1° \ 2°$	primary, secondary (etc.)
$\approx, \cong$	approximately, roughly equal to
$=, \neq$	equals, not equal to
$\equiv, \not\equiv$	equivalent, not equivalent to
$<, >$	smaller than, bigger than
$>>$	much bigger than
$[X]$	concentration of X
Σ	sum
Δ	change
f	function
$\#$	number
∞	infinity, infinite

You should also make up your own abbreviations relevant to the context, e.g. if a lecturer is talking about photosynthesis, you could write 'PS' instead, etc.

Make sure you note down references to books and take special care to ensure accuracy of definitions and numerical examples. If the lecturer repeats or otherwise emphasises a point, highlight (e.g. by underlining) or make a margin note of this – it could come in useful when revising. If there is something you don't understand, ask at the end of the lecture, or make an appointment to discuss the matter if there isn't time to deal with it then. Tutorials may provide an additional forum for discussing course topics.

Lectures delivered by PowerPoint or similar presentation programs

Some students make the mistake of thinking that lectures delivered as computer-based presentations with an accompanying handout or web resource require little or no effort by way of note-taking. While it is true that you may be freed from the need to copy out large diagrams and the basic text may provide structure, you will still need to adapt and add to the lecturer's points. Much of the important detail and crucial emphasis will still be delivered verbally. Furthermore, if you simply listen passively to the lecture, or worse, try to work from the handout alone, it will be far more difficult to understand and remember the content.

If you are not supplied with handouts, you may be given access to the electronic file, so that you can print out the presentation beforehand, perhaps in the '3 slides per page' format that allows space for notes alongside each slide (Fig. 4.3). Scan through this before the lecture if you

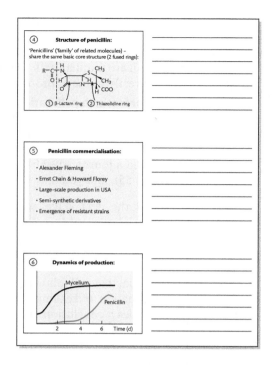

Fig. 4.3 An example of a printout from *PowerPoint* in 'Handouts (3 slides per page)' format.

Sources of abbreviations for note-taking – general abbreviations can be found at http://public.oed.com/how-to-use-the-oed/abbreviations/ whereas specific examples relevant to the biosciences are at http://www.allacronyms.com/biology/ topic'.

can; then, during the presentation, focus on listening to what the lecturer has to say. Note down any extra details, points of emphasis and examples. After lectures, you could also add notes from supplementary reading. The text in presentations can be converted to word processor format if you have access to the electronic file. In *PowerPoint,* this can be achieved from the *Outline* option on the *View* tab, under *Presentation Views > Normal > Outline.* You can copy and paste text between programs in the normal fashion, then modify font size and colour as appropriate.

'Making up' your notes

As soon as possible after each lecture, work through your notes, tidying them up and adding detail where necessary. Add emphasis to any headings you have made, so that the structure is clearer. If you feel it would be more logical for your purposes, change the order. Compare your notes with material in a textbook and correct any inconsistencies. Make notes from, or copy, any useful material you see in textbooks, ready for revision.

Taking notes from books and journal papers

Scanning

Identifying a study goal – reading around your subject is worthwhile and rewarding, but if your time is limited, it is important to identify a purpose for your reading and note-making. Ideally, the goal should relate to specific learning outcomes or assessments.

Good reading skills are vital when taking notes from written sources. When consulting a new source for specific information, the first thing you need to do is to orientate yourself by understanding the text's scope and structure. This is often called 'scanning'.

Validating your sources – in some cases texts and references may be recommended by staff, but in others you may need to evaluate the reliability of the source yourself – for example, by considering whether the author(s) has a lengthy publication record, who he/she works for or, alternatively, whether they might have a biased viewpoint that could skew their analysis.

Definitions

Paraphrasing – This is to quote others' ideas or points by restating them in different words. It is different from summary in that you may include more detail. For example: *Wilson and Brand (2010), based on their studies of the snail population in Arran, take the view that small sample size may be irrelevant when variability is low.*

Quoting – This involves lifting text directly from your source and demonstrating this clearly by using quotation marks (and sometimes italics). If you wish to miss words out, use ellipsis punctuation marks (. . .), as shown in this example: *Wilson and Brand (2010) concluded: 'this indicates that sample size was small in this study. . . but that this was acceptable in view of the low variability encountered.'* Here, the words *'because of financial constraints'* have been omitted as irrelevant to the author's main point. For longer quotes (say 40 words or more) create a separate paragraph of text with a citation (normally at the end).

Summarising – This is taking the essence from a text and expressing it briefly in your own words. For example: *Other researchers defend the use of small sample sizes (e.g. Wilson and Brand, 2010).*

In all cases above, note the citation of a relevant reference, in this case using Harvard style (author names and date, p. 55). Full details of the source should be provided in a reference or bibliography section (p. 56). See also Chapter 10 (especially Box 10.1) in relation to plagiarism.

- **For books,** this might involve a quick inspection of the contents section, a check on how each chapter is constructed, and noting, for example, whether the book has a Glossary that might be useful. Once you are familiar with the structure and layout, then you might either go directly to the appropriate chapter or section, or consult the index.
- **For research publications,** the best idea is to read the Abstract before consulting specific text, figures or tables, as this should detail the essential methods and findings. The paper's layout thereafter will probably involve the same component sections (IMRaD, p. 120), but occasionally in a journal-specific order. Referencing systems may also differ (p. 55), and you may wish to note this.

At the next stage, when you are scanning through the text to find material relevant to your goal, it is useful to have a good understanding of how most writing is structured. Both paragraphs and larger pieces of writing generally include an introduction, a main body and a conclusion. The introductory or 'topic' sentence/paragraph should orientate you to the material that will follow; the main body expands on the topic sentence, with explanation and examples; and the final or 'terminator' sentence/paragraph provides a conclusion and/or a link to the next part.

KEY POINT   Working out how a piece of writing is structured will help you to scan quickly for relevance, digest the content faster and find material of interest as soon as possible.

Depending on your goal, there may be a specific key word or phrase for which you are searching, or you may wish to read paragraphs on a specific topic. As well as an index, most biosciences textbooks use systems of headings and subheadings which facilitate such searches. With digital or online resources, you can press the Control and F keys together to access a 'find' dialog box that will speed this process – however, take care to enter correctly spelled and relevant words, using alternatives if an initial search draws a blank.

Finally, you need to absorb the content and decide what the salient points are (Chapter 10). These will obviously depend on the task you are carrying out. When taking notes, it is vital that you distinguish between your own paraphrasing or summarising of the text and situations where your notes are direct quotes, or you may be guilty of plagiarism (p. 60). If transcribing text word-for-word, always use quotation marks and always note the full citation details of the source (p. 56) at the same time.

Printing *PowerPoint* slides – use the 'Black and White' option on the *Print* menu to avoid wasting ink on printing of coloured backgrounds. If you wish to use colour, remember that slides can be difficult to read if printed in small format. Always print a sample page before printing the whole lecture.

Scanning effectively – you need to stay focused on your key words, otherwise you may be distracted by apparently interesting but peripheral material.

Spotting sequences – writers often number their points (firstly, secondly, thirdly, etc.) and looking for these words in the text can help you skim it quickly.

Making sure you have all the details of a source – when taking notes from a text or journal paper: (a) always take full details (Chapter 9); (b) if copying word-for-word make sure you indicate this using quotes and take special care to ensure you do not alter the original wording.

Skimming

This is a valuable way to gain the maximum amount of information in the minimum amount of time, by reading as little of a text as is required. Essentially, the technique (also termed 'surveying') requires you to look at the organisation of the text, rather than the detail. In a sense, you are trying to see the writer's original plan and the purpose behind each part of the text. Look through the whole of the piece first, to gain an overview of its scope and structure. Headings provide an obvious clue to structure, if present. Next, look for the 'topic sentence' in each paragraph (p. 113), which is often the first. You might then decide that the paragraph contains a definition that is important to note, or it may contain examples, so may not be worth reading in detail for your purpose.

When you have found relevant material, note-taking fulfils the vital purpose of helping you understand and remember the information. If you simply read it, either directly or from a photocopy, you risk accomplishing neither. The act of paraphrasing (using different words to give the same meaning) makes you think about the meaning and forces you to express this for yourself. It is an important active learning technique. A popular method of skimming and note-taking is called the SQ3R technique (Box. 4.1).

> **KEY POINT**   Obtaining information and then understanding it are distinct, sequential parts of the process of learning. As discussed in Chapter 5 (Table 5.1), you must be able to do more than recall facts to succeed.

Methods for finding and evaluating texts and articles are discussed further in Chapters 9 and 10.

Box 4.1   The SQ3R technique for skimming texts

Survey	Gain a quick overview of the contents of the book or chapter, perhaps by rapidly reading the contents page or headings	Recall	Recite to yourself what has been stated every few paragraphs. Write notes of this if appropriate, paraphrasing the text rather than copying it
Question	Ask yourself what the material covers and how precisely it relates to your study objectives	Review	Think about what you have read and/or review your notes as a whole. Consider where it all fits in
Read	Now read the text, paying attention to the ways it addresses your key questions		

Reference

Buzan, T. and Buzan, B. (2009) *The Mind Map Book: Unlock Your Creativity, Boost Your Memory, Change Your Life*. Pearson, Harlow.

[Note that there are several free versions of mind mapping software available on the web, e.g. FreeMind – try using a search engine to find relevant examples and their URLs.]

STUDY EXERCISES

4.1 Experiment with a new note-taking technique. If you haven't tried the pattern or mind mapping methods (Figs 4.1 and 4.2), carry out a trial to see how they work for you. Research the methods first by consulting appropriate books or websites.

4.2 Carry out a 'spring clean' of your desk area and notes. Make a concerted effort to organise your notes and handouts, investing if necessary in files and folders.

This will be especially valuable at the start of a revision period.

4.3 Try out the SQ3R technique. The next time you need to obtain information from a text, compare this method (Box 4.1) with others you may have adopted in the past. Is it faster, and does it aid your ability to recall the information?

Answers to these study exercises are available at www.pearsoned.co.uk/practicalskills.

5   Learning effectively

Be adaptable – you should be prepared to adjust your study methods according to the topic, the learning environment (p. 26) and the approach of your lecturers (p. 27) as well as your personal learning preferences.

There are many different ways of learning and at university you have the freedom to choose which approach to study suits you best. You should tackle this responsibility with an open mind, and be prepared to consider new options. Understanding how you learn best and how you are expected to think about your discipline will help you to improve your approach to study and understand your branch of bioscience at a deeper level.

> **KEY POINT**   At university, you are expected to set your own agenda for learning. There will be timetabled activities, assessments and exam deadlines, but it is your responsibility to decide how you will study and learn, how you will manage your time, and, ultimately, what you will gain from the experience. You should be willing to challenge yourself academically to discover your full potential.

Thinking about thinking

The thinking processes that students are expected to carry out can be presented in a sequence, starting with less complex thought processes and ending with more complex processes, each of which builds on the previous level (see Table 5.1). The first two categories in this ladder apply to gaining basic knowledge and understanding, important when you first encounter a topic. Processes 3 to 6 are those additionally carried out by high-performing university students, with the latter two being especially relevant to final-year students, researchers and professionals. Naturally, the tutors assessing you will want to reward the deepest thinking appropriate for your level of study. This is often signified by the words used in assessment tasks and marking criteria (column four, Table 5.1, and p. 30), and while this is not an exact relationship, being more aware of this agenda can help you to gain more from your studies and appreciate what is being demanded of you.

Aligning your learning with a specific approach – it is important not to read too much into any formal 'diagnosis' of your preferred method of learning. You need to take a flexible approach; no matter how you prefer to learn, many of the outputs assessed at university (for example, exam answers, essays and reports) are in a formal written format and you will need to apply what you have learned to these format. Also, different aspects of your study will require different approaches, irrespective of personal preferences. For example, practical skills are best learned by 'hands-on' activities – learning by doing.

Identifying your personal learning preferences

We do not all learn in the same way. Your preferred approach to learning is simply the one that suits you best for receiving, communicating and understanding information. It therefore involves selections that will help you learn and perform most effectively. There are many different ways of describing such learning strategies, and you may be introduced to specific schemes during your studies. Although methods and terminology may differ among these approaches, it is important to realise that it is the *process* of analysing your learning style that is critical, together with the way you use the information to modify your approach to studying, rather than the specific type of learner that you may identify yourself to be.

Understanding your learning preferences – although the notion that we have 'hard-wired' individual learning 'styles' is strongly contested by many academics and researchers, in practice it can be helpful to reflect on your personal preferences, and how you can make the most of these to enhance your learning.

A useful scheme for describing learning preferences is the **VARK** system devised by Fleming (2001). By answering a short online question-naire, you can 'diagnose' yourself as one of the types shown in Box 5.1, which also summarises important outcomes relating to how information

Table 5.1 A ladder of thinking processes, moving from 'shallower' thought processes (top of table) to 'deeper' levels of thinking (bottom of table). This table is derived from research by Benjamin Bloom and colleagues (Bloom et al., 1956; Anderson *et al.*, 2001). When considering the cue words in typical question instructions, bear in mind that the precise meaning will always depend on the context. For example, while 'describe' is often associated with relatively simple processes of recall, an instruction like 'describe how the human brain works' demands that you exhibit higher-level understanding. Note also that while a 'cue word' is often given at the start of a question/instruction, this is not universally so.

Thinking processes and description (in approximate order of increasing 'depth')	Example in life sciences	Example of typical question structure, with *cue word* highlighted	Other cue words used in question instructions
1. Remembering (knowing facts). If you know information, you can *recall* or *recognise* it. This does not always mean you understand it at a higher level.	You might know the order of bases in a piece of DNA but not understand what this means.	*Describe* the main components of a biological membrane.	• define • list • state • identify
2. Understanding. If you comprehend a fact, you *understand* what it means.	You might know the order of bases in a piece of DNA and understand that they code in triplets for specific amino acids.	*Explain* how membrane components are involved in the accumulation of solutes within living cells.	• distinguish • interpret • outline • illustrate
3. Applying. To apply a fact means that you can *put it to use* in a particular context.	You might be able to take the DNA base sequence and work out the amino acid sequence of the protein for which they code.	Using the Nernst equation, and realistic values for the membrane potential and solute concentrations, *demonstrate* how Na+ ions must be actively transported out of the cells of marine organisms.	• calculate • illustrate • solve • show
4. Analysing. To analyse information means that you are able to *break it down into parts* and show how these components *fit together.*	You might be able to construct a three-dimensional model of a protein derived from the base sequence.	Drawing on information about membrane structure, *defend* the endosymbiotic theory of eukaryote evolution.	• compare • explain • consider • discuss
5. Evaluating. If you evaluate information, you *arrive at a judgement* based on its importance relative to the topic being addressed.	You might be able to comment on theories about how a protein has evolved, by considering the structure of related proteins and relating this to the taxonomic position of their source species.	*Evaluate* the relative importance of passive and active transport in the accumulation of heavy metal salts by the main groups of soil fungi.	• design • integrate • test • create
6. Creating. To create or 'synthesise', you need to be able to *extract relevant facts* from a body of knowledge and use these to *address an issue in a novel way* or *create something new.*	You might be able to work out the function of a protein for which you know the sequence of bases, based on a comparison with other like proteins.	*Devise* an experiment to test the hypothesis that a specific membrane fraction contains a functional ATPase involved in glucose transport.	• review • assess • consider • justify

and concepts can be assimilated, learned and expressed. People show different degrees of alignment with these categories, and Fleming's research indicates that the majority of students are multi-modal learners – that is, falling into more than one category – rather than being only in one grouping. By carrying out an analysis like this, you can become more aware of your personal characteristics and think about whether the methods of studying you currently use are those that are best suited to your learning needs.

Box 5.1 How to diagnose your learning preferences using the VARK scheme

Visit www.vark-learn.com to carry out the online diagnostic test, reflect on whether it is a fair description of your preferences, and think about whether you might change the way you study to improve your performance. None of the outcomes should be regarded as prescriptive – you should mix techniques as you see fit and only use methods that you feel comfortable adopting. Adapted with permission from material produced by Fleming (2001).

Learning approach *Description of learning preference*	Outcomes for your learning, studying and exam technique		
	Advice for taking in information and understanding it	Best methods of studying for effective learning	Ways to cope with exams so you perform better
Visual: *You are interested in colour, layout and design. You probably prefer to learn from visual media or books with diagrams and charts. You tend to add doodles and use highlighters on lecture and revision notes and express ideas and concepts as images.*	Use media incorporating images, diagrams, flowcharts, etc. When constructing notes, employ underlining, different colours and highlighters. Use symbols as much as you can, rather than words. Leave plenty of white space in your notes. Experiment with the 'mind map' style of note-taking (p. 18).	Use similar methods to those described in column two. Reduce lecture notes to pictures. Try to construct your own images to aid understanding, then test your learning by redrawing these from memory.	Plan answers diagrammatically. Recall the images and doodles you used in your notes. Use diagrams in your answers (making sure they are numbered and fully labelled). As part of your revision, turn images into words.
Aural: *You prefer discussing subjects and probably like to attend tutorials and listen to lecturers, rather than read textbooks. Your lecture notes may be poor because you would rather listen than take notes.*	Make sure you attend classes, discussions and tutorials. Note and remember the interesting examples, stories, jokes. Leave spaces in your notes for later recall and 'filling'. Discuss topics with a 'study buddy'. Record lectures (with lecturer's permission) and replay to reinforce your learning.	Expand your notes by talking with others and making additional notes from the textbook. Ask others to listen to you talk about topics. Read your summarised notes aloud to yourself. Record your vocalised notes and listen to them later.	When writing answers, imagine you are talking to an unseen examiner. Speak your answers inside your head. Listen to your voice and write them down. Practise writing answers to previous exam questions.
Read–Write: *You prefer using text in all formats. Your lecture notes are probably good. You tend to like lecturers who use words well and provide lots of information in sentences and notes. In note-taking, you may convert diagrams to text and text to bullet points.*	Focus on note-taking. You may prefer the 'linear' style of note-taking (p. 17). Use the following in your notes: lists; headings; glossaries; and definitions. Expand your notes by adding further information from handouts, textbooks and library readings.	Reduce your notes to lists or headings. Write out and read the lists again and again (silently). Turn actions, diagrams, charts and flowcharts into words. Rewrite the ideas and principles into other words. Organise diagrams and graphs into statements, e.g. 'The trend is . . .'.	Plan and write out exam answers using remembered lists. Arrange your words into hierarchies and bullet points.
Kinesthetic: *You tend to recall by remembering real events and lecturers' 'stories'. You probably prefer field excursions and lab work to theory and like lecturers who give real-life examples. Your lecture notes may be weak because the topics did not seem 'concrete' or 'relevant'.*	Focus on examples that illustrate principles. Concentrate on applied aspects and hands-on approaches, but try to understand the theoretical principles that underpin them. When taking in information, use all your senses – sight, touch, taste, smell, hearing.	Put plenty of examples, pictures and photographs into your notes. Use case studies and applications to help with principles and concepts. Talk through your notes with others. Recall your experience of lectures, tutorials, experiments or field trips.	Write practice answers and paragraphs. Recall examples and things you did in the lab or field trip. Role-play the exam situation in your own room.
Multi-modal: *Your preferences fall into two or more of the above categories. You are able to use these different modes as appropriate.*	If you are diagnosed as having two dominant preferences or several equally dominant preferences, read the study strategies above that apply to each of these. You may find it necessary to use more than one strategy for learning and exam-sitting, feeling less secure with only one.		

Be active in your learning – a common theme in all recommended approaches to effective learning is *activity* on the learner's part. Thus, note-making during lectures requires a greater level of engagement with the material than the more passive action of simply listening; discussing concepts with others involves a deeper degree of thinking than solo thought processes when reading text (see also Box 6.3).

Definition

Take-home message(s) – the main point(s) that a presenter wishes you to remember from what they have said (perhaps a distillation of important facts or a set of conclusions). If a lecture or seminar has been particularly detailed or complex, noting these messages will help you to make sense of it at a later time — they may well prove to be useful for assignments and/or exams.

Lay the foundations for learning – unless you are gifted with photographic recall, it is not easy to commit information to your memory unless you have a conceptual framework of understanding in which to 'place' it. This 'bigger picture' comes from reading, listening and thinking at an early stage in the process.

Learning effectively in different settings

How you learn best depends on your personal characteristics, as noted above, but the approaches you use must also suit the different modes of teaching you encounter.

Lectures

These are designed to impart knowledge and understanding efficiently on the part of the staff, who can teach large classes using this method. However, in some instances they can simply be note-taking exercises for you as a student. This may tempt you to approach such teaching in a 'passive' frame of mind, but this is almost certainly not the best way to learn. To get the most from lectures, you need to:

- **Prepare beforehand** by finding out what the topic will be, what the learning outcomes are, and doing some preliminary reading.
- **Arrive in good time,** sitting where you can see and hear the lecturer, with the right materials for taking notes.
- **Listen attentively,** particularly when the lecturer is stating what he or she plans to cover; when they indicate that you are expected to write things down word-for-word; when a definition is being presented; when key facts are stated (often emphasised by repetition); and especially at the end, when the take-home messages are usually covered.
- *Make useful notes* – these will not be a transcript of the lecturer's words, but rather a personal digest of the key points that have been made (see Chapter 4). Even if the material is already provided as a *PowerPoint* or *Word* file (or equivalent), your aim is to add value by explaining and interpreting for yourself, and thereby to lay the foundations for memorising the content.
- **Ask questions if you do not understand something.** This might not be at the point of delivery, because things may become clearer when you go over your notes, read a textbook or talk about the subject with a fellow student. However, it should be as soon as possible after those options are exhausted. There will be opportunities to enquire at tutorials, via email or online discussion boards or at later lectures or meetings.
- **Above all, do attend lectures,** even if you feel their value may be limited in some cases – do not underestimate the way that looking and listening contribute to understanding and memorisation. Also, 'cold text' on the page rarely has the additional context that a lecture can provide.

Box 5.2 provides some tips for dealing with some of the different teaching styles of lecturers.

Box 5.2 How to accommodate different lecturers' teaching styles

One challenging aspect of lectures is that no two staff members deliver them in the same way. This means that you will need to adjust your listening and note-making approaches to suit. For example:

- It may be **hard to extract facts and meaning** from a lecturer who tells meandering stories or keeps the class amused with anecdotes. You must not lose focus when listening and need to be alert when the key points are mentioned, or use the textbook to fill in the missing material when making up your notes later on.

- You may be **distracted** if the lecturer uses technology in a way that makes you think about the medium rather than the message. By all means enjoy the videos, images and electronic wizardry, but do remember to listen out for the underlying messages and focus your notes on these.

- It may be **difficult to stay attentive** if the lecturer's delivery is quiet, monotonous or mumbled (or all three). Here, you will need to ensure that you sit where you can hear, and listen very carefully for key words and phrases, rather than changes in intonation, that might signal importance or change of emphasis. Do not be afraid to ask the lecturer to speak up if you cannot hear.

- You may **find it difficult to draw out general principles** if the lecturer provides great detail about his or her life in science or current research work. Moreover, the latter may be hard to follow if complex jargon and undefined acronyms are used frequently. The remedy here may be to use the textbook before and after the delivery to create an overarching framework of understanding.

Practicals (lab classes)

In the biosciences, practical work is essential, so that your understanding is grounded in real examples of biological material, processes and equipment. Many practicals follow the 'scientific method' (Chapter 32) and illustrate how to design and carry out observations and experiments. They may help you to develop manipulative skills, as well as those associated with data recording, analysis and reporting. To maximise learning from practicals, you should:

- **prepare as well as you can** by reading through the schedule in advance;
- **make sure that you have all equipment and protective clothing needed;**
- **arrive at the lab or assembly point in good time;**
- **understand and observe safe practice** (Chapter 23);
- **follow the advised procedures;**
- **ask questions when you do not understand;**
- **manage your time during practicals** and use time when waiting for something to happen to start your write-up;
- **take relevant notes;**
- **draw up informal figures and tables as you go along;**
- **complete your write-up as soon as possible after the session.**

These suggestions are discussed in greater detail in later chapters.

Tutorials

These may be designed to fulfil a number of purposes, including:

- **providing guidance,** as part of an advising system;
- **reviewing and testing your understanding** of material covered elsewhere;
- **introducing new material** that is better taught in small groups;

Making the most of lab demonstrators – these staff or postgraduate students are paid to help you, so ensure you get good value from them by listening to their advice and asking them questions. However, do not expect them to hand you answers on a plate – they will probably be under instruction to make you think for yourself.

Getting the most from tutorials – this requires preparation beforehand, for example, by:

- carrying out the recommended reading or tackling the set problems;
- submitting expected work in good time;
- thinking about the issues involved in the subject matter;
- preparing any questions you might have about the topic you have been asked to study;
- reflecting on the potential relevance of the tutorial to your programme of study.

Taking account of tutorial assessment – the tasks prepared for tutorials may be marked, and in some cases your participation will be graded. Consult the course handbook to find out what proportion of the total mark derives from tutorials and what the marking criteria are. Ensure that you maximise the marks obtained from this source, as they may be relatively easy to achieve if you put in the necessary work.

Alternative modes of distance learning – the discussion opposite assumes that e-learning facilities are being used but in some courses the materials are primarily paper-based. If this is the case, then finding the best way to learn may be difficult. You will need to find a method of making the material 'come alive', perhaps by purchasing or borrowing well-illustrated texts or video materials (such as BBC documentaries).

- developing and assessing essay-writing and numerical skills through appropriate tasks;
- promoting interactions among students and with the tutor.

The face-to-face discussions that can occur during tutorials are rarely possible in larger classes and they add value by allowing views and understanding to develop through an exchange of opinion and knowledge. Think of your tutor as a potential ally. He or she may be able to explain things differently from the lecturer, for example, and thereby help you understand any concepts than you find difficult. Never feel embarrassed about asking questions – the tutor is there to provide answers and the chances are that, if you find a subject tricky, so will the rest of the group. Your tutor may also have or assume responsibilities beyond teaching, and be able to help you with subject choices or non-academic issues.

e-Learning and distance learning

These modes of learning can be a mixed blessing. On the one hand, they allow easy access to learning materials at a time of your choosing, while on the other, they assume a personal discipline in your approach to study that may be hard to maintain. Effective learning in this context requires the following:

- **making regular and frequent visits to online modules and portals;**
- **allocating time to study;**
- **paying special attention to announcements, message pages or email systems;**
- **participating in online discussions;**
- **organising materials provided in digital form** such as lists of 'live' links to website addresses;
- **reading the advised materials and more,** whether online, or as texts;
- **making notes as appropriate;**
- **using active learning techniques** (Chapter 6), rather than passively reading the materials provided;
- **observing staff-set milestones** in study, assessment and submission of work;
- **creating personal milestones** in study, self-assessment and preparation of work.

The only person who can judge the effectiveness of your learning at university is you. Only you will know how much effort you put in to your studies and what you are expecting, in terms of a mark or grade; only you will know how comfortable you feel with a particular approach; and only you can respond to the feedback you have been given, in terms of how you will learn for future assessments. Be prepared to change your learning methods if you find that they are unsuccessful or difficult to apply in particular circumstances.

Text references

Anderson, L.W., Krathwohl, D. R. and Bloom, B. S. (2001) *A Taxonomy for Learning, Teaching, and Assessing: A Revision of Bloom's Taxonomy of Educational Objectives.* Allyn & Bacon, London.

Bloom, B., Englehart, M., Furst, E., Hill, W. and Krathwohl, D. (1956) *Taxonomy of Educational Objectives: The Classification of Educational Goals. Handbook I: Cognitive Domain.* Longmans, Green, New York and Toronto.

Fleming, N.D. (2001) *Teaching and Learning Styles: VARK Strategies.* Neil Fleming, Christchurch.

Fleming, N.D. *VARK: A Guide to Learning Styles.* Available: http://www.vark-learn.com/ Last accessed 01/09/15.

Source for further study

McMillan, K.M. and Weyers, J.D.B. (2012) *The Study Skills Book,* 3rd edn. Pearson Education, Harlow.

STUDY EXERCISES

5.1 Visit the VARK-learn website (see Box 5.1). Carry out the short online test there to diagnose your learning preference. Next, review the guidance given on the website for study practices, including taking in and using information for effective learning. Do you feel that any of the suggestions might be useful to you?

5.2 Review your note-taking methods. How well suited are they to your needs? How well suited are they to the lecture styles of the academic staff? Have you captured the important points, or are you merely transcribing exactly what the lecturer says? Where lecturers use *PowerPoint* slides, are you listening for, and capturing in note form, the additional spoken points and examples that the lecturer is adding during the live presentation?

5.3 Think about your thinking. Read through Table 5.1 and consider different thinking processes in relation to: (a) feedback you have received from a previous assignment; or (b) your next assignment and your intended approach to it. Does this reveal that your marks could be suffering because you are not thinking deeply enough or are not capturing your thoughts clearly enough in your writing?

Answers to these study exercises are available at www.pearsoned.co.uk/practicalskills.

Definitions

Formative assessments – these may be mid-term or mid-semester tests and are often in the same format as later exams. They are intended to give you feedback on your performance.

Learning outcomes/objectives – statements of the knowledge, understanding or skills that a learner will be able to demonstrate on successful completion of a module, topic or learning activity.

Summative assessments – these include end-of-year or end-of-module exams. They inform others about the standard of your work. In continuous or 'in-course' assessment, the summative elements are spread out over the course. Sometimes these assessments may involve a formative aspect, if feedback is given.

Example   A set of learning outcomes taken from an introductory lecture on bacterial cell structure.

After this lecture, you will be able to:

- Define the following terms:
  - ✓ prokaryote
  - ✓ eukaryote
  - ✓ envelope
  - ✓ fimbriae
  - ✓ F pilus
  - ✓ plasmid.
- Draw a labelled diagram to illustrate the principal components of a bacterial cell.
- Explain the functions of the major cellular components.
- Demonstrate knowledge of the relative magnitude of bacteria and eukaryotic cells, in terms of typical linear dimensions and volumes.
- Describe the basic process of cell division and give examples of typical timescales for different bacteria, e.g. *Escherichia coli*, *Clostridium perfringens*, *Mycobacterium tuberculosis*.

Where your studies are assessed through an exam, you will need to prepare for this by revising and, because you may have a number of exams in a short period, you will need to be well organised. You will need to align your study effort with the expectations of your lecturers, as explained in the course materials and through the feedback you have been given on previous assignments and exams. Adopting active methods of revision that are suited to your learning style (Chapter 5) can also make a significant difference to your performance.

The role of assessment and feedback in your learning

Your starting point when preparing for assessment should always be the learning outcomes or objectives for each module, topic or learning activity. You will usually find them in your module handbook. They state in clear terms what your tutors expect you to be able to accomplish after participating in each element and reading around the topic. Also of value will be marking/assessment criteria or grade descriptors, which state in general terms what level of attainment is required for your work to reach specific grades. These are more likely to be defined at faculty/college/school/department level and consequently published in appropriate handbooks or websites. Reading learning outcomes and grade descriptors will give you a good idea of what to expect and the level of performance required to reach your personal goals. Relate them to both the material covered (e.g. in lectures and practicals, or online) and past exam papers. Doing this as you study and revise will indicate whether further reading and independent studying is required, and of what type. You will also have a much clearer picture of how you are likely to be assessed.

> **KEY POINT**   Use the learning outcomes for your course (normally published in the handbook) as a fundamental part of your revision planning. These indicate what you will be expected to be able to do after taking part in the course, so exam questions are often closely based on them. Check this by reference to past papers.

There are essentially two types of assessment – formative and summative, although the distinction may not always be clear-cut (see margin). The first way you can learn from formative assessment is to consider the grade you obtained in relation to the work you put in. If this is a disappointment to you, then there must be a mismatch between your understanding of the topic and the marking scheme with that of the marker, or a problem in the writing or presentation of your assignment. If you do not understand the reason for your grade, contact your tutor.

The second way to learn from formative assessment is through the written feedback and notes on your work. These comments may be cryptic, or scribbled hastily, so if you do not understand or cannot read them, ask the tutor who marked the work. Most tutors will be pleased to explain how you could have improved your mark. If you find that the same comments appear frequently, it may be a good idea to seek help from

Using tutors' feedback – it is always worth reading any comments on your work as soon as it is returned. If you do not understand the comments, or are unsure about why you might have lost marks in an assignment, ask for an explanation while the topic is fresh in your mind.

Time management when revising – this is a vital to success and is best achieved by creating a revision timetable (Box 6.2)

Filing lecture notes – make sure your notes are kept neatly and in sequence by using a ring binder system. File the notes in lecture or practical sequence, adding any supplementary notes or photocopies alongside.

Recognise when your concentration is wavering – take a short break from revision when this happens and return to work refreshed and ready to learn. Remember that 20 min is often quoted as a typical limit to a spell of full concentration effort.

your university's academic support unit. Take along examples of your work and representative feedback comments so they can give you the best possible advice. Another suggestion is to ask to see the work of another student who obtained a good mark, and compare it with your own. This will help you judge the standard you should be aiming towards.

Preparing for revision and examinations

Before you start revising, find out as much as you can about each exam, including:

- **its format and duration;**
- **the date and location;**
- **the types of questions;**
- **whether any questions/sections are compulsory;**
- **whether the questions are internally or externally set or assessed;**
- **whether the exam is 'open book', and if so, which texts or notes are allowed.**

Your course tutor is likely to give you details of exam structure and timing well beforehand, so that you can plan your revision; the course handbook and past papers (if available) can provide further useful details (Box 6.1). Always check that the nature of the exam has not changed before you make plans based on past papers.

Organising and using lecture notes, assignments and practical reports

Given their importance as a source of material for revision, you should have sorted out any deficiencies or omissions in your lecture notes and practical reports at an early stage. For example, you may have missed a lecture or practical due to illness, etc., but the exam is likely to assume attendance throughout the year. Ask other students for copies of notes of any sessions missed.

Your practical reports and any assignment work will contain specific comments from the teaching staff, indicating where marks were lost, corrections, mistakes, inadequacies, and so on. Most lecturers are quite happy to discuss such details with students on a one-to-one basis and this information may provide you with 'clues' to the expectations of individual lecturers that may be useful in exams set by the same members of staff. However, you should never 'fish' for specific information on possible exam questions, as this will not be well received.

Revision

Begin early, to avoid last-minute panic. Start in earnest several weeks beforehand, and plan your work carefully.

- **Prepare a revision timetable** – an 'action plan' that gives details of specific topics to be covered (Box 6.2). Find out at an early stage when (and where) your examinations are to be held, and plan your revision around this. Try to keep to your timetable. Time management during this period is as important as keeping to time during the exam itself.

Box 6.1 How to use past exam papers in your revision

Past exam papers are a valuable resource for targeting your revision.

1. **Find out where the past (or sample) exam papers are kept.** Copies may be lodged in your department or the library; or they may be accessible online.

2. **Locate and copy relevant papers for your module(s).** Check with your tutor or course handbook that the style of paper will not change for the next set of examinations.

3. **Analyse the design of the exam paper.** Taking into account the length in weeks of your module, and the different lecturers and/or topics for those weeks, note any patterns that emerge. For example, can you translate weeks of lectures/practicals into numbers of questions or sections of the paper? Consider how this might affect your revision plans and exam tactics, taking into account: (a) any choices or restrictions offered in the paper; and (b) the different types of questions asked (i.e. multiple choice, short-answer or essay).

4. **Examine carefully the style of questions.** Can you identify the expectations of your lecturers? Can you relate the questions to the learning objectives? How much extra reading do they seem to expect? Are the questions fact-based? Do they require a synthesis based on other knowledge? Can you identify different styles for different lecturers? Consider how the answers to these questions might affect your revision effort and exam strategy.

5. **Practise answering questions.** Perhaps with friends, set up your own mock exam when you have done a fair amount of revision, but not too close to the exams. Use a relevant past exam paper; and try not to study it beforehand. You need not attempt all of the paper at one sitting. You will need a quiet room in a place where you will not be interrupted (e.g. a library). Keep close track of time during the mock exam and try to do each question in the length of time you would normally assign to it (see p. 117) – this gives you a feel for the speed of thought and writing required and the scope of answer possible. Mark each other's papers and discuss how each of you interpreted the question and laid out your answers and your individual marking schemes.

6. **Practise writing answer plans and starting answers.** This can save time compared with the 'mock exam' approach. Practise in starting answers can help you get over stalling at the start and wasting valuable time. Writing essay plans gets you used to organising your thoughts quickly and putting your thoughts into a logical sequence.

Box 6.2 How to prepare and use a revision timetable

1. **Make up a grid showing the number of days until your exams are finished.** Divide each day into several sections. If you like revising in large blocks of time, use a.m., p.m. and evening slots, but if you prefer shorter periods, divide each of these in two, or use hourly divisions (see also the table in study exercise 6.2).

2. **Write in your non-revision commitments,** including any time off you plan to allocate and physical activity at frequent intervals. Try to have about one-third or a quarter of the time off in any one day. Plan this in relation to your best times for useful work – for example, some people work best in the mornings, while others prefer evenings. If you wish, use a system where your relaxation time is a bonus to be worked for; this may help you motivate yourself.

3. **Decide on how you wish to subdivide your subjects for revision purposes.** This might be among subjects, according to difficulty (with the hardest getting the most time), or within subjects, according to topics. Make sure there is an adequate balance of time among topics and especially that you do not avoid working on the subject(s) you find least interesting or most difficult.

4. **Allocate the work to the different slots available on your timetable.** You should work backwards from the exams, making sure that you cover every exam topic adequately in the period just before each exam. You may wish to colour-code the subjects.

5. **As you revise, mark off the slots completed** – this has a positive psychological effect and will boost your self-confidence.

6. **After the exams, revisit your timetable** and decide whether you would do anything differently next time.

Question-spotting – avoid taking this approach in an attempt to reduce the amount of time you spend revising. Lecturers are aware that this strategy may be taken and try to ask questions in an unpredictable manner. You may find that you are unable to answer on unexpected topics that you failed to revise. Moreover, if you have a preconceived idea about what will be asked, you may also fail to grasp the nuances of the exact question set, and provide a response lacking in relevance.

- **Study the learning outcomes for each topic** (usually published in the course handbook) to get an idea of what lecturers expect from you.
- **Use past papers as a guide** to the form of the exam and the type of question likely to be asked (Box 6.1).
- **Remember to have several short (five-minute) breaks during each hour of revision** and a longer break every few hours. In any day, try to work for a maximum of three-quarters of the time.
- **Include recreation within your schedule:** there is little point in tiring yourself with too much revision, as this is unlikely to be profitable.
- **Make your revision as active and interesting as possible** (see Box 6.3): the least productive approach is simply to read and reread your notes.
- **Ease back on the revision near the exam:** plan your revision to avoid last-minute cramming and overload fatigue.

Box 6.3 How to revise actively

The following techniques may prove useful in devising an active revision strategy.

- 'Distil' your lecture notes to show the main headings and examples. Prepare revision sheets with details for a particular topic on a single sheet of paper, arranged as a numbered checklist. Wall posters are another useful revision aid.
- Aid recall through effective note-taking: for example, the mind map technique (p. 18), when used to organise ideas, is claimed to enhance recall for some by connecting the material to visual images or linking it to the physical senses.
- Carry out revision checks: it is important to test yourself frequently to ensure that you have retained the information you are revising. Assess what you have learned by taking a blank sheet of paper and writing down all you know on a selected topic. Check your full notes to see if you missed anything out. If you did, go back immediately to a fresh blank sheet and redo the example. Repeat, as required.
- Make up lists of key facts and definitions: these can be a useful starting point for many exam answers. Test yourself repeatedly on these lists, or get a friend to do this. Try to remember *how many* points you need to know in each case – this will help you recall them all during the exam.
- Use mnemonics and acronyms to commit specific factual information to memory. Sometimes, the dafter they are, the better they seem to work.

- Draw diagrams from memory: make sure you can label them fully.
- Try recitation as an alternative to written recall. Talk about your topic to another person, preferably someone in your class. Talk to yourself (in private) if necessary. Explaining something out loud is an excellent test of your understanding.
- Associate facts with images or journeys if you find this method works.
- Use a wide variety of approaches to avoid boredom during revision (e.g. record information on audio tape, use cartoons, or any other method, as long as it is not just reading).
- Form a revision group to share ideas and discuss topics with other students.
- Prepare answers to past papers, e.g. write essays or, if time is limited, write essay plans (see Box 6.1).
- If your subject involves numerical calculations, work through representative problems.
- Make up your own questions: the act of putting yourself in the examiner's mindset by inventing questions can help revision. However, you should not rely on 'question-spotting': this is a risky practice!
- Write key points on Post-it® notes or similar: arrange and rearrange these in clusters or as lists around your study space or room.

> **KEY POINT** When considering assessment questions, look carefully at words used in the instructions. These cues can help you identify what depth is expected in your answer (see Table 5.1). Take special care in multi-part questions, because the first part may require lower-level thinking, while in later parts marks may be awarded for evidence of deeper thinking.

Final preparations – try to get a good night's sleep before an exam. Last-minute cramming will be counterproductive if you are too tired during the exam.

The evening before your exam should be spent in consolidating your material, and checking through summary lists and plans. Avoid introducing new material at this late stage: your aim should be to boost your confidence, putting yourself in the right frame of mind for the exam itself.

Sources for further study

Burns, R. (1997) *The Student's Guide to Passing Exams.* Kogan Page, London.

Hamilton, D. (2003) *Passing Exams: A Guide for Maximum Success and Minimum Stress.* Cengage Learning, Boston.

McMillan, K.M. and Weyers, J.D.B. (2011) *How to Succeed in Exams and Assessments,* 2nd edn. Prentice Hall, London.

McMillan, K.M. and Weyers, J.D.B. (2012) *The Study Skills Book*, 3rd edn. Pearson Education, Harlow.

[Many universities host study skills websites; these can be found using 'study skills', 'revision' or 'exams' as keywords in a search engine.]

STUDY EXERCISES

6.1 Make use of past exam papers. Use the techniques discussed in Box 6.1 to improve your revision strategy: assess their effectiveness in a particular exam, or series of exams.

6.2 Draw up a revision timetable. Use the techniques discussed in Box 6.2 to create a revision timetable for your forthcoming exams. You may wish to use or adapt the arrangement below, either on paper or within a spreadsheet.

6.3 Try out a new active revision technique. Try any or all of the methods mentioned in Box 6.3 when revising. Compare notes with a colleague – which seems to be the most successful technique for you and for the topic you are revising?

A revison timetable planner.

Date	Morning		Lunch	Afternoon		Tea/Dinner	Evening	
	Session 1	Session 2		Session 1	Session 2		Session 1	Session 2

Answers to these study exercises are available at www.pearsoned.co.uk/practicalskills.

Many universities operate a modular system for their biosciences degree courses. This allows greater flexibility in subject choice and accommodates students studying on different degree paths. Modules also break a subject into discrete, easily assimilated elements. They have the advantage of spreading assessment over the academic year, but they can also tempt you to avoid certain difficult subjects or to feel that you can forget about a topic once the module is finished.

> **KEY POINT** You should select your modules with care, mindful of potential degree options and how your transcript and CV will appear to a prospective employer. If you feel you need advice, consult your personal tutor or study advisor.

As you move between levels of the university system, you will be expected to have passed a certain number of modules, as detailed in the progression criteria. These may be expressed using a credit point system. Students are normally allowed two attempts at each module exam and the resits often take place at the end of the summer vacation. If a student does not pass at the second attempt, they may be asked to 'carry' the subject in a subsequent year, and in severe cases of multiple failure, they may be asked to re-take the whole year or even leave the course. Consequently, it is worth finding out about these aspects of your degree. They are usually published in course handbooks.

You are unlikely to have reached this stage in your education without being exposed to the examination process. You may not enjoy being assessed, but you probably want to do well in your course. It is therefore important to understand why and how you are being tested. Identifying and improving the skills required for exam success will allow you to perform to the best of your ability.

Assessed coursework

There is a component of assessed coursework in many modules. This often tests specific skills, and may require you to demonstrate thinking at deeper levels (Table 5.1). The common types of coursework assessment are covered at various points in this book:

- **practical exercises** (Chapters 21–74);
- **essays** (Chapters 17 and 18);
- **numerical problems** (Chapter 72);
- **data analysis** (Chapters 72–74);
- **poster and spoken presentations** (Chapters 15 and 16);
- **literature surveys and reviews** (Chapter 20);
- **project work** (Chapters 19 and 34);
- **problem-based learning** (Box 7.1).

At the start of each year or module, read the course handbook or module guide carefully to find out when any assessed work needs to be submitted.

Box 7.1 Problem-based learning (PBL)

In this teaching method, you are likely to be presented with a 'real-world' problem or issue, often working within a team. As you tackle the problem, you are expected to gain factual knowledge, develop skills and exercise critical thinking (Chapter 10). Because there is a direct and relevant context for your work, and because you have to employ active learning techniques, the knowledge and skills you gain are likely to be more readily assimilated and remembered. This approach also more closely mimics workplace practices. PBL usually proceeds as follows:

1. **You are presented with a problem** (e.g. a case study, a hypothetical patient, a topical issue).

2. **You consider what issues and topics you need to research**, by discussion with others if necessary. You may need to identify where relevant resources can be found (Chapters 9–12).

3. **You then need to rank the issues and topics in importance**, allocating tasks to group members, if appropriate. A structured approach is an important aspect of PBL.

4. **Having carried out the necessary research, you should review the information that has been obtained.** As a result, new issues may need to be explored and, where appropriate, allocated to group members.

5. **You will be asked to produce an outcome, such as a report, diagnosis, seminar presentation or poster.** An outline structure will be required, and for groups, further allocation of tasks will be needed to accomplish this outcome.

If asked to carry out PBL as part of your course, it is important to get off to a good start. At first, the problem may seem unfamiliar. However, once you become involved in the work, you will quickly gain confidence. If working as part of a group, make sure that your group meets as early as possible, that you attend all sessions and that you do the necessary background reading. When working in a team, a degree of self-awareness is necessary regarding your 'natural' role in group situations (Table 3.1). Various methods are used for grading PBL, including written, oral and poster presentations, and this may involve self-assessment and/or peer marking.

Summative and formative aspects of assessment – many assessments have a formative component, providing feedback that is aimed at helping you to improve, as well as a summative component, with a mark that contributes to your final grade.

Definitions

Terms with specialised meaning in university exam contexts:

Diet – a period during which a number of exams take place, according to a published timetable.

Invigilator – a person whose role is to ensure an exam is conducted according to the regulations.

Paper – usually, a set of questions for a single subject.

Question – an element of an exam paper, often couched in terms of an instruction rather than a true question, and often with the marks available stated at the end.

Rubric – the set of instructions at the start of an exam paper.

Note relevant dates in your diary, and use this information to plan your work. Take special note if deadlines for different modules clash, or if they coincide with social or sporting commitments.

KEY POINT If, for some valid reason (e.g. illness), you will be late with an assessment, speak to your tutors as soon as possible. They may be able to take extenuating circumstances into account by not applying a marking penalty. They will let you know what paperwork you may be required to submit to support your claim.

Summative exams – general points

Summative exams (p. 30) normally involve you answering questions without being able to consult other students or your notes. Invigilators are present to ensure appropriate conduct, but departmental representatives may be present for some of the exam. Their role is to sort out any subject-related problems, so if you think something is wrong, ask at the earliest opportunity. It is not unknown for parts of questions to be omitted in error, or for double meanings to arise, for example.

Planning

When preparing for an exam, make a checklist of the items that you will need (see p. 42). On the day of the exam, give yourself sufficient time to arrive at the correct room, without the risk of being late. Double-check the times and places of your exams, both well before the exam, and also on arrival. If you arrive at the exam venue early, you can always rectify a mistake if you find that you have gone to the wrong place.

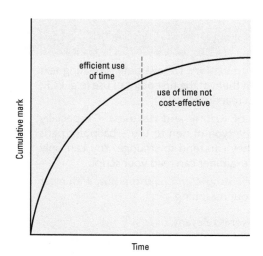

Fig. 7.1 Potential exam marks as a function of time. The marks awarded in a single answer will follow the law of diminishing returns – it will be far more difficult to achieve the final 25% of the available marks than the initial 25%. Do not spend too long on any one question.

Using the question paper – unless this is specifically forbidden, you *should* write on the question paper to plan your strategy, keep to time and organise answers.

Checking your exam answers – look for:

- errors of fact;
- missing information;
- grammatical and spelling errors;
- errors of scale and units;
- errors in calculations.

Adopting different tactics according to the exam – you should adjust your exam strategy (and revision methods) to allow for the differences in question types used in each exam paper.

Tackling the paper

Begin by reading the instructions at the top of the exam paper carefully, so that you do not make any errors based on lack of understanding of the exam structure. Make sure that you confirm:

- **how many questions are set;**
- **how many must be answered;**
- **whether the paper is divided into sections;**
- **whether any parts are compulsory;**
- **what each question/section is worth, as a proportion of the total mark;**
- **whether different questions should be answered in different books.**

Do not be tempted to spend too long on any one question or section: the return in terms of marks will not justify the loss of time from other questions (see Fig. 7.1). Take the first few minutes to read the paper and plan your strategy, before you begin writing. Do not be put off by those who begin immediately; it is almost certain they are producing unplanned work of a poor standard.

Underline the key phrases in the instructions, to reinforce their message. Next, read through the set of questions. If there is a choice, decide on those questions to be answered and decide on the order in which you will tackle them. Quickly prepare a timetable which takes into account the amount of time required to complete each section and which reflects the allocation of marks – there is little point in spending one-quarter of the exam period on a question worth only 5% of the total marks. Use the exam paper to mark the sequence in which the questions will be answered and write the finishing times alongside; refer to this timetable during the exam to keep yourself on course.

Reviewing your answers

At the end of the exam, you should allow some time to check through your script. Make sure your name and/or ID number is on each exam book as required and on all other sheets of paper, including graph paper, even if securely attached to your script, as it is in your interest to ensure that your work does not go astray.

> **KEY POINT** Never leave any exam early. Most exams assess work carried out over several months within a time period of 2–3 h and there is always something constructive you can do with the remaining time to improve your script.

Special considerations for different types of exam question

Essay questions

Essay questions let examiners test the depth of your comprehension and understanding as well as your recall of facts. Essay questions give you plenty of scope to show what you know. They suit those with a good grasp of principles but who perhaps have less ability to recall individual details.

Before you tackle a particular question, you must be sure of what is required in your answer. Ask yourself 'What is the examiner looking for

---

> ### Box 7.2 Writing under exam conditions
>
> Make sure you go into an exam with a strategy for managing the available time.
>
> - Allocate some time (say 5% of the total) to consider which questions to answer and in which order.
> - Share the rest of the time among the questions, according to the marks available. Aim to optimise the marks obtained. A potentially good answer should be allocated slightly more time than one you do not feel so happy about. However, do not concentrate on any one answer (see Fig. 7.1).
> - For each question divide the time into planning, writing and revision phases (see p. 117).
>
> Employ time-saving techniques as much as possible:
>
> - Use spider diagrams (Fig. 17.2) or mind maps (Fig. 4.2) to organise and plan your answer.
> - Use diagrams and tables to save time in making difficult and lengthy explanations, but make sure you refer to each one in the text.
>
> - Use standard abbreviations to save time repeating text but always explain them at the first point of use (e.g. PCR, polymerase chain reaction).
> - Consider speed of writing and neatness – especially when selecting the type of pen to use – ballpoint pens are fastest, but they can tend to smudge. You can only gain marks if the examiner can read your script.
> - Keep your answer simple and to the point, with clear explanations of your reasoning.
>
> Make sure your answer is relevant.
>
> - Do not include irrelevant facts just because you memorised them during revision, as this may do you more harm than good. You must answer the specific question that has been set.
> - Remember that time taken to write irrelevant material is time lost from another question.

Penalties for guessing – if there is a penalty for incorrect answers in a multiple-choice test, the best strategy is *not* to answer questions when you know your answer is a complete guess. Depending on the penalty, it may be beneficial to guess if you can narrow the choice down to two options (but beware false or irrelevant alternatives). However, if there are no such penalties, then you should provide an answer to all questions.

in this particular question?' and then set about providing a *relevant* answer. Consider each word in the question and highlight, underline or circle the key words. Make sure you know the meaning of the terms given in Table 18.1 so that you can provide the appropriate information, where necessary. Spend some time developing a structure for your writing (see Chapter 17). Refer back to the question frequently as you write, to confirm that you are keeping to the subject matter. Box 7.2 gives advice on writing essays under exam conditions.

It is usually a good idea to begin with the question that you are most confident about. This will reassure you before tackling more difficult parts of the paper. If you run out of time, write in note form. Examiners are usually understanding, as long as the main components of the question have been addressed and the intended structure of the answer is clear. Common reasons for poor exam answers in essay-style questions are listed in Box 7.3. Try to avoid making these mistakes.

Multiple-choice and short-answer questions

Multiple-choice questions (MCQs) and short-answer questions (SAQs) are generally used to test the breadth and detail of your knowledge. The various styles that can be encompassed within the SAQ format allow for more demanding questions than MCQs, which may emphasise memory work and specific factual information.

A good approach for MCQ papers is as follows:

1. **First trawl:** read through all of the questions fairly rapidly, noting the 'correct' answer for those you can attempt immediately.

Box 7.3   **Reasons for poor exam answers to essay-style questions**

The following are reasons that lecturers often cite when they give low marks for essay answers:

- **Not answering the exact question set.** Either failing to recognise the specialist terms used in the question, failing to demonstrate an understanding of the terms by not providing definitions, failing to carry out the precise instruction in a question, or failing to address all aspects of the question.

- **Running out of time.** Not matching the time allocated to the extent of the answer. Frequently, this results in spending too long on one question and not enough on the others, or even failing to complete the paper.

- **Not answering all parts** of a multiple-part question, or not recognising that one part (perhaps involving more complex ideas) may carry more marks than another.

- **Failing to provide evidence** to support an answer. Forgetting to state the 'obvious' – either basic facts or definitions.

- **Failing to illustrate an answer appropriately,** either by not including a relevant diagram, or by providing a diagram that does not aid communication, or by not including relevant examples.

- **Incomplete answer(s).** Failing to answer appropriately owing to lack of knowledge.

- **Providing irrelevant evidence** to support an answer. 'Waffling' to fill space.

- **Illegible handwriting.**

- **Poor English,** such that facts and ideas are not expressed clearly.

- **Lack of logic** or structure to the answer.

- **Factual errors,** indicating poor note-taking, poor revision or poor recall.

- **Failing to correct obvious mistakes** by re-reading an answer before submitting the script.

At higher levels, the following aspects are especially important:

- **Not providing enough in-depth information.**

- **Providing a descriptive rather than an analytical answer** – focusing on facts, rather than deeper aspects of a topic (see Table 5.1).

- **Not setting a problem in context,** or not demonstrating a wider understanding of the topic. However, make sure you do not overdo this, or you may risk not answering the question set.

- **Not giving enough evidence of reading around the subject.** Wider reading can be demonstrated by quoting relevant papers and reviews and by giving author names and dates of publication.

- **Not considering both sides of a topic/debate,** or not arriving at a conclusion.

---

2. **Second trawl:** go through the questions again, checking your original answers and answering any other straightforward questions.

3. **Third trawl:** now tackle the difficult questions and those that require longer to answer (e.g. those based on numerical problems).

4. **Final review:** look again at all your answers, checking for obvious mistakes.

This iterative approach is not only efficient, but it also gives scope for other questions to prompt you to recall forgotten information.

When unsure of an answer, the first stage is to rule out options that are clearly absurd or have obviously been placed there to distract you. Next, looking at the remaining options, can you judge between contrasting pairs with alternative answers? Logically, both cannot be correct, so you should see if you can rule one of the pairs out. Watch out, however, in case *both* may be irrelevant to the answer. If the question involves a calculation, try to work this out independently from the answers, so you are not influenced by them.

In SAQ papers, there may be a choice of questions. Choose your options carefully – it may be better to gain half marks for a correct answer to half a question than to provide a largely irrelevant answer that

Answer the question as requested – this is true for all questions, but especially important for SAQs. If the question asks for a diagram, make sure you provide one; if it asks for a specified number of aspects of a topic, try to list this number of points; if there are two or more parts, provide appropriate answers to all aspects. This may seem obvious, but many marks are lost for not following instructions.

apparently covers the whole question but lacks the necessary detail. For the SAQ form of question, few if any marks are given for writing style. Think in 'bullet points' and list the crucial points only. The time for answering SAQ questions may be tight, so get down to work fast, starting with answers that demand remembered facts. Stick to your timetable by moving on to the next question as soon as possible. Strategically, it is probably better to get part-marks for the full number of questions than good marks for only a few.

Extended matching items (EMIs)

EMIs, sometimes called extended matching sets, or extended matching questions, are a more complex and challenging version of the multiple-choice format, where a pool of answers is offered that may be correct in several cases or none at all. Formats differ, but the questions often relate to a complex scenario outlined as a 'vignette'. EMIs are designed to test not only your detailed knowledge but also your deeper understanding and ability to apply this, and are commonly used in disciplines where professional judgements on complex issues must be made.

Exams with an EMI component require you to read carefully, quickly and with understanding, distinguish between relevant and irrelevant information and make accurate connections between statements. As with MCQs, you will require a broad knowledge across your subject, but also a deeper and sometimes quite detailed understanding. An idea of the sorts of topics that might be covered should be apparent from the learning outcomes and case studies covered in the course. Look carefully at any example EMIs or past papers that are offered, as this will help you to develop a strategy for answering.

When answering an EMI paper, first look quickly through the topics and select those that most suit your knowledge and revision. Plan to answer those you are most confident about first. Then, for each individual EMI:

1. **Scan-read the whole question quickly** to get a feel for its scope.
2. **Read both the scenario and the questions more carefully,** and highlight key terms. Some people find it helpful next to try to answer the questions without reference to the options as they will not be put off by distracters.
3. **Work through the list of options ruling them 'in' or 'out'** for each specific question – you should be left with a shortlist of possible answers.
4. **Select the final answers you wish to give,** with reference to the instructions or rules provided.
5. **Make sure you are keeping up with the necessary pace** by setting a target time for completion of each topic. Try to leave some time in your planning to allow you to review and check your answers (p. 37).

You will need to be particularly well focused and work quickly and decisively – time will often be a constraint in EMI exams. If there are no penalty marks, ensure you give a response to each question.

Typical layout for an extended matching items (EMI) question

1. **Description of topic or theme** – use this heading to orientate your thoughts when answering.
2. **Instructions** – this will explain how you should answer and the weighting of marks among the questions. If all the questions are similar, this may appear at the head of the exam paper; if not, read each instruction carefully, as the rules and marks given may change.
3. **A series of options** – you will have to select among these to answer the questions which follow. Some options may be correct more than once, some not at all, and you may or may not be told how many are correct in each case. Each time you answer a question you will have to go through the list carefully and see which apply – slow readers beware!
4. **A scenario (or 'vignette')** – this is a short paragraph that describes the background to the questions. Details here will have a big effect on the answers you give, so you will need to consider the wording very carefully before you answer. It might be a good idea to highlight key words to keep them in your mind.
5. **The questions** – these may be grouped, and may develop in complexity from those demanding a simple factual answer to those requiring a fair amount of reflection and judgement.

Practical and information-processing exams

The prospect of a practical or information-processing exam may cause you more concern than a theory exam. This may be due to a limited experience of practical examinations, or to the fact that practical and observational skills are tested, as well as recall, description and analysis of factual information. Your first thoughts may be that it is not possible to prepare for such exams but, in fact, you can improve your performance by mastering the various practical techniques described in this book.

You may be allowed to take your laboratory reports and other texts into the practical exam. Do not assume that this is a soft option, or that revision is unnecessary: you will not have time to read large sections of your reports or to familiarise yourself with basic principles, etc. The main advantage of 'open book' exams is that you can check specific details of methodology, reducing your reliance on memory, provided you know your way around your practical manual and notes. In all other respects, your revision and preparation for such exams should be similar to theory exams. Make sure you are familiar with all of the practical exercises, including any work carried out in class by your partner (since exams are assessed on individual performance). If necessary, check with the teaching staff to see whether you can be given access to the laboratory to complete any exercises that you have missed.

At the outset of the practical exam, determine or decide on the order in which you will tackle the questions. A question in the latter half of the paper may need to be started early on in the exam period (e.g. an enzyme assay requiring 2-h incubation in a 3-h exam). Such questions are included to test your forward planning and time management skills. You may need to make additional decisions on the allocation of material; e.g. if you are given 30 sterile test tubes, there is little value in designing an experiment that uses 25 of these to answer question 1, only to find that you need at least 15 tubes for subsequent questions.

Make sure you explain your choice of apparatus and experimental design. Calculations should be set out in a stepwise manner, so that credit can be given, even if the final answer is incorrect (see p. 505). If there are any questions that rely on recall of factual information and you are unable to remember specific details, make sure that you describe the item fully, so that you gain credit for observational skills. Alternatively, leave a gap and return to the question at a later stage.

Oral exams and interviews

An oral interview is sometimes a part of final degree exams, representing a chance for the external examiner(s) to meet individual students and to test their abilities directly and interactively. In some departments, orals are used to validate the exam standard or to test students on the borderline between exam grades. Sometimes an interview may form part of an assessment, as with project work or posters. This type of exam is often intimidating – many students say they do not know how to revise for an oral – and many candidates worry that they will be so nervous that they will not be able to do themselves justice.

Preparation is just as important for oral exams as it is for written exams:

- **Think about your earlier performances** – if the oral follows written papers, it may be that you will be asked about questions you did not do

---

Examples These are principal types of question you are likely to encounter in a practical or information-processing exam:

Manipulative exercises Often based on work carried out during your practical course. Tests dexterity, specific techniques (e.g. sterile technique, p. 223).

Spot' tests Short questions requiring an identification, or brief descriptive notes on a specific item (e.g. a prepared slide). Tests knowledge of seen material or ability to transfer this to a new example.

Calculations For example, the preparation of aqueous solutions at particular concentrations (p. 160) or statistical exercises (p. 524). Tests numeracy.

Data analyses May include the preparation and interpretation of graphs (p. 498) and numerical information, from data either obtained during the exam or provided by the examiner. Tests problem-solving skills.

Preparing specimens for examination with a microscope Tests staining technique and light microscopy technique (Chapters 28 and 29).

Interpreting images Sometimes used when it is not possible to provide living specimens, e.g. in relation to electron microscopy. Can test a variety of skills.

Terminology – the oral exams are sometimes known simply as 'orals' or, borrowing Latin, as 'viva voce' (by the living voice) exams or 'vivas'.

so well on. These topics should be revised thoroughly. Be prepared to say how you would approach the questions if given a second chance.

- **Read up a little about the examiner** – he or she may focus their questions in their area of expertise.
- **Get used to giving spoken answers** – it is often difficult to transfer between written and spoken modes. Write down a few questions and get a friend to ask you them, possibly with unscripted follow-up queries.
- **Research and think about topical issues in your subject area** – some examiners will feel this reflects how interested you are in your subject.

Your conduct during the oral exam is important, too:

- **Arrive promptly and wear reasonably smart clothing.** Not to do either might be considered disrespectful by the examiner.
- **Take your time before answering questions.** Even if you think you know the answer immediately, take a little while to check mentally whether you have considered all angles. A considered, logical approach will be more impressive than a quick but ill-considered response.
- **Start answers with the basics, then develop into deeper aspects.** There may be both surface and deeper aspects to a topic and more credit will be given to a student who puts the latter in context.
- **When your answer is finished, stop speaking.** A short, crisp answer is better than a rambling one.
- **If you do not know the answer, say so.** To waffle and talk about irrelevant material is more damaging than admitting that you do not know.
- **Make sure your answer is balanced.** Talk about the evidence and opinions on both sides of a contentious issue.
- **Do not disagree strongly with the examiner.** Politely put your point of view, detailing the evidence behind it. Examiners will be impressed by students who know their own mind and subject area. However, they will expect you to support a position at odds with the conventional viewpoint.
- **Finally, be positive and enthusiastic about your topic.**

Counteracting anxiety before and during exams

Adverse effects of anxiety need to be overcome by anticipation and preparation well in advance (Box 7.4). Exams, with their tight time limits, are especially stressful for perfectionists. To counteract this tendency, focus on the following points during the exam:

- **Do not expect to produce a perfect essay** – this will not be possible in the time available.
- **Do not spend too long planning your answer** once you have an outline plan, get started.
- **Do not spend too much time on the initial parts of an answer** – at the expense of the main message.
- **Concentrate on getting all of the basic points across** – examiners often allocate marks for the main points first, before allocating extra marks for the detail and depth.

Sleeplessness – this is common and does little harm in the short term. Get up, have a snack, do some light reading or other activity, then return to bed. Avoid caffeine (e.g. tea, coffee and cola) for several hours before going to bed.

Lack of appetite – again commonplace. Eat what you can, and take sugary sweets into the exam to keep energy levels up in case you become tired.

Fear of the unknown – it can be a good idea to visit the exam room beforehand, so you can become familiar with the location. By working through the points given in the exam action list on p. 42, you will be confident that nothing has been left out.

Lack of preparation – the remedy, of course, is to adopt an effective approach to your revision (Chapter 6).

Worries about timekeeping – get a *reliable* alarm clock or a new battery for an old one. Arrange for an alarm phone call; ask a friend or relative to make sure you are awake on time. Make reliable travel arrangements, to arrive on time. If your

exam is early in the morning, it may be a good idea to get up early for a few days beforehand.

Blind panic during an exam – explain how you feel to an invigilator. Ask to go for a supervised walk outside. Do some relaxation exercises (see below), then return to your work. If you are having problems with a specific question, it may be appropriate to speak to the departmental representative at the exam, to check that you are not misinterpreting the question.

Feeling tense – shut your eyes, take several slow, deep breaths, do some stretching and relaxing muscle movements. During exams, it can be a good idea to do this between questions, and possibly to have a complete rest for a minute or so. Before exams, try some exercise activity or escape temporarily from your worries by watching TV or a movie.

Running out of time – do not panic when the invigilator says 'five minutes left'. It is surprising how much you can write in this time. Write note-style answers or state the areas you would have covered: you may get some credit.

- **Do not be obsessed with neatness,** either in handwriting, or in the diagrams you draw, but make sure your answers are legible.

- **Do not worry if you forget something.** You can not be expected to know everything. Most marking schemes give a first class grade to work that misses out on up to 30% of the marks available.

After the exam – try to avoid becoming involved in prolonged analyses with other students over the 'ideal' answers to the questions; after all, it is too late to change anything at this stage. Go for a walk, watch TV for a while, or do something else that helps you relax, so that you are ready to face the next exam with confidence.

KEY POINT Everyone worries about exams. Anxiety is a perfectly natural feeling. It works to your advantage, as it helps provide motivation and the adrenaline that can help you 'raise your game' on the day.

There is a lot to be said for treating exams as a game. After all, they are artificial situations contrived to ensure that large numbers of candidates can be assessed together, with little risk of cheating. They have conventions and rules, just like games. If you understand the rationale behind them and follow the rules, this will aid your performance.

Sources for further study

Acres, D. (1998) *Passing Exams Without Anxiety: How to Get Organised, be Prepared and Feel Confident of Success.* How to Books, London.

McMillan, K.M. and Weyers, J.D.B. (2011) *How to Succeed in Exams and Assessments,* 2nd edn. Prentice Hall, London.

O'Brien, D. (2007) *How to Pass Exams: Accelerate Your Learning – Memorise Key Facts – Revise Effectively,* 2nd edn. Baird, Winchester.

[Many universities host study skills websites that cover exam technique; these can be found using 'study skills', 'revision' or 'exams' as keywords in a search engine.]

STUDY EXERCISES

7.1 Analyse your past performances. Think back to past exams and any feedback you received from them. How might you improve your performance? Consider ways in which you might approach the forthcoming exam differently. If you have kept past papers and answers to continuous assessment exercises, look at any specific comments your lecturers may have made.

7.2 Share revision notes with other students. Make a revision plan (see pp. 11 and 32) and then allocate some time to discussing your revision notes with a colleague. Try to learn from his or her approach. Discuss any issues you do not agree upon.

7.3 Plan your exam tactics. Find out from your module handbook or past papers what the format of each paper will be. Confirm this if necessary with staff. Decide how you will tackle each paper, allocating time to each section and to each question within the sections (see p. 117). Write a personal checklist of requirements for the exam (see p. 42).

Answers to these study exercises are available at www.pearsoned.co.uk/practicalskills.

Personal development planning (PDP) and your CV – many PDP schemes (p. 5) also include an element of career planning that may involve creating a draft or generic CV. The PDP process can help you improve the structure and content of your CV, and the language you use within it.

Understanding skills and qualities – it may be helpful to think about how the skills and qualities in Tables 1.1 and 8.1 apply to particular activities during your studies, since this will give them a greater relevance.

Focusing on evidence – it is important to be able to provide specific information that will back up the claims you make under the 'skills and personal qualities' and other sections of your CV. A potential employer will be interested in your level of competence (what you can actually do) and in situations where you have used a skill or demonstrated a particular quality. These aspects can also be mentioned in your covering letter or at interview.

Many students only think about their curriculum vitae immediately before applying for a job. Reflecting this, chapters on preparing a CV are usually placed near the end of texts of this type. Putting this chapter near the beginning of this book emphasises the importance of focusing your thoughts on your CV at an early stage in your studies. There are four main reasons why this can be valuable:

1. Considering your CV and how it will look to a future employer will help you think more deeply about the direction and value of your academic studies.
2. Creating a draft CV will prompt you to assess your skills and personal qualities and how these fit into your career aspirations.
3. Your CV can be used as a record of all the relevant things you have done at university and then, later, will help you to communicate these to a potential employer.
4. Your developing CV can be used when you apply for vacation or part-time employment.

KEY POINT   Developing your skills and qualities needs to be treated as a long-term project. It makes sense to think early about your career aspirations so that you can make the most of opportunities to build up relevant experience. A good focus for such thoughts is your developing curriculum vitae, so it is useful to work on this from a very early stage.

Skills and personal qualities

Skills (sometimes called competences) are generally what you have learned to do and have improved with practice. Table 1.1 summarises some important skills for bioscientists. This list might seem quite daunting, but your tutors will have designed your courses to give you plenty of opportunities to develop your expertise. Personal qualities, on the other hand, are predominantly innate. Examples include honesty, determination and thoroughness (Table 8.1). These qualities need not remain static, however, and can be developed or changed according to your experiences. By consciously deciding to take on new challenges and responsibilities, not only can you develop your personal qualities, but you can also provide supporting evidence for your CV.

Personal qualities and skills are interrelated because your personal qualities can influence the skills you gain. For example, you may become highly proficient at a skill requiring manual dexterity if you are particularly adept with your hands. Being able to transfer your skills is highly important (Chapter 1) – many employers take a long-term view and look for evidence of the adaptability that will allow you to be a flexible employee and one who will continue to develop skills.

Developing your curriculum vitae

The initial stage involves making an audit of the skills and qualities you already have, and thinking about those you might need to develop. Tables 1.1 and 8.1 could form a basis of this self-appraisal. Assessing

Table 8.1 Some positive personal qualities.

Adaptability
Conscientiousness
Curiosity
Determination
Drive
Energy
Enthusiasm
Fitness and health
Flexible approach
Honesty
Innovation
Integrity
Leadership
Logical approach
Motivation
Patience
Performance under stress
Perseverance
Prudence
Quickness of thought
Seeing others' viewpoints
Self-confidence
Self-discipline
Sense of purpose
Shrewd judgement
Social skills (sociability)
Taking initiative
Tenacity
Thoroughness
Tidiness
Tolerance
Unemotional approach
Willingness to take on challenges

Seeing yourself as others see you – you may not recognise all of your personal qualities and you may need someone else to give you an honest appraisal. This could be anyone whose opinion you value: a friend, a member of your family, a tutor, or a careers advisor.

Setting your own agenda – you have the capability to widen your experience and to demonstrate relevant personal qualities through both curricular and extracurricular activities.

Paying attention to the quality of your CV – your potential employer will regard your CV as an example of your very best work and will not be impressed if it is full of mistakes or badly presented, especially if you claim 'good written communication' as a skill!

your skills may be easier than critically analysing your personal characteristics. In judging your qualities, try to take a positive view and avoid being overly modest. It is important to consider personal qualities in a specific context, e.g. 'I have shown that I am trustworthy, acting as treasurer for the Biochemistry Society', as this evidence will form a vital part of your CV and job applications.

If you can identify gaps in your skills, or qualities that you would like to develop, especially in relation to the needs of your intended career, the next step is to think about ways of improving them. This will be reasonably easy in some cases, but may require some creative thinking in others. A relatively simple example would be if you decided to learn a new language or to maintain one you learned at school. There are likely to be many local college and university courses dealing with foreign languages at many different levels, so it would be a straightforward matter to join one of these. A rather more difficult case might be if you wished to demonstrate 'responsibility', because there are no courses available on this. One route to demonstrate this quality might be to put yourself up for election as an officer in a student society or club; another could be to take a leading role in a relevant activity within your community (e.g. voluntary work such as hospital radio). If you already take part in activities like these, your CV should relate them to this context.

Basic CV structures and their presentation

Box 8.1 illustrates the typical parts of a CV and explains the purpose of each part. Employers are more likely to take notice of a well-organised and well-presented CV, in contrast to one that is difficult to read and assimilate. They will expect it to be concise, complete and accurate. There are many ways of presenting information in a CV, and you will be assessed partly on your choices.

- **Order.** There is some flexibility as to the order in which you can present the different parts (see Box 8.1). A chronological approach within sections helps employers gain a picture of your experience.

- **Personality and 'colour'.** Make your CV different by avoiding standard or dull phrasing. Try not to focus solely on academic aspects: you will probably work in a team and the social aspects of teamwork will be enhanced by your outside interests. However, make sure that the reader does not get the impression that these interests dominate your life.

- **Style.** Your CV should reflect *your* personality, but not in such a way that it indicates too idiosyncratic an approach. It is probably better to be formal in both language and presentation, as flippant or chatty expressions will not be well received.

- **Neatness.** Producing a well-presented, word-processed CV is very important. Use a laser quality printer and good-quality paper; avoid poor-quality photocopying at all costs.

- **Layout.** Use headings for different aspects, such as personal details, education, etc. Emphasise words (e.g. with capitals, bold, italics or underlining) sparingly and with the primary aim of making the structure clearer. Remember that careful use of white space is important in design.

Box 8.1 The structure and components of a typical CV and covering letter

There is no right or wrong way to write a CV, and no single format applies. It is probably best to avoid software templates and CV 'wizards' as they can create a bland, standardised result, rather than something that demonstrates your individuality. It is important to focus on the relevant issues, keeping your writing concise, well-structured and well-presented.

You should include the following, with appropriate sub-headings, generally in the order given below:

1. **Personal details.** This section *must* include your full name and date of birth, your address (both home and term-time, with dates, if appropriate) and a contact telephone number at each address. If you have an email account, you might also include this. You need only mention sex if your name could be either male or female.

2. **Education.** Choose either chronological order, or reverse chronological order and make sure you take the same approach in all other sections. Give educational institutions and dates (month, year) and provide more detail for your degree course than for your previous education. Remember to mention any prizes, scholarships or other academic achievements. Include your overall mark for the most recent year of your course, if it seems appropriate. Make sure you explain any gap years.

3. **Work experience.** Include all temporary, part-time, full-time or voluntary jobs. Details include dates, employer, job title and major duties involved.

4. **Skills and personal qualities.** Tables 1.1 and 8.1 give examples of the aspects you might include. Emphasise your strengths, and tailor this section to the specific requirements of the post (the 'job description'): for example, the practical skills you have gained during your degree studies if the post is a biological one, but concentrate on generic transferable skills and personal qualities for other jobs. Provide supporting evidence for your statements in all cases.

5. **Interests and activities.** This is an opportunity to bring out the positive aspects of your personality, and explain their relevance to the post you are applying for. Aim to keep this section short, or it may seem that your social life is more important than your education and work experience. Include up to four separate items, and provide sufficient detail to highlight the *positive* aspects of your interests (e.g. positions of responsibility, working with others, communication, etc.). Use sections 4 and 5 to demonstrate that you have the necessary attributes to fulfil the major requirements of the post.

6. **Referees.** Include the names (and titles), job descriptions, full postal addresses, contact telephone numbers and email addresses of two referees (rarely, some employers may ask for three). It is usual to include your personal tutor or course leader at university (who, among other things, will verify your marks), plus another person – perhaps a current or former employer, or someone who runs a club or society and who knows your personal interests and activities. Unless you have kept in touch with a particular teacher since starting university, it is probably best to choose current contacts rather than those from your previous education.

Some other points to consider:

- Try to avoid jargon and over-complicated phrases in your CV: aim for direct, active words and phrases (see Box 17.1).
- Most employers will expect your CV to be word-processed (and spell-checked). Errors in style, grammar and presentation will count against you, so be sure to check through your final version (and ask a reliable person to second-check it for you).
- Aim for a *maximum* length of two pages, printed single-sided on A4 paper, using a 'formal' font (e.g. Times Roman or Arial) of no less than 12 point for the main text. Always print onto good-quality white paper. Avoid fussy use of colour, borders or fonts.
- Do not try to cram in too much detail. Use a clear and succinct approach with short sentences and lists to improve 'readability' and create structure. Remember that your aim is to catch the eye of your potential employer, who may have many applications to work through.
- It is polite to check that people are willing to act as a referee for you and to provide them with an up-to-date copy of your CV.

Your single-page covering letter should have four major components:

1. **Letterhead.** Include your contact details, the recipient's name and title (if known) and address, plus any job reference number.

2. **Introductory paragraph.** Explain who you are and state the post for which you are applying.

3. **Main message.** This is your opportunity to sell yourself to a potential employer, highlighting particular attributes and experience. Keep it to three or four sentences at the most and relate it to the particular skills and qualities demanded in the job or person specification.

4. **Concluding paragraph.** A brief statement that you look forward to hearing the outcome of your application is sufficient.

Finally, add either 'Yours sincerely' (where the recipient's name is known) or 'Yours faithfully' (in a letter beginning 'Dear Sir or Madam') and then end with your signature.

- **Grammar and proofreading.** Look at your CV carefully before you submit it, as sloppy errors give a very poor impression. Even if you use a spell-checker, some errors may creep in. Ask someone whom you regard as a reliable proofreader to comment on it (many tutors will do this, if asked in advance).
- **Relevance.** If you can, slant your CV towards the job description and the qualifications required (see below). Make sure you provide evidence to back up your assertions about skills, qualities and experience.
- **Accuracy and completeness.** Check that all your dates tally; otherwise, you will seem careless. It is better to be honest about your grades and (for example) a period of unemployment than to cover this up or omit details that an employer will want to know. They may be suspicious if you leave things out.

Adjusting your CV

You should fine-tune your CV for each submission. Employers frequently use a 'person specification' to define the skills and qualities demanded in a job, often under headings such as 'essential' and 'desirable'. This will help you decide whether to apply for a position and it assists the selection panel to filter the applicants. Highlight relevant qualifications as early in your CV as possible. Be selective – don't include every detail about yourself. Emphasise relevant parts and leave out irrelevant details, according to the job. Similarly, your letter of application is not merely a formal document but is also an opportunity for persuasion (Box 8.1). You can use it to state your ambitions and highlight particular qualifications and experience. However, do not go over the top – always keep the letter to a single page.

Creating a generic CV – as you may apply for several jobs, it is useful to construct a CV in electronic format (e.g. as a *Word* file) which includes *all* information of potential relevance. This can then be modified to fit each post. Having a prepared CV on file will reduce the work each time you apply, while modifying this will help you focus on relevant skills and attributes for the particular job.

KEY POINT A well-constructed and relevant CV will not necessarily guarantee you a job, but it may well get you onto the shortlist for interview. A poor-quality CV is a sure route to failure.

Sources for further study

Anon. (2000) *How to Write a Curriculum Vitae.* University of London Careers Service, London.

Anon. *Graduate Prospects Website.* Available: http://www.prospects.ac.uk Last accessed 01/09/15.

[UK Higher Education Careers Services, containing good examples of CVs. and cover letters]

Corfield, R. (2009) *Preparing the Perfect Job Application.* Kogan Page, London.

Jackson, A. and Geckeis, K. (2003) *How to Prepare your Curriculum Vitae.* McGraw-Hill, Chicago.

Purdue University. *Online Writing Lab: Writing the Curriculum vitae.* Available: https://owl.english.purdue.edu/owl/resource/641/01/ Last accessed 01/09/15.

STUDY EXERCISES

8.1 Evaluate your personal attributes. Using Table 8.1, list *five* qualities that you would use to best describe yourself, and cite the evidence you might give to a potential employer to convince them that this was the case. List *five* attributes you could develop, then indicate how you might do this.

8.2 Create a generic CV. Drawing on your school record of achievement, or any CV already prepared, e.g. for a part-time job, create a word-processed generic CV. Save the file in an appropriate (computer) folder and make a back-up copy. Print out a copy for filing.

Periodically update the word-processed version. If appropriate, save different versions to be used in different contexts (e.g. when applying for a vacation job).

8.3 Think about your future career and ask for advice. Make an appointment with one of the advisors in your university's careers service. Ask about career options for graduates with your intended degree, or determine what qualifications or module options might be appropriate for occupations that interest you.

Answers to these study exercises are available at www.pearsoned.co.uk/practicalskills.

Information technology and learning resources

The ability to find scientific information is a skill required for many exercises in your degree programme. You will need to research facts and published findings as part of writing essays, literature reviews and project introductions, and when adding to your lecture notes (p. 19) and revising for exams (p. 31). You must also learn how to follow scientific convention in citing source material as the authority for the statements you have made.

Sources of information

For essays and revision

You are unlikely to use scientific papers (primary sources, p. 61) for these purposes – books and reviews are much more readable. If a lecturer or tutor specifies a particular book, then it should not be difficult to find out where it is shelved in your library, using the computerised index system. Library staff will generally be happy to assist with any queries. If you want to find out which books your library holds on a specified topic, use the system's subject index. You will also be able to search by author or by key words.

There are two main systems used by libraries to classify books: the Dewey Decimal system and the Library of Congress system. Libraries differ in the way they employ these systems, especially by adding further numbers and letters after the standard classification marks, e.g. to signify shelving position or edition number. Enquire at your library for a full explanation of local usage.

The web is an ever-expanding resource for gathering both general and specific information (see Chapters 11 and 12). Sites fall into analogous categories to those in the printed literature: there are sites with original information, sites that review information and bibliographic sites. One problem is that websites may be frequently updated, so information present when you first looked may be altered or even absent when the site is next consulted. Further, very little of the information on the web has been monitored or refereed. Another disadvantage is that the site information may not state the origin of the material, who wrote it or when it was written. This is considered in more detail in Box 11.5.

For literature surveys and project work

You will probably need to consult the primary literature. If you are starting a new research project or writing a report from scratch, you can build up a core of relevant papers by using the following methods:

- **Asking around:** supervisors or their postgraduate students will almost certainly be able to supply you with a reference or two that will start you off.
- **Searching an online database,** either via the Internet (see Chapter 11) or on CD-ROM: these cover very wide areas and are a convenient way to start a reference collection, although a charge is often made for access and sending out a listing of the papers selected (your library may or may not pass this on to you).
- **Consulting the bibliography of other papers in your collection** – an important way of finding the key papers in your field. In effect, you are taking advantage of the fact that another researcher has already done all the hard work.

Browsing in a library – this may turn up interesting material, but remember the books on the shelves are those not currently out on loan. Almost by definition, the latter may be more up-to-date and useful. To find out a library's full holding of books in any subject area, you need to search its catalogue (normally available as an online database). An online search will also bring up electronic versions of textbooks (ebooks).

Example The book *The Selfish Gene* by Richard Dawkins (1976; Oxford University Press) is likely to be classified as follows:

Dewey Decimal system: 591.51

where	591	refers to zoology
	591.5	refers to ecology of animals
	591.51	refers to habits and behaviour patterns

Library of Congress system: QL751

where	Q	refers to science
	QL	refers to zoology
	QL75	refers to animal behaviour
	QL751	refers to general works and treatises.

Researching a new topic – reading reviews (secondary sources, p. 61) can provide you with a useful overview at the start of a new project.

Definitions

Abstracts – shortened versions of papers, often those read at scientific meetings. These may later appear as full papers.

Bibliography – a summary of the published work in a defined subject area.

ebook – a book published online in downloadable form.

ebrary – a commercial service offering ebooks and other online resources.

> **Definitions**
>
> **eJournal** – a journal published online, access consisting of articles structured in the same way as a paper-based journal. A valid username and password may be required for access.
>
> **Journal/periodical/serial** – any publication issued at regular intervals. In biosciences, usually containing papers (articles) describing original research findings and reviews of literature.
>
> **Monograph** – a specialised book covering a single topic.
>
> **Proceedings** – volume compiling written versions of papers read at a scientific meeting on a specific topic.
>
> **Review** – an article in which recent advances in a specific area are outlined and discussed.
>
> **The primary literature** – original research papers, published in specialist scientific periodicals.

> **Web resources** – your university library will provide you with access to a range of web-based databases and information systems. The library Web pages will list these and provide links, which may be worth bookmarking on your Web browser. Resources especially useful to bioscientists include:
>
> - *Web of Science*, including the *Science Citation Index*
> - *IngentaConnect*, including *Ingenta Medline*
> - *CSA Illumina*
> - *PubMed*
> - *ScienceDirect*
> - *Scopus*
> - *Ovid*, including *Cinahl* and *Medline*.
>
> Most of these electronic resources operate on a subscription basis and may require an 'Athens' username and password – for details of how to obtain these consult library staff or your library's website.

- **Referring to 'current awareness' online databases:** these are useful for keeping you up to date with current research; they usually provide a monthly listing of article details (title, authors, source, author address) arranged by subject and cross-referenced by subject and author. Current awareness databases cover a wider range of primary literature than could ever be available in any one library. Examples include: *Current Contents Connect* (ISI), the *Current Advances* series (Elsevier), *Biological Abstracts* and *Zoological Record Archive* (Thomson Scientific). Some online databases also offer a service whereby they will email registered users with updates based on saved search criteria. Consult library staff or your library website to see which of these databases and services are available to you.

- **Using the *Science Citation Index* (SCI):** this is a valuable means of exploring the published literature in a given field, because it lets you see who has cited a given paper; in effect, SCI allows you to move forward through the literature from an existing reference. The Index is available online via ISI *Web of Science*.

For specialised information

You may need to consult reference works, such as encyclopaedias, maps and books providing specialised information. Some of this is now available on the web, or online (consult your library's information service web pages). Three sources worth noting are:

- **The Handbook of Chemistry and Physics** (Haynes, 2014; online via CHEMnetBASE): the Chemical Rubber Company's publication (affectionately known as the 'Rubber Bible') giving all manner of physical constants, radioisotope half-lives, etc.

- **The Merck Index** (O'Neil *et al.*, 2013), which gives useful information about organic chemicals, e.g. solubility, whether poisonous, etc., now also available online.

- **The Geigy Scientific Tables** (8th edition), a series of six volumes (Lentner, 1981; 1984; 1986; 1990; 1992; Lentner *et al.*, 1982) provides a wide range of information centred on biochemistry, e.g. buffer formulae, properties of constituents of living matter.

Obtaining and organising research papers

Obtaining a copy

It is usually more convenient to have personal copies of key research articles in print or electronic format for direct consultation when working in a laboratory or writing. The simplest way of obtaining these is to photocopy the originals or download and/or print off copies online (e.g. as '.pdf' files). For academic purposes, this is normally acceptable within copyright law. If your library does not subscribe to the journal, it may be possible for them to borrow it from a nearby institute or obtain a copy via a national borrowing centre (an 'inter-library loan'). If the latter, you will have to fill in a form giving full bibliographic details of the paper and where it was cited, as well as signing a copyright clearance statement concerning your use of the copy. Alternatively, you might try emailing the communicating author and requesting an electronic copy ('.pdf' file) of the article.

Copyright law – In Europe, copyright regulations were harmonised in 1993 (Directive 93/98/EEC) to allow literary copyright for 70 years after the death of an author and typographical copyright for 25 years after publication. This was implemented in the UK in 1996, where, in addition, the Copyright, Designs and Patents Act (1988) allows the Copyright Licensing Agency to license institutions so that lecturers, students and researchers may take copies for teaching and personal research purposes – no more than a single article per journal issue, one chapter of a book, or extracts to a total of 5% of a book.

Storing printed research papers – these can easily be kept in alphabetical order within filing boxes or drawers, but if your collection is likely to grow large, it will need to be refiled as it outgrows the storage space. You may therefore wish to add an 'accession number' to the record you keep in your database, and file the papers in sequence according to this, as they accumulate. New filing space is only required at the end and you can use the accession numbers to form the basis of a simple cross-referencing system.

Using commercial bibliographic database software to organise your references – for those with large numbers of references in their collection, and who may wish to produce lists of selected references in particular format, e.g. for inclusion in a project report or journal paper, systems like *EndNote*, *Reference Manager* or *ProCite* can reward the investment of time and money required to create a personal reference catalogue. Appropriate bibliographic data must first be entered into fields within a database (some versions assist you to search online databases and upload data from these). The database can then be searched and used to create customised lists of selected references in appropriate citation styles.

Organising papers

Although the number of papers you accumulate in electronic and/or printed formats may be small to start with, it is worth putting some thought into their storage and indexing before your collection becomes disorganised and unmanageable. Few things are more frustrating than not being able to lay your hands on a vital piece of information, and this can seriously disrupt your flow when writing or revising.

Indexing your references

Whether you have obtained a printed copy, have stored downloaded files electronically, or have simply noted the bibliographic details of a reference, you will need to index each resource. This is valuable for the following reasons:

- You will probably need the bibliographic information for creating a reference list for an assignment or report.
- If the index also has database features, this can be useful, allowing you to search for key words or authors.
- If you include an 'accession number' and if you then file printed material sequentially according to this number, it will help you to find the hard copy.
- Depending on the indexing system used, you can add comments about the reference that may be useful at a later time, e.g. when writing an introduction or conclusion.

The simplest way to create an index system is to put the details on reference cards, but database software can be more convenient and faster to sort once the bibliographic information has been entered. If you do not feel that commercial software is appropriate for your needs, consider using a word processor or spreadsheet; their rudimentary database sorting functions (see Chapters 13 and 14) may be all that you require.

If you are likely to store lots of references and other electronic resources digitally, then you should consider carefully how this information is kept, for example by choosing file names that indicate what the file contains and that will facilitate sorting.

Making citations in text

It is particularly important to cite all sources of information used in your work – this will demonstrate to assessors that you have used appropriate source material and it will also avoid any possibility of plagiarism (using another person's work without acknowledgement, p. 61).

There are two main ways of citing articles and creating a bibliography (also referred to as 'references' or 'literature cited').

The Harvard system

For each citation, the author name(s) and the date of publication are given at the relevant point in the text. The bibliography is organised alphabetically, and by date of publication for papers with the same authors. Formats normally adopted are, for example, 'Smith and Jones (2005) stated that . . .' or 'it has been shown that . . . (Smith and Jones, 2005)'. Lists of references within parentheses are separated by semicolons, e.g. '(Smith and Jones, 2005; Jones and Smith, 2009)', normally in order of date of publication. To avoid repetition within the same paragraph, an approach such as 'the investigations of Smith and Jones indicated that . . .' could be used following an initial citation of the paper. Where there are more than two authors it is usual to write 'et al.' (or *et al.* if an italic font is available); this stands for the Latin *et alia* meaning 'and others'. If citing more than one paper with the same authors, put, for example, 'Smith and Jones (2005; 2011)' and if papers by a given set of authors appeared in the same year, letter them (e.g. Smith and Jones, 2012a; 2012b).

The numerical or Vancouver system

Papers are cited via a superscript or bracketed reference number inserted at the appropriate point. Normal format would be, for example: 'DNA sequences[4,5] have shown that . . .' or 'Jones [55,82] has claimed that . . .'. Repeated citations use the number from the first citation. In the true numerical method (e.g. as in *Nature*), numbers are allocated by order of citation in the text, but in the alpha-numerical method (e.g. the *Annual Review* series), the references are first ordered alphabetically in the bibliography, then numbered, and it is this number which is used in the text. Note that with this latter method, adding or removing references is tedious, so the numbering should be done only when the text has been finalised.

KEY POINT   The main advantages of the Harvard system are that the reader might recognise the paper being referred to and that it is easily expanded if extra references are added. The main advantages of the Vancouver system are that it aids text flow and reduces length.

How to list your citations in a bibliography

Whichever citation method is used in the text, comprehensive details are required for the bibliography so that the reader has enough information to find the reference easily. Citations should be listed in alphabetical order with the priority: first author, subsequent author(s), date. Unfortunately, in terms of punctuation and layout, there are almost as many ways of citing papers as there are journals. Your department may specify an exact format for project work; if not, decide on a style and be consistent – if you do not pay attention to the details of citation you may lose marks. Take special care with the following aspects:

- **Authors and editors:** give details of *all* authors and editors in your bibliography, even if given as *et al.* in the text.
- **Abbreviations for journals:** while there are standard abbreviations for the titles of journals (consult library staff), it is a good idea to give the whole title, if possible.
- **Books:** the edition should always be specified as contents may change between editions. Add, for example, '(5th edition)' after the title of the book. You may be asked to give the International Standard Book Number (ISBN), a unique reference number for each book published.
- **Unsigned articles,** e.g. unattributed newspaper articles and instruction manuals: refer to the author(s) in text and bibliography as 'Anon.'.
- **Websites:** there is no widely accepted format at present. You should follow departmental guidelines if these are provided, but, if these are

Thesis:

Smith, A. B. (2012). *Investigations on my Favourite Topic*. PhD thesis, University of Life, Fulchester.

Website:

Jones, A.B. (2011). *My Webpage on My Work in 2011*. Available: http://www.jonesinfo. co.uk/work2011.

Last accessed: 01/04/2012.

Note: if your references are handwritten, you should indicate italics by underlining text and/or numbers.

Finding dates on websites quickly – when visiting a particular page, you can find occurrences of year dates beginning '200' by pressing the *Control* and *F* keys together, entering 200 in the *Find* window that appears, then carrying out a search using the *Find next* command.

Adding web references in other systems – if you are using a different referencing system than Harvard, consult Pears and Shields (2010) or McMillan and Weyers (2012) for further advice on how to cite websites in these systems.

not available, we suggest providing author name(s) and date in the text when using the Harvard system (e.g. Hacker, 2010), while in the bibliography giving the URL details in the following format: Hacker, A. (2010) *University of Anytown Homepage on Aardvarks*. Available: http://www.myserver.ac.uk/homepage. Last accessed: 01/04/2012. In this example, the web page was constructed in 2010, but accessed in April 2012. If no author is identifiable, cite the sponsoring body (e.g. University of Anytown, 2011), and if there is no author or sponsoring body, write Anon.' for 'anonymous', e.g. Anon. (2011), and use Anon, as the 'author' in the bibliography. If the web pages are undated, *either* use the 'last accessed' date for citation and put no date after the author name(s) in the reference list, *or* cite as 'no date' (e.g. Hacker, no date) and leave out a date after the author name(s) in the reference list – you should be consistent whichever option you choose.

- **Unread articles:** you may be forced to refer to a paper via another without having seen it. If possible, refer to another authority who has cited the paper, e.g. '. . . Jones (1990), cited in Smith (2011), claimed that . . .'. Alternatively, you could denote such references in the bibliography by an asterisk and add a short note to explain at the start of the reference list.

- **Personal communications:** information received in a letter, seminar or conversation can be referred to in the text as, for example, '. . . (Smith, pers. comm.)'. These citations are not generally listed in the bibliography of papers, though in a thesis you could give a list of personal communicants and their addresses.

- **Online material:** some papers and articles are published solely online and others are available online ahead of publication in printed form. The item may be given a digital object identifier (DOI), allowing it to be cited and potentially tracked before and after it is allocated to a printed issue (see http://www.doi.org/). DOIs also allow for web page redirection by a central agency, and *CrossRef* (http://www.crossref.org/) is the official DOI registration organisation for scholarly and professional publications. DOIs can be used as 'live' hyperlinks in online articles, or cited in place of the volume and page numbers for the article, with the remainder of the details cited in the usual fashion, e.g. 'Smith. A. and Jones B. (2011) Our latest important research in the form of a web-published article. *Online Biosciences* 8/2011 (p. 781). Published online: 26 March 2011. DOI: 10.1083/mabi.200680019'.

Text references and sources for further study

Anon. *CHEMnetBASE*. Available: http://www. chemnetbase.com
Last accessed 01/09/15.
[Access to the *Handbook of Chemistry and Physics*]

Anon. *ISI Web of Science*. Available: http://wos.mimas.ac.uk
Last accessed 22/5/12.
[Requires Athens password.]

Grix, J. and Watkins, G. (2010) *Information Skills: Finding and Using the Right Resources*. Palgrave Macmillan, Basingstoke.

Haynes, W.M. (ed.) (2014) *CRC Handbook of Chemistry and Physics,* 95th edn. CRC Press, Boca Raton.

Lentner, C. (ed.) (1981) *Geigy Scientific Tables*, Vol. 1: *Units of Measurement, Body Fluids, Composition of the Body. Nutrition*, 8th edn. Ciba-Geigy, Basel.

Lentner, C. (ed.) (1984) *Geigy Scientific Tables,* Vol. 3: *Physical Chemistry, Composition of Blood, Hematology, Somatometric Data,* 8th edn. Ciba-Geigy, Basel.

Lentner, C. (ed.) (1986) *Geigy Scientific Tables,* Vol. 4: *Biochemistry, Metabolism of Xenobiotics, Inborn Errors of Metabolism, Pharmacogenetics, Ecogenetics,* 8th edn. Ciba-Geigy, Basel.

Lentner, C. (ed.) (1990) *Geigy Scientific Tables,* Vol. 5: *Heart and Circulation,* 8th edn. Ciba-Geigy, Basel.

Lentner, C. (ed.) (1992) *Geigy Scientific Tables,* Vol. 6: *Bacteria, Fungi, Protozoa, Helminths,* 8th edn. Ciba-Geigy, Basel.

Lentner, C., Diem, K. and Seldrup, J. (eds) (1982) *Geigy Scientific Tables,* Vol. 2: *Introduction to Statistics, Statistical Tables, Mathematical Formulae,* 8th edn. Ciba-Geigy, Basel.

McMillan, K.M. and Weyers, J.D.B. (2012) *The Study Skills Book* 3rd edn. Pearson Education, Harlow.

Neville, C. (2009) *The Complete Guide to Referencing and Avoiding Plagiarism.* Open University Press, Maidenhead.

O'Neil, M.J., Heckelman, P.E., Koch, C.B. and Roman, K.J. (2013) *Merck Index: An Encyclopedia of Chemicals, Drugs, and Biologicals,* 15th edn. Merck & Co. Inc., Whitehouse Station.

Pears, R. and Shields, G. (2010) *Cite them Right: the Essential Referencing Guide,* 8th edn. Palgrave Macmillan, Basingstoke. Also available as an online resource, see: http://www.citethemrightonline.com/

STUDY EXERCISES

9.1 Test your library skills. This exercise relies on the fact that most university-level libraries serving bioscience departments will take the scientific journal *Nature.* To help you to answer these questions, it may be beneficial to attend a library induction session if you have not done so already. Alternatively, the library's help or enquiry desks may be able to assist you if you are having problems.

(a) First, find out and provide the name of the classification system that your university uses for cataloguing its books and periodicals.

(b) Using your library's cataloguing system (online, preferably), find out the appropriate local classification number for the journal *Nature.*

(c) Where is *Nature* shelved in your library? (Your answer need refer only to most recent issues if some have been archived.)

(d) What is the exact title of the landmark papers in the following two volumes? (i) *Nature* **171,** 737–738 (1953); (ii) *Nature* **408,** 796–815 (2000).

9.2 Explore different methods of citing references. Go to your library and seek out the journal area for biology or life sciences. Choose *three* different journals in your subject area and from a recent edition write down how they would print a typical citation for a multi-author journal paper in the 'references' or 'literature cited' section.

Where used, indicate italicised text with normal underline and bold text with wavy underline. Pay attention to punctuation. Compare these methods with each other, with the methods recommended on pp. 55–6 of this book and with the recommendations your department or your course handbook makes. Are they all the same?

9.3 Make website citations. Use a search engine (pp. 72–3) to find an informative website that covers each of the following:

(a) The use of SI units.

(b) Information about the bacterial culture collection of the Pasteur Institute (l'Institut Pasteur).

(c) The different types of blood cells.

Indicate how you would cite each website at the end of an essay (follow your department's guidelines or use those in this chapter).

9.4 Compare the Harvard and Vancouver methods of citation. Pair up with a partner in your class. Each person should then pick one of the two main methods of citation and consider its pros and cons independently. Meet together and compare your lists. Given the choice, which method would you choose for: (a) a handwritten essay; (b) a word-processed review; (c) an article in an academic journal, and why?

Answers to these study exercises are available at www.pearsoned.co.uk/practicalskills.

10 Evaluating information

Example A web search for the letters 'DNA' (e.g. using the search engine Google, p. 73), will reveal that this acronym appears in several hundred million websites. Not all of these deal with deoxyribonucleic acid – the listed websites include: *Doctors' Net Access*, a web resource for American physicians; the website of the *Dermatology Nurses Association*; and that of *Delta Nu Alpha*, an international organisation for those working in the professions. These examples are easy to identify as irrelevant to research on 'DNA, the molecule', but considering the many websites that cover this topic, which ones might contain the exact information you seek? How valid is the information? Is it biased towards one viewpoint or hypothesis? Does it represent current mainstream thinking on the topic? These are some of the issues that an evaluation of the information sources might deal with.

Definition

Plagiarism – the unacknowledged use of another's work as if it were one's own. In this definition, the concept of 'work' includes ideas, writing, data or inventions and not simply words; and the notion of 'use' does not only mean copy 'word-for-word', but also 'in substance' (i.e. a copy of the ideas involved). Use of another's work is acceptable only if you acknowledge the source (Box 10.1), and you must use quotation marks if you are using the words of another person.

Checking the reliability of information, assessing the relative value of different ideas and thinking critically are skills essential to the scientific approach (p. 200). You will need to develop your abilities to evaluate information in this way because:

- **you will be faced with many sources of information,** from which you will need to select the most appropriate material;
- **you may come across two conflicting sources of evidence** and may have to decide which is the more reliable;
- **the accuracy and validity of a specific fact may be vital to your work;**
- **you may doubt the quality of information** from a particular source;
- **you may wish to check the original source** because you are not sure whether someone else is quoting it correctly.

KEY POINT Evaluating information and thinking critically are regarded as higher order academic skills. The ability to think deeply in this way is greatly valued in the biosciences and will consequently be assessed in coursework and exam questions (see Chapter 7).

The process of evaluating and using information can be broken down into four stages:

1. **Selecting and obtaining material.** How to find sources is covered in Chapter 9. Printed books and journals are important, but if you identify a source of this kind there may be delays in borrowing it or obtaining a photocopy. If the book or journal is available online, then downloading or printing sections or papers will be more convenient and faster. The Internet is often a first port of call if you wish to find something out quickly. However, for many websites, it can be difficult to verify the authenticity of the information given (see Box 11.5).

2. **Assessing the content.** You will need to understand fully what has been written, including any technical terms and jargon used. Establish the relevance of the information to your needs and assure yourself that the data or conclusions have been presented in an unbiased way.

3. **Modifying the information.** In order to use the information, you may need to alter it to suit your needs. This may require you to make comparisons, interpret or summarise. Some sources may require translation. Some data may require mathematical transformation before they are useful. There is a chance of error in any of these processes and also a risk of plagiarism (see Box 10.1).

4. **Analysis.** This may be your own interpretation of the information presented, or an examination of the way the original author has used the information.

KEY POINT Advances in communications and information technology mean that we can now access almost limitless knowledge. Consequently, the ability to evaluate information has become an extremely important skill.

Box 10.1 How to avoid plagiarism and copyright infringement

Plagiarism is defined on p. 59. Examples of plagiarism include the following

- Copying the work of a fellow student (past or present) and passing it off as your own.
- Using 'essay-writing services', such as those on offer on certain websites.
- Copying text or images from a source (book, journal article or website, for instance) and using this within your own work without acknowledgement.
- Quoting others' words without indicating who wrote or said them.
- Copying ideas and concepts from a source without acknowledgement, even if you paraphrase them.

Most students would accept that some of the above can only be described as cheating. However, many students, especially at the start of their studies, are unaware of the academic rule that they must *always* acknowledge the originators of information, ideas and concepts, and that not doing so is regarded as a form of academic dishonesty. If you adopt the appropriate conventions that avoid such accusations, you will achieve higher marks for your work as it will fulfil the marker's expectations for academic writing.

Universities have a range of mechanisms for identifying plagiarism, from employing experienced and vigilant coursework markers and external examiners to analysing students' work using sophisticated software programs. Plagiarism is always punished severely when detected. Penalties may include awarding a mark of zero to all involved – both the copier(s) and the person whose work has been copied (who is regarded as complicit in the crime). Further disciplinary measures may be taken in some instances. In severe cases, such as copying substantive parts of another's work within a thesis, a student may be dismissed from the university.

If you wish to avoid being accused of plagiarism, the remedies are relatively simple:

1. **Make sure the work you present is always your own.** If you have been studying alongside a colleague, or have been discussing how to tackle a particular problem with your peers, make sure you write on your own when working on your assignments.

2. **Never, ever be tempted to 'cut and paste'** from websites or online sources such as word-processed handouts. Read these carefully, decide what the important points are, express these *in your own words* and *provide literature citations to the original sources* (see Chapter 9). In some cases, further investigations may be required to find out details of the original sources. The lecturer's reading list or a book's references may help you here.

3. **Take care when note-taking.** If you decide to copy text word-for-word, make sure you show this clearly in your notes with quotation marks. If you decide to make your own notes based on a source, make sure these are paraphrased, rather than copy phrases from the text. In both cases, write down full details of the source at the appropriate point in your notes.

4. **Place appropriate citations throughout your text, where required.** If you are unsure about when to do this, study reviews and articles in your subject area (see also Chapter 9).

5. **Understand the differences between quoting, summarising and paraphrasing (p. 20).** Apply the relevant conventions for text format: show clearly where you are quoting directly from a source by using quotation marks, and *always* cite your source(s). Try not to rely too much on quotation as this may be regarded as a lack of ability to synthesise your own ideas.

Copyright issues are often associated with plagiarism, and refer to the right to publish (and hence copy) original material, such as text, images and music. Copyright material is indicated by a symbol © and a date (see, for example, p. iv of this book). Literary copyright is the aspect most relevant to students in their academic studies. UK Copyright Law protects authors' rights for life and gives their estates rights for a further 70 years. Publishers have 'typographical copyright' that lasts for 25 years. This means that it is illegal to photocopy, scan or print out copyright material unless you have permission, or unless your copying is limited to an extent that could be considered 'fair dealing'. For educational purposes – private study or research – in a scientific context, this generally means:

- no more than 5% in total of a work;
- one chapter of a book;
- one article per volume of an academic journal;
- 20% of a short book;
- one separate illustration or map.

You may only take one copy within the above limits, may not copy for others, and may not exceed these amounts *even if you own a copy of the original*. These rules also apply to web-based materials, but sometimes you will find sites where the copyright is waived. Some copying may be licensed; you should consult your library's website or helpdesk to see whether it has access to licensed material. Up-to-date copyright information is generally posted close to library and departmental photocopiers.

Evaluating sources of information

One way of assessing the reliability of a piece of scientific information is to think about how it was obtained in the first place. Essentially, facts and ideas originate from someone's research or scholarship, whether they are numerical data, descriptions, concepts or interpretations. Sources are divided into two main types:

1. **Primary sources** – those in which ideas and data are first communicated. The primary literature is generally published in the form of 'papers' (articles) in journals, whether printed or online. These are usually refereed by experts in the academic peer group of the author, and they will check the accuracy and originality of the work and report their opinions back to the editors of the journal. This peer review system helps to maintain reliability, but it is not perfect. Books and, more rarely, websites and articles in magazines and newspapers, can also be primary sources but this depends on the nature of the information published rather than the medium. These sources are not formally refereed, although they may be read by editors and lawyers to check for errors and unsubstantiated or libellous allegations.

2. **Secondary sources** – those which quote, adapt, interpret, translate, develop or otherwise use information drawn from primary sources. It is the act of quoting or paraphrasing that makes the source secondary, rather than the medium. Reviews are examples of secondary scientific sources, and books and magazine articles are often of this type.

When information is modified for use in a secondary source, alterations are likely to occur, whether intentional or unintentional. Most authors do not deliberately set out to change the meaning of the primary source, but they may unwittingly do so, e.g. in changing text to avoid plagiarism or by oversimplification. Others may consciously or unconsciously exert bias in their reporting, for example, by quoting evidence that supports only one side of a debate. Therefore, the closer you can get to the primary source, the more reliable the information is likely to be. On the other hand, modification while creating a secondary source could involve correcting errors, or synthesising ideas and content from multiple sources.

Authorship and provenance

Clearly, much depends on who is writing the source and on what basis (e.g. who paid them?). Consequently, an important way of assessing sources is to investigate the ownership and provenance of the work (who and where it originated from, and why).

Can you identify who wrote the information? If it is signed or there is a 'by-line' showing who wrote it, you might be able to make a judgement on the quality of what you are reading. This may be a simple decision, if you know or can assume that the writer is an authority in the area; otherwise a little research might help (for example, by putting the name into a web search engine). Of course, just because Professor X thinks something does not make it true. However, if you know that this opinion is backed up by years of research and experience, then you might take it a little more seriously than the thoughts of a school pupil. If an author is not cited, effectively nobody is taking responsibility for the content. Could there be a reason for this?

Distinguishing between primary and secondary sources – try the 'IMRaD test'. Many primary sources contain information in the order: Introduction, Materials and Methods, Results and Discussion. If you see this format, and particularly if *data* from an experiment, study or observation are presented, then you are probably reading a primary source.

Example If a journalist wrote an article about a new 'flesh-eating bug' for the *New York Times* that was based on an article in the *British Medical Journal*, the *New York Times* article would be the secondary source, while the *British Medical Journal* article would be the primary source.

Taking account of the changing nature of websites and wikis – by their very nature, these sources may change. This means that it is important to quote accurately from them and to give a 'last accessed' date when citing (see p. 57).

Finding out about authors and provenance – these pieces of information are easy to find in most printed sources and may even be presented just below the title, for convenience. In the case of the web, it may not be so easy to find what you want. Relevant clues can be obtained from 'homepage' links and the header, body and footer information. For example, the domain may be useful, while the use of the tilde symbol (~) in a URL usually indicates a personal, rather than an institutional, website.

Table 10.1 Checklist for assessing information
in science. How reliable is the information you
have been reading? The more 'yes' answers you
can give below, the more trustworthy you can
assume it to be.

Assessing sources
☐ Can you identify the author's name?
☐ Can you determine what relevant qualifications he/she holds?
☐ Can you say who employs the author?
☐ Do you know who paid for the work to be done?
☐ Is this a primary or secondary source?
☐ Is the content original or derived from another source?

Evaluating information
☐ Have you checked a range of sources?
☐ Is the information supported by relevant literature citation?
☐ Is the age of the source likely to be important regarding the accuracy of the information?
☐ Have you focused on the substance of the information presented rather than its packaging?
☐ Is the information fact or opinion?
☐ Have you checked for any logical fallacies in the arguments?
☐ Does the language used indicate anything about the status of the information?
☐ Have the errors associated with any numbers been taken into account?
☐ Have the data been analysed using appropriate statistics?
☐ Are any graphs constructed fairly?

Is the author's place of work cited? This might tell you whether the facts or opinions given are based on an academic study. Is there a company with a vested interest behind the content? If the author works for a public body, there may be publication rules to follow and they may even have to submit their work to a publications committee before it is disseminated.

Evaluating facts and ideas

However reliable the source of a piece of information seems to be, it is probably a good idea to retain a slight degree of scepticism about the facts or ideas involved. Even information from impeccable primary sources may not be perfect – different approaches can give different outcomes, and interpretations can change with time and with further advances in knowledge. Table 10.1 provides a checklist for evaluating sources.

Critically examining facts and ideas is a complex task depending on the particular issues involved, and a number of different general approaches can be applied. You will need to decide which of the following general tips are useful in your specific case:

- **Make cross-referencing checks triangulation** – look at more than one source and compare what is said in each. The cross-referenced sources should be as independent as possible (for example, do not compare a primary source together with a secondary review based on it). If you find that all the sources give a similar picture, then you can be more confident about the reliability of the information.

- **Look at the extent and quality of citations (Chapter 9)** – if references are quoted, these indicate that a certain amount of research has been carried out beforehand, and that the ideas or results are based on genuine scholarship. If you are doubtful about the quality of the work, these references might be worth looking at. How up to date are they? Do they cite independent work, or is the author exclusively quoting their own work, or solely the work of one person?

- **Consider the age of the source** – the fact that a source is old is not necessarily a barrier to truth, but ideas and facts may have altered since the date of publication, and methods may have improved. Can you trace changes through time in the sources available to you? What key events or publications have forced any changes in the conclusions?

- **Try to distinguish fact from opinion** – to what extent has the author supported a given viewpoint? Have relevant facts been quoted, via literature citations or the author's own researches? Are numerical data used to substantiate the points used? Are these reliable and can you verify the information, for example, by looking at the sources cited? Might the author have a reason for putting forward biased evidence to support a personal opinion?

- **Analyse the language used** – words and their use can be very revealing. Subjective wording might indicate a personal opinion rather than an objective conclusion. Propaganda and personal bias might be indicated by absolute terms, such as 'everyone knows . . .'; 'It can be guaranteed that...' or a seemingly one-sided consideration of the evidence. How carefully has the author considered the topic? A less studious approach might be indicated by exaggeration, ambiguity or the use of 'journalese' and slang. Always remember, however, that content should be judged above presentation.

- **Look closely at any numbers** – if the information you are looking at is numerical in form, have statistical errors been taken into consideration, and, where appropriate, quantified? If so, does this help you arrive at a conclusion about how genuine the differences are between important values?
- **Think carefully about any hypothesis-testing statistics used** – are the methods appropriate? Are the underlying hypotheses the right ones? Have the results of any tests been interpreted correctly in arriving at the conclusion? To deal with these matters, you will need at least a basic understanding of the 'statistical approach' and of commonly used techniques (see Chapters 73 and 74).

Interpreting data

Numerical data

Information presented in public, whether as a written publication or a spoken presentation, is rarely in the same form as it was when first obtained. Chapters 30 and 42 deal with processes in which data are recorded and manipulated – take particular care over percentages and proportions (p. 509), while Chapter 73 describes the standard descriptive statistics used to 'encapsulate' large data sets. Chapter 72 covers some relevant mathematical techniques. Sampling (essentially, obtaining representative measurements) is at the heart of many observational and experimental approaches in the biosciences (see Chapters 30 and 32), and analysis of samples is a key component of hypothesis-testing statistics (Chapter 74). Understanding these topics and carrying out the associated study exercises will help you improve your ability to interpret numerical data.

Graphs

Frequently, understanding and analysis in science depend on your ability to interpret data presented in graphical form. Sometimes, graphs may mislead (Chapter 70). This may be unwitting, as in an unconscious effort to favour a 'pet' hypothesis of the author. Graphs may be used to 'sell' a product, e.g. in advertising, or to favour a viewpoint as, perhaps, in politics. Experience in drawing and interpreting graphs will help you spot these flawed presentations, and understanding how graphs can be erroneously presented (Box 70.3) will help you avoid the same pitfalls.

Tables

Tables, especially large ones, can appear as a mass of numbers and thus be more daunting at first sight than graphs. In essence, however, most tables are simpler than most graphs. The construction of tables is dealt with in Chapter 71.

Critical thinking

Critical thinking involves the application of logic to a problem, or case study issue. It requires a wide range of skills. Key processes involved include: acquiring and processing information; creating appropriate hypotheses and formulating conclusions; and acting on the conclusions towards a specific objective.

Analysing a graph – this process can be split into six phases.

1. Considering the context and purpose of the graph.
2. Recognising the type of presentation and examining the axes.
3. Looking closely at the scale on each axis.
4. Examining the data presented (e.g. data points, symbols, curves).
5. Considering errors and statistics associated with the graph.
6. Reaching conclusions based on the above.

These processes are amplified in Chapter 70.

Analysing a table – as with analysing a graph, this process can be split into six phases.

1. Considering the context and purpose.
2. Examining the subheadings to see what information is contained in the rows and columns.
3. Considering the units used and checking any footnotes.
4. Comparing the data values across rows and/or down columns, looking for patterns, trends and unusual values.
5. Taking into account any statistics presented.
6. Reaching conclusions based on the above.

These processes are covered in more detail in Chapter 71.

Learning from examples – as your lecturers introduce you to case studies, you will see how life scientists have applied critical thinking to understand the nature of biomolecules, cells and organisms. Some of your laboratory sessions may mimic the processes involved – observation, hypothesis, experimental design, data gathering, analysis and formulating a conclusion (see Chapter 32). These skills and approaches can be applied in your degree programme, e.g. when writing about a biological issue or carrying out a research project.

'You can prove anything with statistics'– leaving aside the issue that statistical methods deal with probability, not certainty (Chapter 74), it is often possible to analyse and present data in such a way that they support one chosen argument or hypothesis rather than another. Detecting a bias of this kind can be difficult, but the critical thinking skills involved are essential for all life scientists.

Good writing requires good logic – understanding the logic behind what you want to write is a prerequisite for creating high-quality text. Creating a plan for your writing (Chapter 17) will help you to both recognise and organise what you want to say.

> **KEY POINT** Critical thinking needs reliable knowledge, but it requires you to use this appropriately to analyse a problem. It can be contrasted with rote learning – where you might memorise facts without an explicit purpose other than building your knowledge base.

Critical thinking is particularly important in the biosciences, because the subject deals with complex and dynamic systems. These can be difficult to understand for several reasons:

- **they are often multi-faceted, involving many interactions;**
- **it can be difficult to alter one variable in an experiment without producing confounding variables** (see p. 202);
- **many variables may be unmeasured or immeasurable;**
- **heterogeneity (variability) is encountered at all scales** from the molecular scale to the whole organism;
- **perturbation of the system can lead to unexpected ('counter-intuitive') results.**

As a result, conclusions in biological research are seldom clear-cut. Critical thinking allows you to arrive at the most probable conclusion from the results at hand; however, it also involves acknowledging that other conclusions might be possible. It allows you to weigh up these possibilities and find a working hypothesis or explanation, but also to understand that your conclusions are essentially dynamic and might alter when new facts are known. Hypothesis-testing with statistics (Chapter 74) is an important adjunct to critical thinking because it demands the formulation of simple hypotheses and provides rational reasons for making conclusions.

Recognising fallacies in arguments is an important aspect of critical thinking. Philosophers and logicians recognise different forms of argument and many different fallacies in each form. Damer (2013) provides an overview of this wide-ranging and complex topic.

Explaining your thoughts

The context for your evaluation of the literature and the associated critical thinking will normally be an essay, report or similar piece of academic writing (Chapters 18–20). The skills involved in marshalling and explaining your thoughts are regarded as highly important in employment and research.

You may have a very specific remit, as defined in the instruction for the assignment (Chapter 18), or the topic may be open. In either case, your reading around the topic should result in an overarching position or argument you wish to put forward – this is sometimes termed the 'thesis' for your writing. You may be explaining an established viewpoint or creating an original perspective on a topic. For both situations, the same principles apply. When setting out your thoughts, you should:

- **make a clear statement of the issue being considered,** if necessary defining terms and boundaries;
- **consider the issue from different perspectives,** providing evidence for or against different propositions;

Examples
Inductive reasoning

1. The plants growing in a mining spoil tip (Site A) are different from those in nearby land subject to similar climatic conditions (Site B).
2. The spoil tip contains a metal Z known to be toxic to a range of organisms.
3. When metal Z is added to pots containing plants from both sites, the plants common on Site B but not on Site A die; those from Site A all survive.
4. High concentrations of metal Z in the spoil have resulted in the change in species mix owing to differential sensitivity to the metal.

Possible flaw in logic: the concentration of metal Z used at step 3 might have been much higher than that found in the spoil and another potentially toxic metal that is present might not have been tested.

Deductive reasoning

1. A bacterial cell has no membrane-bound nucleus.
2. This cell has no membrane-bound nucleus.
3. Therefore this cell is a bacterium.

Possible flaw in logic: other types of cell may have no membrane-bound nucleus; for example, red blood cells.

- **cite sources of evidence and ideas** that you are evaluating (Chapter 9);
- **ensure your viewpoint is logical and internally consistent;**
- **structure your writing appropriately;** for example, by first considering evidence for a particular view and then evidence against it; or by considering the development of evidence through time;
- **use an academic style of writing,** avoiding personal statements (Box 17.2);
- **arrive at a conclusion,** even if this is that the evidence is not sufficient to allow firm statements to be made.

In the sciences, the norm is to use inductive reasoning – that is, to state observed facts and assumptions at the outset, then draw a logical conclusion based on these. The alternative, deductive reasoning, starts from a general statement, premise or law which is held to be true and then reaches a conclusion by considering facts logically. You should look for these types of argument in texts and papers, and also think about possible flaws in such arguments. Chapter 32 discusses further the use of hypotheses and experiments in biological science.

Text reference

Damer, T.E. (2013) *Attacking Faulty Reasoning: A Practical Guide to Fallacy-Free Arguments,* 7th edn. Wadsworth, Belmont, CA.

Sources for further study

Gilbert, F.S., MacGregor, P.K. and Barnard, C.J. (2011) *Asking Questions in Biology,* 4th edn. Prentice Hall, Harlow.

Smith, A. *Evaluation of Information Sources.* Available: http://www.vuw.ac.nz/staff/alastair_smith/evaln/evaln.htm
Last accessed 01/09/15.
[Part of the Web-based Information Quality Virtual Library giving details of a wide range of sources with criteria for evaluation.]

Van Gelder, T. *Critical Thinking on the Web.* Available: http://www.austhink.org/critical/
Last accessed 01/09/15.
[Includes a useful directory of web resources.]

STUDY EXERCISES

10.1 Distinguish between primary and secondary literature. Based on their titles and any research you can do in your library, determine whether the following journals are primary or secondary sources:

(a) *British Journal of Haematology*

(b) *Proceedings of the National Academy of Sciences*

(c) *Annual Review of Molecular Biology*

(d) *Essays in Biochemistry*

(e) *Microbiological Research*

(f) *Trends in Biotechnology*

(g) *The Lancet*

(h) *Biochemical Journal*

(i) *Scientific American*

(j) *Food Safety News*.

10.2 Consider a controversial issue from both sides. Select a current biological topic being discussed in the newspapers or other media. Relatively controversial issues such as 'genetically modified crops' or 'stem cell research' would be good examples. Next, write out a statement that you might use for a motion in a debate, such as 'Genetic modification of crops is a good thing' or 'Creating chimeric organisms is acceptable if it results in medical advances'. Then write at least five points in support of either side of the argument, which you should organise in tabular form. If you can find more than five points, add these to your table, but for each point that you add to one side, you should add one to the other side.

10.3 Analyse graphic presentations in the media. Many newspapers provide graphic presentations related to current issues, and graphs are frequently used in television news reports. Practise critical thinking skills by determining whether the graphs presented are a fair representation of the facts.

Answers to these study exercises are available at www.pearsoned.co.uk/practicalskills.

Definitions

Browser - a program to display web pages and other Internet resources.

FAQ - Frequently Asked Question; a file or web page giving information on common queries, sometimes used as a file extension (.faq).

FTP - File Transfer Protocol; a mechanism for downloading files.

URL - Uniform Resource Locator; the 'address' for web resources.

Academic use of ICT resources - a range of inappropriate activities will be identified in your university's rules for use of ICT systems. These may include: hacking, spamming, using another person's account, and copyright infringement, as well as broader aspects of behaviour covered by a code of conduct or student charter.

Understanding the technology - you do not need to understand the workings of the Internet to use it – most of it is invisible to the user. To ensure you obtain the right facilities, you may need to know some jargon, such as terms for the speed of data transfer (megabits) and the nature of internet addresses. Setting up a modem and/or local wireless network can be complex, but instructions are usually provided with the hardware. White and Downs (2014) and Gralla (2006) are useful texts if you wish to learn more about computing and the Internet.

Information and communication technology (ICT) is vital in the modern academic world and 'IT literacy' is a core skill for all bioscientists. This involves a wide range of computer-based skills, including:

- **Accessing web pages** using a 'browser' such as Internet Explorer, Firefox, Safari or Chrome.
- **Searching the web for useful information and resources** using a search engine such as Google, or a meta-search engine such as Dogpile.
- **Finding what you need using online databases,** such as library catalogues or complex websites, such as your university's homepage.
- **Downloading, storing and manipulating files.**
- **Communicating via the Internet.**
- **Using e-learning facilities** effectively.
- **Working with 'Office'-type programs and other software** (dealt with in detail in Chapters 12 and 13).

You will probably receive an introduction to your university's networked IT systems and you will be required to follow rules and regulations that are important for the operation of these systems. Whatever your level of experience with PCs and the Internet, you should also follow the basic guidelines shown in Box 11.1. Reminding yourself of these from time to time will reduce your chances of losing data.

The Internet as a global resource

The Internet is a complex network of computer networks; it is loosely organised and no one group organises it or owns it. Instead, many private organisations, universities and government organisations fund and operate discrete parts of it.

The web is the most popular application of the Internet. It allows easy links to information and files which may be located on networked computers across the world. The web enables you to access millions of 'homepages' or 'websites' – the initial point of reference with many individuals, institutions and companies. Besides text and images, these sites may contain 'hypertext links', highlighted words or phrases that take you to another Internet location via a single mouse click.

You can gain access to the Internet either through a network at your university, at most public libraries, at a commercial 'Internet cafe', or from home via a modem connected to a broadband or dial-up internet service provider (e.g. Virgin Media, BT or Sky).

> **KEY POINT**   Most material on the Internet has not been subject to peer review or vetting. Information obtained from the web or posted on newsgroups may be inaccurate, biased or spoof; do not assume that everything you read is true, or even legal.

Online communication

You will be allocated an email account by your university and should use this routinely for communicating with staff and fellow students, rather than using a personal account. You may be asked to use email to submit work as an

---

Box 11.1   Important guidelines for using PCs and networks

Hardware

- Do not drink or smoke around the computer.
- Try not to turn the computer off more than is necessary.
- Never turn off the electricity supply to the machine while in use.
- Switch off the computer and monitor when not in use (saves energy and avoids dangers of 'hijacking').
- Rest your eyes at frequent intervals if working for extended periods at a computer monitor. Consult Health and Safety Executive publications for up-to-date advice on working with display screens (http://www.hse.gov.uk/msd/dse).
- Never try to reformat the hard disk without the help of an expert.

CDs and USB drives

- Protect CDs when not in use by keeping them in holders or boxes.
- Label USB (Universal Serial Bus) drives with your name and return details and consider adding these to a file stored on the drive.
- Try not to touch the surface of CDs, and if they need cleaning, do so carefully with a clean cloth, avoiding scratching. If floppy disks are used, keep these away from sources of magnetism (e.g. speakers).
- Keep disks and USB drives away from moisture, excess heat or cold.
- Do not use disks from others, unless you first check them for viruses.
- Do not insert or remove a disk or USB drive when it is operating (drive light on). Close all files before removing a USB drive and use the *Safely Remove Hardware* feature.
- Try not to leave a disk or USB drive in the computer when you switch off.

File management

- Organise your files in an appropriate set of folders.

- Always use virus-checking programs on copied or imported files before running them.
- Make backups of all important files at frequent intervals (say, every 10 min), e.g. when using a word processor or spreadsheet.
- Periodically clear out redundant files.

Network rules

- Never attempt to 'hack' into other people's files.
- Do not give out any of your passwords to others. Change your password regularly. Make sure it is not a common word, is longer than 8 characters, and includes numerical characters and punctuation symbols, as well as upper and lower case letters.
- Never use network computers to access or provide financial or other personal information: spyware and Trojan programs may intercept your information.
- Never open email attachments without knowing where they came from; always virus-check attachments before opening.
- Remember to log out of the network when finished; others can access your files if you forget to log out.
- Be polite when sending email messages.
- Periodically reorganise your email folder(s). These rapidly become filled with acknowledgements and redundant messages that reduce server efficiency and take up your allocated filespace.
- Do not play games without approval – they can affect the operation of the system.
- If you are setting up your own network, e.g. in your flat, always install up-to-date firewall software, anti-spyware and anti-virus programs.

The Golden Rule – always make backup copies of important files and store them well away from your working copies. Ensure that the same accident cannot happen to both copies.

---

attachment, or you may be asked to use a 'digital drop-box' within the university's e-learning system (Box 11.2). When using email at university, follow conventions, including etiquette, carefully:

- **Check your email account regularly** (daily). Your tutors may wish to send urgent messages to you in this way.
- **Respond promptly to emails.** Even if you are just acknowledging receipt, it is polite to indicate that you have received and understood a message.

Box 11.2 Getting to grips with e-learning

Some key aspects of tackling e-learning are outlined below.

1. **Develop your basic IT skills, if required.** e-learning requires only basic IT skills, such as: use of keyboard and mouse; word processing; file management; browsing and searching. If you feel weak on any of these, seek out additional courses offered by the IT administration or your department.

2. **Visit your e-learning modules regularly.** You should try to get into a routine of doing this on a daily basis at a time that suits you. Staff will present up-to-date information (e.g. lecture room changes) via the 'announcements' section, may post information about assessments, or links to the assessments themselves, and you may wish to provide feedback or look at discussion boards and their threads.

3. **Participate.** e-learning requires an active approach.

   - At the start of each new course, spend some time getting to know what has been provided online to support your learning. As well as valuable resources, this may include crucial information such as learning outcomes (p. 30), dates of submission for coursework and weighting of marks for different elements of the course.

   - If you are allowed to download lecture notes (e.g. in the form of *PowerPoint* presentations), do not think that simply reading through these will be an adequate substitute for attending lectures and making further notes (see p. 17).

   - Do not be tempted to 'lurk' on discussion boards: take part. Ask questions; start new threads; answer points raised by others if you can.

   - Try to gain as much as you can from formative online assessments (p. 30). If these include feedback on your answers, make sure you learn from this and, if you do not understand it, consult your tutors.

   - Learn from the critical descriptions that your lecturers provide of linked websites. These pointers may help you to evaluate such resources for yourself in future.

   - Do not think that you will automatically assimilate information and concepts, just because you are viewing them online. The same principles apply as with printed media: you must apply active learning methods (p. 23).

   - Help your lecturers by providing constructive feedback when they ask for it. You may find this easier to do online than hurriedly filling out a feedback sheet at the end of a session.

4. **Organise files and web links.** Take the time to create a meaningful system of folders and files for downloaded material in tandem with your own coursework files and set up folders on your browser for bookmarked websites (*Favorites* in Internet Explorer).

5. **Take care when submitting coursework.** Make sure you keep a backup of any file you email or submit online and check the version you are sending carefully. Follow instructions carefully, for example regarding file type, or how to use your system's 'digital drop-box'.

Spam, junk mail and phishing - these should be relatively easy to identify, and should never be responded to or forwarded. Some may look 'official' and request personal or financial details (for example, they may pretend to come from your bank, and ask for account details). Never send these details by email or your identity may be used illegally.

- **Be polite.** Email messages can seem abrupt and impersonal. Take care to read your messages through before sending and if you are at all in doubt, do not send your message right away: re-read at a later time and consider how others might view what you say.

- **Consider content carefully.** A useful approach is only to send what you would be happy to hear being read out loud to classmates or family.

- **Take care with language and names when communicating with tutors and other staff.** Slang phrases and text message shorthand are unlikely to be understood. Be friendly without being overfamiliar.

- **Use your university email for academic purposes** this includes discussing coursework with classmates, but not forwarding questionable jokes, potentially offensive images, links to inappropriate websites, etc. In fact, doing so may break regulations and result in disciplinary action.

- **Beware of spam, junk and 'phishing' via email.**

Similar rules apply to discussion boards and the learning management system.

Using newsgroups and online forums - these can be useful to obtain answers to a specific problem: just post a query to the appropriate group and wait for someone to reply. However, always bear in mind that this is the view of an individual person.

The Usenet Newsgroup service is an electronic discussion facility, and there are thousands of newsgroups representing different interests and topics. Any user can contribute to the discussion within a topic by posting their own message; it is like email, but without privacy, since your message becomes available to all other subscribers.

Your university may also make use of social media, including Facebook and Twitter, or dedicated systems such as UCROO – often, the details of these systems will be explained during the initial orientation period, including how to 'follow' the university so that you keep up to date through news feeds.

Internet tools

The specific programs you will use for accessing the Internet will depend on what has been installed locally, on the network you are using, and on your internet service provider. The best way to learn the features of the programs is to try them out, making full use of whatever help services are available.

e-learning systems

Most university departments present their courses through a mixture of face-to-face sessions (e.g. lectures, tutorials, practicals) and online resources (e.g. lecture notes, websites, discussion boards, computerised tests and assessments). This constitutes 'blended learning' on your part, with the online component also being known as e-learning. Some courses are delivered fully online (distance education), or by online study supplemented by block attendance on campus ('intensives' or residential schools).

The e-learning element is usually delivered through an online module within a virtual learning environment (e.g. Blackboard, Moodle). It is important not to neglect the e-learning aspects of your course just because it may not be as rigidly timetabled as your face-to-face sessions. This flexibility is to your advantage, as you can work when it suits you, but it requires discipline on your part. Box 11.2 provides tips for making the most of the e-learning components of your courses. Mobile learning (m-learning), facilitated by wireless devices, is likely to grow in importance in the next few years.

Definition

m-learning - learning delivered via mobile devices, including handheld computers and mobile phones: m-learning occurs independently of the location of the learner.

Definition

Bookmark - a feature of browsers that allows you to save details of websites you have visited. This is termed 'add to favorites' in Internet Explorer. Bookmarks save you the trouble of remembering complex URL names and of typing them into the browser's address window.

Internet browsers

These are software programs that interact with remote server computers around the world to carry out the tasks of requesting, retrieving and displaying the information you require. Many different browsers exist, but the most popular are Internet Explorer, Mozilla Firefox and Google Chrome. These three browsers dominate the market and have plug-ins and add-on programs available that allow, for example, video sequences to be seen online. The standard functions of browsers include:

- accessing web documents;
- following links to other documents;
- printing the current document;
- maintaining a history of visited URLs (including 'bookmarks' for key sites);

- searching for a term in a document;
- viewing images and image maps.

Browsers provide access to millions of websites. Certain sites specialise in providing catalogued links to other sites; these are known as portals and can be of enormous help when searching within a particular area of interest. Your university's library website will almost certainly provide a useful portal to catalogues and search services, often arranged by subject area, and this is often the first port of call for electronic resources; get to know your way around this part of the website as early as possible during your course.

When using a browser program to get to a particular page of information on the web, all you require is the location (the web 'address') of that page, i.e. the URL. Most web page URLs take the form http://, or https:// (secure), followed by the various terms (domains and subdomains) that direct the system to the appropriate site. If you do not have a specific URL in mind but wish to explore appropriate sites, you will need to use a search tool with the browser.

Search tools

With the proliferation of information on the web, one of the main problems is finding the exact information you require. There are a variety of information services that you can use to filter the material on the network. These include:

- **search engines** (Boxes 11.3 and 11.4);
- **meta-search engines;**
- **subject directories;**
- **subject gateways (portals).**

Search engines such as Google (http://www.google.com), Bing (http://www.bing.com) and DuckDuckGo (https://duckduckgo.com) are tools designed to search, gather, index and classify Web-based information. Searching is usually by keyword(s), although specific phrases can be defined. Many search engines offer advanced searching tools such as the use of Boolean operators to specify combinations of keywords to more precisely filter the sites. Box 11.3 provides tips for refining keyword searches, while Box 11.4 provides tips for enhanced searching with Google.

It is important to realise that each search engine will cover at most about 40% of the available sites and if you want to carry out an exhaustive search it is necessary to use several to cover as much of the web as possible. Meta-search engines make this easier. These operate by combining collections of search engines. Examples include Mamma (http://www.mamma.com/) and Dogpile (http://www.dogpile.com/).

Some useful approaches to searching include the following:

- **For a comprehensive search,** use a variety of tools including search engines, meta-search engines and portals or directories.
- **For a complex, finely specified search,** employ Boolean operators and other tools to refine your keywords as fully as possible (Box 11.3). Some search engines allow you to include and exclude terms or restrict by date.

Box 11.3 Useful tips for using search engines

- Keywords should be chosen with care. Try to make them as specific as possible, e.g. search for 'archaebacteria' or 'Archaea', rather than 'bacteria' or 'micro-organisms'.

- Most search engines are case-insensitive. Thus 'Nobel Prize' will return the same number of hits using Google as 'nobel prize'. If in doubt, use lower case throughout.

- Putting keyword phrases in double quotes (e.g. "capillary electrophoresis") will result in a search for sites with the phrase as a whole rather than sites with both (all) parts of the phrase as separate words (i.e. 'capillary' and 'electrophoresis' at different places within a site). This feature allows you to include common words normally excluded in the search, such as 'the'.

- Use multiple words/phrases plus similar words to improve your search, for example 'terrestrial crustacean "land crab" Australia'. If you can, use scientific terms, as you are likely to find more relevant sites, e.g. search for the name of a particular species such as *Coenobita compressus*.

- Adding words preceded with + or − will add or exclude sites with that word present (e.g. 'blood diseases − leukaemia' will search for all blood diseases excluding those mentioning leukaemia). This feature can also be used to include common words normally excluded by the search engine.

- Check that your search terms have the correct spelling, otherwise you may only find sites with the same misspelled word. In some cases, the search engine may prompt you with an alternative (correct) spelling.

- Remember that US and UK spellings may differ (e.g. hemolysis/haemolysis) and a search will only find hits from sites that use the spelling you specify.

- Boolean operators (AND, OR, NOT) can be used with some search engines to specify combinations of keywords to filter the sites identified more precisely (e.g. 'plant NOT engineering' will avoid sites about engineering plant and focus on biological material – though this combination might omit genetic engineering of plants).

- Some search engines allow 'wildcards' to be introduced with the symbol*. For example, this will allow you to specify the root of a word and include all possible endings, as with anthropomorph*, which would find anthropomorphic, anthropomorphism, etc. If the search engine does not allow wildcards (e.g. Google), then you will need to be especially careful with the keywords used, including all possible words of relevance.

- Numbers can be surprisingly useful in search engines. For example, typing in EC 1.1.1.1 will find sites concerned with alcohol dehydrogenase, as this is its Enzyme Commission code number (p. 411). If you know the phone number for a person, institute or company or the ISBN of the book, this can often help you find relevant pages quickly.

- If you arrive at a large site and cannot find the point at which your searched word or phrase appears, press Control and F together and a 'local' search window will appear, allowing you to find the point(s) where it is mentioned.

- **Use 'cascading' searching when available** – this is searching within the results of a previous search.
- **Use advanced search facilities to limit your search,** where possible, to the type of medium you are looking for (e.g. graphics, video), language, sites in a specific country (e.g. UK) or to a subject area (e.g. news only).

However well defined your search is, you will still need to evaluate the information obtained. Chapter 10 covers general aspects of this topic, while Box 11.5 provides specific advice on assessing the quality of information provided on websites.

Directories

A directory is a list of web resources organised by subject. It can usually be browsed and may or may not have a search facility. Directories often contain better-quality information than the lists produced by search engines, as they have been evaluated, often by subject specialists or librarians. Check your library's subject resources for relevant directories.

Box 11.4 Getting the most from Google searches

Google (http://www.google.com) has become the search engine of choice for millions of people, owing to its simplicity and effectiveness. However, you may be able to improve your searches by understanding its default settings and how they can be changed.

- **Download the Google toolbar to your browser.** This is available from the Google homepage and will give you quick access to the Google search facility.

- **Understand how standard operators are used.** For combinations of keywords Google uses the 'minus' operator '−' instead of NOT (exclude) and the 'plus' operator '+' instead of AND (include). Since Google usually ignores small words ('stop words' such as *in* or *the*), use '+' to include them in a search. Where no operator is specified, Google assumes that you are looking for both terms (i.e. '+' is default). If you want to search for alternative words, you can use 'OR' (e.g. *sulphur OR sulfur*). Google does not allow brackets and also ignores most punctuation marks.

- While wildcard truncation of words using '*' is not allowed, **you can use '*' to replace a whole word (or number).** For example, if you type the phrase *"a virus is approximately * nanometres"* (UK spelling) your results will give you results for web pages where the wildcard is replaced by a number.

- **Search for exact wording.** By placing text in double inverted commas (" "), you can ensure that only websites with this exact phrasing will appear at the head of your search results.

- **Search within your results to improve the outcome.** If your first search has produced a large number of results, use the *Search within results* option near the bottom of each page to type in a further word or phrase.

- **Search for words within the title of a web page.** Use the command *intitle:* to find a web page, for example *intitle: "tissue culture"* returns web pages with this phrase in the title (note that phrases must always be in double speech marks, not single quotes).

- **Search within a website.** Use the *site:* command to locate words/phrases on a specific website, for example site.unicef. org *dysentery* returns only those results for this disease on the UNICEF website (unicef.org). Pressing *Control* and *F* when visiting a web page will give you a pop-up search window.

- **Locate definitions, synonyms and spellings.** The operator *define:* enables you to find the meaning of a word. If you are unsure as to the spelling of a word, try each possibility. Google will usually return more results for the correct spelling and will often also prompt you with the correct spelling [*Did you mean . . .?*).

- **Find similar web pages.** Simply click the *Similar pages* option at the end of a Google search result to list other sites (note that these sites will not necessarily include the term(s) searched for).

- **If a web link is unavailable, try the cached (stored) page.** Clicking on *Cached* at the end of a particular result should take you to the stored page, with the additional useful feature that the search term(s) will be highlighted.

- **Use the calculator function.** Simply enter a calculation and press *Enter* to display the result, for example "10 + (2 × 4)' returns 18. The calculator function can also carry out simple interconversion of units, e.g. '2 feet 6 inches in metres' returns 0.762 (see Box 31.1 for interconversion factors between SI and non-SI units).

- **Try out the advanced search features.** In addition to the standard operators, these include the ability to specify the number of results per page (e.g. 50) to reduce the use of the *next* button, language (e.g. English), file format (e.g. for PDF files), recently updated web pages (e.g. past 3 months), usage (e.g. free to use/share).

- **Find non-text material.** These include images, videos and maps – always check that any material you use is not subject to copyright limitations (p. 55).

- **Use Google alerts to keep up to date.** This function (http://www.google.co.uk/alerts) enables you to receive regular updated searches by email.

- **Use Google Scholar to find articles and papers.** Go to http://scholar.google.com/ and type in either the general topic or specific details for a particular article, e.g. author names or words from the title. Results show titles/authors of articles, with links to either the full article, abstract or citation. A useful feature is the *Cited by . . .* link, taking you to those papers that have cited the article in their bibliography and enabling you to carry out **forward citation searching** to locate more recent papers. Also try out the advanced scholar search features to limit your search to a particular author, journal, date or subject area. However, you should note that Google Scholar provides only a basic search facility to easily accessible articles and should not be viewed as a replacement for your library's electronic journal holdings and searching software. For example, if you find the title of a paper via Google Scholar you may then be able to locate the electronic version through your own library's databases, or request it via inter-library loan (p. 54). Another significant limitation is that older (more cited) sources are typically listed first. Use 'Advanced Search' to find sources between years X and Y.

- **Use Google Earth to explore locations.** Download from http://earth.google.com/. This allows you to zoom in on satellite images to find locations.

Box 11.5 How to evaluate information on the web

It is often said that 'you can find anything on the web'. The two main disadvantages of this are, firstly, that you may need to sift through many sources before you find what you are looking for and, secondly, that the sources you find will vary in their quality and validity. *It is important to realise that evaluating sources is a key aspect of using the Internet for academic purposes, and one that you will need to develop during the course of your studies* (see also Chapter 10). The ease with which you can 'point and click' to reach various sources should not make you complacent about evaluating their information content. The following questions can help you to assess the quality of a website – the more times you can answer 'yes', the more credible the source is likely to be, and vice versa:

Authority

- Is the author identified?
- Are the author's qualifications or credentials given?
- Is the owner, publisher or sponsoring organisation identified?
- Is an address given (postal and/or email)?

It is sometimes possible to obtain information on authority from the site's metadata (try the 'View' 'Source' option in Internet Explorer, or look at the URL to see if it gives any clues as to the organisation, e.g. whether the domain name ends in .ac, .edu, .gov or .org, rather than .co or .com).

Content

- Is there any evidence that the information has been peer-reviewed (p. 123), edited or otherwise validated, or is it based on such sources?
- Is the information factual or based on personal opinion?
- Is the factual data original (primary) or derived from other sources (secondary)?
- Are the sources of specific factual information detailed in full (Chapter 9)?
- What is the purpose of the site and who is it aimed at?

- Is the content relevant to the question you are trying to answer?
- Is there any evidence of a potential conflict of interest, or bias? (Is the information comprehensive and balanced, or narrowly focused?)
- Did you find the information via a subject-specific website (e.g. a bioscience gateway such as BIOME), or through a more general source, such as a search engine (e.g. Google)?

The above questions are similar to those that you would use in assessing the value of a printed resource (p. 61), and similar criteria should be applied to web-based information. You should be especially wary of sites containing unattributed factual information or data whose primary source is not given.

Currency

- Is there any indication that the information is up to date?
- Does the site show the date on which it was created, or last updated?

Presentation

- What is your overall impression of how well the site has been put together?
- Are there any grammatical or spelling mistakes?
- Are there links to other websites, to support statements and factual information?

The care with which a site has been constructed can give you an indication of the credibility of the author/organisation. However, while a poorly presented site may cause you to question the credibility of the information, the reverse is not always necessarily true: do not be taken in by a slick, well-presented website – authority, content and currency are *always* more important than presentation.

Using the Internet as a resource

A common way of finding information on the web is by browsing or 'surfing'. However, this can be time-consuming; try to restrict yourself to sites known to be relevant to the topic of interest. Some of the most useful sites are those that provide hypertext links to other locations. Some other resources you can use on the web are:

- **Libraries, publishers and commercial organisations.** Your university library is likely to subscribe to one or more databases providing access to scientific articles; these include Web of Science

Understanding the impermanence of the web - the temporary nature of much of the material on the web is a disadvantage for academic purposes because it may change or even disappear after you have cited it. You may also find it difficult or impossible to find out who authored the material. A case in point are wikis, such as Wikipedia (www.wikipedia.org). This online encyclopaedia has many potential authors and the content may change rapidly as a result of new submissions or edits; nevertheless, it can sometimes be a useful resource for general information about a wide range of topics, though it is not necessarily the best approach for researching scientific assignments.

Using traditional sources - remember that using the Internet to find information is not a substitute for visiting your university library. Internet resources complement rather than replace more traditional printed sources.

Remembering useful websites - create a 'bookmark' (= add a *'favourite'*) for the ones you find of value, to make revisiting easy. This can be done from the menu of your browser program. Make a copy of your bookmark file occasionally, to avoid loss of this information.

Note that URLs may change - make a keyword search using a search engine to find a particular site if the URL information you have does not lead you to an active page. Alternatively, use the 'Way-back Machine' at: http://www.archive.org to locate an earlier version of a URL.

Understanding open access publication- this is a relatively recent movement, pioneered by the Public Library of Science (PLOS). Typically, authors of scholarly articles (or their institutions) pay for all costs required for online publishing, thereby providing free and unrestricted online access to all potential readers (for an example of an open access journal, see: http://www.plosbiology.org).

(http://webofknowledge.com), and Science Direct (http://www.science-direct.com). A password is usually required, especially for off-campus use; consult your library staff for further details. Some scientific database sites give free access, without subscription or password; these include the PubMed website of the National Center for Biotechnology Information (http://www.ncbi.nlm.nih.gov) and the HighWire Press (http://highwire.com). Others allow free searching, but require payment for certain articles, e.g. the Scientific World (http://www.thescientificworld.com/) and Infotrieve (http://www4.infotrieve.com). Publishers such as Pearson and booksellers such as Amazon.com provide online catalogues and ecommerce sites that can be useful sources of information (see http://pearsoned.com and http://www.amazon.com).

- **Online journals and ebooks.** A number of traditional journals have websites. You can keep up to date by visiting the websites of *Nature* (http://www.nature.com), *New Scientist* (http://www.newscientist.com) and *Scientific American* (http://www.sciam.com), or ScienceDirect journals (at http://www.sciencedirect.com/science/journals). Some scientific societies make their journals and other publications available via their websites, e.g. the American Society for Microbiology at http://www.asm.org. Journals solely published in electronic format are also available (e.g. *Molecular Vision,* http://www.molvis.org/molvis) but some require a subscription password for access; check whether your institute is a subscriber. The directory of open access journals (at: http://www.doaj.org/) provides a comprehensive list of all non-subscription (free, open access) journals.

- **Data and images.** Archives of text material, video clips and photographs can be accessed, and much of the material is readily available. The HEA Centre for Bioscience ImageBank (http://heacademy.ac.uk/imagebank/) is a good example. When downloading such material, you should check that you are not breaching copyright and avoid potential plagiarism by giving a full citation of the source if you use such images in an assignment (see p. 59).

- **Biological institutions.** Many culture collections, scientific societies and other biological institutions around the world are now online. Use their sites to obtain specific information about collections, resources, etc. They frequently provide lists of other relevant sites or topics, e.g. the Society for General Microbiology (http://www.sgm.ac.uk/links/).

- **Databases.** In addition to those covering the scientific literature (e.g. Web of Science, or the PubMed service of the US National Library of Medicine), others focus on specific topics (e.g. biodiversity on the Internet, http://www.biodiversity.org.uk/ibs/; or academic employment, http://www.jobs.ac.uk). Examples of nucleic acid databases include the Rutgers University database (http://ndbserver.rutgers.edu) and the FBI CODIS database, used in forensic DNA analysis (http://www.fbi.gov/about-us/lab/biometric-analysis/codis). Chapter 12 provides further guidance on using online databases for the analysis of biomolecular structure (bioinformatics).

Text references

Gralla, P. (2006) *How the Internet Works,* 7th edn. Pearson, Harlow.

White, R. and Downs, T. (2014) *How Computers Work,* 10th edn. Que Publishing, Pearson, Harlow.

Sources for further study

Grassian, E. *Thinking Critically about World Wide Web Resources.* Available: http://www.library.ucla.edu/libraries/college/11605_12337.cfm Last accessed 01/09/15.

Hock, R. (2013) *The Extreme Searcher's Internet Handbook: a Guide for the Serious Searcher,* 4th edn. Cyber Age Books, New Jersey.

STUDY EXERCISES

11.1 Explore the resources of the web using a search engine. Using a search engine, find the answers to the following questions:

(a) Who is Barbara McClintock, and what is she famous for? What prize did she win and when?

(b) What is the common name for *Arabidopsis thaliana*?

(c) What is the postal address of the UK Biochemical Society?

11.2 Compare results from a variety of search engines. First, think of an appropriate biological keyword or phrase (e.g. a species name) and enter this into several search engines. Make sure that you include meta-search engines such as Dogpile. Compare the outcomes to reveal the strengths and weaknesses of the individual search engines. Work with a colleague to compare different searches on a quantitative (i.e. number of hits) and qualitative (quality of hits) basis.

11.3 Organise your bookmarks. Enter the *'Organize favorites'* menu for your preferred browser and create folders with appropriate headings. Move existing bookmarks to these folders and save any new ones appropriately. Doing this will help you find bookmarks more easily, rather than searching through long lists.

Answers to these study exercises are available at www.pearsoned.co.uk/practicalskills.

Bioinformatics is a term used to describe the appli[...] based techniques in the biological sciences and, in particula[...] of nucleic acid sequences (genomics, transcriptomics) and to the prote[...] sequences predicted from nucleic acid sequence data (proteomics). These complex molecules contain a large amount of information within their structures: analysis of this information by visual examination is impractical for all but the smallest polynucleotide or polypeptide fragments.

> **KEY POINT** The only practical approach to understanding the information in complex macromolecules is to use a computer – typically, by comparing sequence data for all or part of a specific biomolecule with that of others in a database.

Table 12.1 gives details of some of the most important databases accessible via the Internet.

A range of programs and packages can be used to analyse nucleic acid and protein sequences. Many programs can be used online, via the Internet, while others can be downloaded, for local operation. In most instances, the information and software are freely available. The large number of different programs available is a reflection of the various algorithms used for analysis, and such programs are likely to be updated in parallel with advances in computer hardware/software. When using a new program, make use of the help facilities to familiarise yourself with the operating principles.

Primary sequence databases

The major databases holding primary sequence information for nucleic acids and proteins are operated by the European Bioinformatics Institute (EBI) of the European Molecular Biology Laboratory, EMBL (at http://www.ebi.ac.uk), the National Center for Biotechnology Information (NCBI) GenBank database (at http://www.ncbi.nlm.nih.gov/genbank/) and the DNA Data Bank of Japan, DDBJ (at http://www.ddbj.nig.ac.jp). These databases hold comprehensive information submitted by researchers and sequencing groups across the world, with collaborative data sharing on a daily basis, as part of the International Nucleotide Sequence Database Collaboration.

Nucleic acid analysis

A typical application would be to investigate whether a nucleic acid sequence obtained, for example, as part of a cloning project (p. 474), has been reported previously. Table 12.1 gives details of URLs for the nucleic acid databases of EMBL-EBI and NCBI GenBank. These databases allow you to:

- **enter a particular nucleic acid sequence and edit the sequence,** if required;
- **find and retrieve a nucleotide sequence from the database** using a keyword, author name, organism name or an accession number, where known;
- **translate a nucleotide sequence into an amino acid sequence and vice versa;**

Definitions

Genome – the entire complement of genetic information (coding and non-coding DNA) of an organism.

Metabolome – all of the metabolites (low molecular mass biomolecules) of an organism.

Proteome – the expressed protein complement of the genome.

Secretome – all of the secreted proteins from an organism.

Transcriptome – the complement of mRNAs transcribed from the genome, weighted by the expression level of each RNA.

Note that terms ending in 'ome' can be used to describe disciplines by changing the ending to 'omics'.

Definition

Chemoinformatics – the application of information technology to chemical data and information, especially in the field of drug design/discovery.

Understanding the scale of primary sequence databases – the EBI-NCBI-DDBJ holds many millions of nucleic acid sequence records from tens of thousands of different organisms.

Example Y10510 is the accession number for a human mRNA sequence for the CD67S protein in the EMBL-EBI database.

able 12.1 Some useful databases and other Internet locations relevant to bioinformatics.

Major nucleic acid sequence databases	
EMBL-EBI European Nucleotide Archive	https://www.ebi.ac.uk/ena
NCBI GenBank	http://www.ncbi.nlm.nih.gov/genbank
Major protein sequence databases	
UniProt (universal protein resource)	http://www.uniprot.org
PIR (protein information resource)	http://pir.georgetown.edu
SWISS-PROT (EBI protein sequence database)	http://ca.expasy.org/
Other bioinformatics databases	
Sanger Institute (links to various databases)	http://www.sanger.ac.uk/
PROSITE (protein motifs)	http://prosite.expasy.org/
PRINTS (protein fingerprints)	http://bioinf.man.ac.uk/dbbrowser/PRINTS/
ExPASy proteomics resource	http://ca.expasy.org/
PDB (Protein Data Bank)	http://www.rcsb.org/pdb
Array Express (Microarrays)	https://www.ebi.ac.uk/arrayexpress
MEROPS (protease/peptidase database)	http://merops.sanger.ac.uk
Transport protein database	http://tcdb.org/
CBS prediction servers (various databases)	http://www.cbs.dtu.dk/services/
PubMed (literature database)	http://www.ncbi.nlm.nih.gov/pubmed

---

Searching primary databases – a search of the EMBL-EBI nucleic acid database can be carried out using BLAST (Fig. 12.1) to locate related sequences.

Definition

Palindrome – a sequence reading the same in both directions, e.g. GACCAG. These are often sites at which restriction enzymes cleave DNA (see p. 458).

Remembering which is which – INTrons (non-coding sequences) INTerrupt coding sequences while EXons are Expressed (translated into an amino acid sequence).

- **search for any resemblance between a particular sequence and sequences within the database,** e.g. by comparing and aligning the sequence for several nucleic acids or proteins, to identify regions of sequence similarity (see Fig. 12.1);

- **carry out phylogeneric analysis,** constructing 'ancestry trees' to show the most likely evolutionary relationships between sequences from various organisms.

Other programs allow you to analyse nucleic acid sequence information in more detail, for example:

- **to assemble/align sequence fragments into contiguous sequences,** also known as 'contigs'. This type of alignment is particularly useful for sets of nucleic acid fragments from a clone library (p. 478);

- **to search for structural and functional 'motifs' within the nucleic acid sequence,** including palindromes, restriction sites (p. 458), likely promoter or control sites and regions that might form secondary structures, such as hairpin loops;

- **to simulate *in vitro* manipulation and cloning procedures,** including the preparation of restriction digests, designing oligonucleotide probes and primers, and simulating PCR (p. 467);

- **to identify likely coding regions (e.g. open reading frames, exons) and non-coding regions (e.g. introns, repeat sequences) in eukaryoric DNA,** 'start' and 'stop' codons, etc. The starting point for this analysis is the creation of a 'six frame translation' of the target sequence (see Fig. 12.2).

```
H. sapiens    atgtatggcattgagaatgaagtcttcctgagccttccatgtatcctcaatgcccgggg
              ||||| ||||| ||||| |||||||||||| || || || || ||||| ||||| |||||
R. norvegicus atgtacggcatcgagaacgaagtcttcctcagtctcccgtgcatccttaatgctcgggg
```

Fig. 12.1 Representative output from a BLAST (basic local alignment search tool) alignment search for DNA sequences from part of the lactate dehydrogenase genes of *Homo sapiens* and *Rattus norvegicus*. Here, identical bases are shown in black, and non-identical bases are shown in green. This region shows 49 identical bases out of 59, i.e. 49 ÷ 59 × 100 = 83% similarity (to the nearest integer).

Fig. 12.2
'Six frame translation' of a DNA sequence. The triplet codons corresponding to each amino acid are given in Table 66.1 (p. 457) and the single letter amino acid code is explained in Table 56.1. Note that the complementary sequence is translated in the reverse direction to the original sequence.

Analysis of proteins and polypeptides

The principal protein sequence databases are SWISS-PROT in Europe, together with PIR in the USA: Table 12.1 gives their current Internet locations. These databases can be searched using a range of programs equivalent to those described above for nucleic acid analysis, allowing the operator to:

- **locate and retrieve a particular protein or amino acid sequence** from the database by keyword, author name, organism name or accession number;
- **enter and edit (cut and paste) a particular amino acid sequence,** or convert a nucleotide sequence into an amino acid sequence, and vice versa;
- **search the database for sequences** with a perfect match, or for similarities in primary or secondary structure at a specified level of homology; such searches may be more informative than those based on DNA homology, since protein sequence information (based on 20 different amino acid residues, p. 382) is more complex than nucleic acid sequence information (based on four different nucleotides, p. 398);
- **compare and align the amino acid sequences of several proteins,** to identify particular regions of similarity and dissimilarity.

Other bioinformatics databases

Pattern recognition databases for proteins

These have been created using primary sequence data, to provide information on the patterns identified within proteins. The representative examples given in Table 12.1 include the PROSITE and PRINTS databases, which identify protein families by their diagnostic 'signature' motifs, and the Expert Protein Analysis System (ExPASy) website, which provides a gateway to the general analysis of protein structure, including

---

Using bioinformatics databases – many of the database websites are rather complex and newcomers may find them a little difficult to navigate, especially if you are just browsing. The best approach is to try out the online tutorials, to get a feel for how the website operates. If the URL changes, use a search engine to locate the new website.

Understanding the basic terminology of protein structure:

- primary structure – sequence of amino acids along a polypeptide;
- secondary structure – regular repeating structural motifs (α-helix and β-sheet), stabilised by hydrogen bonds;
- tertiary structure – overall three-dimensional form of the polypeptide;
- quaternary structure – where more than one polypeptide subunits form a functional complex.

three-dimensional imaging, while the CBS Prediction Servers can be used to identify particular motifs, as well as other aspects of genome analysis.

Many of these 'second-level' databases provide tools and software that can be used for a range of purposes, including:

- **assembling overlapping peptide fragments,** based on their amino acid sequences, e.g. during sequence analysis of an individual protein (p. 385);
- **predicting secondary structure from a primary sequence,** identifying regions most likely to exist as α-helices or β-sheets, etc.;
- **searching for motifs and patterns within a sequence** that might provide information on the origins and/or possible function(s) of the protein;
- **predicting potential sites of post-translational modification;**
- **modelling the likely three-dimensional structure of the protein,** identifying likely membrane-spanning regions, active sites and other tertiary and quaternary features;
- **investigating properties** such as hydropathy, solvent accessibility and antigenicity;
- **predicting and modelling the effects of changes in primary or secondary structure** on protein characteristics and three-dimensional organisation.

Protein-specific databases

Several databases have been created to collect together information on a particular class of proteins – many are devoted to particular enzymes or groups of enzymes. An example of a second-level protein database is that devoted to the analysis of enzymes that degrade, modify or create glycosidic bonds, the carbohydrate-active enzymes (CAZy database, at http://www. cazy.org). This website brings together information on the various families of enzymes that interact with carbohydrates, enabling researchers to compare sequences against known catalytic and carbohydrate-binding modules, arranged by enzyme class and by organism. Links to related websites are also included. Another example is the MEROPS protease database at the Sanger Institute, which provides a structure-based catalogue of proteases and peptidases, giving amino acid and nucleotide sequences, classification and nomenclature, plus links to relevant literature through PubMed. If you are looking for information on a particular enzyme or other protein, it may be worth using a web search engine to see whether a specific database has been created.

Protein structure

The Protein Data Bank (PDB) is the worldwide repository for three-dimensional structures of proteins (http://www.pdb.org/), also serving as a general portal for structural information on biological macromolecules. The database can be searched by identity number (e.g. 1 ALB is the code for an adipocyte lipid-binding protein) or by name (note that this is a US resource, so using the US spelling 'hemoglobin' locates more structures than using the UK spelling). Structures can be viewed in a variety of formats, including those compatible with molecular graphics software such as PyMOL, RasMol and Chime. Additional information includes secondary structure (helices, sheets, hairpins, etc.), crystallographic data, geometry, etc., plus links to related sites. Figure 12.3 shows an example of a typical output – in this instance, for the 3D structure of the enzyme pectate lyase, a virulence factor in some plant pathogenic bacteria.

Hydropathy – hydrophobic interactions within a biomolecule; often important in determining folding and three-dimensional structure.

Bioinformatics in drug design – protein databases provide chemoinformatic information that can be used to locate new therapeutic targets, e.g. when designing novel antimicrobial agents.

Example   PyMOL is a molecular graphics program that enables the user to visualise the structure of biomolecules – it can be obtained from the following website: http://pymol.sourceforge.net/

Fig. 12.3 Representation of the three-dimensional structure of a pectate lyase from *Cellvibrio japonicus*. This is a solid ribbon display, with helices and sheets shown in green. The cleft, in which the plant polysaccharide is bound, is clearly visible on the left of the structure.
Image courtesy of Prof. G. Black, Northumbria University.

Table 12.2 Other Internet locations relevant to bioinformatics.

Databases and resources for particular organisms	
Arabidopsis information resource	http://www.arabidopsis.org
Drosophila database (flybase)	http://flybase.org
Caenorhabditis genomes project	http://www.caenorhabditis.bio.ed.ac.uk.
E. coli genome project	http://www.genome.wisc.edu
Human genome database	http://www.genome.gov/
HIV sequence database	http://www.hiv.lanl.gov/
Plant genome database	http://www.plantgdb.org/
Saccharomyces genome database	http://www.yeastgenome.org/
Other resources	
BioCyc Database Collection (pathways/genomes)	http://biocyc.org
KEGG metabolic pathway database	http://www.genome.jp/kegg/pathway.html
MUSC DNA microarray database	http://proteogenomics.musc.edu/ma/

Other Internet resources

Several websites have been constructed to bring together bioinformatic data and related information for a single organism – many of these sites have been produced as a result of a genome-sequencing project for that particular organism, e.g. the human genome mapping project. In addition to the nucleic acid and/or protein sequence data available in the primary databases, the organism-specific websites provide a broader range of information and resources, often including links to other web locations. As such, they can be a useful means of accessing relevant internet sites: Table 12.2 gives examples, along with the locations of several other relevant websites that may be useful when researching a topic, e.g. as part of an assignment, or when planning a bioinformatics project.

Next-generation sequencing - high throughput techniques (p. 465) have revolutionised the study of genomics, by providing rapid and less expensive approaches than traditional sequencing.

Sources for further study

(Note also the websites given in Tables 12.1 and 12.2.)

Brown, T.A. (2006) *Genomes 3.* Garland Science, New York.
[This book focuses on genomics, in contrast to more traditional molecular biology textbooks, which emphasise individual genes.]

Buffalo, V. (2015) *Bioinformatics Data Skills.* O'Reilly, Cambridge.

Choudhouri, S (2015) *Bioinformatics for Beginners.* Academic Press, New York.

Ghosh, Z. and Mallick, B. (2015) *Bioinformatics Principles and Applications.* Oxford University Press, Oxford.

Korenberg, M.J. (2014) *Microarray Data Analysis.* Humana, New York.

Lesk, A. (2014) *Introduction to Bioinformatics,* 4th edn. Oxford University Press, Oxford.

Pevsner, J. (2009) *Bioinformatics and Functional Genomics.* Wiley, New York.

Ramsden, J. (2015) *Bioinformatics: an Introduction,* 3rd edn. Springer, New York.

Richards, J.E. and Hawley, R.S. (2010) *The Human Genome: a User's Guide,* 3rd edn. Academic Press, London.

STUDY EXERCISES

12.1 Investigate the potential of the Internet for locating information relevant to bioinformatics. Using the web, answer the following questions:

(a) How many nucleotide sequences are currently held by GenBank?

(b) What and where is Hinxton Hall?

(c) Where is the European Molecular Biology Laboratory (EMBL) Mouse Biology Program located?

(d) What is the total number of protein structures currently held by the Protein Data Bank?

(e) What is the full postal address of the US National Center for Biotechnology Information?

12.2 Search GenBank nucleotide database using a DNA sequence. Carry out a BLASTN search of GenBank (at: http://www.ncbi.nlm.nih.gov/BLAST) for the following nucleotide sequence: actgactcata-gagtacccaccgtggtgct. How many 'hits' do you find? What is the length of the sequence showing complete identity?

12.3 Search the NCBI database using a protein sequence. Carry out a BLASTP search (at: http://www.ncbi.nlm.nih.gov/BLAST) for the following amino acid sequence: NVKAAWGKVGAHAGEYGAE (note – p. 382 gives details of the code letters corresponding to each amino acid), (a) What protein is this sequence most likely to correspond to? (b) Which sequence gives the highest alignment score? (c) What is the highest score for a sequence from *Homo sapiens*?

12.4 Convert a DNA base sequence to an amino acid sequence. Using the 'DNA → protein translate tool' of the ExPASy server (at: http://ca.expasy.org/tools/dna.html), carry out a six-frame translation of the following nucleotide sequence: tcagagagaacccaccat. Which of the six frames contains both a start and stop codon and what is the predicted amino acid sequence for this frame?

12.5 Use the Protein Data Bank to investigate the structure of an enzyme. Search the PDB archive (at: http://www.rcsb.org/pdb/) using 'beta-galactosidase' as the keyword in the SearchLite search tool, and locate the entry for the β-galactosidase (= β-glycosidase) enzyme of *Sulfolobus solfataricus*, deposited on 19 September 1996. Use the *'explore'* command to find out: (a) the number of subunits in the holoenzyme; (b) the number of amino acids residues in each chain; (c) the number of helices in each chain. Next, use the *'view structure'* option to look at the 3D configuration (either in quick PDB format, or using *Rasmol*, if this program is available on your PC, or network): (d) save an image as a file and then incorporate it into a *Word* document, as you might in an assignment.

Answers to these study exercises are available at www.pearsoned.co.uk/practicalskills.

13   Using spreadsheets

The spreadsheet is one of the most powerful and flexible computer applications. It can be described as the electronic equivalent of a paper-based longhand calculation, where the sums are carried out automatically. Spreadsheets provide a dynamic method of storing, manipulating and analysing data sets. Advantages of spreadsheets include:

- **Ease and convenience** – especially when complex calculations are repeated on different sets of data.
- **Accuracy** – providing the entry data and cell formulae are correct, the result will be free of calculation errors.
- **Improved presentation** – data can be produced in graphical or tabular form to a very high quality.
- **Integration with other programs** – graphs and tables can be exported to other compatible programs, such as a word processor in the same office suite.
- **Useful tools** – advanced features include hypothesis-testing statistics, database features and macros.

Spreadsheets can be used to:

- **store and manipulate raw data** by removing the drudgery of repeated calculations, allowing easy transformation of data and calculation of statistics;
- **display your data as a graph or chart, for evaluation and/or presentation.** Printouts can be used in practical and project reports;
- **carry out statistical analysis** using in-built procedures or by allowing construction of formulae for specific tasks;
- **model 'what if situations** where the consequences of changes in data can be seen and evaluated.

A typical spreadsheet (Fig. 13.1) is divided into rows (identified by numbers) and columns (identified by alphabetic characters). Each individual

Fig. 13.1 The appearance of a typical spreadsheet, showing cells, rows and columns, toolbars, etc.

Microsoft *Excel* product screenshot reprinted with permission from Microsoft Corporation.

combination of column and row forms a cell which can contain either a data item, a formula or a piece of text. Formulae can include scientific and/or statistical functions and/or a reference to other cells or groups of cells (often called a range). Complex systems of data input and analysis can be constructed. The analysis, in part or complete, can be printed out. New data can be added at any time and the sheet will recalculate automatically. The power a spreadsheet offers is directly related to your ability to create arrays of formulae (models) that are accurate and templates that are easy to use.

Data entry

Spreadsheets have built-in commands which allow you to control the layout of data in the cells (see Fig. 13.2). These include number format, the number of decimal places to be shown (the spreadsheet always calculates using eight or more places), the cell width and the location of the entry within the cell (left, right or centre). An auto-entry facility assists greatly in entering large amounts of data by moving the entry cursor either vertically or horizontally as data are entered. Recalculation default is usually automatic so that when a new data value is entered the entire sheet is recalculated immediately.

The parts of a spreadsheet

Labels

These should be used to identify parts of the spreadsheet, for example, stating what data are contained in a particular column or indicating that a cell's contents represent the end point of a calculation. In *Excel*, it may be useful to use the *Cells > Format > Format Cells* section in the *Home* tab to delimit numerical sections of your spreadsheet. Note that spreadsheet

Data output from analytical instruments – many devices provide output in spreadsheet-compatible form (e.g. a 'comma delimited' file). Once you have uploaded the information into a spreadsheet, you can manipulate, analyse and present it according to your needs. Consult instrument manuals and the spreadsheet help function for details.

Using hidden columns – these are useful for storing intermediate calculations which you do not wish to be displayed on the screen or printout.

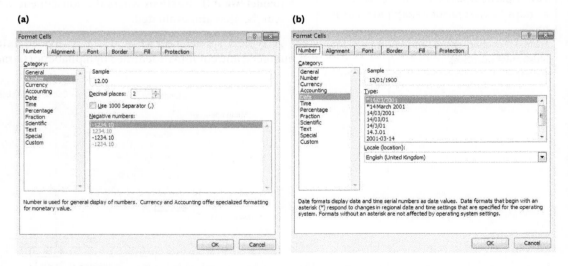

(a) **(b)**

Fig. 13.2 Example of cell formatting options within a Microsoft *Excel* spreadsheet. These menus are accessed via the *Format Cells* option and would apply to all of a range of selected cells. (a) Use of the number formatting option to specify that data will be presented to three decimal places (the underlying data will be held to greater accuracy). (b) Use of the date formatting option to specify that dates will be presented in day/month/year format. (Spreadsheet dates are stored numerically and converted to appropriate formats. This allows a period between two dates to be calculated more easily.) Microsoft *Excel* product screenshots reprinted with permission from Microsoft Corporation.

programs have been designed to make assumptions about the nature of data entry being made. If the first character is a number, then the entry is treated as numerical data; if it is a letter, then it is treated as a text entry; and if it is a specific symbol ('=' in Microsoft *Excel*), then what follows is a formula. If you wish to enter text that starts with a number, then you must type a designated character to show this (a single quote mark in Microsoft *Excel*).

Numbers

You can also enter numbers (values) in cells for use in calculations. Many programs let you enter numbers in more than one way and you must decide which method you prefer. The way you enter the number does not affect the way it is displayed on the screen as this is controlled by the cell format at the point of entry. There are usually special ways to enter data for percentages, currency and scientific notation for very large and small numbers.

Formulae

These are the 'power tools' of the spreadsheet because they do the calculations. A cell can be referred to by its alphanumeric code, e.g. A5 (column A, row 5) and the value contained in that cell manipulated within a formula, e.g. =(A5+10) or =(A5+B22) in another cell. Formulae can include various pre-programmed functions which can refer to a cell, so that if the value of that cell is changed, so is the result of the formula calculation. They may also include branching options through the use of logical operators (IF, TRUE, FALSE, OR, etc.).

Functions

Various functions may be offered — only mathematical and statistical functions will be considered here.

Mathematical functions

Spreadsheets offer a range of options, including trigonometrical functions, angle functions, logarithms and random number functions. Functions are invaluable for transforming sets of data rapidly and can be used in formulae required for more complex analyses. Spreadsheets work with an order of preference of the mathematical operators in much the same way as a standard calculator and this must always be taken into account when operators are used in formulae. They also require a very precise syntax – the program should warn you if you break this.

Statistical functions

Modern spreadsheets incorporate many sophisticated statistical functions, and if these are not appropriate the spreadsheet can be used to carry out the calculations required for most of the statistical tests found in textbooks. The descriptive statistics normally available include:

- **the sum of all data** present in a column, row or block;
- **the minimum and maximum** of a defined range of cells;
- **counts of cells** – a useful operation if you have an unknown or variable number of data values;

Operators and brackets in spreadsheets – the standard mathematical operators ÷ and x are usually replaced by / and * respectively, while ^ signifies 'to the power' (e.g. 10^4 = 10^4). In complex formulae, brackets should be used to separate the elements, otherwise the results may not be what you expect. For example, *Excel* will calculate = A1*B1/C1–D1 differently from (A1*B1)/(C1–D1).

Definition

Function – a pre-programmed code for the transformation of values (mathematical or statistical functions) or selection of text characters (string functions).

Example = sin(A5) is an example of a function in *Excel*. If you write this in a cell, the spreadsheet will calculate the sine of the number in cell A5 (assuming it to be an angle in radians) and write it in the cell. Different programs may use a slightly different syntax.

Working with empty cells – note that these may be given the value 0 by the spreadsheet for certain functions. This may cause errors, e.g. by rendering a minimum value inappropriate. Also, an 'error return' may result for certain functions if the cell content is zero.

- **averages and other statistics describing location;**
- **standard deviations and other statistics describing dispersion.**

A useful function where you have large data sets allows you to create frequency distributions using pre-defined class intervals.

The hypothesis-testing statistical functions are usually reasonably powerful (e.g. *t*-test, ANOVA, regression) and they often return the *probability (P)* of obtaining the test statistic when the null hypothesis (p. 525) is true, so there may be no need to refer to statistical tables. Again, check on the effects of including empty cells within the statistical calculations.

Database functions

Many spreadsheets can be used as simple databases and offer a range of functions to support this, including filtering and sorting options. The rows and columns of the spreadsheet are used as the fields and records of the database (see Chapter 14). For many biological purposes, this form of database is perfectly adequate and should be seriously considered before using a full-feature database product.

Copying

All programs provide a means of copying (replicating) formulae or cell contents when required and this is a very useful feature. This is usually accomplished by 'dragging' a cell's contents to a new range using the mouse. When copying, references to cells may be either relative, changing with the row/column as they are copied, or absolute, remaining a fixed cell reference and not changing as the formulae are copied (Fig. 13.3).

> **KEY POINT** The distinction between relative and absolute cell references is very important and must be understood; it provides one of the most common forms of error when copying formulae.

As an example, in *Excel,* copying is normally *relative* and if you wish a cell reference to be *absolute* when copied, this is done by putting a dollar ($) sign before and after the column reference letter e.g. $C$56.

Naming blocks

When a group of cells (a block) is carrying out a particular function, it is often easier to give the block a name which can then be used in all formulae referring to that block. This powerful feature also allows the spreadsheet to be more readable.

Spreadsheet templates

A template is a pre-constructed spreadsheet containing the formulae required for repeated data analysis. Data are added when they become available, and results are available as soon as the last item is entered. To create a template, the sequence of operations is:

1. **Determine what information/statistics you want to produce.**
2. **Identify the variables you will need to use,** both for original data that will be entered and for any intermediate calculations that might be required.

Statistical calculations – make sure you understand whether the functions you employ apply to populations or samples (see p. 520).

Using text functions – these allow you to manipulate text within your spreadsheet and include functions such as 'find and replace' and alphabetical or numerical 'sort'.

(a)

	Cell	Formula	
Original → cell	A1	=B1+C1	← Original formula
Copied cells	A2	=B2+C2	Copied formulae (relative)
	A3	=B3+C3	
	A4	=B4+C4	

(b)

	Cell	Formula	
Original → cell	A1	=B1/$C$1	← Original formula
Copied cells	A2	=B2/$C$1	Copied formulae (mixed relative and absolute)
	A3	=B3/$C$1	
	A4	=B4/$C$1	

Fig. 13.3 Illustration of relative (a) and absolute (b) copying. In *Excel,* the $ sign before and after the column letter makes the cell reference absolute, as shown in (b).

Using templates – these should contain:
- a data input section;
- data transformation and/or calculation sections;
- a results section, which can include graphics;
- text in the form of headings and annotations;
- a summary section.

Constructing a spreadsheet – start with a simple model and extend it gradually, checking for correct operation as you go.

Preparing graphs using *Excel 2010* – Box 70.2 on p. 490 gives step-by-step guidance for plotted curves, histograms and pie charts.

3. **Set up areas of the spreadsheet for data entry,** calculation of intermediate values (statistical values such as sums of squares, etc.), calculation of final statistics and, if necessary, a summary area.

4. **Establish the format of the numeric data** if this is different from the default values. This can be done globally (affecting the entire spreadsheet) or locally (affecting only a specified part of the spreadsheet).

5. **Establish the column widths** required for the various activities.

6. **Add text (labels) to identify input,** intermediate formulae and output cells. This is valuable in error-tracking and when carrying out further development work. Text can be entered in designated cells, or cells can be annotated using the *New Comment* feature in the *Review* tab.

7. **Enter a test set of values** to use during formula entry: use a fully worked example to check that formulae are working correctly.

8. **Enter the formulae required** to make all the calculations, both intermediate and final. Check that results are correct using the test data.

The spreadsheet is then ready for use. Delete all of the test data values and you have created your template. Save the template to a disk and it is then available for repeated operations.

Graphics display

Most spreadsheets now offer a wide range of graph and chart options which are easy to use, and this represents a way to examine your data sets rapidly and comprehensively. The quality of the final graphics output (to a printer) may be variable but is usually sufficient for data exploration and analysis. Many of the default options are business graphics styles but there are usually histogram, bar, column and pie charts X–Y plotting, line and area graphics options available. Note that some spreadsheet graphics often do not come up to the standards expected for the formal presentation of scientific data, unless you manipulate the option settings appropriately.

Printing spreadsheets

This is usually a straightforward, menu-controlled procedure, made difficult only by the fact that your spreadsheet may be too big to fit on one piece of paper. Try to develop an area of the sheet which contains only the data that you need to print, i.e. perhaps a summary area. Remember that columns and/or rows can usually be hidden for printing purposes and you can control whether the printout is in portrait or landscape mode, and for page size and format. Use a screen preview option, if available, to check your layout before printing. A 'print to fit' option is also available in some programs, making the output fit the page dimensions.

Sources for further study

Frye, C. (2010) *Microsoft Office Home and Student 2010 Step-by-Step.* Microsoft Press, Redmond.

Hart-Davies, G. (2010) *Beginning Microsoft Office 2010.* Apress, New York.

Harvey, G. (2013) *Excel 2013 for Dummies.* Wiley, New York.

Miller, R. (2006) *Point & Click OpenOffice.org* Prentice Hall, New Jersey.

The instructions and tips for these problems assume that you have *Excel* (2003 or later) available. If not, they should be readily modified for most advanced spreadsheet programs. If you have problems with any of the tasks, consult Box 70.2 or try using the program's *Help* facility.

13.1 Create a spreadsheet and graph (introductory).

(i) Copy the information in the table below into a spreadsheet. Name and save the spreadsheet file appropriately.

(ii) From the copied information, create a pie chart using the *Insert* tab and *Pie* (chart) function.

(iii) Adjust the colours selected so the chart will print out in black and white. Save the final version of your spreadsheet. Print the chart out directly from *Excel*.

Relative percentages of the major types of leukocytes in a blood sample

Cell type	Percentage
Neutrophils	60
Eosinophils	4
Basophils	1
Monocytes	7
Lymphocytes	28
Total	100

13.2 Create a spreadsheet and graph (advanced).

(i) Copy the data in the table at the top of the next column into a spreadsheet. Name and save the file appropriately.

My exam timetable

Subject	Date	Time	Paper	Location	Question style
Biochemistry	3 Jun	Morning	1	Great Hall	Multiple choice
Biochemistry	17 Jun	Morning	2	Exam Hall 5	Essay paper
Biochemistry	2 Jun	Afternoon	3	Main Laboratory	Information processing
Genetics	3 Jun	Afternoon	A	Small Hall	Short answer questions
Genetics	14 Jun	Afternoon	B	Exam Hall 5	Essay paper
Cell Biology A	4 Jun	Morning	1	Small Hall	Short answer questions
Cell Biology B	1 Jul	Afternoon	1	Exam Hall 3	Short answer questions
Cell Biology A	13 Jun	Afternoon	2	Exam Hall 5	Essay paper
Cell Biology B	2 Jun	Morning	2	Main Laboratory	Practical exam

Time course of cell population growth

Time (h)	Cell count (per unit volume)
1	2.00
2	6.00
6	50.0
12	6.00×10^3
18	2.20×10^5
24	1.65×10^7

(ii) Use the spreadsheet and chart-making facilities to explore by eye which of the following transformations would result in the best linear fit for these data: reciprocal, square root, cube root or log.

(iii) Add a linear trend-line to the chart for the most appropriate transformation.

(iv) Copy the graph to a file in *Word* and print out.

13.3 Use a spreadsheet as a simple database. Copy the data in the table below into cells within a spreadsheet. Modify the column widths so you can see all of the text on a single screen. Now sort the data in the following ways:

a) by subject, in alphabetical order;

b) by date and then by time of day;

c) by topic, in reverse alphabetical order.

Use the *Hide* function (right-click on the column header in Microsoft *Excel*) so that the information in columns 4 and 6 is not displayed. Find out how to undo this operation.

Reassess your IT skills and knowledge in relation to your course – although you may be familiar with 'office' style software (e.g. Microsoft *Office*, Apple Mac Pages or *Apache OpenOffice)*, you may be required to use features in new or different ways. For example:

- You may be asked to present documents and their components in very specific formats (such as defining fonts, line spacing, paragraph layout, margins, layout of tables and figures).
- You may benefit from using certain editing features to refine your writing (such as word counts, thesaurus, spelling and grammar checker).
- You may need to adapt your methods to longer writing exercises (such as creating an outline structure, writing parts out of sequence, creating a bibliography at the outset and contributing to it as you write).

You will probably be familiar with a range of computer programs used for academic work, and may own a laptop or PC with this software installed. To support your studies, you will also have access to networked PCs and associated programs licensed for your use. However, the specific software and versions may vary between these systems, involving different menu protocols and commands for the same or similar functions. In addition, the specific demands of academic work may involve the use of sophisticated features or even new types of program. This chapter considers some common software applications in relation to relevant tasks. Spreadsheets and presentation software are considered in detail in Chapters 13 and 16.

> **KEY POINT** Even when you are familiar with the basics of a software program, it may be valuable to learn more about advanced features, to ensure that you are working efficiently and effectively.

Word processors

Word processors are generally available as part of an 'office' package with the advantage that they share a common interface in the different components (word processor, spreadsheet, database, etc.), allowing exchange of information (e.g. text, graphics) between applications. Examples include *Pages* (Apple Mac) and Microsoft *Word* (Microsoft Office). Compatible open source software is also available, such as *OpenOffice,* Most word processors have similar general features but differ in operational detail. They offer specific advantages for academic work. For example, you can:

- **refine material many times before submission;**
- **insert material easily, allowing writing to take place in any sequence;**
- **use a spell-checker to check your text;**
- **use a thesaurus when composing your text;**
- **carry out ongoing checks of the word count;**
- **produce high-quality final copies;**
- **reuse part or all of the text in other documents.**

Although use of computers to compose text is now almost universal, it is worth remembering certain disadvantages of this approach:

- **the need for a reliable battery or mains power supply** (important for work in the field);
- **the need to learn specific operational details of the program;**
- **the temptation to make trivial revisions,** hence 'overworking' your text;
- **the risk of losing files or forgetting to save work appropriately.**

> **KEY POINT** It is vital to save your work frequently to a memory stick, hard drive or network drive. This should be done every 10 min or so. If you do not save regularly, you may lose hours or days of work. Most programs can be set to 'autosave' every few minutes – make sure you switch this feature on.

Using online help, tutorials and textbooks – support information is usually provided in one or more of the following ways: as a help facility within the program; as a file on the installation CD; or as an online help support site. If unfamiliar with the software, it may be worthwhile investing in one of the commercial textbooks written to support users of specific programs.

When using a word processor, take full advantage of the differences between word processing and handwriting (which necessarily follows a linear sequence and requires more planning):

- **Simply jot down your initial ideas for a plan,** preferably at paragraph topic level. The order can be altered easily and if a paragraph grows too much it can easily be split.
- **Start writing wherever you wish** and fill in the rest later.
- **Just put down your ideas as you think,** confident in the knowledge that it is the concepts that are important to note; their order and the way you express them can be adjusted later.
- **Do not worry about spelling and use of synonyms** – these can (and should) be checked during a separate scan of your text, using the spell-checker first to correct obvious mistakes, then the thesaurus to change words for style or meaning.
- **Use a draft printout to check (a) for pace and spacing (b) to ensure that words checked for spelling fit the required sense.**

Laying out (formatting) your document

Controlling widow and orphan lines – single lines or pairs of lines at the top or bottom of a page are referred to as 'widows' or 'orphans' respectively and they are generally undesirable in page design. You can control whether they appear in most word processors using the appropriate commands for your program.

The appearance of a document on-screen generally represents what the printout on paper will look like, but this should be checked using the 'print preview' feature before printing large documents, or where the paper size is non-standard. Menu commands can be used to adjust page margins, headers and footers, line spacing and fonts as required. Paragraph indentations should be created using the TAB or key rather than the space bar, but you can also indent or bullet whole blocks of text using special menu options. Although you can reformat your text at any time, it is good practice to enter the relevant details in the *Page Setup* when you start writing: entering them later can lead to problems due to forced reorganisation of the text layout. If you use a particular set of layout formats regularly, e.g. an A4 page with space for a letterhead, create a formatted template that can be called up whenever you start a new document. Advanced users can set up automated styles and formatting features to suit particular requirements (e.g. using the *Home* tab, *Styles* group in Microsoft *Word*).

Editing features

Using a spell-check facility – do not rely on this to spot all errors. Remember that spell-check programs do not correct grammatical errors, nor do they pick up words that are incorrect but that are legitimate words, e.g. 'their' for 'there', or 'form' for 'from'.

Word processors usually have many features designed to make editing documents easy. Most programs allow you to customise the toolbar to add commonly used command 'buttons' – some examples relevant to scientific writing include subscripts and superscripts, word count, A–Z sort and table column formatting. The spell-checkers and thesaurus are self-explanatory, but they generally do not include technical terms and, unless set appropriately, usually suggest US spellings as default. Therefore, where a misspelling is 'incorrectly' indicated (in Microsoft *Word,* by a red wavy underline) you may wish to add the word to the dictionary file.

Grammar checkers are designed to highlight text that may be incorrectly structured (in Microsoft *Word,* this is shown by a green wavy underline). In general, these features are unreliable for scientific writing except when pointing out obvious errors, such as a missing verb. For example, academic style frequently demands the use of impersonal language involving passive tense and this is usually marked up as a matter for you to review, when it is perfectly acceptable. Note that the grammar-check tool is customisable in most programs.

The ability to move blocks of text ('cut and paste') is very valuable when reviewing and editing, and indeed can be said to have changed the nature of writing: it is now easy to write sections out of sequence, as availability of information or personal interest determines. However, there is an associated risk that the planning phase of writing (Chapter 17) is neglected, resulting in weak structure. To counteract this, you should create a plan and use temporary or permanent headings to organise your writing.

An extremely useful editing facility is the 'find' or 'search' procedure, which allows you to scan quickly through a document looking for a specified word, phrase or punctuation (in Microsoft *Word,* available from the *Home* tab, *Editing* group). This is particularly valuable when combined with a replace facility so that, for example, you could replace the word 'test' with 'trial' throughout your document simply and rapidly. *Format Painter* is a convenient function that allows you to copy the format of one section to another, without altering the text (in Microsoft *Word,* available on the *Home* tab, *Clipboard* menu). Most users will be familiar with the 'undo' command to rectify mistakes, but the 'redo' or 'repeat' command is equally valuable where repeated formatting or text entry is required.

Fonts and line spacing

The instructions for assignments will often define specific fonts and line spacing. The terminology involved is explained below.

- **Typeface:** the term for a family of characters of a particular design, each of which is given a particular name. The most commonly used for normal text is Times Roman (as used here for the main text) but many others are widely available, particularly for the better-quality printers. They fall into three broad groups: serif fonts with curves and flourishes at the ends of the characters (e.g. Times Roman); sans serif fonts without such flourishes, providing a clean, modern appearance (e.g. Arial); and decorative fonts used for special purposes only, such as the production of newsletters and notices.

- **Size:** measured in units called points. There are about 28 points per cm. The standard sizes for text are 10, 11 and 12 point, but typefaces are often available up to 72 point or more.

- **Appearance:** many typefaces are available in a variety of styles and weights. Many of these are not designed for use in scientific writing, but for desktop publishing.

- **Spacing:** double or 1.5 line spacing may be specified to allow the marker to add comments or corrections. Character spacing is usually defined by the font type ('proportional' spacing, where each character takes up a different amount of space, being more readable). There are conventions for spacing after each sentence – a double space used to

Language setting – remember to set up your document language before you start compiling text: this will ensure that the spell-check and thesaurus provide the correct spellings by default within your documents.

Quick commands for commonly used functions – most word processors provide several routes to achieve the same end result. These include: standard menus and commands; customisable toolbar commands; and key combinations. Some useful examples of key combinations in Microsoft *Word* and other programs include: *Ctrl+F* for find; *Ctrl+X* for delete, *Ctrl+P* for print, *Ctrl+C* for copy and *Ctrl+ V* for paste.

Presenting your documents – it is good practice not to mix typefaces too much in a formal document; also the font size should not differ greatly for different headings, subheadings and the text.

Preparing final documents – if no specific guidance is provided, use a 12-point proportional serif font, such as Times Roman. At this font size, 1.5 line spacing will generally be sufficient, unless otherwise specified. You can add contrast to headings by using a sans-serif typeface, such as Arial, at 11 point bold. You may wish to add extra blank lines before headings and additional half lines between paragraphs to improve readability by spacing out the text.

be the norm but single is now common. Paragraph indentation may be specified – a common format is not to indent the first paragraph in a section, but to indent thereafter (as here), but increasingly 'blocked' paragraphs are used with no indents.

- **Justification:** this is the term describing the way in which text is aligned vertically. Left justification is normal, but for formal documents, both left and right justification may be used (as here).

Commands for the above features may be accessed from the *Home* menu in Microsoft *Word*.

Table construction

Tables are widely used in the biosciences, but the conventions for their presentation are very specific (see Chapter 71). So, while you can produce basic tables via tabbing or by accessing the standard formats available on most word processor menus, this is rarely appropriate for formal submissions. Your main options here are to:

- **Use the in-built word processor table-constructing commands.** Here, the basic table outline is constructed for you (e.g. four columns by ten rows) and can be adjusted thereafter to fit your presentational needs (e.g. by changing column widths, merging cells or deleting boundary lines).

- **Use a spreadsheet to construct the table and then copy it to the word processor file.** This requires considerably more manipulation than using the word processor directly and is best reserved for special circumstances, such as very large or complex tables of data, and especially where calculations or recalculations are involved and/or the data are already stored as a spreadsheet.

Box 71.2 provides further guidance about how to create such tables to suit your requirements using Microsoft *Office* programs.

Special characters, equations and graphics

Inserting special characters – a wide range of symbols, including Greek letters, and characters from other languages is available within most word processing programs.

You can draw lines and insert other small-scale graphical features such as brackets directly within most word processors, and special characters (e.g. mathematical symbols and Greek characters) should also be available. Word processors can be used to create simple charts and diagrams, but it is better to create graphs using a more flexible and fully featured application (e.g. a spreadsheet, see Box 70.2) then import the final output into the text of a document. Files must be compatible, as with 'office' suites, but if this is so, it is a relatively straightforward procedure. Microsoft *Equation Editor* may be available on your setup to assist with mathematical formulae – consult the help menu for guidance.

Printing

Using the print preview mode – this can reveal errors of several kinds, e.g. spacing between pages, that can prevent you wasting paper and printer ink unnecessarily.

If more than one printer is attached to your PC or network, you will need to specify which one to use from the word processor's print menu. Most printers offer choices as to text and graphics quality, so choose draft (low) quality for all but your final copy since this will save both time and materials. Use a print preview option to show the page layout if it is available. Assuming that you have entered appropriate layout and font commands, printing should then be a straightforward operation. Although document layout settings are modifiable at any time, changing the page, margin and font sizes will cause your text to be rearranged, and this can be frustrating if you have spent time carefully laying out the pages.

Databases

A database is an electronic filing system whose structure is similar to a manual record-card collection. Its collection of records is termed a file. The individual items of information on each record are termed fields. Once the database is constructed, search criteria can be used to view files through various filters according to your requirements. The computerised catalogues in your library are just such a system; you enter the filter requirements in the form of author or subject keywords.

You can use a database to catalogue, search, sort and relate collections of information. The benefits of a computerised database over a manual cardfile system are:

- **the information content is easily amended or updated;**
- **printout of relevant items can be obtained;**
- **it is quick and easy to organise** through sorting and searching/selection criteria, to produce subgroups of relevant records;
- **record displays can easily be redesigned,** allowing flexible methods of presenting records according to interest;
- **relational databases can be combined,** giving the whole system immense flexibility. The older 'flat-file' databases store information in files that can be searched and sorted, but cannot be linked to other databases.

Simple database files can be constructed within spreadsheets using the columns and rows as fields and records respectively. These programs are capable of reasonably sophisticated sorting, searching and filtering operations and are probably sufficient for the types of database you are likely to require as an undergraduate. You may also make use of a bibliographic database specially constructed for that purpose (e.g. *EndNote*, see Chapter 9).

Statistical analysis packages

Statistical packages vary from small programs designed to carry out very specific statistical tasks to large sophisticated packages (e.g. *SYSTAT, SigmaStat, SPSS,* etc.) intended to provide statistical assistance, from experimental design to the analysis of results. Consider the following features when selecting a package:

- **The data entry and editing section** – this should be user-friendly, with options for transforming data.
- **Data exploration options** – these should include descriptive statistics and techniques for exploratory data analysis.
- **Hypothesis-testing techniques** – these should include ANOVA, regression analysis, multi-variate techniques, and parametric and non-parametric statistics.
- **Support for research methods** – the program should provide assistance with experimental design and sampling.
- **Output facilities** – these should be suitable for graphical and tabular formats.

Definition

Relational database – effectively, a set of structured tables of data where information from each array of information is formally related to others. The database can be used to access and construct new data sets without the need to adjust the original tables.

Choosing between a database and a spreadsheet – use a database only after careful consideration. Can the task be done within a spreadsheet? A database program can be complex to set up and usually needs to be updated regularly.

Using spreadsheet statistics functions – before using a specific statistics package, check whether your spreadsheet is capable of carrying out the form of analysis you require, as this can often be the simpler option.

Some programs have very complex data entry systems, limiting the ease of using data in different tests. Ideally, the data entry and storage system should be based on a spreadsheet system, so that subsequent editing and transformation operations are straightforward.

> **KEY POINT**   Make sure that you understand the theoretical basis for your statistical test and the computational techniques involved before using a particular program function.

Graphics and presentation packages

Microsoft *Office* programs or similar open source products can be used to achieve most coursework tasks requiring images, presentation slides or web pages: *PowerPoint* is useful for creating posters (Box 15.1) and for oral presentations (Box 16.1). For more advanced tasks, additional software may be available on your network; for example:

- *SigmaPlot* can produce graphs with floating axes;
- *Adobe Illustrator* is useful for designing complex graphics;
- *Adobe Dreamweaver* enables you to produce high-quality web pages;
- *MindGenius* can be used to produce mind maps.

Important points regarding the use of such packages are:

- **the learning time** required for some of the more complex operations can be considerable;
- **the quality of your printer** will limit the quality of your output;
- **not all files will readily import** into a word processor – you may need to save your work in a particular format. The different types of file are distinguished by the three-character filename extension, e.g. .jpg and .bmp.

> **KEY POINT**   'Default' computer graphics are rarely satisfactory for scientific presentation. Make appropriate changes to suit scientific standards and style. Box 70.1 gives a checklist for graph drawing and Box 70.2 provides guidelines for adapting *Microsoft Excel* output.

Image storage and manipulation

With the widespread use of digital images, programs that facilitate the storage and manipulation of electronic image files have become increasingly important. These programs create a library of your stored images and provide a variety of methods for organising and selecting images. Programs for image manipulation vary widely in capability, cost and associated learning time. *Adobe Photoshop* and *Paint Shop Pro* are two widely available examples. Many are highly sophisticated programs intended for graphic artists. For most scientific purposes, however, only relatively limited functions are required, such as image cropping and resizing.

Presentation using computer packages – though many computer programs enhance presentational aspects of your work, there are occasions when they can make your presentation worse. Take care to avoid the following common pitfalls:

- Default or 'chart wizard' settings for graphs may result in output that is unacceptable for the sciences (see Box 70.2).
- Fonts in labels and legends may not be consistent with other parts of your presentation.
- Some basic programs cannot produce Greek symbols (e.g. μ); do not use 'u' as a substitute. The same applies to scientific notation and superscripts: do not use 14C for ^{14}C, and replace, e.g., 1.4E $+$ 09 with 1.4×10^9. First try cutting and pasting symbols from Microsoft *Word* or, if this fails, leave space and draw the correct symbols by hand on the printout.

Take care when copying images – you may be at risk of committing plagiarism and copyright infringement (Chapter 10).

Sources for further study

Apache *OpenOffice.org.* Available:
http://www.openoffice.org/
Last accessed 01/09/15.
[Website containing information and links for downloading freeware 'office' type programs.]

Fuller, L.U., Cook, K. and Kaufeld, J. (2013) *Access 2013 for Dummies.* Wiley, New York.

Gookin, D. (2013) *Word 2013 for Dummies.* Wiley, New York.

Wang, W. (2013) *Office 2013 for Dummies.* Wiley, New York.

[And similar texts for other packages and release versions.]

STUDY EXERCISES

14.1 Investigate intermediate/advanced *Word* features. The tasks in the following list are likely to be useful in preparing assignments and report writing within the life sciences. Can you carry out all of the tasks? If not, use either a manual or the online *Help* feature to find out how to accomplish them. Tips are given in the answer section.

(a) Sort information in a list into alphabetical order.

(b) Replace a text string word or phrase with a new text string throughout your document.

(c) Replace a text string in normal font with the same text string in italics throughout your document.

(d) Add a 'header' and 'footer' to your document, the former showing the document's title and the latter containing page numbers on the bottom centre of the page.

(e) Adjust the margins of the page to give a 5-cm margin on the left and a 2-cm margin on the right.

(f) Change the type of bullets used in a list from standard (• or ■) to a different form (e.g. –, ◆ or ☑).

(g) Use the 'thesaurus' option to find a different or more suitable word to express your meaning. Try, for example, to find alternatives to the word 'alternative'.

(h) Carry out a spell-check on your document.

(i) Carry out a word count on your document and on a selected part of it.

(j) Open two documents and switch between them.

14.2 Make precise copies of tables. Copy the following tables using word processor software.

Test organism	Results of analysis (units)		
	August	September	October
X			
Y			
Z			

Results of analysis (units)			
Test organism	August	September	October
X			
Y			
Z			

14.3 Investigate what programs and packages are available to you as a student. Test each program with appropriate data, images, etc.

Answers to these study exercises are available at www.pearsoned.co.uk/practicalskills.

Communicating information

A scientific poster is a visual display of the results of an investigation, usually mounted on a rectangular board. Posters are used in undergraduate courses to display project results or assignment work, and at scientific meetings to communicate research findings.

In a written report you can include a reasonable amount of specific detail and the reader can go back and re-read difficult passages. However, if a poster is long-winded or contains too much detail, your reader is likely to lose interest.

> **KEY POINT** A poster session is like a competition – you are competing for the attention of people in a room. Because you need to attract and hold the attention of your audience, make your poster as interesting as possible. Think of it as an advertisement for your work and you will not go far wrong.

Learning from others – look at the various types of posters around your university and elsewhere; the best examples will be visual, not textual, with a clear structure that helps get the key messages across.

Preliminaries

Before considering the content of your poster, you should find out:

- **the linear dimensions of your poster area,** typically up to 1.5 m wide by 1.0 m high;
- **the composition of the poster board and the method of attachment,** whether drawing pins, Velcro tape, or some other form of adhesive; and whether these will be provided – in any case, it is safer to bring your own;
- **the time(s) when the poster should be set up and when you should attend;**
- **the room where the poster session will be held.**

Design

Plan your poster with your audience in mind, as this will dictate the appropriate level for your presentation. Aim to make your poster accessible to a broad audience. Since a poster is a *visual* display, you must pay particular attention to the presentation of information: work that may have taken hours to prepare can be ruined in a few minutes by the ill-considered arrangement of items (Fig. 15.1). Begin by making a draft sketch of the major elements of your poster. It is worth discussing your intended design with someone else, as constructive advice at the draft stage will save a lot of time and effort when you prepare the final version (or consult Sources for further study on p. 103).

Layout

One approach is to divide the poster into several smaller areas, perhaps six or eight in all, and prepare each as a separate item on a piece of card. Alternatively, you can produce a single large poster on one sheet of paper or card and store it inside a protective cardboard tube. However, a single large poster may bend and crease, making it difficult to flatten out. In addition, photographs and text attached to the backing sheet may work loose; a large printed poster with embedded images is an alternative approach.

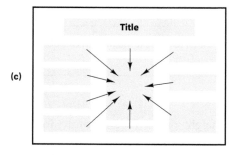

Fig. 15.1 Poster design. (a) An uninspiring design: subunits of equal area, reading left to right, are not recommended. (b) This design is more interesting and the text will be easier to read (column format). (c) An alternative approach, with a central focus and arrows/tapes to guide the reader.

Subdividing your poster means that each smaller area can be prepared on a separate piece of paper or card, of A4 size or slightly larger, making transport and storage easier. It also breaks the reading matter up into smaller pieces, looking less formidable to a potential reader. By using pieces of card of different colours you can provide emphasis for key aspects, or link text with figures or photographs.

You will need to guide your reader through the poster and headings/subheadings will help with this aspect. It may be appropriate to use either a numbering system, with large, clear numbers at the top of each piece of card, or a system of arrows (or thin tapes) to link sections within the poster (see Fig. 15.1). Make sure that the relationship is clear and that the arrows or tapes do not cross.

Title

Your chosen title should be concise (no more than eight words), specific and interesting, to encourage people to read the poster. Make the title large and bold – it should run across the top of your poster, in letters at least 4 cm high, so that it can be read from the other side of the room. Coloured spirit-based marker and block capitals drawn with a ruler work well, as long as your writing is readable and neat (the colour can be used to add emphasis). Alternatively, you can print out each word in large font, using a word processor. Details of authors, together with their addresses (if appropriate), should be given, usually across the top of the poster in somewhat smaller lettering than the title.

Text

Write in short sentences and avoid verbosity. Keep your poster as visual as possible and make effective use of the spaces between the blocks of text. Your final text should be double-spaced and should have a minimum capital letter height of 8 mm (minimum font size 36 point), preferably greater, so that the poster can be read at a distance of 1 m. One method of obtaining text of the required size is to photo-enlarge standard typescript (using a good-quality photocopier), or use a high-quality (laser) printer. It is best to avoid continuous use of text in capitals, since it slows reading and makes the text less interesting to the reader. Also avoid italic, 'balloon' or decorative styles of lettering.

> **KEY POINT**   Keep text to a minimum – aim to have a maximum of 500 words in your poster.

Subtitles and headings

These should have a capital letter height of 15–20 mm, and should be restricted to two or three words. They can be produced by word processor, photo-enlargement, by stencilling or by hand, using pencilled guidelines (but make sure that no pencil marks are visible on your finished poster).

Colour

Consider the overall visual effect of your chosen display, including the relationship between your text, diagrams and the backing board. Colour can be used to highlight key aspects of your poster. However, it is very easy to ruin a poster by the inappropriate choice and application of colour.

Presenting a poster at a formal conference – it can be useful to include your photograph for identification purposes, e.g. in the top right-hand corner of the poster.

Making up your poster – text and graphics printed on good-quality paper can be glued directly onto a contrasting mounting card: use photographic spray mountant or glue stick rather than liquid glue. Trim carefully using a guillotine to give equal margins, parallel with the paper. Photographs should be scanned into an electronic format, or placed in a window mount to avoid the tendency for their corners to curl. Another approach is to trim pages or photographs to their correct size, then encapsulate in plastic film: this gives a highly professional finish and is easy to transport.

Producing composite material for posters – *PowerPoint* is generally more useful than *Word* when you wish to include text, graphics and/or images on the same page. It is possible to use *PowerPoint* to produce a complete poster (Box 15.1), although it can be expensive to have this printed out commercially to A1 or A0 size.

Careful use of two, or at most three, complementary colours and shades will be easier on the eye and should aid comprehension. Colour can be used to link the text with the visual images (e.g. by picking out a colour in a photograph and using the same colour on the mounting board for the accompanying text). For *PowerPoint* posters, careful choice of colours for the various elements will enhance the final product (Box 15.1). Use coloured inks or water-based paints to provide colour in diagrams and figures, as felt pens rarely give satisfactory results.

Content

The typical format is that of a scientific report (see Box 19.1), i.e. with the same headings, but with a considerably reduced content. Keep references within the text to a minimum – interested parties can always ask you for further information. Also note that most posters have a summary/ conclusions section at the end, rather than an abstract.

Introduction

This should give the reader background information on the broad field of study and the aims of your own work. It is vital that this section is as interesting as possible, to capture the interest of your audience. It is often worth listing your objectives as a series of numbered points.

Materials and methods

Keep this short, and describe only the principal techniques used. You might mention any special techniques, or problems of general interest.

Results

Do not present your raw data: use data reduction wherever possible, i.e. figures and simple statistical comparisons. Graphs, diagrams, histograms and pie charts give clear visual images of trends and relationships and should be used in place of data tables (see p. 487). Final copies of all figures should be produced so that the numbers can be read from a distance of 1 m. Each should have a concise title and legend, so that it is self-contained: if appropriate, a series of numbered points can be used to link a diagram with the accompanying text. Where symbols are used, provide a key on each graph (symbol size should be at least 5 mm). Avoid using graphs straight from a written version, e.g. a project report, textbook or a paper, without considering whether they need modification to meet your requirements.

Conclusions

This is where many readers will begin, and they may go no further unless you make this section sufficiently interesting. This part needs to be the strongest section of your poster, summarising the main points. Refer to your figures here to draw the reader into the main part of your poster. A slightly larger or bolder typeface may add emphasis, though too many different typefaces can look messy. For the reference list, a smaller font can be used.

The poster session

A poster display session may be organised as part of the assessment of your coursework, and this usually means those held at scientific meetings and conferences. Staff and fellow students (delegates at conferences) will mill

Presenting at a scientific meeting – never be tempted to spend the minimum amount of time converting a piece of scientific writing into poster format – the least interesting posters are those where the author simply displays pages from a written communication (e.g. a journal article) on a poster board.

Designing the materials and methods section – photographs or diagrams of apparatus can help to break up the text of this section and provide visual interest. It is sometimes worth preparing this section in a smaller typeface.

Keeping graphs and diagrams simple – avoid composite graphs with different scales for the same axis, or with several trend lines (use a maximum of three trend lines per graph).

Listing your conclusions – a series of numbered points is a useful approach, if your findings fit this pattern.

Consider providing a handout – this is a useful way to summarise the main points of your poster, so that your readers have a permanent record of the information you have presented.

Box 15.1 How to create a poster using Microsoft *PowerPoint*

Software such as *PowerPoint* can be used to produce a high-quality poster, providing you have access to a good colour printer. However, you should avoid the standard templates available on the web as they encourage unnecessary uniformity and stifle creativity, leading to a less satisfying end result. The following steps give practical advice on creating a poster as a single *PowerPoint* slide:

1. **Sketch out your plans.** Decide on the main poster elements (images, graphs, tables and text sections) and their relationship with each other and draw out a one-page 'storyboard' (see Fig. 15.1). Think about colours for background, text and graphics (use two or three complementary colours) – dark text on a light background is clearer (high contrast), and uses less ink when printing. Also consider how you will link the elements in sequence, to guide readers through your 'story'.

2. **Get your material ready.** Collect together individual files for pictures, figures and tables. Make any required adjustments to images, graphs or tables before you import them into your poster.

3. **Create a new/blank slide.** Open *PowerPoint*. From the *Home* tab, select *New Slide* and then *Blank*. Then use the *Design* tab > *Page setup* group to select either *Landscape* or *Portrait* orientation and to set the correct page size (use *Width* and *Height* commands, or select a standard size like A4, A3, A2, etc.). Right-click on the slide and select *Ruler* and *Grid and Guides* (to help position elements within the slide – the horizontal and vertical guidelines can be dragged to different positions at later stages, as required). Use the templates (*Themes*) on the top line menu or select an appropriate background style and colour. In general, avoid setting a picture as your background as these tend to detract from the content of the poster. Before going further, save your work. Repeat this frequently and in more than one location (e.g. hard drive and USB memory stick).

4. **Add graphics.** For images, use the *Insert* tab, select *Picture* and browse to *Insert* the correct file. The *Insert, Object* command performs a similar function for *Excel* charts (graphs). Alternatively, use the copy-and-paste functions of complementary software. Once inserted, resize using the sizing handles in one of the corners (for photographs, take care not to alter one dimension relative to the other, or the image will be distorted). To re-position, put the mouse pointer over the image, left-click and hold, then drag to new location. While the *Insert* tab offers standard shapes and other useful features, you should avoid clipart (jaded and overused) and poor-quality images from the web (always use the highest resolution possible) – if you do not have your final images, use blank text boxes to show their position within the poster.

5. **Add text.** Use either the *Insert* tab to select a *Text Box* and place this on your slide, then either type in your text (use the *Enter* key to provide line spacing within the box) or copy-and-paste text from a word-processed file. You will need to consider the font size for the printed poster (e.g. for an A0 poster (size 1189 x 841 mm), a printed font size of 24 point is appropriate for the main text, with larger fonts for headings and titles. If you find things difficult to read on-screen, use the *Zoom* function on the *View* tab (either select a larger percentage in the *Zoom* box on the bottom right toolbar, or hold down the *Control* key and use the mouse wheel to scroll up (magnify) or down (reduce). Use a separate text box for each element of your poster and do not be tempted to type too much text into each box – write in phrases, using bullet points and numbered lists to keep text concise (aim for no more than 50 words per text box). Select appropriate font styles and colours using the *Home* tab, *Font* menu. For a background colour or surrounding line, right-click and use the *Format Text Effects* box and *Format Shape* box commands (line thickness and colour can then be altered). Present supplementary text elements in a smaller font – for example, details of methodology, references cited.

6. **Add boxes, lines and/or arrows** to link elements of the poster and guide the reader (e.g. Fig. 15.1). These features are available from the *Home* tab, *Drawing* group drop-down menu. Note that new inserts are overlaid on older inserts – if this proves to be a problem, select the relevant item and use the *Arrange, Order* functions within the *Drawing* group to change its relative position.

7. **Review your poster.** Get feedback from another student or tutor, e.g. on a small printed version, or use a projector to view your poster without printing (adjust the distance between projector and screen to give the correct size).

8. **Revise and edit your poster.** Revisit your work and remove as much text as possible. Delete any component that is not essential to the message of the poster. Keep graphs simple and clear (p. 487 gives further advice). White space is important in providing structure.

9. **Print the final version.** Use a high-resolution colour printer (this may be costly, so you should wait until you are sure that no further changes are needed).

Note: PowerPoint commands illustrated here may vary among the different versions of this Microsoft Office program.

Coping with questions in assessed poster sessions – you should expect to be asked questions about your poster, and to explain details of figures, methods, etc.

around, looking at the posters and chatting to their authors, who are usually expected to be in attendance. If you stand at the side of your poster throughout the session you are likely to discourage some readers, who may not wish to become involved in a detailed conversation about the poster. Stand nearby. Find something to do – talk to someone else, or browse among the other posters, but remain aware of people reading your poster and be ready to answer any queries they may raise. Do not be too discouraged if you are not asked lots of questions: remember, the poster is meant to be a self-contained, visual story, without need for further explanation.

A poster display will never feel like an oral presentation, where the nervousness beforehand is replaced by a combination of satisfaction and relief as you unwind after the event. However, it can be a very satisfying means of communication, particularly if you follow these guidelines.

Sources for further study

Alley, M. (2007) *The Craft of Scientific Presentations: Critical Steps to Succeed and Critical Errors to Avoid,* 5th edn. Springer-Verlag, New York.

Briscoe, M.H. (1996) *Preparing Scientific Illustrations: A Guide to Better Posters, Presentations and Publications,* 2nd edn. Springer-Verlag, New York.

Davis, M.F. (2004) *Scientific Papers and Presentations,* 2nd edn. Academic Press, New York.

Gosling, P.J. (1999) *Scientist's Guide to Poster Presentations.* Kluwer, New York.

Hess, G., Tosney, K. and Liegel, L. *Creating Effective Poster Presentations.* Available: http://www.ncsu.edu/project/posters/.
Last accessed 01/09/15.

Nicol, A.A.M. (2010) *Displaying Your Findings,* 6th edn. American Psychological Association, Washington.

STUDY EXERCISES

15.1 Design a poster. Working with one or more partners from your year group, decide on a suitable poster topic (perhaps something linked to your current teaching programme). Working individually, make an outline plan of the major elements of the poster, with appropriate subheadings and a brief indication of the content and relative size of each element (including figures, diagrams and images). Exchange draft plans with your partners and arrange a session where you can discuss their merits and disadvantages.

15.2 Prepare a checklist for assessing the quality of a poster presentation. After reading through this chapter, prepare a 10-point checklist of assessment criteria under the heading 'What makes a good poster presentation?' Compare your list with the one that we have provided online – do you agree with our criteria, or do you prefer your own list (and can you justify your preferences)?

15.3 Evaluate the posters in your university. Most universities have a wide range of academic posters on display. Some may cover general topics (e.g. course structures), while others may deal with specific research topics (e.g. poster presentations from past conferences). Consider their good and bad features (if you wish to make this a group exercise, you might compare your evaluation with that of other students in a group discussion session).

Answers to these study exercises are available at www.pearsoned.co.uk/practicalskills.

Opportunities for practising speaking skills – these include:

- answering lecturers' questions;
- contributing in tutorials;
- talking to informal groups;
- giving your views at formal (committee) meetings;
- demonstrating or explaining to other students, e.g. during a practical class;
- asking questions in lectures/seminars;
- answering an examiner's questions in an oral exam.

Learning from experience – use your own experience of good and bad lecturers to shape your performance. Some of the more common errors include:

- speaking too quickly;
- reading from notes or from slides and ignoring the audience;
- inexpressive, impersonal or indistinct speech;
- distracting mannerisms;
- poorly structured material with little emphasis on key information;
- factual information too complex and detailed;
- too few or too many visual aids.

Testing the room – if possible, try to rehearse your talk in the room in which it will be presented. This will help you to make allowance for layout of equipment, lighting, acoustics and sight lines that might affect the way you deliver your talk. It will also put you more at ease on the day, because of the familiarity of the surroundings.

Most students feel very nervous about giving talks. This is natural, since very few people are sufficiently confident and outgoing that they look forward to speaking in public. Additionally, the technical nature of the subject matter may give you cause for concern, especially if you feel that some members of the audience have a greater knowledge than you have, e.g. your tutors. However, this is a fundamental method of scientific communication and an important transferable skill; therefore, it forms an important component of many courses.

The comments in this chapter apply equally to informal talks, e.g. those based on assignments and project work, and to more formal conference presentations. It is hoped that the advice and guidance given below will encourage you to make the most of your opportunities for public speaking, but there is no substitute for practice. Do not expect to find all of the answers from this, or any other, book. Rehearse, and learn from your own experience.

> **KEY POINT** The three 'Rs' of successful public speaking are: reflect – give sufficient thought to all aspects of your presentation, particularly at the planning stage; rehearse – to improve your delivery; revise – modify the content and style of your material in response to your own ideas and to the comments of others.

Preparation

Preliminary information

Begin by marshalling the details needed to plan your presentation, including:

- **the duration of the talk;**
- **whether time for questions is included;**
- **the size and location of the room;**
- **the projection/lighting facilities provided,** and whether pointers or similar aids are available.

It is especially important to find out whether the room has the necessary equipment for digital projection (e.g. PC, projector and screen, black-out curtains or blinds, appropriate lighting) or overhead projection before you prepare your audio-visual aids. If you concentrate only on the spoken part of your presentation at this stage, you are inviting trouble later on. Have a look around the room and try out the equipment at the earliest opportunity, so that you are able to use the lights, projector, etc. with confidence. For digital projection systems, check that you can load/present your material. Box 16.1 gives practical advice on the use of *PowerPoint*.

Audio-visual aids

If you plan to use overhead transparencies, find out whether your department has facilities for their preparation, whether these facilities are available for your use, and the cost of materials. Adopt the following guidelines:

Box 16.1 Tips on preparing and using Microsoft *PowerPoint* slides in a spoken presentation

Microsoft *PowerPoint* can be used to produce high-quality visual aids, assuming a computer and digital projector are available in the room where you intend to speak. The presentation is produced as a series of electronic 'slides' onto which you can insert images, diagrams and text. When creating your slides, bear the following points in mind:

- **Plan the structure of your presentation.** Decide on the main topic areas and sketch out your ideas on paper. Think about what material you will need (e.g. pictures, graphs) and what colours to use for background and text.
- **Choose slide layouts according to purpose.** Once *PowerPoint* is running, from the *Home* tab select *Layout*. You can then add material to each new slide to suit your requirements.
- **Select your background with care.** Many of the preset background templates *(Themes)* available within the *Design* tab are best avoided, since they are overused and fussy, diverting attention from the content of the slides. Conversely, flat, dull backgrounds may seem uninteresting, while brightly coloured backgrounds can be garish and distracting. Choose whether to present your text as a light-coloured font on a dark background (more restful but perhaps less engaging if the room is dark) or a dark-coloured font on a light background (more lively).
- **Use visual images throughout.** Remember the maxim 'a picture is worth ten thousand words'. A presentation composed entirely of text-based slides will be uninteresting: adding images and diagrams will brighten up your talk considerably (use the *Insert* menu, *Picture* option). Images can be taken with a digital camera, scanned in from a printed version or copied and pasted from the web, but you should take care not to break copyright regulations. 'Clipart' is copyright-free, but should be used sparingly as most people will have seen the images before and they are rarely wholly relevant. Diagrams can be made from components created using the *Home* tab *Drawing* group while graphs and tables can be imported from other programs, e.g. *Excel* (Box 15.1 gives further specific practical advice on adding graphics, saving files, etc.).
- **Keep text to a minimum.** Aim for no more than 20 words on a single slide (e.g. four/five lines containing a few words per line). Use headings and subheadings to structure your talk: write only key words or phrases as 'prompts' to remind you to cover a particular point during your talk – never be tempted to type whole sentences as you will then be reduced to reading these from the screen during your presentation, which is boring.
- **Use a large, clear font style.** Use the *Slide Master* option within the *View menu* to set the default font to a non-serif style such as Arial, or Comic Sans MS, and an appropriate colour. Default fonts for headings and bullet points are intentionally large, for clarity. Do not reduce these to anything less than 28 point font size (preferably larger) to cram in more words: if you have too much material, create a new slide and divide up the information.
- **Animate your material.** The *Animation* tab provides a function that enables you to introduce the various elements within a slide, e.g. text can be made to *Appear* one line at a time, to prevent the audience from reading ahead and help maintain their attention.
- **Do not overdo the special effects.** *PowerPoint* has a wide range of features that allow complex slide transitions and animations, additional sounds, etc., but these quickly become irritating to an audience unless they have a specific purpose within your presentation.
- **Always edit your slides before use.** Check through your slides and cut out any unnecessary words, adjust the layout and animation. Remember the maxim 'less is more' – avoid too much text; too many bullet points; too many distracting visual effects or sounds.

When presenting your talk:

- **Work out the basic procedures beforehand.** Practise, to make sure that you know how to move forwards and backwards in your slideshow, turn the screen on and off, hide the mouse pointer, etc.
- **Do not forget to engage your audience.** Despite the technical gadgetry, *you* need to play an active role in the presentation, as explained elsewhere in this chapter.
- **Do not go too fast.** Sometimes, new users tend to deliver their material too quickly: try to speak at a normal pace and practise beforehand.
- **Consider whether to provide a handout.** *PowerPoint* has several options, including some that provide space for notes (e.g. Fig. 4.3). However, a handout should not be your default option, as there is a cost involved.

Note: PowerPoint commands illustrated here vary among the different versions of the Microsoft *Office* program.

Using audio-visual aids – do not let the equipment and computer gadgetry distract you from the essential rules of good speaking (Box 16.2): remember that *you* are the presenter.

- **Keep text to a minimum:** present only the key points, with up to 20 words per slide/transparency.
- **Make sure the text is readable:** try out your material beforehand.
- **Use several simpler figures rather than a single complex graph.**
- **Avoid too much colour on slides:** blue and black are easier to read than red or green.
- **Do not mix slides and transparencies** as this is often distracting.
- **Use spirit-based pens for transparencies:** use alcohol for corrections.
- **Transparencies can be produced from typewritten or printed text using a photocopier,** often giving a better product than pens. Note that you must use special heat-resistant acetate sheets for photocopying.

Electronic presentation software (e.g. *PowerPoint*) can replace these specialist requirements, as long as the necessary facilities are available for your talk (see below).

Audience

You should consider your audience at the earliest stage, since they will determine the appropriate level for your presentation. If you are talking to fellow students you may be able to assume a common level of background knowledge. In contrast, a research lecture given to your department, or a paper at a meeting of a scientific society, will be presented to an audience from a broader range of backgrounds. An oral presentation is not the place for a complex discussion of specialised information: build up your talk from a low level. The speed at which this can be done will vary according to your audience. As long as you are not boring or patronising, you can cover basic information without losing the attention of the more knowledgeable members in your audience.

Pitching your talk at the right level – the general rule should be: 'do not overestimate the background knowledge of your audience'. This sometimes happens in student presentations, where fears about the presence of 'experts' can encourage the speaker to include too much detail, overloading the audience with facts.

Content

While the specific details in your talk will be for you to decide, most spoken presentations share some common features of structure, as described below.

Introductory remarks

It is vital to capture the attention of your audience at the outset. Consequently, you must make sure your opening comments are strong, otherwise your audience will lose interest before you reach the main message. Remember it takes a sentence or two for an audience to establish a relationship with a new speaker. Your opening sentence should be some form of preamble and should not contain any key information. For a formal lecture, you might begin with 'Thank you for that introduction. My talk today is about . . .' then restate the title and acknowledge other contributors, etc. You might show a transparency or slide with the title printed on it, or an introductory photograph, if appropriate. This should provide the necessary settling-in period.

Getting the introduction right – a good idea is to have an initial slide giving your details and the title of your talk, and a second slide telling the audience how your presentation will be structured. Make eye contact with all sections of the audience during the introduction.

After these preliminaries, you should introduce your topic. Begin your story on a strong note – avoid timid or apologetic phrases.

Opening remarks are unlikely to occupy more than 10% of the talk. However, because of their significance, you might reasonably spend up to 25% of your preparation time on them.

What to cover in your introduction – You should:

- explain the structure of your talk;
- set out your aims and objectives;
- explain your approach to the topic.

> **KEY POINT** Make sure you have practised your opening remarks, so that you can deliver the material in a flowing style, with less chance of mistakes.

The main message

This section should include the bulk of your experimental results or literature findings, depending on the type of presentation. Keep details of methods to the minimum needed to explain your data. This is *not* the place for a detailed description of equipment and experimental protocol (unless it is a talk about methodology). Results should be presented in an easily digested format.

> **KEY POINT** Do not expect your audience to cope with large amounts of data; use a maximum of six numbers per slide. Remember that graphs and diagrams are usually better than tables of raw data, since the audience will be able to see the visual trends and relationships in your data (p. 487).

Present summary statistics (Chapter 73) rather than individual results. Show the final results of any analyses in terms of the statistics calculated, and their significance (p. 526), rather than dwelling on details of the procedures used. Figures should not be crowded with unnecessary detail. Every diagram should have a concise title and the symbols and trend lines should be clearly labelled, with an explanatory key where necessary. When presenting graphical data (Chapter 70) always 'introduce' each graph by stating the units for each axis and describing the relationship for each trend line or data set.

> **KEY POINT** Use summary slides at regular intervals, to maintain the flow of the presentation and to emphasise the main points.

Take the audience through your story step-by-step at a reasonable pace. Try not to rush the delivery of your main message due to nervousness. Avoid complex, convoluted story-lines – one of the most distracting things you can do is to fumble backwards through *PowerPoint* slides or overhead transparencies. If you need to use the same diagram or graph more than once then you should make two (or more) copies. In a presentation of experimental results, you should discuss each point as it is raised, in contrast to written text where the results and discussion may be in separate sections. The main message typically occupies approximately 80% of the time allocated to an oral presentation (Fig. 16.1).

Concluding remarks

Having captured the interest of your audience in the introduction and given them the details of your story in the middle section, you must now bring your talk to a conclusion. Do not end weakly, e.g. by running out of steam on the last slide. Provide your audience with a clear 'take-home message', by returning to the key points in your presentation. It is often appropriate to prepare a slide or overhead transparency listing your main conclusions as a numbered series.

Signal the end of your talk by saying 'finally. . .', 'in conclusion. . .', or a similar comment and then finish speaking after that sentence. Your audience will lose interest if you extend your closing remarks beyond this

Allowing time for slides – as a rough guide you should allow at least two minutes per illustration, although some diagrams may need longer, depending on content. Make a note of the halfway point, to help you check timing/pace.

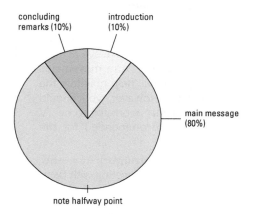

Fig. 16.1 Pie chart showing time allocation for a typical presentation.

Final remarks – make sure you give the audience sufficient time to assimilate your final slide: some of them may wish to write down the key points. Alternatively, you might provide a handout, with a brief outline of the aims of your study and the major conclusions.

Box 16.2 Hints on spoken presentations

In planning the delivery of your talk, bear the following aspects in mind:

- **Using notes.** Many accomplished speakers use abbreviated notes for guidance, rather than reading word-for-word from a prepared script. When writing your talk:

  (a) **Consider preparing your first draft as a full script:** write in spoken English and keep the text simple, to avoid a formal, impersonal style. Your aim should be to *talk* to your audience, not to *read* to them.

  (b) **If necessary, use note-cards with key words and phrases:** it is best to avoid using a full script in the final presentation. As you rehearse and your confidence improves, a set of note-cards may be an appropriate format. Mark the position of slides/key points, etc.: each note-card should contain details of structure as well as content. Your notes should be written/printed in text large enough to be read easily during the presentation (also check that the lecture room has a lectern light or you may have problems reading your notes if the lights are dimmed). Each note-card or sheet should be clearly numbered, so that you do not lose your place.

  (c) **Decide on the layout of your talk:** give each subdivision a heading in your notes, so that your audience is made aware of the structure.

  (d) **Memorise your introductory/closing remarks:** you may prefer to rely on a full written version for these sections, in case your memory fails, or if you suffer 'stage fright'.

  (e) **Using *PowerPoint*** (Box 16.1): here, you can either use the 'notes' option *(View* tab; *Notes Page)*, or you may even prefer to dispense with notes entirely, since the slides will help structure your talk, acting as an *aide-memoire* for your material.

- **Work on your timing.** It is essential that your talk is the right length and the correct pace:

  (a) **Rehearse your presentation:** ask a friend to listen and to comment constructively on those parts that were difficult to follow, to improve your performance.

  (b) **Use 'split times' to pace yourself:** following an initial run-through, add the times at which you should arrive at the key points of your talk to your notes.

These timing marks will help you keep to time during the final presentation.

  (c) **Avoid looking at your wristwatch when speaking:** this sends a negative signal to the audience. Use a wall clock (where available), or take off your watch and put it beside your notes so that you can glance at it without distracting the audience.

- **Consider your image.** Make sure that the image you project is appropriate for the occasion:

  (a) **Think about what to wear:** aim to be respectable without 'dressing up', otherwise your message may be diminished.

  (b) **Maintain a good posture:** it will help your voice projection if you stand upright, rather than slouching, or leaning over a lectern.

  (c) **Deliver your material with expression:** project your voice towards the audience at the back of the room and make sure you look round to make eye contact with all sections of the audience. Arm movements and subdued body language will help maintain the interest of your audience. However, you should avoid extreme gestures (it may work for some TV personalities but it is not recommended for the beginner).

  (d) **Try to identify and control any repetitive mannerisms:** repeated 'empty' words/phrases, fidgeting with pens, keys, etc. will distract your audience. Note-cards held in your hand give you something to focus on, while laser pointers will show up any nervous hand tremors. Practising in front of a mirror may help.

- **Think about questions.** Once again, the best approach is to prepare beforehand:

  (a) **Consider what questions are likely to come up, and prepare brief answers.** However, do not be afraid to say 'I don't know': your audience will appreciate honesty rather than vacillation if you do not have an answer for a particular question.

  (b) **If no questions are asked, you might pose a question yourself** and then ask for opinions from the audience: if you use this approach, you should be prepared to comment briefly if your audience has no suggestions, to avoid the presentation ending in an embarrassing silence.

point. You may add a simple end phrase (for example, 'thank you') as you put your notes into your folder, but do not say 'that's all folks!', or make any similar offhand remark. Finish as strongly and as clearly as you started. Box 16.2 gives further advice on presentation.

Sources for further study

Alley, M. (2007) *The Craft of Scientific Presentations: Critical Steps to Succeed and Critical Errors to Avoid,* 5th edn. Springer-Verlag, New York.

Capp, C.C. and Capp, G.R. (1989) *Basic Oral Communication,* 5th edn. Prentice Hall, Harlow.

Radel, J. *Oral Presentations.* Available: http://people.eku.edu/ritchisong/oralpres.html Last accessed 01/09/15.

STUDY EXERCISES

16.1 Prepare a checklist for assessing the quality of an oral presentation. After reading through this chapter, prepare a 10-point checklist of assessment criteria under the heading 'What makes a good oral presentation?'. Compare your list with the one that we have provided online – do you agree with our criteria, or do you prefer your checklist? (Can you justify your preferences?)

16.2 Evaluate the presentation styles of other speakers. There are many opportunities to assess the strengths and weaknesses of academic 'public speakers', including your lecturers, seminar speakers, presenters of TV documentaries, etc. Decide in advance how you are going to tackle the evaluation (e.g. with a quantitative marking scheme, or a less formal procedure).

16.3 Rehearse a talk and get feedback on your performance. There are a number of approaches you might take, including: (i) recording and reviewing your presentation using a digital movie camera; or (ii) giving your talk to a small group of fellow students and asking them to provide constructive feedback.

Answers to these study exercises are available at www.pearsoned.co.uk/practicalskills.

17 General aspects of scientific writing

Time management – practical advice is given in Chapter 2.

Monday:	morning afternoon evening	Lectures (University) Practical (University) Initial analysis and brainstorming (Home)
Tuesday:	morning afternoon evening	Lectures (University) Locate sources (Library) Background reading (Library)
Wednesday:	morning afternoon evening	Background reading (Library) Squash (Sports hall) Planning (Home)
Thursday:	morning afternoon evening	Lectures (University) Additional reading (Library) Prepare outline (Library)
Friday:	morning afternoon evening	Lab class (University) Write first draft (Home) Complete first draft (Home)
Saturday:	morning afternoon evening	Shopping (Town) Review first draft (Home) Revise first draft (Home)
Sunday:	morning afternoon evening	Free Produce final copy (Home) Proof read and print essay (Home)
Monday:	morning	Final read-through and check Submit essay (deadline midday) Go out with friends

Fig. 17.1 Example timetable for writing a short essay.

Creating an outline – an informal outline can be made simply by indicating the order of sections on a spider diagram (as shown in Fig. 17.2).

Written communication is an essential component of all sciences. Most courses include writing exercises in which you will learn to describe ideas and results accurately, succinctly and in an appropriate style and format. The following features/aspects are common to all forms of scientific writing.

Organising your time

Making a timetable at the outset helps ensure that you give each stage adequate attention and complete the work on time (e.g. Fig. 17.1). To create and use a timetable:

1. **Break down the task into stages.**
2. **Decide on the proportion of the total time each stage should take.**
3. **Set realistic deadlines for completing each stage,** allowing some time for slippage.
4. **Refer to your timetable frequently as you work:** if you fail to meet one of your deadlines, make a serious effort to catch up as soon as possible.

> **KEY POINT** The appropriate allocation of your time to reading, planning, writing and revising will differ according to the task in hand (see Chapters 18–20).

Organising your information and ideas

Before you write, you need to gather and/or think about relevant material (Chapters 9 and 10). You must then decide:

- **what needs to be included and what does not;**
- **in what order it should appear.**

Start by jotting down headings for everything of potential relevance to the topic (this is sometimes called 'brainstorming'). A spider diagram (Fig. 17.2) or a mind map (Fig. 4.2) will help you organise these ideas. The next stage is to create an outline of your text (Fig. 17.3). Outlines are valuable because they:

- **force you to think about and plan the structure;**
- **provide a checklist** so nothing is missed out;
- **ensure the material is balanced in content and length;**
- **help you organise figures and tables,** by showing where they will be used.

> **KEY POINT** A suitable structure is essential to the narrative of your writing, and should be carefully considered at the outset.

In an essay or review, the structure of your writing should help the reader to assimilate and understand your main points. Subdivisions of the topic could simply be related to the physical nature of the subject matter (e.g. levels of organisation of a protein) and should proceed logically (e.g. primary structure then secondary, etc.). A chronological approach is good for evaluation of past work (e.g. the development of the concept of DNA as

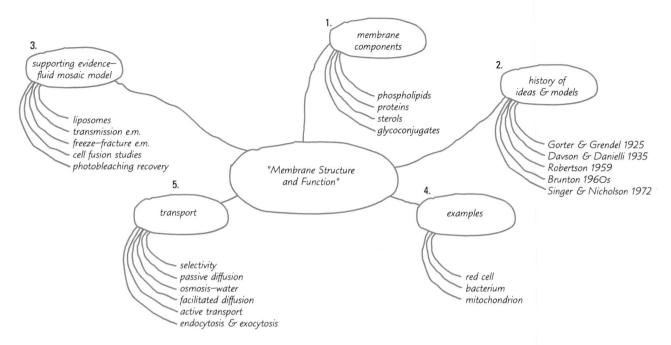

Fig. 17.2 Spider diagram showing how you might 'brainstorm' an essay with the title 'Membrane Structure and Function'. Write out the essay title in full to form the spider's body, and as you think of possible content place headings around this to form its legs. Decide which headings are relevant and which are not and use arrows to note connections between subjects. This may influence your choice of order and may help to make your writing flow because the links between paragraphs will be natural. You can make an informal outline directly on a spider diagram by adding numbers indicating a sequence of paragraphs (as shown). This method is best when you must work quickly, as with an essay written under exam conditions.

Fig. 17.3 Formal outlines. These are useful for a long piece of work where you or the reader might otherwise lose track of the structure. The headings for sections and paragraphs are simply written in sequence with the type of lettering and level of indentation indicating their hierarchy. Two different forms of formal outline are shown, a minimal form (a) and a numbered form (b). Note that the headings used in an outline are often repeated within the essay to emphasise its structure. The content of an outline will depend on the time you have available and the nature of the work, but the most detailed hierarchy you should reasonably include is the subject of each paragraph.

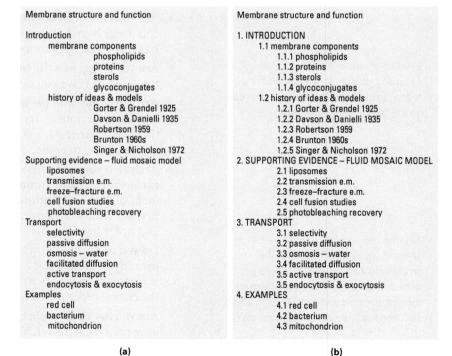

(a) (b)

the hereditary material), whereas a step-by-step comparison might be best for certain exam questions (e.g. 'Discuss the differences between prokaryotes and eukaryotes'). There is little choice about structure for practical and project reports (see p. 121).

Writing

Adopting a scientific style

Your main aim in developing a scientific style should be to put your message across directly and unambiguously. While you can try to achieve this through a set of 'rules' (see Box 17.1), you may find other requirements driving your writing in a contradictory direction. For instance, the need to be accurate and complete may result in text littered with technical terms, and the flow may be continually interrupted by references to the literature. The need to be succinct also affects style and readability through the use of, for example, stacked noun-adjectives (e.g. 'restriction fragment length polymorphism') and acronyms (e.g. 'RFLP'). Finally, style is very much a matter of taste and each tutor, examiner, supervisor or editor will have pet loves and hates which you may have to accommodate. Different assignments will need different styles; Box 17.2 gives further details.

Developing technique

Writing is a skill that can be improved, but not instantly. You should analyse your deficiencies with the help of feedback from your tutors, be prepared to change work habits (e.g. start planning your work more carefully, or find out more about punctuation), and be willing to learn from some of the excellent texts that are available on scientific writing (p. 116).

Getting started

A common problem is 'writer's block' – inactivity or stalling brought on by a variety of causes. If blocked, ask yourself these questions:

- **Are you comfortable with your surroundings?** Make sure you are seated comfortably at a reasonably clear desk and have minimised the possibility of interruptions and distractions.
- **Are you trying to write too soon?** Have you clarified your thoughts on the subject? Have you done enough preliminary reading?
- **Are you happy with the underlying structure of your work?** If you have not made an outline, try this. If you are unhappy because you cannot think of a particular detail at the planning stage, just start writing – it is more likely to come to you while you are thinking of something else.
- **Are you trying to be too clever?** Your first sentence does not have to be earth-shattering in content or particularly smart in style. A short statement of fact or a definition is fine. If there will be time for revision, first get your ideas down on paper and then revise grammar, content and order later.
- **Do you really need to start writing at the beginning?** Try writing the opening remarks after a more straightforward part. For example, with reports of laboratory work, the Materials and Methods section may be the easiest place to start.
- **Are you too tired to work?** Do not try to 'sweat it out' by writing for long periods at a stretch: stop frequently for a rest.

Improving your writing skills – you need to take a long-term view if you wish to improve this aspect of your work. An essential preliminary is to invest in and make full use of a personal reference library (see Box 17.3).

Benefits of talking about your work – discussing your topic with a friend or colleague might bring out ideas or reveal deficiencies in your knowledge.

Writing with a word processor – use the dynamic/interactive features of the word processor (Chapter 14) to help you get started: first make notes on structure and content, then expand these to form a first draft and finally revise/improve the text.

Box 17.1 How to achieve a clear, readable style

Words and phrases

- Choose short, clear words and phrases rather than long ones: e.g. use 'build' rather than 'fabricate'; 'now' rather than 'at the present time'. At certain times, technical terms must be used for precision, but do not use jargon if you do not have to.

- Do not worry too much about repeating words, especially when introducing an alternative might subtly alter your meaning.

- Where appropriate, use the first person to describe your actions ('We decided to'; 'I conclude that'), but not if this is specifically discouraged by your supervisor or department.

- Favour active forms of writing ('the observer completed the survey in ten minutes') rather than a passive style ('the survey was completed by the observer in ten minutes').

- Use tenses consistently. Past tense is always used for materials and methods ('samples were taken from. . .') and for reviewing past work ('Smith (1990) concluded that. . .'). The present tense is used when describing data ('Fig. 1 shows. . .'), for generalisations ('Most authorities agree that. . .') and conclusions ('To conclude, . . .').

- Use statements in parentheses sparingly – they disrupt the reader's attention to your central theme.

- Avoid clichés and colloquialisms – they are usually inappropriate in a scientific context.

Punctuation

- Try to use a variety of types of punctuation to make the text more interesting to read.

- Decide whether you wish to use 'closed' punctuation (frequent commas at the end of clauses) or 'open' punctuation (less frequent punctuation) and be consistent.

- Do not link two sentences with a comma. Use a full stop, this is an example of what *not* to do.

- Pay special attention to apostrophes, using the following rules:

  - To indicate possession, use an apostrophe before an 's' for a singular word (e.g. the rat's temperature was . . .') and after the s for a plural word ending in s (e.g. the rats' temperatures were = the temperatures of the rats were). If the word has a special plural (e.g. woman → women) then use the apostrophe before the s (the women's temperatures were . . .).

  - When contracting words, use an apostrophe (e.g. do not = don't; it's = it is), but remember that contractions are generally *not* used in formal scientific writing.

  - Do *not* use an apostrophe for 'its' as the possessive form of 'it' (e.g. 'the university and its surroundings'). Note that 'it's' is reserved for 'it is'. This is an exception to the general rule and a very common mistake.

  - Never use an apostrophe to indicate plurals. Even for abbreviations, the accepted style is now to omit the apostrophe for the plural (e.g. write 'the ELISAs were . . .').

Sentences

- Do not make them overlong or complicated.

- Introduce variety in structure and length.

- If unhappy with the structure of a sentence, try chopping it into a series of shorter sentences.

Paragraphs

- Get the paragraph length right – five sentences or so. Do *not* submit an essay that consists of a single paragraph, nor one that contains single sentence paragraphs.

- Make sure each paragraph is logical, dealing with a single topic or theme.

- Take care with the first sentence in a paragraph (the 'topic' sentence); this introduces the theme of the paragraph. Further sentences should then develop this theme, e.g. by providing supporting information, examples or contrasting cases.

- Use 'linking' words or phrases to maintain the flow of the text within a paragraph (e.g. 'for example'; 'in contrast'; 'however'; 'on the other hand').

- Make your text more readable by adopting modern layout style. The first paragraph in any section of text is usually *not* indented, but following paragraphs may be (by the equivalent of three character spaces). In addition, the space between paragraphs should be slightly larger than the space between lines. Follow departmental guidelines if these specify a format.

- Group paragraphs in sections under appropriate headings and subheadings to reinforce the structure underlying your writing.

- Think carefully about the first and last paragraphs in any piece of writing: these are often the most important as they respectively set the aims and report the conclusions.

Note: If you are not sure what is meant by any of the terms used here, consult a guide on writing (see sources for further study).

Box 17.2 Using appropriate writing styles for different purposes (with examples)

Note that courses tend to move from assignments that are predominantly descriptive in the early years to a more analytical approach towards the final year (see Chapter 5). Also, different styles may be required in different sections of a write-up, e.g. descriptive for introductory historical aspects, becoming more analytical in later sections.

Descriptive writing

This is the most straightforward style, providing factual information on a particular subject and is most appropriate:

- in essays where you are asked to 'describe' or 'explain' (p. 118)
- when describing the results of a practical exercise, e.g.: The experiment shown in Figure 1 confirmed that enzyme activity was strongly influenced by temperature, as the rate observed at 37°C was more than double that seen at 20°C.'

However, in literature reviews and essays where you are asked to 'discuss' (p. 118) a particular topic, the descriptive approach is mostly inappropriate, as in the following example, where a large amount of specific information from a single scientific paper has been used, without any attempt to highlight the most important points:
'In a study carried out between July and October 2011, a total of 225 sputum samples from patients attending 25 different clinics in England and Wales were screened. Bacteria were isolated from 67.6% of these samples, with 47.42% of the samples giving *Pseudomonas aeruginosa*, 11.76% *Burkholderia cepacia* and 8.59% *Stentrophomonas maltophilia* (Grey and Gray, 2012).'

In the most extreme examples, whole paragraphs or pages of essays may be based on descriptive factual detail from a single source, often with a single citation at the end of the material, as above: such essays often score low marks in essays where evidence of deeper thinking is required (Chapter 5).

Comparative writing

This technique is an important component of academic writing, and it will be important to develop your comparative writing skills as you progress through your course. Its applications include:

- answering essay questions and assignments of the 'compare and contrast' type (p. 118)
- comparing your results with previously published work in the Discussion section of a practical report.

To use this style, first decide on those aspects you wish to compare and then consider the material (e.g. different literature sources) from these aspects – in what ways do they agree or disagree with each other? One approach is to compare/contrast a different aspect in each paragraph. At a practical level, you can use 'linking' words and phrases to help orientate your reader, as you move between aspects where there is agreement and disagreement. These include, for agreement: 'in both cases'; 'in agreement with'; 'is also shown by the study of'; 'similarly'; 'in the same way', and for disagreement: 'however'; 'although'; 'in contrast to'; 'on the other hand'; 'which differs from'. The comparative style is fairly straightforward, once you have decided on the aspects to be compared. The following brief example compares two different studies using this style:
'While Grey and Gray (2011) reported that *Pseudomonas aeruginosa* was present in 47.4% of 225 UK sputum samples, Black and White (2012) showed that 89.1% of sputum samples from 2592 patients were positive for this bacterium.'

Comparative text typically makes use of two or more references per paragraph.

Analytical writing

Typically, this is the most appropriate form of writing for:

- a review of scientific literature on a particular topic;
- an essay where you are asked to 'discuss' (p. 118) different aspects of a particular topic;
- evaluating a number of different published sources within the Discussion of a final-year project report.

By considering the significance of the information provided in the various sources you have read, you will be able to take a more critical approach. Your writing should evaluate the importance of the material in the context of your topic (see also Chapter 10). In analytical writing, you need to demonstrate critical thinking (p. 64) and personal input about the topic in a well-structured text that provides clear messages, presented in a logical order and demonstrating synthesis from a number of sources by appropriate use of citations (p. 55). Detailed information and relevant examples are used only to explain or develop a particular aspect, and not simply as 'padding' to bulk up the essay. The following example show how detail can be used for explanation:
'*Pseudomonas aeruginosa* is often isolated from sputum samples of cystic fibrosis patients: a short-term UK study with a relatively small sample size (225 patients) isolated this bacterium from around half of all samples (Grey and Gray, 2011), while a longer-term study with a far larger sample size (2592 patients) gave an isolation rate of almost 90% (Black and White, 2012).'

Analytical writing is based on a broad range of sources, typically with several citations per paragraph.

Note: do not allow text box to overrun this page.

Box 17.3 How to improve your writing ability by consulting a personal reference library

Using dictionaries

We all know that a dictionary helps with spelling and definitions, but how many of us use one effectively? You should:

- Keep a dictionary beside you when writing and always use it if in any doubt about spelling or definitions.
- Use it to prepare a list of words which you have difficulty in spelling: apart from speeding up the checking process, the act of writing out the words helps commit them to memory.
- Use it to write out a personal glossary of terms. This can help you memorise definitions. From time to time, test yourself.

Not all dictionaries are the same! Ask your tutor or supervisor whether he/she has a preference and why. Try out the *Oxford Advanced Learner's Dictionary*, which is particularly useful because it gives examples of use of all words and helps with grammar, e.g. by indicating which prepositions to use with verbs. Dictionaries of biology tend to be variable in quality, possibly because the subject is so wide and new terms are continually being coined. *Henderson's Dictionary of Biological Terms* is a useful example.

Using a thesaurus

A thesaurus contains lists of words of similar meaning grouped thematically; words of opposite meaning always appear nearby.

- Use a thesaurus to find a more precise and appropriate word to fit your meaning, but check definitions of unfamiliar words with a dictionary.
- Use it to find a word or phrase 'on the tip of your tongue' by looking up a word of similar meaning.
- Use it to increase your vocabulary.

Roget's Thesaurus is the standard. Several publishers produce a combined dictionary and thesaurus.

Using guides for written English

These provide help with the use of words.

- Use guides to solve grammatical problems such as when to use 'shall' or 'will', 'which' or 'that', 'effect' or 'affect', 'can' or 'may', etc.
- Use them for help with the paragraph concept and the correct use of punctuation.
- Use them to learn how to structure writing for different tasks.

Reviewing and revising your text

Detailed review of each draft is strongly advised for all writing, apart from in exams. When using a word processor, this can be a simple process. Where possible, schedule your writing so you can leave each draft to 'settle' for at least a couple of days. When you return to it fresh, you will see more easily where improvements can be made. Try the following structured review process, each stage being covered in a separate scan of your text:

1. **Examine content.** Have you included everything you need to? Is all the material relevant?
2. **Check grammar and spelling.** Can you spot any 'howlers'?
3. **Focus on clarity.** Is the text clear and unambiguous? Does each sentence really say what you want it to say?
4. **Be succinct.** What could be missed out without spoiling the essence of your work? It might help to imagine that an editor has set you the target of reducing the text by 15%.
5. **Improve style.** Could the text read better? Consider the sentence and paragraph structure and the way your text develops to its conclusion.

Reviewing your text – to improve clarity and shorten your text, 'distil' each sentence by taking away unnecessary words and 'condense' words or phrases by choosing a shorter alternative.

Learning from others – ask a colleague to read through your draft and comment on its content and overall structure.

Common errors

These include:

- **Problems over singular and plural words** ('a bacteria is'; 'the results shows')

- **Verbose text** ('One definition that can be employed in this situation is given in the following sentence.')
- **Misconstructed sentences** ('Health and safety regulations should be made aware of . . .')
- **Misuse of punctuation,** especially commas and apostrophes (for examples, see Box 17.1).
- **Poorly constructed paragraphs** (for advice, see Box 17.1).

Sources for further study

Butterfield, J. (ed.) (2004) *Fowler's Modern English Usage,* revised 4th edn. Oxford University Press, Oxford.

Clark, R. *The English Style Book. A Guide to the Writing of Scholarly English.* Available: http://www.literacy.com/stylebook/stylebook.php Last accessed 01/09/15.

Day, R.A. and Sakaduski, N. (2011) *Scientific English: a Guide for Scientists and Other Professionals.* Greenwood, Westport.

Kane, T.S. (2000) *The Oxford Essential Guide to Writing.* Oxford University Press, New York.

[This is excellent for the basics of English – it covers grammar, usage and the construction of sentences and paragraphs.]

Lindsay, D. (2011) *Scientific Writing = Thinking in Words.* CSIRO, Collingwood.

Matthews J.R. and Matthews R.W. (2014) *Successful Scientific Writing: A Step-by-Step Guide for the Biological and Medical Sciences,* 4th edn. Cambridge University Press, Cambridge.

McMillan, K.M. and Weyers, J.D.B. (2009) *The Smarter Study Skills Companion.* Pearson, Harlow.

STUDY EXERCISES

17.1 'Brainstorm' an essay title. Pair up with a partner in your class. Together, pick a suitable essay title from a past exam paper. Using the spider diagram or another technique, individually 'brainstorm' the title. Meet afterwards, compare your ideas and discuss their relative merits and disadvantages.

17.2 Improve your writing technique. From the following checklist, identify the *three* weakest aspects of your writing, either in your own opinion or from essay/assignment feedback:

- grammar;
- paragraph organisation;
- presentation of work;
- punctuation;
- scientific style;
- sentence structure/variety;
- spelling;

- structure and flow;
- vocabulary.

Now either borrow a book from a library or buy a book that deals with your weakest aspects of writing. Read the relevant chapters or sections and for each aspect write down some tips that should help you in future.

17.3 Improve your spelling and vocabulary with two lists. Create a pair of lists and pin these up beside your desk. One should be entitled *Spelling Mistakes* and the other *New Words.* Now, whenever you make a mistake in spelling or have to look up how to spell a word in a dictionary, add the problem word to your spelling list, showing where you made the mistake. Also, whenever you come across a word whose meaning is unclear to you, look it up in a dictionary and write the word and its meaning in the 'new words' vocabulary list.

Answers to these study exercises are available at www.pearsoned.co.uk/practicalskills.

(a)

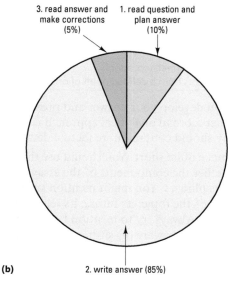

(b)

Fig. 18.1 Typical division of time for an essay written as part of an in-course assessment (a) or under exam conditions (b).

> **Considering essay content** – it is rarely enough simply to lay down facts for the reader – you must analyse them and comment on their significance (see p. 114).

The main functions of an essay are to show how much you understand about a topic and how well you can organise and express your knowledge.

Organising your time

The way you should divide your time when producing an essay depends on whether you are writing it for in-course assessment or under exam conditions (Fig. 18.1). Essays written over a long period with access to books and other resources will probably involve a research element, not only before the planning phase but also when writing (Fig. 18.1(a). For exams, it is assumed that you have revised appropriately (Chapter 6) and essentially have all the information at your fingertips. To keep things uncomplicated, the time allocated for each essay should be divided into three components – planning, writing and reviewing (Fig. 18.(b), and you should adopt time-saving techniques whenever possible (Box 7.2).

Making a plan for your essay

Dissect the meaning of the essay question or title

Read the title very carefully and think about the topic before starting to write. Consider the definitions of each of the important nouns (this can help in approaching the introductory section). Also think about the meaning of the verb(s) used and try to follow each instruction precisely (see Table 18.1). Do not get side-tracked because you know something about one word or phrase in the title: consider the whole title and all its ramifications. If there are two or more parts to the question, make sure you give adequate attention to each part.

Consider possible content and examples

Research content using the methods described in Chapters 9 and 10. If you have time to read several sources, consider their content in relation to the essay title. Can you spot different approaches to the same subject? Which do you prefer as a means of treating the topic in relation to your title? Which examples are most relevant to your case, and why?

Construct an outline

Every essay should have a structure related to its title.

> **KEY POINT** Most marks for essays are lost because the written material is badly organised or is irrelevant. An essay plan, by definition, creates order and, if thought about carefully, should ensure relevance.

Your plan should be written down (but scored through later if written in an exam book). Think about an essay's content in three parts:

1. **The introductory section,** in which you should include definitions and some background information and the context for the topic being considered. You should also tell your reader how you plan to approach the subject.

1. Read the question carefully, and decide exactly what the assessor wants you to achieve in your answer.

2. Make sure you understand the question by considering all aspects – discuss your approach with colleagues or a tutor.

3. Carry out the necessary research (using books, journals, the web), taking appropriate notes. Gain an overview of the topic before getting involved with the details.

4. Always plan your work in outline before you start writing. Check that your plan covers the main points and that it flows logically.

5. Introduce your essay by showing that you understand the topic and stating how you intend to approach it.

6. As you write the main content, ensure it is relevant by continually looking back at the question.

7. Use headings and subheadings to organise and structure your essay.

8. Support your statements with relevant examples, diagrams and references where appropriate.

9. Conclude by summarising the key points of the topic, indicating the present state of knowledge, what we still need to find out and how this might be achieved.

10. Always review your essay before submitting it. Check grammar and spelling and confirm that you have answered *all* aspects of the question.

Using diagrams – give a title and legend to each diagram so that it makes sense in isolation and point out in the text when the reader should consult it (e.g. 'as shown in Fig. 1...' or 'as can be seen in the accompanying diagram,...').

Table 18.1 Instructions often used in essay questions and their meanings. When more than one instruction is given (e.g. compare and contrast; describe and explain), make sure you carry out *both* or you may lose a large proportion of the available marks (see also Table 5.1).

Account for:	give the reasons for
Analyse:	examine in depth and describe the main characteristics of
Assess:	weigh up the elements of and arrive at a conclusion about
Comment:	give an opinion on and provide evidence for your views
Compare:	bring out the similarities between
Contrast:	bring out dissimilarities between
Criticise:	judge the worth of (give both positive and negative aspects)
Define:	explain the exact meaning of
Describe:	use words and diagrams to illustrate
Discuss:	provide evidence or opinions about, arriving at a balanced conclusion
Enumerate:	list in outline form
Evaluate:	weigh up or appraise; find a numerical value for
Explain:	make the meaning of something clear
Illustrate:	use diagrams or examples to make clear
Interpret:	express in simple terms, providing a judgement
Justify:	show that an idea or statement is correct
List:	provide an itemised series of statements about
Outline:	describe the essential parts only, stressing the classification
Prove:	establish the truth of
Relate:	show the connection between
Review:	examine critically, perhaps concentrating on the stages in the development of an idea or method
State:	express clearly
Summarise:	without illustrations, provide a brief account of
Trace:	describe a sequence of events from a defined point of origin

2. **The middle of the essay,** where you develop your answer and provide relevant examples. Decide whether a broad analytical approach is appropriate or whether the essay should contain more factual detail.

3. **The conclusion,** which you can make quite short. You should use this part to summarise and draw together the components of the essay, without merely repeating previous phrases. You might mention such things as: the broader significance of the topic; its future; its relevance to other important areas of biology. Always try to mention both sides of any debate you have touched on, but beware of 'sitting on the fence'.

> **KEY POINT** Use paragraphs to make the essay's structure obvious. Emphasise them with headings and subheadings unless the material beneath the headings would be too short or trivial.

Start writing

- **Never lose track of the importance of content** and its relevance. Repeatedly ask yourself: 'Am I really answering this question?' Never waffle just to increase the length of an essay. Quality, rather than quantity, is important.

- **Illustrate your answer appropriately.** Use examples to make your points clear, but remember that too many similar examples can stifle the flow of an essay. Use diagrams where a written description would be difficult or take too long. Use tables to condense information.

- **For handwritten essays, take care with your handwriting.** You cannot get marks if your writing is illegible. Try to cultivate an open form of handwriting, making the individual letters large and distinct. If there is time, make out a rough draft from which a tidy version can be copied.

Reviewing your answer

Make sure that you leave enough time to:

> Learning from lecturers' and tutors' comments – ask for further explanations if you do not understand a comment or why an essay was less successful than you thought it should have been.

- **re-read the question** to check that you have answered all points;
- **re-read your essay** to check for errors in punctuation, spelling and content. Make any corrections obvious. In an exam, do not panic if you suddenly realise that you have missed a large chunk out as the reader can be redirected to a supplementary paragraph if necessary.

Sources for further study

Good, S. and Jensen, B. (1996) *The Student's Only Survival Guide to Essay Writing*. Orca Book Publishers, Victoria.

McMillan, K.M. and Weyers, J.D.B. (2011) *How to Write Essays and Assignments*, 2nd edn. Pearson Education, London.

Royal Literary Fund. *Writing Essays*. Available: http://www.rlf.org.uk/resource/writing-essays/ Last accessed 01/09/15.
[An accessible, informal guide to essay writing.]

STUDY EXERCISES

18.1 Practise dissecting essay titles. Use past exam papers, or make up questions based on the learning outcomes for your course and your lecture notes. Take each essay title and carefully 'dissect' the wording, working out exactly what you think the assessor expects you to do (see e.g. Table 18.1).

18.2 Write essay plans under self-imposed time limits. Continuing from study exercise 18.1, outline plans for essays from a past exam paper. Allow yourself a maximum of 5 min per outline. Within this time your main goal is to create an essay plan. To do this, you may need to 'brainstorm' the topic. Alternatively, if you allocate 10 min per essay, you may be able to provide more details, e.g. list the examples you could describe.

18.3 Practise reviewing your work carefully. For the next assignment you write, review it fully as part of the writing process. This will require you to finish the first draft about one week before the hand-in date, e.g. by setting yourself an earlier deadline than the submission date. This exercise is best done with a word-processed essay. Do not worry if it is a little over the word limit at this stage.

(a) Print out a copy of the essay. Do not look at it for at least two days after finishing this version.

(b) Review 1: spelling, grammar and sense. Read through the draft critically (try to imagine it had been written by someone else) and correct any obvious errors that strike you. Does the text make sense? Do sentences/ paragraphs flow smoothly?

(c) Review 2: structure and relevance. Consider again the structure of the essay, asking yourself whether you have really answered the question that was set (see study exercise 18.1). Are all the parts in the right order? Is anything missed out? Have you followed precisely the instruction(s) in the title? Are the different parts of the essay linked together well?

(d) Review 3: shorten and improve style. Check the word count. Shorten the essay if required. Look critically at phrasing and, even if the essay is within the word limit, ask yourself whether any of the words are unnecessary or whether the text could be made more concise, more precise or more apt.

Answers to these study exercises are available at www.pearsoned.co.uk/practicalskills.

Practical reports, project reports, theses and scientific papers differ greatly in depth, scope and size, but they all have the same basic structure. Some variation is permitted, however (see Box 19.1), and you should always follow the advice or rules provided by your department.

Additional parts may be specified: for project reports, dissertations and theses, a Title page is often required and a List of Figures and Tables as part of the Contents section. When work is submitted for certain degrees, you may need to include certain declarations and statements made by the student and supervisor. In scientific papers, a list of Key Words is often added following the Abstract: this information may be combined with words in the title for computer cross-referencing systems.

> **KEY POINT** Department, school or faculty regulations may specify a precise format for your report or thesis. Consult a copy of these rules at an early stage and follow them closely, to avoid losing marks.

Practical laboratory reports

These are exercises designed to make you think more deeply about your experiments and to practise and test the skills necessary for writing up research work. Special features are:

- **Introductory material is generally short** and, unless otherwise specified, should outline the aims of the experiment(s) with a minimum of background material.
- **Materials and methods may be provided by your supervisor for practical reports.** If you make changes to this, you should state clearly what you did. With extended projects, your lab notebook (see p. 208) should provide the basis for writing this section.
- **Great attention in assessment will be paid to presentation and analysis of data.** Take special care over graphs (see p. 495 for further advice). Make sure your conclusions are justified by the evidence you present.

Theses and dissertations

These are submitted as part of the examination for a degree following an extended period of research. They act to place on record full details about your experimental work and will normally only be read by those with a direct interest in it – your examiners or colleagues. Note the following:

- **The Introduction and Methods sections are lengthy detailed,** providing the reader with the full context for your research.
- **You are allowed scope to expand on your findings** and to include detail that might otherwise be omitted in a scientific paper.
- **You may have problems with the volume of information that has to be organised.** One method of coping with this is to divide your thesis into chapters, each having the standard format (as in Box 19.1). A General Introduction can be given at the start and a General Discussion at the end. Discuss this with your supervisor.

Typical structure of scientific reports – this usually follows the 'IMRaD' acronym: Introduction, Materials and Methods, Results and Discussion.

Options for discussing data – the main optional variants of the general structure include combining Results and Discussion into a single section and adding a separate Conclusions section.

- The main advantage of a joint Results and Discussion section is that you can link together different experiments, perhaps explaining why a particular result led to a new hypothesis and the next experiment. However, a combined Results and Discussion section may contravene your department's regulations, so you should check before using this approach.
- The main advantage of having a separate *Conclusions* section is to draw together and emphasise the chief points arising from your work, when these may have been 'buried' in an extensive *Discussion* section.

| Box 19.1 | The structure of reports of experimental work |

Undergraduate practical and project reports are generally modelled on this arrangement or a close variant of it, because this is the structure used for nearly all research papers and theses. The more common variations include Results and Discussion combined into a single section and Conclusions appearing separately as a series of points arising from the work. In scientific papers, a list of Key Words (for computer cross-referencing systems) may be included following the Abstract. Acknowledgements may appear after the Contents section, rather than near the end. Department or faculty regulations for producing theses and reports often require a Title page to be inserted at the start and a List of Figures and Tables as part of the Contents section, and may specify declarations and statements to be made by the student and supervisor.

Part (in order)	Contents/purpose	Checklist for reviewing content
Title	Explains what the project was about	❑ Does it explain what the text is about succinctly?
Authors plus their institutions	Explains who did the work and where; also where they can be contacted now	❑ Are all the details correct?
Abstract/Summary	Synopsis of methods, results and conclusion of work described. Allows the reader to grasp quickly the essence of the work	❑ Does it explain why the work was done? ❑ Does it outline the whole of your work and your findings?
List of Contents	Shows the organisation of the text (not required for short papers)	❑ Are all the sections covered? ❑ Are the page numbers correct?
Abbreviations	Lists all the abbreviations used (but not those of SI, chemical elements or standard biochemical terms)	❑ Have they all been explained? ❑ Are they all in the accepted form? ❑ Are they in alphabetical order?
Introduction	Orientates the reader, explains why the work has been done and its context in the literature, why the methods used were chosen, why the experimental organisms were chosen. Indicates the central hypothesis behind the experiments	❑ Does it provide enough background information and cite all the relevant references? ❑ Is it of the correct depth for the readership? ❑ Have all the technical terms been defined? ❑ Have you explained why you investigated the problem? ❑ Have you outlined your aims and objectives? ❑ Have you explained your methodological approach? ❑ Have you stated your hypothesis?
Materials and Methods	Explains how the work was done. Should contain sufficient detail to allow another competent worker to repeat the work	❑ Is each experiment covered and have you avoided unnecessary duplication? ❑ Is there sufficient detail to allow repetition of the work? ❑ Are proper scientific names and authorities given for all organisms? ❑ Have you explained where you got them from? ❑ Are the correct names, sources and grades given for all chemicals?
Results	Displays and describes the data obtained. Should be presented in a form which is easily assimilated (graphs rather than tables, small tables rather than large ones)	❑ Is the sequence of experiments logical? Are the parts adequately linked? ❑ Are the data presented in the clearest possible way? ❑ Have SI units been used properly throughout? ❑ Has adequate statistical analysis been carried out? ❑ Is all the material relevant? ❑ Are the figures and tables all numbered in the order of their appearance? Are their titles appropriate? ❑ Do the figure and table legends provide all the information necessary to interpret the data without reference to the text? ❑ Have you presented the same data more than once?
Discussion/ Conclusions	Discusses the results: their meaning, their importance; compares the results with those of others; suggests what to do next	❑ Have you explained the significance of the results? ❑ Have you compared your data with other published work? ❑ Are your conclusions justified by the data presented?
Acknowledgements	Gives credit to those who helped carry out the work	❑ Have you listed everyone that helped, including any grant-awarding bodies?
Literature Cited (References/ Bibliography)	Lists all references cited in appropriate format: provides enough information to allow the reader to find the reference in a library	❑ Do all the references in the text appear on the list? ❑ Do all the listed references appear in the text? ❑ Do the years of publications and authors match? ❑ Are the journal details complete and in the correct format? ❑ Is the list in alphabetical order, or correct numerical order?

Steps in the production of a practical report or thesis

Choose the observations or the experiments you wish to describe and decide how best to present them

Try to start this process before your lab work ends, because at the stage of reviewing your experiments, a gap may become apparent (e.g. a missing control) and you might still have time to rectify the deficiency. Irrelevant material should be ruthlessly eliminated, at the same time bearing in mind that negative results can be extremely important (see p. 200). Use as many different forms of data presentation as are appropriate, but avoid presenting the same data in more than one form. Relegate large tables of primary data to an appendix and summarise the important points within the main text (with a cross-reference to the appendix). Make sure that the experiments you describe are representative: always state the number of times they were repeated and how consistent your findings were.

Make up plans or outlines for the component parts

The overall structure of practical and project reports is well defined (see Box 19.1), but individual parts will need to be well organised, as with any other form of writing (see Chapter 17).

Write

The Materials and Methods section is often the easiest to write once you have decided what to report. Remember to use the past tense and do not allow results or discussion to creep in. The Results section is the next easiest as it should only involve description. At this stage, you may benefit from jotting down ideas for the Discussion – this may be the hardest part to compose as you need an overview both of your own work and of the relevant literature. It is also liable to become wordy, so try hard to make it succinct. The Introduction should not be too difficult if you have fully understood the aims of the experiments. Write the Abstract and complete the list of references at the end. To assist with the latter, it is a good idea as you write to collate appropriately formatted details of references (Chapter 9) as you progress.

Revise the text

Once your first draft is complete, try to answer all the questions given in Box 19.1. Show your work to your supervisors and learn from their comments. Let a friend or colleague who is unfamiliar with your subject read your text; they may be able to pinpoint obscure wording and show where information or explanation is missing. If writing a thesis, double-check that you are adhering to your institution's thesis regulations.

Prepare the final version

Markers appreciate neatly produced work but a well-presented document will not disguise poor science. If using a word processor, print the final version with the best printer available. Make sure figures are clear and in the correct size and format.

Submit your work

Your department will specify when to submit a thesis or project report, so plan your work carefully to meet this deadline or you may lose marks. Tell your supervisor early of any circumstances that may cause delay. Also, check to see whether any forms are required for late submission, or evidence of extenuating circumstances.

Choosing between graphs and tables – graphs are generally easier for the reader to assimilate, while tables can be used to condense a lot of data into a small space.

Repeating your experiments – remember, if you do an experiment twice, you have repeated it only once.

Presenting your results – remember that the order of results presented in a report need not correspond with the order in which you carried out the experiments: you are expected to rearrange them to provide a logical sequence of findings.

Using the correct tense – always use the past tense to describe the methodology used in your work, since it is now complete. Use the present tense only for generalisations and conclusions.

Oral assessments – there may be an oral exam (*viva voce*) associated with the submission of a thesis or dissertation. The primary aim of the examiners will be to ensure that you understand what you did and why you did it.

Definition

Peer review – the process of evaluation and review of a colleague's work. In scientific communication, a paper is reviewed by two or more expert reviewers ('referees') for comments on quality and significance as a key component of the validation procedure.

Producing a scientific paper

Scientific papers are the means by which research findings are communicated to others. Peer-reviewed papers are published in journals; each covers a well-defined subject area and publishes details of the format they expect.

KEY POINT Peer review is an important component of the process of scientific publication; only those papers whose worth is confirmed by the peer review process will be published.

It would be very unusual for an undergraduate to submit a paper on his or her own – this would normally be done in collaboration with your project supervisor, and only then if your research has satisfied appropriate criteria. However, it is important to understand the process whereby a paper comes into being (Box 19.2), as this can help you understand and interpret the primary literature.

Box 19.2 Steps in producing a scientific paper

Scientific papers are the lifeblood of any science and it is a major landmark in your scientific career to publish your first paper. The main steps in doing this should include the following:

Assessing potential content

The work must be of an appropriate standard to be published and should be 'new, true and meaningful'. Therefore, before starting, the authors need to review their work critically under these headings. The material included in a scientific paper will generally be a subset of the total work done during a project, so it must be carefully selected for relevance to a clear central hypothesis – if the authors will not prune, the referees and editors of the journal certainly will.

Choosing a journal

There are thousands of journals covering biology and each covers a specific area (which may change through time). The main factors in deciding on an appropriate journal are the range of subjects it covers, the quality of its content and the number and geographical distribution of its readers. The choice of journal always dictates the format of a paper since authors must follow to the letter the journal's 'Instructions to Authors'.

Deciding on authorship

In multi-author papers, a contentious issue is often who should appear as an author and in what order they should be cited. Where authors make an equal contribution, an alphabetical order of names may be used. Otherwise, each author should have made a substantial contribution to the paper and should be prepared to defend it in public. Ideally, the order of appearance will reflect the amount of work done rather than seniority. This may not always happen in practice!

Writing

The paper's format will be similar to that shown in Box 19.1 and the process of writing will include outlining, reviewing, etc., as discussed elsewhere in this chapter. Figures must be finished to an appropriate standard and this may involve preparing photographs or digital images of them.

Submitting

When completed, copies of the paper are submitted to the editor of the chosen journal with a simple covering letter. A delay of one to two months usually follows while the manuscript is sent to two or more anonymous referees who will be asked by the editor to check that the paper is novel, scientifically correct and that its length is justified.

Responding to referees' comments

The editor will send on the referees' comments to the authors, who will then have a chance to respond. The editor will decide on the basis of the comments and replies to them whether the paper should be published. Sometimes quite heated correspondence can result if the authors and referees disagree.

Checking proofs and waiting for publication

If a paper is accepted, it will be sent for production. The next the authors see of it is the proofs (first printed version in style of journal), which have to be checked carefully for errors and returned. Eventually, the paper will appear in print, but a delay of six months following acceptance is not unusual. Nowadays, papers are often available electronically, via the web, in PDF format – see p. 57 for advice on how to cite such 'online early' papers using the DOI system.

Sources for further study

Berry, R. (2004) *The Research Project: How to Write it,* 5th edn. Routledge, London.

Davis, M. and Davis K.J. (2012) *Scientific Papers and Presentations: Navigating Scientific Communication in Today's World,* 3rd edn. Academic Press, London.

Day, R.A. and Gastel, B. (2012) *How to Write and Publish a Scientific Paper,* 7th edn. Cambridge University Press, Cambridge.

Hofmann, A.H. (2014) *Scientific Writing, and Communication: Papers, Proposals and Presentations,* 2nd edn. Oxford University Press, Oxford.

Lobban, C. S. and Schefter, M. (1992) *Successful Lab Reports: A Manual for Science Students.* Cambridge University Press, Cambridge.

Luey, B. (2009) *Handbook for Academic Authors,* 5th edn. Cambridge University Press, Cambridge.

McMillan, K.M. and Weyers, J.D.B. (2014) *How to Complete a Successful Research Project.* Pearson Education, London.

Matthews, J.R., Bowen, J.M. and Matthews, R.W. (2007) *Successful Science Writing: a Step-by-Step Guide for the Biological and Medical Sciences,* 3rd edn. Cambridge University Press, Cambridge.

Valiela, I. (2009) *Doing Science: Design, Analysis and Communication of Scientific Research,* 2nd edn. Oxford University Press, Oxford.
[Covers scientific communication, graphical presentations and aspects of statistics.]

STUDY EXERCISES

19.1 Write a formal 'Materials and Methods' section. Adopting the style of a research paper (i.e. past tense, all relevant detail reported such that a competent colleague could repeat your work), write out the Materials and Methods for a practical you have recently carried out. Ask a colleague or tutor to comment on what you have written.

19.2 Describe a set of results in words. Again adopting the style of a research paper, write a paragraph describing the results contained in a particular table or graph. Ask a colleague or tutor to comment on your description, to identify what is missing or unclear.

19.3 Write an abstract for a paper. Pair up with a colleague. Each of you should independently choose a different research paper in a current journal. Copy the paper, but mask over the abstract section, having first counted the words used. Swap papers. Now, working to the same number of words as in the original, read the paper and provide an abstract of its contents. Then compare this with the real abstract. Compare your abstracts.

Answers to these study exercises are available at www.pearsoned.co.uk/practicalskills.

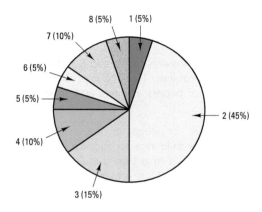

Fig. 20.1 Pie chart showing how you might allocate time for a literature survey:
1. select a topic
2. scan the literature
3. plan the review
4. write first draft
5. leave to settle
6. prepare a structured review of text
7. write final draft
8. produce top copy.

Creating a glossary – one barrier to developing an understanding of a new topic is the jargon used. To overcome this, create your own glossary. You may wish to cross-reference a range of sources to ensure the definitions are reliable and context-specific. Remember to note your sources in case you wish to use the definition within your review.

Using index cards – these can help you organise large numbers of references. Write key points and author information on each card – this helps when considering where the reference fits into the literature. Arrange the cards in subject piles, eliminating irrelevant ones. Order the cards in the sequence in which you wish to write.

The literature survey or review (sometimes termed a 'dissertation') is a specialised form of essay which summarises and reviews the evidence and concepts concerning a particular area of research.

> **KEY POINT** A literature review should *not* be a simple recitation of facts. The best reviews are those which analyse information rather than simply describe it.

Organising your time

Figure 20.1 illustrates how you might divide up your time for writing a literature survey. There are many subdivisions in this chart because of the size of the task: in general, for lengthy tasks, it is best to divide up the work into manageable chunks. Note also that proportionally less time is allocated to writing itself than with an essay. In a literature survey, make sure that you spend adequate time on research and revision.

Selecting a topic

You may have no choice in the topic to be covered, but if you do, carry out your selection as a three-stage process:

1. **Identify a broad subject area that interests you.**
2. **Find and read relevant literature in that area.** Try to gain a broad impression of the field from books and general review articles. Discuss your ideas with your supervisor.
3. **Select a relevant and concise title.** The wording should be considered very carefully as it will define the content expected by the reader. A narrow subject area will cut down on the amount of literature you will be expected to review, but will also restrict the scope of the conclusions you can make (and vice versa for a wide subject area).

Scanning the literature and organising your references

You will need to carry out a thorough investigation of the literature before you start to write. The key issues are as follows:

- **Make a start with relevant literature.** Seek help from your supervisor, who may be willing to supply a few key papers to get you started. Hints on expanding your collection of references are given on p. 53.
- **Assess the relevance and value of each article.** This is the essence of writing a review, but it is difficult unless you already have a good understanding of the field. Try reading earlier reviews in your area and discussing the topic with your supervisor or other academic staff.
- **Clarify your thoughts.** One approach is to use the SPSER method (Box 20.1). Alternatively, subdividing the main topic and assigning your references to these smaller subject areas may help you gain a better overview of the literature.

This method is useful when trying to get to grips with a new and complex subject. The approach helps you to 'deconstruct' the topic and involves considering the task in five discrete stages, given the acronym SPSER. You can note down your thoughts using a table or the 'pattern notes' (p. 17) or 'mind map' (p. 18) approaches, ready for converting into more formal text.

1. **Situation.** Briefly outline the context of the topic and the history of its development. For some complex topics you may need to refer to broader introductory texts or websites to build a foundation for your understanding.

2. **Problem.** State the essential issues, questions or problems that researchers or others are tackling. In a research paper, this might be stated succinctly in the Abstract, but you might also find it at the end of the

Introduction. In some cases this will be clear from the title of the exercise you have been given.

3. **Solutions.** Outline possible answers to questions raised, or ways of tackling the problem. These will probably arise from your reading of papers, reviews and other sources.

4. **Evaluation.** Note positive or negative features of each solution and provide evidence to support your viewpoint. A tabular approach might be useful, listing positives on one side of a vertical line and negatives on the other.

5. **Recommendation.** Arrive at a conclusion by deciding what might/should happen next, outlining your reasoning.

Deciding on structure and content

The general structure and content of a literature survey are described below.

The *Annual Review* series (available in most university libraries) provides good examples of appropriate style for reviews of the biosciences.

Introduction

The introduction should give the general background to the research area, concentrating on its development and importance. You should also make a statement about the scope of your survey; as well as defining the subject matter to be discussed, you may wish to restrict the period being considered.

Defining terms – the introduction is a good place to explain the meaning of the key terms used in your survey or review.

Main body of text

The review itself should discuss the published work in the selected field and may be subdivided into appropriate sections. Within each portion of a review, the approach is usually chronological, with appropriate linking phrases (e.g. 'Following on from this,. . .'; 'Meanwhile, Bloggs (2002) tackled the problem from a different angle. . .'). However, a good review is much more than a chronological list of work done. It should:

- **allow the reader to obtain an overall view of the current state of the research area,** identifying the key areas where knowledge is advancing;
- **show how techniques are developing** and discuss the benefits and disadvantages of using particular organisms or experimental systems;
- **assess the relative worth of different types of evidence** – this is the most important aspect (see Chapter 10). Do not be intimidated from taking a critical approach as the conclusions you may read in the primary literature are not always correct;
- **indicate where there is conflict in findings or theories,** suggesting, if possible, which side has the stronger case;
- **indicate gaps in current knowledge.**

Balancing opposing views – even if you favour one side of a disagreement in the literature, your review should provide a balanced and fair description of all the published views of the topic. Having done this, if you do wish to state a preference, give reasons for your opinion.

You do not need to wait until you have read all the sources available to you before starting to write the main body. Word processors allow you to modify and move pieces of text at any point and it will be useful to write paragraphs about key sources, or groups of related papers, as you read them. Try to create a general plan for your review as soon as possible. Place your draft sections of text under an appropriate set of subheadings that reflects your plan, but be prepared to rearrange these and re-title or re-order sections as you proceed. Not only will working in this way help to clarify your thoughts, but it may help you avoid a last-minute rush of writing near to the submission date.

Conclusions

The conclusions should draw together the threads of the preceding parts and point the way forward, perhaps listing areas of ignorance or where the application of new techniques may lead to advances.

References, etc.

The references or literature cited section should provide full details of all papers referred to in the text (see p. 56). The regulations for your department may also specify a format and position for the title page, list of contents, acknowledgements, etc.

Making citations – a review of literature poses stylistic problems because of the need to cite large numbers of papers; in the *Annual Review* series this is overcome by using numbered references (see p. 56).

Sources for further study

McMillan, K.M. and Weyers, J.D.B. (2012) *The Study Skills Book,* 3rd edn., Pearson Education, Harlow.

McMillan, K.M. and Weyers, J.D.B. (2011) *How to Write Dissertations and Project Reports,* 2nd edn. Prentice Hall, Harlow.

Rudner, L.M. and Schafer, W.D. (1999) How to write a scholarly research report. *Practical Assessment, Research & Evaluation*, 6 (13). Available: http://pareonline.net/getvn.asp?v=6&n=13 Last accessed: 01/09/15.

STUDY EXERCISES

20.1 Summarise the main differences between a review and a scientific paper. From the many subject areas in the *Annual Review* series (find via your library's periodical indexing system), pick one that matches your subject interests, and within this find a review that seems relevant or interesting. Read the review and write down *five* ways in which the writing style and content differ from those seen in primary scientific papers.

20.2 Gather a collection of primary sources for a topic. From the journal section of the library, select an interesting scientific paper published about 5–10 years ago. First, work *back* from the references cited by that paper: can you identify from the text or the article titles which are the most important and relevant to the topic? List *five* of these, using the proper conventions for citing articles in a reference list (see Chapter 9). Note that each of these papers will also cite other articles, always going back in time. Now, using the Science Citation Index or a similar system (e.g. the Web of Science website at http://wos.mimas.ac.uk or Google Scholar at http://scholar.google.com/), work *forward* and find out who has cited your selected article in the time since its publication. Again, list the *five* most important articles found.

20.3 Write a synopsis of a review. Using one of the *Annual Review* series as a source, allow yourself just *five* single-sentence bullet points to summarise the key points reported in a particular review.

Answers to these study exercises are available at www.pearsoned.co.uk/practicalskills.

Fundamental laboratory techniques

All knowledge and theory in science have originated from practical observation and experimentation: this is equally true for disciplines as diverse as microscopy and molecular genetics. Practical work is an important part of most courses and often accounts for a significant proportion of the assessment marks. The abilities developed in practical classes will continue to be useful throughout your course and beyond, some within science and others in any career you choose (see Chapter 1).

Being prepared

> **KEY POINT** You will get the most out of practicals if you prepare well in advance. Do not go into a practical session assuming that everything will be provided, without any input on your part.

The main points to remember are:

- **Read any handouts in advance:** make sure you understand the purpose of the practical and the particular skills involved. Does the practical relate to, or expand upon, a current topic in your lectures? Is there any additional preparatory reading that will help?
- **Take along appropriate textbooks,** to explain aspects in the practical.
- **Consider what safety hazards might be involved,** and any precautions you might need to take, before you begin (p. 143).
- **Listen carefully to any introductory guidance** and note any important points: adjust your schedule/handout as necessary.
- **Organise your bench space during the practical session** – make sure your lab book is adjacent to, but not within, your working area. You will often find it easiest to keep clean items of glassware, etc. on one side of your working space, with used equipment on the other side.
- **Write up your work as soon as possible,** and submit it on time, or you may lose marks.
- **Catch up on any work you have missed** as soon as possible – preferably, before the next practical session.

Ethical and legal aspects

You will need to consider the ethical and legal implications of biological work at several points during your studies:

- Safe working means following a code of safe practice, supported by legislation, alongside a moral obligation to avoid harm to yourself and others, as discussed in Chapter 22.
- Any laboratory work that involves working with humans or animals must be carefully considered.
- Fieldwork can have a legal aspect, e.g. in relation to specimen collection, since legislation may protect particular habitats and/or species. Ethical and human aspects that must all be considered before any work is carried out include: possible damage to fauna and flora; safety of others within your group; implications of your work to others, e.g. landowners, or other users.

Developing practical skills – these will include:

- designing experiments
- observing and measuring
- recording data
- analysing and interpreting data
- reporting/presenting.

Using textbooks in the lab – take this book (or photocopies of relevant pages) along to the relevant classes, so that you can make full use of the information during the practical sessions.

SAFETY NOTE Mobile phones – these should never be used in a lab class, as there is a risk of contamination from hazardous substances. Always switch off your mobile phone before entering a laboratory. Conversely, they are an extremely useful accessory for fieldwork.

Getting to grips with bioethics – in addition to any moral implications of your lab practicals and fieldwork, you may have the opportunity to address broader issues within your course (see Chapter 22). Professional scientists must always consider the consequences of their work, and it is therefore important that you develop your understanding of these issues alongside your academic studies.

Presenting results – while you do not need to be a graphic designer to produce work of a satisfactory standard, presentation and layout are important and you will lose marks for poorly presented work. Chapter 70 gives further practical advice.

Buying and using a calculator – many university departments specify a particular make and model of calculator for students to use from first year onwards. The advantage of this approach is that you can seek help from staff during tutorials and lab classes, if you are unsure of specific functions. Conversely, staff will not be fully familiar with other types of calculator, so may not be able to answer such questions if you are using an alternative make or model.

Using calculators for numerical problems – Chapter 72 gives further advice.

Using inexpensive calculators – many unsophisticated calculators have a restricted display for exponential numbers and do not show the 'power of 10', e.g. displaying 2.4×10^{-5} as 2.4^{-05}, or 2.4E–05, or even 2.4–05.

Presenting graphs and diagrams – ensure these are large enough to be easily read: a common error is to present graphs or diagrams that are too small, with poorly chosen scales (see p. 489).

Basic requirements

Recording practical results

An A4 loose-leaf ring binder offers flexibility, since you can insert laboratory handouts or lined and graph paper at appropriate points. The danger of losing one or more pages from a loose-leaf system is the main drawback. Bound books avoid this problem, although those containing alternating lined/graph or lined/blank pages tend to be wasteful – it is often better to paste sheets of graph paper into a bound book, as required.

A good-quality pencil or propelling pencil is recommended for recording your raw data, making diagrams, etc. as mistakes are easily corrected. Buy a black, spirit-based (permanent) marker for labelling experimental glassware, Petri plates, etc. Fibre-tipped fine line drawing/lettering pens are useful for preparing final versions of hand-drawn graphs and diagrams for assessment purposes. Use a see-through ruler with an undamaged edge for graph drawing, so that you can see data points and information below the ruler as you draw. Alternatively, use graph-drawing software (Chapter 70).

Calculators

These range from basic machines with no pre-programmed functions and only one memory, to sophisticated programmable portable computers with many memories. The following may be helpful when using a calculator:

- **Power sources.** Choose a battery-powered machine, rather than a mains-operated or solar-powered type. You will need one with basic mathematical/scientific operations, including powers, logarithms (p. 510), roots and parentheses (brackets), together with statistical functions such as sample means and standard deviations (Chapter 73).
- **Mode of operation.** Spend some time finding out how your calculator operates, e.g. does it have true algebraic logic ($\sqrt{}$ then number, rather than number then $\sqrt{}$)? How does it deal with (and display) scientific notation and logarithms (p. 510)?
- **Display.** Some calculators will display an entire mathematical operation (e.g. '$2 + 4 = 6$'), while others simply display the last number/operation. The former type may offer advantages in tracing errors.
- **Complexity.** In the early stages, it is usually better to avoid the more complex machines, full of impressive-looking, but often unused, pre-programmed functions – go for more memory, parentheses or statistical functions rather than engineering or mathematical constants. Programmable calculators may be worth considering for more advanced studies. However, it is important to note that such calculators are often unacceptable for exams.

Presenting more advanced practical work

In some practical reports and in project work, you may need to use more sophisticated presentation equipment. Computer-based graphics packages can be useful – choose easily read fonts such as Arial for posters and consider the layout and content carefully. Alternatively, you could use fine line drawing pens and dry-transfer lettering/symbols, although this can be more time-consuming than computer-based systems, e.g. using Microsoft *Excel* (p. 490).

Printing on acetates – standard overhead transparencies are not suitable for use in laser printers or photocopiers: you need to make sure that you use the correct type.

To prepare overhead transparencies for spoken presentations, you can use spirit-based markers and acetate sheets. An alternative approach is to photocopy onto special acetates, or print directly from a computer-based package, using a laser printer and special acetates, or use a PC-based presentation package (e.g. *PowerPoint*) and a suitable projector. Further advice on content and presentation is given in Chapter 16.

Sources for further study

Barnard, C.J., Gilbert, F.S. and MacGregor, P.K. (2011) *Asking Questions in Biology,* 4th edn. Prentice Hall, Harlow.

Bonner, P. and Hargreaves A. (2011) *Basic Bioscience Laboratory Techniques: a Pocket Guide.* Wiley, New York.

Dickey, J.L. (2002) *Laboratory Investigations for Biology,* 2nd edn. Benjamin Cummings, San Francisco.

STUDY EXERCISES

21.1 Consider the value of practical work. Spend a few minutes thinking about the purpose of practical work within a specific part of your course (e.g. a particular first-year module) and then write a list of the six most important points. Compare your list with the generic list that we have provided online, which is based on our experience as lecturers – does it differ much from your list, which is drawn up from a student perspective?

21.2 Make a list of items required for a particular practical exercise. This exercise is likely to be most useful if you can relate it to an appropriate practical session on your course. However, we have given a model list for a tissue dissection as an example.

21.3 Check your calculator skills. Carry out the following mathematical operations, using either a hand-held calculator or a PC with appropriate 'calculator' software.

(a) $5 \times (2 + 6)$

(b) $(8.3 \div [6.4 - 1.9]) \times 24$ (to four significant figures)

(c) $(1 \div 32) \times (5 \div 8)$ (to three significant figures)

(d) $1.2 \times 10^5 + 4.0 \times 10^4$ in scientific notation (see p. 510)

(e) $3.4 \times 10^{-2} - 2.7 \times 10^{-3}$ in 'normal' notation (i.e. conventional notation, not scientific format) and to three decimal places.

(See also the numerical exercises in Chapter 72.)

Answers to these study exercises are available at www.pearsoned.co.uk/practicalskills.

Bioethics is the philosophical study of ethical issues arising from recent advances in biology and medicine, though its scope also extends to cover environmental and global issues. It aims to provide a framework for making decisions, based on specific moral principles that take into account the reasons underlying different choices and the consequences of specific decisions. Particular groups of individuals will often adopt a set of ethical rules and standards as part of their professional code of conduct, e.g. medical ethics.

This chapter aims to provide you with a basic framework that will help you understand the various ethical theories and principles, so that you can consider some of the issues in contemporary bioethics. Without an ethical framework, individuals can only make subjective 'value judgements', based solely on personal opinions and viewpoints, rather than objective analysis.

One starting point is to understand the following fundamental terms:

- **Values** – the various beliefs and views held by individuals and groups, either for or against something; for example, religious, cultural or political values.
- **Morals** – individual or collective beliefs about whether something is 'right' or 'wrong'; for example, is it right to use animals in testing of pharmaceutical products?
- **Ethics** (moral philosophy) – a system of moral values and decision-making, described in terms of principles, rules and standards that are adopted by a specific group of people; for example, research ethics, legal ethics.

Note that sometimes the practical distinction between morals and ethics is less clear-cut than described above, and the terms are occasionally used interchangeably.

Bioscience degree programmes place increasing emphasis on the ethical and social impacts of contemporary scientific advances and on the need for scientists engaged in potentially controversial work to communicate their ideas and decision-making to the general public. Bioscience research raises many moral, ethical and legal dilemmas, requiring difficult choices to be made, e.g. in relation to animal testing of medical products, and students are likely to be asked to reflect on bioethical questions and moral dilemmas, e.g. in group discussions and debates. In discussing such issues, you will also gain experience that will have broader value, in terms of understanding some of the issues linked to the public understanding of science, and of how these issues can be addressed through communication with the public. You will find links and resources addressing this broader topic on the websites of a number of professional scientific organisations such as the Royal Society of London, and the Wellcome Trust in the UK, or the Coalition on the Public Understanding of Science (COPUS) in the USA.

Recognising the difference between morals and ethics – consider a biomedical scientist who believes that murder is fundamentally wrong (immoral), but the scientist's research into a specific disease requires experiments that will lead to the death of some animals, to be carried out under a research ethics code that specifies that the animals should not endure unneccessary pain or suffering. What should the scientist do? You will find that many aspects of bioethics are best considered through specific examples of moral/ethical dilemmas such as this.

Understanding the terminology of issues and dilemmas – these topics are sometimes referred to as ethical, legal and social issues (ELSI).

> **KEY POINT** In contrast to numerical calculations (Chapter 72) there is rarely a single 'right' or 'wrong' answer to bioethical issues. Consequently, it is important to use ethical principles and systems to consider each issue in a logical manner. Your aim should be to develop your skills in critical thinking and reasoning (p. 64), and in explaining how you have arrived at a particular decision on a bioethical issue, with sound evidence and a reasoned argument (pp. 64–5) to support the position that you have taken.

Bioethics can be subdivided into three main areas:

1. **Environmental ethics** – for example, the use of genetically modified organisms (GMOs), the hunting and killing of endangered species.
2. **Animal ethics** – examples include factory farming, the use of transgenic animals and xenotransplantation.
3. **Human ethics** – dealing with medical and social issues, such as human cloning, embryo and stem cell research, genetic testing and the storage of DNA profiles on databases.

Ethical theories

A basic understanding of the various theories of ethics may help you formulate your ideas in relation to specific questions and issues. The major theories include:

- **Utilitarianism** – the notion that it is ethical to choose the action that produces the greatest good (happiness) for the greatest number. For example, the use of atomic bombs against civilians in the Japanese cities of Hiroshima and Nagasaki in World War II was justified in terms of its utilitarian value, in terms of ending the war and preventing Japan's development and use of similar weapons, with potentially greater loss of life on both sides.
- **Deontology** – the theory that states that a particular action is either intrinsically good (right) or bad (wrong), and that such principles are absolute. According to deontological theory, decisions must be based on the actions themselves, rather than on their overall consequences. Such an approach is enshrined in the Judaeo-Christian religious value system (for example, in the ten commandments of the Old Testament). While this approach appears simple to apply, for example, in upholding the principle that 'it is always wrong to lie', it does not allow for particular circumstances, e.g. is it wrong to lie if, in doing so, you could prevent a murder? Perhaps in this example of an ethical dilemma, most people would agree that the greater 'good' of preventing a murder would outweigh the lesser 'good' of not lying, though this then moves towards utilitarianism.
- **Consequentialism** – where decisions are made and choices are selected by considering their likely outcomes and consequences, rather than from any intrinsic viewpoint. From a consequentialist viewpoint, a lie would be the right course of action choice to take if it prevented a murder.
- **Virtue theory** – arising from the philosophy of Plato and Aristotle, this enshrines the notion that making decisions according to established virtues, such as honesty, wisdom, justice and altruism will lead to ethically valid choices. As with consequentialism and utilitarianism,

Definition

Xenotransplantation – surgical transfer of tissues/organs from one species to another, typically discussed in terms of animal-to-human xenotransplantation.

Understanding consequentialism – this theory is probably best summed up by the phrase 'the end justifies the means'.

Examples of current bioethical issues and dilemmas:

- Whether to locate wind farms in the Scottish highlands. (Weigh up the benefits of 'green energy' against those of damage to environmentally sensitive areas.)
- Whether vivisection (using dissection procedures on living animals) can be justified if it may lead to medical advances that reduce human suffering.
- Whether the potential advantages of using GMOs outweigh the possible disadvantages (e.g. the use of 'golden rice', which has been genetically modified to contain genes for vitamin A synthesis, as a means of tackling vitamin A deficiency in some developing countries).

virtue theory is distinct from deontology, since it is not based on rigid absolutes.

- **Objectivism** – the theory that what is right and wrong is intrinsic and applies equally to all people, places and times. The alternative is that morality is subjective, being entirely dependent upon the views of the individual **(subjectivism).**

Ethical principles

The following provides a framework of fundamental principles for considering particular topics and questions from different perspectives, in no particular order:

- **Beneficence** – the obligation to do good. For example, if it is possible to prevent suffering by a particular course of action, then it should be carried out. However, dilemmas arise in particular circumstances when conflicts arise, e.g. is it acceptable to cause suffering to animals used in medical research that aims to prevent suffering to humans in the future?
- **Non-maleficence** – the duty to do no harm. This is enshrined within the Hippocratic oath of medical practitioners, but is not without difficulty in its practical application in specific circumstances. For example, should a medical practitioner continue to maintain someone who is on a life-support machine and who has no reasonable expectation of recovery, if this is causing anguish to the patient's family, and if the practitioner knows that another person is awaiting life support? Here, the practitioner has to weigh up the different aspects of the case, and make a balanced judgement.
- **Justice** – the obligation to treat all people fairly and impartially; for example, lack of discrimination between people on the grounds of race, colour, sex, language, religion, age or other status. This principle is also enshrined in Article 2 of the Universal Declaration of Human Rights (United Nations, 1948).
- **Autonomy** – the duty to allow an individual to make their own choices, without constraints. This principle underlies the notion of 'informed consent' in medical and scientific research.
- **Respect** – the need to show due regard for others; for example, the need to take into account the rights and beliefs of all people equally. This is also an intrinsic component of the UN Universal Declaration of Human Rights. It is also an essential component of the principle of confidentiality of personal information about participants in medical and scientific research (see p. 139).
- **Rationality** – the notion that a particular action or choice should be based on reason and logic. Many scientists would argue that the scientific method itself (pp. 200–201) is an good example of the principle of rationality in action.
- **Honesty** – the obligation to be truthful. This principle also underpins 'informed consent', since individuals cannot make informed choices if they have not been provided with information that is true and complete. It is also a fundamental aspect of scientific research ethics, with scientists having a duty to present their results as they truly are, without falsification or bias. When this principle is breached, the misconduct is often termed 'scientific fraud'.

Examples of scientific fraud – these include the case of Andrew Wakefield, a former doctor who was struck off the UK medical register for 'misleading, dishonest and irresponsible' claims of a link between measles–mumps–rubella (MMR) vaccination and autism, based on seriously flawed research. For more examples, see journalist Ben Goldacre's 'Bad Science' website at: http://www.badscience.net/

- **Precautionary principle** – the notion that it is better not to carry out an action if there is any risk of harm. For example, many people would apply the precautionary principle in deciding that the risks of building nuclear power stations outweigh their potential benefits. It is enshrined in principle 15 of the Rio Declaration (United Nations, 1992) as follows:

> to protect the environment, the precautionary approach shall be widely applied by States according to their capabilities. Where there are threats of serious or irreversible damage, lack of full scientific certainty shall not be used as a reason for postponing cost-effective measures to prevent environmental degradation.

It is also an aspect of the sustainable approach to global issues, such a global warming. While the precautionary principle is widely applied to environmental ethics, it is also relevant to other areas. For example, in medical ethics, what should a doctor recommend where only limited information is available about the possible relationship between exposure to a particular factor and a specific disease?

It is clear from the above listing, that several different principles may apply to a particular bioethical issue, and it will be up to individuals to weigh up the relative merits and disadvantages of particular choices and options in terms of the different outcomes against these principles when faced with an ethical dilemma. Box 22.1 outlines a practical and sequential approach to making ethical decisions.

Box 22.1 A step-wise approach to making ethical decisions

The following text is based on the framework of Shamoo and Resnik (2009):

1. **Define the issue or dilemma.** Typically this will be in the form of a question that requires a choice to be made, for example: 'Should the UK allow euthanasia?' The problem/question will often be given to you as part of a coursework assignment.

2. **Research the literature and gather together relevant information.** This may include (i) finding scholarly papers that consider the ethics of the topic, e.g. medical ethicists who have considered euthanasia from a particular standpoint, (ii) seeking other perspectives, e.g. those of organisations in favour of or opposed to euthanasia.

3. **Work out what options are available.** For example, you might consider under what circumstances euthanasia might be permitted, and how this would be regulated, rather than restricting yourself to a 'yes' or 'no' answer.

4. **Consider the options in relation to ethical theories and principles.** For example, what outcomes would fit with deontological theory, or utilitarianism? How do the principles of non-maleficence, autonomy and respect for the individual apply in relation to euthanasia?

5. **Consider whether anything else should be taken into account.** This step provides an opportunity for a 'second check' on whether you have considered all aspects of the topic, e.g. in relation to euthanasia, have you considered whether the views of relatives would carry any weight? Or whether your decision would apply to an individual suffering from a psychological disorder?

6. **Make your decision** – choose a particular response to the question posed at the outset. In relation to euthanasia, this might be an unqualified 'yes' or 'no', or might take the form of a conditional response, e.g. 'only in the following circumstances. . .'

7. **Explain your choice** – use the evidence gained in steps 2–5 to justify the decision made at step 6. In an assessed coursework assignment, this component of your written text will probably represent the greatest proportion of the marks available (always remember that it is not about reaching the 'right' or 'wrong' answer, but about reaching a logical and reasoned decision, based on interpretation of the issue in relation to sound ethical principles and reasoned discussion, supported by published sources).

Research involving human subjects

As explained on pp. 136–7, a number of ethical principles apply to the conduct of any research involving human subjects. Such principles apply in addition to those of health and safety (Chapter 23). If you carry out any research involving human participants as part of your degree programme, you will need to obtain ethical approval before you begin any practical work. Usually, this will involve submission of a request for ethical approval of your proposed project to your institution's ethics committee.

> **KEY POINT**   For the purposes of most undergraduate research proposals, while a student may be the investigator, the responsibility for following appropriate ethical approval procedures will remain with the supervisor.

Outlined below are the three main aspects that you should consider in all research involving human participants, irrespective of whether the research is 'invasive' (e.g. collecting mouth swabs from participants to study a particular DNA sequence) or 'non-invasive' (e.g. questioning participants about their drinking habits as part of a survey on the effects of alcohol consumption on study patterns and exam performance).

Providing information

All participants should be informed in writing and through oral explanation about the main aspects of the research study, typically through the use of a Participant Information Sheet that should include the following aspects:

- **An invitation to participate in the study,** with an explanation that participation is entirely voluntary and that the person is free to withdraw from the study at any stage.
- **A succinct outline of the purpose of the study,** and its likely outcomes.
- **A summary of what the participant will be expected to do,** and the timelines involved.
- **A statement of any possible benefits or risks involved in participation in the study.**
- **An assurance of confidentiality and anonymity in use of data from the study,** including details of how long the data will be retained within the institution.
- **Information about the funding source(s).**
- **Names and contact details of researchers/supervisors.**

Many institutions provide templates for such information sheets, which helps students to address all of the key aspects of the proposed study. The Participant Information Sheet should be written in plain English so that it can be readily understood. Researchers should also answer any additional questions that participants might have, based on their reading of the information sheet.

Definition

Plain English – a clear, succinct style that aims to be direct and natural, avoiding jargon, difficult language or complex structure. This will allow the meaning of the text to be understood by people from all walks of life.

Seeking informed consent

After reading the Participant Information Sheet, subjects are then requested to complete and sign an Informed Consent Form. As with the Participant Information Sheet, most institutions will provide a suitable template for student use – for broader examples, see the World Health Organization templates at: http://www.who.int/rpc/research_ethics/informed_consent/en/.

Assuring confidentiality

All participants must be assured that their personal details will be protected through processes that ensure anonymity. This means that all data recording and storage must be de-identified, e.g. through the use of randomly assigned codes to individuals, rather than using names. In addition to ethical considerations relating to storage and use of personal information, you must also comply with any legislation relating to data storage and protection, in terms of how it is held and used, the time-limit for maintaining the information and how it will be destroyed at the end of this period.

Box 22.2 provides practical advice that should help ensure that your research study is conducted in accordance with ethical principles.

UK data protection legislation – the Data Protection Act 1998 covers the procedures that must be followed in relation to storage of data, particularly sensitive personal data, such as racial origins, religious beliefs and sexuality (see: http://www.legislation.gov.uk/ukpga/1998/29/contents).

Box 22.2 **A step-wise approach to conducting ethical research**

The following text is based on McMillan and Weyers (2011):

1. **Research your institution's guidelines and procedures for research ethics.** Typically, such material, including general guidelines, templates and forms, will be available through the university website. For discipline-specific information, you may need to consult your departmental website, or seek advice from your supervisor. Investigating these procedures and guidelines at the outset will help you understand (i) the aspects of your own study that require ethical approval and (ii) how these might be addressed.

2. **Consider the ethical dimensions of your project.** Are there any examples of unethical behaviour in the scientific literature or popular media that might impact on the design or implementation of your study? By taking a 'devil's advocate' stance and thinking about your study from different perspectives, you should be able to consider any potential ethical flaws in your proposed experimental design (see also Box 22.1).

3. **Discuss your proposed research with other students and staff.** Such debate will help raise your awareness of any issues that might impact on your proposed study, and

will help you 'defend' your proposal in your application for ethical approval and also in your final report.

4. **Modify the proposed study to avoid ethical problems.** If steps 2 or 3 raise any ethical issues, then it is best to consider how best to modify your experimental design to address these. For example, if self-analysis or peer-group discussion has identified that the research might pose particular risks to specific groups of participants, then you might wish to consider excluding such groups. However, you would also need to consider the implication of such a move to your study outcomes.

5. **Complete all necessary paperwork in advance of your study.** Remember that no experimental work can begin until you have formal written approval from the relevant ethics committee.

6. **Carry out all practical procedures and data collection/storage as required by guidelines and legislation.** Remember that this is an important aspect of your personal development and a key aspect of professional practice that will apply when you graduate. All research scientists are bound by the same requirements to conduct their research within an ethical framework.

Animal welfare legislation in the UK – the Animal Welfare Act 2006 is designed to prevent animal cruelty and directive 2010/63/EU is designed to protect aimals used for scientific purposes. Before any research activities are carried out, they must be approved through a licence issued by the Home Office, under the Animals (Scientific Procedures) Act 1976.

Using animals in teaching and research

In addition to complying with any legislative requirements, such as laws relating to animal welfare, all teaching activities and research studies involving vertebrates and cephalopods (which have been recognised as being capable of feeling pain, suffering distress and suffering lasting harm) must follow the ethical guidelines of the institution, which will include seeking ethical approval in advance of any procedure involving live animals. Such approval must also be sought in relation to killing of animals for teaching or research purposes.

If you are involved in animal experimentation as part of a research project (Chapter 34), you must take care to follow all conditions of the animal licence obtained by your institution – consult your supervisor if you are unsure of any aspect of the regulations.

Text references

McMillan, K. and Weyers, J.D.B. (2011) *The Study Skills Book,* 2nd edn. Pearson Education, Harlow.

Shamoo, A.E. and Resnik, D.B. (2009) *Responsible Conduct of Research,* 2nd edn. Oxford University Press, Oxford.

United Nations (1948) *Universal Declaration of Human Rights: Homepage.* Available: http://www.un.org/ en/documents/udhr/index.shtml Last accessed: 01/09/15.

United Nations (1992) *Rio Declaration: Homepage.* Available: http://www.un.org/documents/ga/ confl51/aconfl5126-lannexl.htm Last accessed: 01/09/15.

Sources for further study

Barry, V. (2011) *Bioethics in a Cultural Context.* Wadsworth Cengage Publishing, Boston.

Bioethics Education Project, BEEP, Bristol University (2011) *Bioethics Homepage.* Available: http://www. beep.ac.uk/ Last accessed: 01/09/15.

Bryant, J., la Velle, L.B. and Searle, J (2005) *Introduction to Bioethics.* Wiley, Chichester.

DeGrazia, D., Mappes, T. and Ballard, J. (2010) *Biomedical Ethics,* 7th edn . McGraw-Hill, New York.

Intute Consortium of UK Universities (2010) *Bioethicsweb.* Available: http://www.intute.ac.uk/ bioethicsweb/ Last accessed: 01/09/15.

National Health Museum, Atlanta (2009) *Bioethics.* Available: http://www.accessexcellence.org/RC/AB/ IE/bioethics.php Last accessed: 01/09/15.

National Institutes of Health (2010) *Bioethics Resources on the Web.* Available: http://bioethics. od.nih.gov/ Last accessed: 01/09/15.

Vaughn, T. (2012) *Bioethics, Principles, Issues and Cases*, 2nd edn . Oxford University Press, Oxford.

STUDY EXERCISES

22.1 Test your understanding of the terminology and theories of ethics. Distinguish between the following pairs of terms:

(a) morals and ethics;

(b) value judgements and ethical decisions;

(c) utilitarianism and deontology;

(d) consequentialism and virtue theory;

(e) objectivism and subjectivism.

22.2 Test your understanding of ethical principles. Explain each of the following terms, in the context of ethics:

(a) beneficence;

(b) non-maleficence;

(c) justice;

(d) autonomy;

(e) precautionary principle.

22.3 Locate web-based information on bioethics posted by relevant organisations and learned societies. This could be important if you are researching a topic with an ethical dimension, e.g. an essay on the pros and cons of genetic modification of farm animals, or if you are considering a research project involving human or animal subjects.

22.4 'Brainstorm' the ethical issues within a research project or topic. Make a 'spider diagram' (p. 111) or 'mind map' (p. 18) with your title or topic at the centre. Then, use the ethical framework described in this chapter to consider the various aspects of the title/topic, and the choices, decisions and actions required. You could carry out this exercise on an individual basis, or as a group exercise (share your spider diagrams or mind maps with other students, and discuss your views and those of others within the group).

Answers to these study exercises are available at www.pearsoned.co.uk/practicalskills.

23   Health and safety

Health and safety law requires academic institutions to provide a working environment that is safe and without risk to health. Where appropriate, training and information on safe working practices must be provided. Students and staff must take reasonable care to ensure the health and safety of themselves and of others, and must not misuse any safety equipment.

> **KEY POINT**   All practical work must be carried out with safety in mind, to minimise the risk of harm to yourself and to others – safety is everyone's responsibility.

Risk assessment

The most widespread approach to safe working practice involves the use of risk assessment, which aims to establish:

1. **The intrinsic chemical, biological and physical hazards,** together with any maximum exposure limits (MELs) or occupational exposure standards (OESs), where appropriate. Chemical manufacturers provide data sheets listing the hazards associated with particular chemical compounds, while pathogenic (disease-causing) microbes are categorised according to their ability to cause illness (p. 225).

2. **The risks involved,** by taking into account the amount of substance to be used, the way in which it will be used and the possible routes of entry into the body (Fig. 23.1).

> **KEY POINT**   It is important to distinguish between the intrinsic hazards of a particular substance and the risks involved in its use in a particular exercise.

3. **The persons at risk,** and the ways in which they might be exposed to hazardous substances, including accidental exposure (spillage).

4. **The steps required to prevent or control exposure.** Ideally, a non-hazardous or less hazardous alternative should be used. If this is not feasible, adequate control measures must be used, e.g. a fume cupboard or other containment system. Personal protective equipment (e.g. lab coats, safety glasses) must be used in addition to such containment measures. A safe means of disposal is also required.

The outcome of the risk assessment process must be recorded and appropriate safety information must be passed on to those at risk. For most practical classes, risk assessments will have been carried out in advance by the person in charge: the information necessary to minimise the risks to students may be given in the practical schedule. Make sure you know how your department provides such information and that you have read the appropriate material before you begin your practical work. You should also pay close attention to the person in charge at the beginning of the practical session, as they may emphasise the major hazards and risks. In project work, you will need to be involved in the risk assessment process along with your supervisor, before you carry out any practical work.

Health and safety legislation – in the UK, the *Health & Safety at Work, etc. Act 1974* provides the main legal framework for health and safety. The *Control of Substances Hazardous to Health (COSHH) Regulations 2002* impose specific legal requirements for risk assessment wherever hazardous chemicals or biological agents are used, with Approved Codes of Practice for the control of hazardous substances, carcinogens and biological agents, including pathogenic microbes.

Definitions

Hazard – the intrinsic ability of a substance or biological agent to cause harm.

Risk – the likelihood that a substance or biological agent might be harmful under specific circumstances.

Distinguishing between hazard and risk– one of the hazards associated with water is drowning. However, the risk of drowning in a few drops of water is negligible.

---

inhalation

ingestion

inoculation or absorption

absorption from spillage

Fig. 23.1 Major routes of entry of harmful substances into the body.

In addition to specific risk assessments, most institutions will have a safety handbook, or equivalent, giving general details of safe working practices, together with the names and telephone numbers of safety personnel, first aiders, hospitals, etc. Make sure you read the relevant sections and follow any instructions given.

Basic rules for laboratory work

- Make sure you know what to do in case of fire, including exit routes, how to raise the alarm, and where to gather on leaving the building. Remember that the most important consideration at all times is human safety: do not attempt to fight a fire unless it is safe to do so.
- All laboratories display notices telling you where to find the first aid kit and who to contact in case of accident or emergency. Report all accidents, even those appearing insignificant – your department will have a formal recording procedure to comply with safety legislation.
- Wear appropriate protective clothing at all times – a clean lab coat (buttoned up), plus safety glasses if there is any risk to the eyes.
- Never smoke, eat or drink in any laboratory, because of the risks of contamination by inhalation or ingestion (Fig. 23.1).
- Never mouth pipette any liquid. Use a pipette filler (see p. 145) or, if appropriate, a pipettor (p. 146).
- Take care when handling glassware – see p. 149 for details.
- Know the warning symbols for specific chemical hazards (see Fig. 23.2).
- Use a fume hood for hazardous chemicals. Make sure that it is working and then open the front only as far as is necessary: many fume hoods are marked with a maximum opening.
- Always use the minimum quantity of any hazardous materials.
- Work in a logical, tidy manner and minimise risks by thinking ahead.
- Always clear up at the end of each session. This is an important aspect of safety, encouraging a responsible attitude towards laboratory work.
- Dispose of waste in appropriate containers. Most labs will have bins for sharps, glassware, hazardous solutions and radioactive waste.

Genetic engineering and molecular genetics

Additional legal constraints apply to practical work involving genetic manipulation. A specific risk assessment must be carried out for any experiment where a cell or organism is modified by genetic engineering techniques (Chapter 68) involving the insertion of DNA into a cell or organism in which it does not normally occur. Before any practical work can be carried out, it must be authorised by the establishment's genetic manipulation safety committee and notified to the relevant authority. Such work must be carried out with appropriate containment, to prevent the accidental release of genetically modified organisms into the environment.

Practicals in molecular genetics will involve some of the techniques of genetic manipulation. Typically, these will be examples of 'self-cloning' where recombinant DNA molecules are constructed from fragments of DNA which naturally occur in that organism; for example, the transformation of laboratory strains of *Escherichia coli* using a pUC plasmid (p. 476).

Fig. 23.2 Warning labels for specific chemical hazards. These appear on suppliers' containers and on tape used to label working vessels.

In the UK, *the Genetically Modified Organisms (Contained Use) Regulations 2000* define the legal requirements for risk assessment and for the notification of work involving genetic manipulation.

The Advisory Committee on Genetic Manipulation provides guidance on the use of genetically manipulated organisms.

The Health and Safety Executive (HSE) has specific responsibility for the operation of these regulations and is the regulatory authority for genetic manipulation in the UK.

Sources for further study

Anon. *Biological Hazards at Work*. Available: http://www.hse.gov.uk/biosafety/information.htm [Provides links to relevant resources] Last accessed: 01/09/15.

Anon. *COSHH Basics*. Available: http://www.hse.gov.uk/coshh/basics.htm Last accessed: 01/09/15. [UK Health and Safety Executive website giving an overview of the principles of risk assessment and management.]

Anon. (2004) *Laboratory Biosafety Manual*, 3rd edn. World Health Organization, Geneva.

Fleming, D.O. and Hunt, D.L. (2006) *Biological Safety: Principles and Practice*. ASM Press, Washington.

Health and Safety Executive (2005) *Control of Substances Hazardous to Health Regulations*. HSE, London.

STUDY EXERCISES

23.1 Test your knowledge of safe working procedures. After reading the appropriate sections of this book, can you remember the following:

   (a) The four main steps involved in the process of risk assessment?

   (b) The major routes of entry of harmful substances into the body?

   (c) The warning labels for the major chemical hazard symbols (either describe them or draw them from memory)?

   (d) The international symbol for a biohazard?

   (e) The international symbol for radioactivity?

23.2 Locate relevant health and safety features in a laboratory. Find each of the following in one of the laboratories used as part of your course (draw a simple location map, if this seems appropriate):

   (a) fire exit(s);

   (b) fire-fighting equipment;

   (c) first aid kit;

   (d) 'sharps' container;

   (e) container for disposal of broken glassware;

   (f) eye wash station (where appropriate).

23.3 Investigate the health and safety procedures in operation at your university. Can you find out the following?

   (a) your university's procedure in case of fire;

   (b) the colour coding for fire extinguishers available in your department and their recommendations for use;

   (c) the accident reporting procedure used in your department;

   (d) your department's Code of Safe Practice relating to a specific aspect of bioscience, e.g. working with micro-organisms.

23.4 Carry out risk assessments for specific chemical hazards. Look up the hazards associated with the use of the following chemicals and list the appropriate protective measures required to minimise risk during use in a lab class:

   (a) formaldehyde solution, used as a preservative for animal tissue, to be used for microscopic examination (Chapter 29);

   (b) acetone, e.g. for use as a solvent for the quantitative analysis of plant pigments (Chapter 47);

   (c) sodium hydroxide, used in solid form to prepare a dilute solution to be used for pH adjustment (Chapter 27).

Answers to these study exercises are available at www.pearsoned.co.uk/practicalskills.

Measuring and dispensing liquids

The equipment you should choose to measure out liquids depends on the volumes being dispensed, the accuracy required and the number of times the job must be done (Table 24.1).

Table 24.1 Criteria for choosing a method for measuring out a liquid.

Method	Best volume range	Accuracy	Usefulness for repetitive measurement
Pasteur pipette	30 μl to 2 mL	Low	Convenient
Conical flask/beaker	25–5000 mL	Very low	Convenient
Measuring cylinder	5–2000 mL	Medium	Convenient
Volumetric flask	5–2000 mL	High	Convenient
Burette	1–100 mL	High	Convenient
Glass pipette	1–100 mL	High	Convenient
Mechanical pipettor	5–1000 μL	High*	Convenient
Syringe	0.5–20 μL	Medium**	Convenient
Microsyringe	0.5–50 μL	High	Convenient
Weighing	Any (depends on accuracy of balance)	Very high	Inconvenient

*If correctly calibrated and used properly (see p. 147).
**Accuracy depends on width of barrel: large volumes are less accurate.

Certain liquids may cause problems:

- **High viscosity liquids** are difficult to dispense: allow time for all the liquid to transfer.
- **Organic solvents** may evaporate rapidly, making measurements inaccurate: work quickly; seal containers without delay.
- **Solutions prone to frothing** (e.g. protein and detergent solutions) are difficult to measure and dispense: avoid forming bubbles due to overagitation; do not transfer quickly.
- **Suspensions** (e.g. cell cultures) may sediment: thoroughly mix them before dispensing.

Pasteur pipettes

Hold correctly during use (Fig. 24.1) – keep the pipette vertical, with the middle fingers gripping the barrel while the thumb and index finger provide controlled pressure on the bulb. Squeeze gently to dispense individual drops. To avoid the risk of cross-contamination, take care not to draw up solution into the bulb or to lie the pipette on its side. Alternatively, use a plastic disposable 'Pastette'.

Measuring cylinders and volumetric flasks

These must be used on a level surface so that the scale is horizontal; you should first fill with solution until just below the desired mark; then fill slowly (e.g. using a Pasteur pipette) until the meniscus is level with the mark. Allow time for the solution to run down the walls of the vessel.

Reading any volumetric scale – make sure your eye is level with the bottom of the liquid's meniscus and take the reading from this point.

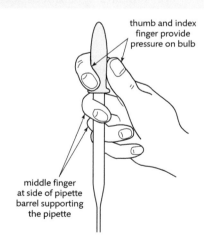

thumb and index finger provide pressure on bulb

middle finger at side of pipette barrel supporting the pipette

Fig. 24.1 How to hold a Pasteur pipette.

(a)   (b)   (c)

Fig. 24.2 Glass pipettes – graduated pipette, reading from zero to shoulder (a); graduated pipette, reading from maximum to tip, by gravity (b); bulb (volumetric) pipette, showing volume (calibration mark to tip, by gravity) above the bulb (c).

Fig. 24.3 A pipettor (autopipettor) – the Gilson Pipetman. The pipettor is designed to fit comfortably in your 'dominant' hand, with all of your fingers wrapped around the upper (coloured) section and your thumb on the pushbutton.

Burettes

These should be mounted vertically on clamp stands – do not overtighten the clamp. First ensure the tap is closed and fill the body with solution using a funnel. Open the tap and allow some liquid to fill the tubing below the tap before first use. Take a meniscus reading, noting the value in your notebook. Dispense the solution via the tap and measure the new meniscus reading. The volume dispensed is the difference between the two readings. Titrations using a burette are usually performed on a magnetic stirrer (p. 156).

Pipettes

These come in various designs, including graduated and bulb (volumetric) pipettes (Fig. 24.2). Take care to check the volume scale before use: some empty from full volume to zero, others from zero to full volume; some scales refer to the shoulder of the tip, others to the tip either by gravity or after blowing out.

> **KEY POINT**   For safety reasons, it is never permissible to mouth pipette – various aids are available such as the Pi-pump.

Pipettors (autopipettors or micropipettes)

These come in two basic types:

- **Air displacement pipettors.** For routine work with dilute aqueous solutions. One of the most widely used examples is the Gilson Pipetman (Fig. 24.3). Box 24.1 gives practical guidance on its use.
- **Positive displacement pipettors.** For non-standard applications, including dispensing viscous, dense or volatile liquids, or certain procedures in molecular genetics, e.g. the PCR (p. 467), where an air displacement pipettor might create aerosols, leading to errors.

Air displacement and positive displacement pipettors may be:

- **Fixed volume:** capable of delivering a single factory-set volume.
- **Adjustable:** where the volume is determined by the operator across a particular range of values.
- **Pre-set:** movable between a limited number of values.
- **Multi-channel:** able to deliver several replicate volumes at the same time.

Whichever type you use, you must ensure that you understand the operating principles of the volume scale and the method for changing the volume delivered – some pipettors are easily misread.

A pipettor must be fitted with the correct disposable tip before use: each manufacturer produces different tips to fit particular models. Specialised tips are available for particular applications e.g. PCR (p. 467).

Syringes

Syringes should be used by placing the tip of the needle in the solution and drawing the plunger up slowly to the required point on the scale. Check the barrel to make sure no air bubbles have been drawn up. Expel slowly and touch the syringe on the edge of the vessel to remove any liquid adhering to

Using a pipettor (autopipettor) to deliver accurate, reproducible volumes of liquid

A pipettor can be used to dispense volumes with accuracy and precision, by following this step-wise procedure:

1. **Select a pipettor that operates over the appropriate range.** Most adjustable pipettors are accurate only over a particular working range and should not be used to deliver volumes below the manufacturer's specifications (minimum volume is usually 10–20% of maximum value). Do not attempt to set the volume above the maximum limit, or the pipettor may be damaged.

2. **Set the volume to be delivered.** In some pipettors, you 'dial up' the required volume. Types like the Gilson Pipetman have a system where the scale (or 'volumeter') consists of three numbers, read from top to bottom of the barrel, and adjusted using the black knurled adjustment ring (Fig. 24.3). This number gives the first three digits of the volume scale and thus can only be understood by establishing the maximum volume of the Pipetman, as shown on the pushbutton on the end of the plunger (Fig. 24.3). The following examples illustrate the principle for two common sizes of Pipetman:

P1000 Pipetman® (maximum volume 1000 μL) if you dial up

```
1
0
0
```
the volume is set at 1000 μL

P20 Pipetman® (maximum volume 20 μL) if you dial up

```
1
0
0
```
the volume is set at 10.0 μL

Note: The Pipetman scale is *not* a percentage one.

3. **Fit a new disposable tip to the end of the barrel.** Make sure that it is the appropriate type for your pipettor and that it is correctly fitted. Press the tip on firmly using a slight twisting motion – if not, you will take up less than the set volume and liquid will drip from the tip during use. Tips are often supplied in boxes, for ease of use: if sterility is important, make sure you use appropriate sterile technique at all times (p. 223). *Never, ever, try to use a pipettor without its disposable tip.*

4. **Check your delivery.** Confirm that the pipettor delivers the correct volume by dispensing volumes of distilled water and weighing on a balance, assuming 1 mg = 1 μL = 1 mm³. The value should be within 1% of the selected volume. For small volumes, measure several 'squirts' together, e.g. 20 'squirts' of 5 μL = 100 mg. If the pipettor is inaccurate (p. 193), giving a biased result (e.g. delivering significantly more or less than the volume set), you can make a temporary correction by adjusting the volumeter scale down or up accordingly (the volume *delivered* is more important than the value *displayed* on the volumeter), or have the pipettor recalibrated. If the pipettor is imprecise (p. 193), delivering a variable amount of liquid each time, it may need to be serviced. After calibration, fit a clean (sterile) tip if necessary.

5. **Draw up the appropriate volume.** Holding the pipettor *vertically*, press down on the plunger/pushbutton until a resistance (spring-loaded stop) is met. Then place the end of the tip in the liquid. Keeping your thumb on the plunger/push-button, release the pressure slowly and evenly: watch the liquid being drawn up into the tip, to confirm that no air bubbles are present. Wait a second or so, to confirm that the liquid has been taken up, then withdraw the end of the tip from the liquid. Inexperienced users often have problems caused by drawing up the liquid too quickly/carelessly. If you accidentally draw liquid into the barrel, seek assistance from your demonstrator or supervisor as the barrel will need to be cleaned before further use.

6. **Make a quick visual check on the liquid in the tip.** Does the volume seem reasonable? (e.g. a 100 μL volume should occupy approximately half the volume of a P200 tip). The liquid will remain in the tip, without dripping, as long as the tip is fitted correctly and the pipettor is not tilted too far from a vertical position.

7. **Deliver the liquid.** Place the end of the tip against the wall of the vessel at a slight angle (10–15° from vertical) and press the plunger/pushbutton slowly and smoothly to the first (spring-loaded) stop. Wait a second or two, to allow any residual liquid to run down the inside of the tip, then press again to the final stop, dispensing any remaining liquid. Remove from the vessel with the plunger/pushbutton still depressed.

8. **Eject the tip.** Press the tip ejector button if present (Fig. 24.3). If the tip is contaminated, eject directly into an appropriate container, e.g. a beaker of disinfectant, for microbiological work, or a labelled container for hazardous solutions (p. 143). For repeat delivery, fit a new tip if necessary and begin again at step 5 above. Always make sure that the tip is ejected before putting a pipettor on the bench.

Example Using Eqn [24.1], weighing 9 g of liquid with a density of 1.2 g mL^{-1} will give a volume of $9 \div 1.2 = 7.5$ mL.

Fig. 24.4 Stabilising pipettor (autopipettor) with the thumb of your second hand (note: thumb is positioned well away from the disposable tip).

the end of the needle. Microsyringes should always be cleaned before and after use by repeatedly drawing up and expelling pure solvent. The dead space in the syringe needle can occupy up to 4% of the nominal syringe volume. A way of avoiding such problems is to fill the dead space with an inert substance (e.g. silicone oil) after sampling. Alternatively, use a syringe where the plunger occupies the needle space (small volumes only).

Balances

These can be used to weigh accurately (p. 157) how much liquid you have dispensed. Convert mass to volume using the equation:

$$\text{mass/density} = \text{volume} \qquad [24.1]$$

Densities of common solvents can be found in Haynes (2014). You will also need to know the liquid's temperature, as density is temperature-dependent.

Holding and storing liquids

Test tubes

These are used for colour tests, small-scale reactions, holding cultures, etc. The tube can be sterilised by heating (p. 223) and maintained in this state with a cap or cotton wool plug.

Beakers

Beakers are used for general purposes, e.g. heating a solvent while the solute dissolves, carrying out a titration, etc.

Conical (Erlenmeyer) flasks

These are used for storage of solutions: their wide base makes them stable, while their small mouth reduces evaporation and is easily sealed.

Bottles and vials

These are used when the solution needs to be sealed for safety, sterility or to prevent evaporation or oxidation. They usually have a screw top or ground glass stopper to prevent evaporation and contamination. Many types are available, including 'bijou', 'McCartney', 'universal' and 'Winkler'.

You should clearly label all stored solutions (see p. 156), including relevant hazard information, preferably marking with hazard warning tape (p. 143). Seal vessels in an appropriate way, e.g. using a stopper or a sealing film such as Parafilm or Nescofilm to prevent evaporation. To avoid degradation store your solution in a fridge, but allow it to reach room temperature before use.

Creating specialised apparatus

Glassware systems incorporating ground glass connections such as Quickfit are useful for setting up combinations of standard glass components, e.g. for chemical reactions. In project work, you may need to adapt standard forms of glassware for a special need. A glassblowing service (often available in chemistry departments) can make special items to order.

Choosing between glass and plastic

Bear in mind the following points:

- **Reactivity.** Plastic vessels often distort at relatively low temperatures; they may be flammable, may dissolve in certain organic solvents and may be affected by prolonged exposure to ultraviolet (UV) light. Some plasticisers

Table 24.2 Typical spectral cut-off values for glass and plastics (λ_{50} = wavelength at which transmission of EMR is reduced to 50%).

Material	λ_{50} (nm)
Routine glassware	340
Pyrex glass	292
Polycarbonate	396
Acrylic	342
Polyester	318
Quartz	220

may leach from vessels and have been shown to have biological activity. Glass may adsorb ions and other molecules and then leach them into solutions, especially in alkaline conditions. Pyrex glass is stronger than ordinary soda glass and can withstand temperatures up to 500 °C.

- **Rigidity and resilience.** Plastic vessels are not recommended where volume is critical as they may distort through time: use class A volumetric glassware for accurate work, e.g. preparing solutions (Chapter 26). Glass vessels are more easily broken than plastic (Box 24.2), which is particularly important for centrifugation (see p. 330).

- **Opacity.** Both glass and plastic absorb light in the UV range of the EMR spectrum (Table 24.2). Quartz should be used where this is important, e.g. in cuvettes for UV spectrophotometry (see p. 310).

- **Disposability.** Plastic items may be cheap enough to make them disposable, an advantage where there is a risk of chemical or microbial contamination.

Cleaning glass and plastic

Take care to avoid the possibility of contamination arising from prior use of chemicals or inadequate rinsing following washing. A thorough rinse with distilled or deionised water immediately before use will remove dust and other deposits and is good practice in quantitative work, but ensure that the rinsing solution is not left in the vessel. 'Strong' basic detergents (e.g. Pyroneg) are good for solubilising acidic deposits. If there is a risk of basic deposits

---

Box 24.2 **Safe working with glass**

Many minor accidents in the laboratory are caused by lack of care with glassware. You should follow these general precautions:

- Always wear safety glasses when there is any risk of glass breakage – e.g. when using low pressures, or heating solutions.

- Take care when attaching tubing to glass tubes and when putting glass tubes into bungs – always hold the tubing and glassware close together, as shown in Fig. 24.5, and wear thick gloves when appropriate.

- Use a 'soft' Bunsen flame when heating glassware – this avoids creating a hot spot, where cracks may start: always use tongs or special heat-resistant gloves when handling hot glassware (never use a rolled-up paper towel).

- Do not use chipped or cracked glassware – it may break under very slight strain and should be disposed of in the broken glassware bin.

- Never carry large glass bottles/flasks by their necks – support them with a hand underneath or, better still, carry them in a basket.

- Do not force bungs too firmly into bottles – they can be extremely difficult to remove. If you need a tight seal, use a screw-top bottle with a rubber or plastic seal.

- Dispose of broken glass thoroughly and carefully – use disposable paper towels and wear thick gloves. Always put pieces of broken glass in the correct bin.

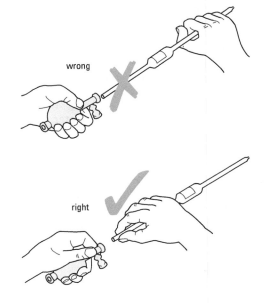

Fig. 24.5 Handling glass pipettes tubing.

remaining, use an acid wash. If there is a risk of contamination from organic deposits, a rinse with Analar grade ethanol is recommended. Glassware can be disinfected by washing with a sodium hypochlorite bleach such as Chloros, with sodium metabisulfite or a blended commercial product such as Virkon – dilute as recommended before use and rinse thoroughly with sterile water after use. Alternatively, to sterilise glassware, heat to at least 121 °C for 15 min in an autoclave or 160 °C for 3 h in an oven.

Text references

Haynes, WM. (ed.) (2014) *CRC Handbook of Chemistry and Physics,* 95th edn. CRC Press, Boca Raton.

Sources for further study

Boyer, R.F. (2011) *Biochemistry Laboratory: Modern Theory and Techniques,* 2nd edn. Prentice Hall, New Jersey.

Henrikson, C., Byrd, L.C. and Hunter, N.W. (2005) *A Laboratory Manual for General, Organic and Biochemistry,* 5th edn. McGraw-Hill, New York.

Ninfa, A.J., Ballou, D.P. and Benore, M. (2010) *Fundamental Laboratory Approaches for Biochemistry and Biotechnology,* 2nd edn. Wiley, New York.

Seidman, L.A. and Moore, C.J. (2008) *Basic Laboratory Methods for Biotechnology: Textbook and Laboratory Reference,* 2nd edn. Benjamin Cummings, San Francisco.

STUDY EXERCISES

24.1 Decide on the appropriate methods and equipment for the following procedures:

(a) Preparing one litre of ethanol at approximately 70% v/v in water for use as a general-purpose reagent.

(b) Adding 10 µL of a sample to the well of an agarose gel during a molecular biology procedure.

(c) Preparing a calibration standard of 100 mL of DNA, to contain 200.0 µg mL^{-1}, for spectrophotometry.

(d) Carrying out a titration curve for a buffer solution.

24.2 Write a protocol for calibrating and using a pipettor. After reading this chapter, prepare a detailed step-wise protocol explaining how to use a pipettor to deliver a specific volume, say of 500 µL (e.g. using a Gilson Pipetman, or an alternative if your department does not use this type). Ask another student to evaluate your protocol and provide you with written feedback – either simply by reading through your protocol, or by trying it out with a pipettor as part of a class exercise (check with a member of staff before you attempt this in a laboratory).

24.3 Determine the accuracy and precision of a pipette. Using the following data for three different models of pipettor, determine which pipettor is most *accurate* and which is most *precise* (check p. 193 if you are unsure of the definitions of these two terms). All three pipettors were set to deliver 1000µL (1.000 mL) and 10 repetitive measurements of the weight of the volume of water in grams delivered were made using a three-place balance:

Model A pipettor:	0.986; 0.971; 0.993; 0.964; 0.983; 0.996; 0.977; 0.969; 0.982; 0.974
Model B pipettor:	1.013; 1.011; 1.010; 1.009; 1.011; 1.010; 1.011; 1.009; 1.011; 1.012
Model C pipettor:	0.985; 1.022; 1.051; 1.067; 0.973; 0.982; 0.894; 1.045; 1.062; 0.928

In your answer, you should support your conclusions with appropriate numerical (statistical) evidence (see Chapter 73 for appropriate measures of location and dispersion).

Answers to these study exercises are available at www.pearsoned.co.uk/practicalskills.

Finding out about chemicals – *The Merck Index* (O'Neil *et al.*, 2013: Fig. 25.1) and the *CRC Handbook of Chemistry and Physics* (Haynes, 2014) are useful sources of information on the physical and biological properties of chemicals, including melting and boiling points, solubility, toxicity, etc.

8671. Sodium Chloride. [7647-14-5] Salt; common salt. ClNa; mol wt 58.44. Cl 60.67%, Na 39.34%. NaCl. The article of commerce is also known as *table salt*, *rock salt* or *sea salt*. Occurs in nature as the mineral *halite*. Produced by mining (rock salt), by evaporation of brine from underground salt deposits and from sea water by solar evaporation: *Faith, Keyes & Clark's Industrial Chemicals*, F. A. Lowenheim, M. K. Moran, Eds. (Wiley-Interscience, New York, 4th ed., 1975) pp 722-730. Toxicity studies: E. M. Boyd, M. N. Shanas, *Arch. Int. Pharmacodyn.* **144**, 86 (1963). Comprehensive monograph: D. W. Kaufmann, *Sodium Chloride*, ACS Monograph Series no. **145** (Reinhold, New York, 1960) 743 pp.

Cubic, white crystals, granules, or powder; colorless and transparent or translucent when in large crystals. d 2.17. The salt of commerce usually contains some calcium and magnesium chlorides which absorb moisture and make it cake. mp 804° and begins to volatilize at a little above this temp. One gram dissolves in 2.8 ml water at 25°, in 2.6 ml boiling water, in 10 ml glycerol; very slightly sol in alcohol. Its soly in water is decreased by HCl. Almost insol in concd HCl. Its aq soln is neutral. pH: 6.7-7.3. d of satd aq soln at 25° is 1.202. A 23% aq soln of sodium chloride freezes at −20.5°C (5°F). LD_{50} orally in rats: 3.75 ±0.43 g/kg (Boyd, Shanas).

Note: **Blusalt**, a brand of sodium chloride contg trace amounts of cobalt, iodine, iron, copper, manganese, zinc is used in farm animals.

Caution: Not generally considered poisonous. Accidental substitution of NaCl for lactose in baby formulas has caused fatal poisoning.

USE: Natural salt is the source of chlorine and of sodium as well as of all, or practically all, their compds, e.g., hydrochloric acid, chlorates, sodium carbonate, hydroxide, etc.; for preserving foods; manuf soap, dyes—to salt them out; in freezing mixtures; for dyeing and printing fabrics, glazing pottery, curing hides; metallurgy of tin and other metals.

THERAP CAT: Electrolyte replenisher; emetic; topical anti-inflammatory.

THERAP CAT (VET): Essential nutrient factor. May be given orally as emetic, stomachic, laxative or to stimulate thirst (prevention of calculi). Intravenously as isotonic solution to raise blood volume, to combat dehydration. Locally as wound irrigant, rectal douche.

Fig. 25.1 Example of typical *Merck Index* entry above showing type of information given for each chemical. From O'Neil *et al.* Reproduced with permission from *The Merck Index*, Fifteenth Edition. (2013). Copyright © 2013 by Merck & Co, Inc., Whitehouse Station, NJ, USA. All rights reserved.

Using chemicals

Safety aspects

In practical classes, the person in charge has a responsibility to inform you of any hazards associated with the use of chemicals. For routine practical procedures, a risk assessment (p. 142) will have been carried out by a member of staff and relevant safety information will be included in the practical schedule: an example is shown in Table 25.1.

In project work, your first duty when using an unfamiliar chemical is to find out about its properties, especially those relating to safety. Your department must provide the relevant information to allow you to do this. If your supervisor has filled out the form, read it carefully before signing. Box 25.1 gives further advice.

> **KEY POINT** Before you use any chemical you must find out whether safety precautions need to be taken, and complete the appropriate forms confirming that you appreciate the risks involved.

Selection

Chemicals are supplied in various degrees of purity and this is always stated on the manufacturer's containers. Suppliers differ in the names given to the grades and there is no conformity in purity standards. Very pure chemicals cost more, sometimes a lot more, and should only be used if the situation demands. If you need to order a chemical, your department will have a defined procedure for doing this.

Preparing solutions

Solutions are usually prepared with respect to their molar concentrations (e.g. mmol L^{-1}, or mol m^{-3}), or mass concentrations (e.g. g L^{-1}, or kg m^{-3}):

Table 25.1 Representative risk assessment information for a practical exercise in molecular biology, involving the isolation of DNA.

Substance	Hazards	Comments
Sodium dodecyl sulfate (SDS)	Irritant	Wear gloves
	Toxic	
Sodium hydroxide (NaOH)	Highly corrosive	Wear gloves
	Severe irritant	
Isopropanol	Highly flammable	No naked flames
	Irritant/corrosive	Wear gloves
	Potential carcinogen	
Phenol	Highly toxic	Use in fume hood
	Causes skin burns	Wear gloves
	Potential carcinogen	
Chloroform	Volatile and toxic	Use in fume hood
	Irritant/corrosive	Wear gloves
	Potential carcinogen	

Box 25.1 Safe working with chemicals

You should always treat chemicals as potentially dangerous, following these general precautions:

- **Do not use any chemical until you have considered the risks involved** – for lab classes, you should carefully read all hazard and risk information provided *before* you start work. In project work, you may need to be involved in the risk assessment process with your supervisor.

- **Wear a lab coat at all times** – the coat should be fully fastened and cleaned appropriately, should any chemical compound be spilled on it. Closed-toe footwear will protect your feet should any spillages occur.

- **Make sure you know where the safety apparatus is kept before you begin working** – this includes eye bath, fire extinguishers and blanket, first aid kit.

- **Wear safety glasses and gloves when working with toxic, irritant or corrosive chemicals, and for any substances where the hazards are not yet fully characterised** – make sure you understand the hazard warning signs (p. 143) along with any specific hazard coding system used in your department. Carry out procedures with solid material in a fume hood.

- **Use aids such as pipette fillers to minimise the risk of contact with hazardous solutions** – these aids are further detailed on p. 146.

- **Never smoke, eat, drink or chew gum in a lab where chemicals are handled** – this will minimise the risk of ingestion.

- **Label all solutions appropriately** – use the appropriate hazard warning information (see pp. 143/223/305).

- **Report all spillages of chemicals/solutions** – make sure that spillages are cleaned up properly.

- **Store hazardous chemicals only in the appropriate locations** – for example, a spark-proof fridge is required for flammable liquids; acids and solvents should not be stored together.

- **Dispose of chemicals in the correct manner** – if unsure, ask a member of staff (do not assume that it is safe to use the lab waste bin or the sink for disposal).

- **Wash hands after any direct contact with chemicals or biochemical material** – always wash your hands at the end of a lab session.

Examples Using Eqn [25.1], 25g of a substance dissolved in 400 mL of water would have a mass concentration of $25 \div 400 = 0.0625$ g mL^{-1} ($\equiv 62.5$ mg mL$^{-1} \equiv 62.5$ g L^{-1}) Using Eqn [25.1], 0.4 mol of a substance dissolved in 0.5 litres of water would have a molar concentration of $0.4 \div 0.5 = 0.8$ mol L^{-1} ($= 800$ mmol L^{-1}).

Solving solubility problems – if your chemical does not dissolve after a reasonable time:

- check the limits of solubility for your compound (see O'Neil *et al.*, 2013);
- check the pH of the solution – solubility often changes with pH, e.g. you may be able to dissolve the compound in an acidic or an alkaline solution.

both can be regarded as an amount of *substance* per unit volume of *solution*, in accordance with the relationship:

$$\text{Concentration} = \frac{\text{amount}}{\text{volume}} \qquad [25.1]$$

The most important aspect of Eqn [25.1] is to recognise clearly the units involved, and to prepare the solution accordingly: for molar concentrations, you will need the relative molecular mass of the compound, so that you can determine the mass of substance required. Further advice on concentrations and interconversion of units is given in Box 26.1.

Box 25.2 shows the steps involved in making up a solution. The concentration you require is likely to be defined by a protocol you are following, and the grade of chemical and supplier may also be specified. Success may depend on using the same source and quality, e.g. with enzyme work. To avoid waste, think carefully about the volume of solution you require, though it is always a good idea to err on the high side because you may spill some or make a mistake when dispensing it. Try to choose one of the standard volumes for vessels, as this will make measuring out easier.

Use distilled or deionised water to make up aqueous solutions and stir to make sure all the chemical is dissolved. Magnetic stirrers are the most convenient means of doing this: carefully drop a clean magnetic stirrer bar ('flea') in the beaker, avoiding splashing; place the beaker centrally on the stirrer plate, switch on the stirrer and gradually increase the speed of stirring. When the crystals or powder have completely dissolved, switch off and retrieve the flea with a magnet or another flea. Take care not to contaminate your solution when you do this, and rinse the flea with distilled water.

Box 25.2   **How to make up an aqueous solution of known concentration from solid material**

1. **Find out or decide the concentration of chemical required** and the degree of purity necessary.

2. **Decide on the volume of solution required.**

3. **Find out the relative molecular mass of the chemical** (M_r). This is the sum of the atomic (elemental) masses of the component elements and can be found on the container. If the chemical is hydrated, i.e. has water molecules associated with it, these must be included when calculating the mass required.

4. **Work out the mass of chemical that will give the concentration desired in the volume required.**
   Suppose your procedure requires you to prepare 250 mL of 0.1 mol L^{-1} NaCl.

   (a) Begin by expressing all volumes in the same units, either millilitres or litres (e.g. 250 mL as 0.25 litres).

   (b) Calculate the number of moles required from Eqn [25.1]: 0.1 = amount (mol) ÷ 0.25.
   By rearrangement, the required number of moles is thus 0.1 × 0.25 = 0.025 mol.

   (c) Convert from mol to g by multiplying by the relative molecular mass (M_r for NaCl = 58.44)

   (d) Therefore, you need to make up 0.025 × 58.44 = 1.461 g to a final volume of 250 mL of solution, using distilled water.

   In some instances, it may be easier to work in SI units, though you must be careful when using exponential numbers (p. 509).
   Suppose your protocol states that you need 100 mL of 10 mmol L^{-1} KCl.

   (a) Start by converting this to 100 × 10^{-6} m³ of 10 mol m^{-3} KCl.

   (b) The required number of mol is thus (100 × 10^{-6}) × (10) = 10^{-3}.

   (c) Each mol of KCl weighs 72.56 g (M_r, the relative molecular mass).

   (d) Therefore you need to make up 72.56 × 10^{-3} g = 72.56 mg KCl to 100 × 10^{-6} m³ (100 mL) with distilled water.

   See Box 26.1 for additional information.

5. **Weigh out the required mass of chemical to an appropriate accuracy.** If the mass is too small to weigh to the desired degree of accuracy, consider the following options:

   (a) Make up a greater volume of solution.

   (b) Make up a stock solution which can be diluted at a later stage (p. 151).

   (c) Weigh the mass first, and calculate what volume to make the solution up to afterwards using Eqn [25.1].

6. **Add the chemical to a beaker or conical flask then add a little less water than the final amount required.** If some of the chemical sticks to the paper, foil or weighing boat, use some of the water to wash it off.

7. **Stir and, if necessary, heat the solution to ensure all the chemical dissolves.** You can determine when this has happened visually by observing the disappearance of the crystals or powder.

8. **If required, check and adjust the pH of the solution when cool** (see p. 170).

9. **Make up the solution to the desired volume.** If the concentration needs to be accurate, use a class A volumetric flask; if a high degree of accuracy is not required, use a measuring cylinder (class B).

   (a) Pour the solution from the beaker into the measuring vessel using a funnel to avoid spillage.

   (b) Make up the volume so that the meniscus comes up to the appropriate measurement line (p. 145). For accurate work, rinse out the original vessel and use this liquid to make up the volume.

10. **Transfer the solution to a reagent bottle or a conical flask and label the vessel clearly.**

'Obstinate' solutions may require heating, but do this only if you know that the chemical will not be damaged at the temperature used. Use a stirrer-heater to keep the solution mixed as you heat it. Allow the solution to cool before you measure volume or pH as these change with temperature.

Stock solutions

Working with chemicals and stock solutions – it is poor laboratory practice to return unused chemicals or stock solution to their original containers after use. They should be discarded, following the appropriate safety procedures.

Stock solutions are valuable when making up a range of solutions containing different concentrations of a reagent or if the solutions have some common ingredients. They also save work if the same solution is used over a prolonged period (e.g. a nutrient solution). The stock solution is

Table 25.2 Use of stock solutions. Suppose you need a set of solutions 10 mL in volume containing differing concentrations of KCl, with and without reagent Q. You decide to make up a stock of KCl at twice the maximum required concentration (50 mmol L^{-1} = 50 mol m^{-3}) and a stock of reagent Q at twice its required concentration. The table shows how you might use these stocks to make up the media you require. Note that the total volumes of stock you require can be calculated from the table (end column).

Stock solutions	Volume of stock required to make required solutions (mL)						Total volume of stock required (mL)
	No KCl plus Q	No KCl minus Q	15 mmol L^{-1} KCl plus Q	15 mmol L^{-1} KCl minus Q	25 mmol L^{-1} KCl plus Q	25 mmol L^{-1} KCl minus Q	
50 mmol L^{-1} KCl	0	0	3	3	5	5	16
[reagent Q] × 2	5	0	5	0	5	0	15
Water	5	10	2	7	0	5	29
Total	10	10	10	10	10	10	60

Storing stock solutions – these are best kept in darkness in a refrigerator to minimize the growth of photoautotrophic microbes.

Making a dilution – use the relationship $[C_1] V_1 = [C_2] V_2$ to determine volume or concentration (see Box 26.1).

Removing a magnetic flea from a volumetric flask – use a strong magnet to bring the flea to the top of the flask, to avoid contamination during removal.

Using the correct volumes for dilutions – it is important to distinguish between the volumes of the various liquids: a one-in-ten dilution is obtained using 1 volume of stock solution plus 9 volumes of diluent (1 + 9 = 10). Note that when this is shown as a ratio, it may represent the initial and final volumes (e.g. 1:10) or, sometimes, the volumes of stock solution and diluent (e.g. 1:9).

more concentrated than the final requirement and is diluted as appropriate when the final solutions are made up. The principle is best illustrated with an example (Table 25.2).

Preparing dilutions

Making a single dilution

You may need to dilute a stock solution to give a particular mass concentration, or molar concentration. Use the following procedure:

1. **Transfer an accurate volume of stock solution to a volumetric flask,** using appropriate equipment (Table 24.1).
2. **Make up to the calibration mark with solvent** – add the last few drops from a pipette or solvent bottle, until the meniscus is level with the calibration mark.
3. **Mix thoroughly,** either by repeated inversion (holding the stopper firmly) or by prolonged stirring, using a magnetic stirrer. Make sure you add the magnetic flea *after* the volume adjustment step.

For routine work using dilute aqueous solutions where the highest degree of accuracy is not required, it may be acceptable to substitute test tubes or conical flasks for volumetric flasks. In such cases, you would calculate the volumes of stock solution and diluent required, with the assumption that the final volume is determined by the sum of the individual volumes of stock and diluent used (e.g. Table 25.2). Thus, a twofold dilution would be prepared using 1 volume of stock solution and 1 volume of diluent. The dilution factor is obtained from the ratio of the initial concentration of the stock solution and the final concentration of the diluted solution. The dilution factor can be used to determine the volumes of stock and diluent required in a particular instance. For example, suppose you wanted to prepare 100 mL of a solution of NaCl at 0.2 mol L^{-1}. Using a stock solution containing 4.0 mol L^{-1} NaCl, the dilution factor is 0.2 ÷ 4.0 = 0.05 = 1/20 (a 20-fold dilution). Therefore, the amount of stock solution required is 1/20th of 100 mL = 5 mL and the amount of diluent needed is 19/20ths of 100 mL = 95 mL.

Preparing a dilution series

Dilution series are used in a wide range of procedures, including the preparation of standard curves for calibration of analytical instruments (Chapter 42), and in microbiology (Chapter 35) and immunoassay

(Chapter 43), where a range of dilutions of a particular sample is often required. A variety of different approaches can be used:

Linear dilution series

Here, the concentrations are separated by an equal amount, e.g. a series containing protein at 0, 0.2, 0.4, 0.6, 0.8, 1.0 µg mL^{-1}. Such a dilution series might be used to prepare a calibration curve for spectrophotometric assay of protein concentration (Box 56.1), or an enzyme assay (p. 411). Use $[C_1]\ V_1 = [C_2]\ V_2$ (p. 162) to determine the amount of stock solution required for each member of the series, with the volume of diluent being determined by subtraction.

Logarithmic dilution series

Here, the concentrations are separated by a constant proportion, often referred to as the step interval.

The most common examples are:

- **Doubling dilutions** – where each concentration is half that of the previous one (twofold step interval, $\log_2$ dilution series). First, make up the most concentrated solution at twice the volume required. Measure out half of this volume into a vessel containing the same volume of diluent, mix thoroughly and repeat, for as many doubling dilutions as are required. The concentrations obtained will be 1/2, 1/4, 1/8, 1/16, etc. times the original (i.e. the dilutions will be two, four, eight and 16-fold, etc.).

- **Decimal dilutions** – where each concentration is one-tenth that of the previous one (tenfold step interval, $\log_{10}$ dilution series). First, make up the most concentrated solution required, with at least a 10% excess. Measure out one-tenth of the volume required into a vessel containing nine times as much diluent, mix thoroughly and repeat. The concentrations obtained will be 1/10, 1/100, 1/1000, etc. times the original (i.e. dilutions of 10^{-1}, 10^{-2}, 10^{-3}, etc.). To calculate the actual concentration of solute, multiply by the appropriate dilution factor.

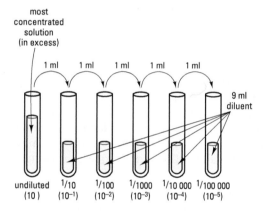

Fig. 25.2 Preparation of a dilution series. The example shown is a decimal dilution series, down to 1/100000 (10^{-5}) of the solution in the first (left-hand) tube. Note that all solutions must be mixed thoroughly before transferring the volume to the next in the series. In microbiology and cell culture, sterile solutions and appropriate aseptic technique will be required (p. 223).

When preparing serial doubling or decimal dilutions, it is often easiest to add the appropriate amount of diluent to several vessels beforehand, as shown in the worked example in Fig. 25.2. When preparing a dilution series, it is essential that all volumes are dispensed accurately, e.g. using calibrated pipettors (p. 147), otherwise any inaccuracies will be compounded, leading to gross errors in the most dilute solutions.

Harmonic dilution series

Here, the concentrations in the series take the values of the reciprocals of successive whole numbers, e.g. 1, 1/2, 1/3, 1/4, 1/5, etc. The individual dilutions are simply achieved by a stepwise increase in the volume of diluent in successive vessels, e.g. by adding 0, 1, 2, 3, 4 and 5 times the volume of diluent to a set of test tubes, then adding a constant unit volume of stock solution to each vessel. Although there is no dilution transfer error between individual dilutions, the main disadvantage is that the series is non-linear, with a step interval that becomes progressively smaller as the series is extended.

Solutions must be thoroughly mixed before measuring out volumes for the next dilution. Use a fresh measuring vessel for each dilution to

Preparing a dilution series using pipettes or pipettors – use a fresh pipette or disposable tip for each dilution, to prevent carry-over of solutions.

avoid contamination, or wash your vessel thoroughly between dilutions. Clearly label the vessel containing each dilution when it is made: it is easy to get confused! When deciding on the volumes required, allow for the aliquot removed when making up the next member in the series. Remember to discard any excess from the last in the series if volumes are critical.

Mixing solutions and suspensions

Various devices may be used, including:

- **Magnetic stirrers and fleas.** Magnetic fleas come in a range of shapes and sizes, and some stirrers have integral heaters. During use, stirrer speed may increase as the instrument warms up.
- **Orbital shakers and shaking water baths.** Use these to provide controlled mixing at a particular temperature, e.g. for long-term incubation and cell growth studies (p. 231).
- **Bottle rollers.** For cell culture work, ensuring gentle, continuous mixing.
- **Vortex mixers.** For vigorous mixing of small volumes of solution, e.g. when preparing a dilution series in test tubes. Take care when adjusting the mixing speed – if the setting is too low, the test tube will vibrate rather than creating a vortex, giving inadequate mixing. If the setting is too high, the test tube may slip from your hand.

SAFETY NOTE Using a vortex mixer with open or capped test tubes – do not vortex too vigorously or liquid will spill from the top of the tube, creating a contamination risk.

Storing chemicals and solutions

Labile chemicals may be stored in a fridge or freezer. Take special care when using chemicals that have been stored at low temperature: the container and its contents must be warmed up to room temperature before use, otherwise water vapour will condense on the chemical. This may render any weighing you do meaningless and it could ruin the chemical. Other chemicals may need to be kept in a desiccator, especially if they are deliquescent (water-absorbing).

SAFETY NOTE Cleaning up chemical spillages – you must always clean up any spillages of chemicals, as you are the only person who knows the risks from the spilled material.

KEY POINT Label all stored chemicals clearly with the following information: the chemical name (if a solution, state solute(s), concentration(s) and pH if measured), plus any relevant hazard warning information, the date made up, and your name.

Separating components of mixtures and solutions

Particulate solids (e.g. soils) can be separated on the basis of size using sieves. These are available in stacking forms which fit on automatic shakers. Sieves with the largest pores are placed at the top and the assembly is shaken for a fixed time until the sample separates. Suspensions of solids in liquids may be separated out by centrifugation (see p. 326) or filtration. Various forms of filter paper are available having different porosities and purities. Vacuum-assisted filtration speeds up the process and is best carried out with a filter funnel attached to a filter flask. Filtration through pre-sterilised membranes with very small pores (e.g. the Millipore type) is an excellent method of sterilising small volumes of solution. Solvents can be removed from solutes by heating, using rotary film evaporation under low pressure and, for water, by freeze drying. The last two are especially useful for heat-labile solutes – refer to the manufacturers' specific instructions for use.

Using balances

Electronic balances with digital readouts are now favoured over mechanical types: they are easy to read and their self-taring feature means the mass of the weighing boat or container can be subtracted automatically before weighing an object. The most common type offers accuracy down to 1 mg over the range 1 mg to 160 g, which is suitable for most biological applications.

To operate a standard self-taring balance:

1. **Check that the balance is level,** using the adjustable feet to centre the bubble in the spirit level (usually at the back of the machine). For accurate work, make sure a draught shield is on the balance.

2. **Place an empty vessel in the middle of the balance pan** and allow the reading to stabilise. *If the object is larger than the pan, take care that no part rests on the body of the balance or the draught shield as this will invalidate the reading.* Press the tare bar to bring the reading to zero.

3. **Place the chemical or object carefully in the vessel** (powdered chemicals should be dispensed with a suitably sized clean spatula). Take care to avoid spillages.

4. **Allow the reading to stabilise and make a note of the value.**

5. **If you add excess chemical, take great care when removing it.** Switch off if you need to clean any deposit accidentally left on or around the balance.

Larger masses should be weighed on a top-loading balance to an appropriate degree of accuracy. Take care to note the limits for the balance: while most have devices to protect against overloading, you may damage the mechanism. In the field, spring or battery operated balances may be preferred. Try to find a place out of the wind to use them.

For extremely small masses, there are electrical balances that can weigh down to 1 μg, but these are very delicate and must be used under supervision.

Measuring and controlling temperature

Heating samples

Care is required when heating samples. Safety glasses should always be worn. Use a thermostatically controlled electric stirrer-heater if possible. If using a Bunsen burner, keep the flame well away from yourself and your clothing (tie back long hair). Use a nonflammable mat beneath a Bunsen to protect the bench. Switch off when no longer required. To light a Bunsen, close the air hole first, then apply a lit match or lighter. Open the air hole if you need a hotter, more concentrated flame: the hottest part of the flame is just above the apex of the blue cone in its centre.

Ovens and drying cabinets may be used to dry specimens or glassware. They are normally thermostatically controlled. If drying organic material for dry weight measurement, do so at about 80 °C to avoid caramelising the sample. Always state the actual temperature used as this affects results. Check that all water has been driven off by weighing until a constant mass is reached.

Weighing – never weigh anything directly onto a balance's pan: you may contaminate it for other users. Use a weighing boat or a small piece of aluminium foil. Otherwise, choose a suitable vessel like a beaker, conical flask or aluminium tray.

Deciding on which balance to use – select a balance that weighs to an appropriate number of decimal places. For example, you should use a top-loading balance weighing to one decimal place for less accurate work. Note that a weight of 6.4 g on such a balance may represent a true value of between 6.350 g and 6.449 g (to three decimal places).

Using a balance – it is poor technique to use a large container to weigh out a small amount of a chemical: you are attempting to make accurate measurements of a small difference between two large numbers. Instead, use a small weighing container.

SAFETY NOTE There is a danger of fire whenever organic material is heated and a danger of scalding from heated liquids.

SAFETY NOTE Heating/cooling glass vessels – take care if heating or cooling glass vessels rapidly as they may break when heat-stressed. Freezing aqueous solutions in thin-walled glass vessels is risky because ice expansion may break the glass.

Understanding thermometer calibration – some are calibrated for use in air, others require partial immersion in liquid and others total immersion – check before use.

SAFETY NOTE If a mercury thermometer is broken, report the spillage, as mercury is a poison.

Example Water vapour can be removed by passing gas over dehydrated $CaCO_3$, and CO_2 may be removed by bubbling through KOH solution.

Fig. 25.3 Parts of a cylinder head regulator. The regulator is normally attached by tightening the attachment nut clockwise; the exception is with cylinders of hydrogen, where the special regulator is tightened *anticlockwise* to avoid the chance of this potentially explosive gas being incorrectly used.

Cooling samples and specimens

Fridges and freezers are used for storing stock solutions and chemicals that would either break down or become contaminated at room temperature. Normal fridge and freezer temperatures are about 4°C and –15°C respectively. Ice baths can be used when reactants must be kept close to 0°C. Most bioscience departments will have a machine which provides flaked ice for use in these baths. If common salt is mixed with ice, temperatures below 0°C can be achieved. A mixture of ethanol and solid CO_2 will provide a temperature of –72°C if required. To freeze a specimen quickly, immerse in liquid nitrogen (–196°C) using tongs and wearing an apron and thick gloves, as splashes will damage your skin. Always work in a well-ventilated room.

Maintaining constant temperature

Thermostatically controlled temperature rooms and incubators can be used to maintain temperature at a desired level. Always check with a thermometer or thermograph that the thermostat is accurate enough for your study. To achieve a controlled temperature on a smaller scale, e.g. for an oxygen electrode (pp. 369–371), use a water bath. These usually incorporate heating elements, a circulating mechanism and a thermostat. Baths for subambient temperatures have a cooling element.

Controlling atmospheric conditions

Gas composition

The atmosphere may be 'scrubbed' of certain gases by passing through a U-tube or Dreschel bottle containing an appropriate chemical or solution.

For accurate control of gas concentrations, use cylinders of pure gas; the contents can be mixed to give specified concentrations by controlling individual flow rates. The cylinder head regulator (Fig. 25.3) allows you to control the pressure (and hence flow rate) of gas; adjust using the controls on the regulator or with spanners of appropriate size. Before use, ensure the regulator outlet tap is off (turn anticlockwise), then switch on at the cylinder (turn clockwise) – the cylinder dial will give you the pressure reading for the cylinder contents. Now switch on at the regulator outlet (turn clockwise) and adjust to desired pressure/ flow setting. To switch off, carry out the above directions in reverse order.

To control dissolved gas composition in liquids, either 'de-gas' under vacuum or bubble with another gas, e.g. when preparing oxygen-free liquids for HPLC (p. 334) by bubbling with nitrogen.

Pressure

Many forms of pump are used to pressurise or provide partial vacuum, usually to force gas or liquid movement. Each has specific instructions for use. Many laboratories are supplied with 'vacuum' (suction) and pressurised air lines that are useful for procedures such as vacuum-assisted filtration. Make sure you switch off the taps after use. Take special care with glass items kept at very low or high pressures. These should be contained within a metal cage to minimise the risk of injury (e.g. a vacuum desiccator, as shown in Fig. 25.4).

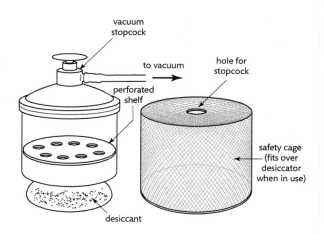

Fig. 25.4 Vacuum desiccator (with mesh safety cage).

Measuring time

Many experiments and observations need to be carefully timed. Large-faced stopclocks allow you to set and follow 'experimental time' and remove the potential difficulties in calculating this from 'real time' on a watch or clock. Some timers incorporate an alarm which you can set to warn when readings or operations must be carried out; 24-h timers are available for controlling light and temperature regimes.

Using a timer – always set the alarm *before* the critical time, so that you have adequate time to react.

Text references

Haynes, W.M. (ed.) (2014) *CRC Handbook of Chemistry and Physics,* 95th edn. CRC Press, Boca Raton.

O'Neil, M.J., Smith, A. and Heckelman, P.E. (2013) *The Merck Index: An Encyclopedia of Chemicals,* *Drugs and Biologicals,* 15th edn. Merck & Co., Inc., New Jersey.

Sources for further study

Boyer, R.F. (2011) *Biochemistry Laboratory: Modern Theory and Techniques,* 2nd edn. Prentice Hall, New Jersey.

Ninfa, A.J., Ballou, D.P and Benore, M. (2009) *Fundamental Laboratory Approaches for Biochemistry and Biotechnology.* Wiley, New York.

Seidman, L.A. and Moore, C.J. (2008) *Basic Laboratory Methods for Biotechnology: Textbook and Laboratory Reference.* 2nd edn. Prentice Hall, New Jersey.

Turgeon, M.L. (2010) *Clinical Laboratory Science: the Basics and Routine Techniques,* 6th edn. Mosby, St Louis.

STUDY EXERCISES

25.1 Practise the calculations involved in preparing specific volumes of aqueous solutions (see also study exercise 26.1). What mass of substance would be required to prepare each of the following (answer in each case to three significant figures):

(a) 100 mL of NaCl at 50 mmol L^{-1} (M_r of NaCl = 58.44)?

(b) 250 mL of mannitol at 0.10 mol L^{-1} (M_r of mannitol = 182.17)?

(c) 200 mL of a bovine serum albumin solution at 800 µg mL^{-1}?

(d) 0.5 litres of $MgCl_2$ prepared from the hexahydrate salt at 22.5 mmol L^{-1} (M_r of $MgCl_2 \cdot 6H_2O$ = 203.30)?

(e) 400 mL of DNA at 20 ng μL^{-1}?

25.2 Practise the calculations involved in preparing dilutions (answer in each case to three significant figures).

(a) If you added 1.0 mL of an aqueous solution of NaCl at 0.4 mol L^{-1} to 9.0 mL of water, what would be the final concentration of NaCl in mmol L^{-1}?

(b) If you added 25 mL of an aqueous solution of DNA at 10 µg mL^{-1} to a 500-mL volumetric flask and made it up to the specified volume with water, what would be the final concentration of DNA, in ng mL^{-1}?

(c) If you added 10 µL of an aqueous solution of sucrose at 200 mmol L^{-1} to a 250-mL volumetric flask and made it up to the specified volume with water, what would be the final concentration of sucrose, in nmol mL^{-1}?

(d) How would you prepare 250 mL of KCl at a final concentration of 20.0 mmol L^{-1} from a solution containing KCl at 0.2 mol L^{-1}?

(e) How would you prepare 1×10^{-3} m^3 of glucose at a final concentration of 50 µmol m^{-3} from a stock solution containing glucose at 20.0 gm^{-3} (M_r of glucose = 180.16)?

25.3 Practise the calculations involved in using stock solutions. Suppose you had the following stock solutions: NaCl 100.0 mmol L^{-1}; KCl 200.0 mmol L^{-1}; $CaCl_2$ 160.0 mmol L^{-1}; glucose 5.0 mmol L^{-1}. Calculate the volumes of each stock solution and the volume of water required to prepare each of the following (answer in each case to three significant figures):

(a) 1.0 mL of a solution containing only KCl at 10.0 mmol L^{-1}.

(b) 50 mL of a solution containing NaCl at 2.5 mmol L^{-1} and glucose at 0.5 mmol L^{-1}.

(c) 100 mL of a solution containing NaCl at 5.0 mmol L^{-1}, KCl at 2.5 mmol L^{-1}, $CaCl_2$ at 40.0 mmol L^{-1} and glucose at 0.25 mmol L^{-1}.

(d) 10 mL of a solution containing NaCl, $CaCl_2$ and KCl, all at 20.0 nmol mL^{-1}.

(e) 25 mL of a solution containing $CaCl_2$ at 8.0 mmol L^{-1}, KCl at 50.0 mmol L^{-1} and glucose at 20.0 nmol mL^{-1}.

Answers to these study exercises are available at www.pearsoned.co.uk/practicalskills.

Definitions

Electrolyte – a substance that dissociates, either fully or partially, in water to give two or more ions.

Mole (of a substance) – the equivalent in mass to relative molecular mass, in grams.

Relative atomic mass (A_r) – the mass of an atom relative to $^{12}C = 12$.

Relative molecular mass (M_r) – the mass of a compound's formula unit relative to $^{12}C = 12$.

Expressing solute concentrations – you should use SI units wherever possible. However, you are likely to meet non-SI concentrations and you must be able to deal with these units too.

Example A 1.0 molar solution of NaCl would contain 58.44 g NaCl (the relative molecular mass) per litre of solution.

A solution is a homogeneous liquid, formed by the addition of solutes to a solvent (usually water in biological systems). The behaviour of solutions is determined by the type of solutes involved and by their proportions, relative to the solvent. Many laboratory exercises involve calculation of concentrations, e.g. when preparing an experimental solution at a particular concentration (p. 153), or when expressing data in terms of solute concentration (p. 152). Make sure that you understand the basic principles set out in this chapter before you tackle such exercises.

Solutes can affect the properties of solutions in several ways, including the following.

Electrolytic dissociation

This occurs where individual molecules of an electrolyte dissociate to give charged particles (ions). For a strong electrolyte, e.g. NaCl, dissociation is essentially complete. In contrast, a weak electrolyte, e.g. acetic acid, will be only partly dissociated, depending upon the pH and temperature of the solution (pp. 169–70).

Osmotic effects

These are the result of solute particles lowering the effective concentration of the solvent (water). These effects are particularly relevant to biological systems since membranes are far more permeable to water than to most solutes. Water moves across biological membranes from the solution with the higher effective water concentration to that with the lower effective water concentration (osmosis).

Ideal/non-ideal behaviour

This occurs because solutions of real substances do not necessarily conform to the theoretical relationships predicted for dilute solutions of so-called ideal solutes. It is often necessary to take account of the non-ideal behaviour of real solutions, especially at high solute concentrations (see Haynes, 2014, and Robinson and Stokes, 2002 for appropriate data).

Concentration

In SI units (p. 198), the concentration of a solute in a solution is expressed in mol m^{-3}, which is convenient for most biological purposes. The concentration of a solute is usually symbolised by square brackets, e.g. [NaCl]. Details of how to prepare a solution using SI and non-SI units are given in Box 25.2.

A number of alternative ways of expressing the relative amounts of solute and solvent are in general use, and you may come across these terms in your practical work or in the literature. These include the following.

Molarity

This is the term used to denote molar concentration, [C], expressed as moles of solute per litre volume of solution (mol L^{-1}). This non-SI term continues to find widespread usage, in part because of the familiarity of working scientists with the term, but also because laboratory glassware is calibrated in millilitres and litres, making the preparation of molar and millimolar solutions relatively straightforward. However, the symbols

> **Box 26.1** Useful procedures for calculations involving molar concentrations
>
> 1. **Preparing a solution of defined molarity.** For a solute of known relative molecular mass, M_r, the following relationship can be applied:
>
> $$[C] = \frac{\text{mass of solute}/\text{relative molecular mass}}{\text{volume of solution}} \quad [26.1]$$
>
> So, if you wanted to make up 200 mL (0.2L) of an aqueous solution of NaCl (M_r 58.44) at a concentration of 500 mmol L^{-1} (0.5 mol L^{-1}), you could calculate the amount of NaCl required by inserting these values into Eqn [26.1]:
>
> $$0.5 = \frac{\text{mass of solute}/58.44}{0.2}$$
>
> which can be rearranged to
>
> $$\text{mass of solute} = 0.5 \times 0.2 \times 58.44 = 5.844\,g$$
>
> The same relationship can be used to calculate the concentration of a solution containing a known amount of a solute, e.g. if 21.1 g of NaCl were made up to a volume of 100 mL (0.1 L), this would give
>
> $$[\text{NaCl}] = \frac{21.1/58.44}{0.1} = 3.61 \text{ mol } L^{-1}$$
>
> 2. **Dilutions and concentrations.** The following relationship is very useful if you are diluting (or concentrating) a solution:
>
> $$[C_1]V_1 = [C_2]V_2 \quad [26.2]$$
>
> where $[C_1]$ and $[C_2]$ are the initial and final concentrations, while V_1 and V_2 are their respective volumes: each pair must be expressed in the same units. Thus, if you wanted to dilute 200 mL of 0.5 mol L^{-1} NaCl to give a final molarity of 0.1 mol L^{-1}, then, by substitution into Eqn [26.2]:
>
> $$0.5 \times 200 = 0.1 \times V_2$$
>
> Thus $V_2 = 1000$ mL (in other words, you would have to add water to 200 mL of 0.5 mol L^{-1} NaCl to give a final volume of 1000 mL to obtain a 0.1 mol L^{-1} solution).
>
> 3. **Interconversion.** A simple way of interconverting amounts and volumes of any particular solution is to divide the amount and volume by a factor of 10^3: thus a molar solution of a substance contains 1 mol L^{-1}, which is equivalent to 1 mmol mL^{-1}, or 1 μmol μL^{-1}, or 1 nmol nL^{-1}, etc. You may find this technique useful when calculating the amount of substance present in a small volume of solution of known concentration, e.g. to calculate the amount of NaCl present in 50 μL of a solution with a concentration (molarity) of 0.5 mol L^{-1} NaCl:
>
> (a) this is equivalent to 0.5 μmol μL^{-1};
>
> (b) therefore 50 μL will contain 50×0.5 μmol = 25 μmol.
>
> Alternatively, you may prefer to convert to primary SI units, for ease of calculation (see Box 25.2).
> The 'unitary method' (p. 511) is an alternative approach to these calculations.

> **Example** A 1.0 molar solution can be expressed as a 1.0 mol L^{-1} solution.

in common use for molar (M) and millimolar (mM) solutions are at odds with the SI system and many people now prefer to use mol L^{-1} and mmol L^{-1} respectively, to avoid confusion. Box 26.1 gives details of some useful approaches to calculations involving molarities.

Molality

This is used to express the concentration of solute relative to the *mass* of solvent, i.e. mol kg^{-1}. Molality is a temperature-independent means of expressing solute concentration, rarely used except when the osmotic properties of a solution are of interest (p. 165).

> **Example** A 0.5 molal solution of NaCl would contain $58.44 \times 0.5 = 29.22$ g NaCl per kg of water.

Per cent composition (% w/w)

This is the solute mass (in g) per 100 g solution. The advantage of this expression is the ease with which a solution can be prepared, since it simply requires each component to be pre-weighed (for water, a volumetric measurement may be used, e.g. using a measuring cylinder) and then mixed together. Similar terms are parts per thousand (‰), e.g. mg g^{-1}, and parts per million (ppm), e.g. μg g^{-1} (see below).

> **Example** A 5% w/w sucrose solution contains 5 g sucrose and 95 g water (= 95 mL water, assuming a density of 1 g mL^{-1}) to give 100 g of solution.

Per cent concentration (% w/v and % v/v)

For solutes added in solid form, this is the number of grams of solute per 100 mL solution. This is more commonly used than per cent composition, since solutions can be accurately prepared by weighing out the required amount of solute and then making this up to a known volume using a volumetric flask. The equivalent expression for liquid solutes is % v/v.

The principal use of mass/mass or mass/volume terms (including $g\ L^{-1}$) is for solutes whose relative molecular mass is unknown (e.g. cellular proteins), or for mixtures of certain classes of substance (e.g. total salt in sea water). You should *never* use the per cent term without specifying how the solution was prepared, i.e. by using the qualifier w/w, w/v or v/v. For mass concentrations, it is simpler to use mass per unit volume, e.g. $mg\ L^{-1}$, $\mu g\ \mu L^{-1}$, etc.

Parts per million (ppm) and parts per billion (ppb) concentration

Ppm is a non-SI weight per volume (w/v) concentration term commonly used in quantitative analysis such as flame photometry, atomic absorption spectroscopy and gas chromatography, where low concentrations of solutes are analysed. The term ppm is equivalent to the expression of concentration as $\mu g\ mL^{-1}$ ($10^{-6}\ g\ mL^{-1}$) and a 1.0 ppm solution of a substance will have a concentration of $1.0\ \mu g\ mL^{-1}$ ($1.0 \times 10^{-6}\ g\ mL^{-1}$).

Parts per billion (ppb) is an extension of this concentration term as $ng\ mL^{-1}$ ($10^{-9}\ g\ mL^{-1}$) and is commonly used to express concentrations of very dilute solutions. For example, the allowable concentration of arsenic in water is 0.05 ppm, more conveniently expressed as 50 ppb. Also note that ppm and ppb are sometimes used as a weight per weight (w/w) term, e.g. $\mu g\ g^{-1}$ and $ng\ g^{-1}$ respectively.

Activity (a)

This is a term used to describe the *effective* concentration of a solute. In dilute solutions, solutes can be considered to behave according to ideal (thermodynamic) principles, i.e. they will have an effective concentration equivalent to the actual concentration. However, in concentrated solutions ($\geq 500\ mol\ m^{-3}$), the behaviour of solutes is often non-ideal, and their effective concentration (activity) will be less than the actual concentration [C]. The ratio between the effective concentration and the actual concentration is called the activity coefficient (γ) where

$$\gamma = \frac{a}{[C]} \qquad\qquad [26.3]$$

Equation [26.3] can be used for SI units ($mol\ m^{-3}$), molarity ($mol\ L^{-1}$) or molality ($mol\ kg^{-1}$). In all cases, γ is a dimensionless term, since a and [C] are expressed in the same units. The activity coefficient of a solute is effectively unity in dilute solution, decreasing as the solute concentration increases (Table 26.1). At high concentrations of certain ionic solutes, γ may increase to become greater than unity.

Example A 5% w/v sucrose solution contains 5 g sucrose in 100 mL of solution. A 5% v/v glycerol solution would contain 5 mL glycerol in 100 mL of solution.

Note that when water is the solvent this is often not specified in the expression, e.g. a 20% v/v ethanol solution contains 20% ethanol made up to 100 mL of solution using water.

Alternative expressions – parts per thousand (%) w/v can be used as an alternative to $g\ L^{-1}$.

Example The concentration of an NaCl solution is stated as 3 ppm. This is equivalent to 3 $\mu g\ mL^{-1}$ (3 $mg\ L^{-1}$). The relative molecular mass of NaCl is 58.44 $g\ mol^{-1}$, so the solution has a concentration of $3 \times 10^{-6} \div 58.44\ mol\ mL^{-1} = 5.13 \times 10^{-8}\ mol\ mL^{-1} = 0.0513\ \mu mol\ mL^{-1} = 51.3\ \mu mol\ L^{-1}$.

Table 26.1 Activity coefficient of NaCl solutions as a function of molality. Data from Robinson and Stokes (2002).

Molality	Activity coefficient at 25 °C
0.1	0.778
0.5	0.681
1.0	0.657
2.0	0.668
4.0	0.783
6.0	0.986

KEY POINT Activity is often the correct expression for theoretical relationships involving solute concentration (e.g. where a property of the solution is dependent on concentration). However, for most practical purposes, it is possible to use the *actual* concentration of a solute rather than the activity, since the difference between the two terms can be ignored for dilute solutions.

> **Example**   A solution of NaCl with a molality of 0.5 mol kg^{-1} has an activity coefficient of 0.681 at 25 °C and a molal activity of $0.5 \times 0.681 = 0.340$ mol kg^{-1}.

> **Examples**   For carbonate ions (CO_3^{2-}), with a relative molecular mass of 60.00 and a valency of 2, the equivalent mass is $60.00/2 = 30.00$ g eq^{-1}.
>
>   For sulfuric acid (H_2SO_4, relative molecular mass 98.08), where 2 hydrogen ions are available, the equivalent mass is $98.08/2 = 49.04$ g eq^{-1}.

> **Example**   A 0.5 N solution of sulfuric acid would contain $0.5 \times 49.04 = 24.52$ g L^{-1}.

> **Example**   Under ideal conditions, 1 mol of NaCl dissolved in water would give 1 mol of Na^+ ions and 1 mol of Cl^- ions, equivalent to a theoretical osmolarity of 2 osmol L^{-1}.

> **Example**   A 1.0 mol kg^{-1} solution of NaCl has an osmotic coefficient of 0.936 at 25 °C and an osmolality of $1.0 \times 2 \times 0.936 = 1.872$ osmol kg^{-1}.

Table 26.2 Osmotic coefficients of NaCl solutions as a function of molality. Data from Robinson and Stokes (2002).

Molality	Osmotic coefficient at 25 °C
0.1	0.932
0.5	0.921
1.0	0.936
2.0	0.983
4.0	1.116
6.0	1.271

The particular use of the term 'water activity' is considered below (p. 165), since it is based on the mole fraction of solvent, rather than the effective concentration of solute.

Equivalent mass (equivalent weight)

Equivalence and normality are outdated terms, although you may come across them in older texts. They apply to certain solutes whose reactions involve the transfer of charged ions, e.g. acids and alkalis (which may be involved in H^+ or OH^- transfer), and electrolytes (which form cations and anions that may take part in further reactions). These two terms take into account the valency of the charged solutes. Thus the equivalent mass of an ion is its relative molecular mass divided by its valency (ignoring the sign), expressed in grams per equivalent (eq) according to the relationship:

$$\text{equivalent mass} = \frac{\text{relative molecular mass}}{\text{valency}} \quad [26.4]$$

For acids and alkalis, the equivalent mass is the mass of substance that will provide 1 mol of either H^+ or OH^- ions in a reaction, obtained by dividing the molecular mass by the number of available ions (n), using n instead of valency as the denominator in Eqn [26.4].

Normality

A 1 normal solution (1 N) is one that contains one equivalent mass of a substance per litre of solution. The general formula is:

$$\text{normality} = \frac{\text{mass of substance per litre}}{\text{equivalent mass}} \quad [26.5]$$

Osmolarity

This non-SI expression is used to describe the number of moles of osmotically active solute particles per litre of solution (osmol L^{-1}). The need for such a term arises because some molecules dissociate to give more than one osmotically active particle in aqueous solution.

Osmolality

This term describes the number of moles of osmotically active solute particles per unit mass of solvent (osmol kg^{-1}). For an ideal solute, the osmolality can be determined by multiplying the molality by n, the number of solute particles produced in solution (e.g. for NaCl, $n = 2$). However, for real (i.e. non-ideal) solutes, a correction factor (the osmotic coefficient, $+\phi$) is used:

$$\text{osmolality} = \text{molality} \times n \times \phi \quad [26.6]$$

If necessary, the osmotic coefficients of a particular solute can be obtained from tables (e.g. Table 26.2): non-ideal behaviour means that φ may have values >1 at high concentrations. Alternatively, the osmolality of a solution can be measured using an osmometer.

Colligative properties and their use in osmometry

Several properties vary in direct proportion to the effective number of osmotically active solute particles per unit mass of solvent and can be used to determine the osmolality of a solution. These colligative properties include freezing point, boiling point and vapour pressure.

An osmometer is an instrument which measures the osmolality of a solution, usually by determining the freezing point depression of the solution in relation to pure water, a technique known as cryoscopic osmometry. A small amount of sample is cooled rapidly and then brought to the freezing point (Fig. 26.1), which is measured by a temperature-sensitive thermistor probe calibrated in mosmol kg^{-1}. An alternative method is used in vapour pressure osmometry, which measures the relative decrease in the vapour pressure produced in the gas phase when a small sample of the solution is equilibrated within a chamber.

Osmotic properties of solutions

Several interrelated terms can be used to describe the osmotic status of a solution. In addition to osmolality, you may come across the following.

Osmotic pressure

This is based on the concept of a membrane permeable to water, but not to solute molecules. For example, if a sucrose solution is placed on one side and pure water on the other, then a passive driving force will be created and water will diffuse across the membrane into the sucrose solution, since the effective water concentration in the sucrose solution will be lower (see Fig. 26.2). The tendency for water to diffuse into the sucrose solution could be counteracted by applying a hydrostatic pressure equivalent to the passive driving force. Thus, the osmotic pressure of a solution is the excess hydrostatic pressure required to prevent the net flow of water into a vessel containing the solution. The SI unit of osmotic pressure is the pascal, Pa ($=$ kg m^{-1} s^{-2}). Older sources may use atmospheres, or bars, and conversion factors are given in Box 31.1. Osmotic pressure and osmolality can be interconverted using the expression 1 osmol kg^{-1} = 2.479 MPa at 25 °C.

The use of osmotic pressure has been criticised as misleading, since a solution does not exhibit an 'osmotic pressure' unless it is placed on the other side of a selectively permeable membrane from pure water.

Water activity (a_w)

This is a term often used to describe the osmotic behaviour of microbial cells. It is a measure of the relative proportion of water in a solution, expressed in terms of its mole fraction, i.e. the ratio of the number of moles of water (n_w) to the total number of moles of all substances (i.e. water and solutes) in solution (n_t), taking into account the molal activity coefficient of the solvent, water (i.e. γ_w):

$$a_w = \gamma_w \frac{n_w}{n_t} \qquad [26.7]$$

The water activity of pure water is unity, decreasing as solutes are added. One disadvantage of a_w is the limited change which occurs in response to a change in solute concentration: a 1.0 mol kg^{-1} solution of NaCl has a water activity of 0.967 (Table 26.3).

Osmolality, osmotic pressure and water activity are measurements based solely on the osmotic properties of a solution, with no regard for any other driving forces, e.g. hydrostatic and gravitational forces. In circumstances where such other forces are important, you will need to measure a variable that takes into account these aspects of water status, namely water potential.

Using an osmometer – it is vital that the sample holder and probe are clean, otherwise small amounts of the previous sample may be carried over, leading to inaccurate measurement.

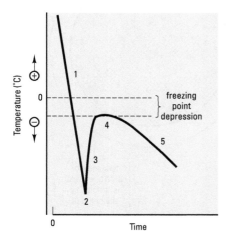

Fig. 26.1 Temperature responses of a cryoscopic osmometer. The response can be subdivided into:
1. initial supercooling
2. initiation of crystallisation
3. crystallisation/freezing
4. plateau, at the freezing point
5. slow temperature decrease.

Example A 1.0 mol kg^{-1} solution of NaCl at 25 °C has an osmolality of 1.872 osmol kg^{-1} and an osmotic pressure of 1.872 $\times$ 2.479 = 4.641 MPa.

Table 26.3 Water activity (a_w) of NaCl solutions as a function of molality. Data from Robinson and Stokes (2002).

Molality	a_w
0.1	0.997
0.5	0.984
1.0	0.967
2.0	0.932
4.0	0.852
6.0	0.760

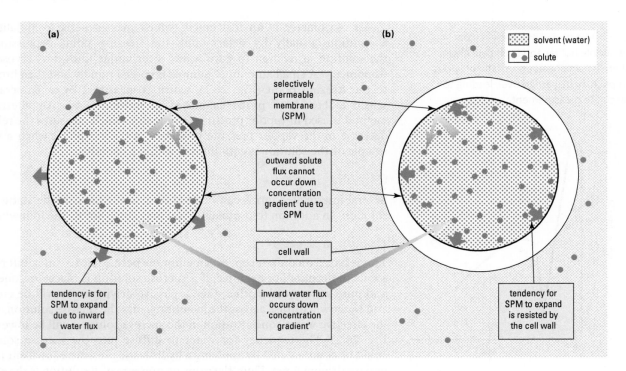

Fig. 26.2 Illustration of forces driving solvent (water) and solute movement across a selectively permeable membrane (SPM). Energetically, both solutes and solvents tend to move down their respective 'concentration gradient' (strictly, down their chemical potential gradient). However, solute molecules cannot leave the model cells illustrated because they cannot pass through the SPM. In the situation illustrated in (a), water will tend to move from outside the cell to within because the solute molecules have effectively 'diluted' the water within the cell (illustrated by the density of point shading), creating a gradient in 'concentration' and, because this molecule is able to pass through the SPM, the result will be an expansion of this model cell (short arrows). The osmotic pressure is the (theoretical) pressure that would need to be applied to prevent this. If the model cell were surrounded by a cell wall, as in (b), this would resist expansion, leading to internal pressurisation (turgor pressure).

Water potential (hydraulic potential) and its applications

Water potential, Ψ_w, is the most appropriate measure of osmotic status in many areas of the biosciences. It is a term derived from the chemical potential of water. It expresses the difference between the chemical potential of water in the test system and that of pure water under standard conditions and has units of pressure (i.e. Pa). It is a more appropriate term than osmotic pressure because it is based on sound theoretical principles and because it can be used to predict the direction of passive movement of water, since water will flow down a gradient of chemical potential (i.e. osmosis occurs from a solution with a higher water potential to one with a lower water potential). Pure water at 20 °C and at 0.1 MPa pressure (i.e. ≈ atmospheric) has a water potential of zero. The addition of solutes will lower the water potential (i.e. make it negative), while the application of pressure, e.g. from hydrostatic or gravitational forces, will raise it (i.e. make it positive).

Often, the two principal components of water potential are referred to as the solute potential, or osmotic potential (Ψ_s, sometimes symbolised as Ψ_π or π) and the hydrostatic pressure potential (Ψ_p) respectively. For a solution at atmospheric pressure, the water potential is due solely to the presence of osmotically active solute molecules (osmotic potential) and

Examples A 1.0 mol kg⁻¹ solution of NaCl has a (negative) water potential of −4.641 MPa.

Pure water at 0.2 MPa pressure (about 0.1 MPa above atmospheric pressure) has a (positive) water potential of 0.1 MPa.

Fig. 26.3 Vapour pressure osmometer.
Courtesy of KNAUER, Germany, www.knauer.net

Measuring water potential - Eqn [26.9] ignores the effects of gravitational forces – for systems where gravitational effects are important, an additional term is required (Nobel, 2005).

may be calculated from the measured osmolality (osmol kg^{-1}) at 25 °C, using the relationship:

$$\Psi_w(MPa) = \Psi_s(MPa) = -2.479 \times osmolality \qquad [26.8]$$

For aquatic microbial cells, e.g. algae, fungi and bacteria, equilibrated in their growth medium at atmospheric pressure, the water potential of the external medium will be equal to the cellular water potential ('isotonic') and the latter can be derived from the measured osmolality of the medium (Eqn [26.8]) by using on osmometer (p. 165). The water potential of such cells can be subdivided into two major parts, the cell solute potential (Ψ_s) and the cell turgor pressure (Ψ_p) as follows:

$$\Psi_w = \Psi_s + \Psi_p \qquad [26.9]$$

To calculate the relative contribution of the osmotic and pressure terms in Eqn [26.9], an estimate of the internal osmolality is required, e.g. by measuring the freezing point depression of expressed intracellular fluid. Once you have values for Ψ_w and Ψ_s, the turgor pressure can be calculated by substitution into Eqn [26.9].

For terrestrial plant cells, the water potential may be determined directly using a vapour pressure osmometer (Fig. 26.3), by placing a sample of the material within the osmometer chamber and allowing it to equilibrate. If Ψ_s of expressed sap is then measured, Ψ_p can be determined from Eqn [26.9].

The van't Hoff relationship can be used to estimate Ψ_s, by summation of the osmotic potentials owing to the major solutes, determined from their concentrations, as:

$$\Psi_s = -RTn\phi[C] \qquad [26.10]$$

where RT is the product of the universal gas constant and absolute temperature (2479 J mol^{-1} at 25 °C), n and ϕ are as previously defined and $[C]$ is expressed in SI terms as mol m^{-3}.

Text references

Haynes, W.M. (ed.) (2014) *CRC Handbook of Chemistry and Physics,* 95th edn. CRC Press, Boca Raton.

Nobel, PS. (2005) *Physicochemical and Environmental Plant Physiology,* 2nd edn. Academic Press, New York.

Robinson, R.A. and Stokes, R.H. (2002) *Electrolyte Solutions.* Dover Publications, New York.

Sources for further study

Burtis, C.A. and Ashwood, E.R. (2014) *Fundamentals of Clinical Chemistry,* 7th edn. Saunders, Philadelphia.

Chapman, C. (1998) *Basic Chemistry for Biology.* McGraw-Hill, New York.

O'Neil, M.J., Smith, A. and Heckelman, P.E. (2006) *The Merck Index: An Encyclopedia of Chemicals, Drugs and Biologicals,* 14th edn. Merck & Co., Inc., Whitehouse Station.

Postma, J.M., Roberts, J. and Hollenberg, J.L. (2009) *Chemistry in the Laboratory,* 7th edn. Freeman, New York.

Seidman, L.A. and Moore, C.J. (2008) *Basic Laboratory Methods for Biotechnology: Textbook and Laboratory Reference,* 2nd edn. Prentice Hall, New Jersey.

Timberlake, K.C. (2013) *Laboratory Manual for General, Organic and Biological Chemistry,* 3rd edn. Prentice Hall, New Jersey.

STUDY EXERCISES

26.1 Practise calculations involving molar concentrations (see also study exercises 25.1 and 25.2). What mass of substance would be required to prepare each of the following aqueous solutions (answer in grams, to three decimal places in each case):

(a) 1 litre of NaCl at a concentration of 1 molar? (M_r of NaCl = 58.44.)

(b) 250 mL of $CaCl_2$ at 100 mmol L^{-1}? (M_r of $CaCl_2$ = 110.99.)

(c) 2.5 L of mannitol at 10 nmol μL^{-1}? (M_r of mannitol = 182.17.)

(d) 400 mL of KCl at 5% w/v?

(e) 250 mL of glucose at 2.50 mol m^{-3}? (M_r of glucose = 180.16.)

26.2 Practise expressing concentrations in different ways. Express all answers to three significant figures:

(a) What is 5 gL^{-1} sucrose, expressed in terms of molarity? (M_r of sucrose = 342.3.)

(b) What is 1.0 mol m^{-3} NaCl, expressed in gL^{-1}? (M_r of NaCl = 58.44.)

(c) What is 5% v/v ethanol, expressed in terms of molarity? (M_r of ethanol = 46.06 and density of ethanol at 25°C = 0.789 g mL^{-1}.)

(d) What is 150 mmol L^{-1} glucose, expressed in terms of per cent concentration (% w/v)? (M_r of glucose = 180.16.)

(e) What is a 1.0 molal solution of KCl, expressed as per cent composition (% w/w)? (M_r of KCl = 74.55.)

26.3 Calculate osmolality and osmotic potentials.

Answer to three significant figures in all cases.

(a) Assuming NaCl, KCl and $CaCl_2$ behave according to ideal thermodynamic principles, what would be the predicted osmolality of a solution containing:

   (i)   NaCl alone, at 50 mmol kg^{-1}?

   (ii) KCl at 200 mmol kg^{-1} and $CaCl_2$ at 40 mmol kg^{-1}?

   (iii) NaCl at 100 mmol kg^{-1}, KCl at 60 mmol kg^{-1} and $CaCl_2$ at 75 mmol kg^{-1} ?

(b) What is the predicted osmotic pressure and osmotic potential of each of the solutions in (a) at 25°C?

Answers to these study exercises are available at www.pearsoned.co.uk/practicalskills.

Definitions

Acid – a compound that acts as a proton donor in aqueous solution.

Alkali – a compound that liberates hydroxyl ions when it dissociates. Since hydroxyl ions are strongly basic, this will reduce the proton concentration.

Ampholyte – a compound that can act as both an acid and a base. Water is an ampholyte since it may dissociate to give a proton and a hydroxyl ion (amphoteric behaviour).

Base – a compound that acts as a proton acceptor in aqueous solution.

Conjugate pair – an acid together with its corresponding base.

SAFETY NOTE Safe working with strong acids or alkalis – these can be highly corrosive; rinse with plenty of water, if spilled.

Table 27.1 Effects of temperature on the ion product of water (K_w), H$^+$ ion concentration and pH at neutrality. Values calculated from Haynes (2014).

Temp. (°C)	K_w (mol^2 L^{-2})	[H$^+$] at neutrality (nmol L^{-1})	pH at neutrality
0	0.11×10^{-14}	33.9	7.47
4	0.17×10^{-14}	40.7	7.39
10	0.29×10^{-14}	53.7	7.27
20	0.68×10^{-14}	83.2	7.08
25	1.01×10^{-14}	100.4	7.00
30	1.47×10^{-14}	120.2	6.92
37	2.39×10^{-14}	154.9	6.81
45	4.02×10^{-14}	199.5	6.70

Example Human plasma has a typical H$^+$ concentration of approximately 0.4×10^{-7} mol L^{-1} (=$10^{-7.4}$ mol L^{-1}), giving a pH of 7.4.

pH is a measure of the amount of hydrogen ions (H$^+$) in a solution: this affects the solubility of many substances and the activity of most biological systems, from individual molecules to whole organisms. It is usual to think of aqueous solutions as containing H$^+$ ions (protons), though protons actually exist in their hydrated form, as hydronium ions (H$_3$O$^+$). The proton concentration of an aqueous solution [H$^+$] is affected by several factors:

- **Ionisation (dissociation) of water,** which liberates protons and hydroxyl ions in equal quantities, according to the reversible relationship:

$$H_2O \rightleftharpoons H^+ + OH^- \qquad [27.1]$$

- **Dissociation of acids,** according to the equation:

$$H-A \rightleftharpoons H^+ + A^- \qquad [27.2]$$

where H–A represents the acid and A$^-$ is the corresponding conjugate base. The dissociation of an acid in water will increase the amount of protons, reducing the amount of hydroxyl ions as water molecules are formed (Eqn [27.1]). The addition of a base (usually, as its salt) to water will decrease the amount of H$^+$, owing to the formation of the conjugate acid (Eqn [27.2]).

- **Dissociation of alkalis,** according to the relationship:

$$X-OH \rightleftharpoons X^+ + OH^- \qquad [27.3]$$

where X–OH represents the undissociated alkali. Since the dissociation of water is reversible (Eqn [27.1]), in an aqueous solution the production of hydroxyl ions will effectively act to 'mop up' protons, lowering the proton concentration.

Many compounds act as acids, bases or alkalis: those that are almost completely ionised in solution are usually called strong acids or bases, while weak acids or bases are only slightly ionised in solution.

In an aqueous solution, most of the water molecules are not ionised. In fact, the extent of ionisation of pure water is constant at any given temperature and is usually expressed in terms of the ion product (or ionisation constant) of water, K_w:

$$K_w = [H^+][OH^-] \qquad [27.4]$$

where [H$^+$] and [OH$^-$] represent the molar concentration (strictly, the activity) of protons and hydroxyl ions in solution, expressed as mol L^{-1}. At 25 °C, the ion product of pure water (Table 27.1) is 10^{-14} mol^2 L^{-2} (i.e. 10^{-8} mol^2 m^{-6}). This means that the concentration of protons in solution will be 10^{-7} mol L^{-1} (10^{-4} mol m^{-3}), with an equivalent concentration of hydroxyl ions (Eqn [27.1]). Since these values are very low and involve negative powers of 10, it is customary to use the pH scale, where:

$$pH = -\log_{10}[H^+] \qquad [27.5]$$

and [H$^+$] is the proton activity in mol L^{-1} (see p. 171).

Measuring pH – the pH of a neutral solution changes with temperature (Table 27.1), owing to the enhanced dissociation of water with increasing temperature. This must be taken into account when measuring the pH of any solution and when interpreting your results.

SAFETY NOTE Preparing a dilute acid solution using concentrated acid – always slowly add the concentrated acid to water, not the reverse, since the strongly exothermic process can trigger a violent reaction with water.

SAFETY NOTE Preparing an alkali solution – typically, the alkali will be in solid form (e.g. NaOH) and addition to water will rapidly raise the temperature of the solution: use only heat-resistant glassware, cooled with water if necessary.

Table 27.2 Properties of some pH indicator dyes.

Dye	Acid–base colour change	Useful pH range
Thymol blue (acid)	red–yellow	1.2–6.8
Bromophenol blue	yellow–blue	1.2–6.8
Congo red	blue–red	3.0–5.2
Bromocresol green	yellow–blue	3.8–5.4
Resazurin	orange–violet	3.8–6.5
Methyl red	red–yellow	4.3–6.1
Litmus	red–blue	4.5–8.3
Bromocresol purple	yellow–purple	5.8–6.8
Bromothymol blue	yellow–blue	6.0–7.6
Neutral red	red–yellow	6.8–8.0
Phenol red	yellow–red	6.8–8.2
Thymol blue (alkaline)	yellow–blue	8.0–9.6
Phenolphthalein	none–red	8.3–10.0

KEY POINT While pH is strictly the negative logarithm (to the base 10) of H^+ activity, in practice H^+ concentration in mol L^{-1} (equivalent to kmol m^{-3} in SI terminology) is most often used in place of activity, since the two are virtually the same, given the limited dissociation of H_2O. The pH scale is not SI: nevertheless, it continues to be used widely in biological science.

The value where an equal amount of H^+ and OH^- ions are present is termed neutrality: at 25°C the pH of pure water at neutrality is 7.0. At this temperature, pH values below 7.0 are acidic while values above 7.0 are alkaline.

Always remember that the pH scale is a *logarithmic* one, not a *linear* one: a solution with a pH of 3.0 is not twice as acidic as a solution of pH 6.0, but one thousand times as acidic (i.e. contains 1000 times the amount of H^+ ions). Therefore, you may need to convert pH values into proton concentrations before you carry out mathematical manipulations (see Box 73.2). For similar reasons, it is important that pH change is expressed in terms of the original and final pH values, rather than simply quoting the difference between the values: a pH change of 0.1 has little meaning unless the initial or final pH is known.

Measuring pH

pH electrodes

Accurate pH measurements can be made using a pH electrode, coupled to a pH meter. The pH electrode is usually a combination electrode, comprising two separate systems: an H^+-sensitive glass electrode and a reference electrode which is unaffected by H^+ ion concentration (Fig. 27.2). When this is immersed in a solution, a pH-dependent voltage between the two electrodes can be measured using a potentiometer. In most cases, the pH electrode assembly (containing the glass and reference electrodes) is connected to a separate pH meter by a cable, although some handheld instruments (pH probes) have the electrodes and meter within the same assembly, often using an H^+-sensitive field effect transistor in place of a glass electrode, to improve durability and portability.

Box 27.1 gives details of the steps involved in making a pH measurement with a glass pH electrode and meter.

pH indicator dyes

These compounds (usually weak acids) change colour in a pH-dependent manner. They may be added in small amounts to a solution, or they can be used in paper strip form. Each indicator dye usually changes colour over a restricted pH range, typically 1–2 pH units (Table 27.2): universal indicator dyes/papers make use of a combination of individual dyes to measure a wider pH range. Dyes are not suitable for accurate pH measurement as they are affected by other components of the solution, including oxidising and reducing agents and salts. However, they are useful for:

- **estimating the approximate pH of a solution;**
- **determining a change in pH,** for example at the end point of a titration or the production of acids during bacterial metabolism;
- **establishing the approximate pH of intracellular compartments,** for example the use of neutral red as a 'vital' stain (p. 184).

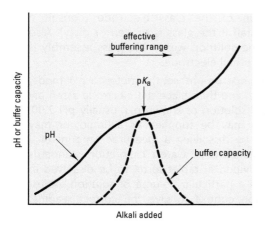

Fig. 27.1 Theoretical pH titration curve for a buffer solution. pH change is lowest and buffer capacity is greatest at the pK_a of the buffer solution.

Buffers

Rather than simply measuring the pH of a solution, you may wish to *control* the pH, e.g. in metabolic experiments, or in a growth medium for cell culture (p. 231). In fact, you should consider whether you need to control pH in any experiment involving a biological system, whether whole organisms, isolated cells, subcellular components or biomolecules. One of the most effective ways to control pH is to use a buffer solution.

A buffer solution is usually a mixture of a weak acid and its conjugate base. Added protons will be neutralised by the anionic base while a reduction in protons, e.g. owing to the addition of hydroxyl ions, will be counterbalanced by dissociation of the acid (Eqn [27.2]); thus the conjugate pair acts as a 'buffer' to pH change. The innate resistance of most biological fluids to pH change is caused by the presence of cellular constituents that act as buffers, e.g. proteins, which have a large number of weakly acidic and basic groups in their amino acid side chains.

Buffer capacity and the effects of pH

The extent of resistance to pH change is called the buffer capacity of a solution. The buffer capacity is measured experimentally at a particular pH by titration against a strong acid or alkali: the resultant curve will be strongly sigmoidal, with a plateau where the buffer capacity is greatest (Fig. 27.1). The mid-point of the plateau represents the pH at which equal quantities of acid and conjugate base are present, and is given the symbol pK_a, which refers to the negative logarithm (to the base 10) of the acid dissociation constant, K_a, where

$$K_a = \frac{[H^+][A^-]}{[HA]} \tag{27.6}$$

By rearranging Eqn [27.6] and taking negative logarithms, we obtain:

$$pH = pK_a + \log_{10} \frac{[A^-]}{[HA]} \tag{27.7}$$

This relationship is known as the Henderson–Hasselbalch equation and it shows that the pH will be equal to the pK_a when the ratio of conjugate base to acid is unity, since the final term in Eqn [27.7] will be zero. Consequently, pK_a is an important factor in determining buffer capacity at a particular pH. In practical terms, a buffer solution will work most effectively at pH values about one unit either side of the pK_a.

Selecting an appropriate buffer

When selecting a buffer, you should be aware of certain limitations to their use. Citric acid and phosphate buffers readily form insoluble complexes with divalent cations, while phosphate can also act as a substrate, activator or inhibitor of certain enzymes. Both of these buffers contain biologically significant quantities of cations, e.g. Na⁺ or K⁺. TRIS (Table 27.3) is often toxic to biological systems: because of its high lipid solubility it can penetrate membranes, uncoupling electron transport reactions in whole cells and isolated organelles. In addition, it is markedly affected by temperature, with a tenfold increase in H⁺ concentration from 4 °C to 37 °C. A number of zwitterionic molecules (having both positive and negative groups) have been introduced to overcome some of the disadvantages of

Box 27.1   Using a glass pH electrode and meter to measure the pH of a solution

The following procedure should be used whenever you make a pH measurement: consult the manufacturer's handbook for specific information, where necessary. Do not be tempted to miss out any of the steps detailed below, particularly those relating to the effects of temperature, or your measurements are likely to be inaccurate.

1. **Stir the test solution thoroughly before you make any measurement:** it is often best to use a magnetic stirrer. Leave the solution for sufficient time to allow equilibration at lab temperature.

2. **Record the temperature of every solution you use,** including all calibration standards and samples, since this will affect K_w, neutrality and pH.

3. **Set the temperature compensator on the meter to the appropriate value.** This control makes an allowance for the effect of temperature on the electrical potential difference recorded by the meter: it does *not* allow for the other temperature-dependent

Fig. 27.2 Measurement of pH using a combination pH electrode and meter. The electrical potential difference recorded by the potentiometer is directly proportional to the pH of the test solution.

effects mentioned elsewhere. Basic instruments have no temperature compensator, and should only be used at a specified temperature, either 20 °C or 25 °C, otherwise they will not give an accurate measurement. More sophisticated systems have automatic temperature compensation.

4. **Rinse the electrode assembly (Fig. 27.2) with distilled water** and gently dab off the excess water onto a clean tissue: check for visible damage or contamination of the glass electrode (consult a member of staff if the glass is broken or dirty). Also check that the solution within the glass assembly is covering the metal electrode.

5. **Calibrate the instrument:** set the meter to 'pH' mode, if appropriate, and then place the electrode assembly in a standard solution of known pH, usually pH 7.00. This solution may be supplied as a liquid, or may be prepared by dissolving a measured amount of a calibration standard in water: calibration standards are often provided in tablet form, to be dissolved in water to give a particular volume of solution. Adjust the calibration control to give the correct reading. Remember that your calibration standards will only give the specified pH at a particular temperature, usually either 20 °C or 25 °C. If you are working at a different temperature, you must establish the actual pH of your calibration standards, either from the supplier, or from literature information.

6. **Remove the electrode assembly from the calibration solution and rinse again with distilled water:** dab off the excess water. Basic instruments have no further calibration steps (single-point calibration), while the more refined pH meters have additional calibration procedures.

   If you are using a basic instrument, you should check that your apparatus is accurate over the appropriate pH range by measuring the pH of another standard whose pH is close to that expected for the test solution. If the standard does not give the expected reading, the instrument is not functioning correctly: consult a member of staff.

   If you are using an instrument with a slope control function, this will allow you to correct for any deviation in electrical potential from that predicted by the theoretical relationship (at 25 °C, a change in pH of 1.00 unit should result in a change in electrical potential of 59.16 mV) by performing a two-point calibration. Having calibrated the instrument at pH 7.00, immerse in a second standard at the same temperature as that of the

(continued)

Box 27.1 (continued)

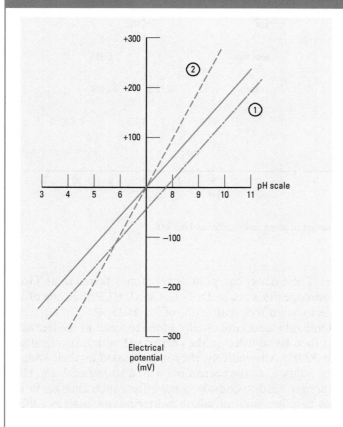

Fig. 27.3 The relationship between electrical potential and pH. The solid line shows the response of a calibrated electrode while the other plots are for instruments requiring calibration: 1 has the correct slope but incorrect isopotential point (calibration control adjustment is needed); 2 has the correct isopotential point but incorrect slope (slope control adjustment is needed).

first standard, usually buffered to either pH 4.00 or pH 9.00, depending upon the expected pH of your samples. Adjust the slope control until the exact value of the second standard is achieved (Fig. 27.3). A pH electrode and meter calibrated using the two-point method will give accurate readings over the pH range from 3 to 11: laboratory pH electrodes are not accurate outside this range, since the theoretical relationship between electrical potential and pH is valid.

7. Once the instrument is calibrated, measure the pH of your solution(s), making sure that the electrode assembly is rinsed thoroughly between measurements. You should be particularly aware of this requirement if your solutions contain organic biological material, e.g. soil, tissue fluids, protein solutions, etc., since these may adhere to the glass electrode and affect the calibration of your instrument. If your electrode becomes contaminated during use, check with a member of staff before cleaning: avoid touching the surface of the glass electrode with abrasive material. Allow sufficient time for the pH reading to stabilise in each solution before taking a measurement: for unbuffered solutions, this may take several minutes, so do not take inaccurate pH readings due to impatience!

8. After use, the electrode assembly must not be allowed to dry out. Most pH electrodes should be stored in a neutral solution of KCl, either by suspending the assembly in a small beaker, or by using an electrode cap filled with the appropriate solution (typically 1.0 mol L^{-1} KCl buffered at pH 7.0). However, many labs simply use distilled water as a storage solution, leading to loss of ions from the interior of the electrode assembly. In practice, this means that pH electrodes stored in distilled water will take far longer to give a stable reading than those stored in KCl.

9. Switch the meter to zero (where appropriate), but do not turn off the power: pH meters give more stable readings if they are left on during normal working hours.
   Problems (and solutions) include: inaccurate and/or unstable pH readings caused by cross-contamination (rinse electrode assembly with distilled water and blot dry between measurements); development of a protein film on the surface of the electrode (soak in 1% w/v pepsin in 0.1 mol L^{-1} HCl for at least an hour); deposition of organic or inorganic contaminants on the glass bulb (use an organic solvent, such as acetone, or a solution of 0.1 mol L^{-1} disodium ethylenediamine tetraacetic acid, respectively); drying out of the internal reference solutions (drain, flush and refill with fresh solution, then allow to equilibrate in 0.1 mol L^{-1} HCl for at least an hour); cracks or chips to the surface of the glass bulb (use a replacement electrode).

Table 27.3 pK_a values at 25 °C and M_r of some acids and bases (upper section) and some large organic zwitterions (lower section) commonly used in buffer solutions. For polyprotic acids, where more than one proton may dissociate, the pK_a values are given for each ionisation step. Only the trivial acronyms of the larger molecules are provided: their full names can be obtained from the catalogues of most chemical suppliers

Acid or base	pK_a value(s)	M_r
Acetic acid	4.8	60.1
Boric acid	9.2	61.8
Citric acid	3.1, 4.8, 5.4	191.2
Glycylglycine	3.1, 8.2	132.1
Phosphoric acid	2.1, 7.1, 12.3	98.0
Phthalic acid	2.9, 5.5	166.1
Succinic acid	4.2, 5.6	118.1
TRIS(base)*	8.1	121.1
CAPS (free acid)	10.4	221.3
CHES (free acid)	9.3	207.3
HEPES (free acid)	7.5	238.3
MES (free acid)	6.1	213.2
MOPS (free acid)	7.2	209.3
PIPES (free acid)	6.8	302.4
TAPS (free acid)	8.4	243.3
TRICINE (free acid)	8.1	179.2

*Note that this compound is hygroscopic and should be stored in a desiccator; also see text regarding its potential toxicity (p. 171).

Table 27.4 Preparation of sodium phosphate buffer solutions for use at 25 °C. Prepare separate stock solutions of (a) disodium hydrogen phosphate and (b) sodium dihydrogen phosphate, both at 200 mol m⁻³. Buffer solutions (at 100 mol m⁻³) are then prepared at the required pH by mixing together the volume of each stock solution shown in the table, then diluting to a final volume of 100 mL using distilled or deionised water

Required pH (at 25 °C)	Volume of stock (a) Na_2HPO_4 (mL)	Volume of stock (b) NaH_2PO_4 (mL)
6.0	6.2	43.8
6.2	9.3	40.7
6.4	13.3	36.7
6.6	18.8	31.2
6.8	24.5	25.5
7.0	30.5	19.5
7.2	36.0	14.0
7.4	40.5	9.5
7.6	43.5	6.5
7.8	45.8	4.2
8.0	47.4	2.6

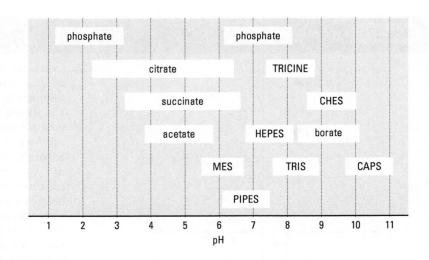

Fig. 27.4 Useful pH ranges of some commonly used buffers.

traditional buffers. These newer compounds are often referred to as 'Good buffers', to acknowledge the work of Dr N.E. Good: HEPES is one of the most useful zwitterionic buffers, with a pK_a of 7.5 at 25 °C.

These zwitterionic substances are usually added to water as the free acid: the solution must then be adjusted to the correct pH with a strong alkali, usually NaOH or KOH. Alternatively, they may be used as their sodium or potassium salts, adjusted to the correct pH with a strong acid, e.g. HCl. Consequently, you may need to consider what effects such changes in ion concentration may have in a solution where zwitterions are used as buffers. In addition, zwitterionic buffers can interfere with protein determinations (e.g. Lowry method, p. 384).

Figure 27.4 shows a number of traditional and zwitterionic buffers and their effective pH ranges. When selecting one of these buffers, aim for a pK_a that is in the direction of the expected pH change (Table 27.3). For example, HEPES buffer would be a better choice of buffer than PIPES for use at pH 7.2 for experimental systems where a pH increase is anticipated, while PIPES would be a better choice where acidification is expected.

Preparation of buffer solutions

Having selected an appropriate buffer, you will need to make up your solution to give the desired pH. You will need to consider two factors:

1. **The ratio of acid and conjugate base required to give the correct pH.**
2. **The amount of buffering required;** buffer capacity depends upon the absolute quantities of acid and base, as well as their relative proportions.

In most instances, buffer solutions are prepared to contain between 10 mmol L⁻¹ and 200 mmol L⁻¹ of the conjugate pair. Although it is possible to calculate the quantities required from first principles using the Henderson–Hasselbalch equation, there are sources that tabulate the amount of substance required to give a particular volume of solution with a specific pH value for a range of buffers (e.g. Anon, 2006, 2011). For traditional buffers, it is customary to mix stock solutions of acidic and basic components in the correct proportions to give the required pH (Table 27.4). For zwitterionic acids, the usual procedure is to add the compound to

Understanding buffering capacity – remember that buffer solutions will only work effectively if they have sufficient buffering capacity to resist the change in pH expected during the course of an experiment. Thus a weak solution of HEPES (e.g. 10 mmol L^{-1}, adjusted to pH 7.0 with NaOH) will not be able to buffer the growth medium of a dense suspension of cells for more than a few minutes.

water, then bring the solution to the required pH by adding a specific amount of strong alkali or acid (obtained from tables). Alternatively, the required pH can be obtained by dropwise addition of alkali or acid, using a meter to check the pH, until the correct value is reached. When preparing solutions of zwitterionic buffers, the acid may be relatively insoluble. Do not wait for it to dissolve fully before adding alkali to change the pH – the addition of alkali will help bring the acid into solution, but make sure it has all dissolved before the desired pH is reached.

Finally, when preparing a buffer solution based on tabulated information, always confirm the pH with a pH meter before use.

Text references

Anon. (2006) *pH Theory and Practice: A Radiometer Analytical Guide.* Available: http://www.radiometer-analytical.com/pdf/ph_theory.pdf Last accessed: 01/09/15.

Anon. (2011) *Buffer Reference Center.* Available: http://www.sigmaaldrich.com/life-science/

Sources for further study

Anon. (2003) *Buffers. A Guide for the Preparation and Use of Buffers in Biological Systems.* Available: http://wolfson.hiji.ac.ie/purification/PDF/Buffers/

core-bioreagents/biological-buffers/learning-center/buffer-reference-center.html Last accessed: 01/09/15.

Haynes, W.M. (ed.) (2014) *CRC Handbook of Chemistry and Physics,* 95th edn. CRC Press, Boca Raton.

Calbiochem_Buffers_Booklet.pdf. Last accessed 0/09/15.

STUDY EXERCISES

27.1 Practise interconverting pH values and proton concentrations. Express all answers to three significant figures.

(a) What is pH 7.4 expressed as [H$^+$] in mol L^{-1}?

(b) What is pH 4.1, expressed as [H$^+$] in mol m^{-3}?

(c) What is the pH of a solution containing H$^+$ at 2×10^{-5} mol L^{-1}?

(d) What is the pH of a solution containing H$^+$ at $10^{-12.5}$ mol L^{-1}?

(e) What is the pH of a solution containing H$^+$ at 2.8×10^{-5} mol m^{-3}?

27.2 Decide on a suitable buffer to use. In the following instances, choose a buffer that would be suitable:

(a) Maintaining the pH at 8.5 during an enzyme assay of a cell-free extract at 25°C.

(b) Keeping a stable pH of 6.5 in an experiment to measure the uptake of radiolabelled glucose by a dense suspension of *E. coli.*

(c) Carrying out an assay of photosynthetic activity at pH 7.2 at temperatures of 10°C, 20°C and 30°C.

(d) Stabilising pH at 5.5 during enzyme extraction, in a solution where you intend to measure total protein concentration at a later stage.

27.3 Practise using the Henderson–Hasselbalch equation. What are the relative proportions of deprotonated (A$^-$) and protonated (HA) forms of each substance at the following pH values:

(a) acetic acid (p$K_a = 4.8$) for use in an experiment at pH 3.8?

(b) boric acid (p$K_a = 9.2$) for use in an experiment at pH 9.5?

(c) HEPES (p$K_a = 7.5$) for use in an experiment at pH 8.1?

Answers to these study exercises are available at www.pearsoned.co.uk/practicalskills.

Many features of interest in biological systems are too small to be seen by the naked eye and can only be observed with a microscope. All microscopes consist of a coordinated system of lenses arranged so that a magnified image of a specimen is seen by the viewer (Figs 28.1 and 28.2). The main differences are the wavelengths of electromagnetic radiation used to produce the image, the nature and arrangement of the lens systems and the methods used to view the image.

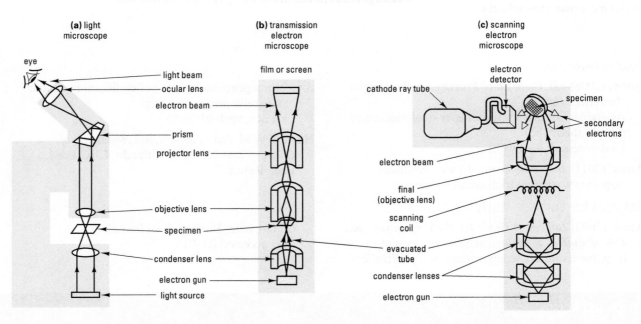

Fig. 28.1 Simplified diagrams of light and electron microscopes. Note that the electron microscopes are drawn upside down to aid comparison with the light microscope.

Table 28.1 Comparison of microscope types. Resolution is that obtained by a skilled user. LM, light microscope; SEM, scanning electron microscope; TEM, transmission electron microscope.

Property	Type of microscope		
	LM	TEM	SEM
Resolution	200 nm	1 nm	10 nm
Depth of focus	Low	Medium	High
Field of view	Good	Limited	Good
Specimen preparation (ease)	Easy	Skilled	Easy
Specimen preparation (speed)	Rapid	Slow	Quite rapid
Relative cost of instrument	Low	High	High

Microscopes allow objects to be viewed with increased resolution and contrast. Resolution is the ability to distinguish between two points on the specimen – the better the resolution, the 'sharper' the image. This is affected by lens design and is inversely related to the wavelength of radiation used. Contrast is the difference in intensity perceived between different parts of an image. This can be enhanced (a) by the use of stains, and (b) by adjusting microscope settings, usually at the expense of resolution.

The three main forms of microscopy are light microscopy, transmission electron microscopy (TEM) and scanning electron microscopy (SEM). Their main properties are compared in Table 28.1 and their suitability for observing cells and organelles is shown in Table 28.2.

Light microscopy

Two forms of the standard light microscope are the binocular (compound) microscope, described in detail in Chapter 29, and the dissecting microscope. These are the instruments most likely to be used in routine practical work. Figure 28.2(a) shows a typical image of cells viewed with a light microscope. In more advanced project work, you may use one or more of the following more sophisticated variants of light microscopy to improve image quality:

Fig. 28.2 Examples of images of a similar specimen (the stomatal complex of *Commelina communis*) obtained using different microscopic techniques: (a) light microscopy (surface view); (b) transmission electron microscopy (transverse section through guard cell pair at mid pore); and (c) scanning electron microscopy (surface view). As an indication of scale, the width of a single guard cell is about 10 μm.

- **Dark field illumination** involves a special condenser which causes reflected and diffracted light from the specimen to be seen against a dark background. The method is particularly useful for near-transparent specimens and for delicate structures like flagella. Care must be taken with the thickness of slides used – air bubbles and dust must be avoided and immersion oil must be used between the dark field condenser and the underside of the slide.

- **Ultraviolet microscopy** uses short-wavelength UV light to increase resolution. Fluorescence microscopy uses radiation at UV wavelengths to make certain fluorescent substances (e.g. chlorophyll or fluorescent dyes which bind to specific cell components, as used in immunofluorescence microscopy) emit light of visible wavelengths. Special light sources, lenses and mountants are required for UV and fluorescence microscopy and filters must be used to prevent damage to users' eyes.

- **Phase contrast microscopy** is useful for increasing contrast when viewing transparent specimens. It is superior to dark field microscopy because a better image of the interior of specimens is obtained. Phase contrast operates by causing constructive and destructive interference effects in the image, visible as increased contrast. Adjustments must be made, using a phase telescope in place of the eyepiece, for each objective lens and a matching phase condenser, and the microscope must be set up carefully to give optimal results.

- **Nomarski or differential interference contrast (DIC) microscopy** gives an image with a three-dimensional quality. However, the relief seen is optical rather than morphological and care should be taken in interpreting the result. One of the advantages of the technique is the extremely limited depth of focus which results: this allows 'optical sectioning' of a specimen.

- **Polarised light microscopy** can be used to reveal the presence and orientation of optically active components within specimens (e.g. starch grains, cellulose fibres), showing them brightly against a dark background.

- **Confocal microscopy** allows three-dimensional views of cells or thick sections. A finely focused laser is used to create electronic images of layered horizontal 'slices', usually after fluorescent staining. Images can be viewed individually or reconstructed to provide a 3D computer-generated image of the whole specimen.

Electron microscopes

Electron microscopes offer an image resolution up to 200 times better than light microscopes (Table 28.1) because they utilise radiation of shorter wavelength in the form of an electron beam. The electrons are produced by a tungsten filament operating in a vacuum and are focused by electromagnets. TEM and SEM differ in the way in which the electron beam interacts with the specimen: in TEM, the beam passes through the specimen (Fig. 28.1(b)), while in SEM the beam is scanned across the specimen and is reflected from the surface (Fig. 28.1(c)). In both cases, the beam must fall on a fluorescent screen before the image can be seen. Permanent images ('electron micrographs') are produced after focusing the beam on photographic film (Figs 28.2(b) and (c)).

Table 28.2 Dimensions of some typical cells and organelles with an indication of suitable forms of microscopy for observing them. LM, light microscope; SEM, scanning electron microscope; TEM, transmission electron microscope. Column 2 data after Rubbi (1994).

Cell or organelle	Approximate diameter or width [µm]	Suitable form of microscopy
Prokaryote cell	0.15–5	LM, SEM, TEM
Eukaryote cell	10–100	LM, SEM, TEM
Fungal hypha	5–20	LM, SEM, TEM
Nucleus	5–25	LM, SEM, TEM
Mitochondrion	1–10	SEM, TEM
Chloroplast	2–8	LM, SEM, TEM
Golgi apparatus	1	SEM, TEM
Lysosome/ peroxisome	0.2–0.5	SEM, TEM
Plant cell wall	0.1–10	LM, SEM, TEM

You are unlikely to use either type of electron microscope as part of undergraduate practical work because of the time required for specimen preparation and the need for detailed training before these complex machines can be operated correctly. However, electron microscopy is extremely important in understanding cellular and subcellular structures and you may be shown electron micrographs, with one or more of the following objectives:

- **to demonstrate cell ultrastructure (TEM);**
- **to show surface features of organisms,** for example, when surface features are coated or when cells are 'freeze-fractured', then coated (SEM);
- **to investigate changes in the number, size, shape and condition of cells and organelles (TEM);**
- **to carry out quantitative studies of cell and organelle disposition (TEM).**

Preparative procedures

Without careful preparation of the material being studied, the biological structures viewed with any type of microscope can be rendered meaningless due to artefacts. Figure 28.3 summarises the processes involved for the main types of microscopy discussed above (see also Chapter 29 for light microscopy).

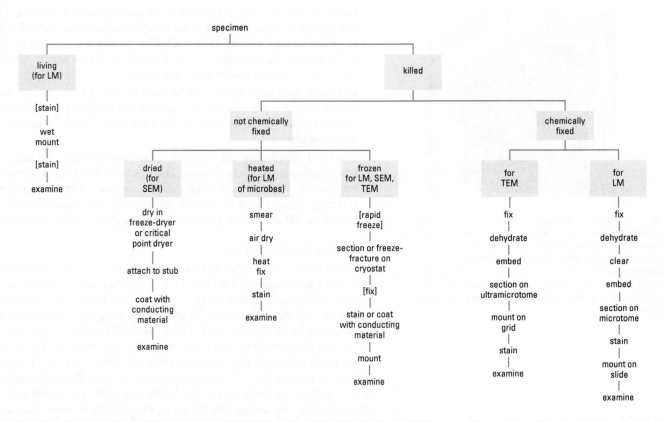

Fig. 28.3 Flowchart of procedures necessary to prepare specimens for different forms of microscopy. Steps enclosed in brackets are optional. LM, light microscope; SEM, scanning electron microscope; TEM, transmission electron microscope.

Text reference

Rubbi, C.P. (1994) *Light Microscopy Essential Data.* Wiley, Chichester.

Sources for further study

Bradbury, S. (1989) *An Introduction to the Optical Microscope.* 2nd edn. Bios, Abingdon.

John Innes Centre *About Microscopy.* Available: https://www.jic.ac.uk/microscopy/about.html Last accessed 01/09/15.
[Covers light microscopy, electron microscopy and scale and provides links to a range of other websites.]

Mertz, J. (2009) *Introduction to Optical Microscopy.* Roberts, Greenwood.

Murphy, D.B. (2012) *Fundamentals of Light Microscopy and Electronic Imaging,* 2nd edn. Wiley-Liss, New York.

STUDY EXERCISES

28.1 Test your microscopy knowledge. Indicate whether the following statements about light microscopy, scanning electron microscopy (SEM) or transmission electron microscopy (TEM) are true or false.

(a) TEM allows you to see at finer resolution than light microscopy.

(b) TEM allows you to see surface features of specimens.

(c) SEM always requires staining of specimens.

(d) The resolution of TEM is about 200 times better than that of light microscopy.

(e) The resolution of a microscope is linked to the wavelength of electromagnetic radiation employed.

(f) The specimen in both TEM and SEM is viewed under near-vacuum conditions.

(g) Specimens for light microscopy can be living or dead.

(h) SEM provides better resolution than TEM.

(i) The depth of focus in light microscopy is greater than that in SEM.

(j) Light microscopy, SEM and TEM all involve the use of a condenser lens within the microscope.

28.2 Fill in the blanks in the following paragraph. Dark field microscopy involves shining reflected and _____ light on the specimen against a dark background. It is particularly useful for _____ specimens. UV microscopy uses short wavelength UV light in order to increase image _____. Phase contrast microscopy utilises constructive and destructive _____ effects to increase image _____. Nomarski microscopy provides a pseudo ___ image, with a very small depth of_____, allowing _____ to be carried out _____ light microscopy allows visualisation of optically active components in the specimen. Confocal microscopy involves the use of a _____ light source and can yield computer-generated 3D images.

28.3 Identify the missing preparative procedures. In each sequence below, one or two steps have been missed out. Using Fig. 28.3, identify the missing procedures.

(a) For light microscopy on a killed and fixed specimen: fix – dehydrate – clear – _____– section –_____– mount – examine.

(b) For light microscopy on a heat-fixed microbial specimen: smear – _____ – heat fix –_____– examine.

(c) For TEM on a killed and fixed specimen: fix – _____ – embed – section – mount – stain – examine.

Answers to these study exercises are available at www.pearsoned.co.uk/practicalskills.

A standard undergraduate binocular microscope consists of three main types of optical unit: eyepieces, objective (Fig. 29.1) and a condenser. These are attached to a stand which holds the specimen on a stage (Fig. 29.2). A monocular microscope is constructed similarly but has one eyepiece lens rather than two.

Setting up a binocular light microscope

Before using any microscope, familiarise yourself with its component parts.

> **KEY POINT** Never assume that the previous person to use your microscope has left it set up correctly: taking account of differences in users' eyes, the microscope needs to be properly set up for each lens combination used.

The procedures outlined below are simplified to allow you to set up microscopes like those of the Olympus CX series (Fig. 29.2). For monocular microscopes, disregard instructions for adjusting eyepiece lenses in (3).

1. **Place the microscope at a convenient position on the bench.** Adjust your seating so that you are comfortable operating the focus and stage controls. Unwind the power cable, plug in and switch on after first ensuring that the lamp setting is at a minimum. Then adjust the lamp setting to about two-thirds of the maximum.

2. **Select a low-power (e.g. × 10) objective.** Make sure that the lens clicks home.

3. **Set the eyepiece (ocular) lenses to your interpupillary distance;** this can usually be read off a scale on the turret. You should now see a single circular field of vision. If you do not, try adjusting in either direction.

4. **Put a prepared slide on the stage.** Examine it first against a light source and note the position, colour and rough size of the specimen. Place the slide (coverslip up) on the stage using the slide holder and, viewing from the side, position it with the stage adjustment controls so that the specimen is illuminated.

5. **Focus the image of the specimen** using first the coarse and then the fine focusing controls (Fig. 29.3). The image will be reversed and upside down compared to that seen when viewing the slide directly.

   (a) If both eyepiece lenses are adjustable, set your interpupillary distance on the scale on each lens. Close your left eye, look through the right eyepiece with your right eye and focus the image with the normal controls. Now close your right eye, look through the left eyepiece with your left eye and focus the image by rotating the eyepiece holder. Take a note of the setting for future use.

   (b) If only the left eyepiece is adjustable, close your left eye, look with the right eye through the static right eyepiece and focus the image with the normal controls. Now close your right eye, look through the left eyepiece with your left eye and focus the image by rotating the eyepiece holder. Take a note of the setting for future use.

Fig. 29.1 Standard features of an objective lens. Most lenses are inscribed to show the details labelled above. The numerical aperture is a measure of the light-gathering power of the lens.

Labels on figure: lens mounting thread; magnification; numerical aperture; lens body; 40x/0.95; 160/0.17; microscope tube length; coverslip thickness (in mm); lens assembly

> **Using binocular eyepieces** – if you do not know your interpupillary distance, ask someone to measure it with a ruler. You should stare at a fixed point in the distance while the measurement is taken in millimetres. Take a note of the value for future use.

> **Issues for spectacle and contact lens wearers** – those who wear glasses can remove them for viewing, as microscope adjustments will accommodate most deficiencies in eyesight (except astigmatism). This is more comfortable and stops the spectacle lenses being scratched by the eyepiece holders. However, it may create difficulties in focusing when drawing diagrams. Those with contact lenses should simply wear them as normal for viewing.

Fig. 29.2 The Olympus binocular microscope model CX41.

- The lamp in the base of the stand (1) supplies light; its brightness is controlled by an on-off switch and voltage control (2). Never use maximum voltage or the life of the bulb will be reduced – a setting two-thirds to three-quarters of maximum should be adequate for most specimens. A field-iris diaphragm may be fitted close to the lamp to control the area of illumination (3).
- The condenser control focuses light from the condenser lens system (4) onto the specimen and projects the specimen's image onto the front lens of the objective. Correctly used, it ensures optimal resolution.
- The condenser-iris diaphragm (5) controls the amount of light entering and leaving the condenser; its aperture can be adjusted using the condenser-iris diaphragm lever (6). Use this to reduce glare and enhance image contrast by cutting down the amount of stray light reaching the objective lens.
- The specimen (normally mounted on a slide) is fitted to a mechanical stage or slide holder (7) using a spring mechanism. Two controls allow you to move the slide in the *x* and *y* planes. Vernier scales on the slide holder can be used to return to the same place on a slide. The fine and coarse focus controls (8) adjust the height of the stage relative to the lens systems. Take care when adjusting the focus controls to avoid hitting the lenses with the stage or slide.
- The objective lens (9) supplies the initial magnified image; it is the most important component of any microscope because its qualities determine resolution, depth of field and optical aberrations. The objective lenses are attached to a revolving nosepiece (10). Take care not to jam the longer lenses onto the stage or slide as you rotate the nosepiece. You should feel a distinct click as each lens is moved into position. The magnification of each objective is written on its side; a normal complement would be ×4, ×10, ×40 and ×100 (oil immersion).
- The eyepiece lens (11) is used to further magnify the image from the objective and to put it in a form and position suitable for viewing. Its magnification is written on the holder (normally ×10). By twisting the holder for one or both of the eyepiece lenses you can adjust their relative heights to take account of optical differences between your eyes. The interpupillary distance scale (12) and adjustment knob allow compensation to be made for differences in the distance between users' pupils.

Photograph supplied by Microscopy, an operating division of KeyMed, with thanks to Olympus KeyMed.

plant cells

(a)   (b)

Fig. 29.3 Importance of correct focus in light microscopy. Stomatal complex of *Commelina communis*, a specimen that is a monolayer of cells approximately 30–50µm thick, (a) Focal plane is on 'internal' walls of the cells; (b) focal plane is on the 'external' walls and stomatal pore. The two images are different and, while it would not be possible to measure the stomatal pore in (a), it would not be possible to see the vacuolar crystals in (b). When looking at specimens, always use the fine focus control to view different focal planes.

SAFETY NOTE Take care when moving microscopes, not only because of the replacement cost if damaged, but also because they weigh several kilograms and could cause injury if dropped. Always carry a microscope using two hands.

Adjusting a microscope with a field–iris diaphragm – adjust this *before* the condenser–iris diaphragm: close it until its image appears in view as a circle of light, if necessary focusing on the edge of the circle with the condenser controls and centring it with the centring screws. Now open it so the whole field is just illuminated.

Using high-power objectives – never remove a slide while a high power objective lens (i.e. ×40 or ×100) is in position. Always turn back to the ×10 first. Having done this, lower the stage and remove the slide.

6. **Close the condenser–iris diaphragm** (aperture–iris diaphragm), then open it to a position such that further opening has no effect on the brightness of the image (the 'threshold of darkening'). The edge of the diaphragm should not be in view. Turn down the lamp if it is too bright.

7. **Focus the condenser.** Place an opaque pointed object (the tip of a mounted needle or a sharp pencil point) on the centre of the light source. Adjust the condenser setting until both the specimen and needle tip/pencil point are in focus together. Check that the condenser–iris diaphragm is just outside the field of view.

8. **For higher magnifications, swing in the relevant objective** (e.g. ×40), carefully checking that there is space for it. Adjust the focus using the fine control only. If the object you wish to view is in the centre of the field with the ×10 objective, it should remain in view (magnified, of course) with the ×40. Adjust the condenser–iris diaphragm and condenser as before – the correct setting for each lens will be different.

9. **When you have finished using the microscope, remove the last slide and clean the stage if necessary.** Turn down the lamp setting to its minimum, then switch off. Clean the eyepiece lenses with lens tissue. Check that the objectives are clean. Unplug the microscope from the mains and wind the cable round the stand and under the stage. Replace the dust cover.

If you have problems in obtaining a satisfactory image, refer to Box 29.1; if this does not help, refer the problem to the class supervisor.

Procedure for observing transparent specimens

Some stained preparations and all colourless objects are difficult to see when the microscope is adjusted as above (Fig. 29.4). Contrast can be improved by closing down the condenser–iris diaphragm. Note that when you do this, diffraction haloes appear round the edges of objects. These obscure the image of the true structure of the specimen and may result in loss of resolution. Nevertheless, an image with increased contrast may be easier to interpret. Figure 29.5 shows several types of non-biological artefacts that you will come to recognise through practical experience.

Box 29.1 Problems in light microscopy and possible solutions

No image; very dark image; image dark and illuminated irregularly

- Microscope not switched on (check plug and base)
- Illumination control at low setting or off
- Objective nosepiece not clicked into place over a lens
- Diaphragm closed down too much or off-centre
- Lamp failure

Image visible and focused but pale and indistinct

- Diaphragm needs to be closed down further (see Fig. 29.4)
- Condenser requires adjustment

Image blurred and cannot be focused

- Dirty objective
- Dirty slide

- Slide upside down
- Slide not completely flat on stage
- Eyepiece lenses not set up properly for user's eyes
- Fine focus at end of travel
- Oil immersion objective in use, without oil

Dust and dirt in field of view

- Eyepiece lenses dirty
- Objective lens dirty
- Slide dirty
- Dirt on lamp glass or upper condenser lens

(a)

(b)

Fig. 29.4 Effect of closing the condenser–iris diaphragm on contrast. Head of human head louse, *Pediculus humanus capitis* (a) with condenser–iris diaphragm open; (b) with condenser–iris diaphragm closed (all other settings the same). Note the difference in detail that can be seen in (b), but also that image (b) is darker. When using the condenser-iris diaphragm in this way you may need to compensate by increasing the light setting.

Using an eyepiece graticule – choose the eyepiece lens corresponding to your stronger eye and check that you have made the correct adjustments to the eyepiece lenses as detailed on p. 180.

Procedure for oil immersion objectives

These provide the highest resolution of which the light microscope is capable. They must be used with immersion oil filling the space between the objective lens and the top of the slide. The oil has the same refractive index as the glass lenses, so loss of light by reflection and refraction at the glass/air interface is reduced. This increases the resolution, brightness and clarity of the image and reduces aberration. Use oil immersion objective(s) as follows:

1. **Check that the object of interest is in the field of view** using e.g. the ×10 or ×40 objective.

2. **Apply a single small droplet of immersion oil** to the illuminated spot on the top of the slide, having first swung the ×40 objective away Never use too much oil: it can run off the slide and mess up the microscope.

3. **Move the high power (×100) oil immersion objective into position** carefully, checking first that there is space for it. Focus on the specimen using the fine control only. You may need a higher brightness setting.

4. **Perform condenser–iris diaphragm and condenser focusing adjustments** as for the other lenses.

5. **When finished, clean the oil immersion lens** by gently wiping it with clean lens tissue. If the slide is a prepared one, wipe the oil off with lens tissue.

You should take great care when working with oil immersion lenses as they are the most expensive to replace. Because the working distance between the lens and coverslip is so short (less than 2 mm), it is easy to damage the lens surface by inadvertently hitting the slide or coverslip surface. You must also remember that they need oil to work properly. If working with an unfamiliar microscope, you can easily recognise oil immersion lenses. Look for a white or black ring on the lens barrel, near the lens, or for 'oil' clearly marked on the barrel.

Stains and staining procedures

Table 29.1 gives a number of widely used stains and their applications for animal, plant and microbial cells and tissues.

Establishing scale and measuring objects

The magnification of a light microscope image is calculated by multiplying the objective magnification by the eyepiece magnification. However, the magnification of the image bears no certain relation to the magnification of any drawing of the image – you may equally well choose to draw the same image 10 mm or 10 cm long. For this reason, *it is essential to add a scale to all your diagrams.* You can provide either a bar giving the estimated size of an object of interest, or a bar of defined length (e.g. 100 μm).

The simplest method of estimating linear dimensions is to compare the size of the image to the diameter of the field of view. You can make a rough estimate of the field diameter by focusing on the millimetre scale of a transparent ruler using the lowest power objective. Estimate the diameter of this field directly, then use the information to work out the field diameters at the higher powers *pro rata*. For example, if the field at an

Table 29.1 A selection of stains for light microscopy.

	Stain	What it stains	Comments
Plant cells	Chlorazol black	Cell walls: black Nuclei: black, yellow or green Suberin: amber	The solvent used (70% ethanol in water or water alone) affects colours developed
	Neutral red	Living cells: pink (pH < 7)	A 'vital' stain used to determine cell viability or to visualise plant protoplasts in plasmolysis experiments; best used at neutral external pH
	Phloroglucinol/HCl	Lignified cell walls: red	Care is required because the acid may damage microscope lenses
	Ruthenium red	Pectins: red	Shows up the middle lamella
	Safranin + Fast green	Nuclei, chromosomes, cuticle and lignin: red Other components: green	Stain in safranin first, then counterstain with fast green (light green will substitute). A differentiation step is required
	Toluidine blue	Lignified cell walls: blue Cellulose cell walls: purple	Best to apply dilute and allow progressive staining to occur
Fungi and bacteria	Giemsa	Bacterial chromosome: purple Bacterial cytoplasm: colourless	Also used in zoology to stain protozoa
	Gram	Gram-positive bacteria: violet/purple Gram-negative bacteria: red/pink Yeasts: violet/purple	See p. 248 for procedure
	Gray	Bacterial flagella: red	Uses toxic chemicals: mercuric chloride and formaldehyde. Leifson's stain is an alternative
	Lactophenol cotton blue	Fungal cytoplasm: blue (hyphal wall unstained)	Shrinkage may occur
	Nigrosin or India ink	Background: grey–black	Negative stains for visualisation of capsules: requires a very thin film
	Shaeffer and Fulton	Bacterial endospores: green Vegetative cells: pink/red	Malachite green is primary stain, heated for 5 min. Counterstained with safranin
	Ziehl–Neelsen	Actinomycetes: red Bacterial endospores: red Other microbes: blue	Requires heat treatment of fuchsin primary stain, decolorisation with ethanol–HCl and a methylene blue counterstain (acid-fast structures remain red)
Animal cells	Azure A/eosin B	Nuclei, RNA: blue Basophilic cells: blue–violet Most other cells: pale blue Muscle cells: pink Necrosing cells: pink Cartilage matrix: red–violet Bone: pink Red blood cells: orange–red Mucins: green–blue/blue–violet	Used in pathology – shows up bacteria as blue; must be fresh; care required over pH: Mann's methyl blue/eosin gives similar results
	Chlorazol black	Chitin: greenish–black Nuclei: black, yellow or green Glycogen: pink or red	Solvent (70% v/v ethanol in water or water alone) affects colours formed
	Iron haematoxylin	Nuclei, chromosomes and red blood cells: black Other structures: grey or blue–black	Good for resolving fine detail; iron alum used as mordant before haematoxylin to differentiate
	Mallory	Nuclei: red Nucleoli: yellow Collagen, mucus: blue Red blood cells: yellow Cytoplasm: pink or yellow	Simple, one stage stain; fades within a year; not to be used with osmium-containing fixatives. Heidenhain's azan gives similar results but does not fade. Cason's one-step Mallory is a rapidly applied stain which is particularly good for connective tissue
	Masson's trichrome	Collagen, mucus: green Cytoplasm: orange or pink	Used as a counterstain after, for example, iron haematoxylin which will have stained nuclei black. Not to be used after osmium fixation
	Mayer's (haemalum and eosin; 'H&E')	Nuclei: blue/purple Cytoplasm: pink	Alum used as mordant for haematoxylin; eosin is the counterstain. To show up collagen, use van Gieson's stain as counterstain

overall magnification of ×40 is 4 mm, at an overall magnification of ×100 it will be: 40/100 × 4 mm = 1.6 mm (1600 μm).

Greater accuracy can be obtained if an eyepiece micrometer (graticule) is used. This carries a fine scale and fits inside an eyepiece lens. The eyepiece micrometer is calibrated using a stage micrometer, basically a slide with a fine scale on it.

Figure 29.6 shows how to calibrate an eyepiece micrometer, along with a worked example. Once you have calibrated your eyepiece micrometer for each objective lens used, you can use it to measure objects: in the example shown in Fig. 29.6, the scale reading is multiplied by 2.65 μm to give the value in micrometres. So, if you measured the width of a human hair at 34 eyepiece micrometer units, then this will be equal to 34 × 2.65 = 90.1 μm. An alternative approach is to put a scale bar on a diagram, e.g. a 100 μm scale bar would be equivalent to the length of almost 38 eyepiece micrometer divisions. You should avoid putting too many significant figures in any estimates of dimensions: there may be quite large errors involved, which could make the implied accuracy misleading (spurious accuracy, p. 193).

Care and maintenance of your microscope

Microscopes are delicate precision instruments. Handle them with care and never force any of the controls. Never touch any of the glass surfaces with anything other than clean, dry lens tissue. Bear in mind that a replacement would be very expensive.

If moving a microscope, hold the stand above the stage with one hand and rest the base of the stand on your other hand. Always keep the microscope vertical (or the eyepieces may fall out). Put the microscope down gently.

(a)

(b)

(c)

(d)

(e)

(f)

Fig. 29.5 Artefacts in light microscopy. Do not mistake any of the following for biological specimens: (a) edge of coverslip; (b) dust on slide; (c) shard of glass; (d) air bubble; (e) grease spots; (f) cellulose fibre.

Fig. 29.6 Calibrating an eyepiece micrometer (graticule). Align two scales and read the number of stage micrometer divisions (shown in colour) for a particular number of eyepiece micrometer divisions (shown in black). In this case 26.5 scale divisions of 0.01 mm are equivalent to 100 eyepiece divisions, so each eyepiece division 0.265 ÷ 100 = 0.00265 mm = 2.65 μm. This is a typical value for a total magnification of ×400 (e.g. ×40 objective and ×10 eyepiece).

Measuring specimens using a dissecting microscope – because of the low magnification, sizes can generally be estimated by comparison with a ruler placed alongside the specimen. If accurate measurements are required, eyepiece graticules can be used (see Fig. 29.6).

Clean lenses by gently wiping with clean, dry lens tissue. Use each piece of tissue once only. Try not to touch lenses with your fingers as oily fingerprints are difficult to clean off. Do not allow any solvent (including water) to come into contact with a lens; sea water is particularly damaging.

The dissecting (stereoscopic) microscope

The dissecting microscope (Fig. 29.7) is a form of stereoscopic microscope used for observations at low total magnification (×4 to ×50) where a large working distance between objectives and stage is required, perhaps because the specimen is not flat or dissecting instruments are to be used. A stereoscopic microscope essentially consists of two separate lens systems, one for each eye. Some instruments incorporate zoom objectives. The eyepiece–objective combinations are inclined at about 15° to each other and the brain resolves the compound image in three dimensions as it does for normal vision. The image is right side up and not reversed, which is ideal for dissections. Specimens are often viewed in a fresh state and need not be placed on a slide – they might be in a Petri dish or on a white tile. Illumination can be from above or below the specimen, as desired.

Most of the instructions for the binocular microscope given above apply equally well to dissecting microscopes, although the latter do not normally have adjustable condensers or diaphragms. With stereoscopic microscopes, make specially sure to adjust the eyepiece lenses to suit your eyes so that you can take full advantage of the stereoscopic effect.

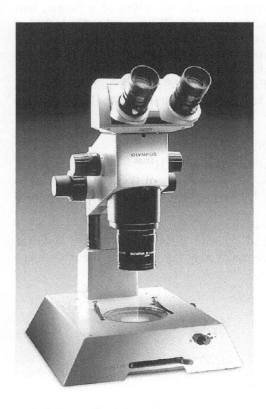

Fig. 29.7 The Olympus Model SZX12 dissecting microscope.
Photograph supplied by Microscopy, an operating division of KeyMed, with thanks to Olympus KeyMed.

Sources for further study

Bradbury, S. and Bracegirdle, B. (1998) *Introduction to Light Microscopy*. Bios Scientific Publishers, Oxford.

Davidson, M.W. and Abramowitz, M. *Nikon Microscopy: Introduction to Microscope Objectives*. Available: http://www.microscopyu.com/articles/optics/objectiveintro.html
Last accessed: 01/09/15.

Olympus Microscopy Resource Centre. *Introduction to Optical Microscopy, Digital Imaging, and Photomicrography*. Available: http://www.olympusmicro.com/primer/
Last accessed: 01/09/15.
[A primer on optical microscopy, including FAQs and links.]

STUDY EXERCISES

29.1 Test your knowledge of the parts of a binocular light microscope. Cover up the legend on the left of Fig. 29.2 with a piece of paper or card. Now identify the parts of the light microscope numbered on the diagram. Check your answers from the legend.

29.2 Identify roles of parts of a binocular light microscope. State briefly the *primary* role of each component of a standard binocular light microscope:

(a) condenser; (b) objective lens; (c) condenser-iris diaphragm; (d) interpupillary distance scale; (e) Vernier scales on the mechanical stage.

29.3 Identify the correct sequence of adjustments when setting up a light microscope.

(a) Focus specimen – set interpupillary distance – adjust condenser-iris diaphragm – make individual eyepiece adjustment – focus condenser – focus specimen.

(b) Make individual eyepiece adjustment – set interpupillary distance – focus condenser – focus specimen – adjust condenser-iris diaphragm – focus specimen.

(c) Set interpupillary distance – focus specimen – make individual eyepiece adjustment – adjust condenser-iris diaphragm – focus condenser – focus specimen.

(d) Make individual eyepiece adjustment – set interpupillary distance – focus specimen – adjust condenser-iris diaphragm – focus condenser – focus specimen.

(e) Focus specimen – make individual eyepiece adjustment – set interpupillary distance – focus condenser – adjust condenser-iris diaphragm.

29.4 Calculate the size of an object seen down a microscope using graticule readings. At 100 × overall magnification, you find that 1 stage micrometer unit of 0.1 mm = 11 eyepiece micrometer units. The length of a parasitic worm on a slide viewed at this magnification appears to be 25 eyepiece units. What is its estimated true length to the nearest μm?

29.5 Select appropriate stains. From Table 29.1, identify stains you could use to help indicate the presence of the following:

(a) glycogen in a liver section; (b) woody (lignified) cells in a plant stem section; (c) a fungal pathogen in a leaf section; (d) living cells in an onion epidermal peel; (e) mucus in a lung section; (f) bacteria in a food sample.

Answers to these study exercises are available at www.pearsoned.co.uk/practicalskills.

The investigative approach

The term 'data' (singular = datum, or data value) refers to items of information, and you will use different types of data from a wide range of sources during your practical work. Consequently, it is important to appreciate the underlying features of data collection and measurement.

Variables

Biological variables (Fig. 30.1) can be classified as follows.

Quantitative variables

These are characteristics whose differing states can be described by means of a number. They are of two basic types:

- **Continuous variables,** such as length; these are usually measured against a numerical scale. Theoretically, they can take any value on the measurement scale. In practice, the number of significant figures of a measurement is directly related to the precision of your measuring system; for example, dimensions measured with Vernier callipers will provide readings of greater precision than a millimetre ruler. Many of the variables measured in biomolecular sciences are continuous and quantitative, e.g. mass, temperature, time, amount of product formed by an enzyme.
- **Discontinuous (discrete) variables;** these are always obtained by counting and therefore the data values must be whole numbers (integers), with no intermediate values.

Ranked variables

These provide data which can be listed in order of magnitude (i.e. ranked). A familiar example is the abundance of an organism in a sample, which is often expressed as a series of ranks, e.g. rare = 1, occasional = 2, frequent = 3, common = 4, and abundant = 5. When such data are given numerical ranks, rather than descriptive terms, they are sometimes called 'semi-quantitative data'. Note that the difference in magnitude between ranks need not be consistent. For example, regardless of whether there was a one-year or a five-year gap between offspring in a family; their ranks in order of birth would be the same.

Qualitative variables (attributes)

These are non-numerical and descriptive; they have no order of preference and, therefore, are not measured on a numerical scale nor ranked in order of magnitude, but are described in terms of categories. Examples include viability (i.e. dead or alive), shape (e.g. round, flat, elongated), presence/absence of a particular biomolecule in a sample (p. 379).

Variables may be independent or dependent. Usually, the variable under the control of the experimenter (e.g. time) is the independent variable, while the variable being measured is the dependent variable (p. 201). Sometimes it is inappropriate to describe variables in this way and they are then referred to as interdependent variables (e.g. the length and breadth of an organism).

The majority of data values are recorded as direct measurements, readings or counts, but there is an important group, called derived (or

---

Working with discontinuous variables – note that while the original data values must be integers, derived data and statistical values do not have to be whole numbers. Thus, it is perfectly acceptable to express the *mean* number of children per family as 2.4.

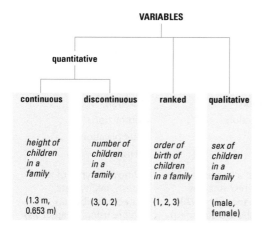

Fig. 30.1 Examples of the different types of variables as used to describe some characteristics of families.

computed), which result from calculations based on two or more data values, e.g. ratios, percentages, indices and rates.

Measurement scales

Variables may be measured on different types of scale:

- **Nominal scale:** this classifies objects into categories based on a descriptive characteristic. It is the only scale suitable for qualitative data.
- **Ordinal scale:** this classifies by rank. There is a logical order in any number scale used.
- **Interval scale:** this is used for quantitative variables. Numbers on an equal unit scale are related to an arbitrary zero point.
- **Ratio scale:** this is similar to the interval scale, except that the zero point now represents an absence of that character (i.e. it is an absolute zero). In contrast to the interval scale, the ratio of two values is meaningful (e.g. a temperature of 200 K is twice that of 100 K).

The measurement scale is important in determining the mathematical and statistical methods used to analyse your data. Table 30.1 presents a summary of the important properties of these scales. Note that you may be

> **Examples** **A nominal scale** for temperature is not feasible, since the relevant descriptive terms can be ranked in order of magnitude.
>
> An **ordinal scale** for temperature measurement might use descriptive terms, ranked in ascending order, e.g. cold = 1, cool = 2, warm = 3, hot = 4.
>
> The **Celsius scale** is an interval scale for temperature measurement, since the arbitrary zero corresponds to the freezing point of water (0°C).
>
> The **Kelvin scale** is a ratio scale for temperature measurement since 0 K represents a temperature of absolute zero (for information, the freezing point of water is 273.15 K on this scale).

Table 30.1 Some important features of scales of measurement.

	Measurement scale			
	Nominal	Ordinal	Interval	Ratio
Type of variable	Qualitative (Ranked)* (Quantitative)*	Ranked (Quantitative)*	Quantitative	Quantitative
Examples	Species Sex Colour	Abundance scales Reproductive condition Optical assessment of colour development	Fahrenheit temperature scale Date (BC/AD)	Kelvin temperature scale Weight Length Response time Most physical measurements
Mathematical properties	Identity	Identity Magnitude	Identity Magnitude Equal intervals	Identity Magnitude Equal intervals True zero point
Mathematical operations possible on data	None	Rank	Rank Addition Subtraction	Rank Addition Subtraction Multiplication Division
Typical statistics used	Only those based on frequency of counts made: contingency tables, frequency distributions, etc. Chi-square test	Non-parametric methods, sign tests. Mann-Whitney U-test	Almost all types of test, t-test, analysis of variance (ANOVA), etc. (check distribution before using, Chapter 74)	Almost all types of test, t-test, ANOVA, etc. (check distribution before using, Chapter 74)

*In some instances (see text for examples).

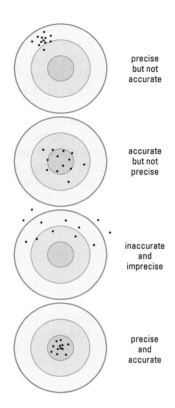

precise
but not
accurate

accurate
but not
precise

inaccurate
and
imprecise

precise
and
accurate

Fig. 30.2 Target' diagrams illustrating precision and accuracy.

able to measure a characteristic in more than one way, or you may be able to convert data collected in one form to a different form. For instance, you might measure light in terms of the photon flux density (p. 307) between particular wavelengths of the EMR spectrum (ratio scale), or simply as 'blue' or 'red' (nominal scale); you could find out the dates of birth of individuals (interval scale) but then use this information to rank them in order of birth (ordinal scale). Where there are no other constraints, you should use a ratio scale to measure a quantitative variable, since this will allow you to use the broadest range of mathematical and statistical procedures (Table 30.1).

Accuracy and precision

Accuracy is the closeness of a measured or derived data value to its true value, while precision is the closeness of repeated measurements to each other (Fig. 30.2). A balance with a fault in it (i.e. a bias, see below) could give precise (i.e. very repeatable) but inaccurate (i.e. untrue) results. Unless there is bias in a measuring system, precision will lead to accuracy and it is precision that is generally the most important practical consideration, if there is no reason to suspect bias. You can investigate the precision of any measuring system by repeated measurements of individual samples.

Absolute accuracy and precision are impossible to achieve, owing to both the limitations of measuring systems for continuous quantitative data and the fact that you are usually working with incomplete data sets (samples). It is particularly important to avoid spurious accuracy in the presentation of results; include only those digits that the accuracy of the measuring system implies. This type of error is common when changing units (e.g. inches to metres) and in derived data, especially when calculators give results to a large number of decimal places. Further advice is given on p. 506.

Bias (systematic error) and consistency

Bias is a systematic or non-random distortion and is one of the most troublesome difficulties in using numerical data. Biases may be associated with incorrectly calibrated instruments, e.g. a faulty pipettor, or with experimental manipulations, e.g. shrinkage during the preservation of a specimen. Bias in measurement can also be subjective, or personal, e.g. an experimenter's preconceived ideas about an 'expected' result.

Bias can be minimised by using a carefully standardised procedure, with fully calibrated instruments. You can investigate bias in 'trial runs' by measuring a single variable in several different ways, to see whether the same result is obtained.

To avoid personal bias, 'blind' measurements should be made where the identity of individual samples is not known to the operator, e.g. using a coding system.

Measurement error

All measurements are subject to error, but the dangers of misinterpretation are reduced by recognising and understanding the likely sources of error and by adopting appropriate protocols and calculation procedures.

A common source of measurement error is carelessness, e.g. reading a scale in the wrong direction or parallax errors. This can be reduced greatly by careful recording and may be detected by repeating the measurement. Other errors arise from faulty or inaccurate equipment, but even a perfectly functioning machine has distinct limits to the accuracy and precision

Minimising errors – determine early in your study what the dominant errors are likely to be and concentrate your time and effort on reducing these.

Working with derived data – special effort should be made to reduce measurement errors because their effects can be magnified when differences, ratios, indices or rates are calculated.

of its measurements. These limits are often quoted in manufacturers' specifications and are applicable when an instrument is new; however, you should allow for some deterioration with age. Further errors are introduced when the subject being studied is open to influences outside your control. Resolving such problems requires appropriate experimental design and representative sampling.

One major influence virtually impossible to eliminate is the effect of the investigation itself: even putting a thermometer in a liquid may change the temperature of the liquid. The very act of measurement may give rise to a confounding variable (p. 202), as discussed in Chapter 32. You should include descriptions of possible sources of errors and estimates of their likely importance in any report. However, do not use 'biological variability' as a catch-all excuse for poor technique or inadequacies in your experimental design.

Sources for further study

Erikson, B.H. and Nosanchuk, T.A. (1982) *Understanding Data,* 2nd edn. Open University Press, Milton Keynes. [A text aimed at social science students but with clear explanations of issues that are generic, including information on analysis of data.]

Rowlett, R. *How Many? A Dictionary of Units of Measurement.* Available: http://www.unc.edu/~rowlett/units/index.html.
Last accessed: 01/09/15.

Shuttleworth, M. *Choosing Scientific Measurements* Available: https://explorable.com/scientific-measurements
Last accessed 01/09/15.

STUDY EXERCISES

30.1 Classify variables. Decide on the type of variables used for the following measures, indicating whether they are quantitative or qualitative, continuous or discontinuous, and the type of scale that would be used.

   (a) Number of organisms in a sample.

   (b) Height of individuals in a sample.

   (c) Eye colour of individuals in a sample.

   (d) Species present in a sample.

   (e) Date of a sample.

   (f) Reproductive status of an animal.

30.2 Investigate types of error. A student weighed a set of standard masses on two electronic balances and obtained the readings shown in the table below. Explain these results in terms of the type of error involved in each case.

Comparison of weights of masses on two balances

	Standard mass (g)				
	10	25	50	100	250
Reading (balance A)	10.050	25.049	50.051	100.048	250.052
Reading (balance B)	10.004	25.011	50.021	100.039	250.102

Answers to these study exercises are available at www.pearsoned.co.uk/practicalskills.

Dimensionless measurements – some quantities can be expressed as dimensionless ratios or logarithms (e.g. pH), and in these cases you do not need to use a qualifying unit.

When describing most measurements, you normally state both a number and a unit (e.g. 'the length is 1.85 metres'). The number expresses the ratio of the measured quantity to a fixed standard, while the unit identifies that standard measure or dimension. Clearly, a single unified system of units is essential for efficient communication of such data within the scientific community. The Système International d'Unités (SI) is the internationally ratified form of the metre-kilogram-second system of measurement and represents the accepted scientific convention for measurements of physical quantities.

Another important reason for adopting consistent units is to simplify complex calculations where you may be dealing with several measured quantities (see p. 505). Although the rules of the SI are complex and the scale of the base units is sometimes inconvenient, to gain the full benefits of the system you should observe its conventions strictly.

The description of measurements in SI involves:

- **seven base units and two supplementary units,** each having a specified abbreviation or symbol (Table 31.1);
- **derived units,** obtained from combinations of base and supplementary units, which may also be given special symbols (Table 31.2);
- **a set of prefixes** to denote multiplication factors of 10^3, used for convenience to express multiples or fractions of units (Table 31.3).

Table 31.1 The base and supplementary SI units.

Measured quantity	Name of SI unit	Symbol
Base units		
Length	metre	m
Mass	kilogram	kg
Amount of substance	mole	mol
Time	second	s
Electric current	ampere	A
Temperature	kelvin	K
Luminous intensity	candela	cd
Supplementary units		
Plane angle	radian	rad
Solid angle	steradian	sr

Table 31.2 Some important derived SI units.

Measured quantity	Name of unit	Symbol	Definition in base units	Alternative in derived units
Energy	joule	J	m^2 kg s^{-2}	N m
Force	newton	N	m kg s^{-2}	J m^{-1}
Pressure	pascal	Pa	kg m^{-1} s^{-2}	N m^{-2}
Power	watt	W	m^2 kg s^{-3}	J s^{-1}
Electric charge	coulomb	C	As	J V^{-1}
Electric potential difference	volt	V	m^2 kg A^{-1} s^{-3}	J C^{-1}
Electric resistance	ohm	Ω	m^2 kg A^{-2} s^{-3}	V A^{-1}
Electric conductance	Siemens	S	s^3 A^2 kg^{-1} m^{-2}	A V^{-1} or Ω^{-1}
Electric capacitance	farad	F	s^4 A^2 kg^{-1} m^{-2}	C V^{-1}
Luminous flux	lumen	lm	cd sr	
Illumination	lux	lx	cd sr m^{-2}	lm m^{-2}
Frequency	hertz	Hz	s^{-1}	
Radioactivity	becquerel	Bq	s^{-1}	
Enzyme activity	katal	kat	mol substrate s^{-1}	

Table 31.3 Prefixes used in the SI.

Multiple	Prefix	Symbol	Multiple	Prefix	Symbol
10^{-3}	milli	m	10^3	kilo	k
10^{-6}	micro	μ	10^6	mega	M
10^{-9}	nano	n	10^9	giga	G
10^{-12}	pico	P	10^{12}	tera	T
10^{-15}	femto	f	10^{15}	peta	P
10^{-18}	atto	a	10^{18}	exa	E
10^{-21}	zepto	z	10^{21}	zetta	Z
10^{-24}	yocto	V	10^{24}	yotta	Y

Example 10 μg is correct, while 10μg and 10 μ g are incorrect. 2.6 mol is right, but 2.6 mols is wrong.

Recommendations for describing measurements in SI units

Basic format

- **Express each measurement as a number separated from its units by a space.** If a prefix is required, no space is left between the prefix and the unit it refers to. Symbols for units are only written in their singular

> **Example**   n stands for nano and N for newtons.

> **Example**   1 982 963.192 309 kg (perhaps better expressed as 1.982 963192 309 Gg).

> **Units combined by multiplication** – an alternative convention is to use a raised dot · between multiplied units to avoid ambiguity. Thus, m ·s indicates metre-second, rather than millisecond (ms).

> **Examples**
> - 10 μm is preferred to 0.00001 m or 0.010mm.
> - 1 mm² = 10^{-6} m² (not one-thousandth of a square metre).
> - 1 dm³ (1 litre) is more properly expressed in SI as 1×10^{-3} m³.
> - The mass of a neutrino is 10^{-36} kg.
> - State as MW m⁻² rather than W mm⁻².

form and do not require full stops to show that they are abbreviated or that they are being multiplied together.

- **Give symbols and prefixes appropriate upper or lower case initial letters** as this may define their meaning. Upper case symbols are named after persons but when written out in full they are not given initial capital letters.

- **Show the decimal sign as a full point on the line.** Some metric countries continue to use the comma for this purpose and you may come across this in the literature: commas should not therefore be used to separate groups of thousands. In numbers that contain many significant figures, you should separate multiples of 10^3 by spaces rather than commas.

Compound expressions for derived units

- **Take care to separate symbols in compound expressions by a space** to avoid the potential for confusion with prefixes. Note, for example, that 200 m s (metre-seconds) is different from 200 ms (milliseconds).

- **Express compound units using negative powers** rather than a solidus (/): for example, write mol m⁻³ rather than mol/m³. The solidus is reserved for separating a descriptive label from its units (see p. 487).

- **Use parentheses to enclose expressions being raised to a power** if this avoids confusion: for example, a photosynthetic rate might be given in mol CO_2 (mol photons)⁻¹ s⁻¹.

- **Where there is a choice, select relevant (natural) combinations of derived and base units:** for example, you might choose units of Pa m⁻¹ to describe a hydrostatic pressure gradient rather than kg m⁻² s⁻², even though these units are equivalent and the measurements are numerically the same.

Use of prefixes

- **Use prefixes to denote multiples of 10^3** (Table 31.3) so that numbers are kept between 0.1 and 1000.

- **Treat a combination of a prefix and a symbol as a single symbol.** Thus, when a modified unit is raised to a power, this refers to the whole unit including the prefix.

- **Avoid the prefixes deci (d) for 10⁻¹ centi (c) for 10⁻², deca (da) for 10 and hecto (h) for 100** as they are not strictly SI.

- **Express very large or small numbers as a number between 1 and 10 multiplied by a power of 10** if they are outside the range of prefixes shown in Table 31.3.

- **Do not use prefixes in the middle of derived units:** they should be attached only to a unit in the numerator (the exception is in the unit for mass, kg).

> **KEY POINT**   For the foreseeable future, you will need to make conversions from other units to SI units, as much of the literature quotes data using imperial, c.g.s. or other systems. You will need to recognise these units and find the conversion factors required. Examples relevant to biology are given in Box 31.1. Table 31.4 provides values of some important physical constants in SI units.

Box 31.1 How to convert values between some redundant units and the SI

Quantity	SI unit/symbol	Old unit/symbol	Multiply number in old unit by this factor for equivalent in SI unit*	Multiply number in SI unit by this factor for equivalent in old unit*
Area	square metre/m^2	acre	$4.046\ 86 \times 10^3$	$0.247\ 105 \times 10^{-3}$
		hectare/ha	10×10^3	0.1×10^{-3}
		square foot/ft^2	$0.092\ 903$	10.7639
		square inch/in^2	645.16×10^{-9}	$1.550\ 00 \times 10^6$
		square yard/yd^2	0.836127	1.19599
Angle	radian/rad	degree/°	$17.453\ 2 \times 10^{-3}$	$57.295\ 8$
Energy	joule/J	erg	0.1×10^{-6}	10×10^6
		kilowatt hour/kWh	3.6×10^6	$0.277\ 778 \times 10^{-6}$
		calorie/cal	4.1868	0.2388
Length	metre/m	Ångstrom/Å	0.1×10^{-9}	10×10^9
		foot/ft	0.3048	3.28084
		inch/in	25.4×10^{-3}	$39.370\ 1$
		mile	1.60934×10^3	$0.621\ 373 \times 10^{-3}$
		yard/yd	0.9144	1.09361
Mass	kilogram/kg	ounce/oz	$28.349\ 5 \times 10^{-3}$	35.2740
		pound/lb	$0.453\ 592$	2.20462
		stone	$6.350\ 29$	$0.157\ 473$
		hundredweight/cwt	$50.802\ 4$	$19.684\ 1 \times 10^{-3}$
		ton (UK)	$1.016\ 05 \times 10^3$	$0.984\ 203 \times 10^{-3}$
Pressure	pascal/Pa	atmosphere/atm	$101\ 325$	$9.869\ 23 \times 10^{-6}$
		bar/b	$100\ 000$	10×10^{-6}
		millimetre of mercury/mmHg	133.322	$7.500\ 64 \times 10^{-3}$
		torr/Torr	133.322	$7.500\ 64 \times 10^{-3}$
Radioactivity	becquerel/Bq	curie/Ci	37×10^9	$27.027\ 0 \times 10^{-12}$
Temperature	kelvin/K	centigrade (Celsius) degree/°C	°C + 273.15	K – 273.15
		Fahrenheit degree/°F	(°F + 459.67) × 5/9	(K × 9/5) – 459.67
Volume	cubic metre/m^3	cubic foot/ft^3	0.0283168	$35.314\ 7$
		cubic inch/in^3	$16.387\ 1 \times 10^{-6}$	$61.023\ 6 \times 10^3$
		cubic yard/yd^3	$0.764\ 555$	$1.307\ 95$
		UK pint/pt	$0.568\ 261 \times 10^{-3}$	$1\ 759.75$
		US pint/liq pt	$0.473\ 176 \times 10^{-3}$	$2\ 113.38$
		UK gallon/gal	$4.546\ 09 \times 10^{-3}$	219.969
		US gallon/gal	$3.785\ 41 \times 10^{-3}$	264.172

*In the case of temperature measurements, use formulae shown.

Table 31.4 Some physical constants in SI terms.

Physical constant	Symbol	Value and units
Avogadro's constant	N_A	$6.022\ 174 \times 10^{23}$ mol^{-1}
Boltzmann's constant	k	$1.380\ 626 \times 10^{-23}$ JK^{-1}
Charge of electron	e	$1.602\ 192 \times 10^{-19}$ C
Gas constant	R	8.31443 J K^{-1} mol^{-1}
Faraday's constant	F	$9.648\ 675 \times 10^4$ C mol^{-1}
Molar volume of ideal gas at STP	V_0	$0.022\ 414$ m^3 mol^{-1}
Speed of light *in vacuo*	c	$2.997\ 924 \times 10^8$ ms^{-1}
Planck's constant	h	$6.626\ 205 \times 10^{-34}$ J s

Using units for volume – in this book, we use L and mL where you would normally find equipment calibrated in that way, but use SI units where this simplifies calculations. In formal scientific writing, constructions such as 1×10^{-6} m³ (= 1 mL) and 1 mm³ (= 1 µL) may be used.

Expressing enzyme activity – the derived SI unit is the katal (kat) which is the amount of enzyme that will transform 1 mol of substrate in 1 s (see Chapter 56).

Converting between concentration units – being able to express concentrations in different units (pp. 161–63) is important as this skill is frequently used when following instructions and interpreting data.

Definition

STP – standard temperature and pressure = 293.15K and 0.101 325 MPa.

Unit oddities – examples of nomenclature that do not fit easily into the SI:

- **Dates:** the UK convention is 10/03/11 for 10th March 2011, but in the USA this would mean 3rd October 2011. Avoid confusion by stating 3 Mar 2011 or making it clear which convention you are adopting in your Materials and Methods or table/figure footnotes/legend.
- **Ohm:** If the symbol Ω is not available in your word processor, then write 'ohm' (not capital O).
- **Computing terminology.** b is used for bit, and B for byte. Mega and giga can be ambiguous as they may refer loosely to 2^{20} or 2^{30} respectively for binary data. The International Electrotechnical Commission has attempted to rationalise this with a new set of prefixes (Rowlett, 2005).

Some implications of SI in bioscience

Volume

The SI unit of volume is the cubic metre, m³, which is rather large for practical purposes. The litre (L) and the millilitre (mL) are technically obsolete, but are widely used and glassware is still calibrated using them. Note also that the US spelling is liter. The litre is often given the symbol L, rather than l to avoid confusion with the number 1 and with the capital letter I in certain font styles.

Mass

The SI unit for mass is the kilogram (kg) rather than the gram (g): this is unusual because the base unit has a prefix applied.

Amount of substance

You should use the mole (mol, i.e. Avogadro's constant, see Table 31.4) to express very large numbers. The mole gives the number of atoms in the atomic mass, a convenient constant. Always specify the elementary unit referred to in other situations (e.g. mol photons m⁻² s⁻¹).

Concentration

The SI unit of concentration, mol m⁻³, is quite convenient for biological systems. It is equivalent to the non-SI term 'millimolar' (mM = mmol l⁻¹) while 'molar' (M = mol l⁻¹) becomes kmol m⁻³. Note that the symbol M in the SI is reserved for mega and hence should not be used for concentrations. If the solvent is not specified, then it is assumed to be water (see Chapter 26).

Time

In general, use the second (s) when reporting physical quantities having a time element (e.g. give photosynthetic rates in mol CO_2 m⁻² s⁻¹). Hours (h), days (d) and years should be used if seconds are clearly absurd (e.g. samples were taken over a 5-year period). Note, however, that you may have to convert these units to seconds when doing calculations.

Temperature

The SI unit is the kelvin, K. The degree Celsius scale has units of the same magnitude, °C, but starts at 273.15 K, the melting point of ice at STP. Temperature is similar to time in that the Celsius scale is in widespread use, but note that conversions to K may be required for calculations. Note also that you *must not* use the degree sign (°) with K and that this symbol must be in upper case to avoid confusion with k for kilo; however, you *should* retain the degree sign with °C to avoid confusion with the coulomb, C.

Light

While the first six base units in Table 31.1 have standards of high precision, the SI base unit for luminous intensity, the candela (cd) and the derived units lm and lx (Table 31.2), are defined in 'human' terms. They are, in fact, based on the spectral responses of the eyes of 52 American soldiers measured in 1923. Clearly, few organisms 'see' light in the same way as this sample of humans. Also, light sources differ in their spectral quality. For these reasons, it is better to use expressions based on energy or photon content (e.g. W m⁻² or mol photons m⁻² s⁻¹) in studies other than those on human vision. Ideally you should specify the photon wavelength spectrum involved (see Chapter 46).

Text reference and sources for further study

Meatyard, B. (ed.) (2009) *Biological Nomenclature: Standard Terms and Expressions Used in the Teaching of Biology*, 4th edn. Institute of Biology, London.

National Institute of Standards and Technology. *The NIST Reference on Constants, Units and Uncertainty: International System of Units (SI)*. Available: http://physics.nist.gov/cuu/Units/index.html/ Last accessed: 01/09/15.

Pennycuick, C.J. (1988) *Conversion Factors: SI Units and Many Others: Over 2100 Conversion Factors for*

Biologists and Mechanical Engineers Arranged in 21 Quick-Reference Tables. The University of Chicago Press, Chicago.

Rowlett, R. (2005) *How Many? A Dictionary of Units of Measurement.* Available: http://www.unc.edu/~rowlett/units Last accessed: 01/09/15.

Tapson, F. *A Dictionary of Units.* Available: http://www.cleavebooks.co.uk/dictunit/dictunit1.htm Last accessed: 01/09/15.

STUDY EXERCISES

31.1 Practise converting between units. Using Box 31.1 as a source, convert the following amounts into the units shown. Give your answers to three significant figures.

(a) 101 000 Pa into atmospheres

(b) One square yard into square millimetres

(c) One UK pint into millilitres

(d) 37 °C into kelvin

(e) 11 stone 6 pounds into kilograms.

31.2 Practise using prefixes appropriately. Simplify the following number/unit combinations using an appropriate prefix so that the number component lies between 0.1 and 1000.

(a) 10000 mm

(b) 0.015 ml

(c) 5×10^9 J

(d) 65 000 ms^{-1}

(e) 0 .000 000 0001 g

31.3 Check units in an equation. The Hagan–Poiseuille equation describes water flux in smooth cylindrical pipes assuming laminar flow. This equation can be expressed as:

$$J_v = \frac{r^2 \times \delta P}{8\eta \times \delta x}$$

where

r = the radius of the cylindrical pipe (m);

η ('eta') = the viscosity of the liquid (Pa s);

δP = the pressure difference across the ends of the pipe (Pa);

δx = the length of the pipe (m).

Verify, by putting relevant units in place of the variables in this equation and simplifying the resulting relationship, that appropriate units for J_v, the mean flow rate, are ms^{-1}.

Answers to these study exercises are available at www.pearsoned.co.uk/practicalskills.

Definitions

Some of the terms used when discussing scientific method may be applied rather loosely in other contexts. For the purposes of this chapter, the following definitions will be assumed.

Hypothesis – an explanation tested in a specific experiment or by a set of observations. Tends to involve a 'small-scale' idea.

Paradigm – theoretical framework so successful and well-confirmed that most research is carried out within its context and does not challenge it – even significant difficulties can be 'shelved' in favour of its retention.

(Scientific) Law – this concept can be summarised as an equation (law) that provides a succinct encapsulation of a system, often in the form of a mathematical relationship. The term is often used in the physical sciences (e.g. 'Beer's Law', p. 310).

Theory – a collection of hypotheses that covers a range of natural phenomena – a 'larger-scale' idea than a hypothesis. Note that a theory may be 'hypothetical', in the sense that it is a tentative explanation.

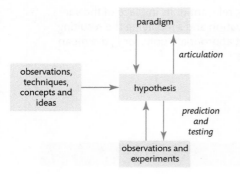

Fig. 32.1 A model of scientific method as used when testing hypotheses on a small scale. Hypotheses can arise as a result of various thought processes on the part of the scientist, and are consistent with the overlying paradigm. Each hypothesis is testable by experiment or observation, leading to its confirmation or rejection. Confirmed hypotheses act to strengthen the status of the paradigm, but rejected ones do not immediately result in the paradigm's replacement.

Science is a body of knowledge based on observation and experiment. Biological scientists attempt to explain life in terms of theories and hypotheses. They make predictions from these hypotheses and test them by experiment or further observations. The philosophy and sociology that underlie this process are complex topics (see, e.g., Chalmers, 2013) and any brief description must involve simplifications.

Figure 32.1 models the scientific process in which you are most likely to be involved – testing hypotheses on a small scale. These represent the sorts of explanation that can give rise to predictions which can be tested by an experiment or a series of observations. For example, you might put forward the hypothesis that the rate of K^+ efflux from a particular cell type is dependent on the intracellular concentration of calcium ions. This might then lead to a prediction that the application of a substance known to decrease the intracellular concentration of calcium ions would reduce K^+ efflux from the cells. An experiment could be set up to test this hypothesis and the results would either confirm or falsify the hypothesis.

If confirmed, a hypothesis is retained with greater confidence. If falsified, it is either rejected outright as false, or modified and retested. Alternatively, it might be decided that the experiment was not a valid test of the hypothesis (perhaps because it was later found that the applied substance could not penetrate the cell membrane to a presumed site of action).

Nearly all scientific research deals with the testing of small-scale hypotheses. These hypotheses operate within a theoretical framework that has proven to be successful (i.e. is confirmed by many experiments and is consistently predictive). This operating model or 'paradigm' is not changed readily, and even if a result appears that seems to challenge the conventional view, it would not be overturned immediately. The conflicting result would be 'shelved' until an explanation was found after further investigation. In the example used above, a relevant paradigm could be the notion that life processes are ultimately chemical in nature.

Although changes in paradigms are rare, they are important, and the scientists who recognise them become famous. For example, a 'paradigm shift' can be said to have occurred when Darwin's ideas about natural selection replaced special creation as an explanation for the origin of species. Generally, however, results from hypothesis-testing tend to support and develop ('articulate') the paradigm, enhancing its relevance and strengthening its status. Thus, research in the area of population genetics has developed and refined Darwinism.

Where do ideas for small-scale hypotheses come from? They arise from one or more thought processes on the part of a scientist:

- **analogy with other systems;**
- **recognition of a pattern;**
- **recognition of departure from a pattern;**
- **invention of new analytical methods;**
- **development of a mathematical model;**
- **intuition;**
- **imagination.**

Recently, it has been recognised that the process of science is not an entirely objective one. For instance, the choice of analogy which led to a new hypothesis might well be subjective, depending on past knowledge or understanding. Also, science is a social activity, where researchers put forward and defend viewpoints against those who hold an opposing view; where groups may work together towards a common goal; and where effort may depend on externally dictated financial opportunities and constraints. As with any other human activity, science is bound to involve an element of subjectivity.

No hypothesis can ever be rejected with certainty. Statistics might allow us to quantify as vanishingly small the probability of an erroneous conclusion, but we are nevertheless left in the position of never being 100% certain that we have rejected all relevant alternative hypotheses, nor 100% certain that our decision to reject some alternative hypotheses was correct. However, despite these problems, experimental science has yielded and continues to yield many important findings.

> **KEY POINT**   The fallibility of scientific 'facts' is essential to grasp. No explanation can ever be 100% certain as it is always possible for a new alternative hypothesis to be generated. Our understanding of biology changes all the time as new observations and methods force old hypotheses to be retested.

Quantitative hypotheses involve a mathematical description of the biological system. They can be formulated concisely by mathematical models. Formulating models is often useful because it forces deeper thought about mechanisms and encourages simplification of the system. A mathematical model:

- **is inherently testable through experiment;**
- **identifies areas where information is lacking or uncertain;**
- **encapsulates many observations;**
- **allows you to predict the behaviour of the system.**

Remember, however, that assumptions and simplifications required to create a model may result in it being unrealistic. Further, the results obtained from any model are only as good as the information put into it.

The terminology of experimentation

In many experiments, the aim is to provide evidence for causality. If x causes y, we expect, repeatedly, to find that a change in x results in a change in y. Hence, the ideal experiment of this kind involves measurement of y, the dependent (measured) variable, at one or more values of x, the independent variable, and subsequent demonstration of some relationship between them. Experiments therefore involve comparisons of the results of treatments – changes in the independent variable as applied to an experimental subject. The change is engineered by the experimenter under controlled conditions.

Subjects given the same treatment are known as replicates (they may be called plots). A block is a grouping of replicates or plots. The blocks are contained in a field, i.e. the whole area (or time) available for the experiment (Fig. 32.2). These terms originated from the statistical analysis of agricultural experiments, but they are now used across all areas of bioscience.

Deciding whether to accept or reject a hypothesis – this is sometimes clear-cut, as in some areas of genetics, where experiments can be set up to result in a binary outcome (Chapter 64). In many other cases, the existence of 'biological variation' means that statistical techniques need to be employed, with implicit acceptance of uncertainty in conclusions. (Chapters 73 and 74; Box 64.2).

Definition

Mathematical model – an algebraic summary of the relationship between the variables in a system.

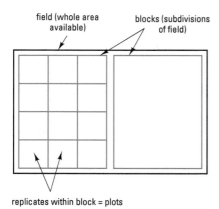

field (whole area available)   blocks (subdivisions of field)

replicates within block = plots

Fig. 32.2 Terminology and physical arrangement of elements in an experiment. Each block should contain the complete range of treatments (treatments may be replicated more than once in each block).

Why you need to control variables in experiments

Interpretation of experiments is seldom clear-cut because uncontrolled variables always change when treatments are given.

Confounding variables

These increase or decrease systematically as the independent variable increases or decreases. Their effects are known as systematic variation. This form of variation can be disentangled from that caused directly by treatments by incorporating appropriate controls in the experiment. A control is really just another treatment where a potentially confounding variable is adjusted so that its effects, if any, can be taken into account. The results from a control may therefore allow an alternative hypothesis to be rejected. There are often many potential controls for any experiment.

The consequence of systematic variation is that you can never be certain that the treatment, and the treatment alone, has caused an observed result. By careful design, you can, however, 'minimise the uncertainty' involved in your conclusion. Methods available include:

- **Ensuring, through experimental design, that the independent variable is the only major factor that changes in any treatment.**
- **Incorporating appropriate controls** to show that potential confounding variables have little or no effect.
- **Selecting experimental subjects randomly** to cancel out systematic variation arising from biased selection.
- **Matching or pairing individuals among treatments** so that differences in response owing to their initial status are eliminated.
- **Arranging subjects and treatments randomly** so that responses to systematic differences in conditions do not influence the results.
- **Ensuring that experimental conditions are uniform** so that responses to systematic differences in conditions are minimised. When attempting this, beware 'edge effects' where subjects on the periphery of the layout receive substantially different conditions from those in the centre.

Nuisance variables

These are uncontrolled variables which cause differences in the value of y independently of the value of x, resulting in random variation. Experimental biology is characterised by the high number of nuisance variables that are present and their relatively great influence on results: biological data tend to have large errors. To reduce and assess the consequences of nuisance variables:

- incorporate replicates to allow random variation to be quantified;
- choose subjects that are as similar as possible;
- control random fluctuations in environmental conditions.

Constraints on experimental design

Box 32.1 outlines the important stages in designing an experiment. At an early stage, you should find out how resources may constrain the design. For example, limits may be set by availability of subjects, cost of treatment, availability of a chemical or bench space. Logistics may be a factor (e.g. time taken to record or analyse data).

Example Suppose you wish to investigate the effect of a metal ion on the growth of a culture. If you add the metal as a salt to the culture and then measure the growth, you will immediately introduce at least two confounding variables, compared with a control that has no salt added. Firstly, you will introduce an anion that may also affect growth in its own right, or in combination with the metal ion; secondly, you will alter the osmotic potential of the medium (see p. 167). Both of these effects could be tested using appropriate controls.

Reducing edge effects – one approach is to incorporate a 'buffer zone' of untreated subjects around the experiment proper.

Evaluating design constraints – a good way to do this is by processing an individual subject through the experimental procedures – this type of 'preliminary run' can help to identify potential difficulties.

| **Box 32.1** | **Checklist for designing and performing an experiment** |

1. Preliminaries

   (a) Read background material and decide on a subject area to investigate.

   (b) Formulate a simple hypothesis to test. It is preferable to have a clear answer to one question than to be uncertain about several questions.

   (c) Decide which dependent variable you are going to measure and how. Is it relevant to the problem? Can you measure it accurately, precisely and without bias?

   (d) Think about and plan the statistical analysis of your results. Will this affect your design?

2. Designing

   (a) Find out the limitations on your resources.

   (b) Choose treatments and conditions that alter the minimum of confounding variables.

   (c) Incorporate as many effective controls as possible.

   (d) Keep the number of replicates as high as is feasible.

   (e) Ensure that the same number of replicates is present in each treatment.

   (f) Use effective randomisation and blocking arrangements.

3. Planning

   (a) List all the materials you will need. Order any chemicals and make up solutions; grow, collect or breed the experimental subjects you require; check equipment is available.

   (b) Organise space and/or time in which to do the experiment.

   (c) Account for the time taken to apply treatments and record results. Make out a timesheet if things will be hectic.

4. Carrying out the experiment

   (a) Record the results and make careful notes of everything you do (see p. 208). Make additional observations to those planned if interesting things happen.

   (b) Repeat the experiment if time and resources allow.

5. Analysing

   (a) Graph data as soon as possible (during the experiment if you can). This will allow you to visualise what has happened and make adjustments to the design, if this seems appropriate (e.g. timing of measurements).

   (b) Carry out the planned statistical analysis.

   (c) Jot down conclusions and new hypotheses arising from the experiment.

Your equipment or facilities may affect design because you cannot regulate conditions as well as you might desire. For example, you may be unable to ensure that temperature and lighting are equal over an experiment laid out in a glasshouse, or you may have to accept a great deal of initial variability in your subjects. This problem is especially acute for experiments carried out under field conditions, or for material collected from the wild.

Using replicates

Replicate results show how variable the response is within treatments. They allow you to compare the differences among treatments in the context of the variability within treatments – you can do this via statistical tests such as analysis of variance (Chapter 74). Larger sample sizes tend to increase the precision of estimates of parameters and increase the chances of showing a significant difference between treatments if one exists. For statistical reasons (weighting, ease of calculation, fitting data to certain tests) it is often best to keep the number of replicates even. Remember that the degree of independence of replicates is important: subsamples cannot act as replicate samples – they tell you about variability in the measurement method but not in the quantity being measured.

Deciding the number of replicates in each treatment – try to maximise the number of replicates in each treatment within the constraints of time and resources available, using the same number of replicates for each treatment.

Pseudo-replication – this term applies to a situation where replicates are used, but they are not fully independent of each other. An example might occur when testing the effect of a chemical on liver function, if several samples from one animal's organ were treated as independent replicates, rather than a sample from the livers of each of several animals. (In practice, an approach which examined both intra-animal and inter-animal variability might be the best way to proceed, using several samples from each of several animals, then analysing the results using ANOVA (Chapter 74).)

If the total number of replicates available for an experiment is limited by resources, you may need to compromise between the number of treatments and the number of replicates per treatment. Statistics can help here, for it is possible to work out the minimum number of replicates you would need to show a certain difference between pairs of means (say 10%) at a specified level of significance (say, $P = 0.05$). For this, you need to obtain a prior estimate of variability within treatments, perhaps from an initial trial run (see Sokal and Rohlf, 2012).

Randomisation of treatments

The two aspects of randomisation you must consider are:

- **positioning of treatments within experimental blocks;**
- **allocation of treatments to the experimental subjects.**

For relatively simple experiments, you can adopt a completely randomised design; here, the position and treatment assigned to any subject are defined randomly. You can draw lots, use a random number generator on a calculator, or use the random number tables which can be found in most books of statistical tables (see Box 32.2).

Box 32.2 How to use random number tables to assign subjects to positions and treatments

This is one method of many that could be used. It requires two sets of *n* random numbers – where *n* is the total number of subjects used.

1. **Number the subjects in any arbitrary order** but in such a way that you know which is which (i.e. mark or tag them).

2. **Decide how treatments will be assigned,** e.g. first five subjects selected treatment A; second five – treatment B, etc.

3. **Use the first set of random numbers in the sequence obtained to identify subjects and allocate them to treatment groups** in order of selection as decided in (2).

4. **Map the positions for subjects in the block or field. Assign numbers to these positions using the second set of random numbers,** working through the positions in some arbitrary order, e.g. top left to bottom right.

5. **Match the original numbers given to subjects with the position numbers.**

To obtain a sequence of random numbers:

1. Decide on the range of random numbers you need.

2. Decide how you wish to sample the random number tables (e.g. row-by-row and top to bottom) and your starting point.

3. Moving in the selected manner, read the sequence of numbers until you come to a group that fits your needs (e.g. in the sequence 978186, 18 represents a number between 1 and 20). Write this down and continue sampling until you get a new number. If a number is repeated, ignore it. Small numbers need to have the appropriate number of zeros preceding (e.g. 5 = 05 for a range in the tens, 21 = 021 for a range in the hundreds).

4. When you come to the last number required, you do not need to sample any more: simply write it down.

Example: You find the following random number sequence in a table and wish to select numbers between 1 and 10 from it.

9059146823	4862925166	1063260345
1277423810	9948040676	6430247598
8357945137	2490145183	5946242208
6588812379	2325701558	3260726568

Working left to right and top to bottom, the order of numbers found is 5, 10, 3, 9, 4, 6, 2, 1, 8, 7 as indicated by coloured type. If the table is sampled by working row-by-row right to left from bottom to top, the order is 6, 10, 7, 2, 9, 3, 4, 8, 1, 6, 1, 9.

> **Example**   If you knew that soil type varied in a graded fashion across a field, you might arrange blocks to be long thin rectangles at right angles to the gradient to ensure conditions within the block were as even as possible.

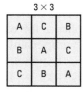

Fig. 32.3 Examples of Latin square arrangements for 3 and 4 treatments. Letters indicate treatments; the number of possible arrangements for each size of square increases greatly as the size increases.

> **Example**   A Latin square format is used for agar bioassays of antibiotics and vitamins (p. 236). Replicates are arranged so that gradients across the plate are cancelled out.

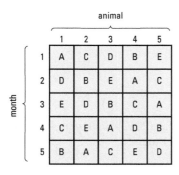

Fig. 32.4 Example of how to use a Latin square design to arrange sequential treatments. The experimenter wishes to test the effect of drugs A–E on weight gain, but only has five animals available. Each animal is fed on a control diet for the first 3 weeks of each month, then on a control diet plus drug for the last week. Weights are taken at start and finish of each treatment. Each animal receives all treatments.

A completely randomised layout has the advantage of simplicity but cannot show how confounding variables alter in space or time. This information can be obtained if you use a blocked design in which the degree of randomisation is restricted. Here, the experimental space or time is divided into blocks, each of which accommodates the complete set of treatments (Fig. 32.2). When analysed appropriately, the results for the blocks can be compared to test for differences in the confounding variables across the field and these effects can be separated out from the effects of the treatments. The size and shape (or timing) of the block you choose is important: besides being able to accommodate the number of replicates desired, the suspected confounding variable should be relatively uniform within the block.

A Latin square is a method of placing treatments so that they appear in a balanced fashion within a square block or field. Treatments appear once in each column and row (see Fig. 32.3), so the effects of confounding variables can be 'cancelled out' in two directions at right angles to each other. This is effective if there is a smooth gradient in some confounding variable over the field. It is less useful if the variable has a patchy distribution, where a randomised block design might be better.

Latin square designs are useful in serial experiments where different treatments are given to the same subjects in a sequence (e.g. Fig. 32.4). A disadvantage of Latin squares is the fact that the number of plots is equal to the number of replicates, so increases in the number of replicates can only be made by the use of further Latin squares.

Pairing and matching subjects

The paired comparison is a special case of blocking used to reduce systematic variation when there are two treatments. Examples of its use are:

- **'Before and after' comparison.** Here, the pairing removes variability arising from the initial state of the subjects, e.g. weight gain of mice on a diet, where the weight gain may depend on the initial weight.
- **Application of a treatment and control to parts of the same subject or to closely related subjects.** This allows comparison without complications arising from different origin of subjects, e.g. drug or placebo given to sibling rats, virus-containing or control solution swabbed on left or right halves of a leaf.
- **Application of treatment and control under shared conditions.** This allows comparison without complications arising from different environments of subjects, e.g. rats in a cage, plants in a pot.

Matched samples represent a restriction on randomisation where you make a balanced selection of subjects for treatments on the basis of some attribute or attributes that may influence results, e.g. age, sex, prior history. The effect of matching should be to 'cancel out' the unwanted source(s) of variation. Disadvantages include the subjective element in choice of character(s) to be balanced, inexact matching of quantitative characteristics, the time matching takes and possible wastage of unmatched subjects.

When analysed statistically, both paired comparisons and matched samples can show up differences between treatments that might otherwise be rejected on the basis of a fully randomised design, but note that the statistical analysis may be different.

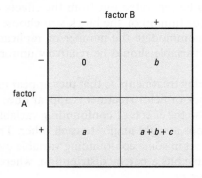

Fig. 32.5 Design of a simple multifactorial experiment. Factors A and B have effects a and b when applied alone. When both are applied together, the effect is denoted by $a + b + c$.

- If $c = 0$, there is no interaction (e.g. $2 + 2 + c = 4$).
- If c is positive, there is a positive interaction (synergism) between A and B (e.g. $2 + 2 + c = 5$).
- If c is negative, there is a negative interaction (antagonism) between A and B (e.g. $2 + 2 + c = 3$).

Reporting results – it is good practice to report how many times your experiments were repeated (in Materials and Methods); in the Results section, you should add a statement saying that the illustrated experiment is representative, or one explaining any differences between results obtained.

Multifactorial experiments

The simplest experiments are those in which one treatment (factor) is applied at a time to the subjects. This approach is likely to give clear-cut answers, but it could be criticised for lacking realism. In particular, it cannot take account of interactions among two or more conditions that are likely to occur in real life. A multifactorial experiment (Fig. 32.5) is an attempt to do this; the interactions among treatments can be analysed by specialised statistics.

Multifactorial experiments are economical on resources because of 'hidden replication'. This arises when two or more treatments are given to a subject because the result acts statistically as a replicate for each treatment. Choice of relevant treatments to combine is important in multifactorial experiments; for instance, an interaction may be present at certain concentrations of a chemical but not at others (perhaps because the response is saturated). It is also important that the measurement scale for the response is consistent, otherwise spurious interactions may occur. Beware when planning a multifactorial experiment that the numbers of replicates do not get out of hand: you may have to restrict the treatments to 'plus' or 'minus' the factor of interest (as in Fig. 32.5).

Repetition of experiments

Even if your experiment is well designed and analysed, you should understand that only limited conclusions can be made. Firstly, what you can say is valid for a particular place and time, with a particular investigator, experimental subject and method of applying treatments. Secondly, if your results were significant at the 5% level of probability (p. 526), there is still an approximately one-in-twenty chance that the results did arise by chance. To guard against these possibilities, it is important that experiments are repeated. Ideally, this would be done by an independent scientist with independent materials. However, it makes sense to repeat work yourself so that you can have full confidence in your conclusions. Many scientists recommend that experiments are done three times in total, but this may not be possible in undergraduate practical classes or project work.

Text references

Chalmers, A.F. (2013) *What is This Thing Called Science?* 4th edn. Open University Press, Buckingham.

Sokal, R.R. and Rohlf, F.J. (2012) *Biometry,* 4th edn. W.H. Freeman, San Francisco.

Sources for further study

Maber, J. (1999) *Data Analysis for Biomolecular Sciences.* Longman, Harlow.

Quinn, G.P. and Keough, M.J. (2002) *Experimental Design and Data Analysis for Biologists.* Cambridge University Press, Cambridge.

Ruxton, G. and Colegrave, N. (2009) *Experimental Design for the Life Sciences.* Oxford University Press, New York.

STUDY EXERCISES

32.1 Consider the application of a Latin square design. Treatments W, X, Y and Z are to be applied to potted plants in a glasshouse where the researcher suspects there may be a slight gradation in temperature and light over an oblong bench that can hold a total of 48 pots. You decide to use an experimental design consisting of three 4 × 4 blocks of plants, each arranged in a different Latin square design (see below). Assign treatments to the locations in the diagram below. Explain why this design will help to eliminate the effects of the confounding variables.

 Suspected temperature gradient

Suspected light gradient

Layout for glasshouse experiment.

32.2 Generate random numbers. Produce a list of 20 random whole numbers between 1 and 5 using a spreadsheet. If using MS *Excel*, investigate the RAND() and INT functions. The RAND() function produces a random number between 0 and 1, so you will need to multiply by a constant factor to scale your final output appropriately. Copy your test formula(e) to several cells to test empirically whether it works.

32.3 Investigate possible interactions. Treatments A, B and C involve tests of three different nutrients on proliferation of cells in a culture of carrot cells. Analyse the results in the table below to determine whether any interactions have occurred between the four possible combinations of treatments. No complex statistical analysis is required, simply a calculation of mean experimental effect (treatment minus controls) in each combined treatment and a comparison with the data for the relevant treatments on their own (in other words, assume that observed differences reflect true underlying differences). Classify the results as 'no interaction', 'antagonism' or 'synergism'.

Results of cell proliferation experiment

Treatment	Replicate (growth of callus in g)				
	1	2	3	4	5
Control	4.8	5.2	5.2	4.8	5.0
A	6.5	7.3	7.0	7.1	7.1
B	7.7	8.3	8.5	7.5	8.0
C	10.7	10.0	9.9	9.8	9.6
A + B	6.6	7.2	6.8	7.2	7.2
A + C	11.9	11.7	12.0	12.0	12.4
B + C	17.3	16.8	17.1	16.6	17.2
A + B + C	12.8	12.9	13.1	13.3	12.9

Answers to these study exercises are available at www.pearsoned.co.uk/practicalskills

33 Making notes of practical work

When carrying out advanced lab work or research projects, you will need to master the important skill of managing data and observations and learn how to keep a reliable record of your studies. This is important for the following reasons:

- **An accurate record helps when using information later,** perhaps for exam purposes or when writing a report.
- **It allows you to practise important skills** such as scientific writing, drawing diagrams, preparing graphs and tables and interpreting results.
- **Analysing and writing up your data as you go along prevents** a backlog at the end of your study time.
- **You can show your work to a future employer** to prove you have developed the skills necessary for writing up properly; in industry, this is vital so that others in your team can interpret and develop your work.

> **KEY POINT** A good set of lab notes should:
> - outline the purpose of your experiment or observation;
> - set down all the information required to describe your materials and methods;
> - record all relevant information about your results or observations and provide a visual representation of the data;
> - note your immediate conclusions and suggestions for further experiments.

Collecting and recording primary data

Individual observations (e.g. laboratory temperature) can be recorded in text form, but tables are the most convenient way to collect large amounts of information. When preparing a table for data collection, you should:

1. **Use a concise title or a numbered code for cross-referencing.**
2. **Decide on the number of variables to be measured and their relationship with each other and lay out the table appropriately:**
   (a) The first column of your table should show values of the independent (controlled) variable, with subsequent columns for the individual (measured) values for each replicate or sample (Fig. 33.1(a)).
   (b) If several variables are measured for the same organism or sample, each should be given a row (Fig. 33.1(b)).
   (c) In time-course studies, put the replicates as columns grouped according to treatment, with the rows relating to different times (Fig 33.1(c)).
3. **Make sure the arrangement reflects the order in which the values will be collected.** Your table should be designed to make the recording process as straightforward as possible, to minimise the possibility of mistakes. For final presentation, a different arrangement may be best (Chapter 71).
4. **Consider whether additional columns are required for subsequent calculations.** Create a separate column for each mathematical manipulation, so the step-by-step calculations are clearly visible. Use software spreadsheets (Chapter 13) if you are manipulating lots of data.

Fig. 33.1 Example layouts of tables for recording data. (a) measurements on replicates at different levels of a controlled variable ('rep' being shorthand for replicate); (b) responses by two organisms to a range of treatments; (c) time course of responses in control and treated specimens. Note the grid-like nature of these tables used for ease of data entry compared with the 'horizontal lines only' style used for formal presentation of tables.

5. **Take sufficient time to record quantitative data accurately.**

6. **Record numerical data to an appropriate number of significant figures,** reflecting the accuracy and precision of your measurement (p. 508). Do not round off data values, as this might affect the subsequent analysis of your data.

7. **Prepare blank recording tables if your experiments or observations will be repeated.**

8. **Explain any unusual data values or observations in a footnote.** Do not rely on your memory.

Recording details of practical and project work

The recommended system is one where you make a dual record.

Primary record

The primary record is made at the bench or in the field. In this, you must concentrate on the detail of materials, methods and results. Include information that would not be used elsewhere, but that might prove useful in error tracing: for example, if you note how a solution was made up (exact volumes and weights used rather than concentration alone), this could reveal whether a miscalculation had been the cause of a rogue result. Note the origin, type and state of the chemicals and organism(s) used. Make rough diagrams to show the arrangement of replicates, equipment, etc.

The basic order of the primary record should mirror that of a research report (see p. 121), including: the title and date; brief introduction; comprehensive materials and methods; the data and short conclusions.

Secondary record

You should make a secondary record which is better organised and presented. This will be used when discussing results with your supervisor, and when writing up a report or thesis, and may be part of your course assessment. Although these notes should retain the essential features of the primary record, they should be more concise and the emphasis should move towards analysis of the experiment. Outline the aims more carefully at the start and link the experiment to others in a series (e.g. 'Following the results of Expt D24, I decided to test whether. . .'). You should present data in an easily digested form, e.g. as tables of means or as summary graphs. Use appropriate statistical tests (Chapter 74) to support your analysis of the results. Secondary records are often kept in digital format, e.g. as a spreadsheet or word-processed document.

The dual method of recording outlined above deals with the inevitable lack of organisation of notes taken at the bench or in the field; these often have to be made rapidly, in awkward positions and in a generally complex environment. Writing a second, neater version forces you to consider again details that might have been overlooked in the primary record and provides a duplicate in case of loss or damage.

If you find it difficult to decide on the amount of detail required in Materials and Methods, the basic ground rule is to record enough information to allow a reasonably competent scientist to repeat your work exactly. You must tread a line between the extremes of pedantic, irrelevant detail and the omission of information essential for a proper interpretation of the data – better perhaps to err on the side of extra detail to begin with. An experienced worker can help you decide which subtle shifts in technique are important (e.g. batch numbers for an important chemical, or when a new stock solution is made up and used). Many important scientific advances have been made because of careful observation and record taking or because coincident data were recorded that did not seem of immediate value.

When creating a primary record, take care not to lose any of the information content of the data: for instance, if you only write down means and not individual values, this may affect your ability to carry out subsequent statistical analyses.

There are numerous ways to reduce the labour of keeping a record. Do not repeat Materials and Methods for a series of similar experiments; use devices such as 'method as for Expt B4'. A photocopy might suffice if the method is derived from a text or article (check with supervisor). To save time, make up and copy a checklist in which details such as chemical batch numbers can be entered.

Using communal records

If working with a research team, you may need to use their communal databases. These avoid duplication of effort and ensure uniformity in techniques. You will be expected to use the databases carefully and to contribute to them properly. They might include:

- **a shared file of common techniques** (e.g. how to make up media or solutions);
- **a set of simplified step-by-step instructions for use of equipment** – manuals are often complex and poorly written, and it may help to redraft them, incorporating any differences in procedure adopted by the group;
- **an alphabetical list of suppliers of equipment and consumables;**
- **a list of chemicals required by the group** and where they are stored;
- **the risk-assessment sheets for dangerous procedures** (p. 142);
- **the record book detailing the use of radioisotopes and their disposal.**

Using photography and imaging to record information

These techniques are primarily for:

- documentation and validation of observations for project work (e.g. of agarose gels used for separation of nucleic acid molecules, p. 459);

Analysing data as soon as possible – always analyse and think about data immediately after collecting them as this may influence your subsequent activities.

- A graphical indication of what has happened can be particularly valuable.
- Carry out statistical analyses before moving on to the next experiment because apparent differences among treatments may not turn out to be statistically significant when tested.
- Write down any conclusions you make while analysing your data: sometimes those that seem obvious at the time of doing the work are forgotten when the time comes to write up a report or thesis.
- Note ideas for further studies as they occur to you – these may prove valuable later. Even if your experiment appears to be a failure, suggestions as to the likely causes might prove useful.

SAFETY NOTE Maintaining and consulting communal lab records – these activities may form a part of the safety requirements for working in a laboratory.

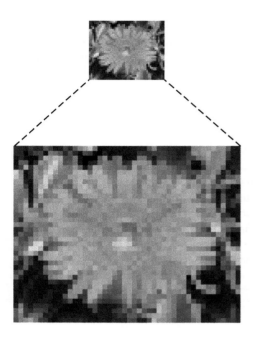

Fig. 33.2 Pixel structure in a digital image.

Understanding the light sensitivity of photographic systems – both digital and film media are rated in terms of their 'speed', measured in either ISO (=ASA) or DIN units. A good range for general purposes is 100 – 160 ISO (21–23 DIN).

Tips for better photography –

- always use a tripod under low-light conditions;
- use electronic flash to provide uniformity of colour balance;
- shadowless lighting of small objects can be obtained using a ring-flash system.

- producing 'hard copy' of microscope images (micrographs);
- preparation of visual material for talks and posters.

Photographs can be taken using two main systems: the digital camera and the film camera. While the basic operation of both types is similar, they differ in their light capturing and image recording systems.

Digital photography and imaging

This is the production of an electronically stored image, produced using a digital camera (still or video/movie). The image is held as a series of discrete picture elements called pixels (Fig. 33.2); the greater the number of pixels per unit surface area, the better the potential resolution of the image and the greater amount of detail captured.

Film photography

Black and white or colour print photography involves capturing images on a film of plastic coated with light-sensitive silver halide salts. When exposed to light and subsequently chemically treated (developed), silver grains are produced to form a negative image, later converted to a positive image during printing. Transparency (slide or positive) films are based upon colour dyes and are effectively grainless. The choice of film is a compromise between speed and resolution: use slower film for fine detail. Faster film is more suitable for moving subjects or when light levels are low, but there is a loss of contrast and detail.

Comparing digital and film systems

Despite advances in digital image technology and reductions in the price of digital cameras and associated memory systems, even the best digital images are of somewhat lower quality (resolution) than high-quality film images. However, digitised images have the advantages of speed and ease of production, since a developing stage is avoided, and are more readily manipulated. Because of the preview facility, they are also useful when it is important to confirm that a successful image has been taken, e.g. when an event is unrepeatable. On the other hand, images stored in magnetic form are not secure from computer failure, or damage from strong magnetic fields (see Box 11.1).

Camera and lens types

For both digital and film systems the primary difference is whether you preview the image through the lens (TTL), often referred to as single lens reflex (SLR), or whether a separate viewfinder is used. Digital cameras also incorporate a monitor which allows you to preview and review images. Lenses are either fixed, i.e. have a set focal length, or zoom, allowing a range of focal lengths, and hence magnification. Some digital systems have a digital zoom feature, often with loss of picture quality with increasing magnification. High-quality images are usually taken with an SLR camera; such systems have better quality lenses and allow greater control over the final image.

Measuring light for photography

The quantity of light can be measured by a photographic light meter, which may be external or built-in. The more light there is, the smaller the lens aperture you can use (larger 'f' number), and the greater will be the depth of field (= depth of focus). By using the camera on a tripod, you can use slow shutter speeds and allow larger 'f ' numbers to be used to maximise depth of focus. The use of an electronic flash system to provide some or all of the lighting makes this even easier, since the effective shutter speed with electronic flash is extremely short (1 ms). Your choice of lighting arrangement will affect the quality of the picture: shadowless lighting is appropriate in some situations, but often shadows help to give three-dimensional form to subjects. In general, the use of more than one light source is advisable to prevent hard shadows.

> **KEY POINT** To use an SLR camera properly, you need to understand the relationship between aperture (f number), shutter speed and depth of field. For moving objects, give fast shutter speed priority by opening the lens aperture (small f number). Where depth of field is required, close the lens aperture (high f number); this will result in slower shutter speeds, so take care to avoid camera shake. Bracket your exposures (make extra exposures at plus and minus 1 f number at same shutter speed) to ensure good results.

Image processing

Digital images are recorded as files on various forms of memory cards and are transferred to a computer for manipulation. For critical work, the rule is to capture images using the largest possible file size, as reductions (e.g. compression of file size) of digital information are possible, but you can never add information.

Digital image files are stored on disk in various formats (e.g. .bmp, .jpg) and can be manipulated using programs such as Adobe Photoshop or Paint Shop Pro. Aspects of image quality such as contrast and colour balance can be adjusted and images can be cropped and resized to suit your application (e.g. small file sizes are recommended for websites). Digital images are valuable for use with word processors and web-authoring software and can be transferred readily between such programs.

Film images require chemical development and are susceptible to irreversible damage by faulty development. The resulting image can only be modified to a limited extent during subsequent printing; this is one of the key reasons for selecting digital image systems.

Photomicrography

This requires the use of a camera system mounted on a microscope. The important operations for successful photomicrography are outlined below:

1. Carefully prepare the object: ensure thorough cleanliness of any slides or coverslips used.

2. Choose the correct film type and any filters required for alteration of colour balance, depending upon the type of light source available.

3. Decide on the magnification to be used: make sure that you know how this relates to the magnification of the negative/transparency. Always include a photograph of a stage micrometer (p. 185) so the final magnification can be calculated and shown on the photograph.

Backing up digital image files – since no physical copy exists, you must remember to download and back up digital image files to appropriate media (hard drive, network drive, CD or USB memory stick) as soon as possible.

Optimising print quality– the quality of your print will depend upon the quality of your image, especially when producing colour prints from film.

Using scales with images– these are vital to the interpretation of biological images; always include information on the magnification used, as well as a scale bar.

Fig. 33.3 Use of white areas for labelling dark prints, illustrated by an SEM micrograph of a stellate hair on a leaf surface.

4. Carefully focus the object onto the film plane: there will be a system-specific method for doing this (consult the manual, as this is a critical aspect).

5. Bracket extra exposures above and below the automatic settings, especially for colour photography.

Printing images

The choice of print type will depend on its purpose. Glossy finish prints generally appear sharper than matt or other finishes. However, if preparing prints for display on a poster, matt/velvet finishes are often preferable, as there is less glare for the viewer. Printing from computers using ink-jet and similar printers requires care, since colours viewed on monitors do not always reproduce on paper as expected – experiment with settings and paper quality to obtain the appropriate results.

Adding scales and labels to images

Having acquired a suitable print, it is often necessary to add information to it:

- When digital images are copied into word processor or presentation software, additional information such as lettering, scales and labels can be added by overlaying a 'text box', or equivalent. You may need to make the background transparent, e.g. using the 'no fill' option in *PowerPoint*. Lettering can be added to printed media using transfer letters (e.g. Letraset).

- Choose a simple font type that will not distract and a size that is legible but not too dominant, e.g. 12 point in final reproduction. Use the same font type/size for related sets of images.

- Choose an appropriate part of the image for the lettering, i.e. a black area for a white letter. If the background is too variable, add a label on which to place the letters (see Fig. 33.3).

Sources for further study

Anon. *Suggestions for Keeping Laboratory Notebooks,* Available: http://otl.stanford.edu/inventors/resources/labnotebooks.html Last accessed: 01/09/15.
[A Stanford University website that looks at the laboratory notebook from the patenting perspective.]

Daney, C. *Photography and Digital Imaging Resources – a Collection of Links.* Available: http://

www.mbay.net/~cgd/photo/photolinks.htm Last accessed: 01/09/15.

Kanare, H.M. (2006) *Writing the Laboratory Notebook.* American Chemical Society, Washington, DC.

Pechenik, J.A.A. (2015) *Short Guide to Writing About Biology,* 9th edn. Longman, Harlow.

STUDY EXERCISES

33.1 **Design a primary data collection sheet for a lab experiment.** Imagine you wish to count and identify the different types of microbes present in water and able to grow on nutrient agar at 25 °C. What might you need to record on your data sheet?

33.2 **Design a secondary record table for the collection and analysis of a set of count data** for the number of bacterial colonies developed in each of ten replicates for each of five different treatments. Assume that you need to calculate means, variances, etc., and then compare the results for each treatment.

Answers to these study exercises are available at www.pearsoned.co.uk/practicalskills.

Research projects are an important component of the final-year syllabus for most degree programmes in the biosciences, while shorter projects may also be carried out during courses in earlier years. Project work can be extremely rewarding, although it does present a number of challenges. The assessment of your project is likely to contribute significantly to your degree grade, so all aspects of this work should be approached in a thorough manner.

Deciding on a topic to study

Assuming you have a choice, this important decision should be researched carefully. Make appointments to visit possible supervisors and ask them for advice on topics that you find interesting. Use library texts and research papers to obtain further background information. Perhaps the most important criterion is whether the topic will sustain your interest over the whole period of the project. Other things to look for include:

- **Opportunities to learn new skills.** Ideally, you should attempt to gain experience and skills that you might be able to 'sell' to a potential employer.
- **Ease of obtaining valid results.** An ideal project provides a means to obtain 'guaranteed' data for your report, but also the chance to extend knowledge by doing genuinely novel research.
- **Assistance.** What help will be available to you during the project? A busy lab with many research students might provide a supportive environment should your potential supervisor be too busy to meet you often; on the other hand, a smaller lab may provide the opportunity for more personal interaction with your supervisor.
- **Impact.** Your project may result in publishable data: discuss this with your prospective supervisor.

Writing a proposal

You may be expected to submit a proposal for your project, especially where it is expected that you should define the precise area of study yourself. The structure for the proposal will probably be provided in the course handbook, or as a form you must complete. Box 34.1 outlines some common features of project proposals and what you should do to complete each part. The aim of formulating a proposal is to ensure that:

- **the project has an appropriate theoretical background;**
- **the objectives you have set are achievable;**
- **the methods chosen are appropriate;**
- **safety and ethical issues have been considered;**
- **sufficient resources are available to complete the work,** such as matching students to available labs and supervisors;
- **you have set yourself a timetable with milestones** on which your progress can be judged;
- **you obtain feedback and suggestions about your plans.**

Obtaining ethical approval – if any aspect of your project involves work with human or animal subjects, then you must obtain the necessary ethical clearance before you begin; consult your department's ethical committee for details.

Using the Internet as an information source for project work – since many university departments and research groups have homepages on the web, searches using relevant key words may indicate where research in your area is currently being carried out. Academics usually respond positively to emailed questions about their area of expertise.

Asking around – one of the best sources of information about supervisors, laboratories and projects is past students. Some of the postgraduates in your department may be products of your own system and they could provide an alternative source of advice.

Box 34.1 How to write a project proposal

Below are listed some common elements of project proposals, with guidance regarding approach and content. Not every part may be included, and the titles for sections may differ – always follow the exact format specified for your course.

Contact details and suggested supervisor(s) – you may be asked whether you have discussed the project with the named supervisor(s).

Proposed title – this should be relatively short, following the style used in research papers. It may change for the final report.

Aims – a general statement of what you plan to achieve.

Objectives – a listing of specific outcomes you expect to fulfil. Typically you will have several specific objectives that all fit within the overall aim.

Brief description of the subject – this section might have an alternative title like Summary, Background, Review of Subject Area or Statement of the Problem to be Addressed. It should contain a brief synopsis of past work, a summary of current ideas and, if relevant, the hypothesis to be tested. In some cases it will consist of a mini-review of the subject area and act as a template for the introduction to your report.

Preliminary bibliography – this will ensure (a) that you have read and understood relevant papers and (b) that you understand how to cite them properly (Chapter 9).

Research methods – this describes how you plan to carry out the investigation. Be quite specific, so the committee can arrive at a valid judgement. Quote references for methods and techniques, where available. Consider use of appropriate controls, experimental design and sampling procedures (Chapters 32 and 33). Check that the results obtained will tell you what you need to know to achieve your objectives and to test your hypothesis.

Resources required – the samples, chemicals, instruments, etc. required to carry out your investigation. Listing these will require quite detailed consideration of the experimental design or field area, as relevant, and a thorough knowledge of methods, such as amounts or volumes of reagents used (Chapters 24–27).

Timetable/plan – a realistic breakdown of the work required, with milestones leading to completion of the project (see Chapter 2 for advice on time management). Always allow time for the unexpected.

Statement or declaration in relation to safety and ethical rules – this is to confirm that you have read and understood relevant issues, have completed relevant forms and processes, and are in a position to proceed with the research. You may be required to attach copies of COSHH/safety forms (Chapter 23) or ethics committee permission (Chapter 22).

Liaising with your supervisor(s) – this is essential if your work is to proceed efficiently. Specific meetings may be timetabled, e.g. to discuss a term's progress, review your work plan or consider a draft introduction. Most supervisors also have an 'open-door' policy, allowing you to air current problems or discuss your results. Prepare well for all meetings: have a list of questions ready before the meeting; provide results in an easily digestible form (but take your lab notebook along); be clear about your future plans for work.

Your proposal outline might be considered by your supervisor or by a formal committee before approval.

Planning your work

As with any lengthy exercise, planning is required to make the best use of the time allocated (p. 10). This is true on a daily basis as well as over the entire period of the project. It is especially important not to underestimate the time it will take to write and produce your thesis (see below). If you wish to benefit from feedback given by your supervisor, you should aim to have drafts in his/her hands in good time. Since a large proportion of marks will be allocated to the report, you should not rush its production.

If your department requires you to write an interim report, look on this as a good opportunity to clarify your thoughts and get some of the time-consuming preparative work out of the way. If not, you should set your own deadlines for producing drafts of the introduction, materials and methods section, etc.

Fig. 34.1 Flowchart showing a recommended sequence of events in carrying out an undergraduate research project. Note the potentially iterative nature of the process, reflecting the scientific method outlined in Chapter 32.

Sources of further information for project work:

- experimental design checklist (Box 32.1);
- advice on recording results (p. 209);
- describing and analysing numerical data (Chapter 73);
- checklist for presenting your final report (Box 19.1).

> **KEY POINT** Project work can be very time-consuming. Try not to neglect other aspects of your course – make sure your lecture notes are up to date and collect relevant supporting information as you go along.

Getting started

Figure 34.1 is a flowchart illustrating how a project might proceed; at the start, do not spend too long reading the literature and working out a lengthy programme of research. Get stuck in and do an experiment. There is no substitute for 'getting your hands dirty' for stimulating new ideas:

- **even a 'failed' experiment will provide some useful information** which may allow you to create a new or modified hypothesis;
- **pilot experiments may point out deficiencies in experimental technique** that will need to be rectified;
- **the experience will help you create a realistic plan of work.**

Designing experiments or sampling procedures

Design of experiments and sampling procedure is covered in Chapter 32. Avoid being too ambitious at the start of your work. It is generally best to work with a simple hypothesis and design your experiments or sampling around this. A small pilot experiment or test sample will highlight potential stumbling blocks, including resource limitations, whether in materials, or time, or both.

Working in a laboratory environment

During your time as a project student, you are effectively a guest in your supervisor's laboratory.

- **Be considerate** – keep your 'area' tidy and offer to do your share of lab duties such as calibrating the pH meter, replenishing stock solutions, distilled water, etc., maintaining cultures, tending plants or animals.
- **Use instruments carefully** – they could be worth more than you think. Careless use may invalidate calibration settings and ruin other people's work as well as your own.
- **Do your homework on techniques you intend to use** – there is less chance of making costly mistakes if you have a good background understanding of the methods you will be using.
- **Always seek advice if you are unsure of what you are doing.**

> **KEY POINT** It is essential that you follow all the safety rules applying to the laboratory or field site. Make sure you know all relevant procedures – if in doubt, ask beforehand.

Keeping notes and analysing your results

Tidy record-keeping is often associated with good research, and you should follow the advice and hints given in Chapter 30. Try to keep copies of all files relating to your project. As you obtain results, you should always calculate, analyse and graph data as soon as you can (see Fig. 34.1). This can reveal aspects that may not be obvious in numerical or readout form. Do not be worried by negative results – these can sometimes be as useful as positive results if they allow you to eliminate hypotheses – and do not be too dispirited if things do not work first time. Thomas Edison's maxim 'Genius is one per cent inspiration and ninety-nine per cent perspiration' certainly applies to research work.

Writing your project report

The structure of scientific reports is dealt with in Chapter 19. The following advice concerns methods of accumulating relevant information.

Introduction

This is a big piece of writing that can be very time-consuming. Therefore, the more work you can do on it early on the better. Allocate some time at the start for library work (without neglecting your lab work) to build up a database of references (Chapter 9). Photocopying can be expensive, but you will find it valuable to have copies of key reviews and references handy when writing away from the library. Discuss proposals for content and structure with your supervisor to ensure your effort is relevant. Leave space at the end for a section on aims and objectives. This is important to orientate readers (including assessors), but you may prefer to finalise the content after the results have been analysed.

Materials and methods

Note as many details as possible *when doing the experiment or making observations.* Do not rely on your memory or hope that the information will still be available when you come to write up. Even if it is, chasing these details can waste valuable time.

Results

Show your supervisor graphed and tabulated versions of your data promptly. These can easily be produced using a spreadsheet (p. 83), but you should seek your supervisor's advice on whether the design and print quality is appropriate to be included in your report. You may wish to use a specialist graphics program to produce publishable-quality graphs and charts: allow some time for learning its idiosyncrasies. If you are producing a poster for assessment (Chapter 15), be sure to mock up the design well in advance. Similarly, think ahead about your needs for any seminar or poster you will present.

Discussion

Because this comes at the end of your report, and some parts can only be written after you have all the results in place, the temptation is to leave the discussion to last. This means that it might be rushed – not a good idea because of the weight attached by assessors to your analysis of data and thoughts about future experiments. It will help greatly if you keep notes of aims, conclusions and ideas for future work as you go along (Fig. 34.1). Another useful tip is to make notes of comparable data and conclusions from the literature as you read papers and reviews.

Acknowledgements

Make a special place in your notebook for noting all those who have helped you carry out the work for use when writing this section of the report.

Brushing up on your IT skills – word processors and spreadsheets are extremely useful when producing a thesis. Chapters 13 and 14 detail key features of these programs. You might benefit from attending courses on the relevant programs or studying manuals or texts so that you can use them more efficiently.

Using drawings and photographs – these can provide valuable records of sampling sites or experimental set-ups and could be useful in your report. Plan ahead and do the relevant work at the time of carrying out your research rather than afterwards.

Using a word processor to record your ideas – remember that you can note down your thoughts and any other important information relevant to the Results and Discussion sections of your project in a file that can then form the basis of your first draft (p. 90); that way, you will not forget to include these points in your final report.

References

Because of the complex formats involved (p. 56), these can be tricky to type. To save time, process them in batches as you go along – bibliographic software (e.g. *EndNote*) can help with organisation of references (p. 55).

> **KEY POINT** Make sure you are absolutely certain about the deadline for submitting your report, and work towards submitting a day or so before it. If you leave things until the last moment, you may find access to printers, photocopiers and binding machines is difficult.

Sources for further study

Ebel, H.F., Bliefert, C. and Russey, W.E. (2004) *The Art of Scientific Writing: from Student Reports to Professional Publications in Chemistry and Related Fields,* 2nd edn. Wiley-VCH, Weinheim, Germany. [Covers record-keeping, notebooks, report writing and dissertations.]

Gilbert, F., McGregor, P. and Barnard, C. (2011) *Asking Questions in Biology,* 4th edn. Benjamin Cummings, San Francisco.

Luck, M. and Wilson, J. (2008) *Student Research Projects: Guidance on Practice in the Biosciences.* HEA Centre for Biosciences, Leeds.

McMillan, K.M. and Weyers, J.D.B. (2014) *How to Complete a Successful Research Project.* Pearson Education, Harlow.

Ruxton, G. and Palgrave, N. (2006) *Experimental Design for the Life Sciences.* Oxford University Press, Oxford.

Sweetnam, D. (2004) *Writing your Dissertation: the Bestselling Guide to Planning, Preparing and Presenting First-Class Work,* 3rd edn. How To Books, Oxford.

STUDY EXERCISES

Note: These exercises assume that you have started a research project, or are about to start one, as part of your studies.

34.1 Prepare a project plan. Make a formal plan for your research project, incorporating any milestones dictated by your department, such as interim reports and final submission dates. Discuss your plan with your supervisor and incorporate his or her comments. Refer back to the plan frequently during your project, to see how well you are meeting your deadlines.

34.2 Resolve to write up your work as you go along. Each time you complete an experiment or observation, write up the materials and methods, analyse the data and draw up the graphs as soon as you can. While you may reject or modify some of these drafts at a later stage, this approach will spread out the majority of the effort and allow time for critical thinking close to the final submission date.

34.3 Devise a computer database for keeping details of your references. Keeping these records up to date will save you a lot of time when writing up. You will need to decide on an appropriate referencing format, or find out about that followed by your department (see Chapter 9).

Answers to these study exercises are available at www.pearsoned.co.uk/practicalskills.

Working with cells and tissues

35 Sterile technique

Sterile technique (aseptic technique) is the name given to the procedures used in cell culture. Although the same general principles apply to all cell types, you are most likely to learn the basic procedures using bacteria and most of the examples given in this section refer to bacterial culture.

Sterile technique serves two main purposes:

1. **To prevent accidental contamination of laboratory cultures** due to microbes from external sources, e.g. skin, clothing or the surrounding environment.

2. **To prevent microbial contamination of laboratory workers,** in this instance you and your fellow students.

> **KEY POINT** *All* microbial cell cultures should be treated as if they contained potentially harmful organisms. Sterile technique forms an important part of safety procedures, and must be followed whenever cell cultures are handled in the laboratory. Avoid any 'hand-to-mouth' operations at all times.

Care is required because:

- **You may accidentally isolate a harmful microbe as a contaminant** when culturing a relatively harmless strain.

- **Some individuals are more susceptible to infection and disease than others** – not everyone exposed to a particular microbe will become ill.

- **Laboratory culture involves purifying and growing large numbers of microbial cells** – this represents a greater risk than small numbers of the original microbe.

- **A microbe may change its characteristics,** perhaps as a result of gene exchange or mutation.

The international biohazard symbol, shown in Fig. 35.1, is used to indicate a significant risk due to a pathogenic microbe (p. 226).

Fig. 35.1 International symbol for a biohazard. Usually red or black on a yellow background, or black on a red background.

Sterilisation procedures

Given the ubiquity of microbes, the only way to achieve a sterile state is by their destruction or removal. Several methods can be used to achieve this objective:

Heat treatment

This is the most widespread form of sterilisation and is used in several basic laboratory procedures, including the following:

- **Red-heat sterilisation.** Achieved by heating metal inoculating loops, forceps, needles, etc. in a Bunsen flame (Fig. 35.2). This is a simple and effective form of sterilisation as no microbe will survive even a brief exposure to a naked flame. Flame sterilisation using alcohol is used for glass rods and spreaders (see below).

- **Dry-heat sterilisation.** Here, a hot-air oven is used at a temperature of at least 160 °C for at least 2 h. This method is used for the routine sterilisation of laboratory glassware. Dry-heat procedures are of little value for items requiring repeated sterilisation during use.

loop handle
wire
blue 'cone' (unburnt gas)
air inlet fully open

Fig. 35.2 'Flaming' a wire loop. Keep the loop in the hottest part of the Bunsen flame (just outside the blue 'cone') until the wire is red-hot.

Fig. 35.3 Autoclave tape – the bottom sample is untreated while the upper sample (with dark diagonal lines) has been autoclaved.

Using a sterile filter – most filters are supplied as pre-sterilised items. Make sure you follow a procedure that does not contaminate the filter on removal from its protective wrapping.

SAFETY NOTE When working with biocides, take care to avoid skin contact or ingestion, as most are toxic and irritant. If you spill biocide on your skin, rinse with plenty of water. If ingested, seek immediate medical help.

Using molten agar – a water bath (at 45–50 °C) can be used to keep an agar-based medium in its molten state after autoclaving. Always dry the outside of the container on removal from the water bath, to reduce the risk of contamination from microbes in the water, e.g. during pour plating (p. 227).

- **Moist-heat sterilisation.** This is the method of choice for many laboratory items, including most fluids, apart from heat-sensitive media. It is also used to decontaminate liquid media and glassware after use. The laboratory autoclave is used for these purposes. Typically, most items will be sterile after 15 min at 121 °C, although large items may require a longer period. The rapid killing action results from the latent heat of condensation of the pressurised steam, released on contact with cool materials in the autoclave. Although special heat-sensitive tape (Fig. 35.3) is sometimes used to check that the autoclave is operating correctly, a better approach is to use spores of *Bacillus stearothermophilus*.

Radiation

Many disposable plastic items used in microbiology and cell biology are sterilised by exposure to UV or ionising radiation. They are supplied commercially in sterile packages, ready for use. Ultraviolet radiation has limited use in the laboratory, while ionising radiation (e.g. γ-rays) requires industrial facilities and cannot be operated on a laboratory scale.

Filtration

Heat-labile solutions (e.g. complex macromolecules, including proteins, antibiotics, serum) are particularly suited to this form of sterilisation. The filters come in a variety of shapes, sizes and materials, usually with a pore size of either 0.2 μm or 0.45 μm. The filtration apparatus and associated equipment is usually sterilised by autoclaving, or by dry heat. Passage of liquid through a sterile filter of pore size 0.2 μm into a sterile vessel is usually sufficient to remove bacteria but not viruses, so filtered liquids are not necessarily virus-free.

Chemical agents

These are known as disinfectants, or biocides, and are most often used for the disposal of contaminated items following laboratory use, e.g. glass slides and pipettes. They are also used to treat spillages. The term 'disinfection' implies destruction of disease-causing bacterial cells, although spores and viruses may not always be destroyed. Remember that disinfectants require time to exert their killing effect – any spillage should be covered with an appropriate disinfectant and left for at least 10 min before mopping up.

Use of laboratory equipment

Working area

One of the most important aspects of good sterile technique is to keep your working area as clean and tidy as possible. Start by clearing all items from your working surface, wipe the bench down with disinfectant and then arrange the items you need for a particular procedure so that they are close at hand, leaving a clear working space in the centre of your bench.

Media

Cells may be cultured in either a liquid medium (broth), or a solidified medium (p. 230). The gelling agent used in most solidified media is agar, a complex polysaccharide from red algae that produces a stiff transparent

Agar (originally agar-agar) – a Malay word meaning jelly.

Working with plastic disposable loops – these are used in many research laboratories: pre-sterilised and suitable for single use, they avoid the hazards of naked flames and the risk of aerosol formation during heating. Discard into a disinfectant solution after use.

Using a Bunsen burner to reduce airborne contamination – working close to the updraught created by a Bunsen flame reduces the likelihood of particles falling from the air into an open vessel.

Using glass pipettes – these are plugged with cotton wool at the top before being autoclaved inside a metal can. Flame the open end of the can on removal of a pipette, to prevent contamination of the remaining pipettes. Autopipettors and sterile disposable tips (p. 146) offer an alternative approach.

Table 35.1 Classification of microbes on the basis of hazard. The following categories are recommended by the UK Advisory Committee on Dangerous Pathogens (ACDP). A similar approach is followed by the World Health Organization.

Hazard group	Comments
1	Unlikely to cause human disease
2	May cause disease: possible hazard to laboratory workers, minimal hazard to community
3	May cause severe disease: may be a serious hazard to laboratory workers, may spread to community
4	Causes severe disease: is a serious hazard to laboratory workers, high risk to community

gel when used at 1–2% (w/v). Agar is used because it is relatively resistant to degradation by most bacteria and because of its rheological properties – an agar medium melts at 98 °C, remaining solid at all temperatures used for routine laboratory culture. Once melted, however, it does not solidify again until the temperature falls to about 44 °C. This means that heat-sensitive constituents (vitamins, blood, cells, etc.) can be added aseptically to the medium after autoclaving.

Inoculating loops

The initial isolation and subsequent transfer of microbes between containers can be achieved by using a sterile inoculating loop. Most teaching laboratories use nichrome wire loops in a metal handle. A wire loop can be repeatedly sterilised by heating the wire, loop downwards and almost vertical, in the hottest part of a Bunsen flame until the whole wire becomes red-hot. Then the loop is removed from the flame to minimise heat transfer to the handle. After cooling for 8–10 s (without touching any other object), it is ready for use.

When re-sterilising a contaminated wire loop in a Bunsen flame after use, do not heat the loop too rapidly, as the sample may spatter, creating an aerosol: it is better to soak the loop for a few minutes in disinfectant than to risk heating a fully charged (contaminated) inoculating loop.

Containers

There is a risk of contamination whenever a sterile bottle, flask or test tube is opened. One method that reduces the chance of airborne contamination is quickly to pass the open mouth of the glass vessel through a flame. This destroys any microbes on the outer surfaces nearest to the mouth of the vessel. In addition, by heating the air within the neck of the vessel, an outwardly directed air flow is established, reducing the likelihood of microbial contamination.

You should flame the mouth of each vessel immediately after opening and then repeat the procedure just before replacing the top. Caps, lids and cotton wool plugs must not be placed on the bench during flaming and sampling: they should be removed and held using the smallest finger of one hand, to minimise the risk of contamination. This also leaves your remaining fingers free to carry out other manipulations. With practice, you will be able to remove the tops from two tubes, flame each tube and transfer material from one to the other while holding one top in each hand.

Laminar-flow cabinets

These are designed to prevent airborne contamination, e.g. when preparing media or subculturing microbes or tissue cultures. Sterile air is produced by passage through a high-efficiency particulate air (HEPA) filter: this is then directed over the working area, either horizontally (towards the operator) or downwards. The operator handles specimens, media, etc., through an opening at the front of the cabinet. Note that standard laminar-flow cabinets do *not* protect the worker from contamination and must not be used with pathogenic microbes: special safety cabinets are used for work with ACDP hazard group 3 and 4 microbes (Table 35.1) and for samples that might contain such pathogens.

Fig. 35.4 Preparation of a streak plate for single colonies. (a) Using a sterile metal loop, take a small sample of the material to be streaked. Distribute the sample over a small sector of the plate (area 1), then flame the loop and allow to cool (approximately 8–10s). (b) Make several small streaks from the initial sector into the adjacent sector (area 2), taking care not to allow the streaks to overlap. Flame the loop and allow to cool. (c) Repeat the procedure for areas 3 and 4, re-sterilising the loop between each step. Finally, make a single, long streak, as shown for area 5. (d) The expected result after incubation at the appropriate temperature (e.g. 37° C for 24h): each step should have diluted the inoculum, giving individual colonies within one or more sectors on the plate. Further subculture of an individual colony should give a pure (clonal) culture.

Microbiological hazards

> **KEY POINT** The most obvious risks when handling microbial cultures are those due to ingestion or entry via a cut in the skin – all cuts should be covered with a plaster or disposable plastic gloves. A less obvious source of hazard is the formation of aerosols of liquid droplets from microbial suspensions, with the risk of inhalation, or surface contamination of other objects.

The following steps will minimise the risk of aerosol formation:

- **Use stoppered tubes when shaking, centrifuging or mixing microbial suspensions.**
- **Pour solutions gently, keeping the difference in height to a minimum.**
- **Discharge pipettes onto the side of the container.**

Other general rules that apply in microbiology laboratories include:

- **Take care with sharp instruments,** including needles and glass Pasteur pipettes.
- **Avoid any 'hand-to-mouth' actions,** e.g. chewing the end of your pencil while thinking.
- **Do not pour waste cultures down the sink** – they must be autoclaved.
- **Put other contaminated items (e.g. slides, pipettes) into disinfectant after use.**
- **Wipe down your bench with disinfectant when practical work is complete.**
- **Always wash your hands before leaving the laboratory.**

Plating methods

Many culture methods make use of a solidified medium within a Petri plate or Petri dish. A variety of techniques can be used to transfer and distribute the organisms prior to incubation. The three most important procedures are described below.

Streak-dilution plate

Streaking a plate for single colonies is one of the most important basic skills in microbiology, since it is used in the initial isolation of a cell culture and in maintaining stock cultures, where a streak-dilution plate with single colonies all of the same type confirms the purity of the strain. A sterile inoculating loop is used to streak the organisms over the surface of the medium, thereby diluting the sample (Fig. 35.4). The aim is to achieve single colonies at some point on the plate: ideally, such colonies are derived from single cells (e.g. in the case of unicellular bacteria, animal and plant cell lines) or from groups of cells of the same species (for example, in filamentous or colonial forms).

> **KEY POINT** Single colonies, containing cells of a single species and derived from a single parental cell, form the basis of all pure culture methods.

Note the following:

- **Keep the lid of the Petri plate as close to the base as possible** to reduce the risk of aerial contamination.

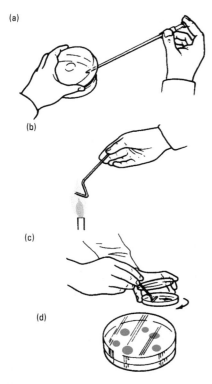

Fig. 35.5 Preparation of a spread plate. (a) Transfer a small volume of cell suspension (0.05–0.5 mL) to the surface of a solidified medium in a Petri plate. (b) Flame sterilise a glass spreader and allow to cool (8–10s). (c) Distribute the liquid over the surface of the plate using the sterile spreader. Ensure an even coverage by rotating the plate as you spread: allow the liquid to be absorbed into the agar medium. Incubate under suitable conditions. (d) After incubation, the microbial colonies should be distributed over the surface of the plate.

- **Allow the loop to glide over the surface of the medium.** Hold the handle at the balance point (near the centre) and use light, sweeping movements, as the agar surface is easily damaged and torn.
- **Work quickly, but carefully.** Do not breathe directly over the exposed agar surface and replace the lid as soon as possible.

Spread plate

This method is used with cells in suspension, either in a liquid growth medium or in an appropriate sterile diluent. It is one method of quantifying the number of viable cells (or colony-forming units) in a sample, after appropriate dilution (pp. 154–6).

An L-shaped glass spreader is sterilised by dipping the end of the spreader in a beaker containing a small amount of 70% v/v alcohol, allowing the excess to drain from the spreader and then igniting the remainder in a Bunsen flame. After cooling, the spreader is used to distribute a known volume of cell suspension across the plate (Fig. 35.5). *There is a significant fire risk associated with this technique,* so take care not to ignite the alcohol in the beaker, e.g. by returning an overheated glass rod to the beaker. The alcohol will burn with a pale blue flame that may be difficult to see, but will readily ignite other materials (e.g. a laboratory coat). Another source of risk comes from small droplets of flaming alcohol shed by an overloaded spreader onto the bench and this is why you *must* drain excess alcohol from the spreader *before* flaming. Many laboratories now provide plastic disposable spreaders for student use, to avoid this risk.

Pour plate

This procedure also uses cells in suspension, but requires molten agar medium, usually in screw-capped bottles containing sufficient medium to prepare a single Petri plate (i.e. 15–20 mL), maintained in a water bath at 45–50 °C. A known volume of cell suspension is mixed with this molten agar, distributing the cells throughout the medium. This is then poured without delay into an empty sterile Petri plate and incubated, giving widely spaced colonies (Fig. 35.6). Furthermore, as most of the colonies are formed within the medium, they are far smaller than those of the surface streak method, allowing higher cell numbers to be counted (e.g. up to 1000 colonies per plate): some workers pour a thin layer of molten agar onto the surface of a pour plate after it has set, to ensure that no surface colonies are produced. Most bacteria and fungi are not killed by brief exposure to temperatures of 45–50 °C, though the procedure may be more damaging to microbes from low temperature conditions, e.g. psychrophilic bacteria.

One disadvantage of the pour plate method is that the typical colony morphology seen in surface-grown cultures will not be observed for those colonies that develop within the agar medium. A further disadvantage is that some of the suspension will be left behind in the screw-capped bottle. This can be avoided by transferring the suspension to the Petri plate, adding the molten agar, then swirling the plate to mix the two liquids. However, even when the plate is swirled repeatedly and in several directions, the liquids are not mixed as evenly as in the former procedure.

Working with phages

Bacterial viruses ('bacteriophages', or simply 'phages') are often used to illustrate the general principles involved in the detection and enumeration of viruses. They also have a role in genome mapping of bacteria (p. 453). Individual

Fig. 35.6 Preparation of a pour plate. (a) Add a known volume of cell suspension (0.05–1.0 mL) to a small bottle of molten agar medium from a 45°C water bath. (b) Mix thoroughly, by rotating between the palms of the hands: do not shake or this will cause frothing of the medium. (c) Pour the mixture into an empty, sterile Petri plate and allow to set. Incubate under suitable conditions. (d) After incubation, the microbial colonies will be distributed throughout the medium: any cells deposited at the surface will give larger, spreading colonies.

phage particles (virions) are too small to be seen by light microscopy, but are detected by their different effects on susceptible host cells.

- **Virulent phages** will infect and replicate within actively growing host cells, causing cell lysis and releasing new infective phages – this 'lytic cycle' takes ≈30 min for T-even phages of *E. coli,* e.g. T4.
- **Temperate phages** are a specialised group, capable either of lytic growth or an alternative cycle, termed lysogeny – the phage becomes latent within a host cell (lysogen), typically by insertion of its genetic information into the host cell genome, becoming a 'prophage'. At a later stage, termed induction, the prophage may enter the lytic cycle. A widely used example is λ phage of *E. coli.*

The lytic cycle can be used to detect and quantify the number of phages in a sample. A known volume of sample is mixed with susceptible bacterial cells in molten soft agar medium (45–50 °C), then poured on top of a plate of the same medium, creating a thin layer of 'top agar'. The upper layer contains only half the normal amount of agar, to allow phages to diffuse through the medium and attach to susceptible cells. On incubation, the bacteria will grow throughout the agar to produce a homogeneous 'lawn' of cells, except in those parts of the plate where a phage particle has infected and lysed the cells to create a clear area, termed a plaque (Fig. 35.7). Each plaque is caused by a single functional phage (i.e. a plaque-forming unit, or PFU). A count of the number of plaques can be used to give the number of phages in a particular sample (e.g. as PFU mL^{-1}), with appropriate correction for dilution and the volume of sample counted in an analogous manner to a bacterial plate count (p. 235). When counting plaques in phage assays you should view them against a black background to make them easier to see: mark each plaque with a spirit-based marker to ensure an accurate count. Temperate phages often produce cloudy plaques, because many of the infected cells will be lysogenised rather than lysed, creating turbidity within the plaque. Samples of material from within the plaque can be used to subculture the phage for further study, perhaps in a broth culture where the phages will cause widespread cell lysis and a decrease in turbidity. Alternatively, phages can be stored by adding chloroform to aqueous suspensions – this will prevent contamination by cellular micro-organisms. A similar approach can be used to detect and count animal or human viruses, using a monolayer of susceptible host cells.

Electron microscopy (EM, pp. 177–8) provides an alternative approach to the detection of viruses, avoiding the requirement for culture of infected host cells, and giving a faster result. However, it requires specialised

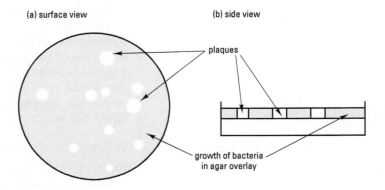

Fig. 35.7 Phage plaques in a 'lawn' of susceptible host bacteria.

equipment and expertise. EM counts are often higher than culture-based methods, for similar reasons to those described for bacteria (p. 233).

Labelling your plates and cultures

Labelling Petri plates – the following information should be recorded on the *base:*

- date
- the growth medium
- your name or initials
- brief details of the experimental treatment.

Petri plates should always be labelled on the *base,* rather than the lid. Restrict your labelling to the outermost region of the plate, to avoid problems when counting colonies, assessing growth, etc. After labelling, Petri plates usually are incubated upside down in a temperature-controlled incubator (often at 37 °C) for an appropriate period (usually 18–72 h). Plates are also usually kept upside down on the lab bench – following incubation, the base (containing medium and microbes) can then be lifted from the lid and examined.

Sources for further study

Anon. (1995) *Advisory Committee on Dangerous Pathogens: Categorisation of Biological Agents According to Hazard and Categories of Containment,* 4th edn. HSE Books, London.

Anon. (2013) *Advisory Committee on Dangerous Pathogens: Approved List of Biological Agents.* Available: http://www.hse.gov.uk/pubns/misc208.pdf Last accessed: 01/09/15.

Collins, C.H., Lyne, P.M., Grange, J.M. and Falkinham, J. (2004) *Collins and Lyne's Microbiological Methods,* 8th edn. Hodder-Arnold, London.

Hawkey, P. and Lewis, D. (2004) *Medical Bacteriology: A Practical Approach,* 2nd edn. Oxford University Press, Oxford.

World Health Organization (2004) *Laboratory Biosafety Manual,* 3rd edn. Available: http://www.who.int/csr/resources/publications/biosafety/en/Biosafety7.pdf Last accessed: 01/09/15.

STUDY EXERCISES

35.1 Decide on the best method of sterilisation. What would be the most appropriate method of sterilisation for the following items?

(a) A box of 100 plastic tips to be used with a pipettor.

(b) A 50-mL batch of blood, for use in 5% v/v blood agar plates.

(c) A 1-litre batch of MacConkey agar.

(d) Ten 5-mL glass pipettes.

(e) A microbiological wire, used for 'stab' cultures.

(f) A 10-mL sample of a heat-sensitive solution of an antibiotic, to be used as a component of a selective isolation medium.

35.2 Find out the biohazard classification (UK ACDP categorisation – Table 35.1) for the following microbes:

(a) *Salmonella typhimurium*

(b) *Leptospira interrogans*

(c) *Shigella dysenteriae* (type 1)

(d) *Escherichia coli* K12

(e) *Escherichia coli* 0157

(f) Human immunodeficiency virus (HIV)

(g) *Cryptococcus neoformans* var. *neoformans*

(h) *Mycobacterium tuberculosis*

(i) *Lactobacillus plantarum*

(j) Marburg virus.

35.3 Consider the advantages and disadvantages of spread plating and pour plating methods. Having read through this chapter, list up to six pros/cons of each plating method and compare your answers either with the list we have provided on the website, or with those of other students, as a group exercise.

Answers to these study exercises are available at www.pearsoned.co.uk/practicalskills.

Definitions

Chemoautotroph – an organism that acquires energy from the oxidation of simple inorganic compounds, fixing CO_2 as a source of carbon (chemosynthetic metabolism).

Heterotroph – an organism that uses complex organic carbon compounds as a source of carbon and energy.

Photoautotroph – an organism that uses light as a source of energy and CO_2 as a carbon source (photosynthetic metabolism).

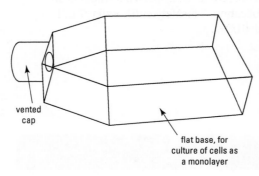

vented
cap

flat base, for
culture of cells as
a monolayer

Fig. 36.1 Plastic flask for animal cell culture – this design provides a large surface area for growth of an adherent monolayer of cells.

Harvesting bacteria from an agar plate – colonies can be harvested using a sterile loop, providing large numbers of cells without the need for centrifugation. The cells are relatively free from components of the growth medium; this is useful if the medium contains substances that interfere with subsequent procedures.

Subculturing – when subculturing microbes from a colony on an agar medium, take your sample from the growing edge, so that viable cells are transferred.

Microbial, animal and plant cell culture methods are based on the same general principles, requiring:

- **a pure culture** (also known as an axenic culture), perhaps isolated as part of an earlier procedure, or from a culture collection;
- **a suitable nutrient medium** to provide the necessary components for growth. This medium must be sterilised before use;
- **satisfactory growth conditions** including temperature, pH, atmospheric requirements, ionic and osmotic conditions;
- **sterile technique** (p. 223) to maintain the culture in pure form.

Heterotrophic animal cells, fungi and many bacteria require appropriate organic compounds as sources of carbon and energy. Non-exacting bacteria can utilise a wide range of compounds and they are often grown in media containing complex natural substances (including meat extract, yeast extract, soil, blood). Animal cells have more stringent growth requirements (p. 259).

Photoautotrophic bacteria, cyanobacteria and algae are grown in a mineral medium containing inorganic ions including chelated iron, with a light source and CO_2 supply. Plant cells may require additional vitamins and hormones (p. 259). For chemoautotrophic bacteria, the light source is replaced by a suitable inorganic energy source, e.g. H_2S for sulfur-oxidising bacteria, NH_3/NH_4^+ for nitrifying bacteria, etc.

Growth on solidified media

Many organisms can be cultured on an agar-based medium (p. 224).

> **KEY POINT** An important benefit of agar-based culture systems is that an individual cell inoculated onto the surface can develop to form a visible colony: this is the basis of most microbial isolation and purification methods, including the streak dilution, spread plate and pour plate procedures (pp. 226–7).

Animal cells are often grown as an adherent monolayer on the surface of a plastic or glass culture vessel (Fig. 36.1), rather than on an agar-based medium (p. 224).

Several types of culture vessel are used:

- **Petri plates** (Petri dishes): usually the pre-sterilised, disposable plastic type, providing a large surface area for growth.
- **Glass bottles or test tubes:** these provide sufficient depth of agar medium for prolonged growth of bacterial and fungal cultures, avoiding problems of dehydration and salt crystallisation. Inoculate aerobes on the surface and anaerobes by stabbing down the centre, into the base (stab culture).
- **Flat-sided bottles:** these are used for animal cell culture, to provide an increased surface area for attachment and allow growth of cells as a surface monolayer. Usually plastic and disposable (Fig. 36.1).

The dynamics of growth are usually studied in liquid culture, apart from certain rapidly growing filamentous fungi, where increases in colony diameter can be measured accurately, e.g. using Vernier callipers.

mouth of flask plugged with cotton wool bung to prevent contamination

wide base provides stability and large surface area for mixing

Fig. 36.2 Conical (Erlenmeyer) flask.

Fig. 36.3 Population growth curve for cells in batch culture (liquid medium).

Example Suppose you counted 2×10^3 cells ($\log_{10} = 3.30$) per unit volume at t_0 and 6.3×10^4 cells ($\log_{10} = 4.80$) after 2 h (t_x).

Substitution into Eqn [36.1] gives [2.303(4.8 − 3.3)] ÷ 2 = 1.727 h^{-1} (or 0.0288 min^{-1}).

Substituting the same values into Eqn [36.2] gives:

[0.301 × 2] ÷ [4.8 − 3.3] = 0.40 h (or 24 min).

Working with logarithms – note that there is no log value for zero, so you cannot plot zero on a log-linear growth curve or on a death curve. Your graph should end at the last measured 'count'.

Growth in liquid media

Many cells, apart from primary cultures of animal cells, can be grown as a homogeneous unicellular suspension in a suitable liquid medium, where growth is usually considered in terms of cell number (population growth) rather than cell size. Most liquid culture systems need agitation, to ensure adequate mixing and to keep the cells in suspension. An Erlenmeyer flask of 100–2000 mL capacity (Fig. 36.2) can be used to grow a batch culture on an orbital shaker, operating at 20–250 cycles per minute. For aerobic organisms, the surface area of such a culture should be as large as possible: restrict the volume of medium to not more than 20% of the flask volume. Larger cultures may need to be gassed with sterile air and mixed using a magnetic stirrer rather than an orbital shaker. The simplest method of air sterilisation is filtration, using glass wool, non-absorbent cotton wool or a commercial filter unit of appropriate pore size (usually 0.2 μm). Air is introduced via a sparger (a glass tube with many small holes, so that small bubbles are produced) near the bottom of the culture vessel to increase the surface area and enhance gas exchange. More complex systems have baffles and paddles to further improve mixing and gas exchange.

Liquid culture systems may be subdivided under two broad headings.

Batch culture

This is the most common approach for routine liquid culture. Cells are inoculated into a sterile vessel containing a fixed amount of growth medium. Your choice of vessel will depend upon the volume of culture required: larger-scale vessels (e.g. 1 litre and above) are often called 'fermenters' or 'bioreactors', particularly in biotechnology. Growth within the vessel usually follows a predictable S-shaped (sinusoidal) curve when plotted in log-linear format (Fig. 36.3), divided into four components:

1. **Lag phase:** the initial period when no increase in cell number is seen. The larger the inoculum of active cells the shorter the lag phase will be, provided the cells are transferred from similar growth conditions.

2. **Log phase,** or exponential phase: where cells are growing at their maximum rate. This may be quantified by the specific growth rate (μ, or k), where:

$$\mu = \frac{2.303 \, (\log N_x - \log N_0)}{(t_x - t_0)} \qquad [36.1]$$

where N_0 is the initial number of cells at time t_0 and N_x is the number of cells at time t_x. For times specified in hours, μ is expressed as h^{-1}.

Prokaryotes grow by binary fission while eukaryotes grow by mitotic cell division; in both cases each cell divides to give two identical offspring. Consequently, the doubling time or generation time (g, or T_2) is:

$$g = \frac{0.301 \, (t_x - t_0)}{\log N_x - \log N_0} \qquad [36.2]$$

Cells grow at different rates, with doubling times ranging from under 20 min for some bacteria to 24 h or more for animal and plant cells. Exponential phase cells are often used in laboratory experiments, since growth and metabolism are nearly uniform.

3. **Stationary phase:** growth decreases as nutrients are depleted and waste products accumulate. Any increase in cell number is offset by death. This phase is usually termed the 'plateau' in animal cell culture.

4. **Decline phase,** or death phase: this is the result of prolonged starvation and toxicity, unless the cells are subcultured. Like growth, death often shows an exponential relationship with time, which can be characterised by a rate (specific death rate), equivalent to that used to express growth or, more often, as the decimal reduction time (d, or T_{90}), the time required to reduce the population by 90%:

$$d = \frac{t_x - t_0}{\log N_x - \log N_0} \qquad [36.3]$$

Some cells undergo rapid autolysis at the end of the stationary period while others show a slower decline.

Batch culture methods can be used to maintain stocks of particular organisms; cells are subcultured onto fresh medium before they enter the decline phase. However, primary cultures of animal cells have a finite life unless transformed to give a continuous cell line, capable of indefinite growth (p. 259).

Continuous culture

This is a method of maintaining cells in exponential growth for an extended period by continuously adding fresh growth medium to a culture vessel of fixed capacity. The new medium replaces nutrients and displaces some of the culture, diluting the remaining cells and allowing further growth.

After inoculating the vessel, the culture is allowed to grow for a short time as a batch culture, until a suitable population size is reached. Then medium is pumped into the vessel: the system is usually set up so that any increase in cell number due to growth will be offset by an equivalent loss due to dilution, i.e. the cell number within the vessel is maintained at a steady state. The cells will be growing at a particular rate (μ), counterbalanced by dilution at an equivalent rate (D):

$$D = \frac{\text{flow rate}}{\text{vessel volume}} \qquad [36.4]$$

where D is expressed per unit time (e.g. h^{-1}). In a chemostat, the growth rate is limited by the availability of some nutrient in the inflowing medium, usually either carbon or nitrogen (see Fig. 36.4). In a turbidostat, the input of medium is controlled by the turbidity of the culture, measured using a photocell. A turbidostat is more complex than a chemostat, with additional equipment and controls.

To determine the specific growth rate (μ, or k) of a continuous culture:

1. Measure the flow of medium through the vessel over a known time interval (e.g. connect a sterile measuring cylinder or similar volumetric device to the outlet), to calculate the flow rate.

2. Divide the flow rate by the vessel volume (Eqn [36.4]) to give the dilution rate (D).

3. This equals the specific growth rate, since $D = \mu$ at steady state.

Example Suppose you counted 5.2×10^5 cells ($\log_{10} = 5.716$) per unit volume at t_0 and 3.7×10^3 cells ($\log_{10} = 3.568$) after 60 min (t_x). Substitution into Eqn [36.3] gives $60 \div [2.148] = 27.9$ min. To the nearest minute, this gives a value for d of 28 min.

---

air filter

reservoir of sterile medium

flow-regulating clamp

air input

air filters

inoculation port

culture vessel

magnetic stirrer   sampling port

reservoir of spent medium

Fig. 36.4 Example of a two-dimensional lab equipment diagram of components of a chemostat.

Example Suppose a continuous culture system of 2000 mL volume had a flow of 600 mL over a period of 40 min (flow rate $600 \div 40 = 15$ mL min^{-1}). Substitution into Eqn [36.4] gives a dilution rate D of $15 \div 2000 = 0.00075$ min^{-1} or $0.00075 \times 60 = 0.45$ h^{-1}.

---

4. If you want to know the doubling time (g), calculate using the relationship:

$$g = \frac{0.693}{\mu} \tag{36.5}$$

(Note that Eqn [36.5] also applies to exponential phase cells in batch culture and is useful for interconverting g and μ.)

Continuous culture systems are more complex to set up than batch cultures. They are prone to contamination, having additional vessels for fresh medium and waste culture: strict aseptic technique is necessary when the medium reservoir is replaced, and during sampling and harvesting. However, they offer several advantages over batch cultures, including the following:

- The physiological state of the cells is more clearly defined, since actively growing cells at the same stage of growth are provided over an extended time period. This is useful for biochemical and physiological studies.
- Monitoring and control can be automated and computerised.
- Modelling can be carried out for biotechnology/fermentation technology.

Measuring growth in cell cultures

The most widely used methods of measuring growth are based on cell number.

Direct microscopic counts

One of the simplest methods is to count the cells in a known volume of medium using a microscope and a counting chamber or haemocytometer (Box 36.1). While this gives a rapid assessment of the total cell number, it does not discriminate between living and dead cells. It is also time-consuming as a large number of cells must be counted for accurate measurement. It may be difficult to distinguish individual cells, e.g. for cells growing as clumps.

Electronic particle counters

These instruments can be used to give a direct (total) count of a suspension of microbial cells. The Coulter counter detects particles, owing to change in electrical resistance when they pass through a small aperture in a glass tube (Fig. 36.5). It gives a rapid count based on a larger number of cells than direct microscopy. It is well-suited for repeat measurements or large sample numbers and can be linked to a microcomputer for data processing. If correctly calibrated, the counter can also measure cell sizes. A major limitation of electronic counters is the lack of discrimination between living cells, dead cells, cell clumps and other particles (e.g. dust). In addition, the instrument must be set up and calibrated by trained personnel. Flow cytometry is a more specialised alternative, since particles can be sorted as well as counted (see Macey, 2010).

Culture-based counting methods

A variety of culture-based techniques can be used to determine the number of microbes in a sample. A major assumption of such methods is that,

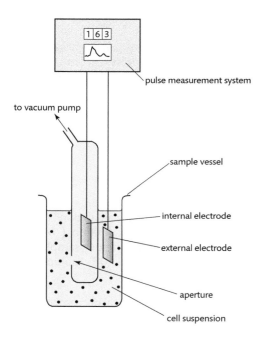

Fig. 36.5 Components of an electronic particle counter. During operation, the cell suspension is drawn through the aperture by the vacuum, creating a 'pulse' of resistance between the two electrodes as each cell passes through the aperture.

to vacuum pump

pulse measurement system

sample vessel

internal electrode

external electrode

aperture

cell suspension

Alternative approaches for measuring growth – these include biomass, dry weight, turbidity (light scattering – nephelometry), uptake of radiolabelled substrates, absorbance or assay of any major cellular component, e.g. protein, nucleic acid, ATP, etc.

Box 36.1   How to use a counting chamber or haemocytometer

A counting chamber is a specially designed slide containing a chamber of known depth with a grid etched onto its lower surface. When a flat coverslip is placed over the chamber, the depth is uniform. Use as follows:

1. **Place the special coverslip over the chamber.** Press the edges firmly, to ensure that the coverslip makes contact with the surface of the slide, but take care that you do not break the slide or coverslip by using too much force. When correctly positioned, you should be able to see interference rings (Newton's rings) at the edge of the coverslip.

2. **Add a small amount of your cell suspension to fill the central space above the grid.** Place on the microscope stage and allow the cells to settle (2–3 min).

3. **Examine the grid microscopically,** using the ×10 objective lens first, since the counting chamber is far thicker than a standard microscope slide. Then switch to the ×40 objective: take care not to scratch the surface of the objective lens, as the special coverslip is thicker than a normal coverslip. For a dense culture, the small squares are used, while the larger squares are used for dilute suspensions. You may need to dilute your suspension if it contains more than 30 cells per small square.

4. **Count the number of cells in several squares:** at least 600 cells should be counted for accurate measurements. Include those cells that cross the upper and left-hand boundaries, but not those that cross the lower and right-hand rulings. A hand tally may be used to aid counting. Motile cells must be immobilised prior to counting (e.g. by killing with a suitable biocide).

5. **Divide the total number of cells (C) by the number of squares counted (S),** to give the mean cell count per square.

6. **Determine the volume (in ml) of liquid corresponding to a single square (V),** e.g. a Petroff–Hausser chamber has small squares of linear dimension 0.2 mm, giving an area of 0.04 mm²; since the depth of the chamber is 0.02 mm, the volume is $0.04 \times 0.02 = 0.0008$ mm³; as there are 1000 mm³ in 1 mL, the volume of a small square is 8×10^{-7}mL; similarly, the volume of a large square (equal to 25 small squares) is 2×10^{-5} mL. Note that other types of counting chamber will have different volumes: check the manufacturer's instructions. For example, the improved Neubauer chamber (Fig. 36.6) has small squares of volume $0.000\,25$ mm³ $= 2.5 \times 10^{-7}$ mL.

7. **Calculate the cell number per ml by dividing the mean cell count per square by the volume of a single square (in mL).**

Fig. 36.6 Haemocytometer grid (Improved Neubauer rulings) viewed microscopically. The large square (delimited by triple etched lines) has a volume of 1/250 mm³ (0.004 mm³ = 4 nL) while each small square (16 contained within the large square) has a volume of 1/4000 mm³ (0.00025 mm³ = 0.25 nL). Note that the boundary line for squares delimited by triple-etched lines is the *middle* line, so this line must be used when counting (count cells straddling the top and left-hand gridlines and ignore those straddling the bottom and right-hand gridlines).

8. **Remember to take account of any dilution of your original suspension** in your final calculation by multiplying by the reciprocal of the dilution (*M*), e.g. if you counted a 1 in 20 dilution of your sample, multiply by 20, or if you diluted to 10^{-5}, multiply by 10^5.

The complete equation for calculating the total microscopic count is:

$$\text{Total cell count (per mL)} = (C \div S \div V) \times M \quad [36.6]$$

e.g. if the mean cell count for a 100-fold dilution of a cell suspension, counted using a Petroff–Hausser chamber, was 12.4 cells in 10 small squares, the total count would be

$$(12.4 \div 10 \div 8 \times 10^{-7}) \times 10^2 = 1.55 \times 10^8 \text{ml}^{-1}$$

A simpler, less accurate approach is to use a known volume of sample under a coverslip of known area on a standard glass slide, counting the number of cells per field of view using a calibrated microscope of known field diameter, then multiplying up to give the cell number per ml.

Box 36.2 **How to make a plate count of bacteria using an agar-based medium**

1. Prepare serial decimal dilutions of the sample in a sterile diluent (pp. 154–5). The most widely used diluents are 0.1% w/v peptone water or 0.9% w/v NaCl, buffered at pH 7.3. Take care that you mix each dilution before making the next one. For soil, food or other solid samples, make the initial decimal dilution by taking 1 g of sample and making this up to 10 mL using a suitable diluent. Gentle shaking or homogenisation may be required for organisms growing in clumps. The number of decimal dilutions required for a particular sample will be governed by your expected count: dilute until the expected number of viable cells is around 100–1000 mL^{-1}.

2. Transfer an appropriate volume (e.g. 0.05–0.5 mL) of the lowest dilution to an agar plate using either the spread plate method or the pour plate procedure (p. 227). At least two, and preferably more, replicate plates should be prepared for each sample. You may also wish to prepare plates for more than one dilution, if you are unsure of the expected number of viable cells.

3. Incubate under suitable conditions for 18–72 h, then count the number of colonies on each replicate plate at the most appropriate dilution. The most accurate results will be obtained for plates containing 30–300 colonies. Mark the base of the plate with a spirit-based pen each time you count a colony. Determine the mean colony count per plate at this dilution (C).

4. Calculate the colony count per mL of that particular dilution by dividing by the volume (in mL) of liquid transferred to each plate (V).

5. Now calculate the count per mL of the original sample by multiplying by the reciprocal of the dilution: this is the multiplication factor (M); e.g. for a dilution of 10^{-3}, the multiplication factor would be 10^3. For soil, food or other solid samples, the count should be expressed per g of sample.

The complete equation for calculating the viable count is:

$$\text{Count per mL (or per g)} = (C \div V) \times M \qquad [36.7]$$

e.g. for a sample with a mean colony count of 5.5 colonies per plate for a volume of 0.05 mL at a dilution of 10^{-7}, the count would be:

$$(5.5 \div 0.05) \times 10^7 = 1.1 \times 10^9 \text{CFU mL}^{-1}$$

The count should be reported as colony-forming units (CFU) per mL, rather than as cells per mL, since a colony may be the product of more than one cell, particularly in filamentous microbes or in organisms with a tendency to aggregate. You should also be aware of the problems associated with counts of zero – these are best recorded as '<1', and you should then apply the appropriate correction factors for dilution and volume to obtain the detection limit. For example, a zero count (<1) of 100 µL of a fivefold dilution gives a detection limit of $(<1 \div 0.1) \times 5 = <50$ CFU mL^{-1}.

Using culture-based methods and selective media – injured cells and stressed microbes may not be able to grow on selective media under standard aerobic conditions unless a non-selective resuscitation stage is added. Alternative approaches involve the addition of scavengers of reactive oxygen species (ROS), e.g. pyruvate.

Definition

CFU – colony-forming unit: a cell or group of cells giving rise to a single colony on a solidified medium.

under suitable conditions, an individual viable microbial cell will be able to multiply and grow to give a visible change in the growth medium, i.e. a colony on an agar-based medium, or turbidity ('cloudiness') in a liquid medium. You are most likely to gain practical experience using bacterial cultures, counted by one or more of the following methods:

- **Spread or pour plate methods** ('plate counts', p. 227). The most widespread approach is to transfer a suitable amount of the sample to an agar medium, incubate under appropriate conditions and then count the resulting colonies (Box 36.2).

- **Membrane filtration.** For bacterial samples where the expected cell number is lower than 10 CFU mL^{-1}, pass the sample through a sterile filter (pore size 0.2 or 0.45 µm). The filter is then incubated on a suitable medium until colonies are produced, giving a count by dividing the mean colony count per filter by the volume of sample filtered.

- **Multiple tube count,** or most probable number (MPN). A bacteriological technique where the sample is diluted and known volumes are transferred to several tubes of liquid medium (typically, five tubes at three volumes), chosen so that there is a low probability of the smallest

Alternative approaches in plate counting – when large numbers of samples have to be counted, a single agar plate can be divided into segments and a single droplet of each dilution placed into the appropriate segment ('Miles and Misra' droplet counting).

Molecular approaches to enumerating microbes – DNA-based methods can be used, including quantitative real-time polymerase chain reaction (p. 471).

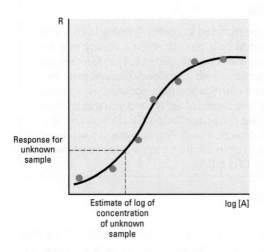

R

Response for unknown sample

Estimate of log of concentration of unknown sample

log [A]

Fig. 36.7 Typical bioassay response curve, showing estimation of an unknown sample. Closed circles represent responses with standard samples. R = response; A = analyte.

Examples   The following are typical bioassays:

● measuring the amount of a plant growth substance in a plant extract using the coleoptile straight growth assay;

● measuring the effect of a drug on the dilation of a person's pupils;

● measuring the mutagenic properties of a chemical compound in the Ames test (see Box 36.3).

volumes containing a viable cell. After incubation, the number of tubes showing growth (turbidity) is compared to tabulated values to give the most probable number (MPN per ml).

The principal advantage of culture-based counting procedures is that dead cells will not be counted. However, for such techniques the incubation conditions and media used may not allow growth of all cells, underestimating the true viable count. Further problems are caused by cell clumping and dilution errors. In addition, such methods require sterile apparatus and media and the incubation period is lengthy before results are obtained. An alternative approach is to use direct microscopy, combined with 'vital' or 'mortal' staining. For example, the direct epifluorescence technique (DEFT) uses acridine orange and UV epifluorescence microscopy to separate living and dead bacteria, while neutral red is a vital stain used for plant cells. Chapter 29 gives examples of vital/mortal stains for other cell types. A further approach is to use DNA-based methods, such as the polymerase chain reaction (PCR, p. 467).

Bioassays and their applications

A bioassay is a method of quantifying a chemical substance (analyte) by measuring its effect on a biological system under controlled conditions. The hypothetical underlying phenomena are summarised by the relationship:

$$A + Rec \rightleftharpoons A\,Rec \rightarrow R \qquad\qquad [36.8]$$

where A is the analyte, Rec the receptor, ARec the analyte–receptor complex and R the response. This relation is analogous to the formation of product from an enzyme–substrate complex and, using similar mathematical arguments to those of enzyme kinetics (p. 416), it can be shown that the expected relationship between [A] and rate of response is hyperbolic (sigmoidal in a log–linear plot like Fig. 36.7). This pattern of response is usually observed in practice if a wide enough range of [A] is tested.

To carry out the assay, the response elicited by the unknown sample is compared to the response obtained for differing concentrations of the substance, as shown in Fig. 36.7. When fitting a curve to standard points and estimating unknowns, the available methods, in order of increasing accuracy, are:

1. fitting by eye;

2. using linear regression on a restricted 'quasi-linear' portion of the assay curve;

3. linearisation followed by regression (e.g. by probit transformation);

4. non-linear regression (e.g. to the Morgan–Mercer–Flodin equation).

In general, bioassay techniques have more potential faults than physicochemical assay techniques. These may include the following:

● **A greater level of variability:** error in the estimate of the unknown compound will result because no two organisms will respond in exactly the same way. Assay curves vary through time, and because they are non-linear, a full standard curve is required each time the assay is carried out.

- **Lack of chemical information:** bioassays provide information about *biological* activity; they say little about the chemical structure of an unknown compound. The presence of a specific compound may need to be confirmed by a physicochemical method (e.g. mass spectrometry).
- **Possibility of interference:** while many bioassays are very specific, it is possible that different chemicals in the extract may influence the results.

Box 36.3 Mutagenicity testing using the Ames test – an example of a widely used bioassay

Chemical carcinogens can be identified by the formation of tumours in laboratory animals exposed to the compound under controlled conditions. However, such animal bioassays are time-consuming and expensive. Dr Bruce Ames and co-workers have shown that most carcinogens are also mutagens, i.e. they will induce mutational changes in DNA. The Ames test makes use of this correlation to provide a simple, rapid and inexpensive bioassay for the initial screening of potential carcinogens. The test makes use of particular strains of *Salmonella typhimurium* with the following characteristics:

- histidine auxotrophy – the tester strains are unable to grow on a minimal medium without added histidine: this characteristic is the result of specific mutational changes to the DNA of these strains, including base substitutions and frame shifts;
- increased cell envelope permeability, to permit access of the test compound to the cell interior;
- defects in excision repair systems and enhanced error-prone repair systems, to reduce the likelihood of DNA repair after treatment with a potential mutagen.

When grown in the presence of a chemical mutagen, the bacteria may revert to prototrophy as a result of back mutations that restore the wild-type phenotype: such revertants grow independently of external histidine and are able to form colonies on minimal medium, unlike the original test strains. The extent of reversion can be used to assess the mutagenic potential of a particular chemical compound. Since many chemicals must be activated *in vivo*, the test incorporates a rat liver homogenate (so-called 'S-9 activator', containing microsomal enzymes) to simulate the metabolic events within the liver. The tester strains are mixed with S-9 activator and a small amount of molten soft agar, then poured as a thin agar overlay (top agar) on a minimal medium plate. The top agar layer contains a very small amount of histidine, to allow the tester strains to divide a few times and express any mutational changes (i.e. prototrophy).

The test can be performed in one of two ways.

1. **Spot test:** a concentrated drop of the test compound is placed at the centre of the plate, either directly on the agar surface, or on a small filter paper disc. The test compound will diffuse into the agar and revertants appear as a ring of colonies around the site of inoculation, as shown in Fig. 36.8. The distance of the ring of colonies from this site provides a measure of the toxicity of the compound, while the number of colonies within the ring gives an indication of the relative mutagenicity of the test substance. The spot test is often carried out in a simplified form, without added S-9 activator, as a rapid preliminary test prior to quantitative analysis.

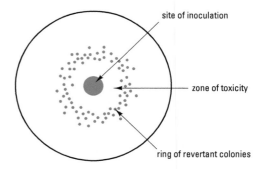

Fig. 36.8 Typical outcome of spot test (Ames test).

2. **Agar incorporation test:** known amounts of the test compound are mixed separately with molten top agar and the other constituents and poured onto separate minimal agar plates. After incubation for 48 h, revertants will appear as evenly dispersed colonies throughout the agar overlay. The number of colonies reflects the relative mutagenicity of the test compounds, with a direct relationship between colony count and the amount of mutagen. Agar incorporation tests can be used to generate dose–response curves similar to that shown in Fig. 36.7. The simplicity, sensitivity and reproducibility of the Ames test has resulted in its widespread use for screening potential carcinogens in many countries, though it is not an infallible test for carcinogenicity.

SAFETY NOTE Correct handling procedures *must* be followed at all times, as the test substances may be carcinogenic – testers should wear gloves and avoid skin contact or ingestion.

Despite these problems, bioassays are still much used. They are 'low-tech' and generally cheap to set up. They often allow detection at very low concentrations. Bioassays also provide the means to assess the biological activity of chemicals and to study changes in sensitivity to a chemical, which physicochemical techniques cannot do. Changes in sensitivity may be evident in the shape of the dose–response curve and its position on the concentration axis.

Bioassays can involve responses of whole organisms or parts of organisms. 'Isolated' responding systems (e.g. excised tissues or cells, Chapter 40) decrease the possibility of interference from other parts of the organism. Disadvantages include disruption of nutrition and wound damage during excision. Isolation can continue down to the molecular level, as in immunoassays (Chapter 43).

Bioassays are the basis for characterising the efficacy of drugs and the toxicity of chemicals. Here, response is often treated as a quantal (all-or-nothing) event. The E_{50} is defined as that concentration of a compound causing 50% of the organisms to respond. Where death is the observed response, the LD_{50} describes the concentration of a chemical that would cause 50% of the test organisms to die within a specified period under a specified set of conditions. Box 36.3 presents details of the Ames test, a widely used bioassay used to assess the mutagenicity of chemicals.

Considerations when setting up a bioassay

- The response should be easily measured and as metabolically 'close' to the initial binding event as possible.
- The experimental conditions should mimic the *in vivo* environment.
- The standards should be chemically identical to the compound being measured and spread over the expected concentration range being tested.
- The samples should be purified if interfering compounds are present and diluted so the response will be on the 'linear' portion of the assay curve.

To check for interference, the bioassay may be standardised against another method (preferably physicochemical). Related compounds known to be present in the analyte solution should be shown to have minimal activity in the bioassay. If an interfering compound is present, this may show up if a known amount of standard is added to sample vials – the result will not be the sum of independently determined results for the standard and sample.

Legal use of bioassays – in the UK, where bioassays involving 'higher' (vertebrate) animals are controlled by the *Animals Scientific Procedures Act (1986)*, they can only be carried out under the direct supervision of a scientist licensed by the Home Office.

Text reference

Macey, M.H. (2010) *Flow Cytometry: Principles and Applications.* Humana Press, London.

Sources for further study

Cann, A.J. (2000) *Virus Culture: a Practical Approach.* Oxford University Press, Oxford.

Davis, J.M. (2002) *Basic Cell Culture: a Practical Approach,* 2nd edn. Oxford University Press, Oxford.

Helgason, C.D. and Miller, C. (2013) *Basic Cell Culture Protocols*, 4th edn. Humana Press, London.

Hewitt, W. and Vincent, S. (1989) *Theory and Application of Microbiological Assay.* Academic Press, New York.

Vinci, V.A. and Parekh, S.R. (eds) (2010) *Handbook of Industrial Cell Culture: Mammalian, Microbial and Plant Cells.* Humana Press, London.

STUDY EXERCISES

36.1 Calculate the specific growth rate and doubling time of cells in culture. What are the specific growth rates and doubling times of the following? Give all answers to three significant figures.

(a) A broth culture of the yeast *Saccharomyces cerevisiae* growing in exponential phase and containing 5.2×10^4 cells at 10 am and 3.4×10^6 cells at 7 pm.

(b) A log phase culture of *E. coli* containing 3.0×10^4 CFU mL^{-1} at the start of the experiment and 6.7×10^7 CFU mL^{-1} at the end of the experiment, 200 minutes later.

(c) An actively growing culture of *Bacillus subtilis* containing 32 bacteria in 25 squares of a haemocytometer chamber at 13.00 hours and 250 bacteria in 25 squares of the same haemocytometer chamber at 15.30 hours.

36.2 Practise the calculations involved in using a haemocytometer to make a direct microscopic count. Express your answers to three significant figures in all cases.

(a) The mean cell count of a yeast suspension per small square of an improved Neubauer counting chamber was 6.42. If the volume of a small square is 2.5×10^{-7} mL, what is the cell count per mL?

(b) The following counts were obtained for bacteria in 20 individual small squares of a Petroff–Hausser counting chamber: 26, 36, 42, 35, 27, 16, 29, 50, 24, 43, 41, 35, 18, 36, 33, 47, 25, 46, 32, 57. If the volume of each small square is 8×10^{-7} mL, what is the cell count per mL?

(c) A 10^{-2} dilution of a dense suspension of *E. coli* was examined microscopically using an improved Neubauer counting chamber, giving a total of 78 cells in a total of 25 small squares. If the volume of each small square is 2.5×10^{-7} mL, what is the cell count per mL of the original (undiluted) suspension?

(d) A dilute suspension of yeast cells was concentrated tenfold by centrifuging 10 mL and resuspending the pellet in 1 mL. The mean cell count of the concentrated yeast suspension per large square of a Petroff–Hausser counting chamber was 6.8. If the volume of the large square is 2×10^{-5} mL, what is the cell count per mL of the original suspension?

36.3 Practise the calculations involved in making a plate count. Express your answers to three significant figures.

(a) The mean spread plate count for 100 μL of a 10^{-5} dilution of a culture of *E. coli* was 54.4 CFU. What is the mean plate count per mL of the original suspension?

(b) Three replicate plates of nutrient agar were each spread with 200 μl of a 10^{-3} dilution of a bacterial suspension, giving colony counts of 34, 40 and 37 after 24 h incubation. What is the mean plate count per mL of the original suspension?

(c) A 20-fold dilution of a yeast suspension was used to prepare four replicate pour plates, each containing 500 μL of this dilution and giving counts of 211, 186, 194 and 202 after incubation. What is the mean plate count per mL of the original suspension?

(d) A sample of 50 g of raw seafood was homogenised and diluted to 2% w/v in sterile saline solution. Three replicate pour plates were prepared using 1 mL of the diluted sample, giving counts for *E. coli* of 35, 41 and 32. What is the average count of *E. coli* per 100 g of raw seafood?

(e) Duplicate samples of 250 mL of river water were filtered through separate sterile membranes of pore size 0.2 μm. The membranes were then transferred to the surface of an agar-based medium and incubated for 48 h at 37°C, giving colony counts of 45 and 32 respectively. What is the mean count per 100 mL of river water?

(f) A sample of bottled drinking water was processed by serial decimal dilution and 500 μL samples of each dilution were pour plated and incubated at 22°C for 72h. The colony counts obtained for the three replicate plates of the 10^{-1} dilution were 28, 32 and 39. Does this water meet the EU regulations for mineral water, which specify a maximum average plate count at 22°C of 100 CFU mL^{-1}?

36.4 Use bioassay data to estimate the amount of analyte in unknown samples. Bioassay data obtained by exposing water fleas (*Daphnia* sp.) to a purified algal toxin are shown in the table below.

(a) Graph the data and use the graph to estimate the toxin contents of two samples obtained from two lochs which gave (i) a heart rate of 130 beats per minute and (ii) a heart rate of 25 beats per minute.

(continued)

(b) Which one of the two estimates would be the least reliable and why?

c) How could you improve the reliability of this estimate if you had more of the sample?

(d) What unmeasured factors might limit the usefulness of this assay?

Responses of *Daphnia* to toxin.

Toxin concentration (mmol L^{-1})	Mean heart rate of *Daphnia* sp. (beats per minute)
1.00×10^{-3}	20
3.16×10^{-4}	23
1.00×10^{-4}	26
3.16×10^{-5}	50
1.00×10^{-5}	180
3.16×10^{-6}	192
1.00×10^{-6}	195

36.5 Use agar diffusion bioassay data to quantify a growth inhibitor in a sample. Agar diffusion bioassay is a simple means of estimating the amount of an inhibitory substance in a test sample, by adding the substance, in aqueous solution, to wells cut in a Petri plate of agar medium seeded with a microbe sensitive to the substance and then incubating the plate at an appropriate growth temperature. Diffusion of the inhibitory substance gives a zone of growth inhibition proportional to $\log_{10}$ of the concentration of substance. In the following example and table (below), the amount of the antibiotic nisin in a test sample can be determined by comparing the degree of growth inhibition of *Bacillus subtilis* for the test solution and a series of standards containing known concentrations of nisin. Prepare a calibration curve and determine the nisin concentration of solution 5 (unknown). Express your answer in µg mL^{-1}, to one decimal place.

Growth inhibition of *B. subtilis* in solutions containing different amounts of nisin.

Solution	Nisin content (ng mL^{-1})	Zone of growth inhibition (mm)
1	40	26.5
2	20	23
3	10	19.5
4	5	17
5	unknown (test solution)	22.5

Answers to these study exercises are available at www.pearsoned.co.uk/practicalskills.

Microbes can be studied by taking samples for analysis, usually in one of the following ways:

- **by direct examination of individual cells** of a particular microbe, e.g. using fluorescence microscopy (p. 177);
- **by isolating/purifying a particular species or related individuals** of a taxonomic group, e.g. the faecal indicator bacterium *Escherichia coli* in sea water;
- **by studying microbial processes,** rather than individual microbes, either *in situ,* or in the laboratory.

The sampling process

Sampling techniques include the use of swabs, sticky tape, strips and agar contact methods for sampling surfaces, bottles for aquatic habitats, plastic bags and corers for soils and sediments. A wide range of complex apparatus is available for accurately sampling water or soil at particular depths.

> **KEY POINT**   An important feature of all microbiological sampling protocols is that the sampling apparatus must be sterile; strict aseptic technique must be used throughout the sampling process (see Chapter 35).

The sampling method must minimise the chance of contamination with microbes from other sources, especially the exterior of the sampling apparatus and the operator. For example, if you are sampling an aquatic habitat, stand downstream of the sampling site. A portable Bunsen burner or spirit lamp can be used to enable sterile technique during field sampling, e.g. while flaming a loop (p. 225). Alternatively, use disposable plastic loops.

Process the sample as quickly as possible, to minimise any changes in microbiological status. As a general guideline, many standard procedures require samples to be analysed within six hours of collection. Changes in aeration, pH and water content may occur after collection. Some microbes are more susceptible to such effects, e.g. anaerobic bacteria may not survive if the sample is exposed to air. Sunlight can also inactivate bacteria; samples should be shielded from direct sunlight during collection and transport to the laboratory.

Soil and water samples are often kept cool (at 0–5°C) during transport to the laboratory. In contrast, some microbes adapted to grow in association with warm-blooded animals may be damaged by low temperatures. An alternative approach is to keep the sample near the ambient sampling temperature using an insulated vessel (e.g. Thermos flask).

Isolation and purification techniques

Several different approaches may be used to obtain microbes in pure culture. The choice of method will depend upon the microbe to be isolated: some organisms are relatively easy to isolate, while others require more involved procedures.

Alternatives to traditional culture-based methods – microbes can now be studied by molecular methods, including the amplification of specific nucleic acid sequences by PCR (p. 467).

Avoiding contamination during sampling – always remember that you are the most important source of contamination of field samples: components of the oral or skin microflora are the most likely contaminants.

Subsampling – to minimise the effects of changes in temperature, aeration and water status during transportation, a primary sample may be returned to the laboratory, where the working sample (subsample) is then taken (e.g. from the centre of a large block of soil).

SAFETY NOTE When working with newly isolated microbes, you should always treat them as potentially harmful (p. 223) until they have been identified (Chapter 38).

Separation methods

Most microbial isolation procedures involve some form of separation to obtain individual microbial cells. The most common approach is to use an agar-based medium for primary isolation, with streak dilution, spread plating or pour plating to produce single colonies, each derived from a single type of microbe (pp. 226–7). It is often necessary to dilute samples before isolation, so that a small number of individual microbial cells is transferred to the growth medium. Strict serial dilution (pp. 154–6) of a known amount of sample is needed for quantitative work.

> **KEY POINT**   If your aim is to isolate a particular microbe, perhaps for further investigation, you will need to subculture individual colonies from the primary isolation plate to establish a pure culture, also known as an axenic culture.

Pure cultures of most microbes can be maintained indefinitely, using sterile technique and microbial culture methods (Chapter 35).
Other separation techniques include:

- **Dilution to extinction.** This involves diluting the sample to such an extent that only one or two microbes are present per millilitre: small volumes of this dilution are then transferred to a liquid growth medium (broth). After incubation, most of the tubes will show no growth, but some tubes will show growth, having been inoculated with a single viable microbe at the outset. This should give a pure culture, though it is wasteful of resources.

- **Sonication/homogenisation.** This is useful for separating individual microbial cells from each other and from inert particles, prior to isolation. However, some decrease in viability is likely.

- **Filtration.** This can be useful where the number of microbes is low. Samples can be passed through a sterile cellulose ester filter (pore size 0.2 μm), which is then incubated on the surface of an appropriate solidified medium. Sieving and filtration techniques are often used in soil microbiology to subdivide a sample on the basis of particle size.

- **Micromanipulation.** It may be possible to separate a microbe from contaminants using a micropipette and dissecting microscope (p. 146 and p. 186). The microbe can then be transferred to an appropriate growth medium, to give a pure culture. However, this is rarely an easy task for the novice.

- **Motility.** Phototactic microbes (including photosynthetic flagellates and motile cyanobacteria) will move towards a light source; heterotrophic flagellate bacteria will move through a filter of appropriate pore size into a nutrient solution, or away from unfavourable conditions (chemotaxis).

Selective and enrichment methods

Selective and enrichment techniques can be considered together, since they both enhance the growth of a particular microbe when compared with its competitors and they are often combined in specific media (Box 37.1).

The difference between selective and enrichment techniques is that the former use growth conditions unfavourable for competitors while the latter provide improved growth conditions for the chosen microbes.

Laboratory incubation under selective conditions will allow the particular microbes to be isolated in pure culture.

Obtaining a pure culture – if a single colony from a primary isolation medium is used to prepare a streak dilution plate and all the colonies on the second plate appear identical, then a pure culture has been established. Otherwise, you cannot assume that your culture is pure and you should repeat the subculture until you have a pure culture.

Using a sonicator – minimise heat damage with short treatment 'bursts' (typically up to 1 min), cooling the sample between bursts, e.g. using ice.

Definitions

Mesophile– a microbe with an optimum growth temperature of 20–45°C (lit. 'middle-loving').

Psychrophile– a microbe with an optimum temperature for growth of <20°C (lit. 'cold-loving').

Psychrotroph– a microbe with an optimum temperature for growth of ≥ 20°C, but capable of growing at lower temperature, typically 0–5°C (lit. 'cold-feeding').

Thermophile– a microbe with an optimum growth temperature of >45°C (lit. 'heat-loving').

Box 37.1 A differential medium for bacterial isolation: example

MacConkey agar is both a selective and a differential medium, useful for the isolation and identification of intestinal Gram-negative bacteria. Each component in the medium has a particular role:

- **Peptone:** (a meat digest) provides a rich source of complex organic nutrients to support the growth of non-exacting bacteria.
- **Bile salts:** toxic to most microbes apart from those growing in the intestinal tract (selective agent).
- **Lactose:** present as an additional, specific carbon source (enrichment agent).
- **Neutral red:** a pH indicator dye, to show the decrease in pH that accompanies the breakdown of lactose.
- **Crystal violet:** selectively inhibits the growth of Gram-positive bacteria. (This is only present in certain formations of MacConkey agar.)

Any intestinal Gram-negative bacterium capable of fermenting lactose will grow on MacConkey agar to produce large red–purple colonies, the red coloration being caused by the neutral red indicator under acidic conditions, while the purple coloration, often accompanied by a metallic sheen, is caused by the precipitation of bile salts and crystal violet at low pH.

In contrast, enteric Gram-negative bacteria unable to metabolise lactose will give colonies with no obvious pigmentation. This differential medium has been particularly useful in medical microbiology, since many enteric bacteria are unable to ferment lactose (e.g. *Salmonella*, *Shigella*) while others metabolise this carbohydrate (e.g. *Escherichia coli*, *Klebsiella* spp.). Colonial morphology (Fig. 38.2) on such a medium can give an experienced bacteriologist important clues to the identity of an organism, e.g. capsulate *Klebsiella* spp. characteristically produce large, convex, mucoid colonies with a weak pink coloration, owing to the fermentation of lactose, while *E. coli* produces smaller, flattened colonies with a stronger red coloration and a metallic sheen.

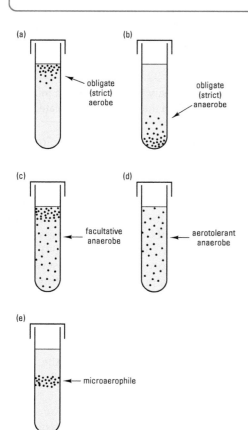

Fig. 37.1 Agar shake tubes. Bacteria are suspended in molten agar at 45–50°C and allowed to cool. The growth pattern after incubation reflects the atmospheric (oxygen) requirements of the bacterium.

> **KEY POINT** Selective methods are based on the use of physicochemical conditions that will permit the growth of a particular group of microbes while inhibiting others. Enrichment techniques encourage the growth of certain bacteria, usually by providing additional nutrients in the growth medium.

Methods based on specific physical conditions include:

- **Temperature.** Psychrophilic and psychrotrophic microbes can be isolated by incubating the growth medium at 4°C, while thermophilic microbes require temperatures above 45°C for isolation. Short-term heat treatment of samples can be used to select for endospore-forming bacteria, e.g. 70–80°C for 5–15 min, prior to isolation.
- **Atmosphere.** Many eukaryotic microbes are obligate aerobes, requiring an adequate supply of oxygen to grow. Bacteria vary in their responses to oxygen: obligate anaerobes are the most demanding, growing only under anaerobic conditions (e.g. in an anaerobic cabinet or jar). Oxygen requirements can be determined using the agar shake tube method as part of the isolation procedure (Fig. 37.1). Some pathogenic bacteria grow best in an atmosphere with a reduced oxygen status and increased CO_2 concentration: such carboxyphilic bacteria (capnophiles) are grown in an incubator where the gas composition can be adjusted.
- **Centrifugation.** This can be used to separate buoyant microbes from their non-buoyant counterparts – on centrifugation, such organisms will collect at the surface while the remaining microbes will sediment. Alternatively, density gradient methods may be used (p. 327). Centrifugation can be combined with repeated washing, to separate microbes from contaminants.
- **Ultraviolet irradiation.** Some microbes are tolerant of UV treatment and can be selected by exposing samples to UV light. However, the survivors may show a greater rate of mutation.

Aerotolerant anaerobe– a microbe that grows by fermentation, but which is insensitive to air/oxygen (in contrast to strict anaerobes, which are typically killed by exposure to air/oxygen).

Capnophile– a microbe that thrives in the presence of high levels of atmospheric carbon dioxide.

Facultative anaerobe– a microbe that grows by aerobic respiration when oxygen is present, switching to fermentation under anaerobic conditions.

- **Illumination.** Samples may be enriched for cyanobacteria and micro-algae by incubation under a suitable light regime. For dilute samples, where the number of photosynthetic microbes is too low to give the sample any visible green coloration, there is a risk of photoinhibition and loss of viability if the irradiance is too high. Such samples need shading during initial growth.

Chemical methods form the mainstay of bacteriological isolation techniques, and various media have been developed for the isolation of specific groups of bacteria. The chemicals involved can be subdivided into the following groups:

- **Selectively toxic substances:** for example, salt-tolerant, Gram-positive cocci can be grown in a medium containing 7.5%w/v NaCl, which prevents the growth of most common heterotrophic bacteria. Several media include dyes as selective agents, particularly against Gram-positive bacteria.

- **Antibiotics:** for example, the use of antibacterial agents (e.g. penicillin, streptomycin, chloramphenicol) in media designed to isolate fungi, or the use of antifungal agents (e.g. cycloheximide, nystatin) in bacterial media. Some antibacterial agents show a narrow spectrum of toxicity and these can be incorporated into selective isolation media for resistant bacteria, e.g. metronidazole for anaerobic bacteria.

- **Nutrients that encourage the growth of certain microbes:** including the addition of a particular carbon source, or specific inorganic nutrients.

- **Substances that affect the pH of the medium:** for example, the use of alkaline peptone water at pH 8.6 for the isolation of *Vibrio* spp.

> **KEY POINT** Note that subcultures from a primary isolation medium must be grown in a non-selective medium, to confirm the purity of the isolate.

Many of the selective and enrichment media used in bacteriology are able to distinguish between different types of bacteria: such media are termed differential media or diagnostic media and they are often used in the preliminary stages of an identification procedure. Box 37.1 gives details of the constituents of MacConkey medium, a selective, differential medium used in clinical microbiology (e.g. for the isolation of certain faecal bacteria), while Table 37.1 gives details of selective agents used. Further details on methods can be found in Collins *et al.* (2004). Note that isolation procedures for a particular microbe often combine several of the techniques described above. For instance, a protocol for isolating food-poisoning bacteria from a foodstuff might involve:

1. **homogenisation** of a known amount of sample in a suitable diluent;
2. **serial decimal dilution;**
3. **separation procedures** using spread or pour plates to quantify the number of bacteria of a particular type present in the foodstuff and provide a count;
4. **selective/enrichment procedures,** e.g. specific media/temperatures/atmospheric conditions, depending on the bacteria to be isolated;
5. **confirmation of identity:** any organism growing on a primary isolation medium would require subculture and further tests, to confirm the preliminary identification (Chapter 38).

Table 37.1 Selective agents in bacteriological media.

Substance	Selective for
Azide salts	*Enterococcus* spp.
Bile salts	Intestinal bacteria
Brilliant green	Gram-negative bacteria
Gentian violet	Gram-negative bacteria
Lauryl sulfate	Gram-negative bacteria
Methyl violet	*Vibrio* spp.
Malachite green	*Mycobacterium*
Polymyxin	*Bacillus* spp.
Sodium selenite	*Salmonella* spp.
Sodium chloride	Halotolerant bacteria
	Staphylococcus aureus
Sodium tetrathionate	*Salmonella* spp.
Trypan blue	*Streptococcus* spp.
Tergitol/surfactant	Intestinal bacteria

Text reference

Collins, C.H., Lyne, P.M. and Grange, J.M. (2004) *Microbiological Methods,* 8th edn. Hodder-Arnold, London.

Sources for further study

Atlas, R.M. (2011) *Handbook of Microbiological Media,* 4th edn. CRC Press, Boca Raton. [Gives details of culture media for a broad range of microbes.]

Eaton, A.D., Clesceri, L.S., Rice, E.W. and Greenberg, A.E. (2005) *Standard Methods for Examination of Water and Wastewater,* 21st edn. American Public Health Association, Washington, DC. [Gives standard US protocols for a range of indicator bacteria.]

Hurst, C.J., Crawford, R.L., Garland, J.L., Lipson, D.A., Mills, A.L. and Stetzenbach, L.D. (2007) *Manual of Environmental Microbiology,* 3rd edn. American Society for Microbiology, Washington, DC.

Levett, P.N. (ed.) (1991) *Anaerobic Microbiology: A Practical Approach.* IRL Press, Oxford. [Provides details of the methods used to isolate and culture anaerobic micro-organisms.]

Pepper, I.L. and Gerba, C.P. (2005) *Environmental Microbiology: A Laboratory Manual,* 2nd edn. Academic Press, New York.

Rochelle, P.A. (2001) *Environmental Molecular Microbiology: Protocols and Applications.* Horizon Scientific Press, Norwich. [Covers molecular approaches to the detection and identification of microbes in environmental samples.]

STUDY EXERCISES

37.1 Plan a collection/sampling strategy for a target group of microbes. What approaches might you take in the following instances?

(a) Collecting representatives of the normal skin microflora.

(b) Sampling psychrotrophic anaerobes from the subsurface mud of an estuary.

(c) Collecting a sample of sea water to enumerate faecal indicator bacteria.

(d) Mapping the microflora on the surface of a leaf.

37.2 Decide on an appropriate isolation procedure for a particular microbe. How might you isolate the following microbes from a sample?

(a) Photosynthetic flagellate algae in a sample of pond water.

(b) Bacteria growing as a biofilm on the surface of sand particles.

(c) Faecal streptococci (enterococci) present at low density (<1 per ml) in a sample of river water.

(d) A distinctively shaped bacterium present at very low numbers in a sample containing a high number of other microbes of different shape but with similar nutritional requirements.

37.3 Investigate the selective basis of microbiological media. Using textbooks on bacteriological methods or the web, research how each of the following media operates, in terms of their selective and diagnostic (differential) features:

(a) Mannitol salt agar for *Staphylococcus aureus.*

(b) Membrane lauryl sulfate broth for coliforms and *E. coli.*

(c) Slanetz and Bartley medium for faecal streptococci (enterococci).

(d) Mannitol pyruvate egg yolk polymyxin (MPYP) medium for *Bacillus cereus.*

Answers to these study exercises are available at www.pearsoned.co.uk/practicalskills.

The changing nature of microbial identification – while most traditional methods are based on microscopy and culture, many contemporary approaches also use the techniques of molecular biology (p. 457) to provide information about nucleic acid composition and sequence.

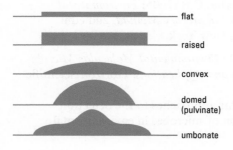

flat
raised
convex
domed (pulvinate)
umbonate

Fig. 38.1 Colony elevations (cross-sectional profile).

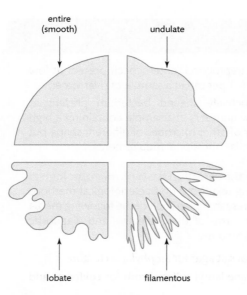

entire (smooth)

undulate

lobate

filamentous

Fig. 38.2 Colony margins (surface view).

SAFETY NOTE Never attempt to smell mould cultures, because of the risk of inhaling large numbers of spores.

Most of the methods described in this chapter were developed for the identification of bacteria, and bacterial examples are used to illustrate the principles involved. While the basic techniques are equally applicable to other types of microbes, the identification systems for some protozoa, fungi and algae rely predominantly on microscopic appearance. Identification of viruses requires electron microscopy or immunological techniques.

KEY POINT   Identification of bacteria is often based on a combination of a number of different features, including growth characteristics, microscopic examination, physiological and biochemical characterisation, and, where necessary, immunological and nucleic acid tests.

Direct observation

Once a microbe has been isolated (Chapter 37) and cultured in the laboratory (Chapter 36), the visual appearance of individual colonies on the surface of a solidified medium may provide useful information. Bacteria typically produce smooth, glistening colonies, varying in diameter from <1 mm to >1 cm. Actinomycete colonies are often <1 cm, with a shrivelled, powdery surface. Filamentous fungi usually grow as large, spreading mycelia composed of microscopic branching filaments (hyphae) with a matt appearance and are identified by microscopy, using the morphological characteristics of their reproductive structures. Yeasts produce small, glistening colonies; identification usually involves microscopy, combined with physiological and biochemical tests similar to those used for bacteria.

Colony characteristics

When measuring colony size, choose a typical colony, well spaced from any others, as colony size is affected by competition for nutrients. The characteristics of a microbial colony on a particular medium include:

- **Size:** some bacteria produce punctiform (pin-point) colonies, with a diameter of less than 1 mm, while motile bacteria may spread over the entire plate.
- **Form:** colonies may be circular, irregular, lenticular (spindle-shaped) or filamentous.
- **Elevation:** colonies may be flat, raised, convex, etc. (see Fig. 38.1).
- **Margin:** the edge of a colony may be entire (smooth) or more distinctive, e.g. undulate, lobate or filamentous (Fig. 38.2).
- **Consistency:** colonies may be viscous (or mucoid), butyrous (of similar consistency to butter) or friable (dry and granular), etc.
- **Colour:** some bacteria produce characteristic pigments. A few pigments are fluorescent under UV light.
- **Optical properties:** colonies may be translucent or opaque.
- **Haemolytic reactions on blood agar:** many pathogenic bacteria produce characteristic zones of haemolysis. Alpha haemolysis is a partial breakdown of the haemoglobin from the erythrocytes, producing a green zone around the colony, while beta haemolysis is the complete destruction of haemoglobin, producing a clear zone.

- **Odour:** some actinomycetes and cyanobacteria produce earthy odours, while certain bacteria and yeasts produce fruity or 'off' odours. However, odour is not a reliable characteristic in bacterial identification and is not recommended on safety grounds.

Microscopic examination

Bacteria are usually observed using an oil immersion objective at a total magnification of ×1000 (p. 183).

Motility

Wet mounts can be prepared by placing a small drop of bacterial suspension on a clean, degreased slide, adding a coverslip and examining the film without delay. For aerobes, areas near air bubbles or by the edge of the coverslip give best results, whereas anaerobes show greatest motility in the centre of the preparation, with rapid loss of motility owing to oxygen toxicity.

Prepare wet mounts using young cultures in exponential growth in a liquid medium (p. 231). It is best to work with cultures grown at 20 or 25 °C, since those grown at 37 °C may not be actively motile on cooling to room temperature. It is essential to distinguish between the following:

- **True swimming motility,** caused by the presence of flagella: bacteria dart around the field of view, changing direction in zigzag, tumbling movements.
- **Brownian motion:** non-motile bacteria show a localised, vibratory, random motion, owing to bombardment of bacterial cells by molecules in the solution.
- **Passive motion,** due to currents within the suspension: all cells will be swept in the same direction at a similar rate of movement.
- **Gliding motility:** a slower, intermittent movement, parallel to the longitudinal axis of the cell, requiring contact with a solid surface.

Cell shape

Bacteria are subdivided into the following groups:

- **Cocci** (singular, coccus): spherical, or almost spherical, cells, sometimes growing in pairs (diplococci), chains or clumps.
- **Rods:** straight, cylindrical cells of variable length, with flattened, tapered or rounded ends; sometimes termed bacilli. Short rods are sometimes called cocco-bacilli.
- **Curved rods:** the curvature varies according to the organism, from short curved rods, sometimes tapered at one end, to spiral shapes.
- **Branched filaments:** characteristic of actinomycete bacteria.

Gram staining

This is the most important differential staining technique in bacteriology (Box 38.1 gives details). It enables us to divide bacteria into two distinct groups, Gram-positive and Gram-negative, according to a particular staining procedure (the technique is given a capital letter, since it is named after its originator, H.C. Gram). The basis of the staining reaction is the different structure of the cell walls of Gram-positive and Gram-negative bacteria. Heat fixation of air-dried bacteria causes some shrinkage, but cells retain their shape: to measure cell dimensions use a chemical fixative.

Gram staining should be carried out using light smears of young, active cultures, since older cultures may give variable results. In particular, certain

---

Using the hanging drop technique – place a drop of bacterial suspension on a coverslip and invert over a cavity slide so that the drop does not make contact with the slide: motile aerobes are best observed at the edge of the droplet, where oxygen is most abundant.

Assessing motility – if you have not seen bacterial motility before, it is worth comparing your unknown bacterium with a positive and a negative control.

Using cell shape in microbial identification – many bacteria are pleomorphic, varying in size and shape according to the growth conditions and the age of the culture: thus, other characteristics are required for identification.

SAFETY NOTE The Gram-staining procedure involves toxic dyes and flammable solvents: avoid skin contact and extinguish any naked flames (e.g. Bunsens).

Assessing the Gram status of an unknown bacterium – if a pure culture gives both Gram-positive and Gram-negative cells, identical in size and shape, it can be regarded as a Gram-positive organism that is demonstrating Gram-variability.

| **Box 38.1** | **Preparation of a heat-fixed, Gram-stained smear** |

Preparation of a heat-fixed smear

The following procedure will provide you with a thin film of bacteria on a microscope slide, for staining.

1. Take a clean microscope slide and pass it through a Bunsen flame twice, to ensure it is free of grease. Allow to cool.

2. Using a sterile inoculating loop, place a single drop of water in the centre of the slide and then mix in a small amount of sample from a single bacterial colony with the drop, until the suspension is slightly turbid. Smear the suspension over the central area of the slide, to form a thin film. For liquid cultures, use a single drop of culture fluid, spread in a similar manner.

3. Allow to air-dry at room temperature, or high above a Bunsen flame: air-drying must proceed gently, or the cells will shrink and become distorted.

4. Fix the air-dried film by passage through a Bunsen flame. Using a slide holder or forceps, pass the slide, film side up, rapidly through the hottest part of the flame (just above the blue cone). The temperature of the slide should be just too hot for comfort on the back of your hand: note that you must not overheat the slide or you may burn yourself (you will also ruin the preparation).

5. Allow to cool: the smear is now ready for staining.

Gram-staining procedure

The version given here is a modification of the Hucker method, since acetone is used to decolorise the smear. Note that some of the staining solutions used are flammable, especially the acetone decolorising solvent: you must make sure that all Bunsens are turned off during staining. The procedure should be carried out with the slides suspended over a sink, using a staining rack.

1. Flood a heat-fixed smear with 2% w/v crystal violet in 20% v/v ethanol: water and leave for 1 min.

2. Pour off the crystal violet and rinse briefly with tap water. Flood with Gram's iodine (2g KI and 1 g I$_2$ in 300 ml water) for 1 min.

3. Rinse briefly with tap water and leave the tap running gently.

4. Tilt the slide and decolorise with acetone for 2–3s: acetone should be added dropwise to the slide until no colour appears in the effluent. This step is critical, since acetone is a powerful decolorising solvent and must not be left in contact with the slide for too long.

5. Immediately immerse the smear in a gentle stream of tap water, to remove the acetone.

6. Pour off the water and counterstain for 10–15 s using 2.5% w/v safranin in 95% v/v ethanol: water.

7. Pour off the counterstain, rinse briefly with tap water, then dry the smear by blotting gently with absorbent paper: all traces of water must be removed before the stained smear is examined microscopically.

8. Place a small drop of immersion oil on the stained smear: examine directly (without a coverslip) using an oil immersion objective (p. 183).

Gram-positive bacteria retain the crystal violet (primary stain) and appear purple, while Gram-negative bacteria are decolorised by acetone and counterstained by the safranin, appearing pink or red when viewed microscopically.

Other decolorising solvents are sometimes used, including ethanol : water, ethyl ether : acetone and acetone : alcohol mixtures. The time of decolorisation must be adjusted depending upon the strength of the solvents used, e.g. 95% v/v ethanol : water is less powerful than acetone, requiring around 30s to decolorise a smear.

Staining bacterial spores – these are relatively impermeable to dyes, and require heat treatment and specialised stains.

Gram-positive bacteria may stain Gram-negative if older cultures are used. This Gram-variability is because of autolytic changes in the cell wall of Gram-positive bacteria. Developing spores are often visible as unstained areas within older vegetative cells of *Bacillus* and *Clostridium*. Other stains are required to demonstrate sub-cellular structures such as capsules or flagella (p. 184).

Basic laboratory tests

At least two simple biochemical tests are usually performed.

Oxidase test

Performing the oxidase test – never use a nichrome wire loop, as this will react with the oxidase reagent, giving a false positive result.

This identifies cytochrome *c* oxidase, an enzyme found in obligate aerobic bacteria. Soak a small piece of filter paper in a fresh solution of 1% (w/v) *N*- *N*- *N'*- *N'*-tetramethyl- *p*-phenylenediamine dihydrochloride on a clean microscope slide. Rub a small amount from the surface of a young, active colony onto the filter paper using a glass rod, a *plastic* loop or a wooden applicator stick: a purple-blue colour within 10 s is a positive result.

Avoiding false negatives – ensure you use sufficient material during oxidase and catalase testing, otherwise you may obtain a false negative result: a clearly visible 'clump' of bacteria should be used.

Carbohydrate utilisation tests and isolation media – many diagnostic agar-based media incorporate one or more specific carbohydrates and pH indicator dyes, thereby providing additional information as part of the isolation procedure.

Understanding biochemical identification – given the variability of different isolates of a single species, a large number of tests must be used to provide a high degree of probability of correctly identifying the isolate. This is often carried out in specialised labs using computer-aided analysis.

Catalase test

This identifies catalase, an enzyme found in obligate aerobes and in most facultative anaerobes, which catalyses the breakdown of hydrogen peroxide into water and oxygen ($2H_2O_2 \rightarrow 2H_2O + O_2$). Transfer a small sample of your unknown bacterium on to a coverslip using a disposable plastic loop or glass rod. Invert onto a drop of hydrogen peroxide: the appearance of bubbles within 30s is a positive reaction. This method minimises the dangers from aerosols formed when gas bubbles burst.

The oxidase and catalase tests effectively allow us to subdivide bacteria on the basis of their oxygen requirements, without using agar shake cultures (p. 243) and overnight incubation, since, in broad terms:

- obligate aerobes will be oxidase-positive and catalase-positive;
- facultative anaerobes will be oxidase-negative and catalase-positive;
- microaerophilic bacteria, aerotolerant anaerobes and strict (obligate) anaerobes will be oxidase-negative and catalase-negative – the latter group will grow only under anaerobic conditions (p. 243).

Once you have reached this stage (colony characteristics, motility, shape, Gram reaction, oxidase and catalase status), it may be possible to make a tentative identification, at least for certain Gram-positive bacteria, at the generic level. To identify Gram-negative bacteria, particularly the oxidase-negative, catalase-positive rods, further tests are required.

Identification tables: further laboratory tests

Bacteria are asexual organisms and strains of the same species may give different results for individual biochemical/physiological tests. This variation is allowed for in identification tables, based on the results of a large number of tests. Identification tables are often used for particular subgroups of bacteria, after Gram staining and basic laboratory tests have been performed: an example is shown in Table 38.1.

A large number of specific biochemical and physiological tests are used in bacterial identification, including:

- **Carbohydrate utilisation tests.** Some bacteria can use a particular carbohydrate as a carbon and energy source. Acidic end products can be identified using a pH indicator dye (p. 170) while CO_2 is detected in liquid culture using an inverted small test tube (Durham tube, Fig. 38.3).

Table 38.1 Identification table for selected Gram-negative rods.

Bacterium	Biochemical test								
	1	2	3	4	5	6	7	8	9
Escherichia coli	V	+	−	+	−	V	V	+	−
Proteus mirabilis	−	−	V	−	+	−	+	−	+
Morganella morganii	−	−	−	−	+	−	+	+	−
Vibrio parahaemolyticus	−	+	V	−	−	+	+	+	−
Salmonella spp.	—	+	V	−	−	+	+	−	+

Key to biochemical tests and symbols:
1. sucrose utilisation
2. mannitol utilisation
3. citrate utilisation
4. β-galactosidase activity
5. urease activity
6. lysine decarboxylase activity
7. ornithine decarboxylase activity
8. indole production
9. H_2S production
+, >90% of strains tested positive
−, <10% of strains tested positive
v, 10–90% of strains tested positive.

Fig. 38.3 Durham tube in carbohydrate utilisation broth. Air within the Durham tube is replaced by broth during the autoclaving procedure.

Aerobic breakdown (via respiration) is termed oxidation, and anaerobic breakdown is known as fermentation. Identification tables usually incorporate tests for several different carbohydrates, e.g. Table 38.1.

- **Enzyme tests.** Most of these incorporate a substance that changes colour if the enzyme is present, e.g. a pH indicator, or a chromogenic substrate.
- **Tests for specific end products of metabolism,** e.g. the production of indole owing to the metabolic breakdown of the amino acid tryptophan, or H_2S from sulfur-containing amino acids.

Identification kits

Some biochemical tests are now supplied in kit form, e.g. the API 20E system incorporates 20 tests within a sterile plastic strip (Fig. 38.4). After inoculation and overnight incubation, the results of the tests are converted into a seven-digit code, for comparison with known bacteria using either a computer program (the Analytical Profile Index), or a reference book. While kit identification systems save time and labour, they are more expensive and less flexible than conventional biochemical tests.

Immunological tests

Tests used in diagnostic microbiology include:

- **Agglutination tests:** based on the reaction between specific antibodies and a particular bacterium (p. 282). These tests are particularly useful for subdividing biochemically similar bacteria.
- **Fluorescent antibody tests:** the reaction between a labelled antibody and a particular bacterium can be visualised using UV microscopy. The direct fluorescent antibody test uses fluorescein isothiocyanate as the label.
- **Enzyme-linked immunoassay tests** using antibodies labelled with a particular enzyme, e.g. the double antibody sandwich ELISA or competitive ELISA tests (p. 285).

While such tests can give specific and accurate confirmation of the identity of a bacterium under controlled laboratory conditions, they are often too expensive and time-consuming for routine identification purposes, especially when large numbers of tests are required.

Typing methods

The identification of bacteria at subspecies level is known as typing: this is usually done in a specialist laboratory, e.g. as part of an epidemiological study to establish the source of an infection. Various methods are used:

- **Antigen typing,** or serotyping, is based on immunological tests.
- **Phage typing** is based on the susceptibility of different strains to certain bacterial viruses (phages).

Molecular approaches to microbial identification – several novel methods of detection and identification are based on nucleic acid techniques, including use of the PCR (p. 467) and Southern blotting (p. 461) to detect particular microbes.

Using antibiotic resistance to identify bacteria – patterns of resistance to antibacterial agents in lab tests can be used to generate an 'antibiogram', giving information that can aid identification, as well as providing data of clinical significance, e.g. for therapy of bacterial infections.

Practical applications of bacterial typing – *E. coli* O157:H7 is a serotype of this bacterium that is capable of causing severe human disease: it can be identified on the basis of an agglutination reaction with an appropriate antiserum. Typing is also discussed on p. 254.

Fig. 38.4 Example of a bacterial identification kit (API 20E). Property of bioMérieux S.A/photo: Andrea Bannuscher.

- **Biotyping** is based on biochemical differences between different strains e.g. enzyme profiles or antibiotic resistance screening ('antibiograms').
- **Bacteriocin typing:** bacteriocins are proteins released by bacteria that inhibit the growth of other members of the same species.

Sources for further study

Barrow, G.I. and Feltham, R.K.A. (2004) *Cowan and Steel's Manual for the Identification of Medical Bacteria,* 3rd edn. Cambridge University Press, Cambridge.
[A standard reference work for the identification of clinically important bacteria.]

Cullimore, D.R. (2010) *Practical Atlas for Bacterial Identification,* 2nd edn. CRC Press, Boca Raton.
[Detailed coverage of practical methods applicable to a wide range of bacteria.]

Fox, A. (ed.) *Journal of Microbiological Methods,* Elsevier, London (available through Science Direct at: http://www.sciencedirect.com/)
[Provides information on novel developments in all aspects of microbiological methods, including microbial culture/isolation and molecular approaches.]

Macfaddin, J.F. (2000) *Biochemical Tests for the Identification of Medical Bacteria,* 3rd edn. Williams & Wilkins, Philadelphia.
[Explains the operating principles underlying most of the biochemical tests in routine use in diagnostic bacteriology.]

Richardson, M. D. and Wamock, D. W. (2012). *Fungal Infections: Diagnosis and Management,* 4th edn. Wiley, New York.

Spencer, J.F.T. and Spencer, A.L.R. (2010) *Environmental Microbiology: Methods and Protocols.* Humana Press, New York.

STUDY EXERCISES

38.1 Describe the colonial characteristics of selected microbes. Either research the features of the following microbes (e.g. via the web), or look at well-isolated individual colonies in the laboratory, following overnight growth on a suitable medium:

a) *Escherichia coli;*

b) *Pseudomonas aerogenes;*

c) *Staphylococcus aureus;*

d) *Streptococcus pneumoniae;*

e) *Bacillus subtilis;*

f) *Saccharomyces cerevisiae.*

38.2 Find out how some of the tests used in microbial identification work. Research the operating principles that underpin the following tests, using microbiological textbooks or the web, and prepare brief notes explaining how each of the following tests works:

a) oxidase test;

b) indole test;

c) β-galactosidase test;

d) urease test;

e) lysine decarboxylase test;

f) H_2S production.

38.3 Identify the following oxidase-negative, catalase-positive. Gram-negative rod-shaped bacteria, using Table 38.1.

a) Positive for citrate, urease, ornithine decarboxylase and H_2S only.

b) Positive for sucrose, mannitol, β-galactosidase, lysine decarboxylase and indole only.

c) Positive for mannitol, citrate, lysine decarboxylase, ornithine decarboxylase and H_2S only.

d) Positive for urease, ornithine decarboxylase and indole only.

e) Positive for mannitol, β-galactosidase and indole only.

Answers to these study exercises are available at www.pearsoned.co.uk/practicalskills.

39 Naming microbes and other organisms

Fig. 39.1 Simplified diagram of the three major domains of the Tree of Life (based on the rRNA sequencing work of Carl Woese).

The use of scientific names is fundamental to all aspects of biological science since it aims to provide a system of identification that is precise, fixed and of universal application. Without such a system, comparative studies would be impossible.

There are two possible bases for such classification:

1. **Phenetic taxonomy,** which involves grouping on the basis of phenotypic similarity, frequently using complex statistical techniques to obtain objective measures of similarity. The characters used have been largely morphological and anatomical, but biochemical, cytological and other characters are increasingly used, especially for microbes (p. 247) where structural characters are few.

2. **Phylogenetic (=phyletic) taxonomy,** which involves grouping on the basis of presumed evolutionary, and therefore genetic, relationships.

These two systems are often broadly similar in outcome, since closely related organisms are usually fairly similar to each other and because judgements of evolutionary relationships are usually themselves based upon similarities. The situation is made more complex by phenomena such as convergent and divergent evolution; phyletic classifications are also liable to subjective bias.

New classifications have been proposed on the basis of studies of biomolecules, e.g. rRNA sequences, that are robust and objective. One such arrangement divides organisms into three major domains: Bacteria, Archaea and Eucarya (Fig. 39.1).

The hierarchical system

Another approach is to recognise six kingdoms if the viruses are included:

1. **Virales:** the viruses. Very simple in structure, obligate parasites of prokaryotic or eukaryotic cells.

2. **Prokaryotae (Monera):** the bacteria and cyanobacteria. Prokaryotic, non-nucleate and lacking membrane-bound organelles, such as: mitochondria, plastids, endoplasmic reticulum, Golgi apparatus.

3. **Protista:** includes algae, protozoa and slime moulds. Eukaryotic and mainly unicellular. A heterogeneous group.

4. **Plantae:** the plants. Eukaryotic, walled, mainly multicellular and usually photoautotrophs.

5. **Fungi:** eukaryotic, usually syncytial, walled and saprotrophic.

6. **Animalia:** the animals. Eukaryotic, multicellular and heterotrophs.

Six other levels of taxa are generally accepted: division/phylum, class, order, family, genus and species.

KEY POINT   Whereas the levels of taxa above genus are rather subjective and vary among authorities, the use of genus and species names are governed by strict, internationally agreed conventions called Codes of Nomenclature.

There are three major Codes of Nomenclature, the Botanical, Zoological and Bacteriological Codes, which operate on similar but not identical principles (Lapage *et al.,* 1992; Wakeham-Dawson *et al.,* 1999; McNeill *et al.,* 2012). Fungi are covered by the same International Code as plants despite the fact that they are not photoautotrophs.

The basis of classification

No simple definition of a species is possible, but there are two generally used definitions:

1. **A group of organisms capable of interbreeding and producing fertile offspring** – this, however, excludes all asexual, parthenogenetic and apomictic forms.

2. **A group of organisms showing a close similarity in phenotypic characteristics** – this would include morphological, anatomical, biochemical, ecological and life history characters.

> **KEY POINT** The basic unit of classification is the species, which represents a group of recognisably similar individuals, clearly distinct from other such groups.

When species are compared, groups of species may show a number of features in common; they are then arranged into larger groupings known as genera (singular genus). This process can be repeated at each taxonomic stage to form a hierarchical system of classification whose different levels are known as taxonomic ranks. The number of levels in this system is arbitrary and based upon practical experience – the seven levels normally used have been found to be sufficient to accommodate the majority of the variation observed in nature.

When a generic or specific name is changed as a result of further study, the former name becomes a synonym; you should always try to use the latest name. Where a generic name has been changed recently, the old name is occasionally given in parentheses to allow easy reference to the old name.

Nomenclature in practice

The scientific name of an organism is effectively a symbol or cipher that removes the need for repeated use of descriptions. It normally comprises two words and is, therefore, called a binomial term; for example, *Escherichia coli*. The name of the genus is followed by a second term that identifies the species, e.g. *Pseudomonas aeruginosa* or *Saccharomyces cerevisiae* (Table 39.1).

The Codes of Nomenclature were established to prevent the ambiguities associated with informal/common names. The Codes require that all

DNA-based definition of a species – in microbiology, members of a bacterial species can be characterised by having a DNA similarity of ≥70% based on nucleic acid hybridisation studies (p. 461).

Writing taxonomic names – always underline (handwritten text) or italicise (word-processed text) generic and species names to avoid confusion: thus bacillus is a descriptive term for rod-shaped bacteria, while <u>Bacillus</u> is a generic name.

Table 39.1 Example of taxonomies for a bacterium and a fungus.

Common name	*Pseudomonas*	Baker's yeast
Kingdom	Monera	Fungi
Division/Phylum	Gracilicutes	Ascomycota
Class	Scotobacteria	Incertae sedis[1]
Order	Pseudomonadales	Saccharomycetales
Family	Pseudomonadaceae	Saccharomycetaceae
Genus	*Pseudomonas*	*Saccharomyces*
Species	*P. aeruginosa*	*S. cerevisiae*

[1] Meaning = 'uncertain taxonomic position'.

> ## Box 39.1 Basic rules for the writing of taxonomic names
>
> - Names of the seven levels of taxa should take lowercase initial letters, e.g. kingdom Fungi.
> - The Latin forms of all taxon names except the specific name take initial capital letters, e.g. 'the Ascomycota ...' but anglicised versions do not, e.g. 'the ascomycetes...'.
> - The names of the higher taxa are all plural, hence 'the Ascomycota are...' while the singular of the anglicised version is used for a single member of that taxon, hence 'an ascomycete is...'.
> - The binomial system gives each species two names, the first being the generic name and the second the specific name, which must never be used by itself. The genus and species names are distinguished from the rest of the text either:
>   - a. by being underlined (when handwritten), e.g. <u>Penicillium notatum</u>; or
>   - b. by being set in italics (in print or on a word processor), e.g. *Penicillium notatum*.
> - The generic name is singular and always takes an initial capital letter. If you use the generic name this implies that the point being made is a generic characteristic unless the specific name is present. Write the generic name in full when first used in a text, e.g. *Escherichia coli*, but subsequent references can be abbreviated to its initial letter, e.g. *E. coli*, unless this will result in confusion with another genus also being considered; for example, with *Escherichia coli* and *Enterococcus faecalis*, these could be abbreviated to *Esch. coli* and *Ent. faecalis*, to avoid any confusion.
> - The abbreviation 'sp.' should be used in place of the specific name if a single unspecified species of a genus is being referred to, e.g. *Bacillus* sp.; it is not underlined or italicised. If more than one unspecified species is meant, then the correct form is 'spp.', e.g. *Bacillus* spp.
> - The name of each species should be followed by the authority: on first usage in formal reports and in titles, the name or names of the person(s) to whom that name is attributed and the date of that description should be quoted. These names may sometimes be abbreviated. If the species was first described under its current generic name, the authority's name, often in abbreviated form, is added. If, however, the species was first described under a different genus, the name of the author of the original description is presented in parentheses, e.g. *Escherichia coli* (Migula) Castellani and Chalmers. The use of authorities should be confined to formal papers, final-year project reports, etc.; they would not normally be used in practical reports, short assignments or examinations.

Example   The full, formal name of baker's yeast is *Saccharomyces cerevisiae* Meyen ex Hansen 1883. After first use, this would be abbreviated to *S. cerevisiae*.

Applications of bacterial typing – typing methods are used widely in epidemiological studies, e.g. tracing a particular type of bacterium responsible for a food-poisoning outbreak, or tracking the development of a particular type of antibiotic-resistant bacterium in a hospital.

scientific names are either Latin or treated as Latin, written in the Latin alphabet and subject to the rules of Latin grammar. Consequently, you must be very precise in your use of such names. In some cases, the Codes stipulate a standardised ending for the names of all taxa of a given taxonomic rank, e.g. fungal and bacterial families end in -aceae. When used in a formal scientific context, you should follow the specific name by the authority on which that name is based, i.e. the name of the person describing that species and the date of the description.

Box 39.1 summarises the basic rules for writing taxonomic names.

Taxa below the rank of species

In microbial classification, use is often made of taxa below the rank of species.

In bacteriology, the use of subspecies is acceptable although a word indicating rank is usually inserted, e.g. *Bacillus subtilis* subsp. *niger*. However, other terms are in widespread use for taxa below the species level, especially in medical microbiology and plant pathology, when a particular strain of bacterium has been identified (pp. 249–50). Subspecies identification is often referred to as typing and the following terms apply:

- **biovar, or biotype:** subdivided according to biochemical characteristics;
- **serovar or serotype:** subdivided by serological methods, using antibodies (see Chapter 43);
- **pathovar:** subdivided according to pathogenicity (ability to cause disease);
- **phagovar or phage type:** subdivided according to susceptibility to particular viruses.

Understanding microbiological terminology – the term *strain* is widely used, particularly in the context of the practice of lodging microbiological strains with culture collections, while the term *isolate* is often used for a pure culture derived from a natural (wild) population. Cell lines (Chapter 40) are often given code names and/or reference numbers.

Many micro-organisms are now referred to by their generic and specific names followed by a culture collection reference number, e.g. *Bacillus subtilis* NCTC 10400, where NCTC stands for the National Collection of Type Cultures and 10400 is the reference number of that strain in the collection.

The special case of viruses

The classification and nomenclature of viruses are less advanced than for cellular organisms and the current nomenclature has been arrived at on a piecemeal, *ad hoc* basis. The International Committee for Virus Taxonomy proposed a unified classification system, dividing viruses into 50 families on the basis of:

- **host preference;**
- **nucleic acid type** (i.e. DNA or RNA);
- **whether the nucleic acid is single- or double-stranded;**
- **the presence or absence of a surrounding envelope.**

Example The virus that causes tobacco mosaic disease belongs to the genus *Tobamovirus* and can be referred to as tobacco mosaic tobamovirus, tobacco mosaic virus or TMV.

Virus family names end in -viridae and genus names in -virus. (Note that these names are *not* latinised and the genus–species binomial is not now approved.) However, this system has not yet been adopted universally and many viruses are still referred to by their trivial names or by code names (sigla), e.g. the bacterial viruses ϕX174, T4, etc. Many of the names used reflect the diseases caused by the virus. Often, a three-letter abbreviation is used, e.g. HIV (for human immunodeficiency virus), TMV (for tobacco mosaic virus).

Text references

Lapage, S.P. *et al.* (eds) (1992) *International Code of Nomenclature of Bacteria (1990 revision).* American Society for Microbiology, Washington, DC.

McNeill, J. *et al.* (2012) *International Code of Nomenclature for Algae, Fungi and Plants: Melbourne Code.* Koeltz Scientific Books, Kóeningstein.

Wakeham-Dawson, A. *et al.* (eds) (1999) *International Code of Zoological Nomenclature,* 4th edn. The International Trust for Zoological Nomenclature, London.

Sources for further study

Buchen-Osmond, C. (2002) *Universal Virus Database of the International Committee on Taxonomy of Viruses.* Available: http://www.ncbi.nlm.nih.gov/ICTVdb/
Last accessed: 01/09/15.
[Includes a catalogue of approved virus names, image gallery, descriptions, etc.]

Institute of Biology (2000) *Biological Nomenclature: Recommendations on Terms, Units and Symbols.* Institute of Biology, London.
[Includes sections on taxonomy and classification of organisms.]

Jeffrey, C. (1989) *Biological Nomenclature,* 3rd edn. Edward Arnold, London.

King A.M.Q. *et al.* (2011) *Virus Taxonomy: IXth Report of the International Committee on Taxonomy of Viruses.* Elsevier, London.

McNeill, J. *et al. International Code of Nomenclature for Algae, Fungi and Plants (Melbourne Code).* available: http://www.iapt-taxon.org/nomen/main.php
Last accessed: 01/09/15.
[Electronic version of the original English text.]

STUDY EXERCISES

39.1 Research full classifications. Provide the full classification of the following species, laid out as in the table in Study exercise 39.2.

(a) The limpet *Patella vulgata*

(b) The great white shark

(c) The giant redwood

(d) The earthworm *Lumbricus terrestris*.

39.2 What is wrong with the hierarchical classifications given in the table below?

Examples of hierarchical classifications

Nautilus	Edible mushroom	*E. coli* bacterium
Animalia	Fungi	Monera (Bacteria)
Mollusca	Basidiomycota	Proteobacteria
Cephalopoda	Basidiomycetes	Enterobacteriales
Nautilidae	Agaricales	Gamma subdivision
Nautilida	Agaricaceae	Enterobacteriaceae
Nautilus	*Agaricus*	*Escherichia*
pompilius	*bisporus*	*coli*

39.3 Compare a variety of current textbooks and Internet sites to discover alternative classification schemes at kingdom level. Make your own notes regarding the *evidence* used to support the alternative kingdom classifications.

Answers to these study exercises are available at www.pearsoned.co.uk/practicalskills.

Working with vertebrate animals and their organs/tissues – remember that procedures must be consistent with the law, i.e. in the UK, the Animals Scientific Procedures Act 1986.

Using human tissues – in the UK, the Human Tissue Act 2004 applies and work is licensed and overseen by the Human Tissue Authority. Note that established cell lines (p. 259) are excluded from the regulations.

Table 40.1 Composition of Ringer's solution (simplified formulation, for amphibians, etc., pH 7.4–7.6).

Compound	Concentration (g L^{-1})
NaCl	6.0
KCl	0.075
CaCl$_2$	0.1
NaHCO$_3$	0.1

While the aim of many studies is to isolate, quantify or characterise individual molecules from a biological system, e.g. the purification of a particular enzyme (p. 403), this is not always the most appropriate course of action. Depending on the purpose of the investigation, it may be more relevant to study the functioning of biomolecules within more complex systems, in order to understand their role in a particular biological process. At one extreme this may be carried out *in vivo*, using whole multicellular organisms (e.g. individual animals or plants), while the other extreme is represented by *in vitro* studies, using subcellular 'cell-free' extracts. Between these two extremes, a range of tissue and cell culture techniques offers some of the biological complexity of the intact organism combined with a degree of experimental control that may not be obtainable *in vivo*. The ethics and costs of whole animal experimentation have provided an additional stimulus to the development of *in vitro* methods, e.g. toxicity tests using mammalian cell culture rather than laboratory animals, while developments in molecular genetics have led to further applications of cell and tissue culture.

Animal tissues and organs

Physiological experiments are carried out using either whole organisms, or a range of animal organs and tissues, including heart, liver, muscle, etc.

KEY POINT A major practical consideration is that the tissue should be studied as soon as possible after the death of the animal, typically under laboratory conditions that mimic the *in vivo* environment as closely as possible.

In most instances, the experiments are relatively short term (<24 h) and the aim is to maintain the tissue in a physiological state similar to that within the living organism. For metabolic studies, the whole organ or a tissue slice (typically 1–10 mm thick) will be bathed in an appropriate perfusion fluid, supplied either by gravity or by peristaltic pump. Practical considerations include the following:

- **Inorganic solute requirements** – the chemical composition of the perfusion fluid is usually chosen to reflect the major inorganic ion requirements of the tissue. For short-term studies, a number of so-called 'physiological salt solutions' may be used, e.g. Ringer's solution, one formulation of which is given in Table 40.1.

- **Oxygen requirements** – it may be necessary to increase the O$_2$ content of the perfusion fluid by bubbling with air, in order to meet the oxygen demand of the innermost parts of the tissue. However, this can lead to oxygen toxicity in the outermost parts and an alternative approach is simply to increase the rate of perfusion.

- **Physicochemical conditions** – including temperature (usually controlled to ±1 °C of normal body temperature), water status (the perfusion fluid and the tissue should be isotonic, p. 166), pH and buffering capacity (e.g. some perfusion fluids have an elevated NaHCO$_3$ concentration, to mimic the buffering capacity of mammalian serum).

Working with plant tissues and organs – the ethical problems associated with animal and human tissues are avoided, and there is a decreased risk of infection to the laboratory worker; practical manipulations are often carried out in horizontal laminar-flow cabinets (p. 225).

Table 40.2 Components of Long Ashton medium (nitrate version).

Stock solution: mass of component required per litre of solution (g)	Volume of stock solution to make 1 litre of medium (mL)
Major nutrients	
KNO_3: 50.60	8
$Ca(NO_3)_2$: 80.25	8
$MgSO_4 \cdot 7H_2O$: 46.00	8
$NaH_2PO_4 \cdot 2H_2O$: 52.00	4
Micronutrients	
FeKEDTA: 3.30	5
$MnSO_4 \cdot 4H_2O$: 2.23	1
$ZnSO_4 \cdot 7H_2O$: 0.29	1
$CuSO_4 \cdot 5H_2O$: 0.25	1
H_3BO_3: 3.10	1
$Na_2MoO_4 \cdot 2H_2O$: 0.12	1
NaCl: 5.85	1
$CoSO_4 \cdot 7H_2O$: 0.056	1

Effects of humans on plants – remember that your exhaled breath will be nearly saturated with water vapour and will contain CO_2 at 3–4% v/v, some 100 times more concentrated than atmospheric CO_2: your breath can thus affect rates of transpiration and photosynthesis.

- **Organic nutrient requirements** – for longer term studies, suitable nutrients will be required: these may be chemically defined additives, e.g. vitamins, amino acids, proteins, etc., or biological fluids such as plasma or serum. Glucose is often added as a carbon and energy source (Freshney, 2010).

Plant tissues and organs

Individual plant components (e.g. leaves, leaf slices and epidermal strips) can be isolated from the main plant body for study under controlled conditions. Since photosynthetic plant parts are autotrophic, they may be maintained *in vitro* for longer than animal organs, given adequate light and CO_2. However, most plant cells are joined via plasmodesmata, and the separation of such connections when the component is removed from the plant often leads to death when these connections are broken.

KEY POINT Plants show wound responses that may affect the metabolic processes under study. For these reasons, the most suitable systems for longer term studies are often whole organs, e.g. whole leaves or entire root systems.

Water culture (hydroponics), e.g. in Long Ashton medium (Table 40.2), is an alternative approach, offering greater control over the root environment (Smith, 2000).

It is best to use vigorous, healthy stock plants and to follow a well-established procedure, taking account of the following:

- **Sterility** – strict attention to sterile technique can be essential to the success of many longer term experiments. Decontamination of plant organs may be especially difficult where specimens are obtained from soil: to achieve this, use a surface wash with disinfectant (e.g. 10% w/v sodium hypochlorite), followed by several rinses with sterile water.

- **Gaseous environment** – in general, the experimental system should be well ventilated. Actively photosynthetic tissues will rapidly deplete the atmospheric CO_2 in a closed vessel: plant parts may also produce physiologically active gases, such as ethylene, especially at wound sites. Turgor loss may occur in isolated plant parts unless a high humidity is maintained.

- **Nutrition** – plant tissues may benefit from a supply of inorganic ions, including K^+, SO_4^{2-}, etc. and may require certain vitamins, micronutrients and plant hormones for prolonged studies.

- **Physicochemical conditions** – light is the most important environmental requirement for green plant parts. At atmospheric CO_2 concentrations, the light compensation point is about 5–10 μmol photons $PARm^{-2}$ s^{-1} (p. 308) and photosynthesis is usually saturated between 500 and 2000 μmol photons $PARm^{-2}$ s^{-1}, depending on the plant type. Light quality and photoperiod (daylength) are also important: fluorescent tubes and incandescent bulbs that mimic the photosynthetic spectrum of sunlight are available (see p. 307), while the photoperiod can be controlled using a timer. The water potential of aqueous media can be adjusted using an impermeant osmoticum such as mannitol, and pH values are often kept close to those of the apoplasm ($\approx$pH 6) using appropriate buffers, if necessary (p. 171).

Definitions

Cell line – a cell culture derived by passage of a primary culture.

Clone – a population of cells derived from a single original cell, i.e. sharing the same genotype.

Confluence/confluent growth – merging of individual cells to form a continuous layer.

Continuous cell line – a culture with the capacity for unlimited multiplication *in vitro*. Sometimes termed an established cell line.

Finite cell line – a culture with a limited capacity for growth *in vitro* (maximum number of cell doublings).

Hybridoma – a hybrid cell produced by fusion of a tumour (myeloma) cell and an antibody-producing B lymphocyte (p. 281). A cell line derived from a single hybridoma will produce a single antibody type (monoclonal).

Immortalisation – conversion of a finite cell line into a continuous cell line.

Passage – an alternative term for subculture.

Primary culture – a cell culture derived from tissue or organ fragments (explants). Primary culture ends on first subculture.

Senescence – the end point in the limited lifespan of a finite cell line, characterised by the lack of proliferation. Conversion to an established (continuous) cell line requires 'escape' from senescence.

Stem cell – an undifferentiated animal cell that can generate new stem cells through mitosis, or can develop and differentiate into a particular specialised cell type.

Transformation – a permanent alteration in the growth characteristics of a finite cell line that may include (i) changes in morphology, (ii) an increased growth rate and/or (iii) the acquisition of an infinite lifespan, often termed immortalisation. Transformation may be spontaneous or may be induced by chemical agents or viruses, and often involves a change in chromosome number.

Cell and tissue cultures

Many of the basic principles involved in culturing animal and plant cells are broadly similar to those described for microbial cell culture (p. 223). Definitions for several key terms are given in the margin.

Applications of cell and tissue culture

The main uses of animal and plant cell culture systems include the following:

- **Experimental model systems in biochemistry, pharmacology and physiology:** cell culture offers certain advantages over whole organism studies, with greater control over environmental conditions and biological variability. The use of genetically defined clones of cells may simplify the analysis of experimental data. Conversely, results obtained with specialised cell-based systems might be unrepresentative of a broader range of cell types and may be more difficult to interpret in terms of the whole organism.

- **Studies of the growth requirements of particular cells:** including studies of the positive effects of growth factors or growth-promoting substances, and the negative effects of xenobiotics or cytotoxic compounds. The use of cell culture in bioassays and mutagenicity testing is considered in Chapter 36.

- **Studies of cell development and differentiation:** including aspects of the cell cycle and gene expression. Cell cultures retaining their ability to differentiate *in vitro* are particularly interesting to researchers, while the lack of differentiation and unlimited growth of many animal cell lines makes them useful models of tumour development.

- **Pathological studies:** including culture of fetal cells for karyotyping, detection of genetic abnormalities, e.g. trisomy, etc.

- **Genetic manipulation:** cell culture techniques have played an essential role in the development of molecular biology, including the production of transgenic animals and plants by techniques such as transfection, etc.

- **Biotechnology:** including the industrial production of therapeutic proteins, vaccines and monoclonal antibodies using large-scale hybridoma culture techniques similar to those used in microbiology (pp. 231–33).

- **Stem cell technology:** increasingly important in biomedical research and development.

> **KEY POINT** One of the main differences between organ and tissue incubation techniques and those used in cell/tissue culture is that the former aim simply to maintain metabolic and physiological activity for a limited period, while the latter provide conditions suitable for cell growth, division and development *in vitro* over an extended timescale, from a few days to several months.

Animal cell culture systems

These may be established either from whole organisms (e.g. chick embryo), discrete organs (e.g. rat liver) or from blood (e.g. lymphocytes), typically using *mild* tissue disruption techniques, e.g. enzyme treatment where necessary.

KEY POINT While, in theory, it is possible to culture nucleated cells from virtually any source, in practice the highest rates of success are most often achieved with young, actively growing tissues.

The principal considerations in animal cell culture are:

SAFETY NOTE Working with tissue cultures safely – do not confuse laminar-flow hoods (p. 225) with biosafety cabinets, as they perform completely different functions. A horizontal laminar-flow hood is designed to minimise contamination of the *culture*, rather than the *worker*.

SAFETY NOTE When sterilising during tissue culture procedures (Box 40.1), take care when using 70% v/v alcohol near a naked flame (e.g. Bunsen or spirit lamp) – it is easily ignited and burns with a weakly visible flame.

Example Dulbecco's modified Eagle's medium contains Ca^{2+}, Fe^{3+}, Mg^{2+}, K^+, Na^+, Cl^-, SO_4^{2-}, PO_4^{2-}, glucose, 20 amino acids, 10 vitamins, inositol and glutathione. Fetal calf serum is usually added at up to 20% v/v.

Using antibiotics in cell culture – a typical antimicrobial supplement might include antibacterial agents, e.g. penicillin and streptomycin, an antifungal agent, e.g. griseofulvin, and an antimycoplasmal agent, e.g. gentamicin. Use sparingly to treat contamination and discontinue once the contaminant has been eradicated.

- **Safety:** it is important to be aware of the potential dangers of infection from cell cultures. Although avian and rodent cells present a reduced risk of disease transmission compared to human cells, all animal cell cultures must be regarded as a potential source of pathogenic microbes, and appropriate sterile technique (Box 40.1) must be used at all times (see also p. 223). Work involving human tissues and cell cultures must be carried out in a biosafety cabinet by trained, experienced personnel – you are unlikely to gain practical experience with such cultures in the early stages of your course. For other cell cultures, you will need to follow the code of safe practice of your department – consult your Departmental Safety Officer if you have any doubts about safe working procedures.

- **Whether to use a primary culture or a cell line:** freshly isolated cells are more likely to reflect the biochemical activities of cells *in vivo*, though they will have a limited lifespan in culture, requiring repeated isolation for longer term projects. Continuous cell lines are more easily cultured and offer the advantage that their growth requirements in culture may be known in some detail, especially for the more widely used cell lines (e.g. BHK, HeLa).

- **The requirements for a solid substratum:** some cells must be attached to a solid surface to grow. Anchorage dependence is a typical feature of primary cultures and finite cell lines – such cultures show density-dependent growth inhibition once the cells have formed a confluent monolayer on the surface of the substratum. An alternative approach is to grow such cells on a particulate support using 'microcarrier' beads. In contrast, many continuous cell lines can be maintained in suspension culture, as individual cells or aggregates.

- **The physicochemical conditions,** including pH (typically 7.2–7.5) and buffering capacity, osmolality (usually 300 ± 20 mosmol kg^{-1}) and temperature (e.g. 35–37 °C for mammalian cells).

- **The requirements of the culture medium:** these will include the provision of inorganic ions (as a balanced salt solution), a carbon/energy source plus other organic nutrients, and in some cases a supplement containing antimicrobial agents to counter the risks of contamination. For example, Dulbecco's modified Eagle's medium is used for many mammalian cell types that grow as adherent monolayers, while suspension cultures of continuous cell lines can be maintained using less stringent media. To support growth, the basal medium is usually supplemented with serum (usually fetal calf serum, at up to 20% v/v), or a chemically defined serum-like supplement containing a mixture of proteins, polypeptides, hormones, lipids and trace components. The *in vitro* level of CO_2 and O_2 must also be considered: many cell cultures are buffered using bicarbonate, and must be maintained in an atmosphere of elevated CO_2, either in a sealed culture vessel or in a CO_2 incubator, to maintain pH balance. In some cases, a pH-sensitive dye (e.g. phenol red, p. 170)

Box 40.1 Sterile technique and its application to animal and plant cell culture

While Chapter 35 gives general advice on the basic principles of sterile technique with microbial cultures, animal and plant cell cultures need additional precautions owing to the complexity of some of the procedures and the likelihood of rapid overgrowth of any contaminating microbes. For routine work on the open bench, you should be aware of the following points.

1. **Consider personal clothing and hygiene:** long hair should be tied back, or retained by a net/cap. Avoid pendulous earrings and similar items. Wash your hands at the outset, to remove any loose skin flakes: surgical gloves may be worn, though there is loss of tactile sensitivity and comfort. Hands/gloves should be swabbed with 70% v/v alcohol to further reduce the risk of contamination. If you have a cold/cough, consider wearing a face mask.

2. **Work in a designated quiet area:** there should be no air currents (avoid open windows) and no 'through traffic' or other activity that might give rise to contamination (e.g. microbes should not be cultured in the same location). The term 'quiet area' also highlights that no talking should occur, to reduce the risk of aerosol contamination from the oral microflora.

3. **Organise your work surface:** clear everything from the bench, swab with 70% v/v alcohol, then position all items around a central area, so you can reach every item without reaching across anything. A well-organised workspace reduces the risk of accidental contact between sterile and non-sterile items. Swab bottles and flasks with 70% v/v alcohol before positioning them at the edge of your workspace. Also, swab your work surface between procedures and at the end of the session.

4. **Work close to a Bunsen flame,** positioned centrally within your workspace where convection currents create an upward airflow, reducing the likelihood of particles falling from the air into an open vessel. Always flame the tops of glass bottles (p. 225), but not plastic items, for 2–3 s both *before* and *after* opening and closing, rotating the bottles during exposure to the flame. Flame glass pipettes, as described on p. 225.

5. **Tilt flasks and bottles during use:** uncapped culture flasks and media bottles are best kept at a shallow angle, to minimise the risk of airborne contamination.

6. **Work without delay, but do not hurry:** always keep in mind that the air contains contaminant microbes and the longer you leave a vessel open to the air, the more likely it is that it will become contaminated.

When preparing to work in a laminar-flow cabinet, the following additional aspects should be considered:

1. **Note whether the airflow is horizontal or vertical:** horizontal flow gives greatest protection to the work area and provides the most stable airflow, while vertical flow cabinets reduce exposure of the operator. A biosafety cabinet should be used for work with potentially hazardous cultures (e.g. human cells/tissues, or with any cells known to be infected with a virus).

2. **Prepare the cabinet:** switch on the cabinet and leave running for at least 10 min, then swab the work surface and other interior surfaces with 70% v/v alcohol. Swab the outsides of bottles, flasks, etc. before bringing them into the cabinet. Arrange items around a crescent-shaped central work area.

3. **Carry out your work with due regard for the airflow within the cabinet:** use of a Bunsen burner is often discouraged, as it disrupts the correct airflow, creates heat and can ignite flammable items within the restricted interior of the cabinet. Always try to keep your hands/arms further away from the sterile airstream than the items within the cabinet, e.g. avoid working *behind* an open vessel in a horizontal flow cabinet and *above* an open vessel in a vertical flow cabinet.

4. **Take particular care when using pipettes:** a laminar-flow hood will have a restricted interior volume, and it is easy for the novice to touch a pipette tip on the hood, or on another item during use. When removing a pipette from its packaging and connecting to a pipette filler (p. 146) you should point the tip of the pipette away from the user and into the airflow, holding it well above the graduated scale markings to avoid contamination. The pipette and filler are then held horizontally until required: as this can be a tricky procedure for the novice, you should not take a pipette out of its container until you are ready to use it.

5. **Work carefully with open bottles and flasks:** caps can be either placed top-down on the work surface, or held in a crooked finger (Fig. 35.6(a)). Since the airflow is sterile, there is less need to work swiftly when compared to the open bench, but you should still recap all containers as soon as you have finished a particular procedure.

Definitions

Callus – an aggregation of undifferentiated plant cells in culture.

Embryogenic callus – tissue with the capacity to differentiate under defined laboratory conditions, typically in response to plant growth regulators in the medium.

Explant – a fragment of tissue used to initiate a culture (the term is also used in animal culture).

Protoplasts – cells lacking their cell walls.

Sphaeroplasts; US spheroplasts – cells with attached fragments of their cell walls: osmotically sensitive.

Totipotency – the ability (of any plant cell) to de-differentiate and redifferentiate into any of the cell types found in the mature plant.

may be incorporated into the growth medium, to provide a visual check on pH status during growth – a colour change indicates acidification of the medium and the growth medium should be renewed.

● **The equipment required:** this may include a laminar-flow hood or biosafety cabinet to reduce the possibility of microbial contamination, suitable culture vessels (typically pre-sterilised, disposable polystyrene dishes, bottles and flasks, treated to create a negatively charged, hydrophilic surface), a supply of high purity water (typically distilled, deionised and carbon filtered), a suitable incubator with temperature control of $\pm0.5°C$ or better, often with CO_2 control and mechanical mixing, and an inverted microscope to examine adherent cell monolayers during growth.

The successive stages of isolation of animal components are shown in Fig. 40.1(a). Box 40.1 gives advice on key aspects of sterile technique in cell culture; additional guidance on the principal practical procedures used in animal cell culture is given in Box 40.2.

Plant tissue and cell culture systems

Plant tissue cultures can be established by growing explants of sterilised tissue on the surface of an agar-based growth medium to give a callus of undifferentiated cells. Initial sterilisation is usually achieved by incubation for 15–20 min in 10% w/v sodium hypochlorite.

Fig. 40.1 Isolation of animal (a) and plant (b) components for *in vitro* study (note decreasing scale, from organisms to cells).

Box 40.2 Practical procedures in animal cell culture

The following activities are most likely to be encountered during the initial stages of your training:

- **Sterile technique (aseptic technique):** your first exercises are likely to involve transferring sterile liquids between vessels using pipettes, preparing solutions/media and making dilutions.

- **Preparing cells for initial culture:** if you are using a cryopreserved (frozen) vial containing a cell line, it will need to be thawed as rapidly as possible, to minimise the chance of intracellular ice crystal damage – thaw by immersion in water at 37°C (if the cells were initially stored in liquid nitrogen, thaw in a lidded container as there is a risk of explosion). The thawed culture should then be diluted slowly with new growth medium. If you are preparing a primary culture, you will first have to disrupt the tissue (Chapter 41 gives further details of methods available).

- **Checking for contamination:** this can be carried out by visual inspection, looking for: (i) a cloudy medium, sometimes with a film of surface growth; (ii) a change in pH; typically a sharp decrease owing to bacterial contamination causes a colour change of the indicator dye (e.g. phenol red turns yellow). Alternatively, examine cultures microscopically, or subculture for specific microbes (Chapter 37).

- **Observing cells:** cells growing as an adherent monolayer can be observed by inverted microscope ($\times 10$ or $\times 40$ objective), looking for typical cell morphology, or cytopathic effects (e.g. rounding off and detachment, irregular shape, or internal vacuolation and granulation), as well as any evidence of microbial growth. Microscopy will also tell you when such cells are reaching confluent growth (p. 259), requiring subculture.

- **Counting cells:** while rough estimates of the growth stage and cell number may be made from direct microscopic examination of cultures, a haemocytometer is required for a fully quantitative count, as detailed in Box 36.1. Adherent cells would first be detached (e.g. using 0.25% w/v trypsin) and counted in suspension. Other quantitative approaches include electronic counting (p. 233), and flow cytometry where other cellular activities can be assessed simultaneously. Alternatively, cellular constituents, e.g. DNA (p. 400) or protein (p. 383) can be assayed.

- **Checking viability:** this is usually carried out by assessing membrane integrity using either dye exclusion (1–2 min in 0.4% w/v trypan blue or 1.0% w/v naphthalene black) or, less frequently, dye uptake (10 min in 0.001% w/v diacetyl fluorescein or 0.5% w/v neutral red), with viability expressed in terms of the percentage of cells unstained or stained, respectively. When combined with haemocytometry (p. 234), the absolute numbers of live and dead cells can be estimated.

- **Feeding an adherent culture:** cells growing as a monolayer may need to have their growth medium replenished before they are ready for subculture, e.g. owing to a drop in pH of the medium, or a deterioration in cell morphology. The original medium should be removed, usually by aspiration using a sterile pipette and suction line, and a replacement volume of fresh medium then added (the new medium should be pre-warmed to 37 °C).

- **Subculturing cells ('passaging'):** typically this is carried out when adherent cells reach confluence, or when a suspension culture almost reaches the maximum population density in stationary phase (p. 231). Cells growing as a monolayer must first be detached from their substrate, typically using a small amount of 0.25% w/v trypsin for up to 15 min (it is important to dilute the trypsin following detachment, to prevent further digestion). Cells in suspension culture can simply be diluted in an appropriate volume of new medium. For both types of cell culture, the reseeding density is important, since over-diluted cultures will grow poorly. For primary cultures, low dilution factors (often termed 'split ratios') of between 1:2 and 1:8 are commonly used, whereas some continuous cell lines can be diluted by up to 1:100 to give a cell density of $10^4 - 10^5$ per mL.

- **Harvesting cells:** gentle centrifugation (at 80–100 g for 10 min) is sufficient to pellet most animal cells, leaving the spent medium as a supernatant, which should be either aspirated, or removed using a pipettor (p. 146). Monolayer cultures would be centrifuged following trypsinisation (see above).

- **Freezing cells (cryopreservation):** cells are first suspended at high density (typically around 10^6 per ml) in a suitable freezing solution, usually growth medium or serum supplemented with a cryoprotectant such as 10% v/v DMSO or 20–30% v/v glycerol, dispensed into small vials ('ampoules'/ 'cryotubes'), and then frozen slowly (ideally, at around 1 °C min^{-1}): one approach is to wrap the ampoules in cotton wool inside a small cardboard tube, then place them in an insulated container in a −70°C freezer for $\geq$6h, then transfer to a liquid nitrogen storage facility. An alternative approach is to use a proprietary system, e.g. 'Mr Frosty' (Nalgene), to provide a controlled rate of cooling.

Embryogenic callus may be induced to differentiate, forming tissues and organs on a medium containing appropriate plant hormones: in many cases, these cultures will develop to form plantlets that can be grown on to mature plants, or encapsulated to produce so-called 'artificial seeds'. This approach can be used to study the conditions necessary for differentiation and development, or to propagate rare plants and other valuable stock (e. g. virus-free stock, or genetically altered plants). Callus derived from anthers can be used to provide haploid cell cultures and haploid plants – these are often useful for experimental genetics and breeding purposes.

Plant cell suspension cultures (Fig. 40.1(b)) are usually obtained by transferring fragments of actively growing callus to a 'shake' flask containing liquid medium in an orbital incubator: gentle agitation causes fragmentation of the callus tissue, to give a suspension culture that will contain individual plant cells and cell aggregates. This culture can be maintained by repeated subculturing of material from the upper layers of the liquid, encouraging the growth of small aggregates. However, in contrast to animal cell suspension cultures, it is rare for plant suspension cultures to be entirely unicellular, owing to the presence of plasmodesmata and plant cell walls. Suspension cultures often require a minimum inoculum size on subculture – an inoculum volume of 10% v/v may be necessary to ensure successful subculture. It may also be helpful to add a small amount of 'conditioned' medium from a previous culture. Commercial-scale suspension cultures are used to produce certain plant pigments and secondary metabolites such as flavourings and high value pharmaceutical compounds.

The growth media used for callus and suspension cultures are more complex than those required for intact plant organs and tissues, with organic nutrient supplements in addition to a balanced salt solution. Such organic supplements often include a major carbon and energy source, e.g. sucrose, plus various vitamins and growth regulators (typically, at least one auxin and a cytokinin), together with undefined components such as yeast extract and hydrolysed casein in some instances. Otherwise, the techniques are broadly similar to those described for microbial systems (pp. 223–28).

Plant protoplasts

Some experimental procedures using plant cells require the enzymatic removal of their cell walls, creating protoplasts that can be manipulated *in vitro*, then maintained under conditions that allow the regeneration of cell walls, with subsequent growth and differentiation to give genetically modified plants. Protoplast isolation often involves pretreatment in a concentrated osmoticum (e.g. sucrose or mannitol, at 300–500 mmol L^{-1}) to plasmolyse the cells, weakening the linkage between cell wall and plasma membrane. Enzymatic treatment is often prolonged, taking several hours in a suitable mixture of enzymes, e.g. cellulase and Macerozyme. The resulting material can be sieved through fine nylon mesh, centrifuged at low speed and then re-suspended, to remove subcellular debris and cell aggregates. The protoplast preparation can be checked for fragments of cell wall using a suitable stain, e.g. 0.1% w/v calcofluor white and a fluorescence microscope.

Using an orbital incubator – continued rapid agitation (>100 r.p.m.) maintains a homogeneous suspension and encourages gas exchange between the culture medium and the atmosphere.

Example   Murashige and Skoog's medium contains a balanced mixture of the principal inorganic ions, plus seven trace element compounds, three vitamins, inositol, glycine and sucrose (at 30 g L^{-1}) as the major carbon source.

Working with protoplasts – remember that all solutions must contain a suitable osmoticum, to prevent bursting. Mannitol is widely used at ≈400 mmol L^{-1}.

Assessing cell and protoplast viability – cytochemical techniques are often used, e.g. exclusion of the mortal stain Trypan blue or Evan's blue. The fluorogenic vital stain fluorescein diacetate is an alternative: it is cleaved by esterases within living cells, liberating fluorescein and giving green-yellow fluorescent cells when viewed by UV fluorescence microscopy.

The fusion of protoplasts from different plants can be used to produce a somatic hybrid: this process can be used to circumvent inter-species reproductive barriers, creating novel plants. Protoplast fusion can be induced by chemical 'fusogens', e.g. using polyethylene glycol (PEG) at high concentration, or by electrofusion (incubation under low alternating current to encourage aggregation, then brief exposure to a high voltage electrical field – typically 1000 V cm^{-1} for 1–2 ms – causing protoplast fusion). Similar techniques can be used with animal cells. The successive stages of isolation of plant components are illustrated in Fig. 40.1(b).

Text references

Freshney, R.I. (2010) *Culture of Animal Cells: A Manual of Basic Technique and Specialised Applications,* 6th edn. Wiley, New York.

Smith, R.H. (2000) *Plant Tissue Culture,* Academic Press, London.

Sources for further study

Anon. (2006) *Fundamental Techniques in Cell Culture . . . a Laboratory Handbook.* Sigma-Aldrich/European Collection of Cell Cultures. Available at: http://www.sigmaaldrich.com/life-science/cell-culture/learning-centre/ecacc-handbook.html. Last accessed: 01/09/15.

Butler, M. (2003) *Animal Cell Culture and Technology: The Basics,* 2nd edn. Garland Science Press, Oxford.

Davey, M.R. and Anthony, P. (2010) *Plant Cell Culture: Essential Methods.* Wiley, New York.

Davis, J.M. (2002) *Basic Cell Culture: a Practical Approach,* 2nd edn. Oxford University Press, Oxford.

Davis, J. (2010) *Animal Cell Culture: Essential Methods.* Wiley, New York.

Loyola-Vargas, V.M. and Vazques-Flota, F. (2005) *Plant Cell Culture Protocols,* 2nd edn. Humana Press, New Jersey.

Newman, K.H., Kumar, A. and Imani, J. (2009) *Plant Cell and Tissue Culture I – A Tool in Biotechnology: Basics and Applications.* Springer, New York.

Reinart, J. and Yeoman, M.M. (2012) *Plant Cell and Tissue Culture: A Laboratory Manual.* Springer, New York.

STUDY EXERCISES

40.1 Calculate the osmolarity of Ringer's solution from its individual constituents. Use the data in Table 40.1 and the information in Chapter 26, and assume that each constituent behaves according to ideal thermodynamic principles. Note the following M_r values: NaCl 58.44; KCl 74.55; CaCl$_2$ 110.99; NaHCO$_3$ 84.01, required to convert the data in Table 40.1 to molar concentrations (see p. 153 if you are unsure about this conversion). Give your final answer in mosmol kg^{-1}, to one decimal place.

40.2 Find out about the availability of cell cultures using the web. Research the culture collections of particular countries via their websites. How useful is each site in terms of features such as: ease of use; access to catalogues; information on individual strains, e.g. culture media; details of costs?

40.3 Research the origins of the code names used for individual cell lines. In many instances, the code names given to individual cell lines are derived from their original source. Use the web to find out the origin of the following descriptors for cell lines:

(a) HeLa; (b) Vero; (c) BHK; (d) CHO; (e) TBY-2.

Answers to these study exercises are available at www.pearsoned.co.uk/practicalskills.

Definitions

Disruption – a process involving structural damage to cells and tissues, to an extent where the biomolecule or complex of interest is released.

Homogenisation – a process where cells and tissues are broken into fragments small enough to create a uniform, stable emulsion (a homogenate).

Homogeniser – a general term for any equipment used to disrupt or homogenise cells and tissues.

Most biological molecules and subcellular complexes must be isolated from their source material to be studied in detail. Unless the biomolecule is already a component of an aqueous medium (e.g. plasma, tissue exudate), the first step will be to disrupt the structure of the appropriate cells or tissues. Following disruption and/or homogenisation, the *in vitro* environment of the homogenate (disrupted tissue) will be very different from that of the intact cell, and it is important that the integrity of the biomolecule or subcellular complex is preserved as far as possible during the isolation procedure.

> **KEY POINT** Disruption may be achieved by chemical, physical or mechanical procedures – the choice of technique(s) required will depend on the intracellular location of the molecule and the nature of the source material.

Types of cell and their susceptibility to disruption

Cell walls are a major obstacle to disruption in many organisms and the technique used for a given application must take into account this aspect. With animal cells, there is no cell wall and the plasma membrane and cytoskeleton are relatively weak: unicellular suspensions of animal cells (e.g. blood cells) can be disrupted by gentle techniques. Animal cells within tissues are more difficult to disrupt, owing to the presence of connective tissue. Muscle requires vigorous techniques, as a result of additional contractile proteins. For plant cells, the plasma membrane is surrounded by a cellulose cell wall, sometimes with additional components, e.g. lignin or waxes. As a result, large shear forces are often required to disrupt plant cells.

Most Gram-positive and Gram-negative bacteria are surrounded by a rigid, protective cell wall. Peptidoglycan is a major structural component and this may be degraded by the enzyme lysozyme, particularly in certain Gram-positive organisms (Fig. 41.1(a)). In contrast, the outer membrane of Gram-negative bacteria (Fig. 41.1(b)) protects against lysozyme: however, prior treatment of Gram-negative bacteria with EDTA destabilises the outer membrane by removing Ca^{2+}, making the cell sensitive to lysozyme. In an osmotically buffered solution, the cells will be converted to sphaeroplasts (p. 262), since components of the Gram-negative envelope will remain after lysozyme treatment. Note also that some Gram-positive bacteria may be insensitive to lysozyme, owing either to a modified peptidoglycan structure or to additional protective outer layers (proteinaceous S layers: Fig. 41.1(a)).

The cell walls of filamentous fungi and yeasts are very robust, containing up to 90% polysaccharide (e.g. chitin, mannan), with embedded protein microfibrils – they are often difficult to disrupt, requiring enzymic digestion and/or mechanical homogenisation. Hopkins (1991) gives details of the various systems available for homogenisation and disruption of biological material.

Homogenising media

The solution used for homogenisation serves several purposes, since it acts as a solvent or suspension medium for the released components; it serves as a cooling medium (since many biological macromolecules

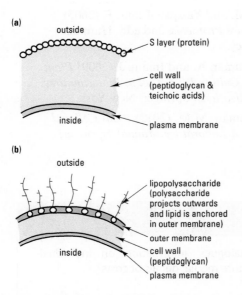

(a)
outside

S layer (protein)

cell wall (peptidoglycan & teichoic acids)

inside

plasma membrane

(b)
outside

lipopolysaccharide (polysaccharide projects outwards and lipid is anchored in outer membrane)

outer membrane

cell wall (peptidoglycan)

inside

plasma membrane

Fig. 41.1 Cell envelopes of (a) Gram-positive and (b) Gram-negative bacteria.

are denatured by heat); and it contains various reagents that may help to preserve the biological integrity of components. A typical medium will contain:

- **Buffer.** This replaces the intracellular buffer systems and is needed to prevent pH changes that might denature proteins. TRIS and phosphate buffers are often used for 'physiological' pH values (7.0–8.0); alternatives include various zwitterionic buffers (Chapter 27).

Using buffers – selecting and preparing an appropriate buffer is covered in more detail on pp. 171–175.

- **Inorganic salts.** The intracellular ionic strength is quite high, so KC1 and NaCl are often included to maintain the ionic strength of the homogenate. However, the total concentration of inorganic salts should be kept below 100 mmol L^{-1}, to avoid thickening of the homogenate owing to solubilisation of structural proteins.

- **EDTA.** This chelates divalent cations and removes metal ions (e.g. Cu^{2+}, Pb^{2+}, Hg^{2+}) that inactivate proteins by binding to thiol groups. In addition, it removes Ca^{2+} which could activate certain proteases, nucleases and lipases in the homogenate.

Using chelating agents – EDTA (ethylenediaminetetraacetic acid) and EGTA (ethylenebis(oxethylenenitrilo)-tetraacetic acid). If Mg^{2+} is an important component of the medium, use EGTA, which does not chelate Mg^{2+}.

- **Sucrose.** This can be used to prevent osmotic lysis of organelles (e.g. mitochondria, lysosomes), and stabilises proteins from hydrophobic intracellular environments by reducing the polarity of the aqueous medium.

- **Mg^{2+}.** This helps to preserve the integrity of membrane systems by counteracting the fixed negative charges of membrane phospholipids.

- **Protease inhibitors** (e.g. phenylmethanesulfonylfluoride, PMSF; L-*trans*-epoxysuccinylleucylamido-(4-guanodino)-butane, E-64; leupeptin). These protect solubilised proteins from digestion by intracellular proteases, mainly released from lysosomes on disruption of the cell. Lysosomal proteases have acid pH optima, another reason for maintaining the pH of the medium close to neutrality.

Extracting proteins from plants – carry out all post-homogenisation procedures as quickly as possible, because plant cell vacuoles may contain phenols that will inactivate proteins when released.

- **Reducing agents** (e.g. 2-mercaptoethanol, dithiothreitol, cysteine at ≈ 1 mmol L^{-1}). These reagents prevent oxidation of certain proteins, particularly those with free thiol groups that may be oxidised to disulfide bonds when released from the cell under aerobic conditions.

- **Detergents** (e.g. Triton X-100, SDS). These cause dissociation of proteins and lipoproteins from the cell membrane, aiding the release of membrane-bound and intracellular components.

Methods of disrupting cells and tissues

Prior to disruption, animal tissues will need to be freed of any visible fat deposits and connective components, and then randomly sliced with fine scissors or a scalpel. Any fibrous and vascular tissue should be removed from plant material. Disruption can be achieved by mechanical and non-mechanical means: the principal applications of various methods are described in Table 41.1 for the major cell and tissue types.

Non-mechanical methods

- **Osmotic shock:** cells are first placed in a hypertonic solution of high osmolality (p. 164), e.g. 20% w/v sucrose, leading to loss of water.

Definitions

Hypertonic – a medium with a more negative water potential, compared with the cells or medium.

Hypotonic – a medium with a less negative water potential, compared with the cells or medium.

Isotonic – a medium with the same water potential (p. 166) as the cells or tissues.

Table 41.1 Summary of techniques for the disruption of tissues and cells – note that safety glasses should be worn for all procedures.

Technique	Suitability	Comments
Non-mechanical methods		
Osmotic shock	Animal soft tissues Some plant cells	Small scale only
Freeze/thaw	Animal soft tissues Some bacteria	Time-consuming; small scale; closed system – suitable for pathogens with appropriate safety measures; some enzymes are cold-labile
Lytic enzymes, e.g. lipases; proteases; pectinase; cellulase	Animal cells Plant cells	Mild and selective; small scale; expensive; enzymes must be removed once lysis is complete
Lysozyme	Some bacteria	Gram-negative bacteria must be pretreated with EDTA. Suitable for some organisms resistant to mechanical disruption
Mechanical methods		
Pestle and mortar + abrasives	Tough tissues	Not suitable for delicate tissues
Ball mills + glass beads	Bacteria and fungi	May cause organelle damage in eukaryotes
Blenders and rotor-stators	Plant and animal tissues	Ineffective for microbes
Homogenisers (glass and Teflon)	Soft, delicate tissues e.g. white blood cells, liver	Glass may shatter – wear safety glasses during use
Solid extrusion (Hughes press)	Tough plant material; bacteria; yeasts	Small scale
Liquid extrusion (French pressure cell)	Microbial cells	Small scale
Ultrasonication	Microbial cells	Cooling required; small scale; may cause damage to organelles, especially in eukaryotic cells.

Using osmotic shock – this approach can be effective for wall-less cells, sphacroplants (p. 262) and protoplants, but is ineffective for cells with robust walls.

On dilution of this solution (e.g. by addition of water or transfer to a hypotonic solution), the cells will burst because of rapid water influx.

- **Freezing and thawing:** causing leakage of intracellular material, following cell wall and membrane damage and internal disruption caused by ice crystal formation.
- **Lytic enzymes:** damaging the cell wall and/or plasma membrane. Cells can then be disrupted by osmotic shock or gentle mechanical treatment.

Mechanical methods

All mechanical procedures for cell disruption generate heat, and this may denature proteins. Therefore it is very important to cool the starting material, the homogenising medium, and, if possible, the homogeniser itself (to ≈4°C). The homogenisation should be carried out in short bursts, and the homogenate should be cooled in an ice bath between each burst. Cooling will also reduce the activity of any degradative enzymes in the homogenate. Ideally, carry out homogenisation in a walk-in cooler (a 'cold room') typically at 4–10°C.

Avoiding protein denaturation during homogenisation – excessive frothing of the homogenate indicates denaturation of proteins (think of whipping egg whites for meringue).

Fig. 41.2 Components of a rotor-stator homogeniser.

Fig. 41.3 Ground-glass homogeniser.

Equipment commonly used includes:

- **Mixers and blenders.** These are similar to domestic liquidisers, with a static vessel and rotating blades. The Waring blender is widely used: it has a stainless steel vessel that will stay cool if pre-chilled. The vessel and blades are designed to maximise turbulence, both disrupting and homogenising tissues and cells.

- **Ball mills** (e.g. Retch mixer mill, Mickle mill). These devices contain glass beads that vibrate and collide with each other and with tissues/cells, leading to disruption.

- **Liquid extrusion devices** (e.g. French pressure cell). Cells are forced from a vessel to the outside, through a very narrow orifice at high pressures ($\approx$100 MPa). The resulting pressure changes are a powerful means of disrupting cells.

- **Solid extrusion** (e.g. Hughes press). Here, a frozen cell paste is forced through a narrow orifice, where the shear forces and the abrasive properties of the ice crystals cause cell disruption.

- **Rotor-stators** (e.g. IKA Ultra-turrax homogeniser). These have a rotor (a set of stainless steel blades) and a stator (a slotted stainless steel cylinder) at the tip of a stainless steel shaft, immersed in the homogenising medium: the arrangement is illustrated in Fig. 41.2. The high speed of the rotor blades causes material in the homogenising fluid to be sucked into the dispersing head, where it is pressed radially through the slots in the stator. Along with the cutting action of the rotor blades, the material is subjected to very high shear and thrust and the resulting turbulence in the gap between rotor and stator gives effective mixing. The vigour of the homogenisation process can be altered by varying the rotor speed setting. Various sizes of rotor-stator are available, with typical diameters in the range 8–65 mm: the smaller sizes are particularly useful for small-scale preparations.

- **Sonicators.** Ultrasonic waves are transmitted to an aqueous suspension of cells via a metal probe. The ultrasound creates bubbles within the liquid and these produce shock waves when they collapse. Successful disruption depends on the correct choice of power and incubation time, together with pH, temperature and ionic strength of the suspension medium, often obtained by trial and error. You can reduce the effects of heating during ultrasonication by using short 'bursts' of power (10–30s), with rests of 30–60s in between, and by keeping your cell suspension on ice during disruption. An ultrasonic water bath provides a more gentle means of disrupting certain types of cells, e.g. some bacterial and animal cells.

- **Homogenisers.** These involve the reciprocating movement of a ground-glass or Teflon pestle within a glass tube (Fig. 41.3). Cells are forced against the walls of the tube, releasing their contents. For glass pestles, the tubes also have ground-glass homogenising surfaces and may have an overflow chamber. The homogeniser can either be hand-operated (e.g. Dounce), or motorised (e.g. Potter-Elvejham). The clearance between the pestle and the tube (range 0.05–0.5mm) must be chosen to suit the particular application.

Cell fractionation and the isolation of organelles

The fractionation and separation of organelles from a cell homogenate by differential centrifugation are described in Chapter 49. Particular organelles can be obtained by appropriate choice of source tissue and homogenisation method, as illustrated in Table 41.2 for the major types of organelle.

Table 41.2 Isolation and fractionation procedures for various organelles.

Stage	Nuclei	Mitochondria	Microsomes	Chloroplasts*
Source	Thymus tissue, which has little cytoplasm, giving high yields	Beef heart, with fat and connective tissue removed, then cubed and minced. Keep at pH 7.5 using TRIS buffer	Rat liver, stored overnight to reduce glycogen content	Spinach leaves, de-ribbed and cut into 1-cm strips
Pretreatment	Rinse with buffered physiological saline. Suspend in homogenising medium	Suspend in 2× volume of ice-cold homogenising medium. Squeeze through muslin	Chop finely with scissors and wash in $2 \times$ volume of homogenising medium	Rinse, then suspend in $3 \times$ volume of prechilled homogenising medium
Homogenising medium	250 mmol L^{-1} sucrose; 10 mmol L^{-1} TRIS/HCI buffer, pH 7.6; 5 mmol L^{-1} $MgCl_2$; 0.2–0.5% v/v Triton X–100	250 mmol L^{-1} sucrose; 10 mmol L^{-1} TRIS/HCI buffer pH 7.7, containing 1 mmol L^{-1} succinic acid and 0.2 mmol L^{-1} EDTA	250 mmol L^{-1} sucrose; 50 mmol L^{-1} TRIS/HCI buffer, pH 7.5; 25 mmol L^{-1} KCl; 5 mmol L^{-1} $MgCl_2$	400 mmol L^{-1} sucrose; 25 mmol L^{-1} HEPES/NaOH buffer at pH 7.6; 2 mmol L^{-1} EDTA
Homogenisation	Waring blender, low speed, 3 min	Bring to pH 7.8 using 2 mol L^{-1} TRIS base: Waring blender, high speed, 15 s: check and adjust pH to 7.8 and repeat blending step, 5s	Potter-Elvejham glass homogeniser with a Teflon pestle – 3×5 min at 800 rpm	Prechilled Waring blender or rotor-stator
Filtration/ centrifugation	Filter through gauze. Spin at 2000 g for 10 min; discard supernatant. Repeat the homogenisation, filtration and centrifugation stages to improve purity of organelles	Spin at 1200 g for 20min: filter supernatant through muslin: centrifuge at 26 000 g for 15 min. Remove and discard upper (lighter) layer of pellet. Resuspend lower layer and rehomogenise ($2 \times$ 5s). Centrifuge at 26000 g for 15 min	Centrifuge at 680 g for 10 min – discard pellet; centrifuge at 10 000 g for 10 min – discard pellet; centrifuge at 100000 g; for 60 min – retain pellet. Resuspend in buffer, pH 8.0 and re-centrifuge at 100 000 g for 60 min	Pass through several layers of muslin (wear gloves); centrifuge at 2500 g; for 60s; resuspend pellet in buffer, pH 7.6; recentrifuge at 2500 g; for 60s. The colour of the supernatants gives a visual indication of chloroplast damage (e.g. if green).
Before use	Resuspend in homogenisation medium without Triton X-100	Resuspend pellet in buffer, pH 7.8 and either use immediately, or store at –20°C overnight	Resuspend in buffer solution at pH 8.0	Resuspend pellet in appropriate incubation medium, containing sucrose, e.g. for CO_2/O_2 studies, p. 430

*An alternative approach is to use plant protoplasts (p. 264) as the starting material, releasing the chloroplasts by gentle lysis – diluting the medium with water.

Text references

Hopkins, T. (1991) Physical and chemical cell disruption for the recovery of intracellular proteins. In: *Purification and Analysis of Recombinant*

Proteins (ed. R. Seetharam and S.K. Sharma), pp. 57–84. Marcel Dekker, New York.

Sources for further study

Anon. *BioLogics Ultrasonic Homogenizers.* Available: http://ultrasonichomogenizer.com Last accessed: 01/09/15. [One of many websites dealing with commercial products.]

Boyer, R. (2011) *Biochemistry Laboratory: Modern Theory and Techniques,* 2nd edn. Prentice Hall, Harlow.

Pryor, P. (2014) *Subcellular Fractionation: A Laboratory Manual.* Cold Spring Harbor Laboratory Press, Cold Spring Harbor.

Simpson, R.J., Adams, P.D. and Golemis, E.A. (2015) *Basic Methods in Protein Purification and Analysis: A Laboratory Manual.* Cold Spring Harbor Laboratory Press, Cold Spring Harbor.

Wilson, K. and Walker, J. (eds) (2010) *Principles and Techniques of Biochemistry and Molecular Biology,* 7th edn. Cambridge University Press, Cambridge. [Also covers a range of other topics, including spectroscopy and chromatography]

STUDY EXERCISES

41.1 Consider suitable disruption procedures for particular types of cell or tissue. After reading this chapter, identify the major problems associated with disrupting each of the following, and list suitable practical procedures to overcome these problems:

(a) Gram-negative bacteria;

(b) skeletal muscle;

(c) plant tissue.

41.2 Select a suitable disruption technique for the preparation of intracellular components. How might the procedure used to prepare a soluble cytoplasmic protein from fresh liver differ from that required for the isolation of intact organelles?

41.3 Test your understanding of the function of homogenisation medium components. The following list provides a number of possible components that might be used in a homogenisation medium:

(a) sucrose;

(b) low ionic strength buffer;

(c) dithiothreitol;

(d) detergent (e.g. Triton X-100).

Select one of these components to include in a medium for preparing:

(i) a membrane protein from liver tissue;

(ii) a mitochondrial suspension from liver tissue;

(iii) an oxygen-sensitive enzyme from pancreatic tissue;

(iv) a protein from the sarcolemma (cytoplasm) of skeletal muscle.

Give a brief explanation for your selection in each case.

Answers to these study exercises are available at www.pearsoned.co.uk/practicalskills.

Analytical techniques

There are many instances where it is necessary to measure the quantity of a test substance using a calibrated procedure. You are most likely to encounter this approach in one or more of the following practical exercises:

- Quantitative spectrophotometric assay of biomolecules (Chapter 47).
- Flame or atomic absorption spectroscopic analysis of metal ions in biological solutions (p. 316).
- Using a chromogenic or fluorogenic substrate to determine the activity of an enzyme (p. 429).
- Quantitative chromatographic analysis, e.g. GC or HPLC (Chapter 51).
- Using a bioassay system (p. 236) to quantify a test substance: examples include immunodiffusion (Chapter 43) and radioimmunoassay (p. 284).

Understanding quantitative measurement - Chapter 30 contains details of the basic principles of valid measurement, while Chapters 43-54 deal with some of the specific analytical techniques used in biomolecular science. The use of internal standards is covered on p. 347.

> **KEY POINT**   In most instances, calibration involves the establishment of a relationship between the measured response (the 'signal') and one or more 'standards' containing a known amount of substance.

Calibrating laboratory apparatus - this is important in relation to validation of equipment, e.g. when determining the accuracy and precision of a pipettor by the weighing method: see p. 147.

In some instances, you can measure a signal because of an inherent property of the substance, e.g. the absorption of UV light by nucleic acids (p. 400), whereas in other cases you will need to react it with another substance to see the result (e.g. molecular weight measurements of DNA fragments after electrophoresis, visualised using ethidium bromide (p. 401), or to produce a measurable response (e.g. the reaction of copper(II) ions and peptide bonds in the biuret assay for proteins, p. 383).

The different types of calibration curve

By preparing a set of solutions (termed 'standards'), each containing either (i) a known *amount* or (ii) a specific *concentration* of the substance, and then measuring the response of each standard solution, the underlying relationship can be established in graphical form as a 'calibration curve', or 'standard curve'. This can then be used to determine either (i) the amount or (ii) the concentration of the substance in one or more test samples. Alternatively, the response can be expressed solely in mathematical terms: an example of this approach is the determination of chlorophyll pigments in plant extracts by measuring absorption at particular wavelengths, and then applying a formula based on previous (published) measurements for purified pigments (p. 312).

There are various types of standard curve: in the simplest cases, the relationship between signal and substance will be linear, or nearly so, and the calibration will be represented best by a straight-line graph (see Box 42.1). In some instances (Fig. 42.1(a)), you will need to transform either the *x* values or *y* values to produce a linear graph (e.g. in radial immunodiffusion bioassay, where the *y* values are squared, p. 283). In other instances, the straight-line relationship may only hold up to a certain value (the 'linear dynamic range') and beyond this point the graph may curve (e.g. in quantitative spectrophotometry, the Beer–Lambert relationship often becomes invalid at high absorbance, giving a curve, (Fig. 42.1(b)). Some calibration curves are sigmoid (Fig. 42.1(c)). Finally,

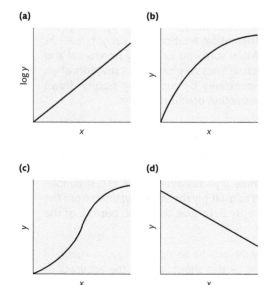

Fig. 42.1 Calibration curves: (a) log-linear; (b) curvilinear; (c) sigmoid, or S-shaped; (d) inverse.

the signal may *decrease* in response to an increase in the substance (Fig. 42.1(d)), e.g. radioimmunoassay, where an inverse sigmoid calibration curve is obtained (p. 284). In some practical classes, you may be told that the relationship is expected to be linear, curvilinear, or whatever, while in others you may have to decide the form of the standard curve as part of the exercise.

Box 42.1 The stages involved in preparing and using a calibration curve

1. **Decide on an appropriate test method** – for example, in a project you may need to research the best approach to the analysis of a particular metabolite in your biological material.

2. **Select either (a) amount or (b) concentration, and an appropriate range and number of standards** – in practical classes, this may be given in your schedule, along with detailed instructions on how to make up the standard solutions. In other cases, you may be expected to work this out from first principles (Chapters 24-26 give worked examples) – aim to have evenly spaced values along the *x* axis.

3. **Prepare your standards very carefully** – due attention to detail is required: for example, you should ensure that you check the calibration of pipettors beforehand, using the weighing method (p. 147). Do not forget the 'zero standard' plus any other controls required, e.g. to test for interference caused by other chemical substances. Your standards should cover the range of values expected in your test samples.

4. **Assay the standards and the unknown (test) samples** – preferably all at the same time, to avoid introducing error due to changes in the sensitivity or drift in the zero setting of the instrument with time. It is a good idea to measure all of your standard solutions at the outset, and then measure your test solutions, checking that the 'zero standard' and 'top standard' give the same values after, say, every six test measurements. If the re-measured standards do not fall within a reasonable margin of the previous value, then you will have to go back and recalibrate the instrument, and repeat the last six test measurements. If your test samples lie outside the range of your standards, you may need to repeat the assay using diluted test samples (extrapolation of your curve may not be valid; see p. 497).

5. **Draw the standard curve, or determine the underlying relationship** – Figure 42.2 gives an example of a typical linear calibration curve, where the spectrophotometric absorbance of a series of standard solutions is related to the amount of substance. When using a spreadsheet (Box 42.2) or a graphics package (Chapter 14), it is often appropriate to use Model I linear regression (p. 534)

Fig. 42.2 Typical calibration curve for spectrophotometric analysis.

to produce a linear trend line (also termed the 'line of best fit') and you can then quote the value of r^2, which is a measure of the 'fit' of the measurements to the line (see p. 534). However, you should take care not to use a linear plot when the underlying relationship is clearly non-linear and you must consider whether the assumptions of the regression analysis are valid (e.g. for transformed data).

6. **Determine the amount or concentration in each unknown sample** – either by reading the appropriate value from the calibration curve or by using the underlying mathematical relationship, i.e. $y = a + bx$ (pp. 510-511). Make sure you draw any horizontal and vertical construction lines very carefully - students often lose marks unnecessarily by submitting poorly drawn construction lines within practical reports.

7. **Correct for dilution or concentration, where appropriate** – for example, if you diluted each test sample by 10-fold, then you would need to multiply by 10 to determine the value for the undiluted test sample. As another example, if you assayed 0.2 mL of test sample, you would need to multiply the value obtained from the calibration curve by 5, to give the value per mL of the original sample.

8. **Quote your test results to an appropriate number of significant figures** – this should reflect the accuracy of the method used (see pp. 506-508), not the size of your calculator's display.

Example   A test tube containing 8 mL of water plus 2 mL of 1% w/v NaCl [$M_r = 58.44$] would have a *mass concentration* of 0.2% w/v NaCl (a fivefold dilution of the original NaCl solution), which can also be expressed as 2 g L^{-1}; in terms of *molar concentration* (p. 161), this would be equivalent to $2 \div 58.44 = 0.0342$ mol L^{-1} (to three significant figures). Expressed in terms of the *amount* of NaCl in the test tube, this would be 0.02 g in *mass*, or $0.02 \div 58.44 = 3.42 \times 10^{-4}$ mol (342 µmol) in *moles*.

Organisations providing national/international standards – these include:

- Laboratory of the Government Chemist (http://www.lgc.co.uk/) for UK standards and European Reference Materials.

- National Institute for Science and Technology (http://www.cstl.nist.gov/), for biochemical/chemical standards in the USA.

- Institute for Reference Materials and Measurements (http://www.irmm.jrc.be/) for European standards, including BCR and ERM materials.

- OIE Biological Standards Commission (http://www.oie.int/bsc/eng/en_bsc.htm), for international animal standards.

Plotting a standard curve – do not force your calibration line to pass through zero if it clearly does not. There is no reason to assume that the zero value is any more accurate than any other reading you have made.

Practical considerations

Amount or concentration?

This first step is often the most confusing for new students. It is vital that you understand the difference between *amount* of substance (e.g. mg, ng), and *concentration* (the amount of substance per unit volume, e.g. mmol L^{-1}, mol m^{-3}, % w/v) before you begin your practical work.

> **KEY POINT**   Essentially, you have to choose whether to work in terms of either (i) the total amount of substance in your assay vessel (e.g. test tube or cuvette) or (ii) the final concentration of the substance in your assay vessel.

The interconversion of amount and concentration is covered in more detail on p. 152. Either way, this is usually plotted on the *x*-(horizontal) axis and the measured response on the *y*-(vertical) axis.

Choice of standards

In your early practical classes, you may be provided with a stock solution (pp. 153–4), from which you then have to prepare a specified number of standard solutions. In such cases, you will need to understand how to use dilutions to achieve the required amounts or concentrations (pp. 154–5). In later work and projects, you may need to prepare your standards from chemical reagents in solid form, where the important considerations are purity and solubility (p. 152). For professional analysis (e.g. in forensic science or clinical biochemistry), it is often important to be able to trace the original standard or stock solution back to national or international standards (Chapter 55) or to certified reference materials.

How many standards are required?

This may be given in your practical schedule, or you may have to decide what is appropriate (e.g. in project work and research). If the form and working range of the standard curve are known in advance, this may influence your choice – for example, linear calibration curves can be established with fewer standards than curvilinear relationships. In some instances, analytical instruments can be calibrated using a single standard solution, often termed a single point 'calibrator'. Replication of each standard solution is a good idea, since it will give you some information on the variability involved in preparing and assaying the standards. Consider whether you should plot mean values on your standard curve, or whether it is better to plot the individual values (if one value appears to be well off the line, you are likely to have made an error, and you may need to check and repeat the measurement).

Preparing your standards

It is extremely important to take the greatest care to measure out all chemicals and liquids very accurately, to achieve the best possible standard curve. The grade of volumetric flask used and temperature of the solution also affect accuracy (grade A apparatus is best). You may also consider what other additives might be required in your standard solutions.

Box 42.2 How to use a spreadsheet (e.g. Microsoft *Excel*) to produce a linear regression plot

Two approaches are possible: either using the trendline feature or using the regression analysis function within the data analysis tool pack. Both make the assumption that the criteria for Model I linear regression are met (see p. 534). In the example shown below, the following simple data set has been used:

Amount (ng)	Absorbance
0	0.00
10	0.19
20	0.37
30	0.56
40	0.63
50	0.78

$y = 0.0155x + 0.0352$
$R^2 = 0.9836$

Fig. 42.3 Calibration curve showing line of best fit and details of linear regression equation.

Using the *Trendline* feature. This quick method provides a line of best fit on an *Excel* chart and can also provide a set of equation values for predictive purposes.

1. **Create a graph (*chart*) of your data.** Enter the data in two columns within your spreadsheet, select the data array (highlight using left mouse button) and then, using the *Insert* tab *Charts* group, select *Scatter*.

2. **Add a trend line.** Right-click on any of the data points on your graph, and select the *Add Trendline* menu. Choose the *Linear* trend line option, but do not click *OK* at this stage. Rather, from the *Trendline Options* menu, select: (i) *Display equation on chart* and (ii) *Display R-squared value on chart*. Now click *Close*. The equation (shown in the form $y = bx + a$) gives the slope (b) and intercept (a) of the line of best fit, while the *R*-squared value (coefficient of determination, p. 534) gives the proportional fit to the line (the closer this value is to 1, the better the fit of the data to the line).

3. **Modify the graph to improve its effectiveness.** For a graph that is to be used elsewhere (e.g. in a lab write-up or project report), adjust the display to remove the default background and gridlines and change the symbol shape (see Box 70.2 for more advice and examples). Right-click on the trend line and use the *Format Trendline > Line Style* menu to adjust the *Width*

of the line. Drag and move the equation panel if you would like to alter its location on the chart, or delete it, having noted the values. Figure 42.3 shows a calibration curve produced in this way for the data presented above.

4. **Use the regression equation to estimate unknown (test) samples.** By rearranging the equation for a straight line and substituting a particular *y* value, you can predict the amount/concentration of substance (*x* value) in a test sample. This is more precise than simply reading the values from the graph using construction lines (Box 42.1). If you are carrying out multiple calculations, the appropriate equation, $x = (y - a)/b$, can be entered into a spreadsheet, for convenience.

Using the *Regression Analysis* tool requires the 'data analysis tool pack' to be loaded beforehand (using the *File* tab, *Options* menu, *Add-Ins*, then select *Analysis ToolPak*) and this provides summary output that contains details of slope, intercept and coefficient of determination along with an analysis of variance (ANOVA) table (Chapter 74 gives further details of these aspects, including an example of output in study exercise 74.4).

Dealing with interfering substances – one approach is to use the method of 'standard additions', where the standards all contain a fixed additional amount of the sample (for more details of this approach see, e.g., Dean, 1997). Internal standards (p. 347) can also be used to detect such problems.

For example, do your test samples have high levels of potentially interfering substances, and should these also be added to your standards? Also consider what controls and blank solutions to prepare.

KEY POINT The validity of your standard curve depends upon careful preparation of standards, especially in relation to accurate dispensing of the volumes of any stock solution and diluting liquid diluent – the results for your test samples can only be as good as your standard curve.

Preparing the calibration curve and determining the amount of the unknown (test) sample(s)

This is described in stepwise fashion in Box 42.1. Check you understand the requirements of graph drawing, especially in relation to plotted curves (p. 489) and the mathematics of straight-line graphs (p. 510). Spreadsheet programs such as Microsoft *Excel* can be used to produce a regression line for a straight-line calibration plot (p. 276). Examples of how to do this are provided in Box 42.2.

Text references

Dean, J.D. (1997) *Atomic Absorption and Plasma Spectroscopy*, 2nd edn. Wiley, Chichester.

[Chapter 1 deals with calibration, and covers the principle of standard additions.]

Sources for further study

Cable, M. (2005) *Calibration: a Technician's Guide.* ISA, Raleigh.

Haris, D.C. (2010) *Quantitative Chemical Analysis*, 8th edn. Freeman, New York.

Miller, J.N. and Miller, J.C. (2010) *Statistics and Chemometrics for Analytical Chemistry*, 6th edn. Prentice Hall, Harlow.
[Gives detailed coverage of calibration methods and the validity of analytical measurements.]

STUDY EXERCISES

42.1 Determine unknowns from a hand-drawn calibration curve. The following data are for a set of calibration standards for Zn, measured by atomic absorption spectrophotometry.

Absorbance measurements for a series of standard solutions containing different amounts of zinc

Zinc concentration ($\mu g\ mL^{-1}$)	Absorbance
0	0.000
1	0.082
2	0.174
3	0.257
4	0.340
5	0.408
6	0.463
7	0.511
8	0.543
9	0.561
10	0.575

Draw a calibration curve by hand using graph paper and estimate the concentration of zinc in the following water samples:

(a) an undiluted sample, giving an absorbance of 0.157;

(b) a 20-fold dilution, giving an absorbance of 0.304;

(c) a fivefold dilution, giving an absorbance of 0.550.

Give your answer to three significant figures in each case.

42.2 Determine unknowns from a calibration curve produced in *Excel*. The following data are for a set of calibration standards for protein content, measured by the Lowry (Folin–Ciocalteau) method (p. 384).

Absorbance measurements for a series of protein standards

Protein (mg per tube)	Absorbance (A_{600})
0.00	0.000
0.02	0.161
0.04	0.284
0.06	0.438
0.08	0.572
0.10	0.762

(continued)

STUDY EXERCISES (continued)

Using PC-based software (e.g. *Excel*), fit a trend line (linear regression) and determine the protein content of solutions with the following absorbances:

a) 0.225   b) 0.465   c) 0.682

Give your answers to three significant figures in each case.

42.3 Identify the errors in a calibration curve. The figure shows a calibration curve of the type that might be submitted in a practical write-up. List the errors and compare your observations with the list given on the website detailed below.

Calibration graph.

Answers to these study exercises are available at www.pearsoned.co.uk/practicalskills.

Antibodies are an important component of the immune system, which protects animals against certain diseases (see Delves *et al*, 2011). They are produced by B lymphocytes in response to foreign macromolecules (antigens). A particular antibody will bind to a site on a specific antigen, forming an antigen–antibody complex (immune complex). Immunological bioassays use the specificity of this interaction for:

- identifying macromolecules, cellular components or whole cells;
- quantifying a particular substance.

Antibody structure

An antibody is a complex globular protein, or immunoglobulin (Ig). While there are several types, IgG is the major soluble antibody in vertebrates and is used in most immunological assays. Its main features are as follows:

- **Shape:** IgG is a Y-shaped molecule (Fig. 43.1), with two antigen-binding sites.
- **Specificity:** variation in amino acid composition at the antigen-binding sites explains the specificity of the antigen–antibody interaction.
- **Flexibility:** each IgG molecule can interact with epitopes which are different distances apart, including those on different antigen molecules.
- **Labelling:** regions other than the antigen-binding sites can be labelled, e.g. using a radioisotope or an enzyme with fluorogenic or chromogenic detection (p. 412).

> **KEY POINT**   The presence of two antigen-binding sites on a single flexible antibody molecule is relevant to many immunological assays, especially the agglutination and precipitation reactions.

Antibody production

Polyclonal antibodies

These are commonly used at undergraduate level. They are produced by repeated injection of antigen into a laboratory animal. After a suitable period (3–4 weeks) blood is removed and allowed to clot, leaving a liquid phase (polyclonal antiserum) containing many different IgG antibodies, resulting in:

- **cross-reaction** with other antigens or haptens;
- **batch variation,** as individual animals produce slightly different antibodies in response to the same antigen;
- **non-specificity,** as the antiserum will contain many other antibodies.

Standardisation of polyclonal antisera is therefore difficult. You may need to assess the amount of cross-reaction, interbatch variation or non-specific binding using appropriate controls, assayed at the same time as the test samples.

Definitions

Antibody – a protein produced in response to an antigen (an *antibody-generating* foreign macromolecule).

Epitope – a site on the antigen that determines its interaction with a particular antibody.

Hapten – a substance that contains at least one epitope, but is too small to induce antibody formation unless it is linked to a macromolecule.

Ligand – a molecule or chemical group that binds to a particular site on another molecule.

Fig. 43.1 Diagrammatic representation of IgG (antibody).

Producing polyclonal antibodies – in the UK, this is controlled by government regulations, since it involves vertebrate animals: personnel must be licensed by the Home Office and must operate in accordance with the Animals Scientific Procedures Act (1986) and with the Code of Practice for the Housing and Care of Animals (2005).

Definition

Monoclonal antibody – an antibody produced by a laboratory-grown clone of either a hybridoma (a fusion of an antibody-producing lymphocyte and a myeloma or lymphoma cell) or a virus-transformed lymphocyte.

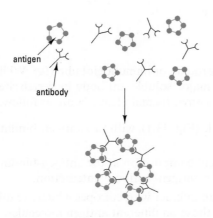

Fig. 43.2 Formation of an antigen–antibody complex (immune complex).

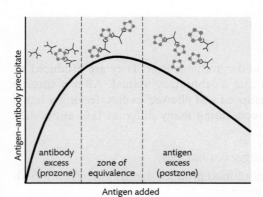

Fig. 43.3 Precipitation curve for an antigen titrated against a fixed amount of antibody.

Monoclonal antibodies

These are specific to a single epitope and are produced from individual clones of cells (hybridomas, p. 259), grown using cell culture techniques (p. 259). Such cultures provide a stable source of antibodies of known, uniform specificity. While monoclonal antibodies are likely to be used increasingly in future years, polyclonal antisera are currently employed for many routine immunological assays.

Agglutination tests

When antibodies interact with a suspension of a particulate antigen, e.g. cells or latex particles, the formation of immune complexes (Fig. 43.2) causes visible clumping, termed agglutination. A positive haemagglutination reaction gives an even 'carpet' of red cells over the base of the tube, while a negative reaction gives a tightly packed 'button' of red cells at the bottom of the tube. Agglutination tests are used in several ways:

- **Microbial identification:** at the species or subspecies level (serotyping), e.g. mixing an unknown bacterium with the appropriate antiserum will cause the cells to agglutinate.
- **Latex agglutination using bound antigens:** by coating soluble (non-particulate) antigens onto microscopic latex spheres, their reaction with a particular antibody can be visualised.
- **Latex agglutination using bound antibodies:** antibodies can be bound to latex microspheres, leaving their antigen-binding sites free to react with soluble antigen.
- **Haemagglutination:** red blood cells can be used as agglutinating particles. However, in some instances, such reactions do not involve antibody interactions (e.g. some animal viruses may haemagglutinate unmodified red blood cells).

Precipitin tests

Immune complexes of antibodies and soluble antigens (or haptens) usually settle out of solution as a visible precipitate: this is termed a precipitin test, or precipitation test. The formation of visible immune complexes in agglutination and precipitation reactions only occurs if antibody and antigen are present in an optimal ratio (Fig. 43.3). It is important to appreciate the shape of this curve: cross-linkage is maximal in the zone of equivalence, decreasing if either component is present in excess. The quantitative precipitin test can be used to measure the antibody content of a solution (Hay *et al.,* 2002 give details). Visual assessment of precipitation reactions forms the basis of several other techniques, described below.

Immunodiffusion assays

These techniques are easier to perform and interpret than the quantitative precipitin test. Precipitation of antibody and antigen occurs within an agarose gel, giving a visible line corresponding to the zone of equivalence (Fig. 43.3). Details of the main techniques are given in Box 43.1.

Box 43.1 How to carry out immunodiffusion assays

The two most widespread approaches are (i) single radial immunodiffusion (Mancini technique) and (ii) double diffusion immunoassay (Ouchterlony technique).

Single radial immunodiffusion (RID) (Mancini technique)

This is used to quantify the amount of antigen in a test solution, as follows:

1. Prepare an agarose gel (1.5% w/v) containing a fixed amount of antibody: allow to set on a glass slide or plate, on a level surface.

2. Cut several circular wells in the gel. These should be of a fixed size between 2 and 4 mm in diameter (see Fig. 43.4(a)). Cut your wells carefully – they should have straight sides and the agarose must not be torn or lifted from the glass plate. All wells should be filled to the top, with a flat meniscus, to ensure identical diffusion characteristics. Non-circular precipitin rings, resulting from poor technique, should not be included in your analysis.

3. Add a known amount of the antigen or test solution to each well.

4. Incubate on a level surface at room temperature in a moist chamber: diffusion of antigen into the gel produces a precipitin ring. This is usually measured after 2–7 d, depending on the molecular mass of the antigen.

5. Examine the plates against a black background (with side illumination), or stain using a protein dye (e.g. Coomassie blue).

6. Measure the diameter of the precipitin ring, e.g. using Vernier callipers.

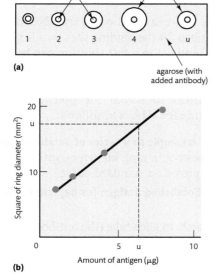

(a)

wells / precipitin rings / agarose (with added antibody)

1   2   3   4   u

(b)

Square of ring diameter (mm²) vs Amount of antigen (µg)

Fig. 43.4 Single radial immunodiffusion (RID). (a) Assay: four standards are shown (wells 1 to 4, each one double the strength of the previous standard), and an unknown (µ), run at the same time. (b) Calibration curve. The unknown contains 6.25 µg of antigen. Note the non-zero intercept of the calibration curve, corresponding to the square of the well diameter: do not force such calibration lines through the origin.

7. Prepare a calibration curve (Chapter 42) from the samples containing known amounts of antigen (Fig. 43.4(b)): the squared diameter of the precipitin ring is directly proportional to the amount of antigen in the well.

8. Quantify the amount of antigen in your test solutions using a calibration curve prepared from standards assayed at the same time.

Double diffusion immunoassay (Ouchterlony technique)

This technique is widely used to detect particular antigens in a test solution, or to look for cross-reaction between different antigens.

1. Prepare an agarose gel (1.5% w/v) on a level glass slide or plate: allow to set.

2. Cut several circular wells in the gel.

3. Add test solutions of antigen or polyclonal antiserum to adjacent wells. Both solutions diffuse outwards, forming visible precipitin lines where antigen and corresponding antibody are present in optimal ratio (Fig. 43.5).

The various reactions between antigen and antiserum are:

- **Identity:** two wells containing the same antigen, or antigens with identical epitopes, will give a fused precipitin line (identical interaction between the antiserum and the test antigens, Fig. 43.5(a)).

- **Non-identity:** where the antiserum contains antibodies to two different antigens, each with its own distinct epitopes, giving two precipitin lines which intersect without any interaction (no cross-reaction, Fig. 43.5(b)).

- **Partial identity:** where two antigens have at least one epitope in common, but where other epitopes are present, giving a fused precipitin line with a spur (cross-reaction, Fig. 43.5(c)).

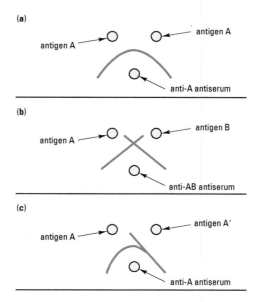

(a)

antigen A / antigen A / anti-A antiserum

(b)

antigen A / antigen B / anti-AB antiserum

(c)

antigen A / antigen A' / anti-A antiserum

Fig. 43.5 Precipitin reactions in double diffusion immunoassay: (a) identity; (b) non-identity; (c) partial identity.

Immunoelectrophoretic assays

These methods combine the precipitin reaction with electrophoretic migration, providing sensitive, rapid assays with increased separation and resolution. The principal techniques follow:

Cross-over electrophoresis (counter-current electrophoresis)

Similar to the Ouchterlony technique, since antigen and antibody are in separate wells. However, the movement of antigen and antibody towards each other is driven by a voltage gradient (p. 349): most antigens migrate towards the anode, while IgG migrates towards the cathode. This method is faster and more sensitive than double immunodiffusion, taking 15–20 min to reach completion.

Quantitative immunoelectrophoresis (Laurell rocket immunoelectrophoresis)

Similar to RID, as the antibody is incorporated into an agarose plate while the antigen is placed in a well. However, a voltage gradient moves the antigen into the gel, usually towards the anode, while the antibody moves towards the cathode, giving a sharply peaked, rocket-shaped precipitin line once equivalence is reached (within 2–10 h). The height of each rocket shape at equivalence is directly proportional to the amount of antigen added to each well. A calibration curve for samples containing a known amount of antigen can be used to quantify the amount of antigen present in test samples (Fig. 43.6).

Radioimmunological methods

These methods use radioisotopes to detect and quantify the antigen–antibody interaction, giving improved sensitivity over agglutination and precipitation methods. The principal techniques follow.

Radioimmunoassay (RIA)

This is based on competition between a radioactively labelled antigen (or hapten) and an unlabelled antigen for the binding sites on a limited amount of antibody. The quantity of antigen in a test solution can be determined using a known amount of radiolabelled antigen and a fixed amount of antibody (Fig. 43.7). As with other immunoassay methods, it is important to perform appropriate controls to screen for potentially interfering compounds. The basic procedure for RIA is as follows:

1. **Add appropriate volumes of a sample to a series of small test tubes.** Prepare a further set of tubes containing known quantities of the substance to be assayed to provide a standard curve.
2. **Add a known amount of radiolabelled antigen (or hapten) to each tube** (sample and standard).
3. **Add a fixed amount of antibody to each tube** (the antibody must be present in limited quantity).
4. **Leave at constant temperature for a fixed time** (usually 24 h), to allow antigen–antibody complexes to form.
5. **Precipitate the antibody and bound antigen using saturated ammonium sulfate, followed by centrifugation.**
6. **Determine the radioactivity of the supernatant or the precipitate** (pp. 299–301).

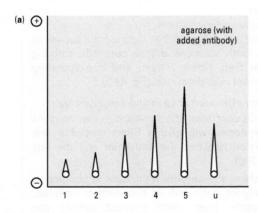

(a) ⊕

agarose (with added antibody)

⊖

1 2 3 4 5 u

(b)

Height of rocket (mm)

20

u

10

0   5   u   10

Amount of antigen (μg)

Fig. 43.6 Laurell rocket immunoelectrophoresis. (a) Assay: precipitin rockets are formed by electrophoresis of five standards of increasing concentration (wells 1 to 5) and an unknown (μ). (b) Calibration curve: the unknown sample contains 7.7 μg of antigen.

Fig. 43.7 Radioimmunoassay calibration curve. Note that the assay is insensitive at very low and very high antigen levels.

SAFETY NOTE When carrying out radio-immunoassays, check health and safety and other requirements for working with radioisotopes with your departmental safety officer.

Chromogenic substrates for enzyme immunoassay – these include tetramethylbenzidine (Box 43.2), which is a highly sensitive substrate used to detect horseradish peroxidase activity, producing a blue-coloured product as a result of peroxidase activity.

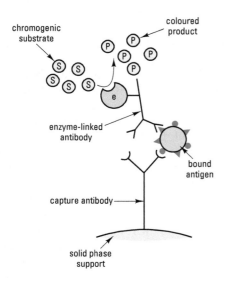

Fig. 43.8 Double antibody sandwich ELISA.

7. **Prepare a calibration curve of radioactivity against $\log_{10}$ antigen** (Fig. 43.7). The curve is most accurate in the central region, so adjust the amount of antigen in your test sample to fall within this range.

Note the following.

- **You must be registered to work with radioactivity**: check with the Departmental Safety Officer, if necessary (p. 304).
- **Measure all volumes as accurately as possible** as the end result depends on the volumetric quantities of unlabelled (sample) antigen, radio-labelled antigen and antibody: an error in any of these reagents will invalidate the assay.
- **Incorporate replicates**, so that errors can be quantified.
- **Seek your supervisor's advice** about fitting a curve to your data: this can be a complex process.

Immunoradiometric assay (IRMA)

This technique uses radiolabelled antibody, rather than antigen, for direct measurement of the amount of antigen (or hapten) in a sample. Most IRMAs are similar to the double antibody sandwich method described below, except that the second antibody is labelled using a radioisotope. The important advantages over RIA are:

- **linear relationship between amount of radioactivity and test antigen;**
- **wider working range for test substance;**
- **improved stability/longer shelf-life.**

Enzyme immunoassays (EIA)

These techniques are also known as enzyme-linked immunosorbent assays (ELISA). They combine the specificity of the antibody–antigen interaction with the sensitivity of enzyme assays using either an antibody or an antigen conjugated (linked) to an enzyme at a site which does not affect the activity of either component. The enzyme is measured by adding an appropriate chromogenic substrate (p. 412), which yields a coloured product. Enzymes offer the following advantages over radioisotopic labels:

- **Increased sensitivity:** a single enzyme molecule can produce many product molecules, amplifying the signal.
- **Simplified assay:** enzyme assays are usually easier than radioisotope assays (p. 411).
- **Improved stability of reagents:** components are generally more stable than their radiolabelled counterparts, giving them a longer shelf-life.
- **No radiological hazard:** no requirement for specialised containment/disposal facilities.
- **Automation is straightforward:** using disposable microtitre plates and an optical scanner.

The principal techniques follow.

Double antibody sandwich ELISA

This is used to detect specific antigens, involving a three-component complex between a capture antibody linked to a solid support, the antigen, and a second, enzyme-linked antibody (Fig. 43.8). This can be used to detect a particular antigen, e.g. a virus in a clinical sample, or to quantify the amount of antigen.

Fig. 43.9 Indirect ELISA.

Fig. 43.10 Competitive ELISA.

Indirect ELISA

This is used for antibody detection, with a specific antigen attached to a solid support. When the appropriate antibody is added, it binds to the antigen and will not be washed away during rinsing. Bound antibody is then detected using an enzyme-linked anti-immunoglobulin, e.g. a rabbit IgG antibody raised against human IgG (Fig. 43.9). One advantage of the indirect assay is that a single enzyme-linked anti-immunoglobulin can be used to detect several different antibodies, since the specificity is provided by the bound antigen.

Competitive ELISA

Here, any antigen present in a test sample competes with added enzyme-labelled antigen for a limited number of binding sites on the capture antibody (Fig. 43.10). Most commercial systems use 96-well microplates (12 columns by 8 rows), where each well is coated with the appropriate antibody. Following addition of known volumes of (i) sample and (ii) enzyme-labelled antigen, the plates are incubated (typically, up to 1 h), then washed thoroughly to remove all unbound material. Bound enzyme-labelled antigen is then detected using a suitable substrate. Quantitative results can be obtained by measuring the absorbance of each well using a spectrophotometric microplate reader: the absorbance at a particular wavelength is *inversely* proportional to the amount of antigen present in the test sample. Calibration standards (p. 276) are required to convert the readings to an amount or concentration. Alternatively, the test can be carried out in positive/negative format. Box 43.2 gives practical details for a sandwich-type ELISA.

Dipstick (chromatographic) immunoassays

Improvements in technology have led to the development of a range of diagnostic kits that can be used either in a medical setting ('near-patient testing') or in the home, including immunoassays for cancer screening and drug testing. The first of these available to the general public was the home pregnancy test kit which measures the hormone human chorionic gonadotrophin (hCG) in urine. The level of this hormone is raised substantially during pregnancy. The principle of this test, which is similar to other home test kits, is that coloured plastic microbeads are conjugated to an anti-hCG monoclonal antibody and then coated onto an absorbent plastic or cellulose strip. When this strip is dipped into urine, any hCG present will bind to the anti-hCG antibody and both hormone–antibody complexes and free antibody will then move up the strip by capillary action until they reach a second hCG-specific antibody that is coupled chemically to the strip so that it cannot move. Any hormone–antibody complex will bind to this 'fixed' antibody (through the hCG component); owing to the presence of the coloured plastic microbeads, a distinct line will then appear at this location on the strip, signifying the presence of the hormone. If no hormone–antibody complex is present, the first antibody will simply continue up the strip until it reaches an antibody raised against the anti-hCG antibody itself, coupled to the strip at a higher point. This will bind any free anti-hCG antibody in the sample and a coloured line will appear at this point. Since there will always be an excess of the original anti-hCG antibody conjugated to the plastic beads, the appearance of two coloured lines means that the hormone is present and the test result is positive, while if

Box 43.2   How to perform an ELISA assay

While the following example is for a sandwich (capture) ELISA assay in microplate format, the same general principles apply to the other types of ELISA.

1. **Prepare the apparatus.** Switch on all equipment required:

    (i)   The microplate reader – used to measure the absorbance of the solution in each well: set the reader to the required wavelength.

    (ii) The microplate washer (where used) – each well must be washed at various stages during the procedure. When using an automated washer, first check that the wash bottle contains sufficient diluent and then test using an old microplate, to check that all wells are being washed correctly. Where required, use a wire needle to clean any blocked wash delivery tubes and repeat. For manual washing, use a wash bottle or multichannel pipettor – make sure you fill each well and empty out all of the wash solution at the end of each wash stage.

    (iii) The computer – this will contain the software required to label the wells, draw the calibration curve and calculate the results for test samples: fill out the ELISA template with details of the assays to be carried out.

2. **Prepare the various solutions to be analysed.** These include:

    (i)   test samples – make sure that each sample is identified with a code that enables you to record what each test well contains;

    (ii) calibrators/standards – including 'cut-off' calibrators and known positive standards;

    (iii) controls – positive and negative controls and blanks.

3. **Coat the wells with the capture antibody.** Typically 100 µL of a solution of the appropriate monoclonal or polyclonal antibody is added to each well and microplates are then incubated overnight at 4°C to allow binding to the well.

4. **Wash the wells.** Transfer the microplate to the washer (or wash manually) – wash six times to remove excess coating (capture) antibody. The final rinse should be programmed so that the washer leaves the wells empty of diluent.

5. **Add blocking solution to each well.** Typically 100 µL of an inert protein solution (bovine serum albumin) is added to each well to block any free binding sites on the well. Microplates are incubated at room temperature for 30 min and then washed, as in step 4.

6. **Add test samples, calibrators and controls to wells.** Typically 100 µL of appropriately diluted test sample, control, etc. is added to each well. Microplates are incubated at room temperature for 90 min and then washed as step 4.

7. **Add the detection antibody to each well.** Add 100 µl of monoclonal or polyclonal detection antibody labelled with a suitable enzyme (e.g. horseradish peroxidase, HRP) to each well. The microplate is then incubated at room temperature for 30–60 min and then washed, as in step 4.

8. **Add chromogenic substrate to each well.** For example, with HRP-labelled antigen add 100 µL of a standard solution of tetramethylbenzidine (TMB) and hydrogen peroxide to each well and reincubate in darkness for 30 min, to allow colour development. The TMB is oxidised in the presence of hydrogen peroxide to produce a blue colour.

SAFETY NOTE TMB and hydrogen peroxide are harmful by inhalation – use a fume hood.

9. **Stop the reactions.** For example, by adding 100 µL of 2 mol L^{-1} sulfuric acid to denature the enzyme. The colour of the oxidised TMB will change from blue to yellow as a result of the pH shift. While the human eye can readily distinguish different shades of blue, it is more difficult to assess different shades of yellow visually, once the reaction has been stopped.

10. **Measure the absorbance of each sample/calibrator/control well.** Transfer the microplate to the reader and assay at an appropriate wavelength: for TMB, use 450 nm.

11. **Interpret the results.** Check that the absorbance values of calibrators and control are within the required range. Then, for each sample, either record the absolute value (convert to concentration or amount, e.g. using a calibration curve, p. 276) or record as 'positive' or 'negative' (based on values for 'cut-off' calibrators), as appropriate.

any uncomplexed antibody binds to a further antibody, raised against it in a different animal (e.g. goat, anti-mouse IgG antibody)

any antigen–antibody complexes bind to a second antibody (double antibody sandwich, see Fig. 43.8)

labelled antibody forms complexes with any antigen present in sample and travels up the dipstick by capillary action

Fig. 43.11 Two-site dipstick immunoassay. For a positive test, coloured bands will appear at positions 1 and 2 while a negative test, as shown here, will give colour in band 2 only (any other colour options indicate a failed test).

only the second coloured line appears, it means that the test is negative (this also serves as a control to confirm that the test is working – lack of any coloured lines indicates that the test has failed). Figure 43.11 shows the general operating principles.

Complement-based assays

Complement is a term applied to a group of around 20 different proteins which are involved in various aspects of immune function, including (i) control of inflammation, (ii) preparation of microbial cells for phagocytosis (opsonisation), (iii) lysis of target cells (e.g. antibody-coated cells) and (iv) activation of leucocytes (for details see Delves *et al.*, 2011). At a practical level, the complement system can be used to detect (i) antibodies, (ii) antigens or (iii) antibody–antigen complexes, often via its lytic action on antibody-coated ('sensitised') cells. The operating principles of a complement fixation assay are illustrated in Fig. 43.12 for antibody detection, with the following steps:

> **Understanding complement-based assays** – the principal benefit from using the complement system is that it operates as an amplifying cascade of reactions, providing higher sensitivity than assays based on agglutination or precipitin reactions. For example, antibodies can be detected at concentrations of less than 1 µg mL^{-1}.

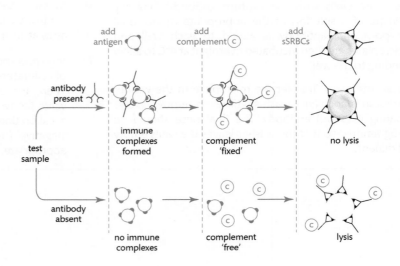

Fig. 43.12 Complement fixation test for antibody. (sSRBCs: sensitised sheep red blood cells.)

1. **Prepare serial doubling dilutions** (pp. 154–5) of a test sample containing an unknown amount of antibody.

2. **Add a standard amount of antigen:** antigen–antibody complexes will be formed in direct proportion to the amount of antibody present in the sample.

3. **Then, add a known amount of complement to the mixture:** if immune complexes (Fig. 43.2) are present, they will 'fix' complement (i.e. the complement components will be consumed and no longer 'free' in solution), whereas if immune complexes are not present, or are at low concentration, complement will remain 'free' in solution, and available for subsequent reaction.

4. **In the final step, add an indicator system for unbound complement** – sensitised sheep red blood cells (sSRBCs) coated with rabbit anti-erythrocyte antibodies are often used. Unfixed complement is detected by lysis of the sensitised erythrocytes. The lowest dilution showing complement fixation (i.e. no erythrocyte lysis) gives the titre of antibody in the test sample.

The test can be modified to test for antigens in a sample by titration against a fixed amount of added antibody. In all cases, suitable controls are required, to check for non-specific reactions and to confirm that immune complexes are not already present in the sample, since these will fix complement and invalidate the titration. Complement fixation tests are useful for many classes of IgG and for IgM, but not for most other antibody types.

Example Complement fixation tests are the basis of the Wasserman test to detect anti-treponemal antibodies in the serum of patients with syphilis.

Text references

Delves, P.J., Martin, S., Burton, D. and Roitt, I. (2011) *Roitt's Essential Immunology,* 12th edn. Blackwell, Oxford.

Hay, F.C., Westwood, O.M.R. and Nelson, P.N. (2002) *Practical Immunology.* Blackwell, Oxford.

Sources for further study

Gosling, J.R.G. (2000) *Immunoassays: a Practical Approach.* Oxford University Press, Oxford.

Lefkovits, I. (1997) *Immunology Methods Manual: The Comprehensive Sourcebook of Techniques.* Academic Press, New York.

Van Emon, J.M. (2006) *Immunoassays and Other Bioanalytical Techniques.* CRC Press, Boca Raton.

Wild, D.G. (2013) *The Immunoassay Handbook,* 4th edn. Elsevier, Amsterdam.

STUDY EXERCISES

43.1 Determine the amount of antigen in a test solution using Laurell rocket immunoelectrophoresis. The figure below represents the results from a rocket immunoelectrophoresis assay for a series of standards. Determine the amount of antigen in a test solution giving a rocket (precipitin line) height of 18.5 mm.

43.2 Determine the amount of antigen in a test solution using single radial immunodiffusion (RID). Using the following data for a series of standards, determine the amount of antigen in a test solution giving a ring diameter of 6.5 mm.

Single radial immunodiffusion results for a series of standards containing 420 μg antigen

Amount of antigen (μg)	Ring diameter (mm)
4	3.5
8	5.5
12	7.0
16	8.0
20	9.0

43.3 Interpret data from a haemagglutination test. In some conditions, such as Hashimoto's disease, autoantibodies to the thyroid protein thyroglobulin are produced in large amounts. The level of these antibodies in serum can be measured by indirect haemagglutination: red blood cells coated with thyroglobulin are mixed with serial doubling dilutions of the test serum. The titre of antithyroglobulin autoantibodies is the lowest dilution at which haemagglutination occurs. The figure below represents the outcome of an antithyroglobulin haemagglutination test on a serum sample. The first well – numbered 1 – represents a 10-fold dilution of the serum sample, with serial doubling dilutions in subsequent wells.

(a) Write a brief explanation of the test results shown in the figure.

(b) Determine the autoantibody titre.

(c) Score the sample as 'positive' if the titre is more dilute than 1 in 5000 (i.e. 1 : >5000) and 'negative' if it is less dilute than 1 in 5000 (i.e. 1: <5000).

Answers to these study exercises are available at www.pearsoned.co.uk/practicalskills.

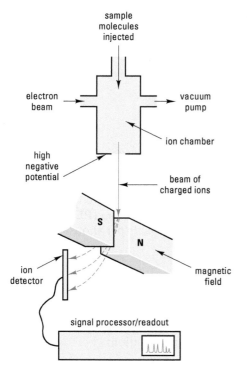

Fig. 44.1 Components of an electron impact mass spectrometer.

Fig. 44.2 Mass spectrum for methanol. m/z = charge ratio.

Stable isotopes of an element all have the same number of protons, but different numbers of neutrons. In contrast to radioisotopes (Chapter 45), they do not spontaneously decay to produce radiation. Stable isotopes have virtually identical gross chemical properties. When two or more stable isotopes of the same element exist in nature, they each contribute to the relative atomic mass (A_r) in direct proportion to their relative abundance.

Given their broad chemical equivalence, stable isotopes have a narrow range of practical applications in biology, and the techniques used to measure them are highly specialised and sensitive; they include mass spectrometry (MS) and nuclear magnetic resonance (NMR) spectrometry.

Measuring stable isotopes by mass spectrometry

Typically, stable isotopes are separated and quantified in terms of their mass-to-charge ratio using some form of mass spectrometry. This technique involves the vaporisation/ionisation of a sample in a gas phase. The ions are accelerated to specific velocities using an electric field and then separated on the basis of their different masses and charges (Fig. 44.1). Each ion of a particular mass-to-charge ratio is detected sequentially with time. The most widely used method in MS is electron impact (EI) ionisation, where the electron source is a heated metal, such as a tungsten filament, subjected to an appropriate potential gradient. The stream of electrons emitted can then interact with elements and molecules in the sample, by either:

- **electron removal** – where an electron in a bond within the sample molecule is 'knocked out' by bombarding electrons, producing a positively charged ion (cation); or

- **electron capture** – where addition of an electron results in the production of a negatively charged ion (anion).

The ions are then subjected to a magnetic field (Fig. 44.1) which results in their detection in proportion to their abundance in the sample. Signal processing and analysis then give a mass spectrum, such as that shown in Fig. 44.2, which can be used to characterise and identify the compound.

For metabolic studies, the technique of isotope ratio mass spectrometry (IRMS) is very useful, since it removes the need to use radioactive isotopes. The technique exploits the ability of MS to distinguish between isotopes such as ^{13}C and ^{12}C. For example, a compound containing ^{13}C will have a greater overall mass than the same compound containing ^{12}C and can therefore be differentiated within the mass spectrum produced as a result of MS. By selectively labelling a key metabolite with a stable isotope that is naturally present at low concentration (e.g. ^{13}C), it is then possible to follow the fate of the metabolite by MS analysis of sequential samples. Typically, the metabolites would be combusted in an oxygen atmosphere, to give gaseous $^{13}CO_2$, followed by exposure of such gases to an EI source (Fig. 44.1); useful metabolic data can be obtained from the isotope ratio – in this instance $^{13}CO_2$ and $^{12}CO_2$.

Using stable carbon isotopes in photosynthetic studies

Measurements of stable carbon isotopes in organic material and in the air passing over photosynthetically active plant material can be used as a probe of photosynthetic physiology. 'Light' and 'heavy' isotopes of carbon (^{12}C and ^{13}C, respectively) have different diffusion rates within the leaf and the key photosynthetic enzyme Rubisco (ribulose bisphosphate carboxylase oxygenase) also discriminates against ^{13}C, leading to changes in the isotope composition of photo-assimilated carbon compared to that of air. By measuring this discrimination, the physiological processes associated with carbon fixation and metabolism can be investigated in a time-integrated manner. Analysis of stable isotope composition makes use of mass spectrometric measurements of the ratio of heavy-to-light isotopes in the sample under investigation and in a predefined standard.

The mass spectrometer compares the mass-to-charge ratio (m/z) of the 44 and 45 masses of $^{12}CO_2$ and $^{13}CO_2$ respectively, to give the isotopic composition of (i) a sample and (ii) a standard, in terms of the ^{13}C molar abundance ratio (R), where:

$$R = \frac{[^{13}CO_2]}{[^{12}CO_2]}$$ [44.1]

These values are then used to calculate the isotopic composition of the sample relative to the standard as a $\delta^{13}C$ value, usually expressed in parts per thousand (% – note that this is a proportion, and *not* a unit – spoken as 'per mil'). The original standard was CO_2 derived from a fossil <u>Be</u>lemnite from the <u>Pee Dee</u> deposits in South Carolina (PDB standard: $R = 0.011\ 24$). The $\delta^{13}C$ value of a sample in parts per thousand is then calculated as:

$$\delta^{13}C = \frac{R_{sample} - R_{standard}}{R_{standard}} \times 1000$$ [44.2]

Isotopic composition values may also be used to calculate discrimination (Δ), which takes into account the isotopic composition of the standard, source and sample, according to the following equation:

$$\Delta = \frac{\delta_a - \delta_p}{1 + \delta_p}$$ [44.3]

where δ_a and δ_p are the carbon isotope discrimination values for air (source) and plant (sample) respectively, expressed in fractional terms (see worked examples in margin). In contrast to $\delta^{13}C$ values which are negative when PDB is used as the standard, values for Δ are positive for plant material, since ^{12}C is favoured over ^{13}C. The advantage in using Δ notation is that it is more straightforward to make comparisons, avoiding the possible confusion involved in discussing differences in negative $\delta^{13}C$ values. Like $\delta^{13}C$ values, Δ values are usually expressed in parts per thousand (%). Values for Δ values in C_3 plants, where Rubisco is the main carboxylating enzyme, usually fall within the range 16–25%, since Rubisco shows substantial discrimination against ^{13}C. In contrast, Δ values for C_4 plants, where PEP carboxylase is the main carboxylating enzyme, are lower – typically in the range 3–7% – since PEP carboxylase shows minimal discrimination against ^{13}C.

Advances in the study of hydrogen, oxygen and nitrogen isotope ratios have further extended the range of tools available to physiologists to

Example Using Eqn [44.1], for air, with $[^{13}CO_2]$ at 0.144 μmol l^{-1} and $[^{12}CO_2]$ at 12.915 μmol l^{-1}, gives a molar abundance ratio, $R = 0.144 \div 12.915 = 0.01115$ (to four significant figures).

Example Using Eqn [44.2], for air, where $R = 0.01115$, $\delta^{13}C = (0.01115 - 0.01124) \div 0.01124 \times 1000 = -8.0\%$ (to one decimal place). $\delta^{13}C$ for plant tissue where Rubisco is the main carboxylating enzyme is *more negative* than $\delta^{13}C$ for C_4 plant tissue, since Rubisco discriminates strongly against ^{13}C (e.g. $R = 0.01095$, $\delta^{13}C = -25.8\%$). $\delta^{13}C$ for C_4 plant tissue where phosphoenolpyruvate (PEP) carboxylase is the main carboxylating enzyme is *less negative* than $\delta^{13}C$ for C_3 plant tissue, since this enzyme shows little discrimination against ^{13}C, and the main effect is due to diffusion within the leaf (e.g. $R = 0.01111$, $\delta^{13}C = -11.6\%$).

Example Using Eqn [44.3], Δ for C_3 plant tissue with a $\delta^{13}C$ of -25.8% is $[-0.008 - (-0.0258)] \div [1 + (-0.0258)] = 0.0182714 = 18.2\%$.
Using Eqn [44.3], Δ for C_4 plant tissue with a $\delta^{13}C$ of -11.5% is $[-0.008 - (-0.0115)] \div [1 + (-0.0115)] = 0.0035407 = 3.5\%$.

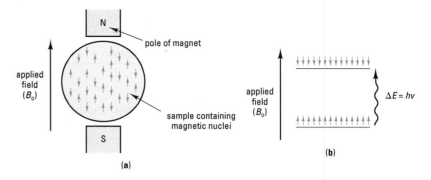

Fig. 44.3 Effect of an applied magnetic field, B_0, on magnetic nuclei. (a) Nuclei in magnetic field have one of two orientations – either with the field or against the field (in the absence of an applied field, the nuclei would have random orientation). (b) Energy diagram for magnetic nuclei in applied magnetic field.

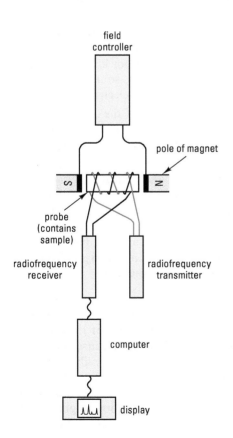

Fig. 44.4 Components of an NMR spectrometer.

investigate areas including plant water relations and nitrogen economy at the individual plant and ecosystem levels. Since the stable carbon isotope ratio of a plant is a broad measure of the overall balance between carbon uptake (photosynthesis) and water loss (transpiration), Δ is found to vary with changes in transpiration rate, with a lower value corresponding to a higher water use efficiency. It has been proposed that concurrent analysis of $^{18}O/^{16}O$ ratios can enable researchers to interpret variations in Δ in terms of water-use efficiency: this can be useful in plant breeding studies, where improved water-use efficiency results in closely coupled ^{13}C and ^{18}O signals.

NMR analysis of stable isotopes

Electromagnetic radiation at radiofrequencies of 1–500 MHz can be used to identify and monitor biological compounds and metabolic processes. This is possible because of differences in the magnetic states of atomic nuclei of different stable isotopes, involving very small transitions in energy level. The atomic nuclei of isotopes of many elements are magnetic, because they are charged and have spin. Typical stable isotopes that are magnetic include ^1H, ^{13}C, ^{14}N, ^{15}N, ^{19}F and ^{31}P. When these nuclei interact with a uniform external magnetic field, they behave like tiny compass needles and align themselves in a direction either (i) parallel or (ii) antiparallel to the external magnetic field. The two orientations have different energies, with the parallel direction having a lower energy than the antiparallel direction (Fig. 44.3). The energy difference between these two levels corresponds to a precise electromagnetic frequency.

When a sample containing an isotope with a magnetic nucleus is placed in a magnetic field and exposed to an appropriate radiofrequency in an NMR spectrometer (Fig. 44.4), transitions between the energy levels of magnetic nuclei will occur when the energy gap and the applied frequency are in *resonance* (i.e. when they are matched exactly). This frequency is a characteristic of the particular isotope and can be used to study the isotope in biological systems. Typically, this is described in terms of the 'chemical shift' of the nuclei involved, which is a field-independent dimensionless value measured against the resonance frequency of a reference compound (for example, tetramethylsilane for ^{1}H and ^{13}C NMR), as:

$$\frac{\text{resonance frequency of sample} \; - \; \text{resonance frequency of reference}}{\text{resonance frequency of the reference}}$$

$$[44.4]$$

Because the difference in frequency (numerator in [44.4]) is very small in relation to the resonance frequency of the reference (denominator in [44.4]), chemical shifts are usually expressed in parts per million (ppm).

The resonance frequency of magnetic nuclei is affected by the presence of shielding owing to small secondary fields generated by electrons in their vicinity, giving rise to different chemical shifts for the same isotope in different parts of a molecule. By convention, the chemical shift is positive if the nucleus is less shielded than the reference compound and negative if it is more shielded. The end result is an NMR spectrum where individual magnetic nuclei produce a 'signal' with peaks representing the different locations within a molecule.

Biomolecular applications of NMR

The sensitivity of NMR has improved dramatically with the development of more powerful magnets. Details of the major uses of various magnetic nuclei are given in Table 44.1. The principal applications include:

- **studies of the structure and function of macromolecules and biological systems,** such as membranes;
- **metabolic investigations on living organisms,** including humans, since NMR can be used to obtain a 'fingerprint' of a particular molecule, and changes in the intensity of spectra can be used for kinetic studies (e.g. Fig. 48.3). This involves use of 'surface coils' as sources of radiation, or placing the organism within the core of an electromagnet;
- **measurement of intracellular pH** by determination of the chemical shift of the phosphate peak, as this changes with pH in a predictable manner;
- **magnetic resonance imaging (MRI),** which is a form of proton NMR that uses a field gradient (as opposed to a uniform field) to produce signals that are translated by computers into anatomical images (Chary and Govil, 2008). Body tissues have different water contents (typically 60–90% w/w). Signals arising from protons in water in different tissues can be used to differentiate between tissues (e.g. 'grey' and 'white' matter in the brain), and between normal tissue and tumours. Fat deposits can be detected, owing to the difference between the ^{1}H signal from fatty acids and that of the water in the surrounding tissue.

^{13}C NMR analysis is particularly useful in photosynthetic studies; for example, in identifying the most abundant organic low molecular weight solutes in plants, and for studying how the concentrations of these solutes are adjusted in response to salt and water stress. The relative insensitivity of this approach, with ^{13}C representing only 1.1% of the naturally occurring carbon atoms in biomolecules, becomes an advantage in such studies. Only those soluble organic molecules that are present in large quantities within a cell will be detected, giving a spectrum that typically shows signals from only one or two solutes. As an example, Fig. 44.5 shows a natural abundance

Using NMR – in contrast to most conventional metabolic studies, NMR is non-invasive and the time course of metabolic reactions can be followed using a single experimental subject or preparation, eliminating variation between samples.

Table 44.1 The relative merits and disadvantages of various magnetic nuclei in biomolecular studies.

Nucleus	Relative sensitivity	Natural abundance	Comments
^1H	100	99.98%	Multiple, but specific, spectral lines are obtained for individual biomolecules. For mixtures, the ubiquitous occurrence of ^1H gives complex, overlapping signals that are often difficult to interpret. Gives a large solvent peak with aqueous samples (can be avoided by using D_2O as solvent). Mainly used for structural studies of pure macromolecules. Essential for MRI.
^{31}P	6.6	100%	Very useful for studies on living systems, with narrow resonance peaks and a wide range of chemical shifts for different molecules. Spectra are simpler and easier to interpret than for ^1H, but are not as distinctive: different compounds may give similar ^{31}P spectra.
			Several important P-containing compounds (including ATP, ADP and inorganic phosphate) can be detected in intact cells – useful in bioenergetic studies.
^{13}C	0.016	1.1%	Gives narrow signals and a wide range of chemical shifts. Resolution is better than for ^1H, and a wide range of organic biomolecules can be detected. Low natural abundance gives low sensitivity, extending the time required to accumulate spectra. However, low natural abundance also means that specific metabolites can be selected for ^{13}C isotope enrichment, allowing particular metabolic pathways to be investigated, e.g. carbon assimilation.

Fig. 44.5 A ^{13}C NMR spectrum of *Himanthalia elongata*, with asterisks showing the peaks corresponding to individual nuclei within the carbohydrate altritol. m = mannitol (Courtesy J. A. Chudek).

^{13}C NMR spectrum of the seaweed *Himanthalia elongata,* which was the first report of the carbohydrate altritol in any living organism (Chudek *et al.,* 1984) – previous chromatographic studies had failed to separate this compound from other isomers, such as mannitol. The relatively low natural abundance of ^{13}C also means that 'labelling' studies can be carried out, e.g. using $^{13}CO_2$, to follow the metabolism of photosynthetically fixed C within the plant.

Text reference

Chary, K.V.R. and Govil, G. (2008) *NMR in Biological Systems: from Molecules to Humans.* Springer, Dordrecht.

Sources for further study

Keller, J. (2010) *Understanding NMR Spectroscopy,* 2nd edn. Wiley, New York.
Smith, R.M. and Butsch, K.L. (2004) *Understanding Mass Spectra: A Basic Approach.* Wiley, New York.

Chudek, J.A., Foster, R., Davison, I.R. and Reed, R.H. (1984) Altritol in the brown alga *Himanthalia elongata, Phytochemistry* 23, 1081–1082.

Westbrook, C. and Roth, C. (2011) *MRI in Practice,* 4th edn. Blackwell, Oxford. [Covers the basics of magnetic resonance imaging in medicine.]

STUDY EXERCISES

44.1 Test your understanding of the terminology of stable isotopes. What do the following terms/acronyms mean? (a) NMR; (b) MRI; (c) EI (d) chemical shift; (e) electron capture.

44.2 Explain the operating principles of a nuclear magnetic resonance (NMR) spectrometer. Make sure that your explanation covers the major components of an NMR spectrometer.

44.3 Calculate the molar abundance ratio ($R = [^{13}CO_2]/[^{12}CO_2]$ and $\delta^{13}C([R_{sample} - R_{standard}]/R_{standard} \times 1000)$ of a sample. Determine R and $\delta^{13}C$ for an air sample with $[^{13}CO_2]$ = 0.152 μmol L^{-1} and $[^{12}CO_2]$ at 12.825 μmol L^{-1}, and PDB standard of R = 0.011 24.

Answers to these study exercises are available at www.pearsoned.co.uk/practicalskills.

The isotopes of a particular element have the same number of protons in the nucleus but different numbers of neutrons, giving them the same proton number (atomic number) but different nucleon numbers (mass number, i.e. number of protons + number of neutrons). Isotopes may be stable (Chapter 44) or radioactive. Radioactive isotopes (radioisotopes) disintegrate spontaneously at random to yield radiation and a decay product.

Radioactive decay

There are three forms of radioactivity (Table 45.1) arising from three main types of nuclear decay:

- **Alpha decay.** This involves the loss of a particle equivalent to a helium nucleus. Alpha (α) particles, being relatively large and positively charged, do not penetrate far in living tissue, but they do cause ionisation damage and this makes them generally unsuitable for tracer studies.

- **Beta decay.** This involves the loss or gain of an electron or its positive counterpart, the positron. There are three subtypes:

  (a) Negatron (β^-) emission: loss of an electron from the nucleus when a neutron transforms into a proton. This is the most important form of decay for radioactive tracers used in biology. Negatron-emitting isotopes of biological importance include ^{3}H, ^{14}C, ^{32}P and ^{35}S.

  (b) Positron (β^+) emission: loss of a positron when a proton transforms into a neutron. This only occurs when sufficient energy is available from the transition and may involve the production of gamma rays when the positron is later annihilated by collision with an electron.

  (c) Electron capture (EC): when a proton 'captures' an electron and transforms into a neutron. This may involve the production of X-rays as electrons 'shuffle' about in the atom (as with ^{125}I) and it frequently involves electron emission.

- **Gamma emission.** Internal transition involves the emission of electromagnetic radiation in the form of gamma (γ) rays from a nucleus in a metastable state and always follows initial alpha or beta decay. Emission of gamma radiation leads to no further change in atomic number or mass.

Note from the above that more than one type of radiation may be emitted when a radioisotope decays. The main radioisotopes used in biology and their properties are listed in Table 45.2.

> **Examples** $^{12}_6$C, $^{13}_6$C and $^{14}_6$C are three of the isotopes of carbon. About 98.9% of naturally occurring carbon is in the stable $^{12}_6$C form. $^{13}_6$C is also a stable isotope but it only occurs at 1.1% natural abundance. Trace amounts of radioactive $^{14}_6$C are found naturally; this is a negatron-emitting radioisotope (see Table 45.2).

> **Examples** ^{226}Ra decays to ^{222}Rn by loss of an alpha particle, as follows:
>
> $$^{226}_{88}\text{Ra} \rightarrow \,^{222}_{86}\text{Rn} + \,^4_2\text{He}^{2+}$$
>
> ^{14}C shows beta decay, as follows:
>
> $$^{14}_6\text{C} \rightarrow \,^{14}_7\text{N} + \beta^-$$
>
> ^{22}Na decays by positron emission, as follows:
>
> $$^{22}_{11}\text{Na} \rightarrow \,^{22}_{10}\text{Ne} + \beta^+$$
>
> ^{55}Fe decays by electron capture and the production of an X-ray, as follows:
>
> $$^{55}_{26}\text{Fe} \rightarrow \,^{55}_{25}\text{Mn} + \text{X}$$
>
> The decay of ^{22}Na by positron emission (β^+) leads to the production of a γ ray when the positron is annihilated on collision with an electron.

Table 45.1 Types of radioactivity and their properties.

Radiation	Range of maximum energies (MeV*)	Penetration range in air (m)	Suitable shielding material
Alpha (α)	4–8	0.025–0.080	Unnecessary
Beta (β)	0.01–3	0.150–16	Plastic (e.g. Perspex)
Gamma (γ)	0.03–3	1.3–13†	Lead

*Note that 1 MeV = 1.6×10^{-13} J.

†Distance at which radiation intensity is reduced to half.

Table 45.2 Properties of some isotopes used commonly in the life sciences. Physical data obtained from Haynes (2014).

Isotope	Emission(s)	Maximum energy (MeV)	Half-life	Main uses	Advantages	Disadvantages
^{3}H	β^-	0.01861	12.3 years	Suitable for labelling organic molecules in wide range of positions at high specific activity	Relatively safe	Low efficiency of detection, high isotope effect, high rate of exchange with environment
^{14}C	β^-	0.156 48	5715 years	Suitable for labelling organic molecules in a wide range of positions	Relatively safe, low rate of exchange with environment	Low specific activity
^{22}Na	β^+ (90%) + γ, EC	2.842 (β^+)	2.60 years	Transport studies	High specific activity	Hazardous
^{32}P	β^-	1.710	14.3 d	Labelling proteins and nucleotides (e.g. DNA)	High specific activity, ease of detection	Short half-life, hazardous
^{33}P	β^-	0.248	25 d	Labelling nucleotides and proteins	Safer than ^{32}P	Moderate half-life
^{35}S	β^-	0.167	87.2 d	Labelling proteins and nucleotides	Low isotope effect	Low specific activity
^{36}Cl	β^-, β^+, EC	0.709 (β^-) 1.142 (β^+, EC)	300 000 years	Transport studies	Low isotope effect	Low specific activity, hazardous
^{125}I	EC + γ	0.178 (EC)	59.9 d	Labelling proteins and nucleotides	High specific activity	Hazardous
^{131}I	β^- + γ	0.971 (β^-)	8.04 d	Labelling proteins and nucleotides	High specific activity	Hazardous

EC, Electron capture.

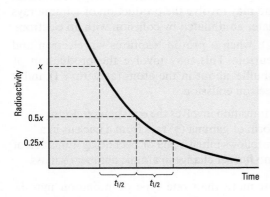

Fig. 45.1 Decay of a radioactive isotope with time. The time taken for the radioactivity to decline from x to $0.5x$ is the same as the time taken for the radioactivity to decline from $0.5x$ to $0.25x$, and so on. This time is the half-life ($t_{1/2}$) of the isotope.

Example For ^{35}S, with a half-life of 87.2 d (Table 45.2), the fraction of radioactivity remaining after 28 d would be worked out as follows: first, from Eqn [45.1], $x = (-0.693 \times 28) \div 87.2 = -0.222522936$, then using Eqn [45.1], $f = e^{-0.222522936} = 0.800496646$ (with appropriate rounding, 80.0% of original activity).

Each radioactive particle or ray carries energy, usually measured in electron volts (eV). The particles or rays emitted by a particular radioisotope exhibit a range of energies, termed an energy spectrum, characterised by the maximum energy of the radiation produced, E_{max} (Table 45.2). The energy spectrum of a particular radioisotope is relevant to the following:

- **Safety:** isotopes with the highest maximum energies will have the greatest penetrating power, requiring appropriate shielding (Table 45.1).
- **Detection:** different instruments vary in their ability to detect isotopes with different energies.
- **Discrimination:** some instruments can distinguish between isotopes, based on the energy spectrum of the radiation produced (p. 301).

The decay of an individual atom (a 'disintegration') occurs at random, but that of a population of atoms occurs in a predictable manner. The radioactivity decays exponentially, having a characteristic half-life ($t_{1/2}$). This is the time taken for the radioactivity to fall from a given value to half that value (Fig. 45.1). The $t_{1/2}$ values of different radioisotopes range from fractions of a second to more than 10^{19} years (see also Table 45.2). If $t_{1/2}$ is very short, as with ^{15}O ($t_{1/2} \approx 2$ min), then it is generally impractical to use the isotope in experiments because you would need to account for the decay during the experiment and counting period.

To calculate the fraction (f) of the original radioactivity left after a particular time (t), use the following relationship:

$$f = e^x, \text{ where } x = -0.693t/t_{1/2} \qquad [45.1]$$

Note that the same units must be used for t and $t_{1/2}$ in this equation.

Measuring radioactivity

The SI unit of radioactivity is the becquerel (Bq), equivalent to one disintegration per second (d.p.s.), but disintegrations per minute (d.p.m.) are also used. The curie (Ci) is a non-SI unit equivalent to the number of disintegrations produced by 1 g of radium (37 GBq). Table 45.3 shows the relationships between these units. In practice, most instruments are not able to detect all of the disintegrations from a particular sample, i.e. their efficiency is less than 100% and the rate of decay may be presented as counts min^{-1} (c.p.m.) or counts s^{-1} (c.p.s.). Most modern instruments correct for background radiation and inefficiencies in counting, converting count data to d.p.m. Alternatively, the results may be presented as the measured count rate, although this is only valid where the efficiency of counting does not vary greatly among samples.

> **KEY POINT** The specific activity is a measure of the quantity of radioactivity present in a known amount of the substance:
>
> $$\text{specific activity} = \frac{\text{radioactivity}\,(\text{Bq, Ci, d.p.m., etc.})}{\text{amount of substance}\,(\text{mol, g, etc.})} \qquad [45.2]$$

This is an important concept in practical work involving radioisotopes, since it allows interconversion of disintegrations (activity) and amount of substance (see Box 45.1).

Two SI units refer to doses of radioactivity and these are used when calculating exposure levels for a particular source. The sievert (Sv) is the amount of radioactivity giving a dose in humans equivalent to 1 gray (Gy) of X-rays: 1 Gy = an energy absorption of 1 J kg^{-1}. The dose received in most biological experiments is a negligible fraction of the maximum permitted exposure limit. Conversion factors from older units are given in Table 45.3.

The most important methods of measuring radioactivity for biological purposes are described below.

The Geiger–Müller (G–M) tube

This operates by detecting radiation when it ionises gas between a pair of electrodes across which a voltage has been applied (Fig. 45.2). You should use a handheld Geiger–Müller tube for routine checking for contamination (although it will not pick up ^{3}H activity).

The scintillation counter

This operates by detecting the scintillations (fluorescence 'flashes') produced when radiation interacts with certain chemicals called fluors (Fig. 45.3). In solid (or external) scintillation counters (often referred to as 'gamma counters') the radioactivity causes scintillations in a crystal of fluorescent material held close to the sample. This method is only suitable for radioisotopes producing penetrating radiation.

Liquid scintillation counters are mainly used for detecting beta decay and they are especially useful in biology. The sample is dispersed or dissolved in a suitable solvent containing the fluor(s) – the 'scintillation cocktail'. The radiation first interacts with the solvent, and the energy from this interaction is passed to the fluors, which produce detectable light. The scintillations are measured by photomultiplier tubes which turn the light pulses into electronic pulses, the magnitude of which is directly related to

Table 45.3 Relationships between units of radioactivity. For abbreviations, see text.

1 Bq = 1 d.p.s	1 Sv = 100 rem
1 Bq = 60 d.p.m.	1 Gy = 100 rad
1 Bq = 27 pCi	1 Gy ≈ 100 roentgen
1 d.p.s. = 1 Bq	1 rem = 0.01 Sv
1 d.p.m. = 0.0167 Bq	1 rad = 0.01 Gy
1 Ci = 37 GBq	1 roentgen ≈ 0.01 Gy
1 mCi = 37 MBq	
1 µCi = 37 kBq	

Example If 0.4 mL of a ^{32}P-labelled DNA solution at a concentration of 50 µmol L^{-1} (amount = 0.4 × 50 ÷ 1000 = 0.02 µmol) gave a count of 2490 d.p.m. (= 41.5 Bq), using Eqn [45.2] this would correspond to a specific activity of 2490 ÷ 0.02 = 124500 d.p.m. µmol^{-1} (or 2075 Bq µmol^{-1}).

Fig. 45.2

Fig. 45.3 Components of a scintillation counter. Note that in most modern instruments, all components are enclosed within a single cabinet.

Box 45.1 How to determine the specific activity of an experimental solution

Suppose you need to make up a certain volume of an experimental solution, to contain a particular amount of radioactivity. For example, 50 mL of a mannitol solution at a concentration of 25 mmol L^{-1}, to contain 5 Bq μL^{-1} – using a manufacturer's stock solution of ^{14}C-labelled mannitol (specific activity = 0.1 Ci mmol^{-1}).

1. **Calculate the total amount of radioactivity in the experimental solution**, in this example 5 × 1000 (to convert μL to mL) × 50 (50 mL required) = 2.5 × 10^5 Bq (i.e. 250 kBq).

2. **Establish the volume of stock radioisotope solution required:** for example, a manufacturer's stock solution of ^{14}C-labelled mannitol contains 50 μCi of radioisotope in 1 ml of 90% v/v ethanol: water. Using Table 45.3, this is equivalent to an activity of 50 × 37 = 1850 kBq. So, the volume of solution required is 250/1850 of the stock volume, i.e. 0.1351 ml (135 μL).

3. **Calculate the amount of non-radioactive substance required** as for any calculation involving concentration (see pp. 153, 162), e.g. 50 mL (0.051) of a 25 mmol L^{-1} (0.025 mol L^{-1}) mannitol (relative molecular mass 182.17) will contain 0.05 × 0.025 × 182.17 = 0.2277 g.

4. **Check the amount of radioactive isotope to be added.** In most cases, this represents a negligible amount of substance, e.g. in this instance, 250 kBq of stock solution at a specific activity of 14.8 × 10^6 kBq mmol^{-1} (converted from 0.4 Ci mmol^{-1} using Table 45.3) is equal to 250/14800000 = 16.89 nmol, equivalent to approximately 3 μg mannitol. This can be ignored in calculating the mannitol concentration of the experimental solution.

5. **Make up the experimental solution** by adding the appropriate amount of non-radioactive substance and the correct volume of stock solution.

6. **Measure the radioactivity in a known volume of the experimental solution.** If you are using an instrument with automatic correction to Bq, your sample should contain the predicted amount of radioactivity, e.g. an accurately dispensed volume of 100 μL of the mannitol solution should give a corrected count of 100 × 5 = 500 Bq (or 500 × 60 = 30 000 d.p.m.).

7. **Note the specific activity of the experimental solution:** in this case, 100 μl (1 × 10^{-4} l) of the mannitol solution at a concentration of 0.025 mol L^{-1} will contain 25 × 10^{-7} mol (2.5 μmol) mannitol. Dividing the radioactivity in this volume (30 000 d.p.m.) by the amount of substance (Eqn [45.2]) gives a specific activity of 30000/2.5 = 12000 d.p.m. μmoL^{-1}, or 12 d.p.m. nmoL^{-1}. This value can be used:

   (a) To assess the accuracy of your protocol for preparing the experimental solution: if the measured activity is substantially different from the predicted value, you may have made an error in making up the solution.

   (b) To determine the counting efficiency of an instrument; by comparing the measured count rate with the value predicted by your calculations.

   (c) To interconvert activity and amount of substance: the most important practical application of specific activity is the conversion of experimental data from counts (activity) into amounts of substance. This is only possible where the substance has not been metabolised or otherwise converted into another form; e.g. a tissue sample incubated in the experimental solution described above with a measured activity of 245 d.p.m. can be converted to nmol mannitol by dividing by the specific activity, expressed in the correct form. Thus 245/12 = 20.417 nmol mannitol.

Correcting for quenching – find out how your instrument corrects for quenching and check the quench indication parameter (QIP) on the printout, which measures the extent of quenching of each sample. Large differences in the QIP would indicate that quenching is variable among samples and might give you cause for concern.

the energy of the original radioactive event. The spectrum of electronic pulses is thus related to the energy spectrum of the radioisotope.

Modern liquid scintillation counters use a series of electronic 'windows' to split the pulse spectrum into two or three components. This may allow more than one isotope to be detected in a single sample, provided their energy spectra are sufficiently different (Fig. 45.4). A complication of this approach is that the energy spectrum can be altered by pigments and chemicals in the sample, which absorb scintillations or interfere with the transfer of energy to the fluor; this is known as quenching (Fig. 45.4). Most instruments have computer-operated quench correction facilities (based on measurements of standards of known activity and energy spectrum) which correct for such changes in counting efficiency.

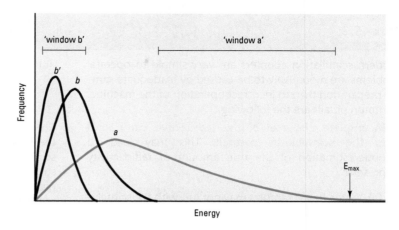

Fig. 45.4 Energy spectra for three radioactive samples, detected using a scintillation counter. Sample *a* is a high energy β-emitter while *b* contains a low energy β-emitter, giving a lower spectral range. Sample *b'* contains the same amount of low energy β-emitter, but with quenching, shifting the spectral distribution to a lower energy band. The counter can be set up to record disintegrations within a selected range (a 'window'). Here, 'window a' could be used to count isotope *a* while 'window b' could give a value of isotope *b*, by applying a correction for the counts due to isotope *a*, based on the results from 'window a'. Dual counting allows experiments to be carried out using two isotopes (double labelling).

Liquid scintillation counting of high energy β-emitters – β-particles with energies greater than 1 MeV can be counted in water (Čerencov radiation), with no requirement for additional fluors (e.g. ^{32}P).

Many liquid scintillation counters treat the first sample as a 'background', subtracting whatever value is obtained from the subsequent measurements as part of the procedure for converting to d.p.m. If not, you will need to subtract the background count from all other samples. Make sure that you use an appropriate background sample, identical in all respects to your radioactive sample but with no added radioisotope, in the correct position within the machine. Check that the background reading is reasonable (15–30 c.p.m. is a reasonable background for most radioisotopes). Tips for preparing samples for liquid scintillation counting are given in Box 45.2.

Gamma-ray (γ-ray) spectrometry

This is a method by which a mixture of γ-ray-emitting radionuclides can be resolved quantitatively by pulse-height analysis. It is based on the fact that pulse heights (voltages) produced by a photomultiplier tube are proportional to the amounts of γ-ray energy arriving at the scintillant or a lithium-drifted germanium detector. The lithium-drifted germanium detector, which is abbreviated to Ge(Li) – pronounced 'jelly' – provides high resolution (narrow peaks), essential in the analysis of complex mixtures, such as in biological material.

Autoradiography

This is a method where photographic film is exposed to the isotope. It is used mainly to locate radioactive tracers in thin sections of an organism or on chromatography papers and gels, but quantitative work is possible. The radiation interacts with the film in a similar way to light, silver grains being formed in the developed film where the particles or rays have passed through. The radiation must have enough energy to penetrate into the film, but if it has too much energy the grain formation may be too distant

Box 45.2 Tips for preparing samples for liquid scintillation counting

Modern scintillation counters are very simple to operate; problems are more likely to be caused by inadequate sample preparation than to incorrect operation of the machine. Common pitfalls are the following:

- **Incomplete dispersal of the radioactive compound in the scintillation cocktail.** This may lead to underestimation of the true amount of radioactivity present:

  (a) Water-based samples may not mix with the scintillation cocktail – change to an emulsifier-based cocktail. Take care to observe the recommended limits, upper and lower, for amounts of water to be added or the cocktail may not emulsify properly.

  (b) Solid specimens may absorb disintegrations or scintillations: extract radiochemicals using an intermediate solvent like ethanol (ideally within the scintillation vial) and then add the cocktail. Tissue-solubilising compounds such as Soluene are effective, particularly for animal material, but extremely toxic, so the manufacturer's instructions must be followed closely. Radioactive compounds on slices of agarose or poly-acrylamide gels may be extracted using a product such as Protosol. Agarose gels can be dissolved in a small volume of boiling water.

  (c) Particulate samples may sediment to the bottom of the scintillation vial – suspend them by forming a gel. This can be done with certain emulsifier-based cocktails by adding a specific amount of water.

- **Chemiluminescence.** This is where a chemical reacts with the fluors in the scintillation cocktail, causing spurious scintillations, a particular risk with solutions containing strong bases or oxidising agents. Symptoms include very high initial counts which decrease through time. Possible remedies are:

  (a) Leave the vials at room temperature for a time before counting. Check with a suitable blank that counts have dropped to an acceptable level.

  (b) Neutralise basic samples with acid (e.g. acetic acid or HCl).

  (c) Use a scintillation cocktail that resists chemiluminescence, such as Hiconicfluor.

  (d) Raise the energy of the lower counts detected to about 8 keV – most chemiluminescence pulses are weak (0–7 keV). This approach is not suitable for weak emitters, e.g. ^3H.

Fig. 45.5 Autoradiograph showing the localisation of a growth factor in epithelial cells of rat kidney tubules (scale bar is 10 urn). The black-silver grains are due to radioactivity from ^{125}I-labelled growth factor.

from the point where the isotope was located to identify precisely the point of origin (e.g. high energy β-emitters). Autoradiography (Fig 45.5) is a relatively specialised method and individual lab protocols should be followed for particular isotopes/ applications.

Biological applications for radioactive isotopes

The main advantages of using radioactive isotopes in biological experiments are:

- **Radioactivity is readily detected.** Methods of detection are sufficiently sensitive to measure extremely small amounts of radioactive substances.

- **Studies can be carried out on intact, living organisms.** If care is taken, minimal disruption of normal conditions will occur when radiolabelled compounds are introduced.

- **Protocols are simple** compared to equivalent methods for chemical analysis.

The main disadvantages are:

- **The 'isotope effect'.** Molecules containing different isotopes of the same atom may react at slightly different rates and behave in slightly different ways from the natural isotope. The isotope effect is more extreme the smaller the atom, and is most important for ^{3}H-labelled compounds of low molecular mass.

- **The possibility of mistaken identity.** The presence of radioactivity does not tell you anything about the compound in which the radioactivity is present: it could be different from the one in which it was applied, due to metabolism or spontaneous breakdown of a ^{14}C-containing organic compound.

The main types of experiments are:

- **Investigations of metabolic pathways:** a radioactively labelled substrate is added (often to an *in vitro* experimental 'system' rather than a whole organism) and samples taken at different time intervals. By identifying the labelled compounds and plotting their appearance through time, an indication of the pathway of metabolism can be obtained.
- **Translocation studies:** radioisotopes are used to follow the fate of molecules within an organism. Uptake and translocation rates can be determined with relative ease.
- **Ecological studies:** radioisotopic tracers provide a convenient method for determining food web interrelationships and for investigating behaviour patterns, while environmental monitoring may involve following the 'spectral signature' of isotopes deliberately or accidentally released.
- **Radiodating:** the age of plant or mineral samples can be determined by measuring the amount of a radioisotope in the sample. The age of the specimen can be found using the half-life by assuming how much was originally incorporated.
- **Mutagenesis and sterilisation:** radioactive sources can be used to induce mutations, particularly in micro-organisms. Gamma emitters of high energy will kill microbes and are used to sterilise equipment such as disposable Petri dishes.
- **Bioassays:** radioisotopes are used in several quantitative detection methods of value to biologists. Radioimmunoassay is described on p. 284. Isotope dilution analysis works on the assumption that introduced radiolabelled molecules will equilibrate with unlabelled molecules present in the specimen. The amount of substance initially present can be worked out from the change in specific activity of the radioisotope when it is diluted by the 'cold' material. A method is required whereby the substance can be purified from the sample and sufficient substance must be present for its mass to be measured accurately.

Lowenthal and Airey (2005) give further details and applications.

Working practices when using radioactive isotopes

By law, undergraduate work with radioactive isotopes must be very closely supervised. In practical classes, the protocols will be clearly outlined, but in project work you may have the opportunity to plan and carry out your own experiments, albeit under supervision. Some of the factors that you should take into account, based on the assumption that your department and laboratory are registered for radioisotope use, are discussed below:

1. **Must you use radioactivity?** If not, it may be a legal requirement that you use an alternative method.

Investigating the metabolic fate of radiolabelled compounds – you may need to separate individual metabolites before counting, e.g. using chromatography (Chapter 50), or electrophoresis (Chapter 52).

Example Carbon dating – living organisms have essentially the same ratio of ^{14}C to ^{12}C as the atmosphere; however, when an organism dies, its ^{14}C/^{12}C falls because the radioactive ^{14}C isotope decays. Since we know the half-life of ^{14}C (5715 years), a sample's ^{14}C/^{12}C ratio will allow us to estimate its age; e.g. if the ratio were exactly 1/8 of that in the atmosphere, the sample is three half-lives old and was formed 17 145 years before present. Such estimates carry an error of the order of 10% and are unreliable for samples older than 50 000 years, for which longer-lived isotopes can be used.

Registering for radioisotope work – in the UK, institutions must be registered for work with specific radioisotopes under the *Radioactive Substances Act (1993)*.

2. **Have you registered for radioactive work?** Usually, all users must register with a local Radiation Protection Supervisor. Details of the project may have to be approved by the appropriate administrator(s). You may have to have a short medical examination before you can start work.

3. **What labelled compound will you use?** Radioactive isotopes must be ordered well in advance through your department's Radiation Protection Supervisor. Aspects that need to be considered include:

   (a) The radionuclide. With many organic compounds this will be confined to ^{3}H and ^{14}C (but see Table 45.2). The involvement of a significant 'isotope effect' may influence this decision (see p. 302).

   (b) The labelling position. This may be a crucial part of a metabolic study. Specifically labelled compounds are normally more expensive than those that are uniformly ('generally') labelled.

   (c) The specific activity. The upper limit for this is defined by the isotope's half-life but, below this, the higher the specific activity, the more expensive the compound.

4. **Are suitable facilities available?** You will need a suitable work area, preferably out of the way of general lab traffic and within a fume cupboard for those cases where volatile radioactive substances are used or may be produced.

In conjunction with your supervisor, decide whether your method of application will introduce enough radioactivity into the system, how you will account for any loss of radioactivity during recovery of the isotope and whether there will be enough activity to count at the end. You should be able to predict approximately the amount of radioactivity in your samples, based on the specific activity of the isotope used, the expected rate of uptake/exchange and the amount of sample to be counted. Use the isotope's specific activity to estimate whether the non-radioactive ('cold') compound introduced with the radiolabelled ('hot') compound may lead to excessive concentrations being administered. Advice for handling data is given in Box 45.1.

Safety and procedural aspects

Make sure the bench surface is one that can be easily decontaminated by washing (e.g. Formica) and always use a disposable surfacing material such as Benchkote. It is good practice to carry out as many operations as possible within a Benchkote-lined plastic tray so that any spillages are contained. You will need a lab coat to be used exclusively for work with radioactivity, safety spectacles and a supply of thin latex or vinyl disposable gloves. Suitable vessels for liquid waste disposal will be required and special plastic bags for solids – make sure you know beforehand the disposal procedures for liquid and solid wastes. Wash your hands after handling a vessel containing a radioactive solution and again before removing your gloves. Gloves should be placed in the appropriate disposal bag as soon as your experimental procedures are complete.

It is important to comply with the following guidelines:

- **Read and obey the local rules for safe usage of radiochemicals.**
- **Maximise the distance between you and the source as much as possible.**
- **Minimise the duration of exposure.**
- **Wear protective clothing** (properly fastened lab coat, safety glasses, gloves) at all times.

Fig. 45.6 Tape showing the international symbol for radioactivity.

- **Use appropriate shielding** at all times (Table 45.1).
- **Monitor your working area for contamination** frequently.
- **Mark all glassware, trays, bench work areas,** etc., with tape incorporating the international symbol for radioactivity (Fig. 45.6).
- **Keep adequate records** of what you have done with a radioisotope – the amount remaining and that disposed of in waste form must agree.
- **Store radiolabelled compounds appropriately** and return them to storage immediately after use.
- **Dispose of waste promptly and with due regard for local rules.**
- **Make the necessary reports about waste disposal,** etc., to your departmental Radiation Protection Supervisor.
- **Clear up** after you have finished each experiment.
- **Wash thoroughly** after using radioactivity.
- **Monitor the work area and your body** when finished.

Text references and sources for further study

L'Annunziata, M.F. (ed.) (2012) *Handbook of Radioactivity Analysis,* 3rd edn. Academic Press, San Diego.

Haynes, W.M. (ed.) (2014) *CRC Handbook of Chemistry and Physics,* 95th edn. CRC Press, Boca Raton.

Lowenthal, G. and Airey, P. (2005) *Practical Applications of Radioisotopes and Nuclear Radiations.* Cambridge University Press, Cambridge.

Slater, J. (ed.) (2002) *Radioisotopes in Biology: a Practical Approach,* 2nd edn. Oxford University Press: Oxford.

STUDY EXERCISES

45.1 Carry out a half-life calculation. A rat dropping found in a pyramid in Egypt had a $^{14}C : {}^{12}C$ ratio that was 57.25% of a modern-day standard. Use this value to estimate the approximate date when the rat visited the pyramid, to the nearest century.

45.2 Practise radioactivity interconversions. Express the following values in the alternative units indicated, with appropriate prefixes as necessary. Answer to three significant figures.

(a) 72 000 d.p.m. as Bq;

(b) 20 μCi as d.p.m.;

(c) 44400 Bq as μCi;

(d) 6.3×10^5 d.p.m. mol^{-1} as Bq g^{-1}, for a compound with a relative molecular mass of 350;

(e) 3108 d.p.m. as pmol, for a sample of a standard where the specific activity is stated as 50 Ci mol^{-1}.

45.3 Use the concept of specific activity in calculations. A researcher wishes to estimate the rate of uptake of the sugar galactose by carrot cells in a suspension culture. She prepares 250 mL of the cell culture medium containing 10^7 cells per mL and unlabelled galactose at a concentration of 5 mmol L^{-1}. She then 'spikes' this with 5 μL (regard this as an insignificant volume) of radioactive standard containing 55 MBq of ^{14}C-labelled galactose (regard as an insignificant concentration). Answer to two significant figures.

(a) Calculate the specific activity of the galactose in the culture solution in Bq mol^{-1}.

(b) If the total cell sample takes up 79.2×10^5 Bq in a 2-h period, calculate the galactose uptake rate in mol s^{-1} cell^{-1}.

Answers to these study exercises are available at www.pearsoned.co.uk/practicalskills.

The measurement of light is directly relevant to several aspects of biology, including: photosynthesis, photomorphogenesis and photoperiodism in plants; perception and thermoregulation in animals; and aspects of aquatic biology.

The nature of light

Light is most strictly defined as that part of the spectrum of electromagnetic radiation detected by the human eye. However, the term is also applied to radiation just outside that visible range (e.g. UV and infrared 'light'). Electromagnetic radiation is emitted by the Sun and by other sources (e.g. an incandescent lamp) and the electromagnetic spectrum is a broad band of radiation, ranging from cosmic rays to radio waves (Fig. 46.1). Most biological experiments involve measurements within the UV, visible and infrared (IR) regions (generally, within the wavelength range 200–1000 nm; see Table 46.1).

Radiation has the characteristics of a particle and of a vibrating wave, travelling in discrete particulate units, or 'packets', termed photons. A quantum is the amount of energy contained within a single photon (it is important not to confuse these two terms, although they are sometimes used interchangeably in the literature). In some circumstances, it is appropriate to measure light in terms of the number of photons, usually expressed directly in moles (6.02×10^{23} photons = 1 mol); older textbooks may use the redundant term Einstein as the unit of measurement, where 1 Einstein = 1 mol photons. Alternatively, the energy content (power) may be measured (e.g. in W m^{-2}). Radiation also behaves as a vibrating electrical and magnetic field moving in a particular direction, with the magnetic and electrical components vibrating perpendicular to one another and perpendicular to the direction of travel.

The wave nature of radiation gives rise to the concepts of wavelength (λ, usually measured in nm), frequency (v, measured in s^{-1}), speed (c, the speed of electromagnetic radiation, which is 3×10^{8} m s^{-1} in a vacuum), and direction. In other words, radiation is a vector quantity, where

$$c = \lambda v \qquad [46.1]$$

Photometric and radiometric measurements

Photometric measurements

These are based on the energy perceived by a 'standard' human eye, with maximum sensitivity in the yellow–green region, around 555 nm (Fig. 46.2). The unit of measurement is the candela, a base unit in the SI system, defined in terms of the visual appearance of a specific quantity of platinum at its freezing point. Derived units are used for the luminous flow (lumen) and luminous flow per unit area (lux). These units were once used in photobiology and you may come across them in older literature. However, it is now recognised that such measurements are of little direct relevance to biologists, including even those who may wish to study visual responses, because they are not based on fundamental physical principles.

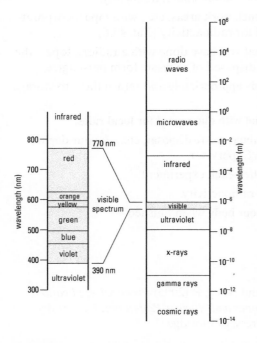

Fig. 46.1 The electromagnetic spectrum. From *Methods for Physical and Chemical Analysis of Fresh Waters*, 2nd edn. International Biological Programme, No. 8, Blackwell Scientific (Golterman, H.L., Clymo, R.S. and Ohnstad, M.A.M., 1978). Reproduced with permission of Blackwell Publishing Ltd.

Table 46.1 Wavelength ranges for UV, visible and IR radiation, compiled from several sources.

Type	Wavelength range (nm)
UVC	100–290*
UVB	290–320
UVA	320–400
Violet	390–450
Blue	450–500
Green	500–560
Yellow	560–600
Orange	600–620
Red	620–770
IR	>770

*Note that UVC is absorbed by the Earth's ozone layer.

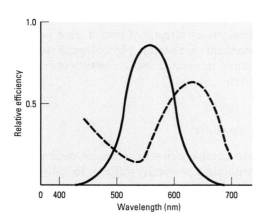

Fig. 46.2 Relative efficiency of vision (solid line) and photosynthesis (dotted line) as a function of wavelength.

Relevance of measurement scale – your choice will depend on relevance to your study. For example, in some environmental studies, you might be interested in total energy input, so irradiance would be a valid measurement; in another type of study, you might be focusing on plant productivity and therefore photosynthetically active radiation (PAR) would be more relevant.

Table 46.2 Approximate conversion factors for a photosynthetic irradiance (PI) of 1 W m^{-2} to photosynthetic photon flux density (PPFD).

Source	PPFD (μmol photons m^{-2} s^{-1})
Sunlight	4.6
'Cool white' fluorescent tube	4.6
Osram 'daylight' fluorescent tube	4.6
Quartz-iodine lamp	5.0
Tungsten bulb	5.0

Source: Lüning, 1981. Reproduced with permission of Blackwell Publishing Ltd.

Radiometric measurements

The radiometric system is based on physical properties of the electromagnetic radiation itself, expressed either as the number of photons, or their energy content. The following terms are used (units of measurement in parentheses):

- **Photon flux** (mol photons s^{-1}) is the number of photons arriving at an object within the specified time interval.
- **Photon exposure** (mol photons m^{-2}) is the total number of photons received by an object, usually expressed per unit surface area.
- **Photon flux density** (mol photons m^{-2}s^{-1}), or PFD, is the most commonly used term to describe the number of photons arriving at a particular surface, expressed per unit surface area and per unit time interval.
- **Photosynthetically active radiation,** or PAR, is radiation within the waveband 400–700 nm, since the photosynthetic pigments (chlorophylls, carotenoids, etc.) show maximum absorption within this band.
- **Photosynthetic photon flux density** (mol photons m^{-2} s^{-1}), or PPFD, is the number of photons within the waveband 400–700 nm arriving at a particular surface, expressed per unit surface area and per unit time interval. Often this term is used interchangeably with PFD.
- **Irradiance** (J m^{-2} s^{-1} = W m^{-2}) is the amount of energy arriving at a surface, expressed per unit surface area and per unit time interval.
- **Photosynthetic irradiance** (W m^{-2}), or **PI,** is the energy of radiation within the waveband 400–700 nm arriving at a surface, expressed per unit surface area and per unit time interval.

Choice of measurement scale

Photon flux density

This is often the most appropriate unit of measurement for biological systems where individual photons are involved in the underlying process, e.g. in photosynthetic studies, where PPFD is measured, since each photochemical reaction involves the absorption of a single photon by a pigment molecule. Most modern light-measuring instruments (radiometers) can measure this quantity, giving a reading in μmol photons m^{-2} s^{-1}.

Irradiance

This is appropriate if you are interested in the energy content of the light, e.g. if you are studying energy balance or thermal effects. Many radiometers measure photosynthetic irradiance within the waveband 400–700 nm, giving a reading in W m^{-2}. It is possible to make an approximate conversion between PPFD and PI measurements, providing the spectral properties of the light source are known (see Table 46.2).

Fig. 46.3 Spectral distribution of energy output from various sources.
Adapted from Golterman *et al.*, 1978. Reproduced with permission of Blackwell Publishing Ltd.

Fig. 46.4 A typical radiometer (quantum) sensor.
Photo courtesy of Skye Instruments Ltd.

Spectral distribution

This can be determined using a spectroradiometer, e.g. to compare different light sources (Fig. 46.3). A spectroradiometer measures irradiance or photon flux density in specific wavebands. This instrument consists of a monochromator (p. 311) to allow separate narrow wavebands (5–25 nm bandwidth) to be measured by a detector; some instruments provide a plot of the spectral characteristics of the source.

Measurements can be instantaneous or integrated over a time period (e.g. a day). Instantaneous measurements are used in physiological studies to record prevailing conditions while integrated measurements are more often used in environmental research.

Using a radiometer ('light meter')

The main components of a radiometer are:

- **receiver (sensor):** either flat-plate, hemispherical or spherical, depending upon requirements. Most incorporate a protective diffuser, to reduce reflection;
- **detector:** either thermoelectric or photoelectric. Some photoelectric detectors suffer from fatigue, with a decreasing response on prolonged exposure: check the manufacturer's handbook for exposure times;
- **processor and readout device** to convert the output from the detector into a visible reading, in analogue or digital form.

Figure. 46.4 shows a typical flat-plate radiation detector with diffusion plate. Modern instruments may incorporate two sets of receiver–detector assemblies to account for sources of radiation over 360°, and data loggers are capable of presenting instantaneous or integrated readings. Box 46.1 gives practical details of the steps involved in using a radiometer. In your write-up, give full details of how the measurement was made, e.g. the type of light source, instrument used, where the sensor was placed, whether an average was calculated, etc.

Box 46.1 Measuring photon flux density or irradiance using a battery-powered radiometer

1. **Check the battery.** Most instruments have a setting that gives a direct readout of battery voltage. Recharge if necessary before use.

2. **Select the appropriate type of measurement** (e.g. photon flux density or irradiance over the PAR waveband, or an alternative range): the simpler instruments have a selection dial for this purpose.

3. **Place the sensor in the correct location and position** for the measurement: it may be appropriate to make several measurements at different positions and take an average.

4. **Choose the most appropriate scale** for the readout device: for needle-type meters, the choice of maximum reading is usually selected by a dial, within the range 0.3 to 30000. Start at a high range and work down until the reading is on the scale. Your final scale should be chosen to provide the most accurate reading, e.g. a reading of 15 µmol photons m^{-2} s^{-1} should be made using the 0–30 scale, rather than a higher range.

5. **Read the value from the meter.** For needle-type instruments there may be two scales, the upper one marked from 0 to 10 and the lower one from 0 to 3: make sure you use the correct one, e.g. a half-scale deflection on the 0–30 scale is 1.5.

6. **Check that the answer is realistic,** e.g. full sunlight has a PPFD of up to 2000 µmol photons m^{-2} s^{-1} (PI $\equiv$ 400 W m^{-2}), though the value will be far lower on a dull or cloudy day, while the PPFD at a distance of 1m from a mercury lamp is around 150 µmol photons m^{-2} s^{-1}, and 50 µmol photons m^{-2} s^{-1} at the same distance from a fluorescent lamp.

Text references

Golterman, H.L., Clymo, R.S. and Ohnstad, M.A.M. (1978) *Methods for Physical and Chemical Analysis of Fresh Waters.* Blackwell Scientific Publications, Oxford.

Sources for further study

Anon. *International Light Homepage.* Available: http://www.intl-lighttech.com Last accessed: 01/09/15.
[Information on products, including several types of detector and meter. Also includes access to a light measurement handbook in .pdf format.]

Boyd, R.W. (1983) *Radiometry and the Detection of Optical Radiation.* Wiley, New York.

National Physics Laboratory *Beginner's Guide to Measurement – Light.*

Lüning, K.J. (1981) 'Light', in C.S. Lobban and M.J. Wynne (eds), *The Biology of Seaweeds,* pp. 326–55, Blackwell Scientific, Oxford.

Available: http://www.npl.co.uk/publications/light/ Last accessed: 01/09/15.
[Gives some background on the development and use of the candela, the SI unit of luminous intensity, and its relationship with human vision.]

Ryer, A.D. (1997) *Light Measurement Handbook.* Available: http://www.intl-light.com/handbook/Last accessed: 01/09/15.
[A commercial site, produced by International Light Technologies, Inc., Peabody, Massachusetts.]

STUDY EXERCISES

46.1 Carry out a calculation involving interconversion of photosynthetic photon flux and photosynthetic irradiance. For sunlight with a photosynthetic photon flux density (PPFD) of 1610mol m^{-2}, what is the total amount of photosynthetic energy (in joules) falling on a leaf of area 45 cm^2 over a 30-min experimental period?

46.2 Compare the photosynthetic photon flux densities (PPFD) of different locations. The following values represent typical light levels for a range of different situations, expressed either as photosynthetic irradiance (PI) or PPFD:

(a) outside, on a sunny day: PI = 300 Wm^{-2};

(b) in a room lit by 'cool white' fluorescent tubes: PPFD = 6.50 nmol photons $cm^{-2} \, s^{-1}$;

(c) under the leaf canopy in a forest: PPFD = 275 μmol photons $m^{-2} \, s^{-1}$;

(d) in a growth cabinet lit by a bank of 'daylight' fluorescent lights: PI = 35.2 W m^{-2};

(e) in a room lit by a tungsten bulb: PI = 1.15 mW cm^{-2}.

Convert all to PPFD in the same units (expressed to three significant figures) and then rank the locations in order of decreasing PPFD.

46.3 Research the various types of commercial radiometers (light meters) available on the web. Make a list of the features that you might want to find out about and see how much information you can find.

Answers to these study exercises are available at www.pearsoned.co.uk/practicalskills.

Using spectroscopy – this technique is valuable for:

- tentatively identifying compounds, by determining their absorption or emission spectra;
- quantifying substances, either singly or in the presence of other compounds, by measuring the signal strength at an appropriate wavelength;
- determining molecular structure;
- following reactions, by measuring the disappearance of a substance, or the appearance of a product as a function of time.

The absorption and emission of electromagnetic radiation of specific energy (wavelength) are characteristic features of many molecules, involving the movement of electrons between different energy states in accordance with the laws of quantum mechanics, and spectroscopic techniques are used to measure and interpret such interactions between molecules and radiation. Electrons in atoms or molecules are distributed at various energy levels, but are mainly at the lowest energy level, usually termed the ground state. When exposed to energy (e.g. from electromagnetic radiation), electrons may be excited to higher energy levels (excited states), with the associated absorption of energy at specific wavelengths giving rise to an absorption spectrum. One quantum of energy is absorbed for a single electron transition from the ground state to an excited state. On the other hand, when an electron returns to its ground state, one quantum of energy is released; this may be dissipated to the surrounding molecules (as heat) or may give rise to an emission spectrum. The energy change (ΔE) for an electron moving between two energy states, E_1 and E_2, is given by the equation:

$$\Delta E = E_1 - E_2 = h\nu \qquad [47.1]$$

where h is the Planck constant (p. 197) and ν is the frequency of the electromagnetic radiation expressed in Hz or s^{-1}. Frequency is related to wavelength (λ, usually expressed in nm) and the speed of electromagnetic radiation, c (p. 197) by the expression:

$$\nu = c \div \lambda \qquad [47.2]$$

UV/visible spectrophotometry

This is a widely used technique for measuring the absorption of radiation in the visible and UV regions of the spectrum. A spectrophotometer is an instrument designed to allow precise measurement at a particular wavelength, while a colorimeter is a simpler instrument, using filters to measure broader wavebands, e.g. light in the green, red or blue regions of the visible spectrum; Jones *et al.* (2015) give details of how to use a colorimeter.

Principles of light absorption

Two fundamental principles govern the absorption of light passing through a solution:

- The absorption of light is exponentially related to the number of molecules of the absorbing solute that are encountered, i.e. the solute concentration [C].
- The absorption of light is exponentially related to the length of the light path through the absorbing solution, l.

These two principles are combined in the Beer–Lambert relationship (sometimes referred to simply as 'Beer's Law'), which is usually expressed in terms of absorbance (A) – the logarithm of the ratio of the incident light (I_0) to the emergent light (I):

$$A = \varepsilon l[C] \qquad [47.3]$$

Definition

Absorbance (A) – this is given by:

$$A = \log_{10}(I_0/I).$$

Usually shown as A_x where 'x' is the wavelength, in nanometres. As an example, for incident light (I_0) = 1.00 and emergent light (I) = 0.16 (expressed in relative terms), $A = \log_{10}(1.00 \div 0.16) = \log_{10} 6.25 = 0.796$ (to three significant figures).

where A is absorbance, ε is a constant for the absorbing substance and the wavelength, termed the absorption coefficient or absorptivity, and $[C]$ is expressed either as mol L^{-1} or g L^{-1} (see p. 152) and l is given in cm.

> **KEY POINT** The Beer–Lambert relationship, expressed in mathematical form in Eqn [47.3], states that there is a direct linear relationship between the concentration of a substance in a solution, $[C]$, and the absorbance of that solution, A.

This relationship is extremely useful, since most spectrophotometers are constructed to give a direct measurement of absorbance (A), sometimes also termed extinction (E), of a solution (older texts may use the outdated term optical density, OD). Note that for substances obeying the Beer–Lambert relationship, A is linearly related to $[C]$. Absorbance at a particular wavelength is often shown as a subscript, e.g. A_{550} represents the absorbance at 550 nm. The proportion of light passing through the solution is known as the transmittance (T), and is calculated as the ratio of the emergent and incident light intensities.

Some instruments have two scales:

1. **An exponential scale** from zero to infinity, measuring absorbance.
2. **A linear scale** from 0 to 100, measuring (per cent) transmittance.

For most practical purposes, the Beer–Lambert relationship applies and you should use the absorbance scale.

UV/visible spectrophotometer

The principal components of a UV/visible spectrophotometer are shown in Fig. 47.1. High intensity tungsten bulbs are used as the light source in basic instruments, capable of operating in the visible region (i.e. 400–700 nm). Deuterium lamps are used for UV spectrophotometry (200–400 nm); these lamps are fitted with quartz envelopes, since glass does not transmit UV radiation.

A major improvement over the simple colorimeter is the use of a diffraction grating to produce a parallel beam of monochromatic light from the (polychromatic) light source. In practice the light emerging from such a monochromator is not of a single wavelength, but is a narrow band of wavelengths. This bandwidth is an important characteristic, since it determines the wavelengths used in absorption measurements – the bandwidth of basic spectrophotometers is around 5–10 nm, while research instruments have bandwidths of less than 1 nm.

Bandwidth is affected by the width of the exit slit (the slit width), since the bandwidth will be reduced by decreasing the slit width. To obtain accurate data at a particular wavelength setting, the narrowest possible slit width should be used. However, decreasing the slit width also reduces the amount of light reaching the detector, decreasing the signal-to-noise ratio. The extent to which the slit width can be reduced depends upon the sensitivity and stability of the detection/amplification system and the presence of stray light.

Most UV/visible spectrophotometers are designed to take cuvettes with an optical path length of 10 mm. Disposable plastic cuvettes are suitable for routine work in the visible range using aqueous and alcohol-based solvents, while glass cuvettes are useful for other organic solvents. Glass cuvettes are manufactured to more exacting standards, so use optically matched glass cuvettes for accurate work, especially at low absorbances (< 0.1), where

---

Definition

Transmittance (T) – this is usually expressed as a percentage, at a particular wavelength, T_x, where

$$T_x = (I/I_0) \times 100 \ (\%).$$

As an example, for incident light (I_0) = 1.00 and emergent light (I) = 0.275 (expressed in relative terms) then transmittance, $T = (0.275 \div 1.00) \times 100 = 27.5\%$.

Fig. 47.1 Components of a UV/visible spectrophotometer.

SAFETY NOTE Working with spectrophotometers – take care not to spill aqueous solutions into the inside of the instrument, owing to the risk of electric shock during use (switch off at mains and seek assistance if this should happen).

Using plastic disposable cuvettes – these are adequate for work in the near-UV region, e.g. for enzyme studies using nicotinamide coenzymes, at 340 nm (p. 312), as well as the visible range.

Examples The molar absorptivity of NADH is 6.22×10^3 L mol^{-1} cm^{-1} at 340 nm. For a test solution giving an absorbance of 0.21 in a cuvette with a light path of 5 mm, using Eqn [47.3] this is equal to a concentration of:

$$0.21 = 6.22 \times 10^3 \times 0.5 \times [C]$$
$$[C] = 0\,000\,0675 \text{ mol L}^{-1}$$
$$\text{(or 67.5 } \mu\text{mol L}^{-1}).$$

The specific absorptivity (10g L^{-1}) of double-stranded DNA is 200 at 260 nm, therefore a solution containing 1g L^{-1} will have an absorbance of 200/10 = 20. For a DNA solution, giving an absorbance of 0.35 in a cuvette with a light path of 1.0 cm, using Eqn [47.3] this is equal to a concentration of:

$$0.35 = 20 \times 1.0\,[C]$$
$$[C] = 0.0175 \text{ g L}^{-1}$$
$$\text{(equivalent to 17.5 } \mu\text{g mL}^{-1}).$$

Chlorophylls *a* and *b* of vascular plants and green algae can be extracted in 90% v/v acetone/water, assayed by measuring the absorbance of the mixed solution at two wavelengths, according to the formulae:

Chlorophyll *a* (mg L^{-1}) =
$11.93\,A_{664} - 1.93\,A_{647}$

Chlorophyll *b* (mg L^{-1}) =
$20.36\,A_{647} - 5.5\,A_{664}$

Note: different equations are required for other solvents.

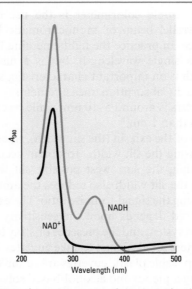

Fig. 47.2 Absorption spectra of nicotinamide adenine dinucleotide in oxidised (NAD$^+$) and reduced (NADH) form. Note the 340 nm absorption peak (A_{340}), used for quantitative work.

any differences in the optical properties of cuvettes for reference and test samples will be pronounced. Glass and plastic absorb UV light and quartz cuvettes must be used at wavelengths below 300 nm.

KEY POINT Before taking a measurement, make sure that cuvettes are clean, unscratched, dry on the outside, filled to the correct level and located in the correct position in their sample holders.

Proteins and nucleic acids in biological samples can accumulate on the inside faces of glass/quartz cuvettes, so remove any deposits using acetone on a cotton bud, or soak overnight in 1 mol L^{-1} nitric acid. Corrosive and hazardous solutions must be used in cuvettes with tightly fitting lids, to prevent damage to the instrument and to reduce the risk of accidental spillage.

Basic instruments use photocells similar to those used in simple colorimeters or photodiode detectors. In many cases, a different photocell must be used at wavelengths above and below 550–600 nm, owing to differences in the sensitivity of such detectors over the visible waveband. The detectors used in more sophisticated instruments give increased sensitivity and stability when compared to photocells.

Digital displays are increasingly used in preference to traditional needle-type meters, as they are not prone to parallax errors and misreading of the absorbance scale. Some digital instruments can be calibrated to give a direct readout of the concentration of the test substance.

KEY POINT Basic spectrophotometers are most accurate within the absorbance range from 0.00 to 1.00 and your calibration standards and test solutions should be prepared to give readings within this range.

Types of UV/visible spectrophotometer

Basic instruments are single beam spectrophotometers in which there is only one light path. The instrument is set to zero absorbance using a blank solution, which is then replaced by the test solution, to obtain an absorbance reading. An alternative approach is used in double beam spectrophotometers, where the light beam from the monochromator is split into two separate beams, one beam passing through the test solution and the other through a reference blank. Absorbance is then measured by an electronic circuit which compares the output from the reference (blank) and sample cuvettes. Double beam spectrophotometry reduces measurement errors caused by fluctuations in output from the light source or changes in the sensitivity of the detection system, since reference and test solutions are measured at the same time (Box 47.1). Recording spectrophotometers are double beam instruments, designed for use with a chart recorder, either by recording the difference in absorbance between reference and test solutions across a predetermined waveband to give an absorption spectrum (Fig. 47.2), or by recording the change in absorbance at a particular wavelength as a function of time (e.g. in an enzyme assay, Chapter 61).

Quantitative spectrophotometric analysis

A single (purified) substance in solution can be quantified using the Beer–Lambert relationship (Eqn [47.3]), provided its absorptivity is known at a particular wavelength (usually at the absorption maximum

---

Box 47.1 How to use a spectrophotometer

1. **Switch on and select the correct lamp** for your measurements (e.g. deuterium for UV, tungsten for visible light).

2. **Allow up to 15 min for the lamp to warm up** and for the instrument to stabilise before use.

3. **Select the appropriate wavelength:** on older instruments a dial is used to adjust the monochromator, while newer machines have microprocessor-controlled wavelength selection.

4. **Select the appropriate detector:** some instruments choose the correct detector automatically (on the basis of the specified wavelength), while others have manual selection.

5. **Choose the correct slit width** (if available): this may be specified in the protocol you are following, or may be chosen on the manufacturer's recommendations.

6. **Insert appropriate reference blank(s):** single beam instruments use a single cuvette, while double beam instruments use two cuvettes (a matched pair for accurate work). The reference blanks should match the test solution in all respects apart from the substance under test, i.e. they should contain all reagents apart from this substance. *Make sure that the cuvettes are positioned correctly, with their polished (transparent) faces in the light path, and that they are accurately located in the cuvette holder(s).*

7. **Check/adjust the 0% transmittance:** most instruments have a control that allows you to zero the detector output in the absence of any light (termed 'dark current' correction). Some microprocessor-controlled instruments carry out this step automatically.

8. **Set the absorbance reading to zero:** usually via a dial, or digital readout.

9. **Analyse your samples:** replace the appropriate reference blank with a test sample, allow the absorbance reading to stabilise (5–10s) and read the absorbance value from the meter/readout device. For absorbance readings greater than 1 (i.e. <10% transmittance), the signal-to-noise ratio is too low for accurate results. Your analysis may require a calibration curve or you may be able to use the Beer–Lambert relationship (Eqn [47.3]) to determine the concentration of test substance in your samples.

10. **Check the scale zero at regular intervals** using a reference blank, e.g. after every 10 samples.

11. **Check the reproducibility of the instrument:** measure the absorbance of a single solution several times during your analysis. It should give the same value.

Problems (and solutions): inaccurate/unstable readings are most often caused by incorrect use of cuvettes, e.g. dirt, fingerprints or test solution on outside of cuvette (wipe the clear faces using a soft tissue before insertion into the cuvette holder and handle only by the opaque faces), condensation (if cold solutions are not allowed to reach room temperature before use), air bubbles (which scatter light and increase the absorbance; tap gently to remove), insufficient solution (causing refraction of light at the meniscus), particulate material in the solution (check for 'cloudiness' in the solution and centrifuge before use, where necessary) or incorrect positioning in light path (locate in correct position).

Measuring high absorbances in colorimetric analysis – if any final solution has an absorbance that is too high to be read with accuracy on your spectrophotometer (i.e. $A > 2$), it is bad practice to dilute the solution so that it can be measured. This dilutes both the sample molecules and the colour reagents to an equal extent. Instead, you should dilute the original sample and reassay.

for the substance, since this will give the greatest sensitivity). The molar absorptivity is the absorbance given by a solution with a concentration of 1 mol L^{-1} (= 1 kmol m^{-3}) of the compound in a light path of 1 cm. The appropriate value may be available from tabulated spectral data (e.g. Anon, 1963), or it can be determined experimentally by measuring the absorbance of known concentrations of the substance (Box 47.1) and plotting a standard curve (see Chapter 42). This should confirm that the relationship is linear over the desired concentration range and the slope of the line will give the molar absorptivity.

The specific absorptivity is the absorbance given by a solution containing 10 g L^{-1} (i.e. 1% w/v) of the compound in a light path of 1 cm. This is useful for substances of unknown molecular weight, e.g. proteins or nucleic acids, where the amount of substance in solution is expressed in terms of its mass, rather than as a molar concentration. For use in Eqn [47.3], the specific absorptivity should be divided by 10 to give the solute concentration in g L^{-1}.

Measuring low absorbances in colorimetric analysis – for accurate readings with highly dilute solutions, use a cuvette with a long optical path length (e.g. 5 cm, rather than 1 cm).

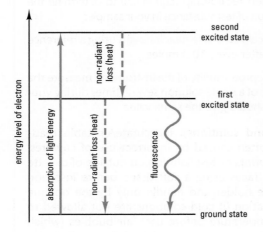

Fig. 47.3 Energy levels and energy transitions in fluorescence.

Fig. 47.4 Fluorescein – a widely used fluorescent label, showing (i) a planar conjugated system of fused rings; (ii) heteroatoms within the conjugated structures; (iii) an electron-donating group; and (iv) an electron-attracting group. In fluorogenic enzyme substrates (p. 412), linkage is usually via one of the two hydroxyl groups, while in fluoroscein-labelled proteins, linkage is via an isothiocyanate group (N≡C≡S) on the lowermost ring (*).

This simple approach cannot be used for mixed samples where several substances have a significant absorption at a particular wavelength. In such cases, it may be possible to estimate the amount of each substance by measuring the absorbance at several wavelengths, e.g. protein estimation in the presence of nucleic acids (pp. 400–401). Further details of spectrophotometric methods of determining the amount of protein in an aqueous sample are given in Chapter 56.

Fluorescence

With most molecules, after electrons are raised to a higher energy level by absorption of electromagnetic radiation, they soon fall back to the ground state by radiationless transfer of energy (heat) to the solvent. However, with some molecules, the events shown in Fig. 47.3 may occur, i.e. electrons may lose only part of their energy by non-radiant routes and the rest may be emitted as electromagnetic radiation, a phenomenon known as fluorescence. Since not all of the energy that was absorbed is emitted (because of non-radiant loss), the wavelength of the fluorescent light is longer than the absorbed light (longer wavelength = lower energy). Thus, a fluorescent molecule has both an absorption spectrum and an emission spectrum. The difference between the excitation wavelength (λ_{ex}) and the emission wavelength (λ_{em}), measured in nm, is known as the Stokes shift, and is fundamental to the sensitivity of fluorescence techniques. The existence of a Stokes shift means that emitted light can be detected against a low background, independently of the excitation wavelength.

Most fluorescent molecules have the following features:

- **A highly conjugated system** (alternating double and single bonds), involving aromatic or heterocyclic rings, usually containing O or N (as heteroatoms).

- **A condensed system of fused rings,** with one or more heteroatoms.

- **Electron-donating groups** such as –OH, –OCH$_3$, –NH$_2$ and –NR$_2$, together with electron-attracting groups elsewhere in the molecule, in conjugation with the electron-donating groups.

- **A rigid, planar structure.**

Figure 47.4 illustrates many of these features for fluorescein, used in a range of biological applications, including nucleic acid detection, fluorescent antibody tests, etc.

Fluorescence spectrophotometry

The principal components of a fluorescence spectrophotometer (fluorimeter) are shown in Figure 47.5. The instrument contains two monochromators, one to select the excitation wavelength and the other to monitor the light emitted, usually at 90° to the incident beam (though light is actually emitted in all directions). As an example, the wavelengths used to measure the highly fluorescent compound aminomethylcoumarin are 388 nm (excitation) and 440 nm (emission). Some examples of biomolecules with intrinsic fluorescence are given in Table 47.1 and the use of fluorogenic enzyme substrates is discussed further in Chapter 61.

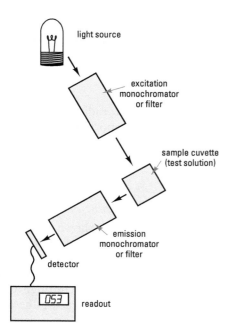

Fig. 47.5 Components of a fluorimeter (fluorescence spectrophotometer). Note that sample cuvettes for fluorimetry must have clear sides all round.

Table 47.1 Examples of compounds with intrinsic fluorescence.

Drugs
Aspirin, morphine, barbiturates, propranolol, ampicillin, tetracyclines

Vitamins
Riboflavin, vitamins A, B6 and E, nicotinamide

Pollutants
Naphthalene, anthracene, benzopyrene

Compared with UV/visible spectrophotometry, fluorescence spectroscopy has certain advantages, including the following:

- **Enhanced sensitivity** (up to 1000-fold), since the emitted light is detected against a background of zero, in contrast to spectrophotometry where small changes in signal are measured against a large 'background' (see Eqn [47.3]).

- **Increased specificity,** because not one but two specific wavelengths are required for a particular compound.

However, there are also certain drawbacks:

- **Not all compounds show intrinsic fluorescence,** limiting its application. However, some non-fluorescent compounds may be coupled to fluorescent dyes, or fluorophores (e.g. proteins may be coupled to fluorescamine).

- **The light emitted can be less than expected due to quenching,** i.e. when substances in the sample (e.g. oxygen) either interfere with energy transfer, or absorb the emitted light (in some instances, the sample molecules may self-quench if they are present at high concentration).

The sensitivity of fluorescence has made it invaluable in techniques in which specific antibodies are linked to a fluorescent dye, including:

- **fluorescence immunoassay (FIA);**

- **immunohistochemistry,** which requires the use of a fluorescence microscope, e.g. using fluorescent antibodies, or fluorescent *in situ* hybridisation (FISH) for nucleic acid detection;

- **flow cytometry** – a fluorescence-activated cell sorter (FACS) uses cell surface protein-specific monoclonal antibodies labelled with fluorescent dyes to separate and enumerate cells such as lymphocytes.

Phosphorescence and luminescence

A phenomenon related to fluorescence is phosphorescence, which is the emission of light following intersystem crossing between electron orbitals (e.g. between excited singlet and triplet states). Light emission in phosphorescence usually continues after the exciting energy is no longer applied and, since more energy is lost in intersystem crossing, the emission wavelengths are generally longer than with fluorescence. Phosphorescence has limited applications in biomolecular sciences.

Luminescence (or chemiluminescence) is another phenomenon in which light is emitted, but here the energy for the initial excitation of electrons is provided by a chemical reaction rather than by electromagnetic radiation. An example is the action of the enzyme luciferase, extracted from fireflies, which catalyses the following reaction:

$$\text{luciferin} + \text{ATP} + O_2 \Rightarrow \text{oxyluciferin} + \text{AMP} + \text{PP}_i + CO_2 + \text{light} \qquad [47.4]$$

The light produced is either yellow–green (560 nm) or red (620 nm). This system can be used in biomolecular analysis of ATP, either to determine ATP concentration in a biological sample, or to follow a coupled reaction (p. 412). Measurement can be performed using the

Fig. 47.6 Components of a flame photometer.

Plotting calibration curves in quantitative analysis – do not force your calibration line to pass through zero if it clearly does not. There is no reason to assume that the zero value is any more accurate than any other reading you have made.

photomultiplier tubes of a scintillation counter (p. 299) to detect the emitted light, with calibration of the output using a series of standards of known ATP content.

Atomic spectroscopy

Atoms of certain metals will absorb and emit radiation of specific wavelengths when heated in a flame, in direct proportion to the number of atoms present. Atomic spectrophotometric techniques measure the absorption or emission of particular wavelengths of UV and visible light, to identify and quantify such metals.

Flame atomic emission spectrophotometry (or flame photometry)

The principal components of a flame photometer are shown in Fig. 47.6. A liquid sample is converted into an aerosol in a nebuliser (atomiser) before being introduced into the flame, where a small proportion (typically fewer than 1 in 10 000) of the atoms will be raised to a higher energy level, releasing this energy as light of a specific wavelength, which is passed through a filter to a photocell detector. Flame photometry is used to measure the alkali metal ions K^+, Na^+ and Ca^{2+} in biological fluids; Box 47.2 gives details of the basic procedure.

Atomic absorption spectrophotometry (or flame absorption spectrophotometry)

This technique is applicable to a broad range of metal ions, including those of Pb, Cu, Zn, etc. It relies on the absorption of light of a specific wavelength by atoms dispersed in a flame. The appropriate wavelength is provided by a cathode lamp, coated with the element to be analysed, focused through the flame and onto the detector. When the sample is introduced

---

When using a flame photometer

- **Allow time for the instrument to stabilise.** Switch on the instrument, light the flame and wait at least 5 min before analysing your solutions.

- **Check for impurities in your reagents.** For example, if you are measuring K^+ in an acid digest of some biological material, check the K^+ content of a reagent blank containing everything except the biological material, processed in exactly the same way as the samples. Subtract this value from your sample values to obtain the true K^+ content. If using acid digestion, always work within a fume hood and wear gloves and safety glasses at all times. Rinse any spillages with a large volume of water.

- **Quantify your samples using a calibration curve** (Chapter 42). Calibration standards should cover the expected concentration range for the test solutions – your calibration curve may be non-linear (especially at concentrations above 1 mmol L^{-1}, i.e. 1 mol m^{-3} in SI units).

- **Analyse all solutions in duplicate,** so that reproducibility can be assessed.

- **Check your calibration.** Make repeated measurements of a standard solution of known concentration after every six or seven samples, to confirm that the instrument calibration is still valid.

- **Consider the possibility of interference.** Other metal atoms may emit light which is detected by the photocell, since the filters cover a wider waveband than the emission line of a particular element. This can be a serious problem if you are trying to measure low concentrations of a particular metal in the presence of high concentrations of other metals (e.g. Na^+ in sea water), or other substances that form complexes with the test metal, suppressing the signal (e.g. phosphate). Atomic absorption spectrophotometry is an alternative approach for such samples.

Practical advantages of atomic absorption spectrophotometry over flame photometry – these include:

- improved sensitivity;
- increased precision;
- decreased interference.

into the flame, it will decrease the light detected in direct proportion to the amount of metal present. Newer variants of this method include flameless atomic absorption spectrophotometry and atomic fluorescence spectrophotometry, both of which are more sensitive than the flame absorption technique.

Text references

Anon. (1963) *Tables of Spectrophotometry Absorption Data for Compounds used for the Colorimetric Detection of Elements (International Union of Pure and Applied Chemistry)*. Butterworth-Heinemann, London.

Jones, A.M., Reed, R.H. and Weyers, J.D.B. (2015) *Practical Skills in Biology,* 6th edn. Pearson, Harlow.

Sources for further study

Anon. *Spectrophotometry Protocols.* Available: http://www.protocol-online.org/prot/General_Laboratory_Techniques/Spectrophotometry/ Last accessed 01/09/15.
[Gives some examples of spectrophotometric analyses of selected biomolecules.]

Boyer, R.F. (2011) *Biochemistry Laboratory: Modern Theory and Techniques,* 2nd edn. Prentice Hall, Harlow.

Gore, M.G. (ed.) (2000) *Spectrophotometry and Spectrofluorimetry: A Practical Approach,* 2nd edn. Oxford University Press, Oxford.

Ninfa, A.J., Ballou, D.P. and Benore, M. (2009) *Fundamental Laboratory Approaches for Biochemistry and Biotechnology,* 2nd edn. Wiley, New York.

Van Dyke, K., van Dyke, C. and Woodfork, K. (2002) *Luminescence Biotechnology: Instruments and Applications.* CRC Press, Boca Raton.

Wilson, K. and Walker, J. (eds.) (2010) *Principles and Techniques of Practical Biochemistry and Molecular Biology,* 7th edn. Cambridge University Press, Cambridge.
[Also covers a range of other topics, including chromatography and centrifugation.]

STUDY EXERCISES

47.1 Write a protocol for using a spectrophotometer. After reading this chapter, prepare a detailed stepwise protocol explaining how to use one of the spectrophotometers in your department. Ask another student or a tutor to evaluate your protocol and provide you with feedback.

47.2 Use the Beer–Lambert relationship in quantitative spectrophotometric analysis. Calculate the following (express your answer to three significant figures):

(a) The concentration of NADH (μmol L^{-1}) in a test solution giving an absorbance at 340 nm (A_{340}) of 0.53 in a cuvette with a path length of 1 cm, based on a molar absorptivity for NADH of 6220 L mol^{-1} cm^{-1} at this wavelength.

(b) The amount of NADH (nmol) in 20 mL of a test solution where $A_{340} = 0.62$ in a cuvette with a path length of 5 cm, based on a molar absorptivity for NADH of 6220 L mol^{-1} cm^{-1} at this wavelength.

(c) The mass concentration (μg mL^{-1}) of double-stranded DNA in a test solution giving an absorbance at 260 nm (A_{260}) of 0.57 in a cuvette of path length 5 mm, based on an absorptivity of 20 L g^{-1} cm^{-1}.

(d) The amount (ng) of double-stranded DNA in a 50 μL subsample from a test solution where $A_{260} = 0.31$ in a cuvette of path length 1 cm, based on an absorptivity of 20 L g^{-1} cm^{-1}.

(e) The chlorophyll a content, in μg (g fresh weight)$^{-1}$, of a plant leaf of fresh weight 1.56 g extracted in 50 mL of 90% acetone, where $A_{664} = 0.182$ at 664 nm and $A_{647} = 0.035$, based on the following equation:

Chlorophyll a (mg L^{-1}) = 11.93 A_{664} − 1.93 A_{647}

(continued)

47.3 Determine the molar absorptivity of a substance in aqueous solution. A solution of *p*-nitrophenol containing 8.8 µg mL^{-1} gave an absorbance of 0.535 at 404 nm in a cuvette of path length 1 cm. What is the molar absorptivity of *p*-nitrophenol at 404 nm, expressed to three significant figures? (Note: M_r of *p*-nitrophenol is 291.27.)

47.4 Determine the concentration of metal ions based on atomic spectroscopy of test and standard solutions. The following data represent a set of calibration standards for K$^+$ in aqueous solution, measured by flame photometry:

Absorbance of standard solutions containing K$^+$ at up to 0.5 mmol L^{-1}

K$^+$ concentration (mmol L^{-1})	Absorbance
0	0.000
0.1	0.155
0.2	0.279
0.3	0.391
0.4	0.537
0.5	0.683

Draw a calibration curve using the above data and use this to estimate the amount of K$^+$ in a test sample prepared by digestion of 0.482 g of tissue in a final volume of 25 mL of solution, giving an absorbance of 0.429 when measured at the same time as the standards shown above. Express your answer in µmol K$^+$ (g tissue)$^{-1}$, to three significant figures. See also study exercise 42.1 for a similar exercise, based on atomic absorption spectrophotometry of Zn.

47.5 Explain the basis of fluorescence. Why is the wavelength of the emitted light longer than that used to excite a fluorescent molecule?

47.6 Account for the difference in design between spectrophotometer and fluorimeter cuvettes. Why does a spectrophotometer cuvette have only two clear sides while a fluorimeter cuvette has four clear sides?

Answers to these study exercises are available at www.pearsoned.co.uk/practicalskills.

Identifying compounds – the techniques described in this chapter can often provide sufficient information to identify a compound with a low probability of error.

Definitions

Spectroscopy – any technique involving the production and subsequent recording of a spectrum of electromagnetic radiation, usually in terms of wavelength or energy.

Spectrometry – any technique involving the measurement of a spectrum, e.g. of electromagnetic radiation, molecular masses.

Wavenumber – the reciprocal of the wavelength (expressed as cm^{-1}): a term used widely in IR spectroscopy, but rarely in other types of analysis.

Understanding the origins of IR and Raman spectra – the IR spectrum is caused by changes in charge displacement in bonds. The Raman spectrum is caused by changes in polarisability in bonds.

The following techniques are unlikely to be encountered in the early stages of your course, though they may be presented as a demonstration. Alternatively, you may be shown spectra for interpretation and analysis. The various types of spectroscopy, and their relationship with the electromagnetic spectrum, are shown in Table 48.1.

Infrared (IR) and Raman spectroscopy

Both of these techniques involve the measurement of frequencies produced by the vibration of chemical bonds (bending and stretching). The IR/Raman region is generally considered to be from 800 to 2500 nm (for near-IR) and up to 16 000 nm (for mid-IR). Near-IR spectroscopy involves recording the spectrum in that region in a manner analogous to UV/visible spectroscopy, and quantitative analysis is possible. However, the most widely used technique is mid-IR spectroscopy, which allows identification of groups or atoms in a sample compound, but is inappropriate for quantitative measurement. A peak at a particular frequency can be identified by reference to libraries or computer databases of IR spectra, e.g. a peak at a wavenumber of 1730–1750 cm^{-1} corresponds to a carbonyl group – ($\sim C = O$), which is present in fatty acids and proteins. The 1400–600 cm^{-1} region is known as the 'fingerprint' region because no two compounds give identical spectra.

The value of IR spectroscopy is greatly enhanced by Fourier transformation (FT), named after the mathematician J.B. Fourier. FT is a procedure for interconverting frequency functions and time or distance functions. In FT-IR, information is obtained from an interferometer, which splits the incident beam so that it passes through both the sample and a reference. When the beam is recombined, interference patterns arise because the two path lengths are different. The interference pattern has the same relationship to a normal spectrum as a hologram has to a picture, and integral computers use FT to convert the pattern into a spectrum in under a minute. The overall result is a greatly enhanced signal-to-noise ratio.

Table 48.1 The electromagnetic spectrum and types of spectroscopy.

Type of radiation	Origin	Wavelength	Type of spectroscopy
γ-rays	Atomic nuclei	<0.1 nm	γ-ray spectroscopy
X-rays	Inner shell electrons	0.1–1.0 nm	X-ray fluorescence (XRF)
Ultraviolet (UV)	Ionisation	10–200 nm	UV spectroscopy
UV/visible	Valency electrons	200–800 nm	UV/visible spectroscopy
Infrared	Molecular vibrations	0.8–25 μm	Infrared spectroscopy (IR) and Raman
Microwaves	Electron spin alignment	400 μm–30 cm	Electron spin resonance (ESR)
Radiowaves	Nuclear spin resonance	>100 cm	Nuclear magnetic resonance (NMR)

Fig. 48.1 IR spectrum for ethanol.

IR spectra

A typical IR spectrum is shown in Fig. 48.1 and you should note the following points:

The use of wavenumber – this is a well-established convention, since high wavenumber = high frequency = high energy = short wavelength. Expression of the IR range, 4000 cm^{-1} to 650 cm^{-1}, is in 'easy' numbers and the high energy is found on the left-hand side of the spectrum. Note that IR spectroscopists often refer to wavenumbers as 'frequencies', e.g. 'the peak of the C＝O stretching 'frequency' is at 1720 cm^{-1}'.

- **The x axis, the wavelength of the radiation, is given in wavenumbers ($\overline{\nu}$), expressed in reciprocal centimetres (cm^{-1}).** You may still see some spectra from old instruments using microns (μ, equivalent to the SI unit 'micrometres', μm, at 1×10^{-6} m) for wavelength.

- **The y axis, expressing the amount of radiation absorbed by the molecule, is usually shown as % transmittance.** When no radiation is absorbed (all is transmitted through the sample) there is 100% transmittance, while 0% transmittance implies that all radiation is absorbed at a particular wavenumber. Since the y axis scale goes from 0 to 100% transmittance, the absorption peaks are displayed *down* from the 100% line; this is *opposite* to most other common spectra.

- **The cells holding the sample usually display imperfections and are not completely transparent to IR radiation, even when empty.** Therefore the base line of the spectrum is rarely set on 100% transmittance and quantitative applications of IR spectroscopy are more complex than for UV/visible spectrophotometry (p. 312).

Applications of IR and Raman spectroscopy

The principal use of IR and Raman spectroscopy is in the identification of drugs (e.g. penicillin), small peptides, pollutants and food contaminants. When an IR spectrometer is coupled to a gas–liquid chromatograph, it can be used for the analysis of drug metabolites.

Using IR – the rapid analysis possible with gaseous samples makes IR ideal for studying CO_2 metabolism in photosynthesis and respiration (p. 430).

One possible future use of IR is for the non-invasive analysis of organic compounds that are of diagnostic significance (e.g. glucose in diabetics). This involves illuminating a small area of skin with IR radiation and using the resultant IR spectrum to quantify a specific compound: an example of such a system is the non-invasive blood glucose monitor described by Malchoff *et al.,* (2002).

Nuclear magnetic resonance (NMR)

Electromagnetic radiation (at radiofrequencies of 1–500 MHz) is used to identify and monitor compounds. This is possible because of differences in the magnetic states of atomic nuclei, involving very small transitions in energy levels. The atomic nuclei of isotopes of many elements are magnetic because they are charged and have spin. Typical magnetic nuclei are ^{1}H, ^{13}C, ^{14}N, ^{15}N, ^{19}F and ^{31}P. When these nuclei interact with a uniform external magnetic field, they behave like tiny compass needles and align themselves in a direction either parallel or antiparallel to the field. The two orientations have different energies, with the parallel direction having a lower energy than the antiparallel (Fig. 44.3). The energy difference between the two levels (ΔE) corresponds to a precise electromagnetic frequency (ν), according to similar quantum principles to those for the excitation of electrons (p. 310). When a sample containing an isotope with a magnetic nucleus is placed in a magnetic field and exposed to an appropriate radiofrequency, transitions between the energy levels of magnetic nuclei will occur when the energy gap and the applied frequency are in *resonance* (i.e. when they are matched exactly). Differences in energy levels, and hence resonance frequencies (ν_0), depend on the magnitude of the applied magnetic field (B_0) and the magnetogyric ratio (λ), according to the equation:

$$\nu_0 = \lambda B_0 / 2\pi \qquad [48.1]$$

The magnetogyric ratio varies from one isotope to another, so NMR is performed at different frequencies for different nuclei at any given value of B_0. The principal components of an NMR spectrometer are shown in Fig. 44.4.

For magnetic nuclei in a given molecule, an NMR spectrum is generated because, in the presence of the applied field, different nuclei experience different local magnetic fields depending on the arrangement of electrons in their vicinity. The effective field (B) at the nucleus can be expressed as:

$$B = B_0(1 - \sigma) \qquad [48.2]$$

where σ (the shielding constant) expresses the contribution of the small secondary field generated by nearby electrons. The magnitude of σ is dependent on the electronic environment of a nucleus, so nuclei in different environments give rise to different resonance frequencies, according to the equation:

$$\nu_0 = \lambda B_0(1 - \sigma)/2\pi \qquad [48.3]$$

The separation of resonance frequencies from a reference value is termed the *chemical shift,* and is expressed in dimensionless terms, as parts per million (ppm). By convention, the chemical shift is positive if the sample nucleus is less shielded than the reference and negative if it is more shielded.

Understanding NMR spectra of biomolecules

Figure 48.2 illustrates the principle underlying this technique – the different chemical environments of the hydrogen atoms in a simple biomolecule such as acetic acid result in two different ^{1}H resonances or 'signals', one corresponding to those of protons in ~CH_3, and the other from that in ~COOH. Furthermore, the relative intensities of the NMR

SAFETY NOTE Working with NMR spectrometers – these instruments can generate very high magnetic fields and can exert strong magnetic forces on materials such as steel items brought close to the magnet (< 1 m). NMR magnets can also destroy data stored in magnetic form – keep items such as memory sticks, laptops and credit cards well away.

Example For an external magnetic field of 2.5 T (Tesla), ΔE for ^{1}H is 6.6×10^{-26}J, and, since $\Delta E = h\nu$, the corresponding frequency (ν) is 100 MHz; for ^{13}C in the same field, ΔE is 1.7×10^{-26}J, and ν is 25 MHz.

Measuring chemical shifts – ppm is not a concentration term in NMR, but is used to reflect the very small frequency changes that occur relative to the reference standard, measured in proportional terms.

Fig. 48.2 ¹H NMR spectrum of acetic acid (CH₃COOH). The relative areas of the two signals are 1:3 and the frequencies (chemical shifts) are expressed in terms of ppm, relative to the reference signal (tetramethyl silane, TMS).

NMR of living tissues – reactions can be studied non-destructively, in 'real time'.

signals, as measured by their areas, are proportional to the number of contributing nuclei, so the relative areas of the peaks due to the protons in ~CH₃ and (undissociated) ~COOH would be 3 : 1. Similarly, with ³¹P NMR of a biologically important molecule such as ATP, there are signals corresponding to the α, β and γ phosphates, the P nuclei of which are in different chemical environments (Fig. 48.3). Thus every molecule that contains one or more magnetic nuclei has its own characteristic NMR fingerprint that may be used for identification and analysis. Spectra such as those shown in Figs 48.2 and 48.3 can be obtained using FT of a large number of individual responses to radiowave pulses. Other factors such as the spin-lattice relaxation time (Jacobson, 2007) can affect signal intensities (peak sizes), and resonances may be split into several lines owing spin–spin coupling (interactions between neighbouring nuclei).

In terms of resolution, narrow signals are obtained only from molecules that are fairly mobile, so most high resolution studies are carried out using solutions. However, since many metabolites in biological samples (cells, tissues, etc.) are in aqueous solution and freely mobile, they can give rise to high resolution spectra, as shown in Fig. 48.3 for the ³¹P-containing compounds, ATP, phosphocreatine and inorganic phosphate. On the other hand, ³¹P signals from immobile molecules such as DNA or phospholipids are very broad, and could be as wide as the whole scale shown in Fig. 48.3.

Electron spin resonance (ESR)

This technique is based on energy transitions of spinning electrons in a magnetic field. As with NMR, the low energy state occurs when the electromagnetic field generated by the spinning electron is parallel to the

Fig. 48.3 Typical ³¹P NMR spectrum from intact skeletal muscle (a) at rest and (b) after a fatiguing series of contractions. In (b) ATP levels are preserved at the expense of phosphocreatinine (PCr). Also, hydrolysis of ATP during contraction results in a large phosphate (Pi) peak which is shifted to the right compared to (a), reflecting the decrease in intracellular pH that accompanies glycolysis.

externally generated field, while the high energy state occurs when the electron-generated field is antiparallel. ESR is very useful for studying metalloproteins and can be used to monitor the activity of such proteins (e.g. cytochrome oxidase) in intact mitochondria or chloroplasts. It can also be used to detect free radicals, for example in irradiated foodstuffs.

Mass spectrometry (MS)

This technique involves the disintegration of organic compounds into fragment ions in a gas phase. These ions are accelerated to specific velocities using an electric field and then separated on the basis of their different masses. Each fragment of a particular mass is detected sequentially with time. The most widely used method in MS is electron impact ionisation (EI), where the electron source is a heated metal, such as a tungsten filament, subjected to an appropriate potential gradient. The stream of emitted electrons may then interact with biomolecules (M) in the sample by either:

1. **electron removal** – where an electron in a bond within the sample molecule is 'knocked out' by bombarding electrons, leaving the bond with only one unpaired electron and resulting in the production of a cationic free radical, i.e. $M + e^- \rightarrow M^{+\bullet} + 2e^-$;

2. **electron capture** – where addition of an extra electron results in the production of an anionic free radical, i.e. $M + e^- \rightarrow M^{-\bullet}$.

Where it is not known whether cations or anions are formed, the radical is given the symbol ¬, i.e. as $M^{\neg\bullet}$. The first ions formed (parent ions) are unstable, and rapidly undergo further disintegration to give smaller fragments and daughter ions, subsequently separated in the mass spectrometer (Fig. 48.4), to give a mass spectrum such as that shown in Fig. 48.5.

For *in vivo* metabolic studies, the technique of isotope ratio mass spectroscopy (IRMS) is useful since it removes the need to use radioactive isotopes. This technique exploits the ability of MS to distinguish between isotopes such as ^{13}C and ^{12}C. For example, compounds containing ^{13}C have a greater mass than the same compound containing ^{12}C and can be differentiated in the mass spectrum. By selectively labelling a key metabolite with a non-radioactive isotope, the fate of this metabolite can be followed by MS analysis of sequential samples. Normally, the materials used are combusted in oxygen to give gases such as $^{13}CO_2$ and $^{15}NO_2$, followed by exposure of such gases to an EI source; useful metabolic data can be obtained by determining the isotope ratio, e.g. of $^{13}CO_2^{\neg\bullet}$ and $^{12}CO_2^{\neg\bullet}$

Pyrolysis-mass spectrometry (PY-MS)

This is another variant of mass spectroscopy that is useful for biomolecular applications. The molecules in a sample are volatilised by heating to a precisely controlled high temperature for a specific time period. The volatile material is then removed by a vacuum, ionised by EI and subjected to MS. Complex mixtures of pyrolysis products are produced, but interpretation is made easier by computer-based multivariate analysis. A useful application of PY-MS is in microbial identification, especially with small samples of organisms that are difficult to culture, e.g. mycobacteria.

> Understanding the terminology of mass spectrometry – since this technique does not involve the production and measurement of electromagnetic spectra and is not based on quantum principles, it should not be referred to as a spectroscopic technique.

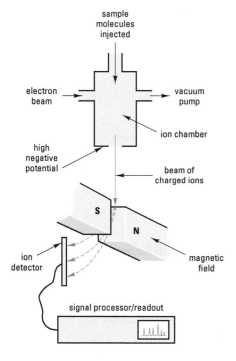

Fig. 48.4 Components of an electron-impact mass spectrometer.

Fig. 48.5 Mass spectrum for methanol.

Example IRMS is used in the diagnosis of gastric ulcers caused by *Helicobacter pylori*. This bacterium produces urease, degrading urea to CO_2 and NH_3. A patient drinks a solution of ^{13}C-labelled urea and, if the organism is present in the stomach, $^{13}CO_2$ will appear at up to 5% v/v in the patient's breath, compared with the expected natural abundance value of 1.1% v/v.

Fast atom bombardment–mass spectrometry (FAB-MS)

This is particularly useful for biomolecular applications because it can be used with aqueous solutions. The solution containing the analyte is mixed with glycerol and applied to a probe which is inserted into a vacuum chamber. The mixture is bombarded with a high velocity stream of atoms (usually argon or xenon) rather than electrons, which induces fragmentation of the biomolecules and allows production of the mass spectrum. One problem with FAB-MS is the suppression effect phenomenon – with a mixed sample, not all components may be equally accessible to the atomic bombardment depending on the way they are distributed in the sample–glycerol mixture. This can be overcome by coupling FAB-MS to techniques that can give high resolution separation of components, e.g. gas–liquid chromatography (GC, p. 335), high-performance liquid chromatography (HPLC, p. 334), or capillary electrophoresis (p. 361). Greaves and Roboz (2013) give further information on biomolecular aspects of MS.

Text references

Greaves, J. and Roboz, J. (2013) *Mass Spectrometry for the Novice*. CRC Press, Boca Raton.

Jacobson, N.E. (2007) *NMR Spectroscopy Explained: Simplified Theory, Applications and Examples for Organic Chemistry and Structural Biology.* Wiley, New York.

Malchoff, C.D., Shoukri, K., Landau, J.I. and Buchert, J.M. (2002) Novel non-invasive blood glucose monitor. *Diabetes Care* **25**: 2268–275.
Available: care.diabetesjournals.org/cgi/content/full/25/12/68
Last accessed: 01/09/15.

Sources for further study

Field, L.D., Sternhall, S. and Kalman, J.R. (2013) *Organic Structures from Spectra*, 5th edn. Wiley, New York.

Gunther, H. (2013) *NMR Spectroscopy: Basic Principles, Concepts and Applications in Chemistry.* Wiley, New York.

Guntzler, H. and Gremlich, H.U. (2002) *IR Spectroscopy: an Introduction.* Wiley-VCH, Mannheim.

Keeler, J. (2010) *Understanding NMR Spectroscopy,* 2nd edn. Wiley, New York.

Larkin, P.J. (2011) *Infrared and Raman Spectroscopy: Principles and Spectral Interpretation.* Elsevier, Amsterdam.

Siuzdak, G. (2006) *The Expanding Role of Mass Spectrometry for Biotechnology,* 2nd edn. MCC Press, New York.
[Written for bioresearchers, it covers principles and applications without extensive use of maths.]

Smith, R.M. and Butsch, K.L. (2004) *Understanding Mass Spectra: a Basic Approach.* Wiley, New York.

STUDY EXERCISES

48.1 Interpret IR signals. The figure shows the IR spectra of acetic acid (ethanoic acid). Using the information provided below, identify the resonance corresponding to (a) the O–H and (b) C=O of the carboxylic acid group in the acetic acid spectrum.

Typical IR absorption ranges for different functional groups

Bond	Location	Wavenumber (cm^{-1})
C–O	esters, alcohols	1000–1300
C=O	ketones, aldehydes, carboxylic acids, esters, amides	1680–1750
O–H	carboxylic acids (H-bonded)	2500–3300
O–H	alcohols (H-bonded)	3230–3350
O–H	free	3580–3670
N–H	amines	3100–3500

48.2 Investigate appropriate reference compounds for biological applications of NMR. Why are reference compounds important when measuring chemical shifts in NMR? Which compounds are suitable references for (a) ^{1}H, (b) ^{13}C and (c) ^{31}P NMR?

48.3 Test your understanding of the NMR terminology. Distinguish between the term ppm as used in NMR and as a concentration term (p. 163).

48.4 Interpret NMR signals from biomolecules in body fluids. A ^{1}H NMR spectrum (range 1–4.5 ppm) of a human urine sample obtained from a healthy adult is shown in the first figure below. The spectrum reveals a large number of components/peaks. The two most intense of these peaks (at approximately 3.05 and 4.05 ppm) correspond to creatinine (C), which is abundant in healthy urine. The molecular structure of creatinine is shown in the second figure below. Explain why the ^{1}H NMR 'fingerprint' of creatinine in urine consists of just two discrete peaks.

Answers to these study exercises are available at www.pearsoned.co.uk/practicalskills.

SAFETY NOTE Working with centrifuges – these instruments generate very high centrifugal forces and should only be used after suitable instruction and training – if in doubt, ask.

Particles suspended in a liquid will move at a rate that depends on:

- **the applied force** – particles in a liquid within a gravitational field, e.g. a stationary test tube, will move in response to the Earth's gravity;
- **the density difference between the particles and the liquid** – particles less dense than the liquid will float upwards while particles denser than the liquid will sink;
- **the size and shape of the particles;**
- **the viscosity of the medium.**

For most biological particles (cells, organelles or molecules) the rate of flotation or sedimentation in response to the Earth's gravity is too slow to be of practical use in separation.

Table 49.1 Relationship between speed (r.p.m.) and acceleration (relative centrifugal field, RCF) for a typical bench centrifuge with an average radius of rotation, $r_{av} = 115$ mm.

r.p.m.	RCF*
500	30
1000	130
1500	290
2000	510
2500	800
3000	1160
3500	1570
4000	2060
4500	2600
5000	3210
5500	3890
6000	4630

*RCF values rounded to nearest 10.

KEY POINT   A centrifuge is an instrument designed to produce a centrifugal force far greater than the Earth's gravity, by spinning the sample about a central axis (Fig. 49.1). Particles of different size, shape or density will thereby sediment at different rates, depending on the speed of rotation and their distance from the central axis.

Fig. 49.1 Principal components of a low-speed bench centrifuge.
Diagram of Fisher Scientific Model MSE Centaur 2, low-speed bench centrifuge.

Examples Suppose you wanted to calculate the RCF of a bench centrifuge with a rotor of r_{av} = 95 mm running at a speed of 3000 r.p.m. Using Eqn [49.1] the RCF would be: $1.118 \times 95 \times (3)^2 = 956\ g$.

You might wish to calculate the speed (r.p.m.) required to produce a relative centrifugal field of 2000 g using a rotor of r_{av} = 85 mm. Using Eqn [49.2] the speed would be: $945.7\sqrt{(2000 \div 85)} = 4587$ r.p.m.

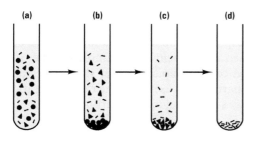

Fig. 49.2 Differential sedimentation. (a) Before centrifugation, the tube contains a mixed suspension of large, medium and small particles of similar density. (b) After low-speed centrifugation, the pellet is predominantly composed of the largest particles. (c) Further high-speed centrifugation of the supernatant will give a second pellet, predominantly composed of medium-sized particles. (d) A final ultracentrifugation step pellets the remaining small particles. Note that all of the pellets apart from the final one will have some degree of cross-contamination.

Fig. 49.3 Density gradients. (a) A continuous (linear) density gradient. (b) A discontinuous (step-wise) density gradient, formed by layering solutions of decreasing density on top of each other. (c) A single-step density barrier, designed to allow selective sedimentation of one type of particle.

How to calculate centrifugal acceleration

The acceleration of a centrifuge is usually expressed as a multiple of the acceleration due to gravity (g = 9.80 m s⁻²), termed the relative centrifugal field (RCF, or 'g value'). The RCF depends on the speed of the rotor (n, in revolutions per minute, r.p.m.) and the radius of rotation (r, in mm) where:

$$\text{RCF} = 1.118r\left(\frac{n}{1000}\right)^2 \tag{49.1}$$

This relationship can be rearranged, to calculate the speed (r.p.m.) for specific values of r and RCF:

$$n = 945.7\sqrt{\left(\frac{\text{RCF}}{r}\right)} \tag{49.2}$$

However, you should note that RCF is not uniform within a centrifuge tube: it is highest near the outside of the rotor (r_{max}) and lowest near the central axis (r_{min}). In practice, it is customary to report the RCF calculated from the average radius of rotation (r_{av}), as shown in Fig. 49.5. It is also worth noting that RCF varies relative to the *square* of the speed: thus the RCF will be doubled by an increase in speed of approximately 41% (Table 49.1).

Centrifugal separation methods

Differential sedimentation (pelleting)

By centrifuging a mixed suspension of particles at a specific RCF for a particular time, the mixture will be separated into a pellet and a supernatant (Fig. 49.2). The successive pelleting of a suspension by spinning for a fixed time at increasing RCF is widely used to separate organelles from cell homogenates. The same principle applies when cells are harvested from a liquid medium.

Density gradient centrifugation

The following techniques use a density gradient, a solution which increases in density from the top to the bottom of a centrifuge tube (Fig. 49.3).

- Rate-zonal centrifugation. By layering a sample onto a shallow preformed density gradient, followed by centrifugation, the larger particles will move faster through the gradient than the smaller ones, forming several distinct zones (bands). This method is time-dependent, and centrifugation *must* be stopped before any band reaches the bottom of the tube (Fig. 49.4).

- Isopycnic centrifugation. This technique separates particles on the basis of their buoyant density. Several substances form density gradients during centrifugation (e.g. sucrose, CsCl, Ficoll, Percoll, Nycodenz). The sample is mixed with the appropriate substance and then centrifuged – particles form bands where their density corresponds to that of the medium (Fig. 49.4). This method requires a steep gradient and sufficient time to allow gradient formation and particle redistribution, but is unaffected by further centrifugation.

Bands within a density gradient can be sampled using a fine Pasteur pipette, or a syringe with a long, fine needle. Alternatively, the tube may be punctured and the contents (fractions) collected dropwise in several tubes. For accurate work, an upward displacement technique can be used (see Ford and Graham, 1991, or Graham, 2001).

(a) rate-zonal centrifugation (shallow gradient) isopycnic centrifugation (steep gradient) **(b)**

Fig. 49.4 Density gradient centrifugation. The central tube shows the position of the sample prior to centrifugation, as a layer on top of the density gradient medium. Note that particles sediment on the basis of size during rate-zonal centrifugation (a), but form bands in order of their densities during isopycnic centrifugation (b). ●, large particles, intermediate density; ▲, medium-sized particles, low density; ▬, small particles, high density.

Density barrier centrifugation

A single-step density barrier (Fig. 49.3(c)) can be used to separate cells from their surrounding fluid, e.g. using a layer of silicone oil adjusted to the correct density using dinonyl phthalate. Blood cell types can be separated using a density barrier of, e.g., Ficoll.

Types of centrifuge and their uses

Low-speed centrifuges

These are bench-top instruments for routine use, with a maximum speed of 3000–6000 r.p.m. and RCF up to 6000 g (Fig. 49.1). They are used to harvest cells, larger organelles (e.g. nuclei, chloroplasts) and coarse precipitates (e.g. antibody–antigen complexes, p. 282). Most modern machines also have a sensor that detects any imbalance when the rotor is spinning and cuts off the power supply (Fig. 49.1). However, some of the older models do not, and must be switched off as soon as any vibration is noticed, to prevent damage to the rotor or harm to the operator. Box 49.1 gives details of operation for a low-speed centrifuge.

Microcentrifuges (microfuges)

These are bench-top machines, capable of rapid acceleration up to 12 000 r.p.m. and 10 000 g. They are used to sediment small sample volumes (up to 1.5 ml) of larger particles (e.g. cells, precipitates) over short timescales (typically, 0.5–15 min). They are particularly useful for the rapid separation of cells from a liquid medium, e.g. silicone oil microcentrifugation.

Continuous flow centrifuges

Useful for harvesting large volumes of cells from their growth medium. During centrifugation, the particles are sedimented as the liquid flows through the rotor.

High-speed centrifuges

These are usually larger, free-standing instruments with a maximum speed of up to 25 000 r.p.m. and RCF up to 60 000 g. They are used for microbial cells, many organelles (e.g. mitochondria, lysosomes) and protein precipitates. They often have a refrigeration system to keep the rotor cool at high speed. You would normally use such instruments only under direct supervision.

Ultracentrifuges

These are the most powerful machines, having maximum speeds in excess of 30 000 r.p.m. and RCF up to 600 000 g, with sophisticated refrigeration and vacuum systems. They are used for smaller organelles (e.g. ribosomes, membrane vesicles) and biological macromolecules. You would not normally use an ultracentrifuge, though your samples may be run by a member of staff.

Rotors

Many centrifuges can be used with tubes of different size and capacity, either by changing the rotor or by using a single rotor with different buckets/adaptors.

- **Swing-out rotors:** sample tubes are placed in buckets which pivot as the rotor accelerates (Fig. 49.5(a)). Swing-out rotors are used on many low-speed centrifuges: their major drawback is their extended path length and the resuspension of pellets due to currents created during deceleration.

Fig. 49.5 Rotors: (a) swing-out rotor; (b) fixed angle rotor; (c) vertical tube rotor.

- **Fixed-angle rotors:** used in many high-speed centrifuges and microcentrifuges (Fig. 49.5(b)). With their shorter path length, fixed rotors are more effective at pelleting particles than swing-out rotors.
- **Vertical tube rotors:** used for isopycnic density gradient centrifugation in high-speed centrifuges and ultracentrifuges (Fig. 49.5(c)). They cannot be used to harvest particles in suspension as a pellet is not formed.

Box 49.1 **How to use a low-speed bench centrifuge**

1. **Choose the appropriate tube size and material for your application**, with caps where necessary. Most low-speed machines have four-place or six-place rotors – use the correct number of samples to *fill* the rotor assembly whenever possible.

2. **Fill the containers to the appropriate level:** do not overfill, or the sample may spill during centrifugation.

3. **It is vital that the rotor is balanced during use.** Therefore, *identical* tubes must be prepared, to be placed opposite each other in the rotor assembly. This is particularly important for density gradient samples, or for samples containing materials of widely differing densities, e.g. soil samples, since the density profile of the tube will change during a run. However, for low-speed work using small amounts of particulate matter in aqueous solution, it is sufficient to counterbalance a sample with a second tube filled with water, or a saline solution of similar density to the sample.

4. **Balance each pair of sample tubes** (plus the corresponding caps, where necessary) to within 0.1 g using a top-pan balance; add liquid dropwise to the lighter tube, until the desired weight is reached. Alternatively, use a set of scales. For small sample volumes (up to 10 mL) added to disposable, lightweight plastic tubes, accurate pipetting of your solution may be sufficient for low-speed use.

5. **For centrifuges with swing-out rotors,** check that each holder/bucket is correctly positioned in its locating slots on the rotor and that it is able to swing freely. All buckets must be in position on a swing-out rotor, even if they do not contain sample tubes – buckets are an integral part of the rotor assembly.

6. **Load the sample tubes into the centrifuge.** Make sure that the outside of the centrifuge tubes, the sample holders and sample chambers are dry: any liquid present will cause an imbalance during centrifugation, in addition to the corrosive damage it may cause to the rotor. For sample holders where rubber cushions are provided, make sure that these are correctly located. Balanced tubes must be placed opposite each other – use a simple code if necessary, to prevent mix-ups.

7. **Bring the centrifuge up to operating speed** by gentle acceleration. Do not exceed the maximum speed for the rotor and tubes used.

8. **If the centrifuge vibrates at any time during use,** switch off and find the source of the problem.

9. Once the rotor has stopped spinning, release the lid and remove all tubes. If any sample has spilled, make sure you clean it up thoroughly using a non-corrosive disinfectant, e.g. Virkon, so that it is ready for the next user.

10. **Close the lid (to prevent the entry of dust) and return all controls to zero.**

Centrifuge tubes

These are manufactured in a range of sizes (from 1.5 mL up to 1000 mL) and materials. The following aspects may influence your choice:

- **Capacity.** This is obviously governed by the volume of your sample. Note that centrifuge tubes must be completely full for certain applications, e.g. for high-speed work.
- **Shape.** Conical-bottomed centrifuge tubes retain pellets more effectively than round-bottomed tubes, while the latter may be more useful for density gradient work.
- **Maximum centrifugal force.** Detailed information is supplied by the manufacturers. Standard Pyrex glass tubes can only be used at low centrifugal force (up to 2000 g).
- **Caps.** Most fixed-angle and vertical tube rotors require tubes to be capped, to prevent leakage during use and to provide support to the tube during centrifugation. For low-speed centrifugation, caps must be used for any hazardous samples. Make sure you use the correct caps for your tubes.
- **Solvent resistance.** Glass tubes are inert, polycarbonate tubes are particularly sensitive to organic solvents (e.g. ethanol, acetone), while polypropylene tubes are more resistant. See manufacturer's guidelines for detailed information.
- **Sterility.** Disposable plastic centrifuge tubes are often supplied in sterile form. Glass and polypropylene tubes can be repeatedly sterilised. Cellulose ester tubes should *not* be autoclaved. Repeated autoclaving of polycarbonate tubes may lead to cracking/stress damage.
- **Opacity.** Glass and polycarbonate tubes are clear, while polypropylene tubes are more opaque.
- **Ability to be pierced.** If you intend to harvest your sample by puncturing the tube wall, cellulose acetate and polypropylene tubes are readily punctured using a syringe needle.

Balancing the rotor

For the safe use of centrifuges, the rotor must be balanced during use (Box 49.1), or the spindle and rotor assembly may be damaged permanently; in severe cases, the rotor may fail and cause a serious accident.

> **KEY POINT** It is vital that you balance your loaded centrifuge tubes before use. As a general rule, balance all sample tubes to within 1% or better, using a top-pan balance or scales. Place balanced tubes opposite each other.

Safe practice

Given their speed of rotation and the extremely high forces generated, centrifuges have the potential to be extremely dangerous, if used incorrectly. You should only use a particular centrifuge if you fully understand the operating principles: if unsure, check with a member of staff. For safety reasons, all centrifuges are manufactured with an armoured casing that should contain any fragments in cases of rotor failure. Machines usually

SAFETY NOTE Working with centrifuge tubes – *never* be tempted to use a tube or bottle which was not designed to fit the machine you are using (e.g. a general-purpose glass test tube, or a screw-capped bottle), or you may damage the centrifuge and cause an accident.

Using microcentrifuge tubes – the integral push-on caps of microcentrifuge tubes must be correctly pushed home before use or they can come off during centrifugation.

SAFETY NOTE Balancing tubes – *never* balance centrifuge tubes 'by eye' – use a balance. Note that a 35-mL tube full of liquid at an RCF of 3000 g has an effective weight greater than a large adult human.

have a safety lock to prevent the motor from being switched on unless the lid is closed and to stop the lid from being opened while the rotor is moving. Do not be tempted to use older machines without a safety lock, or centrifuges where the locking mechanism is damaged/inoperative. Be particularly careful to make sure that hair and clothing are kept well away from moving parts.

Text references

Ford, T.C. and Graham, J.M. (1991) *An Introduction to Centrifugation*. Bios, Oxford.

Graham, J. (2001) *Biological Centrifugation*. Bios, Oxford.

Sources for further study

Bonner, P. and Hargreaves, A. (2011) *Basic Bioscience Laboratory Techniques: a Pocket Guide*. Wiley, New York.

Boyer, R.F. (2011) *Biochemistry Laboratory: Modern Theory and Techniques,* 2nd edn. Benjamin Cummings, San Francisco.

STUDY EXERCISES

49.1 Decide on the type of centrifuge required for a particular application. What centrifuge would you use for each of the following?

(a) Separating mitochondria from a cell homogenate.

(b) Separating yeast cells from their surrounding medium in an experiment to study the uptake of a radiolabelled amino acid as a function of time.

(c) Harvesting cells from a bioreactor containing 25 litres of growth medium.

49.2 Determine centrifugal acceleration for a specific centrifuge. Calculate the centrifugal acceleration (RCF, i.e. the *g* value) of the following (express your answers to three significant figures):

(a) A bench centrifuge with an average radius of rotation of 125 mm operating at 4000 r.p.m.

(b) A bench-top microcentrifuge with an average radius of rotation of 60 mm, operating at 12 000 r.p.m.

(c) An ultracentrifuge with an average radius of rotation of 186 mm, operating at 30 000 r.p.m.

49.3 Determine the speed required to give a particular centrifugal acceleration (*g* value). Calculate the speed required (r.p.m. value) for each of the following (express your answers to three significant figures):

(a) RCF of 1500 *g*, using a bench centrifuge with an average radius of rotation of 95 mm.

(b) RCF of 50 000 *g*, using a high-speed centrifuge with an average radius of rotation of 135 mm.

(c) RCF of 13 000 *g*, using a bench-top microcentrifuge with an average radius of rotation of 5.5 cm.

49.4 Calculate the difference in centrifugal acceleration (*g* value) between the top, middle and bottom of a centrifuge tube. Assuming that the minimum, average and maximum radial distances of a centrifuge tube in a swing-out rotor of a bench centrifuge operating at 5000 r.p.m. are 45 mm, 70 mm and 95 mm respectively, what are the corresponding *g* values at the top, middle and bottom of the tube when the centrifuge is operating? (Express your answer to three significant figures.)

Answers to these study exercises are available at www.pearsoned.co.uk/practicalskills.

50 Chromatography – separation methods

Chromatography is used to separate the individual constituents within a sample on the basis of differences in their physical characteristics, e.g. molecular size, shape, charge, volatility, solubility and/or adsorptivity. The essential components of a chromatographic system are as follows:

- **A stationary phase:** either a solid, a gel or an immobilised liquid, held by a support matrix.
- **A chromatographic bed:** the stationary phase may be packed into a glass or metal column, spread as a thin layer on a sheet of glass or plastic, or adsorbed on cellulose fibres (paper).
- **A mobile phase:** either a liquid or a gas which acts as a solvent, carrying the sample through the stationary phase and eluting from the chromatographic bed.
- **A delivery system:** to pass the mobile phase through the chromatographic bed.
- **A detection system:** to monitor the test substances (Chapter 51).

Individual substances interact with the stationary phase to different extents as they are carried through the system, enabling separation to be achieved.

> **KEY POINT** In a chromatographic system, those substances which interact strongly with the stationary phase will be retarded to the greatest extent while those which show little interaction will pass through with minimal delay, leading to differences in distances travelled or elution times.

Types of chromatographic system

Chromatographic systems can be categorised according to the form of the chromatographic bed, the nature of the mobile and stationary phases and the method of separation.

Thin-layer chromatography (TLC) and paper chromatography

Here, you apply the sample as a single spot near one end of the sheet, by microsyringe or microcapillary. This sheet is allowed to dry fully, then it is transferred to a glass tank containing a shallow layer of solvent (Fig. 50.1). Remove the sheet when the solvent front has travelled across 80–90% of its length.

You can express movement of an individual substance in terms of its relative frontal mobility, or R_F value, where:

$$R_F = \frac{\text{distance moved by substance}}{\text{distance moved by solvent}}\qquad[50.1]$$

Alternatively, you may express movement with respect to a standard of known mobility, as R_X, where:

$$R_X = \frac{\text{distance moved by test substance}}{\text{distance moved by standard}}\qquad[50.2]$$

glass tank with lid

solvent front

direction of travel of sample components

point of application of samples

solvent (mobile phase)

glass plate with silica gel layer

Fig. 50.1 Components of a TLC system.

The R_F (or R_X) value is a constant for a particular substance and solvent system (under standard conditions) and closely reflects the partitioning of the substance between the stationary and mobile phases. Tabulated values are available for a range of biological molecules and solvents (e.g. Stahl, 1969, or Touchstone, 1992). However, you should analyse one or more reference compounds on the same sheet as your unknown sample, to check their R_F values.

Column chromatography

Here, you pack a glass column with the appropriate stationary phase and equilibrate the mobile phase by passage through the column, either by gravity (Fig. 50.2), or using a low-pressure peristaltic pump. You can then introduce the sample to the top of the column, to form a discrete band of material. This is then flushed through the column by the mobile phase. If the individual substances have different rates of migration, they will separate within the column, eluting at different times as the mobile phase travels through the column.

You can detect eluted substances by collecting the mobile phase as it elutes from the column in a series of tubes (discontinuous monitoring), either manually or with an automatic fraction collector. Fractions of 2–5% of the bed volume are usually collected and analysed, e.g. by chemical assay. You can now construct an elution profile (or chromatogram) by plotting the amount of substance against either time, elution volume or fraction number, which should give a symmetrical peak for each substance (Fig. 50.3).

You can express the migration of a particular substance at a given flow rate in terms of its retention time (t), or elution volume (V_e). The separation efficiency of a column is measured by its ability to distinguish between two similar substances, assessed in terms of:

- **selectivity** (α), measured using the following equation, which takes into account the retention times of the two peaks (i.e. t_a and t_b), plus the column dead time (t_0):

$$\alpha = \frac{t_b - t_0}{t_a - t_0} \qquad [50.3]$$

The column dead time is the time it takes for an unretained compound to pass through the column without any interaction with the stationary phase.

- **resolution** (R), quantified in terms of the retention time and the base width (W) of each peak:

$$R = \frac{2(t_a - t_b)}{W_a + W_b} \qquad [50.4]$$

where the subscripts a and b refer to substances a and b, respectively (Fig. 50.3). For most practical purposes, R values of 1 or more are satisfactory, corresponding to 98% peak separation for symmetrical peaks.

Non-symmetrical peaks may result from column overloading, coelution of solutes, poor packing of the stationary phase or interactions between the substances and the support material.

Using a TLC system – it is essential that you allow the solvent to pre-equilibrate in the chromatography tank for at least 2 h before use, to saturate the atmosphere with vapour. Deliver drops of sample with a blunt-ended microsyringe. Make sure you know exactly where each sample is applied, so that R_F values can be calculated.

Fig. 50.2 Equipment for column chromatography (gravity feed system).

Fig. 50.3 Peak characteristics in a chromatographic separation, i.e. a chromatogram. For symbols, see Eqn [50.4].

High-performance liquid chromatography (HPLC)

Column chromatography originally used large 'soft' stationary phases that required low-pressure flow of the mobile phase to avoid compression; separations were usually time-consuming and of low resolution ('low performance'). Subsequently, the production of small, incompressible, homogeneous particulate support materials and high-pressure pumps with reliable, steady flow rates have enabled high-performance systems to be developed. These systems operate at pressures up to 10 MPa, forcing the mobile phase through the column at a high flow rate to give rapid separation with reduced band broadening, owing to smaller particle size.

HPLC columns are usually made of stainless steel, and all components, valves, etc., are manufactured from materials that can withstand the high pressures involved. The two main systems are:

- **isocratic separation:** a single solvent (or solvent mixture) is used throughout the analysis;
- **gradient elution separation:** the composition of the mobile phase is altered using a microprocessor-controlled gradient programmer, which mixes appropriate amounts of two different substances to produce the required gradient.

Most HPLC systems are linked to a continuous monitoring detector of high sensitivity, e.g. proteins may be detected spectrophotometrically by monitoring the absorbance of the eluent at 280 nm as it passes through a flow cell (cuvette). Other detectors can be used to measure changes in fluorescence, current or potential (Chapter 51). Most detection systems are non-destructive, which means that you can collect eluent with an automatic fraction collector for further study (Fig. 50.4).

The speed and sensitivity of HPLC make this the method of choice for the separation of many small molecules of biological interest, normally using reverse-phase partition chromatography (p. 335). Separation of macromolecules (especially proteins and nucleic acids) usually requires 'biocompatible' systems in which stainless steel components are replaced by titanium, glass or fluoroplastics, using lower pressures to avoid denaturation, e.g. the Pharmacia FPLC system. Such separations are carried out using ion-exchange, gel permeation and/or hydrophobic interaction chromatography (p. 337).

Using HPLC – this is a versatile form of chromatography, used with a wide variety of stationary and mobile phases, to separate individual compounds of a particular class of molecules on the basis of size, polarity, solubility or adsorption characteristics.

Fig. 50.4 Components of an HPLC system.

Fig. 50.5 Components of a GC system.

Applications of gas chromatography – GC is used to separate volatile, non-polar compounds: substances with polar groups must be converted to less polar derivatives prior to analysis, in order to prevent adsorption on the column, resulting in poor resolution and peak tailing.

Fig. 50.6 Adsorption chromatography (polar stationary phase).

Fig. 50.7 Liquid–liquid partition chromatography, e.g. reverse-phase HPLC.

Gas chromatography (GC)

Modern GC uses capillary chromatography columns (internal diameter 0.1–0.5 mm) up to 50 m in length (Fig. 50.5). The stationary phase is generally a cross-linked silicone polymer, coated as a thin film on the inner wall of the capillary: at normal operating temperatures, this behaves in a similar manner to a liquid film but is far more robust. The mobile phase ('carrier gas') is usually nitrogen or helium. Selective separation is achieved as a result of the differential partitioning of individual compounds between the carrier gas and silicone polymer phases. The separation of most biomolecules is influenced by the temperature of the column, which may be constant during the analysis ('isothermal' – usually 50–250 °C) or, more commonly, may increase in a pre-programmed manner (e.g. from 50 °C to 250 °C at 10 °C per min). Samples are injected onto the 'top' of the column, through a sample injection port containing a gas-tight septum. The output from the column can be monitored by flame ionisation, electron capture or thermal conductivity (Chapter 51).

Spectrometric detection systems include mass spectrometry (GC-MS) and infrared spectroscopy (GC-IR). GC can only be used with samples capable of volatilisation at the operating temperature of the column, e.g. short chain fatty acids. Other substances may need to be chemically modified to produce more volatile compounds, e.g. long chain saturated fatty acids (Chapter 57) are usually analysed as methyl esters while monosaccharides (Chapter 58) are converted to their trimethylsilyl derivatives.

Separation methods

Adsorption chromatography

This is a form of solid–liquid chromatography. The stationary phase is a porous, finely divided solid which adsorbs molecules of the test substance on its surface owing to dipole–dipole interactions, hydrogen bonding and/ or van der Waals interactions (Fig. 50.6). The range of adsorbents is limited, e.g. polystyrene-based resins (for non-polar molecules), silica, aluminium oxide and calcium phosphate (for polar molecules). Most adsorbents must be activated by heating to 110–120°C before use, since their adsorptive capacity is significantly decreased in the presence of bound water. Adsorption chromatography can be carried out in column or thin-layer form, using a wide range of organic solvents.

Partition chromatography

This is based on the partitioning of a substance between two liquid phases, in this instance the stationary and mobile phases. Substances that are more soluble in the mobile phase will pass rapidly through the system while those which favour the stationary phase will be retarded (Fig. 50.7). In normal phase partition chromatography the stationary phase is a polar solvent, usually water, supported by a solid matrix (e.g. cellulose fibres in paper chromatography) and the mobile phase is an immiscible, non-polar organic solvent. For reverse-phase partition chromatography the stationary phase is a non-polar solvent (e.g. a C18-hydrocarbon, such as octadecylsilane) which is chemically bonded to a porous support matrix (e.g. silica), while the mobile phase can be chosen from a wide range of polar solvents, usually water or an aqueous buffered solution containing one or more organic solvents, e.g. acetonitrile. Solutes interact with the stationary phase through non-polar interactions and so the *least* polar solutes elute last from the column. Solute retention and separation are controlled by changing the stationary phase is a cation exchange resin

sample cations displace mobile counterions (e.g. H+) and are retained on the column

anionic/neutral solutes are not retained

flow of mobile phase

Fig. 50.8 Ion-exchange chromatography (cation exchanger).

small molecules have access to the whole pore space, taking longest to pass through the column

medium-sized molecules may have access to part of the pore space

stationary phase is a microporous gel

large molecules are totally excluded from the pores, passing rapidly through the column

flow of mobile phase

Fig. 50.9 Gel permeation chromatography.

composition of the mobile phase (e.g. % v/v acetonitrile). Reverse-phase high-performance liquid chromatography (RP-HPLC) is used to separate a broad range of non-polar, polar and ionic biomolecules, including peptides, proteins, oligosaccharides and vitamins for ionic and ionisable solutes.

Ion-exchange chromatography (IEC)

Here, separations are carried out using a column packed with a porous matrix which has a large number of ionised groups on its surfaces, i.e. the stationary phase is an ion-exchange resin. The groups may be cation or anion exchangers, depending upon their affinity for positive or negative ions. The net charge on a particular resin depends on the pK_a of the ionisable groups and the pH of the solution, in accordance with the Henderson–Hasselbalch equation (p. 171).

For most practical applications, you should select the ion-exchange resin and buffer pH so that the test substances are strongly bound by electrostatic attraction to the ion-exchange resin on passage through the system, while the other components of the sample are rapidly eluted (Fig. 50.8). You can then elute the bound components by raising the salt concentration of the mobile phase, either step-wise or as a continuous gradient, so that exchange of ions of the same charge occurs at oppositely charged sites on the stationary phase. Weakly bound sample molecules will elute first, while more strongly bound molecules will elute at a higher concentration. Computer-controlled gradient formers are available: if two or more components cannot be resolved using a linear salt gradient, an adapted gradient can be used in which the rate of change in salt concentration is decreased over the range where these components are expected to elute. To maximise resolution in IEC (and HIC) keep your columns as short as possible. Once the sample components have been separated, they should be eluted as quickly as possible from the column to avoid band broadening resulting from diffusion of sample ions in the mobile phase.

IEC can be used to separate mixtures of a wide range of ionic biomolecules, including amino acids, peptides, proteins and nucleotides. Electrophoresis (Chapters 52 and 53) is an alternative means of separating charged biomolecules.

Gel permeation chromatography (GPC) or gel filtration

Here, the stationary phase is in the form of beads of a cross-linked gel containing pores of a discrete size (Fig. 50.9). The size of the pores is controlled so that at the molecular level, the pores act as 'gates' that will exclude large molecules and admit smaller ones. However, this gating effect is not an all or nothing phenomenon: molecules of intermediate size partly enter the pores. A column packed with such beads will have within it two effective volumes that are potentially available to sample molecules in the mobile phase, i.e. V_i, the volume surrounding the beads, and V_{ii}, the volume within the pores. If a sample is placed at the top of such a column, the mobile phase will carry the sample components down the column, but at different rates according to their molecular size. A very large molecule will have access to all of V_i but to none of V_{ii}, and will therefore elute in the minimum possible volume (the 'void volume', or V_o, equivalent to V_i). A very small molecule will have access to all of V_i and all of V_{ii}, and therefore it has to pass through the total liquid volume of the column (V_t, equivalent to $V_i + V_{ii}$) before it emerges. Molecules of intermediate size have access to all of V_i but only part of V_{ii}, and will elute at a volume between V_o and V_t, in order of decreasing size depending on their access to V_{ii}.

Table 50.1 Fractionation ranges of selected gel permeation chromatography media.

M_r	Medium
50–1000	Sephadex G15
	Bio-gel P-2
1000–5000	Sephadex G-25
1500–30 000	Sephadex G-50
	Bio-gel P-10
4000–150 000	Sephadex G-100
5000–250 000	Sephadex G-200
20 000–1 500 000	Sephacryl S 300
60 000–20 000 000	Sepharose 4B

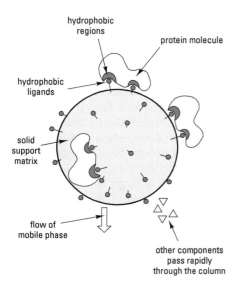

Fig. 50.10 Hydrophobic interaction chromatography. Hydrophobic interactions between ligands and hydrophobic amino acid residues release 'structured' water, making the interactions thermodynamically favourable.

Using HIC for purification – HIC can be used immediately after salt precipitation, when the salt concentration is high. If the desired biomolecule is then eluted using a gradient of reducing salt concentration, it may be possible to follow HIC directly by IEC (which requires an initial buffer of low ionic strength) without changing the buffer.

Cross-linked dextrans (e.g. Sephadex), agarose (e.g. Sepharose) and Polyacrylamide (e.g. Bio-gel) can be used to separate mixtures of macromolecules, particularly enzymes, antibodies and other globular proteins (see Table 50.1 for details). Selectivity in GPC is solely dependent on the stationary phase, with the mobile phase being used solely to transport the sample components through the column. Thus, it is possible to estimate the molecular mass of a sample component by calibrating a given column using molecules of known molecular mass and similar shape. A plot of elution volume (V_e) against $\log_{10}$ molecular mass is approximately linear. A further application of GPC is the general separation of low molecular mass and high molecular mass components, e.g. 'desalting' a protein extract by passage through a Sephadex G-25 column is faster and more efficient than dialysis.

Hydrophobic interaction chromatography (HIC)

This technique is used to separate proteins, and exploits the fact that many proteins have hydrophobic sites, with hydrophobic amino acid residues (e.g. leucine, isoleucine, valine, phenylalanine) on their surfaces. Proteins differ in the nature and extent of these hydrophobic regions. The underlying principle is similar to that of RP-HPLC in that it involves hydrophobic interactions between the sample components and a non-polar stationary phase. However, RP-HPLC is only useful for analytical separations of proteins where retention of biological activity is not required, and is unsuitable for separation of native proteins for several reasons:

- the stationary phase in RP-HPLC columns tends to be densely packed with hydrophobic groups, leading to tight protein binding, possibly through multi-site attachment;
- the use of polar organic solvents may lead to protein denaturation;
- the high pressures used to obtain rapid flow rates in HPLC may also denature proteins.

The groups used on HIC stationary phases are both less densely packed and less hydrophobic than those used in RP-HPLC, and this results in milder adsorption of proteins (octyl or phenyl groups are commonly used in HIC, rather than octadecyl groups). Furthermore, retention and elution can be achieved using aqueous solutions so that an individual protein can be isolated with its 3D conformation intact.

Separations are based on interactions between the three components of the system, i.e. the hydrophobic stationary phase, the hydrophobic sample molecules and the aqueous stationary phase. In an aqueous environment, hydrophobic groups tend to associate, and this results in certain proteins binding to the stationary phase, where the strength of binding is related to the degree of hydrophobicity of the protein (Fig. 50.10). This tendency is promoted by the presence of certain salts, most commonly ammonium sulfate, that produce 'salting out' effects. Elution of components can be achieved by a variety of means:

- reducing the ammonium sulfate concentration of the mobile phase, to decrease the 'salting out' effect;
- changing the salt in the mobile phase to one that does not promote salting out;

- including non-ionic detergents (e.g. Triton X-100) to reduce hydrophobic interactions;
- including aliphatic alcohols, reducing the polarity of the mobile phase; changing the pH;
- reducing the temperature.

Affinity chromatography (AC)

Affinity chromatography enables biomolecules to be purified on the basis of their biological specificity rather than by differences in physicochemical properties, and a high degree of purification (>1000-fold) can be expected. It is especially useful for isolating small quantities of material from large amounts of contaminating substances. The technique involves the immobilisation of a complementary binding substance (the ligand) onto a solid matrix in such a way that the specific binding affinity of the ligand is preserved. When a biological sample is applied to a column packed with this affinity support matrix, the molecule of interest will bind specifically to the ligand, while contaminating substances will be washed through with buffer (Fig. 50.11). Elution of the desired molecule can be achieved by changing the pH or ionic strength of the buffer, to weaken the non-covalent interactions between the molecule and the ligand, or by the addition of other substances that have greater affinity for the ligand.

In AC, the ligand must show specific, but reversible, interaction with the molecule to be purified. It must also contain a reactive functional group, independent of the biospecific site, that will allow covalent attachment to the matrix. The support matrix must be free of non-specific adsorption effects and have sufficient reactive functional groups for the attachment of ligands. Agarose is an ideal support matrix for use in AC. If the ligand is small and the molecule to be purified is large, binding may be restricted owing to the proximity of the matrix surface. This problem may be overcome by the introduction of a 'spacer arm' (e.g. a hexane group) between the ligand and the matrix (Fig. 50.11).

A potential disadvantage of the specificity of AC is that a new ligand may have to be sought for each individual separation – a potentially time-consuming and expensive process. It may be more practical to use 'group-specific adsorbents' which contain ligands that have an affinity for a class of biochemically related substances. Examples of group-specific adsorbents include the following:

Fig. 50.11 Affinity chromatography.

ligand
spacer arm
complementary molecule bound by specific interaction with ligand
solid support matrix
flow of mobile phase
all other solutes are eluted

Examples Biospecific molecules used in affinity chromatography include:
- enzymes and inhibitors/cofactors/substrates;
- hormones and receptor proteins;
- antibodies and antigens;
- complementary base sequences in DNA and RNA (p. 399).

- **Lectins,** which are a group of proteins produced from moulds, plants and animals, that bind reversibly with specific sugar residues. They are very useful for the purification of sugar-containing macro-molecules such as glycoproteins, serum lipoproteins and membrane proteins such as receptors. Different lectins show different specificities, e.g. concanavalin A and lentil lectin bind to sugars having –OH groups at C-3, C-4 and C-5 (i.e. mannose and glucose), while wheatgerm lectin binds to *N*-acetyl glucosamine residues. Once bound, substances can be resolved by eluting the column with a salt gradient, or by including free sugars to act as competitive binding agents. Lectins are sometimes called agglutinins because of their ability to agglutinate different types of erythrocyte (e.g. A, B and O) by binding to the specific receptors on their surfaces.

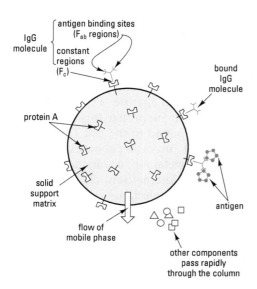

Fig. 50.12 Affinity chromatography using protein A as a ligand to bind IgG antibodies via the F_c region.

Fig. 50.13 Covalent chromatography. A protein with a free surface thiol group reduces the ligand disulfide, releasing the blocking group and forming a new disulfide.

- **Protein A,** which is a commercially available surface protein from the bacterium *Staphylococcus aureus* that has a specific binding capacity for the constant (F_c) region of IgG-type antibodies from most mammalian species (Fig. 50.12). As well as being useful in the purification of IgG, protein A can also be used for the isolation of other molecules, as long as an antibody can be raised against them. This is possible because interaction of protein A with IgG occurs via the F_c segment of immunoglobulins of this class, irrespective of their F_{ab} components (see Fig. 43.1). Immobilisation of a specific IgG antibody onto a protein A-agarose column produces an affinity support that can be used to purify the desired antigen.

- **Immobilised dyes,** which can be used to purify a wide range of enzymes and proteins. One of the most widely used dyes is Cibacron Blue F3G-A, available as an affinity support as Blue Sepharose. This dye has some structural similarities to nucleotide cofactors such as NAD^+ and $NADP^+$, and is useful for the purification of enzymes that require such cofactors. Chemical modification of such dyes increases their specificity, and it is likely that computer modelling will play a role in the future development of new dye-based affinity supports. Dyes are also used in the related non-chromatographic technique of affinity precipitation. This makes use of two dye molecules linked via a spacer molecule; when this 'bis-ligand' is added to a solution containing the molecule to be purified, the ligand specifically binds two desired molecules, forming an insoluble complex. Alternatively, a heterobifunctional ligand can be used – a combination of a specific ligand and a group that can be used to initiate precipitation once the desired molecule has bound to the ligand. Such interactions are equivalent to the precipitin reaction in immunology (p. 282).

- **Poly(U)-agarose** is an affinity support that can be used for the isolation of mRNA because of the biospecific hybridisation of poly(U) with the poly(A) 'tail' sequence characteristic of mRNAs. It can also be used to isolate proteins and enzymes that bind to RNA, such as reverse transcriptases.

Covalent chromatography

This is a variant of affinity chromatography which involves the formation of fairly strong, but reversible, covalent bonds between the affinity support and the molecule to be purified. One type of covalent chromatography is used for the purification of proteins containing thiol (–SH) groups (Fig. 50.13). If the sample is applied to a column containing matrix-attached disulfide 2'-pyridyl groups, thiol-containing proteins displace the 2'-pyridyl group on the support and become immobilised via disulfide bridges. Elution of bound components can be achieved by including thiol reagents such as cysteine, glutathione, 2-mercaptoethanol or dithiothreitol (DTT) in the mobile phase. Another form of covalent chromatography uses immobilised boronic acid, which binds certain carbohydrate groups (e.g. in glycoproteins).

Immobilised metal affinity chromatography (IMAC)

This exploits the ability of certain metal ions, especially Ni^{2+}, Cu^{2+}, Zn^{2+}, Hg^{2+} and Cd^{2+}, to bind to proteins by forming coordination complexes with imidazole groups of histidine, indole groups of tryptophan, or thiol groups of cysteine residues. The metal ion may be immobilised by chelation with iminodiacetic acid covalently bound to agarose. Proteins that bind to the metal ions can be eluted using free metal-binding ligands (e.g. amino acids) in the mobile phase.

A very useful application of IMAC is in the purification of 'histidine-tagged' recombinant proteins, e.g. using pET plasmid vectors (see p. 476). The technique exploits the affinity of histidine-containing proteins for a chelating gel, typically Ni^{2+} immobilised on agarose. The histidine tag present at the *N*- or *C*-terminus of a recombinant protein forms complexes with the transition metal and is therefore bound to the gel. The proteins can be desorbed (eluted) from the gel with increasing concentration of a substrate with an affinity for the chelated metal ion, e.g. imidazole, using either a gradient or a step-wise approach. Alternatively, chelating agents (e.g. EDTA) or low pH can be used for protein elution.

Optimising chromatographic separations

In any chromatographic technique, sample components leave the column at different elution volumes/different times, and are then monitored by a suitable detector (Chapter 51). The responses of the detector are recorded on a chart or a screen in the form of a *chromatogram*. Ideally the sample biomolecules will be completely separated and detection of components will result in a series of discrete individual peaks corresponding to each type of biomolecule (e.g. Fig. 50.3). However, to minimise the possibility of overlapping peaks, or of peaks composed of more than one component, it is important to maximise the separation efficiency of the technique, which depends on:

- **selectivity,** as measured by the relative retention times of the two components, or by the volume of mobile phase between the peak maxima of the two components after they have passed through the column; this depends on the ability of the chromatographic method to separate two components with similar properties;
- **band-broadening properties of the chromatographic system,** which influence the width of the peaks; these are mainly due to the effects of diffusion.

The separation efficiency, or resolution of two adjacent components, can be defined in terms of selectivity and peak broadening, using Eqn [50.4]. In practical terms, good resolution is achieved when there is a large 'distance' (either time or volume) between peak maxima, and the peaks are as narrow as possible. The resolution of components is also affected by the relative amount of each substance: for systems showing low resolution, it can be difficult to resolve small amounts of a particular component in the presence of larger amounts of a second component. If you cannot obtain the desired results from a poorly resolved chromatogram, other chromatographic conditions, or even different methods, should be tried in an attempt to improve resolution. For liquid chromatography, changes in the following factors may improve resolution:

- **Stationary phase particle size** – the smaller the particle, the greater the area available for partitioning between the mobile phase and the stationary phase. This partly accounts for the high resolution observed with HPLC and FPLC compared with low-pressure methods.
- **The slope of the salt gradient** in eluting IEC or HIC columns, e.g. using computer-controlled adapted gradients (p. 405).

Learning from experience – if you are unable to separate your biomolecule using a particular method, do not regard this as a failure, but instead, think about what this tells you about either the substance(s) or the sample.

In low-pressure liquid chromatography, the flow rate of the mobile phase must be optimised because this influences two band-broadening effects which are dependent on diffusion of sample molecules, i.e. (i) the flow rate must be slow enough to allow effective partitioning between the mobile phase and the stationary phase, and (ii) it must be fast enough to ensure that there is minimal diffusion along the column once the molecules have been separated. To allow for these opposing influences, a compromise flow rate must be used. If you prepare your own columns, they must be packed correctly, with no channels present that might result in uneven flow and eddy diffusion.

Text references

Stahl, E. (1969) *Thin Layer Chromatography – a Laboratory Handbook,* 2nd edn. Springer-Verlag, Berlin.

Touchstone, J.C. (1992) *Practice of Thin Layer Chromatography.* Wiley, New York.

Sources for further study

Bliesner, D.M. (2006) *Validating Chromatographic Methods: a Practical Guide.* Wiley, New York.

Boyer, R.F. (2011) *Biochemistry Laboratory: Modern Theory and Techniques,* 2nd edn. Benjamin Cummings, San Francisco.
[Also covers other topics, including centrifugation and spectroscopy]

Cazes, J. (2009) *Encyclopedia of Chromatography,* 3rd edn. CRC Press, Boca Raton.

McMaster, M. (2005) *LC-MS – a Practical User's Guide.* Wiley, New York.
[Covers principles and practice of HPLC systems and MS detectors.]

McMaster, M. (2009) *HPLC: a Practical User's Guide.* Wiley, New York.

McNair, H.M. and Miller, J.M. (2009) *Basic Gas Chromatography,* 2nd edn. Wiley, New York.

Meyer, V. (2010) *Practical High-Performance Liquid Chromatography.* Wiley, New York.

Miller, J.M. (2005) *Chromatography: Concepts and Contrasts,* 2nd edn. Wiley, New York.

Wellings, D.A. (2006) *A Practical Handbook of Preparative HPLC.* Elsevier, Amsterdam.
[Gives advice on column preparation and process optimisation.]

Wilson, K. and Walker, J. (eds) (2010) *Principles and Techniques of Practical Biochemistry and Molecular Biology,* 7th edn. Cambridge University Press, Cambridge.
[Also covers a range of other topics, including spectroscopy and centrifugation.]

STUDY EXERCISES

50.1 Calculate R_F and R_X values from a chromatogram. The figure represents the separation of three pigments by thin-layer chromatography.

(a) What is the R_F value of each pigment?

(b) What is the mobility of pigments B and C, relative to pigment A (R_A)?

Express both answers to three significant figures.

Thin-layer chromatographic separation of a mixture of three pigments (A, B and C).

50.2 Calculate the resolution and selectivity of two components from a chromatogram. Two compounds were separated by column chromatography, giving retention times of 4 min 30 s for A and 6 min 12s for B, while a compound that was completely excluded from the stationary phase was eluted in 1 min 35s. The base width of peak A was 40 s and the base width of peak B was 44s. Calculate (a) the selectivity and (b) the resolution for these two compounds (express all answers to three significant figures).

50.3 Predict chromatographic behaviour on ion-exchange columns using protein titration curves. The variation of net charge with pH for two proteins, A and B, is shown in the figure.

(a) What is the isoelectric point (pI) of each protein?

(b) If a mixture of these proteins was passed down an anion exchange column (e.g. Q-Sepharose) equilibrated with buffer at pH 7.0, how would each protein behave in terms of elution and retention?

(c) If the proteins were bound to a cation exchanger (e.g. S-Sepharose) at pH 5.5, predict the order in which the proteins would elute using a salt gradient.

50.4 Test your understanding of the action of ammonium sulfate on proteins. Explain why mobile phases containing a concentration of ammonium sulfate promote the binding of many proteins to HIC columns.

Once the individual components in a sample mixture have been separated by a given chromatographic technique (Chapter 50), a suitable detection system is required to monitor and record the elution of the components of the mixture as they pass through a 'flow cell'.

KEY POINT The most appropriate detector depends on the type of chromatography and the application: ideally, the detector should show high sensitivity, a low detection limit and minimal noise or drift. These terms are defined in Chapter 55.

Liquid chromatography detectors

UV/visible detectors

These are widely used and have the advantages of versatility, sensitivity and stability. Such detectors are of two types: fixed wavelength and variable wavelength. Fixed wavelength detectors are simple to use, with low operating costs. They usually contain a mercury lamp as a light source, emitting at several wavelengths between 254 and 578 nm; a particular wavelength is selected using suitable cut-off filters. The most frequently used wavelengths for analysis of biomolecules are 254 nm (for nucleic acids) and 280 nm (for proteins). Variable wavelength detectors use a deuterium lamp and a continuously adjustable monochromator for wavelengths of 190–600 nm. For both types of detector, sensitivity is in the absorbance range 0.001–1.0 (down to $\cong 1$ ng), with noise levels as low as 4×10^{-5}. Note that sensitivity is partly influenced by the path length of the flow cell (typically 10 mm). Monitoring at short wavelength UV (e.g. below 240 nm) may give increased sensitivity but decreased specificity, since many biological molecules absorb in this range. Additional problems with short wavelength UV detection include instrument instability, giving a variable baseline, and absorption by components of the eluting buffer (e.g. TRIS, which absorbs at 206 nm).

An important development in chromatographic monitoring is diode array detection (DAD). The incident light comprises the whole spectrum of light from the source, which is passed through a diffraction grating and the diffracted light detected by an array of photodiodes. A typical DAD can measure the absorbance of each sample component at 1–10-nm intervals over the range 190–600 nm. This gives an absorbance spectrum for each eluting substance which may be used to identify the compound and give some indication as to its purity. An example of a three-dimensional diode array spectrum is shown in Fig. 51.1.

Fluorescence detectors

Many biomolecules, including some vitamins, nucleotides and porphyrins, show natural fluorescence (Table 47.1), or can be made to fluoresce by pre-column or post-column derivatisation with a fluorophore. Fluorescence detection is more sensitive than UV/visible detection (p. 412), and may allow analysis in the picogram range. A fluorescence detector consists of a light source (e.g. a xenon lamp), a diffraction grating to supply light at the excitation wavelength and a photomultiplier to monitor the emitted

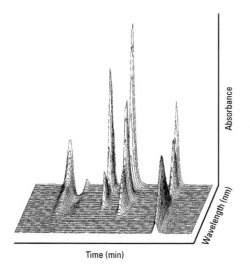

Fig. 51.1 Diode array detector absorption spectra of the eluent from an HPLC separation of a mixture of four steroids, taken every 15 s.

electrical output

platinum anode

platinum cathode

flame

jet assembly

air input

FID body

hydrogen input

gas flow from column

Fig. 51.2 Components of a flame ionisation detector (FID).

light (usually arranged to be at right angles to the excitation beam). The use of instruments with a laser light source can give an extremely narrow excitation waveband, and increased sensitivity and specificity.

Electrochemical detectors

These offer very high sensitivity and specificity, with the possibility of detection of femtogram amounts of electroactive compounds such as catecholamines, vitamins, thiols, purines, ascorbate and uric acid. The two main types of detector, amperometric and coulometric, operate on similar principles, i.e. by measuring the change in current or potential as sample components pass between two electrodes within the flow cell. One of these electrodes acts as a reference (or counter) electrode (e.g. calomel electrode), while the other – the working electrode – is held at a voltage that is high enough to cause either oxidation or reduction of sample molecules. In the oxidative mode, the working electrode is usually glassy carbon, while in reductive mode a mercury electrode is used. In either case, a current flow between the electrodes is induced and detected.

Gas chromatography detectors

The most commonly used detectors for GC analysis of biomolecules are as follows.

Flame ionisation detector (FID)

This is a widely used detector, being particularly useful for the analysis of a broad range of organic biomolecules. It involves passing the exit gas stream from the column through a hydrogen flame that has a potential of >100 V applied across it (Fig. 51.2). Most organic compounds, on passage through this flame, produce ions and electrons that create a small current across the electrodes, and this is amplified for measurement purposes. The FID is very sensitive (down to ≈ 0.1 pg), with a linear response over a wide concentration range. One drawback is that the sample is destroyed during analysis.

Thermal conductivity detector (TCD)

This simple detector is based on changes in the thermal conductivity of the gas stream brought about by the presence of separated sample molecules. The detector elements are two electrically heated platinum wires, one in a chamber through which only the carrier gas flows (the reference detector cell), and the other in a chamber that takes the gas flow from the column (the sample detector cell). In the presence of a constant gas flow, the temperature of the wires (and therefore their electrical resistance) is dependent on the thermal conductivity of the gas. Analytes in the gas stream are detected by temperature-dependent changes in resistance that are dependent on the thermal conductivity of each separated molecule; the size of the signal is directly related to concentration of the analyte.

The advantages of TCD include its applicability to a wide range of organic and inorganic biomolecules and its non-destructive nature, since the sample can be collected for further study. Its major limitation is its low sensitivity (down to ≈10 ng), compared with other systems.

Fig. 51.3 Components of an electron capture detector (ECD).

Chromatographic systems used in conjunction with mass spectrometry – this approach has helped to identify and characterise hundreds of components present in complex biological systems, including flavour molecules in food, drug metabolites and pollutants in water.

Electron capture detector (ECD)

This highly sensitive detector (Fig. 51.3) is useful for the detection of certain compounds with electronegative functional groups, e.g. halogens, peroxides and quinones. The gas stream from the column passes over a β-emitter (p. 297) such as ^{63}Ni, which provides electrons that cause ionisation of the carrier gas (e.g. nitrogen). When carrier gas alone is passing the β-emitter, its ionisation results in a constant current flowing between two electrodes placed in the gas flow. However, when electron-capturing sample molecules are present in the gas flow, a decrease in current is detected. An example of the application of the ECD is in detecting and quantifying chlorinated pesticides.

Interfacing GC or HPLC with mass spectrometry (MS)

Mass spectrometry (p. 323) used in conjunction with chromatographic methods can provide a powerful tool for identifying the components of complex mixtures, e.g. aqueous pollutants. The procedure requires computer control of instruments and data storage/analysis (Fig. 51.4). One drawback is the limited capacity of the mass spectrometer – owing to its vacuum requirements – compared with the volume of material leaving the chromatography column. For capillary GC, the relatively small output can be fed directly into the ionisation chamber of the mass spectrometer. For packed column GC, a 'jet separator' is used to remove most of the carrier gas from the molecules to be analysed before they enter the spectrometer. Similarly, in HPLC, devices have been developed for solving the problem of large solvent volumes, e.g. by splitting the eluent from the column so that only a small fraction reaches the mass spectrometer.

The computer-generated outputs from the mass spectrometer are similar to chromatograms obtained from other methods, and show peaks corresponding to the elution of particular components. However, it is then possible to select an individual peak and obtain a mass spectrum for the component in that peak to aid in its identification (p. 323).

Coupling capillary GC columns with FT-IR spectrometers (pp. 319–20) provides another powerful means of separating and identifying compounds in complex biological mixtures.

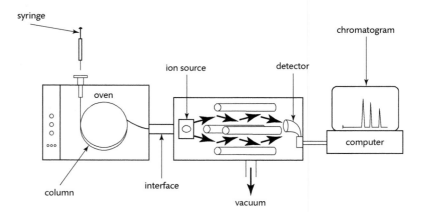

Fig. 51.4 Schematic diagram of a GC-MS instrument.

Recording and interpreting chromatograms

Recording detector output

For analytical purposes, the detector output is usually connected to a computer-based data acquisition and analysis system. This consists of a personal computer (PC) with data acquisition hardware to convert the analogue detector signal to digital format, plus software to control the data acquisition process, store the signal information and display the resulting chromatogram. The software will also detect peaks and calculate their retention times and sizes (areas) for quantitative analysis. The software often incorporates functions to control the chromatographic equipment, enabling automatic operation. In sophisticated systems, the detector output may be compared with that from a 'library' of chromatograms for known compounds, to suggest possible identities of unknown sample peaks.

In simpler chromatographic systems, you may need to use a chart recorder for detector output. Two important settings must be considered before using a chart recorder:

1. **The baseline reading** – this should be set only after a suitable quantity of mobile phase has passed through the column (prior to injection of the sample) and stability is established. The chart recorder is usually set a little above the edge of the chart paper grid, to allow for any baseline drift.

2. **The detector range** – this must be set to ensure that the largest peaks do not go off the top of the chart. Adjustment may be based on the expected quantity of analyte, or by trial-and-error process. Use the maximum sensitivity that gives intact peaks. If peaks are still too large, you may need to reduce the amount of sample used, or prepare and analyse a diluted sample.

Interpretation of chromatograms

Make sure you know the direction of the horizontal axis (usually, either volume or time) – it may run from right to left or vice versa – and make a note of the detector sensitivity on the vertical axis.

Ideally, the baseline should be 'flat' between peaks, but it may drift up or down owing to a number of factors, including:

- **air bubbles (in liquid chromatography);** if the buffers used in the mobile phase are not effectively degassed, air bubbles may build up in the flow cell of the detector, leading to a gradual upward drift of the baseline, followed by a sharp fall when the accumulated air is released. Small air bubbles that do not become trapped may give spurious small peaks.

- **changes in the composition of the mobile phase** (e.g. in gradient elution, p. 334);

- **tailing of material from previous peaks;**

- **carry-over of material from previous samples;** this can be avoided by efficient cleaning of columns between runs – allow sufficient time for the previous sample to pass through the column before you introduce the next sample;

- **loss of stationary phase from the column** (column 'bleed'), caused by extreme elution conditions.

> **Interpreting chromatograms** – *never* assume that a single peak is a guarantee of purity: there may be more than one compound with the same chromatographic characteristics.

> **Using a fraction collector** – make sure you can relate individual fractions to the position of peaks on the chromatogram. Most fraction collectors send a signal to the recorder each time a fraction is changed.

> **Avoiding problems with air bubbles in liquid chromatography** – always ensure that buffers are effectively degassed by vacuum treatment before use, and regularly clean the flow cell of the detector.

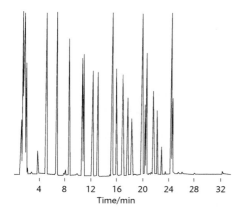

Fig. 51.5 A multicomponent chromatogram. Separation of many compounds, some that are well resolved, e.g. peaks at 12–13 min and others that are not, e.g. peaks at 24–25 min.

Using an external standard – samples and standards should be analysed more than once, to confirm the reproducibility of the technique.

Using an internal standard – you should add the internal standard to the sample at the first stage in the extraction procedure, so that any loss or degradation of test substance during purification is accompanied by an equivalent change in the internal standard, as long as the extraction characteristics of the internal standard and the test substance are very similar.

Quantifying biomolecules – note that quantitative chromatographic analysis often requires assumptions about the *identity* of separated components (p. 379) and that further techniques may be required to provide information about the *nature* of the biomolecules present, e.g. mass spectrometry (see Chapter 48).

A peak close to the origin is likely to be caused by non-retained sample molecules, flowing at the same rate as the mobile phase, or to artefacts, e.g. air (GC) or solvent (HPLC) in the sample. Whatever its origin, this peak can be used to measure the column dead time or column void volume (p. 333). No peaks from genuine sample components should appear before this peak.

Peaks can be denoted on the basis of their elution volume (used mainly in liquid chromatography) or their retention times, as in Fig. 51.5 (mainly in GC). If the peaks are not narrow and symmetrical, they may contain more than one component. Where peaks are more curved on the trailing side compared with the leading side, this may indicate too great an association between the component and the stationary phase, or may result from overloading of the column.

Quantitative analysis

Most detectors and chemical assay systems give a linear response with increasing amounts of the test substance over a given 'working range' of concentration. Alternative ways of converting the measured response to an amount of substance are as follows:

- **External standardisation:** this is applicable where the sample volume is sufficiently precise to give reproducible results (e.g. HPLC, column chromatography). You measure the peak areas (or heights) of known amounts of the substance to give a calibration factor or calibration curve which can be used to calculate the amount of test substance in the sample.

- **Internal standardisation:** where you add a known amount of a reference substance (not originally present in the sample) to the sample, to give an additional peak in the elution profile. You determine the response of the detector to the test and reference substances by analysing a standard containing known amounts of both substances, to provide a response factor (r), where

$$r = \frac{\text{peak area (or height) of test substance}}{\text{peak area (or height) of reference substance}} \qquad [51.1]$$

Use this response factor to quantify the amount of test substance (Q_t) in a sample containing a known amount of the reference substance (Q_r), from the relationship:

$$Q_t = \frac{\text{peak area (or height) of test substance}}{\text{peak area (or height) of reference substance}} \times (Q_r/r) \qquad [51.2]$$

Internal standardisation should be the method of choice wherever possible, since it is unaffected by small variations in sample volume (e.g. for GC microsyringe injection). The internal standard should be chemically similar to the test substance(s) and must give a peak that is distinct from all other substances in the sample. An additional advantage of an internal standard which is chemically related to the test substance is that it may show up problems caused by changes in detector response, incomplete derivatisation, etc. A disadvantage is that it may be difficult to fit an internal standard peak into a complex chromatogram.

Sources for further study

Bouchonnet, S. (2013) *Introduction to GC-MS Coupling*. CRC Press, Boca Raton.

Cazes, J. (2009) *Encyclopedia of Chromatography*, 3rd edn. CRC Press, Boca Raton.

Grob, R.L. and Barry, E.F. (2004) *Modern Practice of Gas Chromatography*, 4th edn. Wiley, New York.

McKay, P. *An Introduction to Chromatography*. Available: http://www.accessexcellence.org/LC/SS/ chromatography_background.php Last accessed 01/09/15.

McNair, H.M. and Miller, J.M. (2009) *Basic Gas Chromatography*. Wiley, New York.

Miller, J.M. (2005) *Chromatography: Concepts and Contrasts*, 2nd edn. Wiley, New York.

Scott, R.P.W (1996) *Chromatography Detectors: Design, Function and Operation*. Marcel Dekker, New York.

Sparkman, O.D., Penton, Z. and Kitson, F.G. (2011) *Gas Chromatography and Mass Spectrometry: A Practical Guide*, 2nd edn. Academic Press, Oxford.

STUDY EXERCISES

51.1 Test your knowledge of detector terminology. Explain what the following acronyms stand for:

(a) FID;

(b) TCD;

(c) ECD;

(d) DAD.

51.2 Check your knowledge of liquid chromatography detectors. Make a list of the various major types of liquid chromatography detectors in order, from highest to lowest sensitivity. Which of these methods is most versatile and why?

51.3 Consider how fluorescence detection can be applied to non-fluorescent molecules. Fluoresence detection of natural molecules is limited to molecules such as porphyrins, nucleotides and some vitamins, e.g. riboflavin. What procedures can convert a non-fluorescent molecule into a form suitable for sensitive fluorescence detection? How does this increase the selectivity of the analysis?

51.4 Calculate the amount of substance in a chromatographic separation using an internal standard. The chromatogram shown in the figure on the right represents the separation of three carbohydrates by gas chromatography. Peak A corresponds to mannitol, peak B to sucrose and peak C to the internal standard, trehalose, at 1.50 mg. The results shown in the table were obtained for the retention times and areas of the three peaks.

Gas–liquid chromatographic separation of three carbohydrates (A, B and C). Note that the plotter shows peaks A and B to be off-scale; however, the instrument still gives a valid area measurement for these two peaks (see table).

Retention times and peak areas for carbohydrates A, B and C.

Peak	Retention time (min)	Area (relative)
A	3.92	2060
B	5.82	1898
C	6.03	604

Given a response factor (*r*) of 1.26 for mannitol and 0.92 for sucrose (relative to the internal standard), determine the amount of A and B in the sample (express your answer to three significant figures).

KEY POINT Electrophoresis is a separation technique based on the movement of charged molecules in an electric field. Dissimilar molecules move at different rates and the components of a mixture will be separated when an electric field is applied. It is a widely used technique, particularly for the analysis of complex mixtures or for the verification of purity (homogeneity) of isolated biomolecules.

Understanding electrophoresis – this is, in essence, an incomplete form of electrolysis (p. 369), since the applied electrical field is switched off well before sample molecules reach the electrodes.

While electrophoresis is mostly used for the separation of charged macromolecules, techniques are available for high resolution separations of small molecules such as amino acids (for example, by capillary electrophoresis, p. 361). This chapter deals mainly with the electrophoretic separation of proteins, which has many applications, including clinical diagnosis. However, the principles apply equally to other molecules (separation of nucleic acids is considered in Chapter 66).

The electrophoretic mobility of a charged molecule depends on the following:

- **Net charge** – negatively charged molecules (anions) migrate towards the anode (+), while positively charged molecules (cations) migrate towards the cathode (−); highly charged molecules move faster towards the electrode of opposite charge than those with lesser charge.
- **Size** – frictional resistance exerted on molecules moving in a solution means that smaller molecules migrate faster than large molecules.
- **Shape** – the effect of friction also means that the shape of the molecule will affect mobility, e.g. globular proteins compared with fibrous proteins, linear DNA compared with circular DNA.
- **Electrical field strength** – mobility increases with increasing electrical potential (voltage), but there are practical limitations to using high voltages, especially because of heating effects.

Definitions

Electrical field strength – a measure of the magnitude of an electrical field, usually expressed in terms of electrical potential difference per unit length, as V m^{-1}.

Electrophoretic mobility – the rate of migration of a particular type of molecule in response to an applied electrical field.

The combined influence of net charge and size means that mobility (μ) is determined by the charge: density ratio or the charge: mass ratio, according to the formula:

$$\mu = \frac{qE}{r}$$

[52.1]

where q is the net charge on the molecule, r is the molecular radius and E is the field strength.

Table 52.1 pK$_a$ values of ionisable groups in selected amino acid residues of proteins.

Group/residue	pK$_a$*
Terminal carboxyl	3.1
Aspartic acid	4.4
Glutamic acid	4.4
Histidine	6.5
Terminal amino	8.0
Cysteine	8.5
Tyrosine	10.0
Lysine	10.0
Arginine	12.0

*Note that these are typical values: the pK$_a$ will change with temperature and ionic strength. Acidic residues will tend to be negatively charged at pH values above their pK$_a$, while basic residues will tend to be positively charged below their pK$_a$.

Electrophoresis and the separation of proteins

The net charge of a sample molecule determines its direction of movement and significantly affects its mobility. The net charge of a protein molecule is pH-dependent, and is determined by the relative numbers of positively and negatively charged amino acid side chains at a given pH (Table 52.1). The degree of ionisation of each amino acid side chain is pH-dependent, resulting in a variation of net charge on the protein at different pH values (Fig. 52.1). Since an individual protein will have a unique content of ionisable amino acids, each protein will have a characteristic 'titration curve' when net charge is plotted against pH. Thus, electrophoresis is

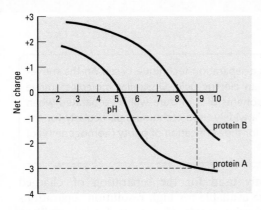

Fig. 52.1 Titration curves for two proteins, A and B, containing different proportions of acidic and basic amino acid residues.

Fig. 52.2 Apparatus for vertical slab electrophoresis (components move downwards from wells, through the gel matrix).

Minimising diffusion – make the sample zone as narrow as possible, and fix and stain protein bands as soon as possible after the run.

Definition

Ohm's law $V = IR$, where V = voltage, I = current and R = resistance.

Optimising electrophoresis – attempting to minimise heat production using very low currents is not practical, since it leads to long separation times, and therefore to increased diffusion.

---

always carried out at *constant* pH and a suitable buffer must be present along with the sample to maintain that pH (Chapter 27). If the proteins shown in Fig. 52.1 were subjected to electrophoresis at pH 9.0, and if the proteins were of similar size and shape, then the rate at which protein A (net charge, –3) migrates towards the anode would be faster than that for protein B (net charge, –1). Separation of proteins is usually carried out at alkaline pH, where most proteins carry a net negative charge.

Basic apparatus

Most types of electrophoresis using supporting media (described below) are simple to carry out and the apparatus can be easily constructed, although inexpensive equipment is commercially available. High resolution techniques such as 2D-electrophoresis and capillary electrophoresis require more sophisticated equipment, both for separation and analysis (p. 361).

Simple electrophoretic separations can be performed either vertically (Fig. 52.2) or horizontally (Fig. 66.2). The electrodes are normally made of platinum wire, each in its own buffer compartment. In vertical electrophoresis, the buffer solution forms the electrical contact between the electrodes and the supporting medium in which the sample separation takes place. In horizontal electrophoresis, electrical contact can be made by buffer-soaked paper 'wicks' dipping in the buffer reservoir and laid upon the supporting medium. The buffer reservoir normally contains a divider acting as a barrier to diffusion (but not to electrical current), so that localised pH changes which occur in the region of the electrodes (as a result of electrolysis, p. 369) are not transmitted to the supporting medium or the sample. Individual samples are spotted onto a solid supporting medium containing buffer or are applied to 'wells' formed in the supporting medium. The power pack used for most types of electrophoresis should be capable of delivering ≈500 V and ≈100 mA.

Using a supporting medium

The effects of convection currents (resulting from the heating effect of the applied field) and the diffusion of molecules within the buffer solution can be minimised by carrying out the electrophoresis in a porous supporting medium. This contains buffer electrolytes and the sample is added in a discrete location or zone. When the electrical field is applied, individual sample molecules remain in sharp zones as they migrate at different rates. After separation, post-electrophoretic diffusion of selected biomolecules (e.g. proteins) can be avoided by 'fixing' them in position on the supporting medium, e.g. using trichloroacetic acid (TCA).

The heat generated during electrophoresis is proportional to the square of the applied current and to the electrical resistance of the medium: even when a supporting medium is used, heat production will lead to zone broadening by increasing the rate of diffusion of sample components and buffer ions. Heat denaturation of sample proteins may also occur, resulting in loss of biological activity, e.g. with enzymes (p. 415). Another problem is that heat will reduce buffer viscosity, leading to a decrease in resistance. If the electrophoresis is run at constant voltage, Ohm's law dictates that, as resistance falls, the current will increase, leading to further heat production. This can be avoided by using a power pack that provides constant power. In practice, most electrophoresis equipment incorporates a cooling device; even so, distortions of an electrophoretic zone from the ideal 'sharp, linear band' can often be explained by inefficient heat dissipation.

Types of supporting media

These can be subdivided into:

- **inert media** – these provide physical support and minimise convection: separation is based on charge density only (e.g. cellulose acetate);
- **porous media** – these introduce molecular sieving as an additional effect: their pore size is of the same order as the size of molecules being separated, restricting the movement of larger molecules relative to smaller ones. Thus, separation depends on both the charge density and the size of the molecule.

With some supporting media, e.g. cellulose acetate, a phenomenon called electro-endosmosis or electro-osmotic flow (EOF) occurs. This is because of the presence of negatively charged groups on the surface of the supporting medium, attracting cations in the electrophoresis buffer solution and creating an electrical double layer. The cations are hydrated (surrounded by water molecules) and when the electric field is applied, they are attracted towards the cathode, creating a flow of solvent that opposes the direction of migration of anionic biomolecules towards the anode. The EOF can be so great that weakly anionic biomolecules (e.g. antibodies) may be carried towards the cathode.

Where necessary, EOF can be avoided by using supporting media such as agarose or Polyacrylamide (p. 352), but it is not always a hindrance to electrophoretic separation. Indeed, the phenomenon of EOF is used in the high resolution technique of capillary electrophoresis (p. 361).

Cellulose acetate

Acetylation of the hydroxyl groups of cellulose produces a less hydrophilic structure than cellulose in the form of paper: as a result it holds less water and diffusion is reduced, with a corresponding increase in resolution. Cellulose acetate is often used in the electrophoretic separation of plasma proteins in clinical diagnosis – it can be carried out quickly ($\approx$45 min) and its resolution is adequate to detect gross differences in various types of protein (e.g. paraproteins in myeloma). Cellulose acetate has a fairly uniform pore structure and the pores are large enough to allow unrestricted passage of all but the largest of molecules as they migrate through the medium.

Agarose

Agarose is the neutral, linear polysaccharide component of agar (from seaweed), consisting of repeating galactose and 3,6-anhydrogalactose subunits (Fig. 52.3). Powdered agarose is mixed with electrophoresis buffer at concentrations of 0.5–3.0% w/v, boiled until the mixture becomes clear, poured onto a glass plate, then allowed to cool until it forms a gel. Gelation is due to the formation of hydrogen bonds both between and within the agarose polymers, resulting in the formation of pores. The pore size depends on the agarose concentration. Low concentrations produce gels with large pores relative to the size of proteins, allowing them to migrate relatively unhindered through the gel, as determined by their individual charge densities. Low concentrations of agarose gel are suitable for techniques such as immunoelectrophoresis (p. 284) and isoelectric focusing (p. 360), where charge is the main basis of separation. The smaller pores produced by higher concentrations of agarose may result in molecular sieving.

Definition

Electro-osmotic flow – the osmotically driven mass flow of water resulting from the movement of ions in an electrophoretic system.

Handling cellulose acetate – the fragile strips must be carefully handled, avoiding touching the flat surfaces with your fingers.

Fig. 52.3 Structure of agarose. Additional sulfate and pyruvyl groups are attached at selected hydroxyls in the polymer.

How to carry out agarose gel electrophoresis of DNA

1. **Prepare the gel.** Typically, a small volume (10–20 mL) of buffer plus agarose is heated gently until the powder dissolves – take care not to overheat, or it will boil over. Nowadays, gels are often cast with a small amount of a visualising dye, such as SYBR®Safe.

2. **Prepare the samples.** A small amount of sucrose or glycerol is usually added, to increase the density of the sample. A water-soluble anionic 'tracking' dye (e.g. bromophenol blue or xylene cyanol) is also added to each sample, so that migration can be followed visually.

3. **Load the samples onto the gel.** Individual samples are added to the preformed wells using a pipettor (the sample should be retained within the well owing to its higher density, compared with the buffer solution). The volume of sample added to each well is small – typically less than 25 µL, so a very steady hand and careful dispensing are needed to pipette each sample accurately.

4. **Load the DNA markers onto the gel.** Typically, these standards of known size are added to the first and last wells of the gel; after electrophoresis, the relative positions of the bands of known size can be used to

prepare a calibration curve (p. 275), usually by plotting $\log_{10}$ size (length) against distance travelled.

5. **Carry out ('run') the electrophoresis.** DNA separation is usually carried out at 100–150 V for 30–60 min (see manufacturer's instructions for specific details, according to which 'power pack' you are using); the gel should be run until the 'tracking' dye has migrated across 80% of the gel.

6. **Examine the result.** If you have used a visualising dye within the agarose, then you simply transfer the gel to a UV transilluminator and look for 'bands' of fluorescence corresponding to each DNA fragment.

> **SAFETY NOTE – always wear suitable UV-filtering safety glasses when working with UV radiation to protect your eyes.** Use a digital camera to photograph your gel. Alternatively, a dedicated image capture system can be used, e.g. GelDoc®.

7. **Extract any DNA bands of interest.** If a particular band is required for further study, the piece of gel containing that band can be cut from the gel using either a clean scalpel or a specialised gel band cutter.

When agarose gels are used for the separation of DNA, the large fragment size means that molecular sieving is observed, even with low concentration gels. This is the basis of the electrophoretic separation of nucleic acids (Box 52.1; see also Chapter 66).

Polyacrylamide

Polyacrylamide gel electrophoresis (PAGE) plays a major role in protein analysis, both for one-dimensional and two-dimensional separations. The gel is formed by polymerising acrylamide monomer into long chains and cross-linking these chains using *N, N*'-methylene bisacrylamide (often abbreviated to 'bis'). The process is shown in Fig. 52.4. In most protocols, polymerisation is initiated by free radicals produced by ammonium persulfate in the presence of *N, N, N*', *N*'-tetramethylethylenediamine (TEMED). The photodecomposition of riboflavin can also be used as a source of free radicals.

The formation of Polyacrylamide from its acrylamide monomers is extremely reproducible under standard conditions, and electrophoretic separations are correspondingly precise. The pore size, and hence the extent of molecular sieving, depends on the total concentration of monomer (%T), i.e. acrylamide plus bisacrylamide in a fixed ratio. This means that pores in the gel can be 'tailored' to suit the size of biomolecule to be separated: gels containing 3% acrylamide have large pores and are used in methods where molecular sieving should be avoided (e.g. in isoelectric focusing, p. 360), while higher concentrations of acrylamide (5–30%T) introduce molecular sieving to various degrees depending on the size of the

> **Advantages of Polyacrylamide gels** – in addition to their versatility in terms of pore size, these gels are chemically inert, stable over a wide range of pH, ionic strength and temperature, and transparent.

> **SAFETY NOTE Preparing gels for PAGE** – both acrylamide and bisacrylamide are extremely potent neurotoxins, so you must wear plastic gloves when handling solutions containing these reagents. Although the polymerised gel is nontoxic, it is still advisable to wear gloves when handling the gel, because some monomer may still be present.

Fig. 52.4 Reactions involved in the formation of Polyacrylamide gels.

(a)

(b)

Fig. 52.5 Determination of relative molecular mass (M_r) of proteins by SDS-PAGE: (a) gel samples: 1, cytochrome c; 2, myoglobin; 3, γ-globulin; 4, carbonic anhydrase; 5, ovalbumin; 6, albumin; 7, transferrin; 8, mixture of samples 1–7 (b) plot of log M_r against distance travelled through gel.

Photo (a) courtesy of Pharmacia Biotech.

sample components, i.e. with 30% acrylamide gels, molecules as small as M_r 2000 may be subject to molecular sieving. Gels of $< 2.5\%$ are necessary for molecular sieving of molecules of $M_r > 10^6$, but such gels are almost fluid and require 0.5% agarose to make them solid. Note that a gel of 3% will separate DNA by molecular sieving, owing to the large size of the nucleic acid molecules (p. 459).

Before you embark on a particular PAGE protein separation, you will need to think about the general strategy for that separation and make certain choices, including whether to use the following:

- **Rod or slab gels** – flat slab gels are formed between glass plates, using plastic spacers 0.75–1.5 mm thick: rod gels are made in narrow bore tubes. For most separations using several samples, a slab gel saves time because up to 25 samples can be separated under identical conditions in a single gel, while rod gels can only be used for individual samples. Rectangular slab gels are also easier to read, by densitometry, and photograph. However, rod gels are useful in preliminary separations, for determining a suitable pH and gel concentration, and for applications where the gel is sliced to extract and assay proteins of interest.

- **Dissociating or non-dissociating conditions** – the most widely used PAGE protein separation technique uses an ionic detergent, usually sodium dodecyl sulfate (SDS), which dissociates proteins into their individual polypeptide subunits and gives a uniform net charge along each denatured polypeptide. This technique, known as SDS-PAGE, requires only microgram amounts of sample and is quick and easy to carry out. On the other hand, if it is necessary to preserve the native protein conformation and biological activity, non-dissociating conditions are used, i.e. no SDS is added. In SDS-PAGE the sample protein is normally heated to 100°C for 2 min, in buffer containing 1% (w/v) SDS and 1% (w/v) 2-mercaptoethanol, the latter to cleave any disulfide bonds. The resultant polypeptides bind to SDS in a constant weight ratio, with 1.4 g of SDS per gram of protein. As a result, the intrinsic net charge of each polypeptide is 'swamped' by the negative charge imposed by SDS, and there is a uniform negative charge per unit length of polypeptide. Since the polypeptides now have identical charge densities, when they are subject to PAGE (with SDS present) using a gel of appropriate pore size, molecular sieving will occur and they will migrate strictly according to polypeptide size. This not only gives effective separation, but the molecular mass of a given polypeptide can be determined by comparing its mobility to polypeptides of known molecular mass run under the same conditions (Fig. 52.5). Several manufacturers (e.g. Pharmacia, Sigma) supply molecular mass standard kits which may include polypeptides of M_r 11 700 to 212 000 (Table 52.2), together with details of their preparation and use. Where necessary, the treated sample can be concentrated by ultrafiltration and the buffer composition can be altered by diafiltration (p. 409).

- **Continuous or discontinuous buffer systems** – a continuous system is where the same buffer ions are present in the sample, gel and buffer reservoirs, all at the same pH. The sample is loaded directly onto a gel (the 'separating gel' or 'resolving gel') that has pores small enough to introduce molecular sieving. In contrast, discontinuous systems have different buffers in the gel compared to the reservoirs, both in terms of buffer ions and pH. The sample is loaded onto a large-pore

---

Preparing Polyacrylamide gels – most solutions used for gel preparation can be made in advance, but the ammonium persulfate solution must be prepared immediately before use.

Selecting and using gels in 2D electrophoresis – a rod gel may be used for the first dimension and a slab gel for the second dimension.

Table 52.2 Molecular masses of standard proteins used in electrophoresis.

Protein	M_r	$\log_{10}M_r$
Cytochrome c	11 700	4.068
Myoglobin	17 200	4.236
γ-globulin (light chain)	23 500	4.371
Carbonic anhydrase	29 000	4.462
Ovalbumin	43 000	4.634
γ-globulin (heavy chain)	50 000	4.699
Human albumin	68 000	4.832
Transferrin	77 000	4.886
Myosin (heavy chain)	212 000	5.326

Following the progress of PAGE – add bromophenol blue solution (0.002% w/v) to the sample in the ratio 1:25 (dye: sample). This highly ionic, small M_r dye migrates with the electrophoretic front.

Choosing a buffer system for PAGE – discontinuous systems are more time-consuming to prepare, but have the advantage over continuous systems in that relatively large volumes of dilute sample can be used and good resolution is still obtained.

Fig. 52.6 Apparatus for discontinuous electrophoresis.

'stacking gel', previously polymerised on top of a small-pore separating gel (Fig. 52.6). The individual proteins in the sample concentrate into very narrow zones during their migration through the large-pore gel and stack up according to their charge densities, prior to separation in the small-pore gel, giving enhanced results compared with continuous systems.

- **pH and buffer for the separation** – PAGE can be carried out between pH 3 and 10. The pH is not critical for continuous SDS-PAGE, since SDS-treated polypeptides are negatively charged over a wide pH range. However, when using non-dissociating systems, the pH can be critical, particularly if the biological activity of the molecules is to be retained.

- **Gel concentration** – at one extreme, a gel with a very high percentage T might totally exclude the sample components, while a gel with a very low percentage T might lead to all SDS-treated proteins migrating at the same rate, i.e. with the electrophoretic front. A sensible approach is to set up a series of rod gels in the range 5–15% T and observe the separation and resolution obtained. Alternatively, a *gradient gel* can be used, in which the percentage T increases, and hence pore size decreases, in the direction of protein migration. A useful gradient for a preliminary experiment would be 5–20% T. Such gels are able to resolve protein mixtures with a wide range of molecular masses. Furthermore, as proteins migrate into regions of ever-decreasing pore size, the movement of the leading edge of a zone will become increasingly restricted. This allows the trailing edge to catch up, resulting in considerable zone sharpening. Gradient formers are fairly simple to make, and are commercially available.

Practical details of the preparation of PAGE and SDS-PAGE gels are given in Table 52.3, while Box 52.2 gives the procedure for SDS-PAGE separation of proteins (see Westermeier, 2004, or Gersten, 1996 for further details).

Choosing a pH for protein electrophoresis – many proteins have isoelectric points in the range pH 4–7 and in response to electrophoresis with buffers in the region pH 8.0–9.5, most proteins will migrate towards the anode.

Separating protein mixtures – for high resolution, a combination of dissociating and discontinuous PAGE in slab gels is the system of choice.

What to do if your Polyacrylamide gels fail to set – polymerisation is inhibited by oxygen, so solutions should be degassed, and the surfaces of the polymerisation mixture exposed to air should be overlaid with water; if your gels still do not polymerise, the most common cause is the use of 'old' ammonium persulfate stock solution. If low pH buffers are used, polymerisation may be delayed because TEMED is required in the free base form.

Table 52.3 Preparation of gels for PAGE and SDS-PAGE. The gel solutions are made by mixing the components in the proportions and in the order shown. Figures are ml of each solution required to give the stated % gel strength.

Solution (added in order shown)	PAGE			SDS-PAGE		
	3.5% gel (T = 3.6%)	5% gel (T = 5.1%)	7.5% gel (T = 7.7%)	5% gel (T = 5.1%)	7.5% gel (T = 7.7%)	10% gel (T = 10.2%)
1. Distilled water	19.3	14.9	7.5	14.9	7.5	–
2. TRIS-glycine buffer, pH 8.9, 0.1 mol L^{-1}	33.0	33.0	33.0	–	–	–
3. Imidazole buffer, pH 7.0, 0.1 mol L^{-1} plus 0.2% w/v SDS	–	–	–	33.0	33.0	33.0
4. Acrylamide solution 22.2% w/v and 0.6% w/v bis	10.4	14.8	22.2	14.8	22.2	29.7
5. Ammonium persulfate solution, 0.15% w/v	3.2	3.2	3.2	3.2	3.2	3.2
6. TEMED	0.1	0.1	0.1	0.1	0.1	0.1
Final volume (mL)	66.0	66.0	66.0	66.0	66.0	66.0

How to carry out SDS-PAGE for protein separation

1. **Prepare the gel.** Nowadays, many laboratories use pre-cast gels, bought from a manufacturer (e.g. Bio-Rad®). If you are preparing your own gel, you will need to follow the protocol very carefully. Typically, the correct proportions of acrylamide, bisacrylamide and SDS are mixed together and degassed, under vacuum. Then ammonium persulfate and tetramethylethylenediamine (TEMED) are added to trigger the polymerisation. Once the latter two constituents are added, the gel should be poured immediately into the casting tray (including the well former 'comb').

2. **Prepare the samples.** The protein sample is mixed with a buffer solution containing SDS (to bind to the dissociated proteins), plus dithiothreitol (DTT) (to cleave disulfide bonds in the proteins) and a 'tracker' dye (e.g. bromophenol blue), then heated for 5 min at 95°C, to disrupt the tertiary structure and 'linearise' the polypeptide chains.

3. **Load the samples onto the gel.** Individual samples are added to the wells using a pipettor (p. 146). The volume of sample added to each well is typically less than 100 µL, so a very steady hand and careful dispensing are needed to pipette each sample accurately (steady the pipettor using both hands if this helps). To optimise the separation of proteins, the volume added should be kept as small as possible.

4. **Load the molecular mass standards.** Nowadays, many labs use 'rainbow' markers, with a wide range of proteins of known molecular mass, each of which is stained a different colour, to enable estimation of the molecular masses of unknown proteins, by visual comparison.

5. **Run the electrophoresis.** The gel is positioned with the well/samples closest to the cathode (negative electrode), since they will move towards the anode during electrophoresis as a result of their negative charge. Protein separation is typically carried out at 80–100V for 1–2h (see manufacturer's instructions for specific details, according to which 'power pack' you are using): the gel should be run until the 'tracking' dye has migrated across 80% of the gel. Higher voltages give faster separation, but poorer separation of protein 'bands', and may cause denaturation of proteins due to heating.

6. **Fix and stain the gel.** One of the most widely used approaches uses Coomassie Brilliant Blue stain. Typically, this involves immersing the gel in 0.25% w/v Coomassie Blue for 1 h then destaining overnight in a methanol/acetic acid solution. A safer, water-based alternative can be bought from a manufacturer (e.g. Bio-Rad Bio-Safe Coomassie Stain), which allows the gel to be destained in water overnight. After destaining, separated proteins are visible as blue bands against an unstained background. For higher resolution, a silver stain can be used instead of Coomassie Brilliant Blue.

7. **Examine the results.** After destaining, separated proteins are visible as blue bands against an unstained background (Fig. 52.5(a)). The position of these bands can be compared with the molecular mass standards to determine the size of each band in the test sample. For greater accuracy, the distance moved by the molecular mass standards can be plotted against $\log_{10}$ relative molecular mass and the resulting relationship (Fig. 52.5(b)) used to determine the size of unknown bands.

Post-electrophoretic procedures – handling of the supporting medium, staining and analysis

For protein electrophoresis, the following stages are appropriate: details for nucleic acids are given on pp. 459–60.

Handling

All types of supporting medium should be handled carefully: wearing gloves is advisable, for safety and to avoid transfer of proteins from the skin. Cellulose acetate strips should be immediately transferred to a fixing and staining solution. Agarose gels should be dried quickly before staining. Polyacrylamide gels in vertical slabs must be freed carefully from one of the glass plates in which they are formed, taking care to lever the glass at a point well away from the part of the gel containing the wells: once free, the gels should be immediately transferred to fixing or staining solution.

Rod gels are recovered by a process called 'rimming', a technique that takes a little practice to master. To remove the gel from the tube, hold the tube in one hand and, using a syringe with a long blunt needle, squirt water between the gel and the inner wall of the tube: while squirting, move the needle gently up and down and rotate the tube. If the gel does not become free, perform the same procedure at the other end of the tube, until the gel is released. If necessary, a Pasteur pipette bulb applied to one end can be used to push the gel out of the tube.

Fixing, staining and destaining

To prevent separated proteins from diffusing, they are usually fixed in position. Proteins on cellulose acetate strips can be fixed and stained using Ponceau S (0.2% w/v in a solution of 3%v/v aqueous trichloracetic acid, TCA).

Fixing is also required for most types of gel electrophoresis: 3% v/v TCA is often used. The most widely used stain for protein separations in gels is Coomassie Blue R-250 (where R = reddish hue): the detection limit is ≈ 0.2 µg and staining is quantitative up to 20 µg for some proteins. It is normal for background staining of the gel to occur, and removal of background colour ('destaining') can be achieved either by diffusion or electrophoresis. To destain by diffusion, transfer the gel to isopropanol: acetic acid: water (12.5 : 10: 77.5 v/v/v) and allow to stand for 48 h, or change the solution several times to speed up the staining process. Electrophoretic destaining can remove Coomassie Blue, which is anionic: stained gels are placed between porous plastic sheets with electrodes on each side, and the tank is filled with 7% acetic acid. Passing a current of up to 1.0 A through the gel for around 30–40 min. should result in substantial destaining.

If you need greater sensitivity (e.g. for nanogram to femtogram amounts), or when using high resolution techniques such as 2D-electrophoresis (p. 361), silver staining can be used. Depending on the protocol chosen and the proteins being stained, the silver technique can be fivefold to 200–fold more sensitive than Coomassie Blue. The method involves a fixation step (e.g. with TCA), followed by exposure to silver nitrate solution and development of the stain. The silver ions are thought to react with basic and thiol groups in proteins, and subsequent reduction (e.g. by formaldehyde at alkaline pH, or by photodevelopment) leads to deposits of silver in the protein bands. Most proteins stain brown or black, but lipoproteins may stain blue, and some glycoproteins stain yellow or red. Some proteins lacking in amino acids with reducing groups (e.g. those lacking cysteine residues) may stain negatively, i.e. the bands are more transparent than the background staining of the gel. Although many protocols have been published, silver stain kits are commercially available, e.g. from Bio-Rad.

Although silver staining has clear advantages in terms of sensitivity, for routine work it is more laborious and expensive to carry out than the Coomassie Blue method. It also requires high purity water, otherwise significant background staining occurs. Another feature is that the staining can be non-specific, since DNA and polysaccharides may stain on the same gel as proteins.

Other methods for the detection of separated components include:

- **autoradiography** for proteins labelled with ^{32}P or ^{125}I (p. 301);
- **fluorescence** for proteins pre-labelled with fluorescent dyes (p. 383);
- **periodic acid-Schiff (PAS) stain** using dansyl hydrazine for glycoproteins.

Handling gels – avoid touching gels with paper as it sticks readily and is difficult to remove without tearing.

Avoiding overloaded gels and band distortion – determine the protein concentration of the sample beforehand. Around 100 µg of a complex mixture, or 1–10 µg of an individual component, will be sufficient, but bear in mind that underloading may result in bands being too faint to be detected.

Optimising resolution – keep the sample volume as small as possible; methods for concentrating protein solutions are considered on pp. 406–9. For vertical slab gels and for rod gels, include 10% w/v sucrose or glycerol to increase density and allow buffer solution to be overlaid on the sample without dilution.

Understanding the terminology of high purity water – this is usually produced by reverse osmosis and its purity is often expressed with respect to its resistivity: ultrapure water has a resistivity of 18 MΩ cm^{-1} (informally referred to as 'eighteen megaohm water').

Blotting

The term 'blotting' refers to the transfer of separated proteins from the gel matrix to a thin sheet such as nitrocellulose membrane (commercially available from, for example, Millipore, Amersham). The proteins bind to this membrane, and are immobilised. Blotting of proteins is usually achieved by electrophoretic transfer, and this process is normally referred to as Western blotting (see also Southern blotting and Northern blotting for DNA and RNA, respectively, pp. 461–2). Its major advantage is that the immobilised proteins on the surface of the membrane are readily accessible to detection reagents, and staining and destaining can be achieved in less than 5 min. Use of labelled antibodies to detect specific proteins (immunoblotting) can take less than 6h. In addition, it is easy to dry and store Western blots for long periods, for further analysis.

Detection of enzymes

If you need to detect enzyme activity you should use a non-denaturing gel. The gel matrix will hinder diffusion of the enzyme, but will allow access to the small molecular weight substrates, co-factors and dyes necessary to localise enzyme activity *in situ*. Most methods for detecting enzymes on gels are modifications of protocols originally developed by histochemists, e.g.:

- **NAD^+-requiring oxidoreductases** can be detected by incubating the gels with substrate, NAD^+ and a solution of a tetrazolium salt which, when reduced, forms an insoluble coloured formazan dye;

- **transferases** and isomerases can be detected by coupling their reactions to an oxidoreductase-requiring reaction, visualised as described above;

- **hydrolases** can be detected using appropriate chromogenic or fluorogenic substrates;

- **phosphatases** can be detected by precipitating any phosphate released from the substrate with Ca^{2+}.

For more details, see Manchenko (2002).

Recording and quantification of results

A number of expensive, dedicated instruments are available for the analysis of gels, e.g. laser densitometers. Alternatively, gel scanning attachments can be purchased for standard spectrophotometers, allowing measurement and recording of the absorbance of the Coomassie Blue stained bands at 560–575 nm: for instruments connected to a computer, quantification of individual components can be achieved by integrating the areas under the peaks.

You can photograph gels using a conventional camera (fine grain film), digital camera, or using a photocopier. Alternatively, a dedicated image capture and analysis system may be used, e.g. GelDoc. The gel should be placed on a white glass transilluminator. A red filter will increase contrast with bands stained with Coomassie Blue. If the gels themselves need to be retained, they can be preserved in 7% acetic acid. Alternatively, they can be dried using a commercially available gel dryer.

Origins of 'Western' blotting – following the description of 'Southern' blotting of DNA by Dr Ed Southern (p. 461), other points of the compass have been used to describe other forms of blotting, with 'Western' blotting for proteins.

Example   For detection of lactate dehydrogenase (LDH): when a solution containing lactate, NAD^+, phenazine methasulfate (PMS) and methyl thiazolyl tetrazolium (MTT) is added to a gel containing LDH, a series of redox reactions occurs in the enzyme-containing regions, starting with the oxidation of lactate to pyruvate, and proceeding via NAD^+, PMS and MTT to the eventual reduction of MTT to formazan dye. After incubation at 37°C in the dark (since MTT is light-sensitive), LDH is detected by the appearance of blue–black bands on the gel.

Detecting enzymes in gels – minimise zone spreading by incorporating substrates, etc. in a thin agarose indicator gel poured over the separating gel.

Measuring peak areas – a valid 'low-tech' alternative to computer-based systems is to cut out and weigh peaks from a recorder chart.

Text references

Gersten, D. (1996) *Gel Electrophoresis: Proteins* (Essential Techniques Series). Wiley, New York.

Manchenko, G.P. (2002) *Handbook of Detection of Enzymes on Electrophoretic Gels,* 2nd edn. CRC Press, Boca Raton.

Westermeier, R. (2004) *Electrophoresis in Practice: A Guide to Methods and Applications of DNA and Protein Separations,* 4th edn. Wiley VCH, Berlin.

Sources for further study

Anon. *The American Electrophoresis Society Homepage,* Available: http://www.aesociety.org Last accessed 01/09/15.
[Includes details on a variety of electrophoretic techniques.]

Anon. *British Society for Proteomics Research Homepage.* Available: http://www.bspr.org/ Last accessed 01/09/15.
[Formerly the British Electrophoresis Society, includes links to other relevant websites and databases.]

Garcia, C.D., Chumbimuni-Torres, K.Y. and Carrilho, E. (2013) *Capillary Electrophoresis and Microchip Capillary Electrophoresis: Principles, Applications and Limitations.* Wiley, New York.

Rabilloud, T. (2004) *Proteome Research: Two-dimensional Gel Electrophoresis and Identification Methods.* Springer, Heidelberg.

Tietz, D. (1998) *Nucleic Acid Electrophoresis.* Springer-Verlag, Berlin.

STUDY EXERCISES

52.1 Find out why the net charge on a protein molecule varies with pH. Identify the amino acids primarily responsible for determining the net charge on a protein molecule and draw simple diagrams to represent the ionisation of their side chains, indicating how you would expect these side chains to be charged at acid, neutral and alkaline pH values.

52.2 Explain the function of the various reagents used in SDS-PAGE analysis of proteins. What is the function of each of the following SDS-PAGE reagents? (a) acrylamide; (b) *N,N'*-methylene bisacrylamide; (c) ammonium persulfate; (d) *N,N,N',N'*-tetramethylethylenediamine; (e) sodium dodecyl sulfate; (f) 2-mercaptoethanol; (g) buffer; (h) Coomassie Blue R-250; (i) bromophenol blue.

52.3 Test your knowledge of 'blotting' terminology. What is Western blotting, and how does it differ from Northern and Southern blotting?

52.4 Determine the M_r of subunits of a protein by SDS-PAGE. A number of proteins of known molecular weight were analysed by SDS-PAGE and the results are shown in the table. In the same experiment, a sample of purified lactate dehydrogenase (LDH) was treated in identical fashion and run, producing a band with a relative mobility of 0.77. (a) What is the M_r of the polypeptide produced from the LDH sample?

Relative mobilities of molecular weight standards by SDS-PAGE.

Protein	M_r	Relative mobility
Myoglobin	17 200	1.00
Carbonic anhydrase	29 000	0.84
Ovalbumin	43 000	0.67
Human albumin	68 000	0.51
Transferrin	77 000	0.46

(b) If the M_r of native LDH is 136 000, what do you conclude about the subunit structure of this enzyme?

52.5 Find out about 'troubleshooting' problems with gels. Investigate the pitfalls of working with and polymerising gels.

Answers to these study exercises are available at www.pearsoned.co.uk/practicalskills.

Using electrophoresis in proteomics – note that many of the techniques described in this chapter are employed to generate the information used in bioinformatic analysis (Chapter 12).

Although the resolution obtained using the basic electrophoretic techniques described in Chapter 52 is adequate for many bimolecular applications, a number of advanced techniques are available that give very high resolution with very small amounts of sample material.

Isoelectric focusing (IEF)

In contrast to electrophoresis, which is carried out at constant pH, IEF is carried out using a pH gradient. The gradient is formed using small molecular mass ampholytes, which are analogues and homologues of polyamino-, polycarboxylic acids that collectively have a range of isoelectric points (pI values) between pH 3 and 10. The mixture of ampholytes (p. 169), either in a gel or in free solution, is placed between the anode in acid solution (e.g. H_3PO_4), and the cathode in alkaline solution (e.g. NaOH). When an electric field is applied, each ampholyte migrates to its own pI and forms a stable pH gradient which will persist for as long as the field is applied. When a protein sample is applied to this gradient, separation is achieved, since individual proteins will migrate to their isoelectric points. The net charge on the protein when first applied will depend on the specific 'titration curve' for that protein (Fig. 53.1). As an example, consider two proteins, X and Y, having pI values of pH 5 and pH 8 respectively, which are placed together on the gradient at pH6 (Fig. 53.2). At that pH, protein X will have a net negative charge, and will migrate towards the anode, progressively losing charge until it reaches its pI (pH 5) and stops migrating. Protein Y will have a net positive charge at pH 6, and so will migrate towards the cathode until it reaches its pI (pH8).

Using a Polyacrylamide gel as a supporting medium and a narrow pH gradient, proteins differing in pI by 0.01 units can be separated. Even greater resolution is possible in free solution (e.g. in capillary electrophoresis, p. 361). Such resolution is possible because protein molecules that diffuse away from the pI will acquire a net charge (negative at increased pH, positive at decreased pH) and immediately be focused back to their pI. This focusing effect will continue for as long as the electric field is applied.

A useful variant of IEF is in obtaining *titration curves* for proteins. A pH gradient is set up, and the sample applied in a line at a right angle to the gradient. The net charge on a given protein will vary according to its position on the gradient – when electrophoresis is carried out at right angles to the pH gradient, the protein will migrate at a velocity and direction governed by that charge. When stained, each protein will appear as a continuous curved line, corresponding to its titration curve (Fig. 53.3). This technique can be usefully performed during protein purification, prior to ion-exchange chromatography (p. 336): by obtaining the titration curve for a protein of interest and those of major contaminants, the mobile phase pH that gives optimal separation can be selected.

In IEF, it is important that electro-osmotic flow (EOF, p. 351) is avoided, as this would affect the ability of the proteins to remain stationary at their pIs. For gel IEF, Polyacrylamide minimises EOF, while capillary IEF uses narrow bore tubing with an internal polymer coating (p. 364).

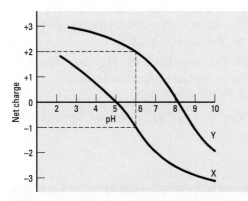

Fig. 53.1 Titration curves for two proteins, X and Y.

Fig. 53.2 The migration of two proteins, X and Y, in response to a pH gradient.

Two-dimensional electrophoresis

The most commonly used version of this high resolution technique involves separating proteins by charge in one dimension using IEF in Polyacrylamide gel, followed by separation by molecular mass in the second dimension using denaturing SDS-PAGE (p. 354). The technique allows up to 1000 proteins to be separated from a single sample. Typically, the first dimension IEF run (pH 3–10) is carried out on gel strips of length 7–24 cm. Strips are run at a voltage of 500–3500 V for 1.5 h, then at 3500 V for a further 4 h. Gel strips can then be used immediately, or frozen until required.

It is common for the second-dimension SDS-PAGE separation to be carried out on a discontinuous slab gel 0.5–1.5 mm thick, which includes a low percentage T stacking gel and a separating gel with an exponential gradient of 10–16%T. The separating gel can be prepared in advance, but the stacking gel must be formed shortly before addition of the rod gel from the one dimensional run.

After equilibration with the buffer used in SDS-PAGE, the IEF gel strip is loaded onto the 2D gel (still between the glass plates in which it was formed) and sealed in position using acrylamide or agarose. Before the sealing gel sets, a well should be formed in it at one end to allow addition of molecular mass markers. The second-dimension is run at 100–200 V until the dye front is ≈1 cm from the bottom edge of the slab. After running, the gel is processed for the detection of polypeptides, e.g. using Coomassie Blue or silver stain. Analysis of the complex patterns that result from 2D electrophoresis requires computer-aided gel scanners to acquire and process data from a gel image, such as that shown in Fig. 53.4. These systems can compare, adjust and match up patterns from several gels, allowing both accurate identification of spots and quantification of individual proteins. Allowance is made for the slight variations in patterns found in different runs, using internal references ('landmarks'), which are either added standard proteins or particular spots known to be present in all samples.

Capillary electrophoresis (CE)

This technique combines the high resolving power of electrophoresis with the speed and versatility of HPLC (p. 334). The technique largely overcomes the major problem of carrying out electrophoresis without a supporting medium, i.e. poor resolution caused by convection currents and diffusion. A capillary tube has a high surface area: volume ratio, and consequently the heat generated as a result of the applied electric current is rapidly dissipated. A further advantage is that very small sample volumes (5–10 nl) can be analysed. The versatility of CE is demonstrated by its use in the separation of a range of biomolecules, e.g. amino acids, proteins, nucleic acids, drugs, vitamins, organic acids and inorganic ions; CE can even separate neutral species, e.g. steroids, aromatic hydrocarbons (see Weinberger, 2000).

The components of a typical CE apparatus are shown in Fig. 53.5. The capillary is made of fused silica and externally coated with a polymer for mechanical strength. The internal diameter is usually 25–50 μm, a compromise between efficient heat dissipation and the need for a light path that is not too short for detection using UV/visible spectrophotometry. A gap in the polymer coating provides a window for detection purposes.

Avoiding streaking in 2D electrophoresis – check carefully to ensure that the sample contains no particulate material (e.g. from protein aggregation); filter or centrifuge before use.

Maximising resolution in 2D electrophoresis of proteins – try to minimise nucleic acid contamination of your sample (see Chapter 60), as they may interact with polypeptides/proteins, affecting their movement in the gel.

Freezing gel strips – be sure to mark the identity and orientation of each gel strip before freezing, e.g. by inserting a fine wire into one end of the strip. Note that if urea is used in the gel strip, it will form crystals on freezing.

Fig. 53.3 Titration curves of bovine muscle proteins produced by electrofocusing-electrophoresis Photo courtesy of Pharmacia Biotech.

low pH ←———— IEF ————→ high pH

SDS – PAGE

Fig. 53.4 Two-dimensional separation of proteins from 100x concentrated urine (2.5 µg total protein; silver stain).
Courtesy of T. Marshall and K.M. Williams.

Fig. 53.5 Components of a capillary electrophoresis system.

Samples are injected into the capillary by a variety of means, e.g. electrophoretic loading or displacement. In the former, the inlet end of the capillary is immersed in the sample and a pulse of high voltage is applied. The displacement method involves forcing the sample into the capillary, either by applying pressure in the sample vial using an inert gas, or by introducing a vacuum at the outlet. The detectors used in CE are similar to those used in chromatography (p. 343), e.g. UV/visible spectrophotometric systems. Fluorescence detection is more sensitive, but this may require sample derivatisation. Electrochemical and conductivity detection is also used in some applications, e.g. conductivity detection of inorganic cations such as Na^+ and K^+.

Electro-osmotic flow (EOF), described on p. 351, is essential to the most commonly used types of CE. The existence of EOF in the capillary is the result of the net negative charge on the fused silica surface at pH values over 3.0. The resulting solvent flow towards the cathode is greater than the attraction of anions towards the anode, so they will flow towards the cathode (note that the detector is situated at the cathodic end of the capillary). The greater the net negative charge on an anion, the greater is its resistance to the EOF and the lower its mobility. Separated components migrate towards the cathode in the order: (1) cations, (2) neutral species, (3) anions.

Capillary zone electrophoresis (CZE)

This is the most widely used form of CE, and is based on electrophoresis in free solution and EOF, as discussed above. Separations are caused by the charge: mass ratio of the sample components, and the technique can be used for almost any type of charged molecule, and is especially useful for peptide separation and confirmation of purity.

Micellar electrokinetic chromatography (MEKC)

This technique involves the principles of both electrophoresis and chromatography. Its main strength is that it can be used for the separation of neutral molecules as well as charged ones. This is achieved by including surfactants (e.g. SDS, Triton X-100) in the electrophoresis buffer at concentrations that promote the formation of spherical micelles, with a hydrophobic interior and a charged, hydrophilic surface. When an electric field is applied, these micelles will tend to migrate with or against the EOF, depending on their surface charge. Anionic surfactants like SDS are attracted by the anode, but if the pH of the buffer is high enough to ensure that the EOF is faster than the migration velocity of the micelles, the net migration is in the direction of the EOF, i.e. towards the cathode. During this migration, sample components partition between the buffer and the micelles (acting as a pseudo-stationary phase); this may involve both hydrophobic and electrostatic interactions. For neutral species it is only the partitioning effect that is involved in separation; the more hydrophobic a sample molecule, the more it will interact with the micelle, and the longer will be its migration time, since the micelle resists the EOF. The versatility of MEKC enables it to be used for separations of biomolecules as diverse as amino acids and polycyclic hydrocarbons. MEKC is also known as micellar electrokinetic capillary chromatography (MECC).

Chiral capillary electrophoresis (CCE)

Resolution of a pair of chiral enantiomers (optical isomers) represents one of the biggest challenges for separation science, because each member of the pair will have identical physicochemical properties. CE offers an effective method of separating enantiomers by inducing a 'chiral selector' in the electrophoresis medium. The most commonly used chiral selectors are cyclodextrins such as the highly sulfated cyclodextrins (HSCDs) (Fig. 53.6). As the enantiomers migrate along the capillary, one will tend to interact more strongly than the other and its mobility will be reduced relative to the other. Figure 53.7 shows separation of the R and S forms of amphetamine using 5% HSCD in the electrophoresis buffer. The R-form has greater affinity for the HSCD used, so its retention time on the capillary is longer than that of the S-form.

Note that HSCDs can also be used in HPLC, but CCE is more effective, with shorter development times and lower reagent costs.

Capillary gel electrophoresis (CGE)

The underlying principle of this technique is directly comparable with that of conventional PAGE, i.e. the capillary contains a polymer that acts as a molecular sieve. As charged sample molecules migrate through the polymer network, larger molecules are hindered to a greater extent than

Distinguishing between stereoisomers – the R and S convention involves prioritising atoms or groups bonded to an asymmetric carbon atom in order of their atomic number. With the smallest atom or group pointing away from you, note the size of the remaining three groups or atoms. If the configuration in order of increasing size is clockwise, this is termed the R-configuration (L. *rectus*); if the order is anticlockwise, this is the S-configuration (L. *sinister*). Note that the older, alternative terminology of 'D' (L. *dextro*) and 'L' (L. *laevo*) isomers is widely used in the life sciences (p. 382), but the two terminologies do not always coincide (i.e. R is not always D, and S is not always L).

Fig. 53.6 General structure of highly sulfated cyclodextrins (HSCDs). The central cavity of the cyclodextrin (shaded green) can interact differently with the R and S isomers of a biomolecule, enabling separation in CE.
Copyright © 1999–2002 Beckman Coulter, Inc. http://www.beckmancoulter.com/products/splashpage/chira!38/default.asp/

Fig. 53.7 CE separation of R and S enantiomers of amphetamine using Beckman Coulter highly sulfated gamma cyclodextrin.
Copyright © 1999–2002 Beekman Coulter, Inc. http://www.beckmancoulter.com/products/splashpage/chira!38/default.asp/

smaller ones and will tend to move more slowly. CGE differs from CZE and MEKC in that the inner surface of the capillary is polymer-coated to prevent EOF; this means that for most applications (e.g. polypeptide or oligonucleotide separations) sample components will migrate towards the anode at a rate determined by their size. The technique also differs from conventional PAGE in that a 'polymer network' is used rather than a gel: the polymer network may be Polyacrylamide or agarose.

CGE offers the following advantages over conventional electrophoresis:

- efficient heat dissipation means that a high electrical field can be applied, giving shorter separation times;
- detection of the separated components as they move towards the anodic end of the capillary (e.g. using a UV/visible detector) means that staining is unnecessary;
- automation is feasible.

Capillary isoelectric focusing (CIEF)

This is used mainly for protein separation. Here, the principles of IEF are valid as long as EOF is prevented by using capillaries that are polymer-coated on their inner surface. Sample components migrate to their isoelectric points and become stationary. Once separated (<10 min), the components must be mobilised so that they flow past the detector. This is achieved by changing the NaOH solution in the cathodic reservoir with an NaOH/NaCl solution. When the electric field is reapplied, Cl^- enters the capillary, causing a decrease in pH at the cathodic end and the subsequent migration of sample components.

Choosing a detector for capillary electrophoresis – most types of HPLC detector are suitable for CE and related applications (see Chapter 51).

Text references

Weinberger, R. (2000) *Practical Capillary Electrophoresis.* Academic Press, New York.

Sources for further study

Anon. *The American Electrophoresis Society Homepage.* Available: http://www.aesociety.org Last accessed 01/09/15. [Includes details on electrophoretic techniques, including IEF and CE.]

Anon. *Cyclodextrin Resource.* Available: http://www.cyclodex.com/index.html Last accessed 01/09/15. [Information on cyclodextrin structure, applications, etc.]

Costello, M. (2013) *Electrophoresis: Theory and Practice.* Callisto, Berlin.

Garcia, C.D., Chumbimuni-Torres, K.Y. and Carilho, E. (2013) *Capillary Electrophoresis and Microchip*

Capillary Electrophoresis: Principles, Applications and Limitations. Wiley, New York.

Schmitt-Kopplin, P. (2010) *Capillary Electrophoresis: Methods and Protocols.* Humana, New York.

Strege, M.A. and Lagu, A.L. (2004) *Capillary Electrophoresis of Proteins and Peptides.* Humana Press, New Jersey.

Westermeier, R. (2016) *Electrophoresis in Practice: A Guide to Methods and Applications of DNA and Protein Separations,* 5th edn. Blackwell Verlag, Berlin. [Covers the principles underlying IEF.]

STUDY EXERCISES

53.1 Consider the requirements for sample application in PAGE and IEF. Explain why in PAGE the sample is applied in a discrete narrow band, usually at the cathodic end of the gel, while in IEF the sample can be applied at any point along the length of the gel without concern about location or narrowness of the sample zone.

53.2 Choose a suitable CE technique for a range of applications. Which variant of CE would you consider most suitable for the following applications? (a) Separation of small peptides of similar size; (b) separation of oligonucleotides of different lengths; (c) separation of proteins with different isoelectric points; (d) a sample of urine tested for several anabolic steroids.

53.3 Check the similarities and differences between electrophoresis and isoelectric focusing. Complete the following table, indicating (i) whether each of the factors listed is relevant to each technique and (ii) briefly explain why, in each case.

Potential factors influencing electrophoresis (E) and isoelectric focusing (IEF).

Factor	E	IEF
Dependence upon ionisable groups in proteins		
Requirement for the application of an electric field		
Dependence upon differences in net charge between sample molecules		
Dependence upon differences in pI values between sample molecules		
Requirement for buffer at constant pH		
Requirement for a pH gradient		
Requirement for ampholytes to be included in solution		
Use of supporting medium with molecular sieving		
Treatment of samples with SDS is possible		
Diffusion of separated proteins is significant		

53.4 Identify proteins by 2D electrophoresis. The figure represents the separation of a mixture of 10 proteins by 2D electrophoresis. Using the information in the table, identify the individual proteins A–J.

M_r and pI values of selected protein subunits.

Protein	Subunit M_r	Subunit pI
Nerve growth factor	13 300	9.3
Ribonuclease	13 700	7.8
Haemoglobin	16 000	7.0
β-lactoglobulin	17 500	5.2
Ceramide trihexosidase	22 000	3.0
Trypsinogen	24 500	9.3
Triose phosphate isomerase	27 000	5.7
Galactokinase	27 000	5.7
Arginase	30 000	9.2
Glycerol-3-phosphate dehydrogenase	34 000	6.4
Alcohol dehydrogenase	35 000	5.4
Deoxyribonuclease II	38 000	10.2
Aldolase	40 000	5.2
Pepsinogen	41 000	3.7
Hexokinase	51 000	5.3
Lipoxidase	54 000	5.7
Catalase	57 500	5.4
Alkaline phosphatase	69 000	4.4
Acetylcholine esterase	70 000	4.5
Glyceraldehyde-3-phosphate dehydrogenase	72 000	8.5

Answers to these study exercises are available at www.pearsoned.co.uk/practicalskills.

Definitions

Oxidation – loss of electrons by an atom or molecule (or gain of O atoms, loss of H atoms, increase in positive charge, or decrease in negative charge).

Reduction – gain of electrons by an atom or molecule (or loss of O atoms, gain of H atoms, decrease in positive charge or increase in negative charge).

Fig. 54.1 A simple galvanic electrochemical cell. The KCl salt bridge allows migration of ions between the two compartments but prevents mixing of the two solutions.

Definition

Galvanic cell – an electrochemical cell in which reactions occur spontaneously at the electrodes when they are connected externally by a conductor, producing electrical energy.

Electrochemical methods are used to quantify a broad range of different biomolecules, including ions, gases, metabolites, drugs and hormones.

> **KEY POINT** The basis of all electrochemical analysis is the transfer of electrons from one atom or molecule to another atom or molecule in an obligately coupled oxidation–reduction reaction (a redox reaction).

It is convenient to separate such redox reactions into two half-reactions and, by convention, each is written as:

$$\text{oxidised form} + \text{electron(s) (ne}^-) \underset{oxidation}{\overset{reduction}{\rightleftharpoons}} \text{reduced form} \qquad [54.1]$$

You should note that the half-reaction is reversible: by applying suitable conditions, reduction *or* oxidation can take place. As an example, a simple redox reaction occurs when metallic zinc (Zn) is placed in a solution containing copper ions (Cu^{2+}), as follows:

$$Cu^{2+} + Zn \rightarrow Cu + Zn^{2+} \qquad [54.2]$$

The half-reactions are (i) $Cu^{2+} + 2e^- \rightarrow Cu$ and (ii) $Zn^{2+} + 2e^- \rightarrow Zn$. The oxidising power of (i) is greater than that of (ii), so in a coupled system the latter half-reaction proceeds in the opposite direction to that shown above, i.e. as $Zn - 2e^- \rightarrow Zn^{2+}$. When Zn and Cu electrodes are placed in separate solutions containing their ions, and connected electrically (Fig. 54.1), electrons will flow from the Zn electrode to Cu^{2+} via the Cu electrode owing to the difference in oxidising power of the two half-reactions.

By convention, the electrode potential of any half-reaction is expressed relative to that of a standard hydrogen electrode (half-reaction $2H^+ + 2e^- \rightarrow H_2$) and is called the standard electrode potential, E^o. Table 54.1 shows the values of E^o for selected half-reactions. With any pair of half-reactions from this series, electrons will flow from that having the lowest electrode potential to that of the highest. E^o is determined at pH = 0. It is often more appropriate to express standard electrode potentials at pH 7 for biological systems, and the symbol $E^{o'}$ is used: in all circumstances, it is important that the pH is clearly stated.

The arrangement shown in Fig. 54.1 represents a simple galvanic cell where two electrodes serve as the interfaces between a chemical system and an electrical system. For analytical purposes, the magnitude of the potential (voltage) or the current produced by an electrochemical cell is related to the concentration (strictly, the activity, *a*, p. 163) of a particular chemical species. Electrochemical methods offer the following advantages:

- **excellent detection limits**, and wide operating range (10^{-1}–10^{-8} mol L^{-1});
- **measurements may be made on very small volumes** (μL), allowing small amounts (pmol) of sample to be measured in some cases;
- **miniature electrochemical sensors can be used** for certain *in vivo* measurements, e.g. pH, glucose, oxygen content.

Table 54.1 Standard electrode potentials* (E°) for selected half-reactions.

Half-reaction	E° at 25° C (V)
$Cl_2 + 2e^- \rightleftharpoons 2Cl^-$	+1.36
$O_2 + 4H^+ + 4e^- \rightleftharpoons 2H_2O$	+1.23
$Br_2 + 2e^- \rightleftharpoons 2Br^-$	+1.09
$Ag^+ + e^- \rightleftharpoons Ag$	+0.80
$Fe^{3+} + e^- \rightleftharpoons Fe^{2+}$	+0.77
$I_3^- + 2e^- \rightleftharpoons 3I^-$	+0.54
$Cu^{2+} + 2e^- \rightleftharpoons Cu$	+0.34
$Hg_2Cl_2 + 2e^- \rightleftharpoons 2Hg + 2Cl^-$	+0.27
$AgCl + e^- \rightleftharpoons Ag + Cl^-$	+0.22
$Ag(S_2O_3)_2^{3-} + e^- \rightleftharpoons Ag + 2S_2O_3^{2-}$	+0.01
$2H^+ + 2e^- \rightleftharpoons H_2$	+0.00
$AgI + e^- \rightleftharpoons Ag + I$	−0.15
$PbSO_4 + 2e^- \rightleftharpoons Pb + SO_4^{2-}$	−0.35
$Cd^{2+} + 2e^- \rightleftharpoons Cd$	−0.40
$Zn^{2+} + 2e^- \rightleftharpoons Zn$	−0.76

*From Milazzo et al. (1978). Reproduced with permission of John Wiley & Sons Ltd.

Using a calomel electrode – always ensure that the KCl solution is saturated by checking that KCl crystals are present.

Fig. 54.2 Components of a potentiometric cell.

Potentiometry and ion-selective electrodes

Operating principles

These systems involve galvanic cells (p. 366) and are based on measurement of the potential (voltage) difference between two electrodes in solution when no net current flows between them: no net electrochemical reaction occurs and measurements are made under equilibrium conditions. These systems include methods for measuring pH, ions and gases such as CO_2 and NH_3. A typical potentiometric cell is shown in Fig. 54.2. It contains two electrodes:

1. **A 'sensing' electrode,** the half-cell potential of which responds to changes in the activity (concentration) of the substance to be measured; the most common type of indicator electrodes are ion-selective electrodes (ISE).

2. **A 'reference' electrode,** the potential of which does not change, forming the second half of the cell.

To assay a particular analyte, the potential difference between these electrodes is measured by a millivolt meter (e.g. a standard pH meter). Reference electrodes for potentiometry are of three main types:

1. **The standard hydrogen electrode,** which is the reference half-cell electrode, defined as 0.0 V at all temperatures, against which values of E° are expressed. H_2 gas at 1 atmosphere pressure is bubbled over a platinum electrode immersed in an acid solution with an activity of unity. This electrode is rarely used for analytical work, since it is unstable and other reference electrodes are easier to construct and use.

2. **The calomel electrode** (Fig. 54.3), which consists of a paste of mercury covered by a coat of calomel (Hg_2Cl_2), immersed in a saturated solution of KCl. The half-reaction: $Hg_2Cl_2 + 2e^- \rightarrow 2Hg + 2Cl^-$ gives a stable standard electrode potential of +0.24 V.

3. **The silver/silver chloride electrode.** This is a silver wire coated with AgCl and immersed in a solution of constant chloride concentration. The half-reaction: $AgCl + e^- \rightarrow Ag + Cl^-$ gives a stable, standard electrode potential of +0.20 V.

> **KEY POINT** Ion-selective electrodes (ISEs) are based on measurement of a potential across a membrane which is selective for a particular analyte.

An ISE consists of a membrane, an internal reference electrode and an internal reference electrolyte of fixed activity. The ISE is immersed in a sample solution that contains the analyte of interest, along with a reference electrode. The membrane is chosen to have a specific affinity for a particular ion, and if the activity of this ion in the sample differs from that in the reference electrolyte, a potential develops across the membrane that is dependent on the ratio of these activities. Since the potentials of the two reference electrodes (internal and external) are fixed, and the internal electrolyte is of constant activity, the measured potential, E, is dependent on the membrane potential and is given by the Nernst equation (Robinson and Stokes, 2002) as:

$$E = K + 2.303\frac{RT}{zF}\log[a] \qquad [54.3]$$

Fig. 54.3 A calomel reference electrode.

Fig. 54.4 Underlying principles of a gas-sensing electrode.

where K represents a constant potential which is dependent on the reference electrode, z represents the net charge on the analyte, [a] represents the activity of analyte in the sample and all other symbols and constants have their usual meaning (p. 197). For a series of standards of known activity, a plot of E against log [a] should be linear over the working range of the electrode, with a slope of 2.303 RT/zF (0.059 V at 25 °C). Although ISEs strictly measure *activity*, the potential differences can be approximated to concentration as long as (i) the analyte is in dilute solution (p. 161), (ii) the ionic strength of the calibration standards matches that of the sample, e.g. by adding appropriate amounts of a high ionic strength solution to the standards, and (iii) the effect of binding to sample macromolecules (e.g. proteins, nucleic acids) is minimal. Potentiometric measurements on undiluted biological fluids, e.g. K$^+$ and Na$^+$ levels in plasma, tissue fluids or urine, are likely to give lower values than flame emission spectrophotometry, since the latter procedure measures total ion levels, rather than just those in aqueous solution.

All of the various types of membrane used in ISEs operate by incorporating the ion to be analysed into the membrane, with the accompanying establishment of a membrane potential. The scope of electrochemical analysis has been extended to measuring gases and non-ionic compounds by combining ISEs with gas-permeable membranes, enzymes and even immobilised bacteria or tissues.

Glass membrane electrodes

The most widely used ISE is the glass membrane electrode for pH measurement (p. 169). The membrane is thin glass (50 μm wall thickness) made of silica which contains some Na$^+$. When the membrane is soaked in water, a thin hydrated layer is formed on the surface in which negative oxide groups (Si-O$^-$) in the glass act as ion exchange sites. If the electrode is placed in an acid solution, H$^+$ exchanges with Na$^+$ in the hydrated layer, producing an external surface potential: in alkaline solution, H$^+$ moves out of the membrane in exchange for Na$^+$. Since the inner surface potential is kept constant by exposure to a fixed activity of H$^+$, a consistent, accurate potentiometric response is observed over a wide pH range. Glass electrodes for other cations (e.g. Na$^+$, NH$_4^+$) have been developed by changing the composition of the glass, so that it is predominantly sensitive to the particular analyte, though the specificity of such electrodes is not absolute. The operating principles and maintenance of such electrodes are broadly similar to those for pH electrodes (p. 169).

Gas-sensing glass electrodes

Here, an ISE in contact with a thin external layer of aqueous electrolyte (the 'filling solution') is kept close to the glass membrane by an additional, outer membrane that is selectively permeable to the gas of interest. The arrangement for a CO$_2$ electrode is shown in Fig. 54.4: in this case the outer membrane is made of CO$_2$-permeable silicone rubber. When CO$_2$ gas in the sample selectively diffuses across the membrane and dissolves in the filling solution (in this case an aqueous NaHCO$_3$/NaCl mixture), a change in pH occurs owing to the shift in the equilibrium:

$$CO_2 + H_2O \rightleftharpoons H_2CO_3 \rightleftharpoons H^+ + HCO_3^- \qquad [54.4]$$

The pH change is 'sensed' by the internal ion-selective pH electrode, and its response is proportional to the partial pressure of CO$_2$ of the solution

internal solution

Fig. 54.5 Underlying principles of a liquid membrane ion-selective electrode. A^+ = analyte; C = neutral carrier ionophore; E_m = surface potential; membrane potential = E_m(internal) − E_m(external).

Definition

Ionophore – a compound that enhances membrane permeability to a specific ion: an ionophore may be incorporated into an ISE to detect that ion.

Definitions

Electrolysis – a non-spontaneous chemical reaction resulting from the application of a potential to an electrode.

Electrolytic cell – an electrochemical cell in which reactions are driven by the application of an external voltage greater than the spontaneous potential of the cell.

Commercial oxygen electrodes – the system developed by Leyland C. Clark is used in the Rank Brothers oxygen electrode, forming the basis of many practical exercises in the biosciences.

(pCO_2). A similar principle operates in the NH_3 electrode, where a Teflon membrane is used, and the filling solution is NH_4Cl.

Liquid and polymer membrane electrodes

In this type of ISE, the liquid is a water-insoluble viscous solvent containing a soluble ionophore, i.e. an organic ion-exchanger, or a neutral carrier molecule, that is specific for the analyte of interest. When this liquid is soaked into a thin membrane such as cellulose acetate, it becomes effectively immobilised. The arrangement of analyte (A^+) and ionophore in relation to this membrane is shown in Fig. 54.5. The potential on the inner surface of the membrane is kept constant by maintaining a constant activity of A^+ in the internal solution, so the potential change measured is that which results from A^+ in the sample interacting with the ionophore in the outer surface of the membrane.

A relevant example of a suitable ionophore is the antibiotic valinomycin (p. 425), which specifically binds K^+. Other ionophores have been developed for measurement of, e.g., NH_4^+, Ca^{2+}, Cl^-. In addition, electrodes have been developed for organic species by using specific ion-pairing reagents in the membrane that interact with ionic forms of the organic compound, e.g. with drugs such as 5,5-diphenylhydantoin.

Solid-state membrane electrodes

These contain membranes made from single crystals or pressed pellets of salts of the analyte. The membrane material must show some permeability to ions and must be virtually insoluble in water. Examples include:

- the fluoride electrode, which uses LaF_3 impregnated with Eu^{2+} (the latter to increase permeability to F^-). A membrane potential is set up when F^- in the sample solution enters spaces in the crystal lattice;
- the chloride electrode, which uses a pressed-pellet membrane of Ag_2S and AgCl.

Voltammetric methods

Voltammetric methods are based on measurements made using an electrochemical cell in which electrolysis is occurring. Voltammetry, sometimes also called amperometry, involves the use of a potential applied between two electrodes (the working electrode and the reference electrode) to cause oxidation or reduction of an electroactive analyte. The loss or gain of electrons at an electrode surface causes current to flow, and the size of the current (usually measured in mA or μA) is directly proportional to the concentration of the electroactive analyte. The materials used for the working electrode must be good conductors and electrochemically inert, so that they simply transfer electrons to and from species in solution. Suitable materials include Pt, Au, Hg and glassy carbon.

Two widely used devices that operate on the voltammetric principle are the oxygen electrode and the glucose electrode. These are sometimes referred to as amperometric sensors.

Oxygen electrodes

The Clark (Rank) oxygen electrode

These instruments measure oxygen in solution using the polarographic principle, i.e. by monitoring the current flowing between two electrodes

Fig. 54.6 Transverse section through a Clark (Rank) oxygen electrode.

Using polytetrafluoroethylene (PTFE) membranes – these are readily permeable to many small molecules and are used for monitoring of dissolved gases in aqueous systems.

when a voltage is applied. The most widespread electrode is the Clark cell type (Fig. 54.6), manufactured by Rank Bros, Cambridge, UK, which is suitable for measuring O_2 concentrations in cell, organelle and enzyme suspensions. Pt and Au electrodes are in contact with a solution of electrolyte (normally saturated KCl). The electrodes are separated from the medium by a Teflon membrane, permeable to O_2. When a potential is applied across the electrodes, this generates a current proportional to the O_2 concentration (Wilson and Walker, 2010). The reactions can be summed up as:

$$4Ag \rightarrow 4Ag^+ + 4e^-$$ (at silver anode)

$$O_2 + 2e^- + 2H^+ \rightleftharpoons H_2O_2$$ (in electrolyte solution; O_2 replenished by diffusion from test solution)

$$H_2O_2 + 2e^- + 2H^+ \rightleftharpoons 2H_2O$$ (at platinum cathode)

Setting up and using a Clark (Rank) oxygen electrode

Box 54.1 describes the steps involved: if you are setting up from scratch, perform steps 1–13; if a satisfactory membrane is already in place, start at step 7. Box 54.2 gives advice on how to use a chart recorder with an oxygen electrode, including how to calculate rates of oxygen consumption/production, using data for oxygen saturation of air-equilibrated water shown in Table 54.2.

The temperature of the incubation vessel should be controlled by passing water (e.g. from a water bath) through the outer chamber (Fig. 54.6). Cells or organelles may be present in the solution added to the incubation chamber or can be added via the hole in the stopper using a syringe, as can chemicals such as inhibitors (Box 54.2). Take care not to introduce air bubbles, and remove any that appear by gently raising and lowering the stopper. The electrode can be used repeatedly, providing the membrane is satisfactory:

Table 54.2 Oxygen saturation values for distilled water and sea water at standard atmospheric pressure and a range of temperatures (derived from Green and Carritt, 1967).

Temperature (°C)	Distilled water mol m^{-3} (mmol L^{-1})	mg L^{-1} (p.p.m.)	ml L^{-1}	Sea water mol m^{-3} (mmol L^{-1})	mg L^{-1} (p.p.m.)	ml L^{-1}
0	0.460	14.7	10.3	0.359	11.5	8.04
2	0.435	13.9	9.75	0.342	10.9	7.65
4	0.413	13.2	9.24	0.326	10.4	7.30
6	0.392	12.5	8.78	0.311	9.95	6.97
8	0.373	11.9	8.35	0.298	9.54	6.66
10	0.355	11.4	7.95	0.285	9.12	6.37
12	0.339	10.8	7.59	0.273	8.74	6.11
14	0.324	10.4	7.25	0.261	8.35	5.85
16	0.310	9.92	6.94	0.251	8.03	5.62
18	0.297	9.50	6.65	0.241	7.71	5.40
20	0.285	9.12	6.38	0.232	7.42	5.19
22	0.274	8.77	6.12	0.224	7.17	5.00
24	0.263	8.42	5.89	0.215	6.88	4.82
26	0.253	8.1	5.67	0.208	6.66	4.64
28	0.244	7.81	5.46	0.200	6.40	4.48
30	0.235	7.52	5.27	0.193	6.18	4.32
37	0.211	6.75	4.71	0.174	5.57	3.88

Notes: mol m^{-3} ≡ mmol L^{-1} ≡ μmol mL^{-1} ≡ nmol μL^{-1}. Tabulated values assume atmospheric pressure = 101.3 kPa (/760 mmHg); for more accurate work, a correction for any deviation can be made by multiplying the appropriate figures from the table by the ratio of the real pressure to the assumed pressure.

remove solutions carefully (e.g. using a pipette, or vacuum line and trap). Keep water in the chamber when not in use. Replace the membrane if:

- the reading becomes noisy;
- the electrode will not zero after adding sodium dithionite;
- the response becomes too slow (check by switching off the in-built stirrer – oxygen concentration should drop rapidly as the available oxygen is consumed).

> **KEY POINT** Successful operation of a Clark (Rank) oxygen electrode requires an intact Teflon membrane and clean electrodes. When replacing the Teflon membrane, you should always check that the electrode surfaces are clean (shiny); if necessary, use a mild abrasive paste to remove any oxidised material.

Box 54.1 How to set up a Clark (Rank) oxygen electrode

1. Detach the base of the incubation vessel (see Fig. 54.6) by unscrewing the locking ring.

2. Add enough saturated KCl to cover the electrodes.

3. Cut a 1-mm square hole in the centre of a 10×10-mm square of lens tissue and place this on the KCl solution so that the hole is over the central platinum cathode.

4. Cut a 10×10-mm square of Teflon membrane and place over the lens tissue; seal by gently lowering the incubation vessel and tightening the locking ring, making sure that the rubber O-ring is correctly positioned over the membrane.

   (a) Do not overtighten the locking ring.
   (b) Take care not to trap air bubbles beneath the membrane.
   (c) Make sure that the membrane does not become twisted.

5. Clamp the electrode over the magnetic stirrer base using the clamping screw.

6. Connect the electrode leads to the polarising unit/recording device (silver anode to positive, platinum cathode to negative). You can either use a digital readout system on the control unit or the output can be passed to a chart recorder, giving a readout of oxygen status as a function of time (see Box 54.2). Check that the polarising voltage is set to 0.60 V and adjust, if necessary, using the 'polarising voltage' control (use a small screwdriver to adjust).

7. Add air-saturated experimental solution and a small Teflon-coated magnetic stirrer bar to the chamber. The volume of the incubation chamber can be adjusted by moving the locking nut on the stopper. To adjust, add the appropriate volume of liquid to the chamber (9.5 mL) using a pipette, insert the stirrer bar and stopper and screw the locking nut on the stopper until the solution just fills the incubation chamber.

8. Gently push the stopper (volume adjuster) into position, making sure that no air bubbles are trapped in the chamber, and switch on the stirrer. Adjust the rate of stirring using the 'stirrer' control, if required.

9. Set the zero: remove the stopper and add a few crystals of sodium dithionite – the dithionite ions ($S_2O_4^{2-}$) will be oxidised to sulfite (SO_3^{2-}) and sulfate ions (SO_4^{2-}), thereby consuming all of the oxygen in the solution within 3–5 min (an alternative is to bubble N_2 through the solution for >10 min). Once the reading has stabilised, if this is not at zero, then adjust using the 'set zero' control (in practice, the zero is usually quite stable and often needs no adjustment).

10. Adjust the sensitivity – rinse out the dithionite solution thoroughly/repeatedly, then replace with air-equilibrated water and allow the reading to stabilise (this may take 5–10 min.). Then adjust the 'sensitivity' control of the electrode system to set the oxygen saturated value at an appropriate point (for a controller with a digital readout, this is best set to 100, and then the values represent percentage oxygen saturation). To check that the instrument is working correctly, switch off the stirrer for a few moments – the reading should drop quickly as oxygen is consumed at the electrode surface and should then rise rapidly when the stirrer is switched back on.

11. Rinse the incubation chamber thoroughly and add fresh experimental solution. Make sure that all traces of sodium dithionite are removed.

12. Carry out your experiment.

13. Remove the solution and check the calibration. If the reading for air-equilibrated water is different, the electrode's sensitivity or the temperature may have changed and you may need to recalibrate and repeat the measurement.

Box 54.2 How to convert a chart recorder trace to a rate of O_2 consumption or production

Having set up your oxygen electrode (Box 54.1) or probe, the most common experiments are those in which you measure the rate of oxygen consumption (respiration) or production (photosynthesis) in a fixed volume of solution (e.g. in the chamber of a Rank (Clark) oxygen electrode). Typically, this involves attaching the electrode to a chart recorder and following the change in oxygen concentration with time. The principal steps are as follows.

1. Calculate the amount of oxygen in the electrode chamber – multiply the appropriate oxygen-saturation concentration, from Table 54.2, by the volume of the chamber. For example, if you are working in mmol L^{-1} (= μmol mL^{-1}) then at 20°C, air-equilibrated (oxygen-saturated) distilled water contains 0.285 μmol mL^{-1}. For an electrode chamber of volume 5 mL, there will be $0.285 \times 5 = 1.425$ μmol oxygen in the water within the chamber at saturation.

2. Set the zero on the recorder – following the addition of a small amount of sodium dithionite to the chamber, to remove oxygen (Box 54.1), the trace is monitored until it has stabilised. Then adjust the 'zero' or 'back off' control of the recorder until the chart trace is set to zero on the chart paper: this is marked as 'A' on Fig. 54.8.

3. Set the oxygen-saturated reading on the recorder – rinse out the dithionite solution thoroughly/repeatedly, replace with air-equilibrated water and allow the reading to stabilise (this may take 3–5 min). Then adjust the 'sensitivity' control of the chart recorder to set the oxygen-saturated value at an appropriate point on the chart – for respiration measurements, this can be close to the full width of the chart paper, whereas for photosynthesis measurements it is more usual to set it closer to the mid-point of the chart paper, to allow measurements over 100% to be recorded. In the example shown in Fig. 54.8, where the chart paper is calibrated in millimetre divisions, the oxygen-saturated reading is set at 80 divisions (80 mm), marked as 'B'. Note that the recorder pen can be lifted between readings, and during rinsing of the chamber, to reduce the amount of 'noise' on the chart.

4. Calculate the amount of oxygen equivalent to a single division of the chart paper – divide the amount of oxygen in the chamber by the number of divisions (mm) to give the amount per division (e.g. $1.425 \div 80 = 0.0178125$ μmol (= 17.8125 nmol per division).

5. Carry out your experiment – the example shown in Fig. 54.8 represents the consumption of oxygen during respiration of a suspension of yeast cells, marked as 'C'. Continue the recording until you are satisfied that the rate has remained stable for at least 5 min. In many experiments, you will be adding substances that change the rate (e.g. metabolic inhibitors, or substrates) – this can be done by injecting a solution containing the substance through the small hole in the stopper of the electrode (keep the volume of this solution to <1% of the total volume of the chamber, to minimise the effect of the added solution).

6. Convert the gradient of the trace into a rate of oxygen consumed or produced – using a transparent ruler, draw the line of best fit through that part of the trace showing a stable relationship (see Fig. 54.8). Use this line of best fit to work out the gradient of the trace, by converting from linear dimensions to an amount per unit time; e.g. the gradient shown in Fig. 54.8 is equivalent to a decrease of 24 chart divisions (mm) in 3 cm. Using the value calculated in step 4, the amount of oxygen consumed is thus $17.8125 \times 24 = 427.5$ nmol. For a chart speed of 0.2 cm min^{-1}, this represents the change occurring over a time interval of $3 \div 0.2 = 15$ min. Thus the rate of oxygen consumed is $427.5 \div 15 = 28.5$ nmol min^{-1}.

7. Express the rate on an appropriate basis – in most instances, this will be either per cell (e.g. determined using a haemocytometer, p. 234) or per unit mass of protein (p. 383), chlorophyll (p. 312) or an equivalent marker. For the example shown in Fig. 54.8, if there were 2.4×10^7 yeast cells per ml of suspension, then in the 5-ml chamber of the electrode there would be a total of $2.4 \times 10^7 \times 5 = 1.2 \times 10^8$ yeast cells, giving a rate of $28.5 \div (1.2 \times 10^8) = 2.375 \times 10^{-7}$ nmol min^{-1} $cell^{-1}$, probably better expressed as 237.5 amol min^{-1} $cell^{-1}$. Note that when inhibitors/substrates are used, you should always wait until a new stable rate has been achieved (for 5 min) before using the trace to determine the new rate of oxygen consumption/production.

Oxygen probes

Clark-type oxygen electrodes are also available in probe form for immersion in the test solution (Fig. 54.7), e.g. for field studies, allowing direct measurement of oxygen status *in situ*, in contrast to chemical assays.

Definition

Biosensor – a device for measuring a substance, combining the selectivity of a biological reaction with that of a sensing electrode.

KEY POINT Note that the solution must be stirred during measurement, to replenish the oxygen consumed by the electrode, owing to the 'boundary layer' effect: gentle movement of the probe during measurement provides sufficient agitation.

Fig. 54.7 A Clark-type oxygen probe.

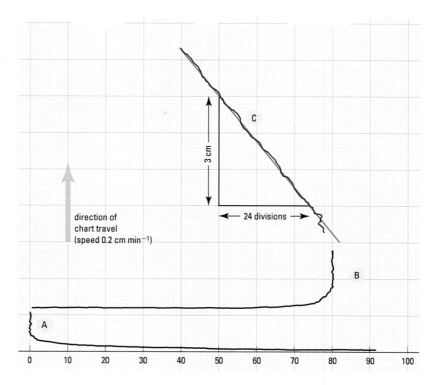

Fig. 54.8 Chart recorder output for a Clark (Rank) oxygen electrode. A = zero oxygen; B = oxygen saturation; C = respiration of yeast.

Fig. 54.9 Underlying principles of a glucose electrode.

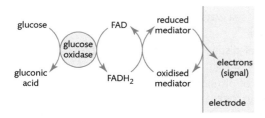

Fig. 54.10 Operating principles of a glucose biosensor. In the ExacTech system, the mediator is a ferrocene derivative.

Glucose electrodes

These are simple types of biosensor; a simple design is shown in Fig. 54.9. It consists of a Pt electrode, overlaid by two membranes. Sandwiched between these membranes is a layer of the immobilised enzyme glucose oxidase. The outer membrane is glucose-permeable and allows glucose in the sample to diffuse through to the glucose oxidase layer, where it is converted to gluconic acid and H_2O_2. The inner membrane is selectively permeable to H_2O_2, which is oxidised to O_2 at the surface of the Pt electrode. The current arising from this release of electrons is proportional to the glucose concentration in the sample within the range 10^{-7}–10^{-3} mol L^{-1}. The efficiency of a biosensor can be increased by including an electrode surface that is capable of capturing electrons directly. This is the basis of recently developed blood glucose meters for diabetics such as the ExacTech meter (Medisense-Abbott Laboratories). Here a pinprick blood sample is applied to a test strip containing glucose oxidase; if glucose is present, a series of reactions occurs as shown in Fig. 54.10.

The electrons produced are donated to the electrode surface, producing an initial current that is proportional to the initial rate at which glucose is oxidised which, in turn, is proportional to the glucose concentration in the blood sample. The future design of biosensors is likely to involve microchip technology in which the biocatalyst and the transducer are even more intimately linked (for more details, see Eggins, 2002).

Electrochemical detectors used in chromatography operate by voltammetric principles and currents are produced as the mobile phase flows over electrodes set at a fixed potential: to achieve maximum sensitivity, this potential must be set at a level that allows electrochemical reactions to occur in all analytes of interest.

Coulometric methods

Here, the charge required to electrolyse a sample completely is measured: the time required to titrate an analyte is measured at constant current and related to the amount of analyte using Faraday's law. There are few biological applications of this technique, though it is sometimes used for determination of Cl^- in serum and body fluids.

Text references

Eggins, B.R. (2002) *Chemical Sensors and Biosensors.* Wiley, New York.

Green, E.J. and Carritt, D.E. (1967) New tables for oxygen saturation of seawater, *Journal of Marine Research*, 25, 140–7.

Milazzo, G., Caroli, S. and Sharma, V.K. (1978) *Tables of Standard Electrode Potentials.* Wiley, London.

Sources for further study

Anon. *Rank Dissolved Oxygen Electrode.* Available: http://www.rankbrothers.co.uk/prod1.htm Last accessed 01/09/15.
[Homepage of the manufacturing company – contains technical manuals and contact information.]

Knopf, G.K. and Bassi, A.S. (2006) *Smart Biosensor Technology.* Taylor & Francis, New York.

Ozkan, S.A., Kauffmann, J-M. and Zuman, P. (2013) *Electroanalysis in Biomedical and Pharmaceutical*

Robinson, R.A. and Stokes, R.H. (2002) *Electrolyte Solutions.* Dover Publications, New York.

Wilson, K. and Walker, J. (eds) (2010) *Principles and Techniques of Practical Biochemistry and Molecular Biology,* 7th edn. Cambridge University Press, Cambridge.
[Also covers a range of other topics, including chromatography, spectroscopy and centrifugation.]

Sciences: Voltammetry, Amperometry, Biosensors, Applications. Springer, New York.

Sanders, G.H.W, Compton, R.G. and Compton, S. (1996) *Electrode Potentials.* Oxford University Press, Oxford.
[Covers the underlying theory, plus applications relating to ion-selective electrodes.]

STUDY EXERCISES

54.1 Test your understanding of the circumstances under which ion-selective electrodes can be used. List the three main assumptions underlying the measurement of ion concentration using an ion-selective electrode.

54.2 Test your knowledge of biosensor design. List the two main functional components of a biosensor such as the glucose electrode. Research a specific application – note that it should be different from the ones used in this chapter (e.g. using the web).

54.3 Calculate the oxygen content of specified volumes of water at defined temperatures. Using the information in Table 54.2 and assuming

air equilibration at 101.3 kPa, what is the amount of oxygen in each of the following (give all answers to three significant figures):

(a) 4 mL of distilled water at 20 °C (express your answer in µmol);

(b) 20 mL of sea water at 12 °C (express your answer in µmol);

(c) 10 mL of distilled water at 15 °C (express your answer in µmol);

(d) 250 mL of distilled water at 37 °C (express your answer in mg);

(e) 200 mL of sea water at 25 °C (express your answer in mL).

(continued)

54.4 **Calculate respiration using data from an oxygen electrode readout.** An oxygen electrode was set up to contain 5.0 mL of air-equilibrated distilled water at 30 °C, giving a full-scale reading on a chart recorder (200 divisions on the chart recorder paper), while a zero reading gave no deflection (0 divisions on the chart recorder paper). A yeast cell suspension (containing 4.1×10^9 cells in total) was added to the same electrode system along with a suitable growth medium, giving an approximately linear trace on the chart recorder with a slope of -25 units per cm, at a chart speed of 0.5 cm min^{-1}. What is the respiratory oxygen consumption in nmol min^{-1} (10^9 cells)$^{-1}$? (Give your answer to three significant figures.)

54.5 **Determine net photosynthetic rate from oxygen electrode readings.** The data below represent oxygen electrode readings taken every minute from 0 min for an illuminated suspension of a cyanobacterium in a 5.0-mL chamber of an oxygen electrode at 20 °C: 70.4, 73.6, 75.3, 78.5, 83.6, 85.9, 88.2, 91.4, 94.1. The system was calibrated to read 100.0 for air-equilibrated distilled water and 0.0 for anaerobic water, in effect making the scale read in terms of percentage oxygen saturation. The chlorophyll a content of an equivalent amount of cyanobacterial suspension to that used in the electrode chamber was measured at 2.13 µg mL^{-1}. What is the net photosynthetic rate of the cyanobacterial suspension, in µmol min^{-1} (mg chlorophyll a)$^{-1}$? (Express your answer to three significant figures.)

Answers to these study exercises are available at www.pearsoned.co.uk/practicalskills.

Assaying biomolecules and studying metabolism

55 Analysis of biomolecules: fundamental principles

Definitions

Accuracy – the closeness of an individual measurement, or a mean value based on a number of measurements, to the true value.

Detection limit – the minimum concentration of an analyte that can be detected at a particular confidence level.

Drift – 'baseline' movement in a particular direction: drift can be a problem between analyses (e.g. using a spectrophotometer for colorimetric analysis), or for a single analysis (e.g. when separating biomolecules by chromatography).

Noise – random fluctuations in a continuously monitored signal.

Precision – the extent of mutual agreement between replicate data values for an individual sample.

Quality assurance – procedures to monitor, document and audit a process against a set of criteria, usually defined externally (e.g. ISO 9000), to confirm that the process meets the specified requirements.

Quality control – internal procedures designed to ensure that the performance of a process meets a particular standard, e.g. analysis of a test sample containing a known amount of a substance, to confirm that the measurement falls within an acceptable range of the true value.

Range of measurement – the range of values from the detection limit to the upper concentration at which the technique becomes inaccurate or imprecise.

Replicate – repeated measurement.

Selectivity – the extent to which a method is free from interference owing to other substances in the sample.

Sensitivity – the ability to discriminate between small differences in analyte concentration.

Validation – the process of checking the accuracy and precision of a particular analytical method in relation to specific standards, using an appropriate reference material containing a known amount of analyte.

Biochemical analysis involves the characterisation of biological components within a sample using appropriate laboratory techniques. Most analytical methods rely on one or more chemical or physical properties of the test substance (the analyte) for detection and/or measurement. There are two principal approaches:

1. **Qualitative analysis** – where a sample is assayed to determine whether a biomolecule is present or absent. As an example, a blood sample might be analysed for a particular drug or a specific antibody (p. 281), or a bacterial cell might be probed for a nucleic acid sequence (p. 462).

2. **Quantitative analysis** – where the quantity of a particular biomolecule in a sample is determined, either as an amount (e.g. as grams, or moles) or in terms of its concentration in the sample (e.g. as $g\ L^{-1}$, or $mol\ m^{-3}$). For example, a blood sample might be analysed to determine its pH ($-\log_{10} [H^+]$), alcohol concentration in $mg\ mL^{-1}$, or glucose concentration in $mmol\ L^{-1}$. Skoog *et al.* (2006) give details of methods.

Your choice of approach will be determined by the purpose of the investigation and by the level of accuracy and precision required (p. 193). Many of the basic quantitative methods described in Chapters 56–60 rely on chemical reactions of the analyte and involve assumptions about the nature of the test substance and the lack of interfering compounds in the sample: such assumptions are unlikely to be wholly valid at all times. If you need to make more exacting measurements of a particular analyte, it may be necessary to separate it from the other components in the sample, e.g. using chromatography (Chapter 50), ultracentrifugation (Chapter 49), or electrophoresis (Chapter 52), and then identify the separated components, e.g. using spectroscopic methods (Chapter 47). However, each stage in the separation and purification procedure may introduce further errors and/or loss of sample, as described in detail for protein purification in Chapter 60.

> **KEY POINT** In general, you should aim to use the simplest procedure that satisfies the purpose of your investigation – there is little value in using a complex, time-consuming or costly analytical procedure in a study where a high degree of accuracy is not required.

Identification of any chemical substance requires specific information about its molecular structure. For some chemicals and techniques, this is straightforward, as very specific signals are produced, e.g. absorption spectra of metal ions. In cases where the probability of interference by other compounds is minimal, a purification step may not be necessary. There may be difficulties in identifying organic compounds owing to the possibility of isomers providing similar signals. An additional complication in biological and environmental samples is that these compounds are always present in complex mixtures, so purification is generally essential. Consequently, an element of uncertainty is present in any analytical technique.

Most of the routine methods based on chemical analysis are destructive, since the analyte is usually converted to another substance which is then

Assaying biomolecules and studying metabolism **379**

Evaluating a new method – results from a novel technique can be compared with an established 'standardised' technique by measuring the same set of samples by each method and analysing the results by correlation (p. 534).

Table 55.1 Some examples of colorimetric assays

Analyte	Reagent/wavelength
Amino acids	ninhydrin/540 nm (p. 382)
Proteins	biuret/520 nm (p. 383)
	Folin–Ciocalteau/600 nm (p. 384)
Carbohydrates	anthrone/625 nm (p. 394)
Reducing sugars	dinitrosalicylate/540 nm (p. 395)
DNA	diphenylamine/600 nm (p. 401)
RNA	orcinol/660 nm (p. 401)

Interpreting results from 'spiked' samples – remember that such procedures tell you nothing about the extraction efficiency of biomolecules from a particular sample, e.g. during homogenisation (Chapter 41).

Criteria for the selection of a particular analytical method:

- the required level of accuracy and precision;
- the number of samples to be analysed;
- the amount of each sample available for analysis;
- the physical form of the samples;
- the expected concentration range of the analyte in the samples;
- the sensitivity and detection limit of the technique;
- the likelihood of interfering substances;
- the speed of analysis;
- the ease and convenience of the procedure;
- the skill required by the operator;
- the cost and availability of the equipment.

assayed, e.g. in colorimetric assays of the major types of biomolecules (Table 55.1). In contrast, many of the analytical methods based on physical properties are non-destructive (e.g. the intrinsic absorption and emission of electromagnetic radiation in spectrophotometry, p. 311, or nuclear magnetic resonance techniques, p. 321). Non-destructive methods are often preferred, as they allow the further characterisation of a particular sample. Most biological methods are destructive: bioassays are often sensitive to interference and require validation (p. 236).

Validity and quantitative analysis

Before using a particular procedure, you should consider its possible limitations in terms of:

- **measurement errors, and their likely magnitude:** these might include processing errors (e.g. in preparing solutions and making dilutions), instrumental errors (e.g. a pH electrode that has not been set up correctly), calibration errors (e.g. converting a digital readout analyte concentration) and errors because of the presence of interfering substances;

- **sampling errors:** these may occur if the material used for analysis is not representative, e.g. owing to biological differences between the individual organisms used in the sampling procedure (p. 514).

Replication will allow you to make quantitative estimates of several potential sources of error; for example, repeated measurements of the same sample can provide information on the precision of the analytical method, e.g. by calculating the coefficient of variation (p. 520), while measurements of several different samples can provide information on biological and sample variability, e.g. by calculating the standard deviation (Chapter 73). Analyses of different sets of samples at different times (e.g. on different days) can provide information on 'between batch' variability, as opposed to 'within batch' variability (based on a single set of analytical data).

The reliability of a particular method can be assessed by measuring 'standards' (sometimes termed 'controls'). These are often prepared in the laboratory by adding a known amount of analyte to a real sample (this is often termed 'spiking' a sample), or by preparing an artificial sample containing a known amount of analyte along with other relevant components (e.g. the major sample constituents and possible interfering substances). In many instances, several standards (including a 'blank' or 'zero') are assayed to construct a 'standard curve', which is then used to convert sample measurements to amounts of analyte (see Chapter 42 for details). Such standard curves form the basis of many routine laboratory assays: while hand-drawn linear calibration curves are sufficient for basic assays, more complex curves are often fitted to a particular mathematical function using a computer program (e.g. bioassays, p. 236). Standards can also be used to check the calibration of a particular method; a mean value based on repeated measurements of an individual 'standard' can be compared with the true value using a modified t-test (p. 531, Eqn [74.1]), in which there is only one standard error term, i.e. that associated with the measured values.

Validation of a particular method can be important in certain circumstances, e.g. in a forensic science or a clinical biochemistry laboratory, where particular results can have important implications. Such laboratories

Evaluating a new method – results from a novel technique can be compared with an established 'standard' technique by assaying the same samples by both methods and analysing the results by correlation (p. 533).

operate strict validation procedures, including: (i) adherence to standard operating procedures for each analytical method; (ii) calibration of assays using certified reference materials containing a known amount of analyte and traceable to a national reference laboratory; (iii) effective systems for internal quality control and external quality assurance (p. 379); (iv) detailed record-keeping, covering all aspects of the analysis and recording of results. Although such rigour is not required for routine analysis, the general principles of standardisation, calibration, assessment of performance and record-keeping are equally valid for all analytical work.

Text reference

Skoog, D.A., Holler, F.J. and Crouch, S.R. (2006) *Instrumental Analysis Principles*. Brooks-Cole, Pacific Grove.

Sources for further study

Hage, D.S. and Carr, J.R. (2010) *Analytical Chemistry and Quantitative Analysis*. Pearson Education, Harlow.

Venn, R.F. (2008) *Principles and Practice of Bioanalysis,* 2nd edn. CRC Press, Boca Raton.

Wilson, K. and Walker, J. (eds) (2010) *Principles and Techniques of Practical Biochemistry,* 7th edn. Cambridge University Press, Cambridge.

STUDY EXERCISES

55.1 Check your understanding of the fundamental principles of biomolecular analysis. Distinguish between each of the following pairs of terms:

(a) qualitative analysis and quantitative analysis;

(b) sensitivity and selectivity;

(c) accuracy and precision;

(d) validation and replication;

(e) noise and drift.

55.2 Research sources of calibration standards and reference materials. Using the web, find a supplier of a validated reference standard for the following:

(a) cortisol in serum;

(b) cotinine in urine.

55.3 Compare equipment specifications. The table (alongside) gives operational details for two glucose meters, A and B.

Which meter has:

(a) the lower detection limit?

(b) the wider linear dynamic range?

(c) the greater accuracy?

(d) the better level of precision?

(e) the smaller sample volume?

Performance of two glucose meters

Aspect	Meter A	Meter B
Units of measurement	mmol L^{-1}	mg dL^{-1}
Detection limit	0.10	1.00
Linearity	to 20.0	to 250.0
Accuracy	±0.050	±1.50
Relative standard deviation (%)	3.5	2.5
Sample volume	40 μL	0.05 mL

Answers to these study exercises are available at www.pearsoned.co.uk/practicalskills.

Fig. 56.1 Structure of α-amino acids. R = side chain (see Table 56.1 for examples).

Table 56.1 The 20 amino acids incorporated into protein, grouped according to their side chains.

Name	Three-letter code	Capital letter code
Aliphatic side chains		
Glycine	gly	G
Alanine	ala	A
Valine	val	V
Leucine	leu	L
Isoleucine	ile	I
Aromatic side chains		
Phenylalanine	phe	F
Tyrosine	tyr	Y
Tryptophan	trp	W
S-containing side chains		
Cysteine	cys	C
Methionine	met	M
Side chains with –OH groups*		
Serine	ser	S
Threonine	thr	T
Basic side chains		
Histidine	his	H
Lysine	lys	K
Arginine	arg	R
Acidic side chains		
Aspartate	asp	D
Glutamate	glu	E
Amide side chains		
Asparagine	asn	N
Glutamine	gln	Q
Cyclic structure (imino acid)		
Proline	pro	P

*tyr also has an OH group.

Definitions

Conjugated protein – a protein containing a non-amino acid component (the prosthetic group), in addition to a polypeptide component (the apoprotein).

Simple protein – a protein composed entirely of amino acid residues.

Proteins are linear polymers, formed by the linkage of α-amino acids (Fig. 56.1) via peptide bonds to create polypeptide chains. The 20 amino acids most commonly incorporated into proteins are listed in Table 56.1. Others may be found when a protein is hydrolysed, owing to post-translational modification (e.g. hydroxyproline).

> **KEY POINT** In all amino acids apart from glycine, the α-carbon atom is asymmetrical, resulting in stereoisomers. Most life scientists use the D- and L- nomenclature (Fischer convention) when referring to stereoisomers of amino acids, although the R- and S- convention (p. 363) can be a useful alternative when considering enzymatic reaction mechanisms.

While most of the chemical properties of stereoisomers are identical, enzymes (including those involved in protein synthesis) can distinguish between them and, with the exception of a few bacterial components, only the L-isomers of amino acids are normally used in protein synthesis.

Amino acids and their properties

There is no agreed formal classification of amino acids but, in terms of protein structure and function, the side chains of amino acid residues can be grouped into:

- **those that carry a charge** (e.g. asp, glu, lys, arg, his) – they may participate in electrostatic interactions and charge–dipole (hydrophilic) interactions;
- **those that contain O, N or S** (e.g. ser, met, thr, asn, gln, cys, tyr) – they can participate in hydrogen bonding and hydrophilic interactions;
- **those that are hydrocarbon in nature** (e.g. ala, leu, val, ile, phe, trp) – they may participate in hydrophobic interactions;
- **cysteine** – this amino acid can form intra- and inter-polypeptide disulfide bridges.

Detection and quantification of amino acids

The primary amino group of amino acids will react with ninhydrin to give a purple-coloured product – this reaction can be used for qualitative assay, e.g. to detect individual amino acids in chromatography (p. 332), or for quantitative colorimetric assay by measuring the absorbance at 570 nm using a spectrophotometer (p. 313). Note that different amino acids give different amounts of coloured product on reaction with ninhydrin, so careful standardisation is needed. The secondary amino groups of the cyclic imino acids hydroxyproline and proline give yellow products with ninhydrin and are assayed at 440 nm.

Proteins and their properties

The diversity of protein structure must be taken into account when considering a suitable analytical method. Globular proteins (e.g. albumin, antibodies) are relatively soluble in dilute salt solutions, while fibrous proteins (e.g. collagens, keratin) are typically insoluble, unless hydrolysed (e.g. using 6 mol 1^{-1} HCl). Conjugated proteins (e.g. metalloproteins, glycoproteins, nucleoproteins) also include non-amino acid components, which may influence the choice of analytical method.

Definitions

Oligopeptide – a very short chain, typically containing 6–10 amino acid residues.

Peptide – a general term for a chain of amino acids linked by peptide bonds, often applied to shorter polymers of up to 50 amino acid residues.

Polypeptide – a single chain of amino acid residues, typically applied to longer polymers of over 50 amino acid residues.

Chemical properties of proteins and peptides

The features most commonly exploited for quantitative analysis are:

- **the peptide bond,** e.g. in the biuret reaction (Box 56.1);
- **the phenolic group of tyr and the indole group of trp,** which react with the oxidising agents phosphotungstic and phosphomolybdic acids in the Folin–Ciocalteau reagent to produce a blue colour. This is combined with the biuret reaction in the Lowry method (Box 56.1);
- **dye binding to hydrophobic regions,** e.g. Bradford assay (Box 56.1);
- **the primary amino groups of lys residues, the guanidino group of arg residues and the N-terminal amino acid residue** will react with ninhydrin, allowing colorimetric assay similar to that described above for isolated amino acids. However, a more sensitive method uses the reaction of these groups with fluorescamine to give an intensely fluorescent product, which you can measure by spectrofluorimetry (p. 314).

Most assays for proteins and peptides do not give absolute values, but require standard solutions, containing appropriate amounts of a particular

---

Box 56.1 Methods of determining the amount of protein/peptide in an aqueous solution

For all of the following methods, the amounts are appropriate for semi-micro cuvettes (1.5 mL volume, path length 1 cm). Appropriate controls (blanks) must be analysed, to assess possible interference (e.g. due to buffers, etc.).

Biuret method

This is based on the specific reaction between cupric ions (Cu^{2+}) in alkaline solution and two adjacent peptide bonds, as found in proteins and peptides. As such, it is not significantly affected by differences in amino acid composition.

1. Prepare protein standards over an appropriate range (typically, between 1 and 10mg mL^{-1}).

2. Add 1 mL of each standard solution to separate test tubes. Prepare a reagent blank, using 1 mL of distilled water, or an appropriate solution.

3. Add 1 mL of each unknown solution to separate test tubes.

4. Add 1 mL of biuret reagent (1.5 g CuSO$_4$ 5H$_2$O, 6.0 g sodium potassium tartrate in 300 mL of 10 w/v NaOH) to all standard and unknown tubes and to the reagent blank.

5. Incubate at 37 °C for 15 min.

6. Read the absorbance of each solution at 520 nm against the reagent blank. The violet colour is stable for several hours.

The main limitation of the biuret method is its lack of sensitivity – it is unsuitable for solutions with a protein content of less than 1 mg mL^{-1}.

Direct measurement of UV absorbance (Warburg-Christian method)

Proteins and peptides absorb EMR maximally at 280nm (owing to the presence of aromatic amino acids) and this forms the basis of the method. The principal advantages of this approach are its simplicity and the fact that the assay is non-destructive. The most common interfering substances are nucleic acids, which can be assessed by measuring the absorbance at 260 nm: a pure solution of protein will have a ratio of absorption (A_{260}/A_{280}) of approximately 0.6, decreasing with increasing nucleic acid contamination. Note also that any free aromatic amino acids in your solution will absorb at 280 nm, leading to an overestimation of protein content. The simplest procedure, which includes a correction for small amounts of nucleic acid, is as follows (use quartz cuvettes throughout).

1. Measure the absorbance of your solution at 280 nm (A_{280}): if A_{280} is greater than 1, dilute by an appropriate amount and remeasure (see p. 313).

2. Repeat at 260 nm (A_{260}).

3. Estimate the approximate protein concentration of the solution using the following relationship:

$$[\text{protein}]\ \text{mg mL}^{-1} = 1.45A_{280} - 0.74A_{260} \qquad [56.1]$$

This equation is based on the work of Warburg and Christian (1942) for enolase. For other proteins, it should not be used for quantitative work, since it gives only a rough approximation of the amount present, due to variations in aromatic amino acid composition.

continued

Box 56.1 (continued)

Lowry (Folin–Ciocalteau) method

This is a colorimetric assay, based on a combination of the biuret method, described above, and the oxidation of tyrosine and tryptophan residues with Folin–Ciocalteau reagent to give a blue–purple colour. The method is extremely sensitive (down to a protein/peptide content of 20 µg mL^{-1}), but is subject to interference from a wide range of non-protein substances, including many organic buffers (e.g. TRIS, HEPES), EDTA, urea and certain sugars. The choice of an appropriate standard is important, as the intensity of colour produced for a particular protein/peptide is dependent on the amount of aromatic amino acids present.

1. Prepare protein standards within an appropriate range for your samples (the method can be used from 0.02 to 1.00 mg mL^{-1}).

2. Add 1 mL of each standard solution to separate test tubes. Prepare a reagent blank, using 1 mL of distilled water, or an appropriate solution.

3. Add 1 mL of each of your unknown solutions to separate test tubes.

4. Then, add 5 mL of 'alkaline solution' (prepared by mixing 2% w/v Na$_2$CO$_3$ in 0.1 mol L^{-1} NaOH, 1% w/v aqueous CuSO$_4$ and 2% w/v aqueous NaK tartrate in the ratio 100:1:1. Mix thoroughly and allow to stand for at least 10 min.

5. Add 0.5 mL of Folin–Ciocalteau reagent (commercial reagent, diluted 1:1 with distilled water on the day of use). Mix rapidly and thoroughly and then allow to stand for 30 min.

6. Read the absorbance of each sample at 600 nm.

Dye-binding (Bradford) method

Coomassie brilliant blue combines with proteins and peptides to give a dye–protein complex with an absorption maximum of 595 nm. This provides a simple and sensitive means of measuring protein content, with few interferences. However, the formation of dye–protein complex is affected by the number of basic amino acids within a protein, so the choice of an appropriate standard is important. The method is sensitive down to a protein content of approximately 5 µg mL^{-1} but the relationship between absorbance and concentration is often non-linear, particularly at high protein content.

1. Prepare protein standards over an appropriate range (between 5 and 100 µg mL^{-1}).

2. Add 100 µl of each standard solution to separate test tubes. Prepare a reagent blank, using 100 µL of distilled water, or an appropriate solution (note that these small volumes must be accurately dispensed, e.g. using a calibrated pipettor, p. 147).

3. Add 100 µL of your unknown solutions to separate test tubes.

4. Add 5.0 mL of Coomassie brilliant blue G250 solution (0.1 g L^{-1}).

5. Mix and incubate for at least 5 min: read the absorbance of each solution at 595 nm.

Other methods are less widely used. They include determination of the total amount of nitrogen in solution (e.g. using the Kjeldahl technique) and calculating the protein content, assuming a nitrogen content of 16%. An alternative approach is to precipitate the protein (e.g. using trichloroacetic acid, tannic acid or salicylic acid) and then measure the turbidity of the resulting precipitate (using a nephelometer, or a spectrophotometer, p. 313).

Advice on preparing standard (calibration) curves – see p. 276 for practical guidance.

Exploiting the biological properties of proteins and peptides – relevant analytical methods include immunoassay (p. 281), enzymatic analysis (p. 411) and affinity chromatography (p. 338), based on specific biological interactions.

protein, to be analysed at the same time, so that a standard curve can be constructed. Bovine serum albumin (BSA) is commonly used as a protein standard. However, you may need an alternative standard if the protein you are assaying has an amino acid composition that is markedly different from that of BSA, depending on your chosen method.

Physical properties of proteins and peptides

The characteristics that can be used for separation and analysis include:

- **absorbance of aromatic amino acid residues** (phe, tyr and trp) at 280 nm (Warburg–Christian method, Box 56.1);

- **prosthetic groups of conjugated proteins** – these often have characteristic absorption maxima, e.g. the haem group of haemoglobin absorbs strongly at 415 nm, allowing quantitative assay;

- **density** – most proteins have a density of 1.33 kg L^{-1}, but lipoproteins have a lower density. This can be used to separate lipoproteins from other classes of protein, or to subdivide the various classes of lipoproteins;
- **net charge** – proteins and peptides differ in the types and number of amino acids with ionisable groups in their side chains. The ionisation of these side chains is pH-dependent, resulting in variation of net charge with pH (p. 349). This property is exploited in the techniques of electrophoresis (Chapter 52), isoelectric focusing (Chapter 53) and ion-exchange chromatography (Chapter 50);
- **water solubility and surface hydrophobicity** – these are exploited in salt fractionation (p. 406) and hydrophobic interaction chromatography (p. 337).

Determining the primary structure of a protein

The term 'primary structure' refers to the linear sequence of amino acid residues along the polypeptide: ultimately this determines the 3D shape of the molecule and hence its biological properties.

Amino acid analysis

Once a particular protein has been purified (Chapter 60), the first step in determining its primary structure is to hydrolyse the polypeptide and then determine the constituent amino acids. This can be achieved by several techniques, including thin-layer chromatography and high-performance liquid chromatography (Chapter 50). Often a dedicated amino acid analyser is used for quantitative assays: a polystyrene resin-based cation exchange column is used to separate the amino acids on the basis of ion-exchange and hydrophobic interactions.

Sequence analysis

Having obtained the amino acid composition of the protein, the next step is to determine the order of the amino acid residues along the polypeptide chain. This may be achieved by the following process.

1. **Cleavage of the polypeptide** at specific peptide bonds, resulting in smaller fragments (peptides); e.g. using cyanogen bromide (CNBr), which cleaves peptide bonds formed by the carboxyl group of met, or by proteolytic enzymes, e.g. trypsin (cleaving after lys and arg residues) and chymotrypsin (cleaving after phe, tyr and trp).
2. **Separation of the peptides,** e.g. by column chromatography.
3. **Determination of the sequence of each peptide** by a process called Edman degradation. This involves the selective removal of the N-terminal amino acid by treating the fragments with phenyl isothiocyanate ($C_6H_5N{=}C{=}S$) followed by acid hydrolysis; the substituted phenylthiohydantoin formed can then be identified by chromatography. The process can be repeated as many as 50 times, releasing the N-terminal residue on each occasion, and allowing the sequence of each fragment to be established.
4. **Matching of the sequenced, overlapping peptide fragments,** to determine the overall sequence of amino acid residues in the original protein.

Definitions

Glycoproteins – compounds consisting of proteins covalently attached to carbohydrate residues in post-translational modifications of the protein constituents.

Lipoproteins – globular, micelle-like particles consisting of a non-polar core of triacyl-glyceryl and cholesteryl esters surrounded by a coating of proteins, phospholipids and cholesterol: involved in the transport of lipids in blood.

Purifying a protein – the practical procedures involved in the purification of a particular protein from a biological sample are considered in detail in Chapter 60.

SAFETY NOTE Hydrolysis of proteins – this is usually achieved using 6 mol L^{-1} HCl at 110°C for 24–72h: care is required when working with hot acids.

Investigating the secondary and tertiary structure of a protein – computer-based molecular modelling programs can be used to predict the 3D conformation of a protein from information based on the primary structure (Chapter 12).

Text reference

Warburg, O. and Christian, W. (1942) *Isolierung und Kristallisation des Garungsferments Enolase.* Biochemische Zeitschrift, **310,** pp. 384–421.

Sources for further study

Hughes, A.B. (2012) *Amino Acids, Peptides and Proteins in Organic Chemistry: Analysis and Function of Amino Acids and Peptides.* Wiley, New York.

Lesk, A.M. (2010) *Introduction to Protein Science: Architecture, Function and Genomics,* 2nd edn. Oxford University Press, Oxford.

White, J.S. and White, D.C. (2002) *Proteins, Peptides and Amino Acids Sourcebook.* Humana Press, New York.

Whitford, D. (2005) *Proteins: Structure and Function.* Wiley, Chichester.

STUDY EXERCISES

56.1 Test your knowledge of the basis of protein assays. Match up each of the following methods with the corresponding underlying reaction: Method: (a) Warburg–Christian; (b) Bradford; (c) Lowry (Folin–Ciocalteau); (d) Biuret. Reaction: (i) Coomassie blue binding; (ii) copper(II) ion–peptide bond interaction; (iii) absorption of UV radiation by aromatic amino acids; (iv) a combination of the copper(II) ion–peptide bond reaction and the oxidation of tyrosine and tryptophan residues.

56.2 Determine the amount of amino acid in samples, using a 'standard curve' (see also study exercise 42.2 for an equivalent exercise for protein assay). The following data are for a set of calibration standards, measured using the ninhydrin method, which is non-linear at high amino acid concentrations.

Absorbance of calibration standards.

Amino acid concentration in assay tube (μg mL⁻¹) (total assay volume 5 mL)	Absorbance of standard (A_{570})
0.0	0.000
2.0	0.151
4.0	0.305
6.0	0.481
8.0	0.623
10.0	0.751
12.0	0.866
14.0	0.935
16.0	0.977
18.0	1.012
20.0	1.028

(a) Draw a calibration curve by hand using graph paper and estimate the concentration of amino acid in three different test samples giving the following A_{570} values: (i) 0.341; (ii) 0.686 (iii) 0.969.

(b) Assuming that these A_{570} values were obtained using the following dilutions and sample volumes in a total assay volume of 5 ml for all samples and standards alike, calculate the amino acid concentration of each of the three original samples, consisting of: (i) 100 μL of undiluted sample; (ii) 50 μL of a 10-fold dilution of sample; (iii) 80 μL of a 25-fold dilution of sample. Give all answers in mg mL⁻¹, to one decimal place.

56.3 Determine the amino acid sequence of a polypeptide from cleavage fragment data. The following data represent the results for a polypeptide of unknown amino acid sequence, when a subsample of the polypeptide was digested to completion with trypsin while another subsample was cleaved with cyanogen bromide (CNBr). In both cases, four peptide fragments were produced. These peptide fragments produced were then subjected to Edman degradation, and the following results were obtained for each peptide, in order of successive rounds of Edman degradation.

Tryptic digestion products: fragment (i): val asn lys; fragment (ii): ala gly met ser arg; fragment (iii): trp phe met ala ala; fragment (iv): his gly met ala glu lys.

CNBr cleavage products: fragment (i): ala ala; fragment (ii): his gly met; fragment (iii): ser arg trp phe met; fragment (iv): ala glu lys val asn lys ala gly met. What is the sequence of the intact polypeptide?

Answers to these study exercises are available at www.pearsoned.co.uk/practicalskills.

57 Assaying lipids

Table 57.1 Some examples of fatty acids.

No. of carbon atoms	Systematic name	Trivial name
Saturated fatty acids (no C=C bonds)		
12	n-Dodecanoic	Lauric
14	n-Tetradecanoic	Mystiric
16	n-Hexadecanoic	Palmitic
18	n-Octadecanoic	Stearic
20	n-Eicosanoic	Arachidic
22	n-Docosanoic	Behenic
Mono-unsaturated fatty acids (one C=C bond)		
12	cis-9-Dodecenoic	Lauroleic
14	cis-9-Tetradecenoic	Myristoleic
16	cis-9-Hexadecenoic	Palmitoleic
18	cis-9-Octadecenoic	Oleic
20	cis-9-Eicosenoic	Gadoleic
22	cis-9-Docosenoic	Erucic

Note: palmitic, stearic, palmitoleic and oleic acids are quantitatively the most common fatty acids in the majority of organisms. Major polyunsaturated fatty acids include linoleic acid (C_{18}, two double bonds), linolenic acid (C_{18}, three double bonds) and arachidonic acid (C_{20}, four double bonds).

Fig. 57.1 General structure of a neutral lipid: the structure of a triacylglycerol. R = remainder of fatty acid.

The term 'lipid' is used to describe a broad group of compounds with a wide variety of chemical structures, physical properties and biological functions. Lipids in biological systems are often combined with either proteins (e.g. in lipoproteins) or polysaccharides (e.g. in lipopolysaccharides). This chapter provides only a brief outline of the techniques used in lipid analysis (for further details, see Gunstone and Harwood, 2007).

> **KEY POINT** The defining feature of lipids is their relative insolubility in water: consequently, they are extracted from biological material using organic solvents, e.g. acetone, ether and chloroform. Because of their diversity and complexity, they are often referred to by more than one name, and the various types of non-systematic names can be confusing.

Biological lipids are often subdivided into two main types, each of which contains fatty acids as a major structural component.

Simple or neutral lipids

These are esters of fatty acids and an alcohol (e.g. Fig. 57.1). Fatty acids are straight-chain carboxylic acids, typically with an even number of carbon atoms and chain lengths of C_{12}–C_{22}, which may be saturated or unsaturated (Table 57.1). The greater the chain length and the fewer the number of double bonds, the higher the melting point, making most long chain saturated fatty acids solids at room temperature. Glycerol is the most common alcohol found in simple lipids, though higher M_r alcohols occur in waxes, and cyclic alcohols (sterols, e.g. cholesterol) occur in bile acids, steroid hormones and vitamins (e.g. vitamin K). While glycerol is a liquid at room temperature, cholesterol remains solid up to 150 °C.

Neutral fats (triglycerides or triacylglycerol) are esters of fatty acids and glycerol, as shown in Fig. 57.1. Many animal triglycerides ('fats') contain mainly saturated fatty acids and are solids at room temperature, while plant triglycerides ('oils') often have shorter chain lengths and a greater degree of unsaturation and are liquids at room temperature. Waxes are esters of fatty acids with alcohols of higher M_r than glycerol. The major biological functions of simple lipids include: (i) energy storage, e.g. oils in plant seeds, (ii) insulation, e.g. subcutaneous fat deposits in whales, and (iii) waterproofing, e.g. waxes in the cuticles of plant leaves.

Complex, compound or polar lipids

These are acyl esters of glycerol, or the amino alcohol sphingosine, that also include a hydrophilic group (e.g. a phosphoryl or carbohydrate group). They are often described in terms of this hydrophilic group, e.g. phospholipids contain a phosphoryl group while glycolipids contain a carbohydrate group.

> **KEY POINT** An important feature of complex lipids is their amphipathic nature, i.e. each molecule has a polar (hydrophilic) and a non-polar (hydrophobic) region.

Fig. 57.2 General structure of a phospholipid. R = remainder of fatty acid, X = hydrophilic group, e.g. $-CH_2CH_2NH_3^+$ in phosphatidyl ethanolamine and $-CH_2CH_2N(CH_3)_3^+$ in phosphatidyl choline.

SAFETY NOTE Working with solvents – chloroform and benzene must be used with care owing to their high toxicity: they are often replaced by dichloromethane and toluene respectively. All mixing and pouring steps should be carried out in a spark-free fume cupboard. Note that ethers may form explosive peroxides on prolonged storage. Lipid extracts extracted in flammable solvents must be stored in a spark-proof refrigerator, not in routine lab fridges.

Definition

Emulsion – a colloidal mixture where one liquid is dispersed (but not dissolved) in another liquid, e.g. lipids often aggregate to form micelles in aqueous solutions.

The two major types of complex lipid follow.

1. **Phospholipids** – the most common types (phosphoglycerides) are based on phosphatidic acid, with two fatty acids esterified to glycerol. Most phospholipids also contain a hydrophilic amino alcohol or a similar group, attached to the phosphoryl group (Fig. 57.2). The principal phospholipid classes are: (i) phosphatidyl cholines (or lecithins), which form stable emulsions with water and dissolve completely on addition of bile salts – these are insoluble in acetone, a feature that enables lecithins to be separated from most other lipids; (ii) phosphatidyl ethanolamines (or cephalins) – unlike lecithins, they are insoluble in ethanol and methanol; (iii) phosphatidyl serine; (iv) phosphatidyl inositol; and (v) plasmalogens.

2. **Sphingolipids** – these incorporate the amino dialcohol, sphingosine, rather than glycerol. Fatty acids are linked to sphingosine via an amide bond to form ceramides, which include: (i) cerebrosides (glycosphingolipids); (ii) sulfatides (sulfated cerebrosides); (iii) gangliosides (glycosphingolipids containing sialic acid residues); and (iv) phosphosphingolipids, including sphingomyelins, which are esters of a ceramide and phosphoryl choline.

Complex lipids have important roles in biological membranes: phospholipids are major structural components while sphingolipids are involved in cell–cell recognition and similar membrane features, e.g. glycosphingolipids are determinants of human ABO blood groups.

Extraction and analysis of lipids

Solvent extraction

Lipid extraction is carried out in a suitable organic solvent, using a homogeniser to disrupt cell and tissue structure (Chapter 41). There are two possible approaches:

1. **Total extraction of all lipids, followed by separation of the different lipid classes.** A commonly used solvent mixture is methanol: chloroform: water (2:1:0.5, v/v/v). Adding an equal volume of aqueous 1%w/v NaCl solution to the extract results in the formation of two layers: the lipids are present in the lower layer, while the upper aqueous layer contains other biomolecules (e.g. proteins). Hexane: isopropanol (3:2, v/v) can be used instead of chloroform: methanol, since it is less toxic and it dissolves very little non-lipid material.

2. **Selective lipid extraction.** Neutral lipids, e.g. within storage tissue, can be extracted by relatively non-polar solvents including hexane, diethyl ether and chloroform. Extraction of membrane lipids (e.g. phospholipids) requires disruption of the membrane using more polar solvents (e.g. methanol or ethanol), with selective precipitation by adding cold, dry acetone. Glycolipids can be extracted using acetone.

The principal disadvantages of solvent extraction methods include: (i) the requirement for large volumes of potentially hazardous solvents; and (ii) the possible formation of an emulsion, with incomplete extraction of component lipids.

Adsorption chromatography

Silica gel, octadecylsilane-bonded silica or ion-exchange resins can be used to bind solvent-extracted lipids by a combination of polar, ionic and van der Waals forces. In practice, a glass column is packed with a slurry of adsorbent in an appropriate organic solvent, and the lipid extract (dissolved in the same solvent) is applied to the top of the column. Lipids can then be selectively eluted; a mixture can be broadly separated into neutral lipids, glycolipids and phospholipids using solvents of increasing polarity, e.g. chloroform $\rightarrow$ acetone $\rightarrow$ methanol. It is possible to further separate the lipid subfractions on silica gel columns, as follows.

- **Neutral lipids** can be separated and eluted using hexane containing increasing proportions of diethyl ether ($0 \rightarrow 100\%$ v/v) in the order: hydrocarbons, cholesterol esters, triacylglycerols, free fatty acids, cholesterol, diacylglycerols, then monoacylglycerols.

- **Glycolipids and sulfolipids** can be separated by first eluting the glycolipids with chloroform: acetone (1:1 v/v), then using acetone to elute the sulfolipids.

- **Phospholipids** can be eluted using chloroform containing increasing proportions of methanol ($5\% \rightarrow 50\%$ v/v) in the order: phosphatidic acid, phosphatidyl ethanolamine, phosphatidyl serine, phosphatidyl choline, phosphatidyl inositol, sphingomyelin.

Thin-layer chromatography (TLC) of lipids

This can be used to separate lipid mixtures into their constituents, or to quantify particular lipids, as part of an analytical procedure. The principle of the technique is described in Chapter 50.

Silica gel G60 is the most frequently used stationary phase, acting as a polar absorbent. When the separation is carried out with a non-polar mobile phase, non-polar lipids will migrate more rapidly (i.e. they will have high R_F values, p. 332), while polar lipids will migrate more slowly. By increasing the polarity of the solvent, the R_F values of the polar lipids can be increased. Your choice of solvent will depend on the lipids in the extract:

- **For neutral lipids,** a typical solvent system is hexane: diethyl ether: glacial acetic acid at $80:20:2$ (v/v/v), separating in the following order of decreasing R_F: steryl esters, wax esters, fatty acids, methyl esters, triacylglycerols, fatty acids, fatty alcohols, sterols, 1,2-diacylglycerols, monoacylglycerols.

- **For polar lipids,** silica gel H is preferred, as silica gel G prevents the separation of acidic phospholipids. Most solvent systems for polar lipids are based on chloroform: methanol: water, e.g. at $65:25:4$ (v/v/v). In this system, the relative order of migration is: mono-galactosyldiacylglycerol, cerebrosides, phosphatidic acid, cardiolipin, lysophosphatidyl ethanolamine, phosphatidyl ethanolamine and digalactosyldiacylglycerol, sulfatides, phosphatidyl choline, phosphatidyl inositol, sphingomyelin and phosphatidyl serine.

KEY POINT No single solvent system will completely separate all lipid components in a single TLC procedure. However, 2D TLC may separate up to 200 individual lipid constituents from a sample.

Lipids can also be separated by reversed-phase TLC (RP-TLC), where the silica gel is made non-polar (e.g. by silanisation), and highly polar solvents are used as the mobile phase: here, the polar lipids will have the highest R_F values and the non-polar lipids the lowest R_F values. Lipids separated by TLC can be located by staining. Several stains are nonspecific, enabling almost all types of lipid to be visualised, while others will locate particular lipid classes. Staining can be carried out by immersion of the TLC plate in the stain, or by spraying. Non-specific staining methods include:

- **spraying with iodine solution** (e.g. 1–3% w/v in chloroform) – most lipids appear as brown spots on a yellow background, though glycolipids stain weakly by this method;

- **treatment with a strong oxidising agent followed by charring** (e.g. spray with 5%v/v sulfuric acid in ethanol, followed by heating in an oven at 180 °C for 30–60 min) – lipids appear as black deposits; the detection limit is 1–2 µg. Scanning densitometry can provide quantitative information;

- **fluorescent stains** (e.g. the widely used 2′,7′-dichlorofluorescein, DCF, at 0.1–0.2%w/v in ethanol). Under UV, lipids appear as bright yellow spots on a yellow–green background; the limit of detection is 5 µg. Other fluorescent stains such as 1-anilo-8-naphthalene sulfonate (ANS, 0.1% w/v in water) can detect nanogram quantities of lipid.

Quantitative assay of lipids and their components

While TLC can be used to quantify particular lipids, it is more common to assay the compounds released on hydrolysis of simple or complex lipids, namely the alcohols, fatty acids or other components, rather than the native lipid. Alkali hydrolysis of lipids containing fatty acids results in the formation of a soap (i.e. saponification), e.g. the incubation of tripalmitin with KOH yields potassium palmitate and glycerol. Acid hydrolysis of triacylglycerols releases 'free' fatty acids and glycerol.

Basic information about the relative size of the fatty acid component of oils and fats is given by the saponification value, determined by titration against 0.8 mol L^{-1} KOH; the lower the saponification value, the higher the M_r of the fatty acids. The degree of unsaturation of the fatty acids is given by the iodine number; the higher the iodine number, the greater the content of unsaturated fatty acids.

To obtain the iodine number for a particular fat, the free iodine remaining after reaction is titrated against 0.1 mol L^{-1} sodium thiosulfate ($Na_2S_2O_3$) using a trace amount of starch as an indicator, giving a titration volume for the test solution, V_t. A blank containing no fat is also titrated, to establish the volume of sodium thiosulfate required to titrate the initial free iodine, V_o. This allows the amount of iodine that reacts with the fat to be calculated according to the formula:

$$\text{iodine number (in g)} = \frac{1.27(V_0 - V_t)}{m} \quad [57.1]$$

where m is the mass of test fat (in g), and V_0 and V_t are expressed in millilitres.

Measurement of glycerol content

Glycerolipids can be quantified by measuring the glycerol released on hydrolysis. A widely used method involves a coupled enzyme assay (p. 412), with the following reactions:

$$\text{triacylglycerol.} \xrightarrow{\text{\textit{lipase}}} \text{glycerol} + \text{fatty acids} \qquad [57.2]$$

$$\text{glycerol} + \text{ATP} \xrightleftharpoons{\text{\textit{glycerol kinase, Mg}}^{2+}} \text{glycerol-3-phosphate} + \text{ADP} \qquad [57.3]$$

$$\text{ADP} + \text{phosphoenolpyruvate} \xrightleftharpoons{\text{\textit{pyruvate kinase}}} \text{ATP} + \text{pyruvate} \qquad [57.4]$$

$$\text{pyruvate} + \text{NADH} + \text{H}^+ \xrightleftharpoons{\text{\textit{lactate dehydrogenase}}} \text{lactate} + \text{NAD}^+ \qquad [57.5]$$

The glycerol concentration is determined by measuring the decrease in absorbance of NADH at 340 nm (A_{340}), compared to that of a blank with no added lipase.

Measurement of cholesterol content

Cholesterol in cholesterol esters can be estimated after hydrolysis to free cholesterol using cholesterol esterase. The subsequent assay is as follows:

$$\text{cholesterol} + \text{O}_2 \xrightarrow{\text{\textit{cholesterol oxidase}}} \text{cholest-4-en-3-one} + \text{H}_2\text{O}_2 \qquad [57.6]$$

The hydrogen peroxide produced as a result of the action of cholesterol oxidase can be measured amperometrically or colorimetrically, via the peroxidase-catalysed reaction:

$$2\text{H}_2\text{O}_2 + \text{phenol} + \text{4-aminoantipyrene} \xrightarrow{\text{\textit{peroxidase}}} \text{quinoneimine} + 4\text{H}_2\text{O} \quad [57.7]$$

This is the basis of many commercially available cholesterol testing kits. An alternative approach is to measure the cholest-4-en-3-one directly, via its absorbance maximum at 240 nm.

Gas chromatography (GC) of lipids and lipid components

This technique is used for the quantitative and qualitative analysis of a broad range of lipids. Volatile lipids may be analysed without modification, while non-volatile lipids must first be converted to a more volatile form, either by degradation (e.g. phospholipids), or derivatisation (p. 335). The most effective GC columns are support-coated open tubular (SCOT) capillary columns, with a thin film (0.1–10 μm) of the stationary phase coated onto the internal wall (p. 334). The choice of stationary phase depends on the components to be separated, e.g. some non-polar stationary phases cannot resolve methyl esters of saturated and monounsaturated fatty acids. Many stationary phases are based on silicone greases or polysiloxane, ranging from the non-polar dimethyl polysiloxanes (e.g. OV-101) to the polar trifluoropropyl methyl polysiloxanes (e.g. OV-210). Other polar stationary phases are based on polyethylene glycol (e.g. Carbowax 20M).

Understanding coupled enzyme assays based on NADH/NAD$^+$ interconversion – these are explained in more detail on p. 412, while the procedure required to convert changes in A_{340} to [NADH] is given on p. 312.

Alternative approaches to measuring H_2O_2 – the hydrogen peroxide produced from the oxidation of cholesterol in Eqn [57.6] can be converted to water and oxygen by the enzyme catalase ($2H_2O_2 \rightarrow 2H_2O + O_2$) and the oxygen produced can be quantified, for example, using an oxygen electrode (p. 370).

Example Fatty acids are derivatised to their methyl esters by boiling for 2–3 min with boron trifluoride (BF_3) solution (14%w/v in methanol) prior to GC analysis to increase their volatility and stability in the gas phase. The separated components are normally detected by flame ionisation (p. 344).

Text references

Gunstone, F.D. and Harwood, J.L. (2007) *The Lipid Handbook,* 3rd edn. CRC Press, Boca Raton.

Sources for further study

Anon. *Cyberlipid Center Homepage.* Available: http://www.cyberlipid.org Last accessed 01/09/15.

Byrdwell, W.C. (2006) *Modem Methods for Lipid Analysis by Liquid Chromatography/Mass Spectrometry and Related Techniques.* American Oil Chemists Society, Boulder.

Mossoba, M.M., Kramer, J.K.G., Brenna, J.T. and McDonald, R.E. (2006) *Lipid Analysis and Lipidomics: New Techniques and Applications.* American Oil Chemists Society, Boulder.

Ridgway, N. and McLeod, R (2015) *Biochemistry of Lipids, Lipoproteins and Membranes,* 6th edn. Elsevier, Amsterdam.

STUDY EXERCISES

57.1 Consider how to avoid oxidative rancidity. What practical steps could you take to avoid lipid oxidation during extraction and analysis?

57.2 Use the web to discover more about lipid structure and analysis. Visit the Cyberlipid Center website (at http://www.cyberlipid.org/index.htm), a resource giving details of a wide range of lipid structures, including triglycerides, phospholipids and glycolipids: (a) identify the main lipid components of beeswax and (b) suggest which separation methods might be appropriate for the analysis of these components.

57.3 Derive the equation for calculation of the iodine number of a lipid. Given that the following reactions (i)-(iii) are involved in the volumetric analysis for calculating the iodine number of a lipid, show how Eqn [57.1] is derived:

(i). $-CH\!=\!CH- + IC1 \rightarrow -(I)CH\!-\!CH(C1)-$

(ii). $IC1 + KI \rightarrow HC1 + I_2$

(iii). $I_2 + 2Na_2S_2O_3 \rightarrow 2NaI + Na_2S_4O_6$

57.4 Interpret a GC chromatogram for a lipid mixture. The chromatogram shown in the figure below represents a GC analysis of triacylglycerols in butter fat. There are 15 peaks, corresponding to lipids with even numbers of acyl carbon atoms between 26 and 54. How would you expect the lipids of different hydrocarbon lengths to be associated with the retention times of the different peaks? (Give a brief explanation of the reasoning underlying your expectation.)

Answers to these study exercises are available at www.pearsoned.co.uk/practicalskills.

These are compounds with a formula based on $C_x(H_2O)_x$. They play a key role in energy metabolism, and are essential constituents of cell walls and membranes. They may exist individually or as heteropolymers, e.g. linked to protein in glycoproteins (where carbohydrates form the minor component), or proteoglycans (where they form the major component).

> **KEY POINT** The identification and quantitative analysis of carbohydrates in biological samples can be difficult, owing to structural and physico-chemical similarities between related compounds. Several routine analytical methods cannot distinguish between isomeric forms of a particular carbohydrate.

Fig. 58.1 Ring and straight chain forms of glucose. Note that β-D-glucose has H and OH groups reversed at C-1 in the pyranose form.

Monosaccharides

The simplest carbohydrates are the monosaccharide sugars, which are polyhydroxy aldehydes and polyhydroxy ketones (so-called aldoses and ketoses), typically with three to seven carbon atoms per molecule. The common names end with the suffix '-ose', e.g. glucose (Fig. 58.1), which contains six carbon atoms and is a hexose. The simplest aldose is the three-carbon compound (triose) glyceraldehyde (Fig. 58.2(a)), and the other aldoses can be considered to be derived from glyceraldehyde by the addition of successive secondary alcohol groups (H–C–OH). In a similar manner, all ketoses can be considered to be structurally related to the triose dihydroxyacetone (Fig. 58.2(b)).

Fig. 58.2 Structure of (a) glyceraldehyde and (b) dihydroxyacetone.

Monosaccharides are assigned as D or L isomers according to a convention based on D-glyceraldehyde as the reference compound, with the carbon atoms numbered from the end of the chain containing the reactive group. Most of the carbohydrates found in biological systems are D isomers. While trioses and tetroses exist in linear form, pentoses and larger monosaccharides can be represented either as a linear structure (Fischer form) or as a ring structure (Haworth form), as in Fig. 58.1. The cyclic structure results from the reaction of the carbonyl group (aldehyde or ketone) at one end of the molecule with a hydroxyl group at the other end of the chain, forming a hemiacetal or hemiketal, as shown for glucose (Fig. 58.1) and fructose (Fig. 58.3).

The formation of a hemiacetal or hemiketal creates another asymmetric carbon atom, so that two ring forms exist – one with the –OH group on this asymmetric carbon positioned below the plane of the ring (α) and another with the –OH group above the plane of the ring (β). These different isomeric forms are called anomers and are particularly difficult to separate by chromatographic methods. The six-membered ring structure shown for glucose has a structure similar to that of pyran and is termed glucopyranose. A few monosaccharides have a five-membered ring structure similar to that of furan, e.g. fructofuranose.

Fig. 58.3 Ring and straight chain forms of fructose. Note that β-D-fructose has OH and CH_2OH groups reversed at C-2 in the furanose form.

While the open chain form is present in very small amounts in aqueous solution, it has a very reactive carbonyl group which is responsible for the reducing properties of sugars: several analytical methods are based on this feature (see p. 394).

Studying the biological roles of glycosides – these compounds are often formed in plants during the detoxification of certain compounds, or to control plant hormone activity.

Fig. 58.4 Formation of a glycosidic link (α (1 $\rightarrow$ 4)) in the disaccharide maltose.

Fig. 58.5 The repeating structure of cellulose (β (1 $\rightarrow$ 4) glycosidic links), $n \approx 5000$.

SAFETY NOTE Using the anthrone method – note that this involves hot H_2SO_4 – wear gloves, carry out the heating step in a fume hood and rinse all spillages thoroughly with excess water.

Glycosides

These are formed when a covalent bond, or glycosidic link, is created between the hemiacetal or hemiketal group of a carbohydrate (e.g. a monosaccharide) and the hydroxyl group of a second compound (e.g. a polyhydroxy alcohol, or another monosaccharide). If the sugars are not joined via their reactive groups, they will still show reducing properties, as in the disaccharide maltose (Fig. 58.4). In contrast, sucrose is a non-reducing disaccharide, since the hemiacetal and hemiketal groups of glucose and fructose are involved in the formation of the glycosidic link.

Polymeric carbohydrates have important biological roles – oligosaccharides contain up to 10 sugars, while polysaccharides are larger polymers with M_r of up to several million, e.g. glycogen, amylopectin, cellulose. The structure of polysaccharides is not fully defined, in contrast to proteins (Chapter 56) and nucleic acids (Chapter 59). Polysaccharide structure is the result of the separate actions of a number of biosynthetic enzymes, producing a range of molecules that may vary in the number of monosaccharide residues and the types of glycosidic bond. The terminology used to describe the glycosidic links in such compounds denotes the anomer involved (α or β) and the C atoms involved in the link, e.g. α(1 $\rightarrow$ 4) in maltose (Fig. 58.4), β(1 $\rightarrow$ 4) in cellulose (Fig. 58.5).

Extraction and analysis of carbohydrates

While most low M_r carbohydrates are soluble in water, ethanol: water (80% v/v) is more often used, since polysaccharides and other biological macromolecules are insoluble in aqueous ethanol. In contrast, polysaccharides are more diverse, and the isolation procedures vary greatly, e.g. boiling water, mild acid or mild alkali can be used to solubilise storage polysaccharides, e.g. starch, while more vigorous treatment is required for structural polysaccharides, e.g. 24% w/v KOH for cellulose. Among the techniques used to purify extracted polysaccharides are gel permeation chromatography (p. 336) and ultracentrifugation (p. 328).

Identification and quantification of carbohydrates

This can be achieved by a variety of procedures, detailed below.

Chemical methods

Several monosaccharide assay methods are based on the reductive capacity of the aldehyde or ketone groups (Table 58.1). A widely used method for quantitative analysis is that based on reduction of 3,5-dinitrosalicylate. This method is also suitable for glycosides, provided the reducing carbonyl groups are not involved in the glycosidic links. Certain polysaccharides react with iodine in acid solution to form coloured complexes: starch gives a blue colour, while glycogen gives a red–brown colour. The anthrone method (typically using 0.1 g L^{-1} anthrone in H_2SO_4, at 100 °C for 10 min, then assayed at 630 nm) is an alternative assay for estimating total carbohydrate content. Careful choice of calibration standards is required for quantitative work – try to use a standard that matches the likely composition of the samples. While such chemical methods can give a general indication of the relative amount of carbohydrate in a sample, they can provide little useful information on the types of carbohydrate present.

Table 58.1 Methods for carbohydrate analysis.

Method	Principle	Comments
Chemical assay		
Benedict's test (and Fehling's test)	Reduction of Cu^{2+} to Cu^+ in presence of reducing sugar; alkaline solution plus heat results in formation of Cu_2O; solution turns from blue, through yellow, to red	Usually presence/absence test; quantitative assay involves measurement of Cu_2O formed
Dinitrosalicylate (DNS)	Reduction of DNS (yellow) to orange–red derivative; alkaline solution plus heat (100 °C, 10 min)	Quantitative: read at 540 nm
Enzymatic assay		
Glucose oxidase (coupled reaction)	$\beta\text{-D-glucose} + O_2 \xrightarrow{\ glucose\ oxidase\ } \text{gluconic acid} + H_2O_2$ $H_2O_2 + \text{reduced dye} \xrightarrow{\ peroxidase\ } H_2O + \text{oxidised dye}$	Mutarotation allows reaction to reach completion; hydrogen peroxide formed may be measured using peroxidase; ABTS* is a suitable dye Assay at 437 nm

*ABTS = 2,2′-azino-di-[3-ethylbenzthiazoline]-6-sulfonate.

Enzymatic methods

These offer a higher degree of specificity in monosaccharide assay, and may allow differentiation between the various stereoisomeric and anomeric forms. The glucose-specific method based on glucose oxidase shown in Table 58.1 also forms the basis of the glucose electrode (see Chapter 54). Hydrogen peroxide may be assayed using peroxidase and a suitable chromogenic substrate, or by electrochemical methods, e.g. using an amperometric sensor. Alternatively, the consumption of oxygen in the initial reaction can be measured using an oxygen electrode (p. 369). Glycosidases can be used to hydrolyse specific disaccharides or polysaccharides into their constituent monosaccharides, which can then be identified and quantified, e.g. α-glucosidase, which hydrolyses the $\alpha(1 \rightarrow 4)$ linkage between the glucose residues in maltose (Fig. 58.4). However, you should note that it is rare for such enzymes to show absolute specificity for a particular substrate, so you should be alert to the possibility of interference owing to related compounds in the sample, or to impurities in the enzyme preparation.

> **Other applications of glycosidase assays** – some microbial identification schemes are based on the detection of specific glycosidase enzymes, e.g. β-glucuronidase for *E. coli* (p. 413).

Chromatographic methods

Traditional methods include paper and thin-layer chromatographic procedures. Suitable supports for TLC include microcrystalline cellulose (in which the sugars partition between the mobile phase and the cellulose-bound water complex) and silica gel. A wide range of traditional solvent systems can be used (Stahl, 1965, gives details) and your choice of mobile phase will depend upon the expected composition of the mixture, e.g. cellulose with ethyl acetate: pyridine: water (100:35:25 v/v/v) gives a good separation of pentoses and hexoses, and will resolve glucose and galactose, as well as some disaccharides. Staining methods usually exploit the reducing properties of carbohydrates. Of the high resolution techniques, HPLC (p. 334) is the preferred method for the analysis of simple monosaccharide mixtures, and for oligosaccharide analysis and purification. Ion-exchange columns are often used for these purposes, with refractive index, or electrochemical, detection of separated components (Chapter 50). GC is more suitable for complex monosaccharide mixtures,

> **Measuring carbohydrate migration in TLC systems** – the distance migrated by glucose is taken as a reference ($R_F = 100$), and the migration of other carbohydrates is given as the R_G value, where R_G is:
>
> $$\frac{\text{distance moved by carbohydrate}}{\text{distance moved by glucose}} \times 100$$

Complex heteropolymers that contain carbohydrates – these include:

- lignins – structural polymers in plants: composed of amino acids (phenyl-alanine and tyrosine) together with carbohydrates, and extremely resistant to hydrolysis;

- peptidoglycan – found in the bacterial cell wall: composed of amino acids and carbohydrates, providing strength and rigidity;

- proteoglycans – found in animal/human connective tissues: composed of a core protein with many carbohydrate side chains attached, responsible for hydration, lubrication, resistance to compressive forces and mediation of cellular interactions.

and can analyse subnanomolar amounts of carbohydrates and their derivatives, e.g. polyols, including glycerol (p. 387). However, a preliminary step is necessary to produce volatile derivatives of the carbohydrates in the mixture, e.g. methylation or, more often, trimethylsilylation (adding hexamethyldisilazane and trimethylchlorosilane at 2:1 v/v at room temperature rapidly produces trimethylsilyl ethers – TMS derivatives). Efficient separations can be obtained with a non-polar stationary phase, e.g. methylpolysiloxy gum (OV-1): the use of SCOT columns (p. 391) can enable over 30 components to be resolved. A disadvantage of GC methods is that the carbohydrates are first converted to a derivative form, then quantified by destructive means, preventing further analysis.

Capillary electrophoresis

Capillary electrophoresis (Chapter 53) is a powerful technique for the separation of carbohydrates, though it is usual to modify uncharged carbohydrates by reductive amination to form primary amines to allow their separation within the capillary.

Characterisation of polysaccharides

Polysaccharides are characterised according to the relative proportions of the constituent sugar residues, the various types of glycosidic links and the M_r. To investigate the composition of a given type of polysaccharide, its glycosidic links must be hydrolysed (e.g. by heating with concentrated HCl at 60 °C for 30 min), followed by separation, identification and quantification of the individual components. The position of glycosidic linkages can be determined by methylation of all free hydroxyl groups of the polysaccharide followed by complete hydrolysis to give a mixture of partially methylated monosaccharides. These are then reduced to alditols, and acetylated. The partially methylated alditol acetates can be identified by GC-mass spectrometry (p. 335), allowing the types of glycosidic links to be deduced from the positions of the acetylated groups. The α and β configuration of the glycosidic linkages can be determined by enzymic assay or by NMR spectroscopy (p. 321).

Text references and sources for further study

Cui, S.W. (2006) *Food Carbohydrates: Chemistry, Physical Properties and Applications.* CRC Press, Boca Raton.

Davis, B.G. and Fairbanks, A.J. (2002) *Carbohydrate Chemistry.* Oxford University Press, Oxford. [A primer on carbohydrate chemistry]

Nielsen, S.S. (2010) *Food Analysis Laboratory Manual,* 2nd edn. Springer, New York.

Packer, N.H. and Karlsson, N.G. (2009) *Glycomics: Methods and Protocols.* Humana, New York.

Stahl, E. (1965) *Thin Layer Chromatography – a Laboratory Handbook.* Springer, Berlin.

Volpi, N. (2010) *Capillary Electrophoresis: from Monosaccharides to Complex Polysaccharides.* Humana, New York.

STUDY EXERCISES

58.1 Check your understanding of the terminology of carbohydrate structure. For each of the following terms: (i) provide a definition; and (ii) give an example of a representative compound:

(a) aldose;

(b) pentose;

(c) hemiacetal;

(d) pyranoside;

(e) β anomer.

58.2 Identify sugars by thin-layer chromatography (TLC). A typical TLC separation of sugars on silica gel is shown in the figure opposite.

(a) Identify the sugars present in the hydrolysed raffinose sample. Check the structure of raffinose, to see whether your findings are consistent with the monosaccharide constituents.

(b) If the relative frontal mobility of unhydrolysed raffinose relative to glucose (R_G, Chapter 50) is known to be 0.34, and the distance migrated by glucose is 35 mm, how far (in mm) would you expect the raffinose to have migrated?

58.3 Consider the operating principles underlying chemical methods for quantitative assay of carbohydrates. Why is the dinitrosalicylate (DNS) method unsuitable for the assay of the disaccharide sucrose, yet it is suitable for measuring its constituent monosaccharides, i.e. glucose and fructose? Is DNS unsuitable for all disaccharides?

Thin-layer chromatographic separation of sugars: lane 1, lactose; lane 2, glucose; lane 3, galactose; lane 4, xylose; lane 5, ribose; lane 6, fructose; lane 7, arabinose; lane 8, hydrolysed raffinose (dotted line indicates origin and arrow indicates direction of solvent flow).

Answers to these study exercises are available at www.pearsoned.co.uk/practicalskills.

Fig. 59.1 Nitrogenous bases in nucleic acids.

Fig. 59.2 A nucleoside triphosphate – deoxyadenosine 5′ triphosphate (dATP).

Nucleic acids are nitrogen-containing compounds of high M_r, often found within nucleic acid–protein (nucleoprotein) complexes in cells. The two main groups of nucleic acids follow.

1. **Deoxyribonucleic acid (DNA)** – found in chromosomes; the principal molecule responsible for the storage and transfer of genetic information (Chapter 64).

2. **Ribonucleic acid (RNA)** – involved with the DNA-directed synthesis of proteins in cells. Three principal types of RNA exist: messenger RNA (mRNA), ribosomal RNA (rRNA) and transfer RNA (tRNA). In some viruses, RNA acts as the genetic material. In eukaryotes, mRNA molecules initially synthesised in the nucleus from genomic DNA (nascent mRNA) will contain several sequences – called introns – that are not transcribed into protein. These are successively excised in the nucleus, leaving only coding sequences (exons) in the mRNA that migrates to the cytoplasm, to be translated at the ribosome.

> **KEY POINT**   Nucleic acids are important in the transmission of information within cells, and one of the most important aspects of nucleic acid analysis is to decipher the coded information within these molecules (see also Chapters 12 and 66).

The structure of nucleic acids

Nucleic acids are polymers of nucleotides (polynucleotides), where each nucleotide consists of:

- **a nitrogenous base,** of which there are five main types. Two have a purine ring structure, i.e. adenine (A) and guanine (G), and three have a pyrimidine ring, i.e. thymine (T), uracil (U) and cytosine (C), as shown in Fig. 59.1. Their carbon atoms are numbered C-1, C-2, etc.

- **a pentose sugar,** which is ribose in RNA and deoxyribose in DNA. The carbon atoms are denoted as C-1′, C-2′, etc. and deoxyribose has no hydroxyl group on C-2′ (Fig. 59.2). The C-1′ of the sugar is linked either to the N-9 of a purine or the N-1 of a pyrimidine;

- **a phosphate group,** which links with the sugars to form the sugar–phosphate backbone of the polynucleotide chain.

A compound with sugar and base only is called a nucleoside (Fig. 59.2) and the specific names given to the various nucleosides and nucleotides are listed in Table 59.1. The individual nucleotides within nucleic acids are linked by phosphodiester bonds between the 3′ and 5′ positions of the sugars (Fig. 59.3).

> **KEY POINT**   RNA and DNA differ both in the nature of the pentose sugar residue and in their base composition: both types contain adenine, guanine and cytosine, but RNA contains uracil while DNA contains thymine.

Table 59.1 Nomenclature of nucleosides and nucleotides.

Base	Nucleoside	Nucleotide
Adenine	Adenosine	Adenylic acid
Guanine	Guanosine	Guanylic acid
Uracil*	Uridine	Uridylic acid
Cytosine	Cytidine	Cytidylic acid
Thymine†	Thymidine	Thymidylic acid

*In RNA.
†In DNA

Minimising damage to chromosomes – chromosomes vary in size from 0.3 to 200 megabase pairs (Mb), so some breaks in DNA inevitably occur during manipulation. Shear effects can be minimised by using wide-mouthed pipettes, gentle mixing, by avoiding rotamixing and by precipitating DNA with ethanol at –20°C.

Fig. 59.3 Linkage of nucleotides in nucleic acids.

Working with precipitated DNA – present in sufficient quantity, this can be spooled from solution by winding it around a glass rod.

Differences also exist in the conformation of the two types of nucleic acid. DNA typically exists as two interwoven helical polynucleotide chains, with their structure stabilised by hydrogen bonds between matching base pairs on the adjacent strands: A always pairs with T (two hydrogen bonds), and G with C (three hydrogen bonds). This complementarity is important since it stabilises the DNA duplex (double helix) and provides the basis for replication and transcription (Chapter 66). While most double-stranded (ds) DNA molecules are in this form, i.e. as a double helix, some viral DNA is single-stranded (ss). Intact DNA molecules are very large indeed, with high M_r values (e.g. 10^9). In the main, RNA is single-stranded and in the form of a gentle right-handed helix stabilised by base-stacking interactions, although some sections of RNA (i.e. tRNA) have regions of self-complementarity, leading to base pairing. Typical values for M_r of RNA range from 10^4 for tRNA to 10^6 for other types.

Extraction and purification of nucleic acids

The types of nucleic acid most commonly isolated are chromosomal DNA, plasmid DNA and mRNA. Irrespective of the source, extraction and purification involve the following stages, in sequence:

- **disruption of cells** to release their contents;
- **removal of non-nucleic acid components** (e.g. protein), leaving DNA and/or RNA;
- **concentration of the remaining nucleic acids.**

DNA isolation procedures

Specific details for the preparation of plasmid DNA from bacterial cells are given in Chapter 68. For other sources of DNA, such as mammalian tissue or plant material, the following steps are required:

1. **Homogenisation** – tissues can be disrupted by the methods described in detail in Chapter 41, e.g. by lysis in a buffered solution containing the detergent sodium dodecyl sulfate (SDS) or Triton X-100.

2. **Enzymic removal of protein and RNA** – using proteinase (e.g. proteinase K at 0.1 mg ml⁻¹) and ribonuclease (typically at 0.1 µg mL⁻¹) for 1–2h.

3. **Phenol–chloroform extraction** – to remove any remaining traces of contaminating protein.

4. **Precipitation of nucleic acids** – usually by adding twice the volume of ethanol.

5. **Solubilisation** in an appropriate volume of buffer (pH 7.5) – ribonuclease is often added to remove any traces of contaminating RNA.

Density gradient centrifugation (p. 327) is an alternative approach to the separation of DNA from contaminating RNA, as described below.

RNA isolation procedures

Each mammalian cell contains about 10 pg of RNA, made up of rRNA (80–85%), tRNA (10–15%) and mRNA (1–5%). While rRNA and tRNA components are of discrete sizes, mRNA is heterogeneous and varies in length from several hundred to several thousand nucleotides.

Avoiding contamination of glassware by RNases – autoclave all glassware before use, to denature RNases. These enzymes are present in skin secretions: use gloves at all times and use plastic-ware wherever possible.

Avoiding RNase degradation of RNA – endogenous RNases can be inhibited by including diethyl pyrocarbonate (DEP) (at 0.1%v/v g ml^{-1}) in solutions used for RNA extraction.

Purification of mRNA – following whole cell extraction of RNA, mRNA can be separated from other types by affinity chromatography (p. 338) using poly (U)-Sepharose which binds to the poly (A) 'tail' sequence at the 3´ end of mRNA molecules.

Measuring nucleic acids by spectrophotometry – ideally, the nucleic acid extracts should be prepared to give A_{260} values of between 0.10 and 0.50, for maximum accuracy and precision.

Fig. 59.4 Ethidium bromide (EtBr), a fluorescent molecule used for the detection and assay of DNA.

> **KEY POINT** RNA is more difficult to purify than DNA, partly because of degradation during the extraction process because of the action of contaminating ribonucleases, and partly because the rigorous treatment required to dissociate the RNA from protein in ribosomes may fragment the polyribonucleotide strands.

RNA can be prepared either from the cytoplasm of cells (to give mainly rRNA, tRNA and fully processed mRNA), or from whole cells, in which case nascent mRNA from the nucleus will also be present (p. 403).

- Preparation of cytoplasmic RNA involves lysis of cells or protoplasts with a hypotonic buffer, leaving the nuclei intact. Cell debris and nuclei are then removed by centrifugation (Chapter 49), and sodium dodecyl sulfate (SDS) is added to the supernatant (the cytoplasmic fraction) to inhibit ribonuclease. Proteinase K can be added to release rRNA from ribosomes. Phenol–chloroform extraction removes contaminating proteins, as for DNA preparation, and the RNA present in the aqueous phase can then be precipitated by addition of twice the volume of ethanol.

- Preparation of whole-cell RNA requires more vigorous cell lysis, e.g. with a solution containing 6mol L^{-1} guanidinium chloride and 2-mercaptoethanol: this effectively denatures any ribonuclease present. Caesium chloride is added to the extract to give a final concentration of 2.4mol L^{-1}, and the solution is processed by density gradient centrifugation (Chapter 49) at 100 000 g for 18h, using a cushion of 5.7mol L^{-1} CsCl. DNA and protein remain in the upper layers of CsCl, while RNA forms a pellet at the base of the tube. The RNA pellet is redissolved in buffer, and precipitated in cold ethanol.

Separating nucleic acids

Electrophoresis is the principal method used for separating nucleic acids. At alkaline pH values, linear DNA and RNA molecules have a uniform net negative charge per unit length owing to the charge on the phosphoryl group of the backbone. Electrophoresis using a supporting medium that acts as a molecular sieve (e.g. agarose or Polyacrylamide, pp. 351–2) enables DNA fragments or RNA molecules to be separated on the basis of their relative sizes (see Chapter 52 for further details).

Quantitative analysis of nucleic acids

Measuring nucleic acid content

The concentration of reasonably pure samples of DNA or RNA can be measured by spectrophotometry (p. 312). In contrast, measurement of the nucleic acid content of whole cell or tissue homogenates requires chemical methods, since the homogenates will contain many interfering substances. The principles involved in each technique are as follows:

- **Spectrophotometry** – DNA and RNA both show absorption maxima at ≈260 nm, owing to the conjugated double bonds present in their constituent bases. At 260 nm, an A_{260} value of 1.0 is given by a 50 μg mL^{-1} solution of dsDNA, or a 40 μg mL^{-1} solution of ssRNA. If the absorbance at 280 nm is also measured, protein contamination can be quantified. Pure nucleic acids give A_{260}/A_{280} ratios of 1.8–2.0, and a value below 1.6 indicates significant protein contamination.

SAFETY NOTE Working with ethidium bromide – this compound is highly toxic and mutagenic. Avoid skin contact (wear gloves) and avoid ingestion. Use a safe method of disposal (e.g. adsorb from solution using an appropriate adsorbant, e.g. activated charcoal). Less toxic alternatives include SYBR Safe and Gel Red.

Fig. 59.5 The diphenylamine reaction for assay of DNA.

Fig. 59.6 The orcinol reaction for assay of RNA.

Further purification steps are required for contaminated samples, e.g. by repeating the phenol–chloroform extraction step. RNA contamination of a DNA preparation is indicated if A_{260} decreases when the sample is treated with 2.5 μL of RNase at 20 μg μL^{-1}. DNA contamination of an RNA preparation might be suspected if the sample is very viscous, and this can be confirmed by electrophoresis.

- **Spectrofluorimetry** – this is the best approach for samples where the DNA concentration is too low to allow direct assay by the spectrophotometric method described above. The traditional method uses the fluorescent dye ethidium bromide (Fig. 59.4), which binds to dsDNA by insertion between stacked base pairs, a phenomenon termed 'intercalation'. The fluorescence of ethidium bromide is enhanced 25-fold when it interacts with dsDNA. Since ssDNA gives no significant enhancement of fluorescence, dsDNA can be quantified in the presence of denatured DNA. The concentration of dsDNA in solution, $[dsDNA]_x$, can be calculated by comparing its fluorescence (excitation, 525 nm; emission, 590 nm) with that of a standard of known concentration, $[dsDNA]_{std}$, using the relationship:

$$[dsDNA]_x = \frac{[dsDNA]_{std} \times \text{fluorescence of unknown}}{\text{fluorescence of standard}} \quad [59.1]$$

- **Chemical methods** – these are mostly based on colorimetric reactions with the pentose groups of nucleic acids. The total DNA concentration can be measured by the diphenylamine reaction (Fig. 59.5), which is specific for 2-deoxypentoses. The diphenylamine reaction involves heating 2 mL of DNA solution with 4 mL of freshly prepared diphenylamine reagent (diphenylamine, at 10 gL^{-1} in glacial acetic acid, plus 25 mL concentrated sulfuric acid) for 10 min in a boiling water bath. The acids cleave some of the phosphodiester bonds, and hydrolyse the glycosidic links between the deoxyribose and purines. Deoxyribose residues are converted to ω-hydroxylaevulinyl aldehyde (Fig. 59.5), which reacts with the diphenylamine to produce a blue pigment, assayed at 600 nm. By constructing a DNA standard curve ranging from 0 to 400 μg mL^{-1}, the DNA concentration of the unknown can be determined.

  RNA concentration can be measured by the orcinol reaction, which is a general assay for pentoses. The orcinol reaction involves heating 2 mL of RNA solution with 3 mL of orcinol reagent (prepared by dissolving 1 g of $FeCl_3 \cdot 6H_2O$ in 1 litre of concentrated HCl, and adding 35 mL of 6%w/v orcinol in ethanol) in a boiling water bath for 20 min. The acid cleaves some phosphodiester bonds and hydrolyses the glycosidic links between the ribose and purines. The hot acid also converts the ribose to furfural (Fig. 59.6), which reacts with orcinol in the presence of ferric ions to produce green-coloured compounds, assayed at 660 nm. A standard curve for RNA ranging from 0 to 400 μg mL^{-1} is used to determine the RNA concentration of the unknown. The orcinol reaction is less specific than the diphenylamine reaction, as deoxyribose reacts to some extent and DNA gives about 10% of the colour given by the same concentration of RNA. If the DNA concentration of the extract is known (e.g. from a diphenylamine assay), the contribution of DNA to A_{660} can be measured for a standard prepared to have the same DNA concentration as the sample: the A_{660} due to DNA can then be subtracted from the result of the orcinol reaction, and the remaining A_{660} value is then used to determine RNA concentration from the RNA standard curve.

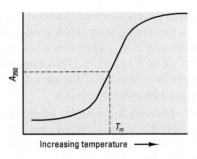

Fig. 59.7 Effect of temperature on the absorbance (A_{260}) of DNA in solution. T_{m} is the temperature for the mid-point of the absorbance change.

Assaying the relative proportions of base pairs in DNA

Double-stranded DNA has a lower molar absorptivity at 260 nm than single-stranded DNA, owing to electron interactions between the stacked base pairs and hydrogen bonding between the complementary bases of dsDNA. When a dsDNA solution is heated slowly, there is little change in absorbance until a temperature is reached where the hydrogen bonds are broken and the DNA strands separate. At this so-called 'melting temperature' (T_{m}), the A_{260} value increases sharply (Fig. 59.7). This temperature-dependent increase in absorbance is often referred to as the 'hyperchromic effect'. The actual value of the T_{m} (and hence the stability of dsDNA) is dependent on the base pair composition of the DNA being studied: GC base pairs have three double bonds while AT base pairs have two, and the higher the % GC content, the higher the value of the T_{m}. The length of the molecule also affects the T_{m}, since short dsDNA molecules do not show a hyperchromic effect. Studies on a number of DNA samples have shown the following relationship, for GC contents between 30 and 70%:

$$\%GC = 2.44(T_{m} - 69.3) \qquad [59.2]$$

The relative proportion of AT base pairs can then be calculated by subtracting the %GC value from 100, giving %AT.

Sources for further study

(See also Chapter 12 for nucleic acid databases and related information.)

Blackburn, G.M., Gait, M.G, Loakes, D. and Williams, D.M. (2005) *Nucleic Acids in Chemistry and Biology,* 3rd edn. Royal Society of Chemistry, London.

Kerr, J.T. and Birch, L. (eds) (2008) *Essentials of Nucleic Acid Analysis: A Robust Approach.* Royal Society of Chemistry, London.

Neidle, S. (2007) *Principles of Nucleic Acid Structure,* 2nd edn. Academic Press, Oxford.

Rapley, R. (2000) *The Nucleic Acids Protocols Handbook.* Humana Press, New Jersey.

STUDY EXERCISES

59.1 Calculate the concentration of DNA in a sample using a fluorescence assay (see also exercises 47.2c and 47.2d, for equivalent exercises using UV spectrophotometry). A volume of 10 μL of a DNA standard solution (of concentration 50 μg mL^{-1}) was added to 3 mL of an ethidium bromide solution (containing 0.5 mgL^{-1} ethidium bromide, buffered at pH 7.5) in a fluorimeter cuvette, mixed, then read in a spectrofluorimeter using an excitation wavelength of 525 nm and an emission wavelength of 590 nm, giving a relative fluorescence intensity (RFI) of 58. This was repeated using 10 μL of a DNA solution of unknown concentration instead of the DNA standard solution, giving an RFI of 32. What is the DNA concentration of the unknown? (Express your answer to three significant figures.)

59.2 Investigate the binding properties of ethidium bromide. Find out: (i) why ethidium bromide fluorescence is enhanced by DNA; and (ii) under what circumstances ethidium bromide can also bind to RNA.

59.3 Consider how to avoid problems in DNA spectrofluorometric analysis caused by the presence of RNA in a sample. What steps could you take to ensure that all the fluorescence obtained from a cell extract was caused by the reaction of ethidium bromide with DNA rather than from DNA plus contaminating RNA?

Answers to these study exercises are available at www.pearsoned.co.uk/practicalskills.

Much of our current knowledge about metabolic and physiological events has been gained from *in vitro* studies of purified proteins. Such studies range from investigations into the kinetics and regulation of enzymes (Chapter 61) to the determination of the structure of a protein and its relationship to function. In addition, certain purified proteins have a role as therapeutic agents, e.g. Factor VIII in the treatment of haemophilia.

> **KEY POINT**   The purification of most proteins involves a series of procedures, based on differential solubility and/or chromatography, that selectively separate the protein of interest from other proteins and materials. It is rare for purification to be a one-step process.

Deciding on your objectives

Purity, yield and cost are the major considerations, and the relative importance of these factors will depend on the purpose of the purification. For studies on the biological activity of a protein (e.g. an enzyme), only microgram amounts may be required and, as long as there are no interfering substances present, 100% purity may not be necessary. For structural studies, milligram amounts may be needed and the protein must be pure. A protein produced for industrial applications (e.g. an enzyme for use in starch degradation) needs to be produced in large amounts (grams or kilograms), but purity is not usually essential. In contrast, commercially produced proteins for therapeutic use need to be free of any significant contamination.

The cost of the purification will depend on the nature of the source material, the number of purification steps required and the price of materials for the separation techniques employed. Inevitably, some protein will be lost at each stage of the purification procedure.

> **KEY POINT**   You should aim to use the smallest number of purification steps that will give you the yield and purity required for your application.

Preliminary considerations
Investigating what is known about the protein

Rather than approaching every purification step on a 'trial and error' basis, try to find out as much as you can about the physical and biological properties of the desired protein before you begin your practical work. Even if the protein to be purified is novel, it is likely that similar proteins will be described in the literature. Information on the isoelectric point (pI) and M_r of both the protein and of the likely major contaminants, is particularly useful for ion-exchange and gel permeation chromatography (p. 336). Knowledge of other factors such as metal ion and co-enzyme requirements, presence of thiol groups and known inhibitors can indicate useful chromatographic steps, and may allow you to take steps to preserve the tertiary structure and biological activity of the desired protein during the purification process.

Protein purification and proteomics – an important aspect of proteomics (p. 77) is the characterisation of individual proteins, using some of the techniques described in this chapter.

Preventing loss of protein during purification – silicone-coated glass containers prevent the adsorption of proteins to the surface of a glass vessel.

Definition

Isoelectric point (pI) – the pH value at which a protein has a net charge of zero, i.e. where positive and negative charges on amino acid side chains are balanced.

Choosing the source material

If you have a choice of starting material, use a convenient source in which the protein of interest is abundant. This can vary from animal tissues (e.g. heart, kidney or liver) obtained from an abattoir or from laboratory animals, to plant material, or to microbial cells from a laboratory culture.

Homogenisation and solubilisation

Unless the protein is extracellular, the cells in the source material need to be disrupted and homogenised by one of the methods described in Chapter 41. Following homogenisation, proteins from the cytosol or extracellular fluid will normally be present in a soluble form, but membrane-bound proteins and those within organelles will require further treatment. Isolation of organelles by differential centrifugation (p. 327) will provide a degree of purification, while membrane-bound proteins will need organic solvents or detergents to render them soluble. Once the protein of interest is in a soluble form, any particulate material in the extract should be removed, e.g. by centrifugation or filtration.

Preserving enzyme activity during purification

Once an enzyme has been released from its intracellular environment, it will be exposed to potentially adverse conditions that may result in denaturation and permanent inactivation, e.g. a lower pH, an oxidising environment or exposure to lysosomal proteases. During homogenisation, and in some or all of the purification steps, buffers should contain reagents that can counteract these potentially damaging effects (see Chapter 41). Fewer precautions may be needed with extracellular enzymes. For most proteins, the initial procedures should be carried out at 4 °C to minimise the risk of proteolysis.

Devising a strategy for protein purification

Any successful protein purification scheme will exploit the unique properties of the desired protein in terms of its size, net charge, hydrophobic nature, biological activity, etc. The chromatographic techniques that separate on the basis of these properties are detailed in Chapter 50. However, the *order* in which the various purification steps are carried out needs some thought, and each stage needs to be considered in relation to the following factors:

- **Capacity** – the amount of material (volume or concentration) that the technique can handle. High capacity techniques such as ammonium sulfate precipitation (see later) and ion-exchange chromatography (p. 336) should be used at an early stage, to reduce the sample volume.
- **Resolution** – the efficiency of separating one component from another. At later stages of the purification, the sample volume will be considerably reduced, but any contaminating proteins may well have very similar properties to those of the protein of interest. Therefore, a high resolution (but low capacity) technique will be required, such as covalent chromatography or immobilised metal affinity chromatography (IMAC).
- **Yield** – the amount of protein recovered at each step, expressed as a percentage of the initial amount (p. 406). For example, yields of >80% can be obtained with ammonium sulfate precipitation. With certain types of affinity chromatography, where harsh elution conditions need

Table 60.1 Principal causes of decreased yield in protein purification.

Cause	Possible solution
Denaturation	Include EDTA or reducing agents in buffers; avoid extreme temperatures
Inhibition	Check buffer composition for possible inhibitors
Proteolysis	Include protease inhibitors in buffers
Non-elution	Alter salt concentration or pH of the eluting buffer
Co-factor loss	Recombine fractions on a trial-and-error basis

Calculating total protein content – multiply the protein concentration by the total volume, making sure that the units are consistent, e.g. mg mL^{-1} × mL. A similar procedure is required to determine the total amount of enzyme.

Using immunoassays – note that inactive or denatured forms of the protein may be detected along with the biologically active form.

Example For an enzyme extract having a specific activity of 250.2 U mg^{-1} compared to an initial specific activity of 45.5 U mg^{-1}, using Eqn [60.1] gives a purification factor of 250.2 ÷ 45.5 = 5.5-fold (to one decimal place).

to be employed, yields may be quite low (e.g. <20%). Some of the other possible reasons for a decreased yield are considered in Table 60.1.

An ideal purification procedure will have the following features:

1. **an initial step that has a high capacity,** but not necessarily a high resolution;
2. **a series of chromatographic steps,** each of which exploits a different property of the protein to increase purity;
3. **a minimum number of manipulations** and changes in conditions (e.g. in buffer composition) between steps.

The following series of steps would meet the above conditions:

Step 1 **Ammonium sulfate precipitation** – a high capacity, low resolution technique that yields a protein solution with a greatly reduced volume and a high concentration of ammonium sulfate.

Step 2 **Hydrophobic interaction chromatography (HIC)** – an absorption technique (p. 337) using a starting buffer with a high ammonium sulfate concentration. During development of the column, the ionic strength of the mobile phase is reduced in a gradient elution.

Step 3 **Ion-exchange chromatography (IEC)** – another absorption technique (p. 336). Here the starting buffer has a low ionic strength, and a gradient of increasing ionic strength is used to separate components.

Step 4 **Gel permeation chromatography (GPC)** – a low capacity method (p. 336) where separation is independent of the composition of the mobile phase.

The sample from step 1 may be able to be applied directly to the HIC column, and that from step 2 may be used for IEC without changing the buffer. A concentration step (pp. 406–407) would be required before step 4. The above scheme is very much an idealised approach: in practice, buffers may need to be changed by dialysis or ultrafiltration, as described later. However, the principle of using the minimum number of manipulations to obtain the desired purification remains valid.

Monitoring purification

At each step, the separated material is usually collected as a series of 'fractions'. Each fraction must be assayed for the protein of interest, and fractions containing that protein are pooled prior to the next step. The assay performed will depend on the properties of the protein of interest, but specific enzyme assays or immunoassays are most commonly used. The protein concentration (p. 383) and the volume of the pooled fraction must be determined, and these values, together with those obtained for the amount of the protein of interest, are used to determine the purity and yield after each step. For enzymes, the biological activity is measured and used to calculate the specific activity (the enzyme activity per unit mass of protein) as described on p. 411. By determining the specific activity of the enzyme at each step, the degree of purification, or purification factor (*n*-fold purification), can be obtained from this relationship:

$$\text{Purification factor} = \frac{\text{specific activity after a particular step}}{\text{specific activity of initial sample}} \qquad [60.1]$$

Table 60.2 Example of a record of purification for an enzyme.

Step	Procedure	Total protein (mg)	Total enzyme (U)*	Specific activity (U mg⁻¹)	Purification factor (*n*-fold)	Enzyme yield (%)
Initial sample	–	210	1984	9.45	–	100.0
Step 1	'Salting out'	112	1740	15.54	1.6	87.7
Step 2	HIC	30	1701	56.70	6.0	85.7
Step 3	IEC	6.0	1604	267.33	28.3	80.8
Step 4	GPC	3.0	1550	516.66	54.7	78.1

*U = unit (p. 411).

Example For an enzyme extract having a total activity of 8521 U compared to an initial total activity of 9580 U, using Eqn [60.2] gives a yield of [8521 ÷ 9580] × 100 = 88.9% (to one decimal place).

Assaying enzymes – if possible, use a preliminary screening technique to detect active fractions prior to quantitative assay, e.g. using tetrazolium dyes in microtitre plates to detect oxidoreductases (p. 358).

Measuring the mass of proteins – this is sometimes expressed in daltons (Da), or kilodaltons (kDa), where 1 dalton = 1 atomic mass unit. M_r is an alternative, numerically equivalent, expression (p. 162).

Increased purification usually represents a decrease in total protein relative to the biological activity of the protein of interest, though in some instances it may reflect the loss of an inhibitor during a purification step.

Calculation of the yield of enzyme at each step is straightforward, since:

$$\text{Yield} = \frac{\text{total enzyme activity after a particular step}}{\text{total enzyme activity in initial sample}} \times 100 \ (\%) \qquad [60.2]$$

Note that the yield equation uses the *total amount* of enzyme and is therefore unaffected by the volume of the solutions involved. You should make a record of the progress of your purification procedure at each step, as in Table 60.2.

Monitoring progress using electrophoresis

Polyacrylamide gel electrophoresis (PAGE) of subsamples from each step will give an indication of the number of proteins still present: ideally, a pure protein will give only one band after silver staining (p. 352); however, this stain is so sensitive that even trace impurities can be detected in what you expect to be a pure sample, so do not be dismayed by the appearance of the gel after staining.

Carrying out SDS-PAGE and isoelectric focusing (or, ideally, 2D-PAGE, p. 361) on subsamples also gives information on the M_r and pI of any remaining contaminants, and this can be taken into account when planning the next chromatographic step.

Differential solubility separation techniques

Often the volume and protein concentration of the initial soluble extract will be quite high. Application of a differential solubility technique at this stage results in the precipitation of selected proteins. These can be recovered by filtration or centrifugation, washed and then resuspended in an appropriate buffer. This will reduce the sample volume and may give a small degree of purification, making the sample more suitable for subsequent chromatographic steps.

Ammonium sulfate precipitation ('salting out')

This is the most widely used differential solubility technique, having the advantage that most precipitated enzymes are not permanently denatured, and can be redissolved with restoration of activity. Precipitation depends on the existence of hydrophobic 'patches' on the surface of proteins, inducing

Table 60.3 Amount of $(NH_4)_2SO_4$ $(g\ L^{-1})$ required for a particular percentage saturation.

Final concentration (%) →	20	30	40	50	60	70	80	90	100
Initial concentration (%) ↓	Ammonium sulfate added (g L⁻¹)								
0	107	166	229	295	366	442	523	611	707
10	54	111	171	236	305	379	458	545	636
20	–	56	115	177	244	316	392	476	565
30		–	57	119	184	253	328	408	495
40			–	59	122	190	262	340	424
50				–	61	127	197	272	353
60					–	63	131	204	283
70						–	66	136	212
80							–	68	141
90								–	71

a reorganisation of water molecules in their vicinity. When ammonium sulfate is added to the extract, it dissolves to give ions that become hydrated, leaving fewer water molecules in association with the protein. As a result, the hydrophobic patches become 'exposed', and hydrophobic interactions between different protein molecules lead to their aggregation and precipitation. The basis of fractionation in this method is that, as the salt concentration of the extract is increased, proteins with larger or more abundant hydrophobic patches will precipitate before those with smaller or fewer patches.

Unless you can obtain information from the literature about the $(NH_4)_2SO_4$ concentration that will precipitate your target protein, fractionation is done on a trial-and-error basis. The $(NH_4)_2SO_4$ concentration is expressed in terms of the percentage saturation value: Table 60.3 shows the amount of $(NH_4)_2SO_4$ required to give 20–100% saturation. For each percentage saturation chosen, the $(NH_4)_2SO_4$ salt should be added slowly while stirring, and the mixture left at 4 °C for 1 h before centrifuging at 3000 g for 40 min. For an effective separation, start with the maximum percentage saturation that does not precipitate the protein of interest, then increase the percentage saturation by the minimum amount that will then precipitate it. The proteins precipitated between any two values of percentage saturation (say between 30 and 50%) are referred to as a 'cut'. An alternative approach is to add saturated $(NH_4)_2SO_4$ solution – the volume (V_a) to be added to an initial volume of solution (V_i) with an initial saturation S_i, to give a final saturation S_f, is given by the equation:

$$V_a = \frac{V_i(S_f - S_i)}{1 - S_f} \qquad [60.3]$$

where S_i and S_f are expressed as fractional saturation, e.g. $S_f = 0.5 = 50\%$ saturation, and both volumes are expressed in the same terms, e.g. millilitres.

Precipitation by changing pH

Proteins are least soluble at their isoelectric points because, at that pH, there is no longer the repulsion that occurs between positively or negatively charged protein molecules at physiological pH values. If the precipitated proteins are required for further purification, it is essential that the protein of interest is not irreversibly denatured. The method is probably best employed by precipitating contaminating proteins, leaving the desired protein in solution. Use citric acid for <pH 3, acetic acid for <pH 4 and sodium carbonate or ethanolamine for >pH 8.

Heat denaturation

Exposure of most proteins to high temperatures disrupts their conformation through effects on non-covalent interactions such as hydrogen bonds and van der Waals forces. However, different proteins are denatured, and hence precipitated, at different temperatures, and this can provide a basis for the separation of some heat-stable proteins. By incubating small aliquots (≈1 mL) of extract for 1 min at a range of temperatures between 45 and 65 °C, it is possible to determine the temperature that gives maximum precipitation of contaminating protein with minimal inactivation of the desired protein.

Solvent and polymer precipitation methods

Organic solvents (e.g. acetone, ethanol) cause precipitation of proteins by lowering the dielectric constant of the solution. Performing the precipitation at 0 °C minimises permanent denaturation. Step-wise concentration (%v/v) increments are used, giving 'cuts' of precipitated proteins, as with ammonium sulfate precipitation.

Organic polymers, particularly polyethylene glycol (PEG), also lower the dielectric constant, but at lower concentrations than with acetone or ethanol. The most commonly used PEG preparations have M_r values of 6000 or 20 000, and these can be removed from the sample by ultrafiltration. PEG precipitation does not involve salts, so it may be a useful preliminary step prior to ion-exchange chromatography, which starts with low salt buffer. Also, since the techniques of PEG and ammonium sulfate precipitation involve different principles, they can be used sequentially.

Concentration by ultrafiltration

This involves forcing water and small molecules through a semi-permeable membrane using high pressure or centrifugation. A range of membranes with 'nominal' molecular weight cut-offs between 500 and 300 000 are commercially available (e.g. Amicon, Millipore), with pore sizes of 0.1–10 μm. Concentration of small samples (<5mL) can be achieved using either a membrane backed by an absorbent pad (e.g. Minicon, Fig. 60.1, available with M_r cut-off from 5000 to 30 000), or by using a centrifugal concentrator (e.g. Vivaspin, Centricon). Larger volumes (up to 400 mL) can be concentrated using a stirred ultrafiltration chamber (an ultrafiltration 'cell') where the liquid is forced through the membrane using nitrogen or an inert gas.

Ultrafiltration not only concentrates the sample, but also may give a degree of purification. It can also be used to change the buffer composition by diafiltration (p. 409). Note that molecular weight cut-off values are quoted for globular proteins – fibrous proteins of higher M_r may pass through the ultrafiltration membrane.

Removing salts and changing the buffer

Although you should aim to use the minimum number of manipulations to obtain the desired purification, it may be necessary to remove salts or

Example The HPII catalase of stationary phase *E. coli* remains intact when heated to 55 °C for 15 min, aiding its separation from other proteins.

Definition

Dielectric constant ($\mathcal{E}$) – a dimensionless measure of the screening effect on the force (F) between two charges (q_1, and q_2) owing to the presence of solvent, from the equation $F = (kq_1q_2) \div (\mathcal{E}r^2)$, where r is the distance between the two particles and k is a constant. The dielectric constant for water is high (≈80), while most organic solvents have lower values, in the range 1–10.

(a)

- sample in well
- absorbent pad
- membrane
- impermeable seal

wall of vessel

(b)

- absorbent pad now contains water from sample
- membrane
- concentrated sample in well
- impermeable seal

Fig. 60.1 Minicon ultrafiltration system (a) with sample added, (b) after ultrafiltration.

to change the buffer before the next step will work effectively (e.g. when carrying out IEC, the ionic strength or the pH of the sample may need to be changed before the target protein will bind to the column). Several methods are available, including the following:

- **Dialysis:** the sample is placed in a bag consisting of semi-permeable membrane (e.g. Visking tubing) and placed in at least 20 volumes of the required buffer. The membrane allows molecules of $M_r <$ 20 000 to pass freely, while retaining larger molecules (note that dialysis is not suitable for small proteins). The small molecules diffuse through the membrane until the osmotic pressure between the sample and the dialysis buffer is equalised. Several changes of dialysis buffer may be required to obtain the desired buffer conditions within the sample. Normally dialysis is carried out overnight at 4 °C. Efficient stirring is required – use a magnetic stirrer and stirrer bar.

- **Diafiltration:** quicker than dialysis and more suitable for larger volumes. It involves addition of the desired buffer to the sample solution, followed by ultrafiltration. Several buffer addition and ultrafiltration steps may be necessary to obtain the desired sample conditions.

- **Gel permeation chromatography** (gel filtration): a gel filtration medium of small pore size (e.g. Sephadex G-25) is used to prepare a column of ≈5 times the volume of the sample. When the sample passes through the column, the large protein molecules will elute with the void volume, while salt ions are retained. This method is only suitable for small volumes, and it results in dilution of the sample.

Protein crystallisation

Much of the information on protein structure that is now available on protein sequence databases (see Table 12.1) has been obtained by X-ray crystallographic studies on protein crystals. To obtain protein crystals, the protein should be purified to at least 90–95% and its concentration in solution should be 10–20 mg mL⁻¹. The method most widely used for crystallisation is the hanging drop vapour diffusion technique (Fig. 60.2). Here, a small droplet of the protein sample (usually about 1–2 μL) is mixed with a crystallisation reagent on a siliconised glass coverslip and inverted over a microtitre plate well containing 0.5–1.0 mL of the same crystallisation reagent. The presence of protein in the droplet means that the initial concentration of crystallisation reagent in the droplet is less than that in the well. As a result, water will diffuse from the droplet into the vapour phase until an equilibrium is achieved between the droplet, the vapour phase and the solution in the well. During this equilibration, the protein is concentrated within the droplet, enhancing the formation of protein crystals. The essential components of a crystallisation reagent are: (i) protein precipitant, and (ii) buffer. The specific reagents differ, depending on the type of protein to be crystallised. Commercially available reagents include the PEG/Ion Screen reagents (Hampton Research, USA) and the Clear Strategy Screen reagents (Molecular Dimensions, UK). An image of protein crystals produced by this method is shown in Fig. 60.3.

Using Visking tubing – this must be boiled before use to ensure a uniform pore size and to remove heavy metal contaminants.

Avoiding protein precipitation during dialysis or gel filtration – make sure your buffer pH is either above or below the pI of the proteins.

Fig. 60.2 The hanging drop technique for protein crystallisation.

Fig. 60.3 Image of protein crystals produced by the hanging drop method

Image courtesy of S. Charnock.

Sources for further study

Ahmed, H. (2016) *Principles and Reactions of Protein Extraction, Purification and Characterisation,* 2nd edn. CRC Press, Boca Raton.

Coligan, J.E., Dunn, B.M., Speicher, D.W. and Wingfield, P.T. (2003) *Short Protocols in Protein Science.* Wiley, New York.
[Provides details of over 500 specific methods.]

Hunte, C, von Jagow, G. and Schägger, H. (2003) *Membrane Protein Purification and Crystallisation: a Practical Guide,* 2nd edn. Academic Press, San Diego.
[Gives specific procedures for isolation of membrane proteins.]

Rosenberg, I.M. (2005) *Protein Analysis and Purification: Benchtop Techniques,* 2nd edn. Springer-Verlag, Berlin.

Scopes, R.K. (2010), *Protein Purification: Principles and Practice,* 3rd edn. Springer, New York.

Simpson, R.J., Adams, P.D. and Golemis, E.A. (2008) *Basic Methods in Protein Purification and Analysis: a Laboratory Manual.* Cold Spring Harbor Laboratory Press, New York.

Wilkins, M.R., Appel, R.D., Williams, K.L. and Hochstrasser, D.F. (eds) (2008) *Proteome Research: Concepts Technology and Application (Principles and Practice),* 2nd edn. Springer, New York.

STUDY EXERCISES

60.1 Practise ammonium sulfate fractionation calculations. For a mixture of three proteins, A, B and C, in 100 mL of aqueous solution, you are provided with the information that protein A precipitates at an $(NH_4)_2SO_4$ concentration of 40%, and that protein B precipitates in the 40–60% 'cut', while protein C is not precipitated by 60% $(NH_4)_2SO_4$. Determine how much solid $(NH_4)_2SO_4$ (in grams) would be needed to achieve 40% saturation and, once protein A is removed, how much $(NH_4)_2SO_4$ will then need to be added to the solution to reach 60% saturation. What will be the fate of protein C in this procedure?

60.2 Select a chromatographic technique for use with a particular protein. Using the information given in Chapter 50, which affinity techniques would be most suitable for the following proteins: (a) an enzyme that uses $NAD^+/NADH$ as a co-factor, (b) a genetically engineered protein with a polyhistidine tag at its *N*-terminal end, (c) a glycoprotein?

60.3 Calculate yield and relative purification of an enzyme. The table shows representative data for an enzyme purification involving five stages. Calculate the yield and *n*-fold purification (relative to the initial values) at each stage. Give all answers to three significant figures.

Volume, protein concentration and enzyme activity at several stages during a protein purification procedure.

Stage	Total volume of extract (mL)	Enzyme activity (U mL^{-1})	Protein concentration (mg mL^{-1})
1	78	1.2	4.8
2	70	1.2	1.7
3	10	6.4	2.5
4	5	10.8	1.4
5	0.6	65.7	3.4

Answers to these study exercises are available at www.pearsoned.co.uk/practicalskills.

Enzymes are globular proteins that increase the rate of specific biochemical reactions. Each enzyme operates on a limited number of substrates of similar structure to generate products under well-defined conditions of concentration, pH, temperature, etc. In metabolism, groups of enzymes work together in sequential pathways to carry out complex molecular transformations, e.g. the multi-reaction conversion of glucose to lactate in glycolysis.

> **Example**   Enzyme EC 1.1.1.1 is usually known by its trivial name, alcohol dehydrogenase.

> **KEY POINT**   Enzymes are categorised according to the chemical reactions they catalyse, leading to a four-figure Enzyme Commission (EC) code number and a systematic name for each enzyme. Most enzymes also have a recommended trivial name, often denoted by the suffix 'ase'.

Measuring enzyme reactions

Activity

This is measured in terms of the rate of enzyme reaction. Activity may be expressed directly as amount of substrate utilised per unit time (e.g. nmol min^{-1}, etc.), or in terms of the non-SI international unit (U, or sometimes IU), defined as the amount of enzyme which will convert 1 µmol of substrate to product(s) in 1 min under specified conditions. However, the recommended (SI) unit of enzyme activity is the katal (kat), which is the amount of enzyme that will convert 1 mol of substrate to product(s) in 1 s under optimal conditions, determined from the following equation:

> **Example**   An enzyme preparation that converted 295 µmol of substrate to product in 15 min would have an activity, expressed in terms of non-SI units (U) of: $295 \div 15 = 19.7$ U (to three significant figures). Using Eqn [61.1], the same preparation would have an activity, expressed in katals, of: $295 \times 10^{-6} \div 900 = 3.28 \times 10^{-7} = 328$ nkat (to three significant figures).

$$\text{enzyme activity(kat)} = \frac{\text{substrate converted(mol)}}{\text{time (s)}} \qquad [61.1]$$

This unit is relatively large (1 kat $= 6 \times 10^7$ U) so SI prefixes are often used, e.g. nkat or pkat (p. 195). Note that the units involve amount of substrate (mol), not concentration (mol L^{-1} or mol m^{-3}).

For enzymes with macromolecular substrates of unknown molecular weight (e.g. deoxyribonuclease, amylase), activity can be expressed as the mass of substrate consumed (e.g. ng DNA min^{-1}), or amount of product formed (e.g. nmol glucose min^{-1}). You must ensure that your units clearly specify the substrate or product used, especially when the enzyme transformations involve different numbers of substrate or product molecules. Specific activity, expressed in terms of the amount or mass of substrate or product, is useful for comparing the purity of different enzyme preparations (e.g. p. 405).

> **Example**   The hydrolysis of one molecule of maltose to give two glucose molecules by α-glucosidase means that enzyme activity specified in terms of substrate consumption (nmol maltose) would be half the value expressed with respect to product formation (nmol glucose).

The turnover number of an enzyme is the amount of substrate (mol) converted to product in 1 s by 1 mol of enzyme operating under optimum conditions. In practice, this requires information on the molecular weight of the enzyme, the amount of enzyme present and its maximum activity.

Types of assay

The rate of substrate utilisation or product formation must be measured under controlled conditions, using some characteristic which changes in direct proportion to the concentration of the test substance.

> **Definition**
>
> **Specific activity** – enzyme activity (e.g. kat, U, ng min^{-1}) expressed per unit mass of protein present (e.g. mg, µg).

Spectrophotometric assays

Many substrates and products absorb visible or UV light, and the change in absorbance at a particular wavelength provides a convenient assay method (p. 312). In other cases, a product may be measured by a colorimetric chemical reaction.

Several assays are based on interconversion of the nicotinamide adenine dinucleotide co-enzymes NAD^+ or $NADP^+$ which are reversibly reduced in many enzymic reactions. The reduced form (either NADH or NADPH) can be detected at 340 nm, where the oxidised form has negligible absorbance (p. 312). An alternative approach is to use a coupled enzyme assay, where a product of the test enzyme is used as a substrate for a second enzyme reaction which involves oxidation/reduction of nucleotide coenzymes. Such assays are particularly useful for continuous monitoring of enzyme activity and for reactions where the product from the test substance is too low to detect by other methods, since coupled assays are more sensitive. Note that the reaction of interest (test enzyme) must be the rate-limiting process, not the indicator reaction (second enzyme).

Radioisotopic assays

These are useful where the substrate and product can be easily separated, e.g. in decarboxylase assays using a ^{14}C-labelled substrate, where gaseous $^{14}CO_2$ is produced.

Electrochemical assays

Enzyme reactions involving acids and bases can be monitored using a pH electrode (p. 172), though the change in pH will also affect the activity of the test enzyme. An alternative approach is to measure the amount of acid or alkali required to maintain a constant preselected pH in a pH-stat.

An oxygen electrode can be used if O_2 is a substrate or a product (p. 370). Other ion-specific electrodes can monitor ammonia, nitrate, etc.

Chromogenic and fluorogenic enzyme substrates

Artificial substrates that generate either (i) a coloured product (i.e. they are chromogens) or (ii) a fluorescent product (fluorogens) are used widely for the detection and quantification of enzymes, especially when no suitable spectrophotometric assay is available for the natural product or the natural substrate. In most cases, these artificial substrates are analogues of the natural substrates, composed of a 'core molecule' that is either fluorescent (p. 314) or coloured, coupled to another group by a covalent bond. The addition of this group to the core molecule usually either causes a decrease in fluorescence and a shift in the excitation and emission signal to longer wavelengths, or converts the coloured compound to a non-coloured form. The covalent bond between core molecule and added group is recognised and cleaved by the target enzyme to liberate the core molecule, thereby generating either a fluorescent or a coloured product that can be measured fluorimetrically (p. 314) or spectrophotometrically (p. 310). The principle is best illustrated using a specific example of each type of substrate (see Fig. 61.1). Such substrates serve as a sensitive and specific means of detecting individual enzymes in the presence of a range of other biomolecules. Applications include:

- **biochemical identification of bacteria,** on the basis of enzyme profiles (p. 250; see also Fig. 61.1);

Example Phosphoenolpyruvate carboxylase (PEP carboxylase) can be assayed by coupling to malate dehydrogenase (MDH): the PEP carboxylase reaction converts phosphoenolpyruvate to oxaloacetate, which is then oxidised to malate by MDH (present in excess), with a stoichiometric (1:1) reduction of NAD^+. The coupled assay is monitored spectrophotometrically as an increase in A_{340} against time (p. 312).

Working with radioisotopes – this is covered in Chapter 45.

Understanding fluorescence – the basis of fluorescence is explained on p. 314.

Fig. 61.1 (a) The chromogenic substrate *ortho*-nitrophenyl-β-D-galactoside (ONP-β-GAL); (b) the fluorogenic substrate 4-methylumbelliferyl-β-D-glucuronide (MU-β-GUR). Both substrates show minimal colour and/or fluorescence owing to the coupling of the core molecule to a carbohydrate group. ONP-β-GAL is cleaved by the enzyme β-galactosidase to liberate ONP, which is yellow (usually assayed spectrophotometrically at 420 nm). In contrast, the cleavage of MU-β-GUR by the enzyme β-glucuronidase liberates MU, which is strongly fluorescent under UV light (e.g. excitation wavelength 360 nm, emission wavelength 440 nm). These two substrates are used to detect and differentiate between coliform bacteria (which contain β-galactosidase) and *Escherichia coli* (which contains both β-galactosidase and β-glucuronidase), e.g. in the IDEXX Colilert system for water analysis (see: http://www.idexx.com/water/colilert/).

- **enzyme-linked immunoassays,** e.g. ELISA (p. 285);
- **nucleic acid hybridisation and blotting** (p. 461);
- **detection of specific enzymes** in living cells;
- **quantitative assay for enzyme activity** – e.g. fluorogenic substrates for HIV protease.

Note that care is required for quantitative fluorimetric assay, since impurities in the enzyme preparation can lead to background fluoresence or to quenching (reduction) of the signal: appropriate controls must be run at the same time as test samples. A wide range of chromogenic core molecules (chromophores) are used in artificial substrates (Table 61.1), including: nitrophenols (e.g. ONP, Fig. 61.1); nitroanilines (e.g. Z-arginine-*para*-nitroanilide derivatives for assaying trypsin activity); indoxyl substrates (e.g. 5-bromo-4-chloro-3-indolyl-β-D-galactoside, also known as X-GAL, which is used widely to detect the activity of the *lacZ* gene in molecular biology, p. 476), along with derivatives of alizarin and 3,3′,5,5′-tetramethylbenzidine (Fig. 61.2). Similarly, several different fluorophores are commercially available, including coumarin derivatives (e.g. 4-methylumbelliferone, MU, Fig. 61.1, and 7-amino-4-methylcoumarin, AMC, which fluoresce blue), fluorescein (Fig. 47.4) and resorufin (Table 61.1). For further details see James (1994).

Example Detection of enzymes that degrade β-lactam antibiotics in resistant bacteria can be carried out using the chromogenic substrate nitrocefin, which changes from pale yellow to red when hydrolysed by β-lactamases.

Table 61.1 Core molecules used in chromogenic and fluorogenic enzyme substrates.

Core molecule	Reaction
Alizarin	Red colour
Aminomethyl coumarin	Blue fluorescence
Fluorescein	Green fluorescence
Indoxyl (and derivatives)	Blue colour (and others)
Methylumbelliferone	Blue fluorescence
Nitroaniline	Yellow colour
Ortho-nitrophenol	Yellow colour
Tetramethylbenzidine	Blue (yellow at low pH)
Rhodamine	Red fluorescence

Methods of monitoring substrate utilisation/product formation

Continuous assays (kinetic assays)

The change in substrate or product is monitored as a function of time, to provide a progress curve for the reaction (Fig. 61.3). These curves start off in a near-linear manner, decreasing in slope as the reaction proceeds and

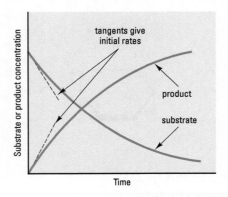

Fig. 61.2 The chromophore 3,3′,5,5′- tetramethylbenzidine (TMB), used in peroxide assays, e.g. in ELISA (p. 287).

> **SAFETY NOTE** – TMB is harmful if swallowed or absorbed through the skin – avoid ingestion and skin contact.

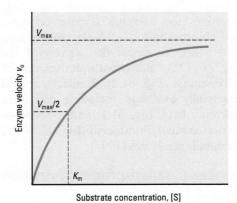

Fig. 61.3 Enzyme reaction progress curve: substrate utilisation/product formation as a function of time.

substrate is used up. The initial velocity of the reaction (v_0) is obtained by drawing a tangent to the curve at zero time and measuring its slope. Continuous monitoring can be used when the test substance can be assayed rapidly (and non-destructively), e.g. using a chromogenic substrate. Reaction rate analysers allow simultaneous addition of reactant(s) or enzyme, mixing and measurement of absorbance: this enables the initial rate to be determined accurately.

Discontinuous assays (fixed time assays)

It is sometimes necessary to measure the amount of substrate consumed or product formed after a fixed time period, e.g. where the test substance is assayed by a (destructive) colorimetric chemical method. It is vital that the time period is kept as short as possible, with the change in substrate concentration limited to around 10%, so that the assay is within the linear part of the progress curve (Fig. 61.3). A continuous assay may be carried out as a preliminary step, to determine whether the reaction is approximately linear over the time period to be used in the fixed time assay.

Enzyme kinetics

For most enzymes, when the initial reaction rate (v_0) using a fixed amount of an enzyme is plotted as a function of the concentration of a single substrate [S] with all other substrates present in excess, a rectangular hyperbola is obtained (Fig. 61.4). At low substrate concentrations v_0 is directly proportional to [S], with a decreasing response as substrate concentration is increased until saturation is achieved. The shape of this plot can be described by a mathematical relationship, known as the Michaelis–Menten equation:

$$v_0 = \frac{V_{max}[S]}{K_m + [S]} \qquad [61.2]$$

This equation makes use of two kinetic constants:

- V_{max}, the maximum velocity of the reaction (at infinite substrate concentration).
- K_m, the Michaelis constant, is the substrate concentration where $v_0 = \frac{1}{2}V_{max}$.

V_{max} is a function of the amount of enzyme and is the appropriate rate to use when determining the specific activity of a purified enzyme. The Michaelis constant is expressed in terms of substrate concentration (mol L^{-1}) and is independent of enzyme concentration. K_m is derived from the individual rate constants of the reaction – for example, with a single substrate (S) and single product (P) enzymic reaction, the process can be described as follows:

$$E + S \underset{k_2}{\overset{k_1}{\rightleftharpoons}} ES \overset{k_3}{\rightarrow} E + P \qquad [61.3]$$

where E = enzyme, ES = enzyme–substrate complex, and k_1, k_2 and k_3 are rate constants. The Michaelis constant can be expressed as:

$$K_m = \frac{k_2 + k_3}{k_1} \qquad [61.4]$$

Fig. 61.4 Effect of substrate concentration on enzyme activity.

(a)

(b)

(c)

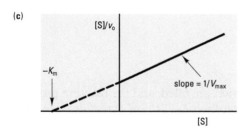

Fig. 61.5 Graphical transformations for determining the kinetic constants of an enzyme. (a) Lineweaver–Burk plot. (b) Eadie–Hofstee plot. (c) Hanes–Woolf plot.

Example The EMBOSS free, open-source software package (at http://emboss.source-forge.net/) can be used to find the K_m and V_{max} of an enzyme, using Michaelis–Menten and Hanes–Woolf plots.

For many enzymes, $k_3 \ll k_2$, and Eqn [61.4] simplifies to:

$$K_m = \frac{k_2}{k_1} = \frac{[E]\,[S]}{[ES]} \tag{61.5}$$

When this applies, K_m provides a measure of the affinity of an enzyme for the substrate, and this is an important characteristic of each particular enzyme. Values for mammalian enzymes usually fall within the range 10^{-2}–10^{-5} mol L^{-1}. Thus an enzyme with a large K_m usually has a low affinity for its substrate, while an enzyme with a small K_m usually has a high affinity. K_m values can be used to select a substrate concentration that will give maximum reaction velocity (p. 416) or to compare the affinities of different substrates for a given enzyme, or the same substrate with different enzymes.

Your first step in determining the kinetic constants for a particular enzyme is to measure the rate of reaction at several substrate concentrations, as in Fig. 61.4. There are various ways to obtain K_m and V_{max} from such data, mostly involving drawing a graph representing a linear transformation of Eqn [61.2].

- **The Lineweaver–Burk plot:** a graph of the reciprocal of the reaction rate ($1/v_0$) against the reciprocal of the substrate concentration ($1/[S]$) gives $-1/K_m$ as the intercept of the x axis and $1/V_{max}$ as the intercept of the y axis (Fig. 61.5(a)). The slope of the plot is most affected by the least accurate values, i.e. those measured at low substrate concentration.

- **The Eadie–Hofstee plot:** v_0 against $v_0/[S]$, where the intercept on the y axis gives V_{max} and the slope equals $-K_m$ (Fig. 61.5(b)).

- **The Hanes–Woolf plot:** $[S]/v_0$ against $[S]$, giving $-K_m$ as the intercept of the x axis and $1/V_{max}$ from the slope (Fig. 61.5(c)).

- **The direct linear plot** (Isenbhal and Cornish) is another option that may give more accurate estimates of kinetic constants than the other plots.

There are several computer packages that will plot the above relationships and calculate the kinetic constants from a given set of data using linear regression analysis (p. 534). While the Eadie–Hofstee and Hanes–Woolf plots distribute the data points more evenly than the Lineweaver–Burk plot, the best approach to such data is to use non-linear regression on untransformed data. This is usually outside the scope of the simpler computer programs, though tailor-made commercial packages can carry out such analyses. Note also that some enzymes, particularly those involved in the control of metabolism, do not show Michaelis–Menten kinetics.

Factors affecting enzyme activity

If you want to measure the maximum rate of a particular enzyme reaction, you will need to optimise the following.

Co-factors

Many enzymes require appropriate concentrations of specific co-factors for maximum activity. These are subdivided into co-enzymes (soluble, low molecular weight organic compounds that are actively involved in catalysis by accepting or donating specific chemical groups, i.e. they are co-substrates of the enzyme; examples include NAD^+ and ADP); and activators (inorganic metal ions, required for maximal activity, e.g. Mg^{2+}, K^+).

Removing a bottle of freeze-dried enzyme from a freezer or fridge – do not open it until it has been warmed to room temperature or water may condense on the contents – this will make weighing inaccurate and may lead to loss of enzyme activity.

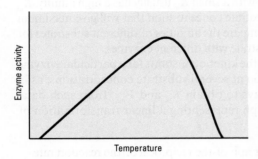

Fig. 61.6 Effect of temperature on enzyme activity.

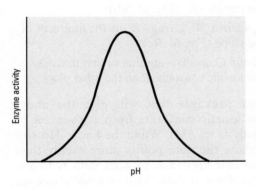

Fig. 61.7 Effect of pH on enzyme activity.

Definitions

Inhibition – the reduction in enzyme activity caused by the presence of another compound (an inhibitor).

Negative feedback – a process within a metabolic pathway where the product of a reaction inhibits an enzyme earlier in the pathway, resulting in a reduction in product formation.

Temperature

Enzyme activity increases with temperature, until an optimum is reached. Above this point, activity decreases as a result of protein denaturation (Fig. 61.6). Note that the optimum temperature for enzyme *activity* may not be the same as that for maximum *stability* (enzymes are usually stored at temperatures near to or below 0 °C, to maximise stability).

pH

Enzymes work best at a particular pH, because of changes in ionisation of the substrates or of the amino acid residues within the enzyme (Fig. 61.7). Most enzyme assays are performed in buffer solutions (p. 171), to prevent changes in pH during the assay. Note that some enzymes have different pH optima for different substrates.

Substrate concentration

A substrate must be present in excess to ensure maximum reaction velocity. The K_m of an enzyme can be used to predict the substrate concentration required for the enzyme to operate at or near its maximum rate – this occurs when all active sites of the enzyme are filled. Using the Michaelis–Menten equation, the fraction of active sites filled (f_{ES}) for a given reaction velocity, v_0, is given by:

$$f_{ES} = \frac{v_0}{V_{max}} = \frac{[S_o]}{[S_o] + K_m}$$ [61.6]

When $[S_o] = K_m$, 50% of active sites are filled, and the reaction proceeds at $\frac{1}{2}V_{max}$. When $[S_o]$ is 10-fold greater than K_m, 91% of active sites are filled. When $[S_o]$ is 100-fold greater than K_m, 99% of active sites are filled and the reaction proceeds at 99% of V_{max}.

> **KEY POINT** Note that temperature and pH optima are dependent upon reaction conditions, including co-factor and substrate concentrations – you should therefore specify the experimental conditions under which such optima are determined.

Enzyme inhibition

It is important to investigate the inhibition of enzyme activity by specific molecules and ions because:

- enzyme inhibition is an important control mechanism in biological systems (e.g. negative feedback by a product);
- many drugs act by inhibiting enzymes;
- the action of many toxins can be explained by enzyme inhibition.

In terms of their interaction with enzymes, inhibitors can be involved in either reversible or irreversible reactions, and the inhibition can be competitive, non-competitive or uncompetitive, depending on whether inhibition can be reduced by an increase in the concentration of the natural substrate.

Irreversible inhibitors

These are substances, usually not of biological origin, which react covalently with an enzyme (E), preventing substrate binding or catalysis, e.g. iodoacetamide, which binds to thiol groups in enzymes as follows:

$$E—CH_2—SH + ICH_2CONH_2 \rightarrow E—CH_2—S—CH_2CONH_2 + HI \qquad [61.7]$$

The thiol group may be within the active site, forming part of the binding and/or the catalytic sites, or it may be further from the active site and affect the 3D conformation of the enzyme.

The toxicity of heavy metal ions (e.g. Hg^{2+}) is largely because of their irreversible effects on enzyme activity. Other examples include the inhibition of acetylcholinesterase by organophosphate pesticides and nerve agents. In such cases, it is often an active site serine residue that is affected, e.g. with diisopropylfluorophosphate (DFP):

$$E—CH_2OH + F—\overset{\overset{R}{|}}{\underset{\underset{R}{|}}{P}}{=}O \rightarrow E—CH_2O—\overset{\overset{R}{|}}{\underset{\underset{R}{|}}{P}}{=}O + HF \qquad [61.8]$$

where R = isopropyl residue.

Reversible inhibitors

These inhibitors do not react covalently with an enzyme, but show rapid reversible binding and dissociation. The velocity of the enzyme-catalysed reaction is reduced by the formation of enzyme–inhibitor (EI) or enzyme–substrate–inhibitor (ESI) complexes. Reversible inhibitors are subdivided according to their effects on K_m and V_{max}:

- **Competitive inhibitors bind reversibly to groups in the active site.** Often the inhibitor resembles the substrate, and the occupation of the active site by an inhibitor molecule prevents a substrate molecule from binding to the same active site (Fig. 61.8(a)). The enzyme (E) can bind to the substrate (S), to form an ES complex, or to the inhibitor (I), to form EI, but cannot bind to both (ESI). A competitive inhibitor lowers the rate of catalysis by reducing the proportion of enzyme molecules bound to the substrate. It is possible to reverse competitive inhibition by increasing the substrate concentration; V_{max} is unaffected, but K_m increases (see p. 418). An example of competitive inhibition is the inhibition of succinate dehydrogenase by malonate, an analogue of the natural substrate, succinate.

- **Non-competitive reversible inhibition occurs when an inhibitor binds to an enzyme whether or not the active site is occupied by the substrate.** This type of inhibition often involves natural inhibitors of enzymes, and is important in the control of metabolism. The inhibitor binds to a site other than the active site (Fig. 61.8(b)). K_m is unchanged because the affinity of substrate molecules that bind to any uninhibited enzyme is unaffected, but V_{max} decreases owing to the concentration of active enzyme molecules being effectively reduced by the presence of inhibitor.

- **Uncompetitive inhibition, where the inhibitor binds only after the substrate has bound to the enzyme** (Fig. 61.8(c)). It is most common in reactions involving more than one substrate.

Using enzyme inhibition kinetics to identify the type of inhibitor

Enzyme kinetics can be used to distinguish between the different forms of inhibition and to provide quantitative information on the effectiveness of various inhibitors, by allowing the dissociation constant of the

Determining the effects of an inhibitor on enzyme kinetics – measure enzyme activity at $\geqslant 6$ different substrate concentrations, both in the presence and absence of inhibitor (see Fig. 61.9).

Fig. 61.8 Representation of: (a) competitive inhibition, where the inhibitor (I) binds to the same site on the enzyme (E) as the substrate (S); (b) simple linear non-competitive inhibition; (c) uncompetitive inhibition (P = products).

(a)

(b)

(c)

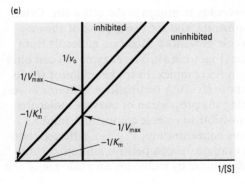

Fig. 61.9 Lineweaver–Burk plots showing the effect of (a) competitive inhibition, (b) simple linear non-competitive inhibition, (c) uncompetitive inhibition.

enzyme-inhibitor complex, K_i, to be determined. K_i is an expression that relates to the strength of binding of the inhibitor to the enzyme.

Competitive inhibition

With inhibitor present, the enzyme can react with the substrate (Eqn [61.3]), but can also react reversibly with the inhibitor (I), to give an inactive enzyme–inhibitor complex (EI), as follows:

$$E + I \underset{k_2'}{\overset{k_1'}{\rightleftharpoons}} EI \qquad [61.9]$$

As a result, the presence of I decreases the amount of free enzyme [E] available for interaction with S, i.e. $[E] = [E_T] - [ES] - [EI]$, where E_T is the total amount of enzyme present. The concentration of the EI complex depends on the concentration of free inhibitor and on the dissociation constant, K_i, where:

$$K_i = \frac{[E][I]}{[EI]} = \frac{k_2'}{k_1'} \qquad [61.10]$$

The Michaelis–Menten equation (Eqn [61.2]) must be modified to account for the presence of inhibitor and for a competitive inhibitor, and results in the Lineweaver–Burk plot shown in Fig. 61.9(a). The lines in the presence and absence of inhibitor intersect on the ordinate (y axis), indicating that, with competitive inhibition, the inhibitory effect disappears at high substrate concentration (when $1/[S] = 0$, $[S] = \infty$). The value of K_i can be calculated from knowledge of the K_m values obtained in the presence and absence of inhibitor (intercepts on the x axis), and the following relationship applies:

$$\text{slope}_{\text{inhibited}} = \text{slope}_{\text{uninhibited}} \frac{(1 + [I])}{K_i} \qquad [61.11]$$

Non-competitive inhibition

In this case, not only can I bind E, but I can also bind to the ES complex to give an inactive ESI complex, i.e. $E + I \rightleftharpoons EI$ and $ES + I \rightleftharpoons ESI$. In the simplest case, the binding of S has no effect on the binding of I, and vice versa, so the dissociation constant of ESI is the same as that of ES, i.e. the K_m. A typical Lineweaver–Burk plot for a non-competitive inhibitor is shown in Fig. 61.9(b). This indicates that this type of inhibitor decreases V_{max} but does not affect K_m. Effectively, this means that the inhibitor removes a certain fraction of active enzyme from operation, no matter what the concentration of substrate. The V_{max} changes by a factor of $(1 + [I])/K_i$, and K_i can be obtained by comparison of the slopes obtained with the uninhibited enzyme and the inhibited enzyme, as described above for a competitive inhibitor.

Uncompetitive inhibition

An uncompetitive inhibitor leads to a Lineweaver–Burk plot as shown in Fig. 61.9(c). Parallel lines are obtained, and both K_m and V_{max} are affected. Here K_i pertains to the dissociation of I from ESI. Uncompetitive inhibitors are fairly rare.

Regulatory enzymes

In cell metabolism, groups of enzymes work together in sequential pathways to carry out a given metabolic process. In such enzyme systems, the reaction product of the first enzyme becomes the substrate for the next. Most of the enzymes in each system obey Michaelis–Menten kinetics. However, in each system there is at least one enzyme – often the first in the sequence – that sets the rate of the overall sequence (the flux through the pathway) because it catalyses the slowest or rate-limiting step. These regulatory enzymes exhibit increased or decreased catalytic activity in response to certain signals. By the action of such regulatory enzymes, the rate of each metabolic sequence is constantly adjusted to meet changes in the cell's demands for energy and biosynthesis.

The activities of regulatory enzymes are altered by non-covalent binding of various types of signal molecules, which are generally small metabolites or co-factors termed 'modulators'. Such enzymes are generally described as 'allosteric'.

Properties of allosteric enzymes

The following general characteristics enable allosteric enzymes to be identified:

- **Feedback inhibition:** in some multi-enzyme systems the regulatory enzyme is specifically inhibited by the end product of the pathway whenever the end product increases in excess of the cell's needs. The end product does not bind to the active site but binds to another, specific, site termed the regulatory site. This binding is non-covalent and readily reversible. Thus if the concentration of the end product decreases, the rate of enzyme activity increases.

- **Modulators for allosteric enzymes may be either inhibitory or stimulatory.** An activator is often the substrate itself, and regulatory enzymes for which substrate and modulator are identical are called homotropic. When a modulator is a molecule other than the substrate, the enzyme is heterotropic. Some enzymes have two or more modulators.

- **Each enzyme molecule has one or more regulatory or allosteric sites for binding the modulator.** Just as the enzyme's active site is specific for its substrate, the allosteric site is specific for its modulator. Enzymes with several modulators generally have different specific binding sites for each. In homotropic enzymes the active site and regulatory site are the same.

- **These enzymes are generally larger and more complex than simple enzymes and normally consist of two or more subunits.** A useful technique for preliminary determination of subunit structure is SDS-PAGE (Chapter 52).

- **Michaelis–Menten kinetics are not followed.** When v_0 is plotted against [S] a sigmoid saturation curve normally results (Fig. 61.10), rather than the hyperbolic curve shown by non-regulatory enzymes (Fig. 61.4). With sigmoidal kinetics, although a value of [S] can be determined at which v_0 is half maximal, this value is not equivalent to K_m. Instead, the symbol $[S]_{0.5}$, or $K_{0.5}$, is often used to represent the substrate concentration giving half maximum velocity of

---

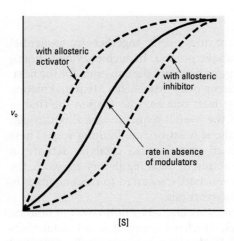

Fig. 61.10 Allosteric effects in enzyme kinetics. The graph shows the rate of enzyme reaction (v_o) against substrate concentration [S].

Example The binding of adrenaline to skeletal muscle receptors triggers reactions that culminate in phosphorylation of a serine residue in glycogen phosphorylase: this activates the enzyme, increasing glycogen breakdown.

the reaction catalysed by an allosteric enzyme. Sigmoid kinetic behaviour generally reflects cooperative interactions between multiple protein subunits, i.e. changes in the structure of one subunit result in structural changes in adjacent subunits that affect substrate binding. In general, allosteric activators cause the curve to become more nearly hyperbolic (Fig. 61.10), with a decrease in $K_{0.5}$, but no change in V_{max}, and therefore v_0 is higher for any value of substrate concentration. However, some allosteric enzymes respond to an activator by an increase in V_{max} with little change in $K_{0.5}$. In contrast, an allosteric inhibitor may produce a more sigmoidal curve with an increase in $K_{0.5}$ (Fig. 61.10).

Covalently modified regulatory enzymes

Some enzymes are regulated by covalent modification, e.g. by phosphorylation of the hydroxyl groups of specific serine, threonine or tyrosine residues. Depending on the enzyme, this may result in activation or inactivation: in most cases, the non-phosphorylated form of the enzyme is relatively inactive. Often these modifications are under hormonal control. Some regulatory enzymes are influenced both by covalent modification and by allosteric effects.

Text reference

James, A.L. (1994) Enzymes in taxonomy and diagnostic bacteriology. In M. Goodfellow and A.G. McDonnell (eds) *Chemical Methods in Bacterial Systematics*, pp. 471–92. Wiley, London.

Sources for further study

Anon. *International Union of Biochemistry and Molecular Biology.* Available: http://www.iubmb.unibe.ch/ Last accessed 01/09/15.
[Links to enzyme nomenclature database and other relevant websites.]

Bisswanger, H. (2008) *Enzyme Kinetics: Principles and Methods.* Wiley, New York.

Cook, P.F. and Cleland, W.W. (2006) *Enzyme Kinetics and Mechanisms.* Garland Science, Oxford.

Cornish-Bowden, A. (2012) *Fundamentals of Enzyme Kinetics*, 4th edn. Wiley, New York.

Eisethal, R. and Danson, M. (2002) *Enzyme Assays: A Practical Approach,* 2nd edn. Oxford University Press, Oxford.

Marangoni, A.G. (2002) *Enzyme Kinetics: a Modern Approach.* Wiley, New York.

Purich, D. (2010) *Enzyme Kinetics, Catalysis and Control: a Reference of Theory and Best-Practice Methods.* Elsevier, Amsterdam.

Taylor, K.B. (2010) *Enzyme Kinetics and Mechanisms.* Springer, New York.
[Deals with steady-state enzyme kinetics.]

STUDY EXERCISES

61.1 Determine enzyme activity and purity (specific activity). An extract of α-glucosidase containing 2.5 mg of protein was incubated with p-nitrophenol-α-D-glucose at 1.0 mmol L^{-1} and assayed by following the liberation of p-nitrophenol at 404 nm. This gave a progress curve with an initial velocity equal to an absorbance change of 0.25 min^{-1}. Given that the conversion of 10 nmol of substrate to product results in an absorbance change of 0.122:

(a) determine the amount of product formed in nmol min^{-1};

(b) calculate the enzyme activity in kat;

(c) determine the specific activity, in terms of (i) nmol min^{-1} (mg protein)$^{-1}$ and (ii) kat (mg protein)$^{-1}$.

(Quote all answers to three significant figures.)

61.2 Estimate K_m and V_{max} for an enzyme. The table below shows representative data for enzyme activity as a function of substrate concentration. Use a suitable graphical method to determine the kinetic constants K_m and V_{max} of the enzyme. (Give your answers to three significant figures.)

Enzyme activity at various substrate concentrations.

Substrate concentration (μmol L^{-1})	Enzyme activity (U)
0	0
5	16.2
7	20.5
10	28.1
15	32.7
25	39.2
50	56.7
100	68.2

61.3 Identify the type of enzyme inhibition shown by therapeutic drugs. In the treatment of acquired immunodeficiency syndrome (AIDS), a possible mode of therapy is to inhibit the reverse transcriptase (RT) of the human immunodeficiency virus (HIV), which is required for the retrovirus to be propagated by RNA-directed DNA synthesis. In the figure, one of the substrates for RT is thymidine (a); two drugs, AZT (b) and HBY097 (c) are known to inhibit HIV RT.

Look at the structures and predict the type of inhibition (i.e. competitive or non-competitive) likely to be shown by each drug. Outline an experiment that would enable you to confirm the type of inhibition by investigating enzyme kinetics and explain how you would interpret the results.

Answers to these study exercises are available at www.pearsoned.co.uk/practicalskills.

Fig. 62.1 Time course of (a) tracer influx and
(b) tracer efflux (A_t). A_α = tracer activity at
equilibrium.

Example Suppose 5 mL of a red cell
suspension containing 2 mg protein mL^{-1}
accumulated 1050 Bq of radiolabelled solute
of specific activity 210 Bq $nmol^{-1}$ in 8 min.
The net rate of solute accumulation over
this time period, expressed in nmol mg
$protein^{-1}$ min^{-1}, would be
$1050 \div 210 \div 10 \div 8 = 0.0625$
($\equiv 62.5$ pmol mg $protein^{-1}$ min^{-1}).

All living cells have a variety of processes whereby solutes move
across membranes, including those processes involved in the uptake
of nutrients and the loss of metabolic waste products at the plasma
membrane, and the intracellular transport of substances across
internal membranes. Membrane transport processes serve to regulate
the intracellular environment, creating suitable conditions for internal
metabolism. In microbes, the plasma membrane plays the principal role
in the selective movement of biomolecules into, or out of, the cell. In
multicellular animals and plants, membrane transport processes in the
outermost epithelial layers are especially important in determining the
solute composition of tissues and organs. Individual organs often have
specialised transport functions, e.g. in animals, nutrient absorption by
intestinal epithelial cells and urea excretion via the proximal tubular
epithelial cells of the kidney or, in plants, the uptake of nutrients by
root hair cells.

Measuring solute transport

Usually, the rate of movement of a solute in a particular biological system
is determined using a radioisotope-labelled (p. 297) or a stable isotope-
labelled solute (p. 323) as a 'tracer'. Typically, in solute uptake (influx),
the labelled solute is added to the external medium and its accumulation
is then monitored by assaying samples of cells or tissue after known time
intervals, enabling you to plot the accumulation of tracer as a function
of time. Loss of solute (efflux) is studied using cells or tissues pre-loaded
with labelled solute and then transferred to a tracer-free incubation
medium: samples of the medium can be assayed after known time
intervals, allowing the time course to be followed without destruction
of the biological material. For microbial cell suspensions, filtration or
silicone oil microcentrifugation (p. 328) may be used to separate the cells
from their incubation medium.

> **KEY POINT** Determination of the relationship between the amount of
> tracer and the amount of solute (the 'specific activity', p. 300) enables the
> uptake or loss of solute to be expressed in terms of the net amount of solute
> transported per unit biomass (e.g. per mg protein, or per cell) per unit time.
> If the relationship between biomass and membrane area is known, the net
> uptake or loss of solute can be expressed per m^2 surface area per unit time.

Short-term tracer studies will give approximate values for the
unidirectional flux rate from the labelled side of the membrane: longer
term studies may show more complex kinetics owing to two-way tracer
movement or to secondary transport within the cells.

For non-metabolised solutes that are accumulated to particular levels
within individual cells, e.g. inorganic ions such as K^+, Na^+, Cl^-, the time
course for tracer accumulation or loss will often show exponential
kinetics, as the transport process allows the tracer to equilibrate
across the cell membrane (Fig. 62.1). Such exponential plots can be
transformed to give linear plots using the relationship $\ln (1 - A_t/A_\alpha)$
for influx, where A_α is the activity at equilibrium and A_t is the activity
after time = t, while a plot of ln A_t/A_α should be linear when the cell

---

Example For a tissue with a surface area of 24 000 m² per m³, an internal K⁺ concentration of 230 mol m⁻³, and an influx rate constant of 0.000036 s⁻¹, substitution into Eqn [62.1] gives an influx of: (0.000036 × 230) ÷ 24 000 = 3.45 × 10⁻⁷ mol m⁻² s⁻¹ (345 nmol m⁻² s⁻¹).

Example For a tissue with a surface area of 24 000 m² per m³, an internal K⁺ concentration of 230 mol m⁻³, and an influx rate constant of 0.000036 s⁻¹, substitution into Eqn [62.1] gives an influx of: $(0.000036 \times 230) \div 24\,000 = 3.45 \times 10^{-7}$ mol m⁻² s⁻¹ (345 nmol m⁻² s⁻¹).

Example 3-O-methyl glucose and 2-deoxyglucose are non-metabolised analogues of glucose, while methyl-β-D-thiogalactoside is a non-metabolised analogue of lactose.

Definition

Fick's first law of diffusion – each solute diffuses in a direction that eliminates its concentration gradient and at a rate proportional to the size of the gradient.

Example For a solute with a flux of 9×10^{-6} mol m⁻² s⁻¹, an internal concentration of 250 mol m⁻³ and an external concentration of 15 mol m⁻³, rearranging Eqn [62.2] gives $P = (9 \times 10^{-6}) \div (250 - 15) = 3.83 \times 10^{-8}$ m s⁻¹ (to three significant figures).

---

behaves as a single compartment with respect to tracer exchange. The slope of such a plot gives a rate constant for exchange (k) that can be used to quantify the unidirectional solute movement, either influx or efflux (ϕ) from the following equation:

$$\phi = \frac{k[C_i]}{M_a} \qquad [62.1]$$

where $[C_i]$ is the internal solute concentration, determined either from tracer equilibration or by chemical assay (p. 379), and M_a is the membrane surface area, expressed per unit volume. For some cells and tissues, an initial, short-term component is observed, owing to tracer exchange between the extracellular space and the bulk medium – this can be subtracted from the longer term component, allowing correction for extracellular tracer content.

Metabolised solutes introduce additional complexity, since they do not simply equilibrate across membranes – for example, glucose may be used as a source of carbon or energy once transported into a cell. Transport studies of metabolised solutes may need to account for such processes, e.g. by separating and assaying the various labelled metabolites. A simpler alternative is to use a non-metabolised analogue of the solute, enabling transmembrane movement to be studied in the absence of metabolism and simplifying the interpretation of tracer studies.

Identifying the transport mechanism for a particular solute

Many studies are carried out to provide information on the nature of the transport process. Solute movement is often subdivided into a number of categories described below.

Simple diffusion – passive transport

In accordance with Fick's first law of diffusion, an uncharged molecule can move through the lipid bilayer of a membrane with a net flux, J, determined by its lipid solubility and by the difference in concentration (strictly, the difference in activity) across the membrane, according to the relationship:

$$J = P\,([C_o] - [C_i]) \qquad [62.2]$$

where P is the permeability coefficient for the substance while $[C_o]$ and $[C_i]$ are external and internal solute concentrations, respectively. The permeability coefficients of various solutes correlate with their partition coefficients between non-polar organic solvents and water, reflecting their relative solubility in the lipid phase. At a practical level, the flux rate calculated from Eqn [62.2] will apply only to the initial state, since diffusion will reduce the concentration gradient and hence the flux rate.

At a practical level, a number of characteristics can be used to recognise simple diffusion, including:

- **the measured rates of transport are consistent with the lipid permeabilities of artificial lipid bilayers;**
- **the rate of transport is directly proportional to the concentration gradient** (Eqn [62.2]), giving a straight-line relationship with no evidence of saturation, in contrast to *all* other transport systems;
- **movement is insensitive to inhibitors,** including structural analogues of the solute, metabolic inhibitors and chemical inactivators;

- **passive diffusion lacks specificity** – chemically similar solutes with comparable lipid solubilities will show similar rates of transmembrane movement.

Simple diffusion can account for the movement of only a limited number of low M_r, uncharged substances, including O_2, CO_2, NH_3 and non-polar hydrocarbons. There is negligible transbilayer diffusion of all other polar and ionic solutes, owing to their hydrophilic, lipophobic nature and to the presence of a surrounding shell of water molecules, preventing movement through the hydrophobic interior of the membrane.

> **KEY POINT** All solutes not transported across membranes by simple diffusion are moved by protein-mediated transporters, often termed 'permeases', 'porters' or 'translocators'.

Facilitated diffusion – passive transport

Here, the transmembrane movement is the result of specific membrane proteins (often termed 'uniporters'). While net movement is energetically 'downhill', as in simple diffusion, the rate of movement is more rapid than would be predicted from Eqn [62.2], since the individual uniporter molecules act as solute-specific channels across the membrane. Examples include glucose transport across the erythrocyte membrane, maltodextrin transport across the outer membrane of Gram-negative bacteria and the stretch-activated ion channels in stomatal membranes. The diagnostic features of facilitated diffusion include the following:

- **The measured rates of transport are far greater than can be accounted for by simple diffusion and Fick's first law:** for example, the erythrocyte glucose system operates at more than 10 000 times the rate predicted from permeability coefficients of synthetic lipid bilayers.
- **The system exhibits specificity,** transporting the solute but not chemically related compounds with a different 3D shape.
- **The rate of transport should show saturation kinetics similar to those observed for enzymes** (Fig. 62.2), with a hyperbolic relationship between the rate of transport (v) and solute concentration [C] according to the relationship:

$$v = \frac{V_{max}[C]}{K_s + [C]} \qquad [62.3]$$

where V_{max} is the maximum rate of transport and K_s the half saturation constant, is a measure of the affinity of the transport system for the solute – a low K_s generally indicates a high-affinity transport system while a high K_s shows that the transport system has a low affinity for the solute.

- **Movement is sensitive to competitive inhibitors** – the presence of structurally similar compounds will reduce the rate of solute transport. Competitive inhibitors decrease the affinity of the permease (increased K_s), but V_{max} is not affected (Fig. 62.2).
- **Transport is susceptible to inactivation by non-competitive inhibitors that act as protein denaturants,** e.g. heavy metal salts. Such irreversible inactivation confirms that the transport process is mediated by a protein. At low concentrations of inhibitor, only some of the permease molecules will be inactivated, giving a decreased V_{max} but unchanged K_s (Fig. 62.2).

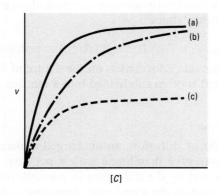

Fig. 62.2 Relationship between the rate of transport (v) and solute concentration [C] for (a) uninhibited 'control', (b) with a competitive inhibitor, (c) with a non-competitive inhibitor.

Example For K$^+$ (a monovalent cation) in a cell with an internal concentration of 450 mol m^{-3} in an experimental solution at an external concentration of 10 mol m^{-3}, substitution into Eqn [62.4] with a value for the gas constant, R = 8.31443 J K^{-1} mol^{-1} and Faraday's constant, F = 9.648675 × 10^4 J V^{-1} mol^{-1} at 20 °C (T = 293.15 K) gives a transmembrane equilibrium potential, E_n = 2.303 [(8.31443 × 293.15) ÷ (+1 × 9.648675 × 10^4)] log (10 ÷ 450) = −0.0961782 V (−96.2 mV to three significant figures).

Example A membrane potential of −60 mV (interior negative) can account for a tenfold passive accumulation of a monovalent cation and a tenfold passive exclusion of a monovalent anion.

Measuring membrane potentials – depending on the cell type, this can be carried out using microelectrodes, lipophilic cations or fluorescent dyes (see p. 436).

If facilitated diffusion of an uncharged solute occurs, then net solute movement will occur down the transmembrane concentration gradient. However, for charged solutes (ions), net movement will depend on the transmembrane electrical potential (the membrane potential), as well as the concentration gradient. The transmembrane equilibrium potential for a particular ion, E_n, can be calculated from the Nernst equation, as:

$$E_n = 2.303 \frac{RT}{zF} \log \frac{[C_o]}{[C_i]}$$ [62.4]

where all symbols have their usual meanings (p. 197; see also Chapter 54). A comparison of the equilibrium potential and the membrane potential will show whether the ion is in electrochemical equilibrium across the membrane, or whether it has been transported against the electrochemical potential gradient.

KEY POINT The transmembrane electrical potential, responsible for the facilitated diffusion of ions, is usually established as a direct result of the action of one or more ion-pumping, active transport systems, e.g. H$^+$-ATPases.

Ionophores (Table 62.1) can be viewed as simple models of facilitated diffusion systems: they can be used to study the effects of an increase in the permeability of membranes to a single solute or group of solutes, to create transmembrane movements of ions, or to monitor specific ions in solution.

Table 62.1 Ionophores and inhibitors of transmembrane solute movement*.

Ionophores	Effect
valinomycin	electrogenic carrier: increased permeability to K$^+$ and Rb$^+$
gramicidin A	electrogenic channel: uniport of monovalent ions, particularly H$^+$
monensin	electroneutral carrier: exchange of Na$^+$ or Li$^+$ for H$^+$
nigericin	electroneutral carrier: exchange of K$^+$ or Rb$^+$ for H$^+$
A-23187	electroneutral carrier: exchange of divalent cations (e.g. Ca^{++}, Mg^{++}) for 2H$^+$
2,4-dinitrophenol (DNP)	electrogenic carrier: increased H$^+$ permeability (protonophore)
carbonylcyanide m-chlorophenyl-hydrazone (CCCP)	electrogenic carrier: increased H$^+$ permeability (protonophore)
carbonylcyanide p-trifluoromethoxy-phenylhydrazone (FCCP)	electrogenic carrier: increased H$^+$ permeability (protonophore)
nonactin	increased permeability of monovalent cations
amphotericin B	anion channel (weak selectivity)
nystatin	anion channel (weak selectivity)
Transport inhibitors†	**Effect**
oligomycin	inhibition of F$_0$F$_1$-type H$^+$-ATPases (mitochondria, bacteria, chloroplasts)
ouabain	inhibition of Na$^+$-K$^+$-ATPases (mammalian cell plasma membrane)
digitoxigenin	inhibition of Na$^+$-K$^+$-ATPases (mammalian cell plasma membrane)
vanadate	inhibition of E$_1$E$_2$-type ATPases (plant plasma membrane)
N,N'-dicyclohexylcarbodiimide (DCCD)	inhibition of F$_0$F$_1$ and V-type ATPases (mitochondria and plant tonoplast)
N-ethylmaleimide (NEM)	inhibition of plant tonoplast V-type ATPases
phlorizin	inhibition of Na$^+$-glucose symporters (intestinal epithelial cells)
furosemide	inhibition of co-transporters (mammalian cell membrane)
cytochalasin B	inhibition of glucose transporters (mammalian cell membrane)
amiloride	inhibition of Na$^+$/H$^+$ antiporters (mammalian and plant cells)
4,4′-diisothiocyano-2,2′-stilbene-disulfonate (DIDS)	inhibition of mammalian anion antiporters

*$Safety note:$ all of these inhibitors are highly toxic and must be handled carefully, observing the appropriate safety precautions.
†Details of electron transport inhibitors are given in Chapter 63.

Active transport

The principal criterion to establish active transport is that movement of the solute occurs against its electrochemical potential gradient. The energy for such 'uphill' transport can be provided by one of two mechanisms:

1. **Primary active transport.** The movement of solute molecules is coupled to the hydrolysis of ATP, via a membrane-bound solute-specific ATPase, or to a redox chain. For example, the Na^+–K^+ ATPase of mammalian cells is a primary active transporter, moving 3 Na^+ ions outwards and 2 K^+ ions inwards for every molecule of ATP hydrolysed, i.e. it is *electrogenic,* with a net charge separation across the membrane. Some of the most important primary active transport processes are those which create a transmembrane H^+ gradient, or proton-motive force (Δp), either via an H^+-translocating ATPase or an electron transport chain (p. 436). In many membranes, Δp provides the major driving force for the movement of other solutes (see Nicholls and Ferguson, 2013, for details). Some Gram-negative bacteria use a variation of phosphoryl-driven active transport to accumulate certain sugars – this is known as 'group translocation' since it involves the simultaneous phosphorylation of each sugar molecule as it is transported. Because the cell membrane is impermeable to sugar phosphates, the bacterial 'phosphotransferase' system can concentrate these solutes within cells.

2. **Secondary active transport, or secondary transport.** Here, the movement of solute is coupled to the 'downhill' movement of another solute, either via a 'symporter' (co-transporter) that simultaneously transports the two solutes in the same direction, or an 'antiporter' (counter-transporter) that simultaneously moves the two solutes in opposite directions across the membrane (Fig. 62.3). These systems are 'active' only in the sense that the 'downhill' movement of the other solute can only occur after primary active transport has established a transmembrane solute gradient (e.g. for H^+ or Na^+). Thus, lactose uptake in bacterial cells is coupled to the transmembrane proton gradient, via the lactose–H^+ symporter, while Na^+ efflux from plant cells is mediated by an Na^+–H^+ antiporter. In both instances, the H^+ gradient is created by a primary active transport system that pumps protons out of the cell, generating a proton-motive force that can be utilised by secondary transport systems. Similarly, the Na^+–K^+ primary active transport system creates an Na^+ gradient that can be utilised for nutrient uptake in mammalian cells, e.g. via the Na^+– glucose symporter (a secondary transporter). Primary and secondary transport systems can be envisaged as operating as a 'circuit' for the movement of ions across membranes (Fig. 62.3).

In addition to the characteristics of specificity, saturation and inhibition shared by facilitated diffusion, active transport systems can be identified by the following features:

- **The transmembrane distribution of the solute is not consistent with Eqn [62.2]** (where $J = 0$ at equilibrium, i.e. $[C_0] = [C_i]$) for an uncharged solute, or Eqn [62.4] for an ion.

- **The active component will be sensitive to metabolic inhibitors,** e.g. ATPase inhibitors, for primary active transport, or specific inhibitors of secondary transporters (see Table 62.1).

Definition

Proton-motive force – an electrochemical gradient of protons (H^+) generated by active transport of H^+ across a membrane: the gradient can then be used to carry out work, e.g. solute transport or ATP synthesis.

membrane

uniport **(a)**

symport **(b)**

antiport **(c)**

Fig. 62.3 Diagrammatic representation of various membrane transport systems, indicating the direction of movement of transported solute molecules.

Interpreting inhibitor experiments – while short-term effects may be unique to a particular inhibitor, in the longer term more general, non-specific effects, are likely. For example, inhibitors of primary active transport will eventually affect secondary transport systems, owing to their effects on ion gradients and membrane potentials.

• **For secondary transport systems, dissipation of the primary ion gradient using a suitable ionophore will inhibit transport of the solute.** Conversely, artificially generated ion gradients can be used to drive solute movement, either in whole cells or in membrane vesicle preparations.

Studying membrane transport using patch-clamp techniques

In this technique, part of a membrane (a 'patch') is sealed to the end of a heat-polished micropipette of diameter ≈ 1 µm and the flow of current across the membrane patch is then measured for different bathing solutions and/or transmembrane voltages, set at particular values ('clamped') by a feedback amplifier. The size and orientation of the membrane patch can be controlled, as shown in Fig. 62.4. Since the membrane patch can be relatively small, the operation of ion-transporting proteins can be seen as changes in current flow, owing to the opening or closing of individual membrane 'channels' (voltage-sensitive transporters). While the physiological factors controlling the *in vivo* operation of such channels remain unclear, patch clamping provides a powerful means of studying channel-mediated ion movements *in vitro*, enabling investigation of:

• the voltage required to produce 'channel' opening;

• the membrane conductance under conditions where the ion concentrations on both sides of the membrane are known;

• temporal effects, including opening and closing times, frequency of opening, etc.;

• the ion selectivity of particular 'channels';

• the effects of inhibitors, ionophores, etc.

When whole cells are used, the flow of ions through primary active ion transport systems (ion 'pumps') can be studied, since changes in current flow represent the sum of many transporters, acting together. Patch-clamp systems have provided a considerable amount of information on the transport properties of individual membranes, e.g. the plasma membrane and tonoplast of plant cells (see Yeo and Flowers, 2007, for details).

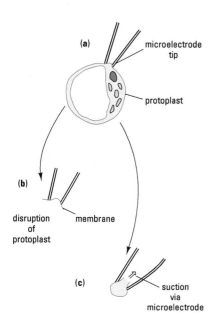

Fig. 62.4 Patch-clamp techniques: (a) whole plant cell, (b) inside-out membrane patch, (c) outside-out membrane patch.

Text references and sources for further study

Baldwin, S.A. (2000) *Membrane Transport: A Practical Approach.* Oxford University Press, Oxford.

Bernhardt, I. and Ellory, J.C. (2003) *Red Cell Membrane Transport in Health and Disease.* Springer-Verlag, Heidelberg.

Blatt, M. (2003) *Membrane Transport in Plants. Annual Plant Reviews,* Vol. 15. Blackwell, Oxford.

Nicholls, D.G. and Ferguson, S.J. (2013) *Bioenergetics,* 4th edn. Academic Press, London.

Prior, R. (2014) *Membrane Function,* 6th edn. Prior Educational Resources, Trail.
[A basic introductory guide.]

Yan, Q. (2003) *Membrane Transporters: Methods and Protocols.* Humana Press, New York.

Yan, Q. (2010) *Membrane Transporters in Drug Discovery and Development: Methods and Protocols.* Humana, New York.

Yeo, A.R. and Flowers, T.J. (2007) *Plant Solute Transport.* Blackwell, Oxford.

STUDY EXERCISES

62.1 Calculate flux rates of various solutes based on data from tracer studies. Using the following data for rate constant (k), internal concentration ($[C_i]$) and membrane surface area (M_a), calculate the unidirectional flux of each solute (express your answers to three significant figures, in all of the study exercises in this chapter).

(a) Solute uptake for a cell where $k = 0.0001$ s^{-1}, $[C_i] = 150$ mol m^{-3} and $M_a = 50\,000$ m^2 m^{-3}.

(b) K$^+$ efflux from a protozoan, where $k = -1.65 \times 10^{-4}$ s^{-1}, $[C_i] = 106$ mol m^{-3} and $M_a = 8.25 \times 10^4$ m^2 m^{-3}.

(c) Na$^+$ uptake into a bacterial cell, where $k = 4.75 \times 10^{-5}$ s^{-1}, $[C_i] = 10.5$ mmol dm^{-3} and $M_a = 1.66 \times 10^{-9}$ m^2 dm^{-3}.

62.2 Determine passive transmembrane fluxes of uncharged solutes. What is the direction and magnitude of the net passive transmembrane flux for each of the following cases, all involving uncharged solutes?

(a) Movement of glycerol in a halotolerant bacterial cell with an internal concentration of 10 mol m^{-3}, an external concentration of 1.5 mol m^{-3} and a permeability coefficient of 2×10^{-7} m s^{-1}.

(b) Glucose movement in an animal cell with an internal concentration of 8.65 mol m^{-3}, an external concentration of 0.03 mol m^{-3} and a permeability coefficient of 1.6×10^{-9} m s^{-1}.

(c) Methanol flow in a fungal cell with an internal concentration of 0.25 mmol dm^{-3}, an external concentration of 3.45 mmol dm^{-3} and a permeability coefficient of 2.6×10^{-3} cm s^{-1}.

62.3 Interpret kinetic parameters of transport systems. Details of the kinetic parameters for the transport of an amino acid into cells of three different types are given in the table.

Kinetic parameters for amino acid uptake into three different cell types.

Cell type	V_{max}	K_s
A	15.2 pmol dm^{-3} s^{-1}	1.24×10^{-6} pmol dm^{-3}
B	9.6 nmol m^{-3} s^{-1}	6.54×10^5 nmol m^{-3}
C	84×10^{-11} mol m^{-3} s^{-1}	2.54×10^{-4} mol m^{-3}

(a) Which cell type has the highest maximum rate of transport?

(b) Which cell type has the highest affinity for the amino acid?

(c) Which cell type gives the fastest rate of uptake at an amino acid concentration of 1×10^4 mol m^{-3}?

62.4 Compare calculated transmembrane equilibrium potentials with the membrane potential to decide whether an ion has been actively transported. At 25 °C, a marine algal cell has a measured plasma membrane potential of –56 mV and the following internal (cytoplasmic) and external ion concentrations.

Internal and external ion concentrations for a marine alga.

Ion	Internal concentration (mol m^{-3})	External concentration (mol m^{-3})
K$^+$	197	10
Na$^+$	8.4	225
Cl$^-$	27.3	235

(a) What is the transmembrane equilibrium potential for each of these ions?

(b) Which ions appear to be in electrochemical equilibrium and which appear to be actively transported (and in what direction)?

Fig. 63.1 Non-cyclic electron transport from H_2O to $NADP^+$ in plant photosynthesis (Z scheme) showing the principal components: PS II, photosystem II; PQ, plastoquinone; cyt, cytochrome; PC, plastocyanin; PS I, photosystem I; FD, ferredoxin; $NADP^+$, nicotinamide adenine dinucleotide phosphate.

Measuring photosynthetic pigments – equations for spectrophotometric analysis of chlorophylls are given on p. 312.

Definitions

C_3 plants – those in which phosphoglyceric acid is the first stable product.

C_4 plants – those in which malate or aspartate are the principal products, via oxaloacetic acid, in a reaction catalysed by phosphoenolpyruvate (PEP) carboxylase.

CAM plants – those in which malic acid is produced, via oxaloacetic acid, by PEP carboxylase activity at night, with little or no net assimilation of CO_2 during the day (crassulacean acid metabolism).

Techniques for investigating photosynthesis and respiration are considered together, since:

- **Gas exchange processes in these two processes may be studied using similar techniques,** and interpreted in similar stoichiometric terms, e.g. by relating changes in O_2 or CO_2 to those of other reactants or products.
- **Membrane-bound electron transport systems are involved in the production of a transmembrane H^+ gradient** that drives ATP synthesis via an ATPase embedded in the same membrane. Measurement of electron transport in photosynthesis and respiration often makes use of redox dyes as artificial electron donors or acceptors, with spectrophotometric analysis of the reactions.
- **Some electron transport inhibitors can affect both photosynthesis and respiration,** while others are specific to one or the other process.

> **KEY POINT** Photosynthesis and respiration are complex, multi-stage metabolic pathways involving a large number of cellular components: they may be studied as a whole, or as isolated components and individual metabolic reactions, to gain a deeper insight into the underlying processes.

Photosynthesis

This is the conversion of light energy to chemical energy by photoautotrophic plants and bacteria. The overall process is often summarised in terms of the synthesis of carbohydrate, $(CH_2O)_n$, driven by the energy within photons of light (hv), as:

$$nCO_2 + 2nH_2O^* + 4nhv \rightarrow (CH_2O)_n + nO_2^* + nH_2O \qquad [63.1]$$

where the asterisks show that all of the O_2 is derived from the photolysis of water. Carbohydrate synthesis occurs as a result of two distinct processes, namely (i) light reactions and (ii) dark reactions. In the former, light energy is absorbed by pigments within membrane-bound reaction centres (photosystems). The water-splitting light reactions generate electrons, protons and molecular oxygen. Electron and proton movement via a series of membrane-bound redox carriers leads to the reduction of $NADP^+$ and the generation of ATP (photophosphorylation), driven by a transmembrane H^+ gradient, or proton-motive force (p. 438). The light reactions of plant-type photosynthesis are often represented by the 'Z scheme' (Fig. 63.1).

The dark reactions use the products of the light reactions to 'fix' CO_2, via a series of soluble enzymes known as the Calvin cycle, or reductive pentose phosphate cycle. The primary carboxylating enzyme in temperate green plants (C_3 plants) and cyanobacteria is ribulose bisphosphate carboxylase ('Rubisco'). This enzyme catalyses the addition of CO_2 to ribulose bisphosphate (C_5), producing two molecules of phosphoglyceric acid (C_3) which are then reduced to triose phosphate (C_3) by enzymes that utilise the NADPH and ATP generated by the light reactions. Thus both light and dark reactions operate in a coupled manner when photosynthetically active cells are illuminated with visible light.

Definitions

Gross photosynthesis – the rate of CO_2 uptake or O_2 production, allowing for gas exchange due to respiratory and photorespiratory activity.

Net photosynthesis – the net rate of CO_2 uptake or O_2 production, including respiratory and photorespiratory gas exchange.

Correcting for photorespiration and respiration – measuring the rate of O_2 consumption or CO_2 production in the first few minutes after the plant material has been transferred from the light to darkness will give you the best estimate of photorespiratory and respiratory gas exchange. Use this rate to convert net photosynthesis to gross photosynthesis.

Preventing oxygen supersaturation in oxygen electrode studies – lower the O_2 content of your experimental solutions by bubbling them with N_2 before use, so that the O_2 evolved during the experiment remains in solution.

Fig. 63.2 A portable infrared gas analyser, used to measure photosynthetic CO_2 uptake (leaf cuvette/probe shown in left hand, with analysis unit and user interface below).
Courtesy of PP Systems, Amesbury, MA, USA
(http:// www.ppsystems.com).

Measuring photosynthetic activity

Equation [63.1] shows that the rate of photosynthesis can be measured in terms of the amount of O_2 evolved or CO_2 fixed. However, several other metabolic processes may lead to concurrent changes in O_2 and/or CO_2 status, particularly:

- **Photorespiration** – light-dependent O_2 uptake, due to the oxygenase activity of Rubisco, leading to the production of glycolic acid (C_2) and the subsequent release of CO_2 due to the operation of a 'scavenging' pathway where two molecules of glycolate are converted to one of phosphoglyceric acid (C_3). The relative rates of photosynthesis and photorespiration depend upon the concentration of O_2 and CO_2 at the active site of Rubisco – in C_3 plants, photorespiration is highest under conditions of high temperature, high light and low water availability, while C_4 plants show low levels of photorespiration.

- **Respiration** – generation of ATP via carbohydrate oxidation, as in non-photosynthetic cells (p. 436). In most photoautotrophs, respiratory activity is likely to be lower during the day than night. However, plant tissues often respond to wounding or environmental stress by *increasing* their rate of respiration – this may be significant if you are measuring photosynthetic activity in tissue fragments, isolated cells or protoplasts. If you wish to calculate gross rates of photosynthesis, you must make appropriate allowances for both respiration and photorespiration.

Measurement of oxygen production

This is most often used with aquatic systems, including algae and photosynthetic bacteria, higher plant cells, protoplasts, chloroplast suspensions or isolated thylakoids. Oxygen in solution can be determined by end-point chemical analysis (e.g. the Winkler method or, more conveniently, by continuous monitoring using an oxygen electrode and chart recorder, to give the rate of net O_2 production (Chapter 54) under various conditions, e.g. light, temperature, etc. For photosynthesis–irradiance (P–I) curves using an O_2 electrode, you should note that the electrode assembly can act as a lens, so the light within the chamber can be higher than that measured at the outside surface. The rate of gross photosynthesis is obtained by correcting for O_2 uptake when the electrode assembly is transferred to darkness (e.g. using a thick black cloth).

Measurement of carbon dioxide uptake

For higher plant studies, this is most easily achieved using an infrared gas analyser, or IRGA (Fig. 63.2). Most modern instruments are portable and generally incorporate: (i) a cuvette, with a transparent window that attaches to a whole leaf or a known area of leaf providing a gas-tight seal, and within which the air is stirred by a fan; (ii) a gas supply system that allows control over input gases; (iii) miniaturised infrared analysers to detect differences in CO_2 and H_2O content of input and output gas streams; (iv) systems for measuring (and possibly controlling) other environmental variables, e.g. light and temperature; (v) an on-board microprocessor for calculating results as rates of photosynthesis and transpiration and estimating the leaf's internal pCO_2 (C_i) and for exporting these and other data (e.g. time, leaf temperature, photosynthetic photon flux density) to PCs. Portable IRGAs are particularly

Fig. 63.3 Example of a photosynthesis–irradiance (*P–I*) curve for a typical plant leaf. Such a curve might be obtained using an IRGA, manipulating *I* in steps down from high values and measuring *P* rates when stable (see Parsons *et al.*, 1997). Typical values for estimated parameters might be: dark respiration rate, -4 µmol CO_2 m^{-2} s^{-1}; light compensation point, 30 µmol photons m^{-2} s^{-1}; maximum photosynthetic efficiency, 8 photons per CO_2 molecule fixed (the theoretical maximum is about 20); maximum photosynthetic rate, 20 µmol CO_2 m^{-2} s^{-1}, which might be achieved at *I* = 1000 µmol photons m^{-2} s^{-1} (full sunlight is about 1800 µmol photons m^{-2} s^{-1}). Various methods are used to fit lines to data points and thereby estimate the above parameters – there remains controversy as to which is the best mathematical function to use for this.

Maximum photosynthetic rate: phase where *P* reaches an asymptote and systems are effectively saturated with light (**photosynthetic capacity**)

Photo-inhibition: phase in high light where *P* may be reduced in some cases

Maximum photosynthetic efficiency: where slope of curve is steepest, measured in mol CO_2 assimilated per photon received

Light compensation point: where *P* is 0 and photosynthesis balances respiration

Low light phase where photosynthesis exceeds respiration and *P* is negative: (at *I* = 0, an estimate of the **dark respiration rate**)

Definitions

Compensation point – PFD or pCO_2 at which net photosynthesis = 0, i.e. where gross photosynthesis, respiration and photorespiration are balanced: from the intercept of a *P–I* or *P–C*$_i$ curve (*x* axis).

Photosynthetic efficiency – the rate of photosynthesis when limited either by (i) by supply of photons or (ii) by supply of CO_2, determined from the initial gradient of a *P–I* or *P–C*$_i$ curve respectively.

Photosynthetic capacity – the rate of photosynthesis under saturating photon flux density (PFD) or CO_2 supply, determined from the upper asymptote of a *P–I* or *P–C*$_i$ curve respectively.

Interpreting ^{14}C fixation data – short-term studies of a few minutes may estimate gross photosynthesis, while longer-term studies will give a rate closer to net photosynthesis, since some of the fixed ^{14}C will be respired.

Studying the release of photo-assimilated ^{14}C in aquatic photoautotrophs – the loss of glycolate and other metabolites can be quantified by acidifying a known amount of medium, to drive off unfixed ^{14}C, then counting directly, or after an appropriate separation procedure.

---

useful for the rapid construction of photosynthesis–irradiance (*P–I*) and photosynthesis–pCO_2 (*P–Ci*) curves used to estimate photosynthetic efficiency, photosynthetic capacity and compensation points (Fig. 63.3).

Using carbon isotopes

Radioisotopic ^{14}C and stable ^{13}C are used to study photosynthesis. Radiocarbon (^{14}C) can be supplied as $^{14}CO_2$ for gas exchange studies or, more readily, as $H^{14}CO_3^-$ for studies in aqueous systems, since there will be an interconversion of soluble CO_2, HCO_3^- and $H_2CO_3^{2-}$. In aqueous systems, the plant material is incubated in medium containing ^{14}C–labelled bicarbonate for a known time period, then removed and prepared for liquid scintillation counting (p. 299). Microalgal cells and photosynthetic bacteria can be separated from the experimental medium by filtration, or silicone oil microcentrifugation (p. 328) – mild acid treatment (e.g. 50 mmol L^{-1} HCl) will ensure that any unfixed ^{14}C is released as $^{14}CO_2$. To express the results in terms of the amount of C assimilated, you will need to calculate (i) the total inorganic C content of the medium, i.e. $CO_2 + HCO_3^- + H_2CO_3^{2-}$, obtained from pH and alkalinity measurements or by IR spectroscopy of the CO_2 produced when a known amount of the medium is acidified and (ii) the relationship between the total inorganic C content and the amount of ^{14}C tracer added (i.e. the specific activity of the experimental solution, p. 419).

One of the advantages of studying the photoassimilation of ^{14}C is that the radiotracer can be used to follow the fate of fixed carbon, by separating and fractionating the various cellular components prior to counting, e.g. by sequential solvent extraction of the plant material in 80% ethanol (low molecular weight solutes) then boiling water (polysaccharides), leaving a residual fraction (structural polysaccharides, proteins and nucleic acids). More sophisticated separation techniques must be used to quantify the amount of radioactivity within individual biomolecules, e.g. using column chromatography (p. 333) and autoradiography (p. 301), or 2-D thin layer chromatography (p. 332). High specific activity $H^{14}CO_3^-$ and short incubation times of a few seconds are required to study the early stages in carbon photoassimilation.

The use of the stable carbon isotopes ^{12}C and ^{13}C in photosynthetic studies is covered in detail within Chapter 63, including example calculations of the relative abundance of each stable isotope, and $\delta^{13}C$.

Chlorophyll fluorescence

This can be used to provide a non-destructive indication of photosynthetic function. When light energy is absorbed by the light-harvesting apparatus of green plants, several reactions compete for the deactivation of excited chlorophyll molecules. Principally, there are three possible outcomes: (i) the light energy may be trapped and used to reduce photosystem II (PS II) and drive photochemical reactions (Fig. 63.1); (ii) the energy may be lost as heat; or (iii) the energy may be re-emitted as fluorescence, measured at wavelengths around 685 nm using a purpose-built chlorophyll fluorometer. Since these three processes are mutually exclusive, measurements of chlorophyll fluorescence can provide information about changes in photochemistry and heat dissipation, since the maximum potential fluorescence yield is 'quenched' (reduced) by (i) photochemical and (ii) non-photochemical processes. Most measurements are made using pulse-amplified modulated (PAM) fluorometers (Fig. 63.4), where the light source used to generate the chlorophyll fluorescence can be turned on and off at high frequency (pulse modulated), enabling the chlorophyll fluorescence signal to be separated from all other light, and therefore allowing measurements to be made in light-adapted leaves under 'natural' conditions.

To interpret fluorescence changes, the relative contribution of the two quenching processes must be established. The starting point is to consider what happens when a strong light is shone on a dark-adapted leaf: there is a rapid rise in chlorophyll fluorescence to a maximum value, as photosynthetic electron transport is rapidly saturated and no further electrons can be accepted by PS II. If the light remains on, the chlorophyll fluorescence falls progressively, due to photochemical processes (e.g. activation of carbon-fixing enzymes) and non-photochemical processes (dissipation of chlorophyll energy as heat). By measuring changes in chlorophyll fluorescence, the relative contributions of these two processes can be quantified. At a practical level, this is achieved by first shining a single flash of high-intensity light (the saturating pulse) onto a leaf in the dark-adapted state and the resulting maximum fluorescence signal, F_m (see Fig. 63.5 for detail) is then used to determine the potential maximum quantum yield of PS II as:

$$\text{Maximum quantum yield PS II} = \frac{F_m - F_0}{F_m} \qquad [63.2]$$

The leaf is then allowed to adapt to ambient light, and the saturation pulse is then reapplied: the reduction in maximum fluorescence (to $F_m{}'$) is the result of non-photochemical quenching (NPQ), i.e. dissipation of energy as heat (Fig. 63.5). This can be determined from the relationship:

$$\text{NPQ} = \frac{F_m - F_{m'}}{F_{m'}} \qquad [63.3]$$

NPQ processes protect the protein components of the photosynthetic apparatus from oxidative damage due to excess light. The remaining

Fig. 63.4 A pulse-amplified modulated fluorometer (PAM), used to measure chlorophyll fluorescence (probe shown in left hand, attached to leaf, with analysis/recording unit on shoulder strap).
Courtesy of Heinz Walz GmbH, Effeltrich, Germany (http://www.walz.com).

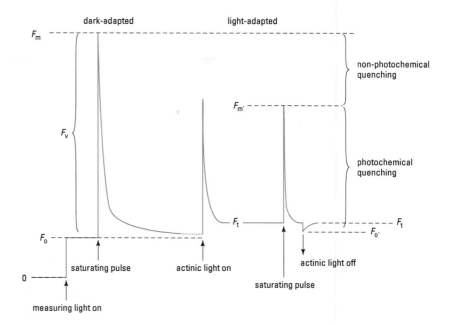

Fig. 63.5 Representation of chlorophyll fluorescence measurement using the saturation pulse method (for details, see text).

Measuring photosynthetic quotients – in studies where both O_2 and CO_2 are measured, the photosynthetic quotient (PQ) can be determined from the relationship: PQ = O_2 evolved ÷ CO_2 consumed. In the simplest case, where fixed carbon accumulates as carbohydrate, a value of 1 should be obtained. However, the PQ may vary, depending on the amount of carbon incorporated into fats, proteins, etc. and the utilisation of photosynthetic energy for other metabolic processes and growth, e.g. NO_3^- assimilation, carbon storage.

fraction is equivalent to photochemical quenching (qP), a measure of the proportion of PS II centres that are 'open', calculated from the relationship:

$$qP = \frac{F_{m'} - F_t}{F_{m'} - F_{0'}}$$ [63.4]

The quantum yield of PS II ($\Phi_{PS\,II}$), which is a measure of the efficiency of PS II photochemistry, can be determined as:

$$\Phi_{PS\,II} = \frac{F_{m'} - F_t}{F_{m'}}$$ [63.5]

The light and dark reactions of photosynthesis are coordinated, since electron transport generates the ATP and NADPH required for CO_2 fixation. For example, a reduction in the rate of CO_2 assimilation due to a metabolic inhibitor will be mirrored by a decrease in $\Phi_{PS\,II}$ and, in many cases, by an increase in NPQ as the light intensity is increased to levels in excess of those required for photosynthesis.

Studying photosynthetic electron transport

Robert Hill first showed that isolated chloroplasts can generate O_2 in the absence of CO_2 fixation, as long as they are provided with an artificial electron acceptor that intercepts electrons from the photosynthetic electron transport chain. The net result of this 'Hill reaction' is the photolysis of water and the reduction of the Hill acceptor (A), as follows:

$$2nH_2O + 4nA \rightarrow 4nAH + nO_2$$ [63.6]

Measuring the Hill reaction using DCPIP – since the dye will revert to its oxidised (blue) form as soon as the chloroplasts are removed from the light, you must measure the absorbance as quickly as possible.

The practical significance of the Hill reaction is that it allows the photochemical reactions of the electron transport system to be studied independently of the dark reactions of photosynthesis. Many of the Hill acceptors are redox dyes that show changes in absorbance as they are reduced, e.g. ferricyanide ($Fe(CN)_6^{3-}$), which accepts electrons from the PS I complex, or 2,6-dichlorophenolindophenol (DCPIP), which intercepts

Table 63.1 Some inhibitors of photosynthetic electron transport

Inhibitor – target site
Hydroxylamine (NH_2OH) – photolysis of H_2O
DCMU (3(3,4-dichlorophenyl)-1,1-dimethylurea) – electron transport from PS II to plastoquinone
DBMIB (2,5-dibromo-3-methyl-6-isopropyl-*p*-benzoquinone) - electron flow from plastoquinone to cytochrome *f*
Methyl viologen (Paraquat) – electron flow from PS I to $NADP^+$
Atrazine (2-chloro-4-(2-propylamino)-6-ethylamine-5-triazine) – electron flow to plastoquinone
HOQNO (2-heptyl-4-hydroxyquinoline-*N*-oxide) – electron flow between quinones and cytochromes
DSPD (disalicylidinepropanediamine) - electron flow via ferredoxin

electrons from the electron transport chain between PS II and PS I (Fig. 63.1). The reduction of these artificial electron acceptors can be followed spectrophotometrically, allowing the Hill reaction to be quantified. Alternatively, an O_2 electrode can be used to follow O_2 evolution from PS II in the presence of various artificial electron acceptors. Redox mediators can be used to investigate the following aspects of photosynthetic electron flow:

- **the activity of PS II and PS I operating in series** can be studied in whole chloroplasts using a suitable terminal acceptor, e.g. ferricyanide;

- **the activity of PS II can be studied in whole chloroplasts** using DCPIP (membrane-permeable), or in fragmented chloroplasts using ferricyanide;

- **the activity of PS I can be measured if PS II is blocked by the herbicide diuron (DCMU)** and if electron flow is maintained by the addition of an artificial electron donor, e.g. ascorbate plus DCPIP. Note that this cannot be measured by O_2 evolution, since PS II is inoperative – methyl viologen can be used as an electron acceptor from PS I and is rapidly reoxidised, consuming O_2 in a reaction which can be measured using an O_2 electrode (as O_2 *uptake*, rather than O_2 *production*);

- **the sites of action of inhibitors (Table 63.1) can be determined** by measuring their effects on the various components of the electron transport system, measured using redox dyes or the methyl viologen/O_2 electrode system.

Respiration

At the cellular level, respiration can be considered as the oxidation of organic compounds coupled to the production of so-called high-energy intermediates, such as ATP. The overall process is often represented as:

$$C_6H_{12}O_6 + 12O_2^* + 6H_2O + nADP + nP_i \rightarrow$$
$$6CO_2 + 12H_2O^* + nATP + nH_2O \qquad [63.7]$$

the asterisks showing that O_2 is converted to water, and where n has a *maximum* value of 38. The process can be divided into three principal stages, namely (i) glycolysis, (ii) the tricarboxylic acid (TCA) cycle and (iii) oxidative phosphorylation. The glycolytic pathway operates in the cytosol and results in the partial breakdown of glucose (C_6) to two molecules of pyruvate (C_3) plus two molecules of NADH and two molecules of ATP. The mitochondrial TCA cycle involves the sequential dismantling of pyruvate to CO_2 (via an intermediate step involving decarboxylation to acetyl-coA), producing one molecule of $FADH_2$, four molecules of NADH and one of GTP for each pyruvate. The final stage occurs at the inner mitochondrial membrane – the movement of electrons and protons along the respiratory transport chain from reductant to O_2 as the terminal electron acceptor creates a transmembrane H^+ gradient that leads to the net synthesis of ATP, as summarised in Fig. 63.6. Thus, as in photosynthesis, respiration involves

Fig. 63.6 Proton flow in mitochondrial oxidative phosphorylation: the creation of a trans-membrane H^+ gradient due to the respiratory electron transport chain drives ATP synthesis via a membrane-bound ATPase.

Fig. 63.7 Representative data for O_2 uptake by a suspension of mitochondria in response to additions of substrate and ADP.

Interpreting respiratory quotients – the complete oxidation of carbohydrates should give values close to 1.0, in agreement with Eqn [63.8], while the oxidation of fats will give values close to 0.7 and protein oxidation will produce values of about 0.8.

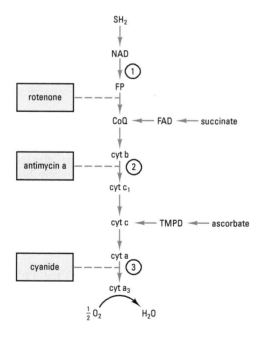

Fig. 63.8 The mitochondrial electron transport system, showing the sites of action of various inhibitors (dotted lines) and the three sites of ATP synthesis (numbered 1–3). S, substrate; FP, flavoprotein; CoQ, coenzyme Q; cyt, cytochrome; TMPD, tetramethylphenylenediamine.

enzyme-catalysed interconversion of organic compounds plus membrane-bound electron transport reactions that are coupled to ATP synthesis.

The enzymic reactions of glycolysis and the TCA cycle can be studied using ^{14}C-labelled intermediates and pulse-chase experiments, using techniques similar to those used to investigate the dark reactions of photosynthesis (p. 431), or by purification and characterisation of the individual enzymes. Oxidative phosphorylation can be studied using techniques appropriate for electron transport reactions.

Measuring the respiratory activity of cells and tissues

The overall rate of cellular respiration is usually measured in terms of the amount of O_2 consumed by a known amount of material in a given time, e.g. as $\mu mol\ O_2\ mg\ protein^{-1}\ min^{-1}$.

The principal techniques include:

- **Manometry** – this traditional approach involves measuring either the pressure change (Warburg manometer) or volume change (Gilson manometer) as gases are produced or consumed during respiration. Manometry is relatively insensitive and may be subject to large measurement errors: the principal application is in studying the relationship between the amounts of O_2 consumed and CO_2 evolved in terms of the respiratory quotient (RQ) for a particular substrate, from the relationship:

$$RQ = \frac{CO_2\ evolved}{O_2\ consumed} \qquad [63.8]$$

The calculations involved in determining CO_2 production or O_2 consumption are complex and are specific to individual instruments – manufacturer's guidelines should be followed carefully.

- **Oxygen electrode studies** – this has largely replaced manometry, as the apparatus is more versatile and less complex to set up and interpret, giving a continuous readout of O_2 status via a chart recorder. Net O_2 evolution can be measured using whole cells, mitochondrial suspensions or submitochondrial preparations.

Using an oxygen electrode to study respiratory electron transport

Intact mitochondria suspended in an isotonic medium will show little respiratory activity unless supplied with (i) a suitable electron donor or substrate (e.g. NADH), (ii) ADP and (iii) P_i: this is termed respiratory control and the mitochondria are said to be tightly coupled, since there is a close link between electron transport, O_2 consumption and phosphorylation. The extent of this coupling can be determined using an O_2 electrode (p. 369) to measure O_2 consumption in the presence of substrate, P_i and ADP, with that in the absence of ADP (Fig. 63.7), as:

$$respiratory\ control\ ratio = \frac{rate\ of\ O_2\ consumption\ with\ ADP}{rate\ of\ O_2\ consumption\ without\ ADP} \qquad [63.9]$$

Freshly prepared, tightly coupled mitochondria should have a respiratory control ratio of ≥ 4.

To investigate the relationship between the number of ATP molecules produced per substrate molecule, substrate and P_i are supplied in excess and the O_2 uptake produced by a known amount of ADP is measured

(a)

(b)

Fig. 63.9 Representative data for O_2 uptake by mitochondria in response to the addition of (a) electron transport inhibitors and substrates, or (b) inhibitors of ATP synthesis and uncouplers. Steeper slopes indicate faster rates of O_2 uptake.

using an O_2 electrode (Fig. 63.8), allowing the P/O ($\equiv$ADP/O) ratio for a particular substrate to be calculated as:

$$\text{P/O ratio} = \frac{\mu\text{mol ADP added}}{2 \times \mu\text{mol } O_2 \text{ consumed}} \quad\quad [63.10]$$

Note that the amount of O_2 is multiplied by 2, since O_2 is a diatomic molecule. The P/O ratio is determined by the site at which electrons are transferred to the respiratory chain (Fig. 63.9). Thus NADH should give a P/O ratio of 3, while succinate and $FADH_2$ should have values of 2. Artificial electron donors to cytochrome c have P/O ratios of 1, e.g. ascorbate/tetramethylphenylenediamine (TMPD). However, measured P/O ratios are always less than these values, due to partial uncoupling and the action of ATPases.

The effects of inhibitors of respiratory electron transport on the rate of oxygen consumption of mitochondria can be used to determine their site of action (Fig. 63.8). Thus, an inhibitor of electron flow will prevent O_2 uptake while an artificial substrate that donates electrons at a location beyond the site of inhibition will restore O_2 uptake, as shown in Fig. 63.9.

'Uncouplers' act by increasing the permeability of membranes to protons, causing the dissipation of the transmembrane H^+ gradient. Addition of an uncoupler increases electron flow along the respiratory chain in the absence of phosphorylation, i.e. there is loss of respiratory control. Uncouplers can reverse the effects of inhibitors such as oligomycin, whose target is the mitochondrial ATPase. Oligomycin prevents the return of H^+ to the interior of the mitochondrion, thereby inhibiting respiratory electron transport and O_2 uptake – the addition of an uncoupler (e.g. CCCP) reverses this inhibition.

KEY POINT When working with inhibitors, make sure you treat all of these as potentially toxic, observing appropriate safety precautions (Chapter 23), and that you know the procedure to be followed in case of accident or spillage.

Measuring the components of the proton-motive force in chloroplasts and mitochondria

According to chemiosmotic principles, the transmembrane electro-chemical potential gradient of protons drives ATP synthesis in photophosphorylation and oxidative phosphorylation via a membrane-bound ATPase (Fig. 63.6). This gradient is often expressed as the proton-motive force, or PMF (Δp), expressed in mV as:

$$\Delta p = E_m - 59(\text{pH}_i - \text{pH}_0) \quad\quad [63.11]$$

at 25 °C, where E_m is the transmembrane electrical potential (mV) and pH_i and pH_o are internal and external pH values. The individual components of the proton-motive force can be determined by a variety of methods:

- **Transmembrane electrical potential**, by the distribution of a suitable radiolabelled lipophilic cation, e.g. tetraphenylphosphonium (TPP^+),

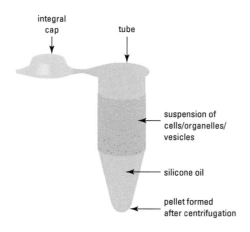

Fig. 63.10 Silicone oil microcentrifugation system: cells, organelles or membrane vesicles are pelleted below the silicone oil layer as a result of their higher density, while the less dense suspension medium remains about the silicone oil layer on centrifugation.

in accordance with the Nernst equation; by the measurement of K^+ or Rb^+ in the presence of the ionophore valinomycin; or by the quenching of a fluorescent dye (e.g. cyanine or oxanol dyes).

● **Transmembrane pH gradient** from the equilibrium distribution of a radiolabelled weak permeant acid, e.g. 5,5-dimethyl-2,4-oxazolidinedione (DMO); by quenching of a fluorescent pH probe, e.g. 9-aminoacridine; or by ^{31}P-NMR.

For radiolabelled probes, organelles or membrane vesicles can be separated from the bathing medium by silicone oil microcentrifugation (Fig. 63.10), with suitable correction for carry-over of external medium, e.g. using a membrane-impermeant solute, labelled with a second radioisotope. Care is required in such double-labelled experiments, as carry-over can represent a significant component, leading to substantial measurement errors if uncorrected.

Typical measurements for Δp across energy-transducing membranes are 180–200 mV (inside-negative): in mitochondria, E_m is the principal component while ΔpH represents the largest component of Δp in thylakoid membranes.

Text references

Maxwell, K. and Johnson, G.N. (2000) Chlorophyll fluorescence – a practical guide. *Journal of Experimental Botany,* **51**, 659–68.

Parsons, R., Weyers, J.D.B., Lawson, T. and Godber, I. (1997) Rapid and straightforward estimates of photosynthetic characteristics using a portable gas exchange system. *Photosynthetica,* **34**, 265–279.

Sources for further study

Lambers, H., Chapin, F.S. and Pons, T.L. (2008) *Plant Physiological Ecology,* 2nd edn. Springer, New York.

Nobel, P.S. (2009) *Physicochemical and Environmental Plant Physiology,* 4th edn. Academic Press, Oxford.

Ridge, I. (ed.) (2002) *Plants.* Oxford University Press, Oxford.
[Chapter 2 is a good general introduction to the physiology of photosynthesis.]

Taiz, L. and Zeiger, E. (2014) *Plant Physiology,* 6th edn. Sinauer, Sunderland, USA.
[Provides a detailed treatment of photosynthesis.]

West, J.B. (2011) *Respiratory Physiology: The Essentials,* 9th edn. Lippincott, Williams and Wilkins, Philadelphia.
[Provides a comprehensive guide to the human respiratory system.]

Wilson, K. and Walker, J. (eds) (2010) *Principles and Techniques of Practical Biochemistry and Molecular Biology,* 7th edn. Cambridge University Press, Cambridge.
[Also covers other topics, including spectroscopy and centrifugation.]

STUDY EXERCISES

(See also Study exercises 54.4 and 54.5 for calculations of respiratory oxygen consumption and photosynthetic oxygen evolution based on oxygen electrode measurements.)

63.1 **Check your understanding of photosynthesis and respiration.** Having read through this chapter, distinguish between each of the following pairs of terms:

(a) net photosynthesis and gross photosynthesis;

(b) C_3 photosynthesis and C_4 photosynthesis;

(c) the oxygenase and carboxylase functions of Rubisco;

(d) coupling and uncoupling;

(e) respiratory quotient and respiratory control ratio.

63.2 **Calculate rates of photosynthetic carbon fixation from ^{14}C data.** A sample of the photosynthetic cyanobacterium *Spirulina platensis* containing 0.29 mg chlorophyll *a* (chl a) was incubated in the light for 3 min in 10 mL of an aqueous solution containing inorganic carbon (HCO_3^-) at a concentration of 12 mmol L^{-1} and at a specific activity of 12 Bq pmol^{-1} ($\equiv$ 0.72 d.p.m. nmol^{-1}) At the end of this period, the cyanobacterial cells were separated from the medium and assayed by liquid scintillation counting, giving a quench-corrected value of 2808 d.p.m. What is the photosynthetic rate, expressed in terms of µmol C fixed per mg chl a per minute? (Express your answer to three significant figures.)

63.3 **Interpret results from an experiment using DCPIP to study photosynthetic electron transport.** The following table shows data for the change in A_{600} of a chloroplast suspension incubated with the Hill acceptor DCPIP, plus different levels of the herbicide simazine.

A_{600} measurements for a chloroplast suspension incubated with DCPIP and various levels of simazine

Time (s)	Control (no simazine)	Simazine at:		
		1.0 nmol L^{-1}	10.0 nmol L^{-1}	100.0 nmol L^{-1}
0	0.462	0.461	0.464	0.462
30	0.372	0.394	0.41	0.436
60	0.320	0.345	0.371	0.413
90	0.284	0.315	0.343	0.397
120	0.253	0.284	0.317	0.385

(a) What is the reason for the rapid decrease in A_{600} in the control sample, with no added simazine?

(b) How do you account for the fact that increasing amounts of simazine lead to a reduction in the extent of change in A_{600} with time?

(c) Plot a graph of A_{600} against time to show the composite data from the table. Use this to determine the initial rate of change in A_{600}. Then, calculate the percentage inhibition at each simazine concentration. Finally, plot percentage inhibition against simazine concentration, to determine the concentration causing 50% inhibition (I_{50}); give your answer to three significant figures.

63.4 **Calculate the P/O ratio for a respiring mitochondrial suspension.** An oxygen electrode chamber containing a mitochondrial suspension (0.5 mL), NADH (0.5 mL) and buffer (4 mL) showed a fall in oxygen concentration from 218 to 174 µ mol L^{-1} on addition of 0.05 mL of a 25 mmol L^{-1} ADP solution. What was the P/O ratio over this period?

63.5 **Calculate mitochondrial proton-motive force.** A suspension of mitochondria in buffer at an external pH of 7.2 had a measured internal (matrix) pH of 8.6 and a measured transmembrane potential (inner membrane) of −85.6 mV. Calculate the proton-motive force across the inner membrane (express your answer to three significant figures). What is your interpretation of the calculated value?

Answers to these study exercises are available at www.pearsoned.co.uk/practicalskills

Genetics

Gregor Mendel, an Austrian monk, made pioneering studies of the genetics of eukaryotic organisms in the middle of the nineteenth century. He made crosses between different forms of flowering plants. Through careful examination and numerical analysis of the observable characteristics, or phenotype, of the parents and their progeny, Mendel was able to deduce much about their genetic characteristics, or genotype. The principles derived from these experiments explain the basis of heredity, and hence underpin our understanding of sexual reproduction, biodiversity and evolution. Mendelian genetics is concerned primarily with the transmission of genetic information, as opposed to molecular genetics, which deals with the molecular details of the genome and techniques for altering genes (see Chapters 66–68) and genome analysis (genomics) (see Chapter 12).

> **KEY POINT**   A common initial stumbling block in genetics is terminology. In many cases the definitions are interdependent, so your success in this subject depends on your grasp of all the definitions and underlying ideas explained below.

Important terms and concepts

Each character in the phenotype is controlled by the organism's genes, the basic units of inheritance. Each gene includes the 'genetic blueprint' (DNA) that usually defines the amino acid sequence for a specific polypeptide or protein – often an enzyme or a structural protein. The protein gives rise to the phenotype through its activity in metabolism or its contribution to the organism's structure. The full complement of genes in an individual is known as its genome. Individual genes can exist in different forms, each of which generally leads to a different form of the protein for which it codes. These different gene forms are known as alleles.

In eukaryotes, the genes are located in a particular sequence on chromosomes within the nucleus. The number of chromosomes per cell is characteristic for each organism (its chromosome number, n). For example, the chromosome number for man is 23. In cells of most 'higher' organisms, there are two of each of the chromosomes (2n). This is known as the diploid state. As a result of the process of meiosis, which precedes reproduction, special haploid cells are formed (gametes) that contain only one of each chromosome (1n). In sexual reproduction, haploid gametes from two individuals fuse to form a zygote, a diploid cell with a new genome, which gives rise to a new individual through the process of mitosis. Cell numbers are increased by this process, producing genetically identical cells.

Organisms vary in the span of the diploid and haploid phases. In some 'lower' organisms, the haploid phase is the longer-lasting form; in most 'higher' organisms, the diploid phase is dominant. A life-cycle diagram can be used to show how the phases are organised and what life forms are involved in each case.

Definitions

Genotype – an individual's genetic make-up, i.e. the organism's genes.

Phenotype – the observable characteristics of an individual organism; the consequence of the underlying genotype and its interaction with the environment.

Essential vocabulary for Mendelian genetics – make sure you know what the following terms and symbols mean:

- chromosome, sex chromosome, autosome;
- gene, allele;
- dominant, recessive, lethal;
- haploid, diploid, gamete, zygote;
- heterozygous, homozygous;
- P, F1, F2, ♀, ♂.

Genomes and chromosome numbers – remember that mitochondria and chloroplasts contain DNA molecules, but these are not included when calculating the chromosome number.

Definitions

Meiosis – division of a diploid cell that results in haploid daughter cells carrying half the original number of chromosomes. Occurs during gamete (sperm and egg) formation.

Mitosis – division of a cell into two new cells, each with the same chromosome number. Occurs in somatic cells, e.g. during growth, development, repair, replacement.

Example   In the garden pea, *Pisum sativum*, studied by Mendel, the yellow seed allele Y was found to be dominant over the green seed allele y. A cross of YY × yy genotypes would give rise to Yy in all of the F_1 generation, all of which would thus have the yellow seed phenotype. If the F_1 generation were interbred, this Yy × Yy cross would lead to progeny in the next, F_2, generation with the genotypes YY:Yy:yy in the expected ratio 1:2:1. The expected phenotype ratio would be 3:1 for yellow:green seed.

Since each diploid individual carries two of each chromosome, it has two copies of each gene in every cell. The number of alleles of each character present depends on whether the relevant genes are the same – the homozygous state – or whether they are different – the heterozygous state. Hence, while there may be many alleles for any given gene, an individual could have two at most, and might have only one if it were homozygous. This will depend on the alleles present in the parental gametes that fused when the zygote was formed. Offspring that inherit a different combination of alleles at the two loci compared with their parents are known as *recombinants*.

The basis of Mendel's experiments and of many exercises in genetics are *crosses*, where individuals showing particular phenotypes are mated and the phenotypes of the offspring, or F_1 generation, are studied (see Box 64.1). If homozygous individuals carrying alternative alleles for a character are crossed, one of the alleles may be *dominant*, and all the F_1 generation will show that character in their phenotype. The character not evident is said to be *recessive*.

In describing crosses, geneticists denote each character with a letter of the alphabet, a capital letter being used for the dominant allele and lower case for the recessive. Taking, for example, a gene with dominant and recessive forms A and a, respectively, there are three possibilities for each individual: it can be (a) homozygous recessive aa; (b) homozygous dominant AA; or (c) heterozygous Aa.

The reasons for dominance might relate to the activity of an enzyme coded by the relevant gene; for example, Mendel's yellow pea allele is dominant because the gene involved codes for the breakdown of chlorophyll. In the homozygous recessive case, none of the functional enzyme will be present, chlorophyll breakdown cannot occur and the seeds remain green. Not all alleles exhibit dominance in this form. In some cases, the heterozygous state results in a third phenotype (incomplete or partial dominance); in others, the heterozygous individual expresses both genotypes (codominance). Another possible situation is epistasis, in which one gene affects the expression of another gene that is independently inherited. Note also that many genes, such as those coding for human blood groups, have multiple alleles.

KEY POINT   Genetics problems may well involve one of the 'standard' crosses shown in Box 64.1. Before tackling the problem, try to analyse the information provided to see if it fits one of these types of cross. Fig. 64.1 is a flowchart detailing the steps you should take in answering genetics problems.

Unless otherwise stated or obvious from the evidence, you should assume that the genes being considered in any given case are on separate chromosomes. This is important because it means that they will assort independently during meiosis. Thus, the fact that an allele of gene A is present in any individual will not influence the possibility of an allele of gene B being present. This allows you to apply simple probability in predicting the genetic make-up of the offspring of any cross (see p. 445).

Where genes are present on the same chromosome, they are said to be linked genes, and thus it would appear that they would not be able to assort independently during metaphase 1 of meiosis, when homologous chromosomes are independently orientated. However,

read the problem carefully

↓

decide the dominant-recessive relationship between alleles, if any; decide whether any genes are sex-linked

↓

assign appropriate symbols to the alleles

↓

decide what type of cross is involved (see Box 64.1)

↓

use a Punnett square or probability calculation to work out expected progeny ratios

↓

if required, analyse the results using a χ^2 test (see Box 64.2)

Fig. 64.1 Flowchart for tackling problems in Mendelian genetics.

Pedigree notation:

○   normal female

□   normal male

● ■   female or male displaying
an inherited condition

○──□   mating

   offspring

Example of simple family pedigree:

This diagram shows the offspring of a normal male and female. The two daughters are normal, but the son displays the inherited condition.

Fig. 64.2 Pedigree notation and family trees.

Denoting linked genes – these are often shown diagrammatically, with a double line indicating the chromosome pair. For example, the two possible linkages for the genotype AaBb would be shown as:

$$\frac{A\ b}{a\ B}\quad \text{or} \quad \frac{A\ B}{a\ b}$$

although physically attached to each other, they may become separated when crossing-over occurs between the homologous chromosomes at an early stage of meiosis. Exchange of genetic information between homologous chromosomes is called recombination. Linkage can be detected from a cross between individuals heterozygous and homozygous recessive for the relevant genes, e.g. AaBb × aabb. If the genes A and B are on different chromosomes, we expect the ratio of AaBb, aabb, Aabb and aaBb to be $1:1:1:1$ in the F_1. However, if the dominant alleles of both genes occur on the same chromosome, the last two combinations will occur, but rarely. Just how rarely depends on how far apart they lie on the relevant chromosome – the further apart, the more likely it is that crossing-over will occur. This is the basis of chromosome mapping (see Box 64.1).

Another complication you will come across is sex-linked genes. These occur on one of the X or Y chromosomes that control sex. Because one or other of the sexes – depending on the organism – is determined as XX and the other as XY (see Box 64.1), this means that rare recessive genes carried on the X chromosome may be expressed in XY individuals. Sex-linked genes are sometimes obvious from differences in the frequencies of phenotypes in male and female offspring. Pedigree charts (Fig. 64.2) are codified family trees that are often used to show the inheritance and expression of sex-linked characteristics through various generations.

Analysis of crosses

There are two basic ways of working out the results of crosses from known or assumed genotypes:

1. **The Punnett square method** provides a good visual indication of potential combinations of gametes for a given cross. Lay out your Punnett squares consistently as shown in Fig. 64.3. Then group together the like genotypes to work out the genotype ratio and proceed to work out the corresponding phenotype ratio if required.

2. **Probability calculations** are based on the fact that the chance of a number of independent events occurring is equal to the probabilities of each event occurring, multiplied together. Thus if the probability P of

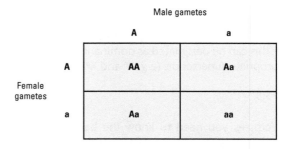

Fig. 64.3 Layout for a simple Punnett square for the cross Aa × Aa. The genotypic ratios for this cross are $1:2:1$ for AA:Aa:aa, and the phenotypic ratio would be $3:1$ for characteristic A to characteristic a. In this simple Punnett square, the allele frequencies are treated as equal ($f = 0.5$); if different from this, the probability of genotypes in each combination will be the relevant frequencies multiplied together.

Box 64.1 Types of cross and what you can (and cannot) learn from them

Monohybrid cross – the simplest form of cross, considering two alleles of a single gene.

Example: AA × aa

If only the parental *phenotypes* are known, you cannot always deduce the parental genotypes from the phenotype ratio in the F_1. An individual of dominant phenotype in the F_1 could arise from a homozygous dominant or heterozygous genotype. However, crossing the F_1 generation with themselves to give an F_2 generation may provide useful information from the phenotype ratios that are found.

Dihybrid cross – a cross involving two genes, each with two alleles.

Example: AaBB × AaBb

As with a monohybrid cross, you cannot always deduce the parental genotype from the phenotype ratio in the F_1 generation alone.

Test cross – a cross of an unknown genotype with a homozygous recessive.

Example: AABb × aabb

A test cross is one between an individual dominant for A and B with one recessive for both genes. The progeny will all be dominant for A, revealing the homozygous nature of the parent for this gene, but the progeny phenotypes will be split approximately 1:1 dominant to recessive for gene B, revealing the heterozygous nature of the parent for this gene. This type of cross reveals the unknown parental genotype in the proportions of phenotypes in the F_1.

Sex-linked cross – a cross involving a gene carried on the X chromosome; this can be designated as dominant or recessive using appropriate superscripts (e.g. X^A and X^a).

Example: $X^A X^a × X^A Y$

In sex-linked crosses you need to know the basis of sex determination in the species concerned – which of XX or XY is male (for example, the former in birds, butterflies and moths, the latter in mammals and *Drosophila*). The expected ratios of phenotypes in the offspring will depend on this. Note that the recessive genotype a will be expressed in the $X^a Y$ case.

Crosses with linked genes – genes are linked if they are on the same chromosome. This is revealed from a cross between individuals heterozygous and homozygous recessive for the relevant genes.

Example 1 (genes on separate chromosomes): expected offspring frequency from the cross AaBb × aabb is AaBb, aabb, Aabb and aaBb in the ratio 1:1:1:1.

Example 2 (linked genes): the frequencies of the last two combinations in example 1 might be skewed according to the direction of parental linkage.

Chromosome mapping uses the frequency of crossing-over of linked genes to estimate their distance apart on the chromosome on the basis that crossing-over is more likely the further apart are the genes. So-called 'map units' are calculated on the following basis:

$$\frac{\text{no. of recombinant progeny}}{\text{total no. of progeny}} \times 100 = \%\ \text{crossing-over}$$

[64.1]

By convention, 1% crossing-over = 1 map unit (centimorgan, cM). The order of a number of genes can be worked out from their relative distances from each other. Thus, if genes A and B are 12 map units apart, while A and B are respectively 5 and 7 map units from C, the assumed order on the chromosome is ACB (Fig. 64.4).

Fig. 64.4 Genetic map showing relative positions of genes A, B and C.

Using probability calculations – this can be simpler and faster than Punnett squares when two or more genes are considered.

Carrying out genetic crosses – typically these are performed using organisms that have a large number of offspring, to even out random variation. Other useful attributes include short generation time, ease of maintenance and wide range of mutations readily observed in the phenotype.

Limitation of the Chi² test – note that the formula cited in Box 64.2 is valid only if expected numbers are greater than 5.

Table 64.1 Values of Chi² *(χ²)* for which $P = 0.05$. The value for $(n - 1)$ degrees of freedom (d.f.) should be used, where n = the number of categories (= phenotypes) considered (normally fewer than 4 in genetics problems). If χ^2 is less than or equal to this value, accept the null hypothesis that the observed values arose by chance; if χ^2 is greater than this value, reject the null hypothesis and conclude that the difference between the observed and expected values is statistically significant.

Degrees of freedom	χ^2 value for which $P = 0.05$
1	3.84
2	5.99
3	7.82
4	9.49

a child being a boy is 0.5 and the probability of the child of particular parents being blue-eyed is 0.5, then the probability of that couple having a blue-eyed son is $0.5 \times 0.5 = 0.25$, and that of having two blue-eyed boys is:

$$P = (0.5 \times 0.5) \times (0.5 \times 0.5) = 0.0625$$

How do you decide whether the results of an experimental cross fit your expectation from theory as calculated above? This is not easy, because of the element of chance in fertilisation. Thus, while you might expect to see a 3:1 phenotype ratio of progeny for a given cross, in 500 offspring you might actually observe a ratio of 379:121, which is a ratio of just over 3.13:1. Can you conclude that this is significantly different from 3:1 in the context of random error? The answer to this problem comes from statistics. However, the answer is not certain, and your conclusion will be based on a balance of probabilities (see Box 64.2 and Table 64.1).

Box 64.2 **Example of a Chi² $(χ^2)$ test**

This test allows you to assess the difference between observed (O) and expected (E) values and is extremely useful in biology. It is particularly valuable in determining whether progeny phenotype ratios fit your assumptions about their genotypes. The operation of the test is best illustrated by the use of an example. Assume that your null hypothesis (see p. 525) is that the phenotypic ratio is 3:1 and you observe that in 500 offspring the phenotype ratio is 379:121 whereas the expected ratio is 375:125.

Start the test by calculating the test statistic $χ^2$. The general formula for calculating $χ^2$ is:

$$\chi^2 = \Sigma \frac{(O - E)^2}{E} \tag{64.2}$$

In this example, this works out as:

$$\chi^2 = \frac{(379 - 375)^2}{375} + \frac{(121 - 125)^2}{125} = \frac{16}{375} + \frac{16}{125} = 0.171$$

The probability associated with this value can be obtained from $χ^2$ tables for $(n - 1)$ degrees of freedom (d.f.), where n = the number of categories = number of phenotypes considered. Here the d.f. value is $2 - 1 = 1$. Since the $χ^2$ value of 0.171 is lower than the tabulated value for 1 d.f. (3.84, Table 64.1), we therefore accept the null hypothesis and conclude that the difference between observed and expected results is not significant (since $P > 0.05$). Had $χ^2$ been ≥ 3.84, then $P \leq 0.05$ and we would have rejected the null hypothesis and concluded that the difference was significant, i.e. that the progeny phenotype did not fit the expected ratio.

You can carry out Chi² calculations using the CHITEST function in Microsoft *Excel*. The *Help* function within that program contains useful guidance.

The Hardy–Weinberg Principle – this was named after its first, independent protagonists, Godfrey H. Hardy and Wilhelm Weinberg. It is sometimes referred to as the Hardy–Weinberg equilibrium, theorem, law or rule.

Population genetics

Population genetics is largely concerned with the frequencies of alleles in a population and how these may change in time. The Hardy–Weinberg Principle states that the frequency of alleles f remains the same between generations, unless influenced by some outside factor(s).

To understand why this is the case, consider alleles H and h for a particular gene, which exist in the breeding population at frequencies p and q, respectively. If the individuals carrying these alleles interbreed randomly, then the expected phenotype and allele ratios in the F_1 generation can be calculated simply as:

$$f(\text{HH}) = p^2;$$
$$f(\text{Hh}) = 2pq; \quad \text{and}$$
$$f(\text{hh}) = q^2$$

If you wish to confirm this, lay out a Punnett square with appropriate frequencies for each allele. Now, by summation, the frequency of H in the $F_1 = p^2 + pq$ (a similar calculation can be made for allele h); and since in this example there are only two alleles, $p + q = 1$ and so $q = (1 - p)$. Substituting $(1 - p)$ for q, the frequency of H in the F_1 is thus:

$$p^2 + p(1 - p) = p^2 + p - p^2 = p \qquad [64.3]$$

i.e. the frequency of the allele is unchanged between generations. A similar relationship exists for the other alleles.

The Hardy–Weinberg Principle holds so long as the following criteria are satisfied:

1. **random mating** – so that no factors influence each individual's choice of a mate;

2. **large population size** – so that the laws of probability will apply;

3. **no mutation** – so that no new alleles are formed;

4. **no emigration, immigration or isolation** – so that there is no interchange of genes with other populations nor isolation of genes within the population;

5. **no natural selection** – so that no alleles have a reproductive advantage over others.

Population geneticists use the Hardy–Weinberg Principle and Eqn [64.3] to gain an idea of the rate of evolution and the influences on evolution. By ensuring that criteria 1–4 hold, if there are any changes in allele frequency between generations, then the rate of change of allele frequencies indicates the rate of evolutionary change (natural selection).

Example Cystic fibrosis occurs in 0.04% of Caucasian babies. If this condition results from a double recessive allele aa, then, following the Hardy–Weinberg Principle and Eqn [64.3], $q^2 = 0.0004$ (0.04% expressed as a fraction of 1) and so $q = \sqrt{0.0004} = 0.02$, or 2%. Since $p = (1 - q)$, p = 0.98, or 98%. The frequency of carriers of cystic fibrosis in the Caucasian population (people having the alleles Aa) is given by $2pq$. From the above, $2pq = 2(0.02 \times 0.98) = 0.0392$. Hence 3.92% of the Caucasian population are carriers (roughly one in 25).

Sources for further study

Blumberg, R.B. *MendelWeb*. Available: http://www.mendelweb.org/
Last accessed 01/09/15.
[A resource for students interested in the origins of classical genetics.]

Brooker, R. (2014) *Genetics: Analysis and Principles*, 5th edn. McGraw-Hill, Burr Hill.

Fletcher, H., Hickey, I. and Winter, P. (2012) *Instant Notes in Genetics*, 3rd edn. BIOS, Oxford.

Hamilton, M. (2009) *Population Genetics*. Wiley, New York.

Klug, WS. and Cummings, M.R. (2012) *Essentials of Genetics*, 8th edn. Prentice-Hall, Upper Saddle River, NJ.

Lewis, R. (2009) *Human Genetics: Concepts and Applications*. McGraw-Hill, New York.

STUDY EXERCISES

64.1 Use a Punnett square to predict the outcome of a cross. Lay out a Punnett square for a cross between genotypes RrOO × RrOo, where R is a semi-dominant gene for flower colour such that RR = red, Rr = pink and rr = white; and O is a dominant gene for corolla shape such that OO = closed corolla, Oo = closed corolla and oo = open corolla. From the Punnett square, derive both the genotypic and phenotypic ratios for the cross.

64.2 Use probability to predict the outcome of a cross. Two hazel-eyed parents are heterozygous for the eye-colour gene B. When expressed as bb, the individual is blue-eyed. Mum's hair is (genuine) blonde but Dad's is mousy-brown. In this case she is double recessive (mm) for the hair-colour gene M and he is heterozygous (Mm). What is the probability that they will have a blue-eyed, blonde daughter?

64.3 Work out a likely genetic scenario for a given set of results. Five tailless female mice were crossed with normal males (with tails). There were 31 normal mice and 28 tailless mice in the F_1 progeny. When pairs of tailless mice from the F_1 generation were crossed, their (F_2) progeny were as follows: normal, 27; tailless, 55; dead on birth, but tailless, 30. In each case the ratio of males to females was roughly 50:50. Provide a logical explanation for these results.

64.4 Predict parental genotypes from the results of crosses involving sex-linked genes. A red-eye gene is known to be sex-linked in *Drosophila*; that is, the alleles R (red-eyed) or r (white-eyed) are carried on the X chromosome, while the Y chromosome does not carry these eye-colour alleles. In the fruit-fly, XX = female and XY = male. Predict the possible parental genotypes from the following F_1 progeny ratios:

(a) 35 red-eyed female, 17 red-eyed male, 19 white-eyed male;

(b) 27 red-eyed male, 29 red-eyed female;

(c) 19 red-eyed female, 18 white-eyed female, 22 red-eyed male, 21 white-eyed male.

64.5 Carry out a χ^2 (Chi^2) test. A geneticist expects the results of a test cross to be in the phenotype ratio 1:2:1. He observes 548 progeny from his cross in the ratio 125:303:120. What should he conclude?

Answers to these study exercises are available at www.pearsoned.co.uk/practicalskills.

Definitions

Copy number – the average number of copies of a particular cellular molecule.

Merozygote – a cell containing two copies of a part of its genome, i.e. a partial diploid (sometimes also termed a merodiploid).

Phage – a bacterial virus (bacteriophage).

Prophage – a bacterial virus genome integrated within the genome of a host bacterial cell.

Transposon – a section of DNA coding for its own movement from one genomic location to another and carrying other genes in addition to those coding for transposition.

Using and interpreting standard nomenclature in bacterial genetics – different forms of three-letter abbreviations are used for:

- **phenotypic features** – non-italic text, with superscripts where appropriate, e.g. Lac⁺;
- **genotypic features** – lower case italic or underlined text, with individual letters to denote individual genes, e.g. *lacZ*;
- **gene products** (polypeptides and proteins) – the non-italicised equivalent of the abbreviation for the gene and with a capitalised first letter, e.g. the LacZ protein, which is a β-galactosidase and is the product of the *lacZ* gene in Lac⁺; cells.
- **transposon** – inactivated genes show the transposon after the individual gene, separated by two colons, e.g. *lacZ*::Tn6.

Several abbreviations may be combined, e.g. a single strain might have the phenotype Ampʳ Lac⁺ Trp⁻.

In eukaryotic organisms, genetic reassortment usually involves the fusion of two haploid gametes to form a zygote and a new generation (p. 441).

KEY POINT Bacterial genetics is very different from eukaryotic genetics (Chapter 64), due to the nature of the bacterial genome, which typically consists of a single chromosome plus none, one or several types of plasmid and/or bacteriophage (phage), depending on the particular strain of bacterium.

In most bacteria, the chromosome is usually a covalently closed circular DNA molecule, carrying genes for essential metabolic functions and structural components. As a consequence, bacteria can be regarded as haploid organisms. Plasmids are additional 'mini-chromosomes', typically coding for non-essential features, e.g. antibiotic resistance, heavy metal tolerance. They are often present at a higher copy number than the chromosome and may also carry genes within mobile transposable elements (transposons). A single bacterium may contain more than one type of plasmid (though not if they are closely related plasmids, i.e. from the same incompatibility group). Plasmids can be introduced into a bacterial cell by conjugation (p. 452) or by transformation (p. 478). Bacteria can be 'cured' of their plasmids by chemical treatment, e.g. using acridine dyes that interfere with replication, or by growth under particular conditions, e.g. at high temperatures, where plasmid replication may be unable to keep up with cell division.

A phage may replicate inside a bacterial cell or, in selected instances (temperate phages, p. 228), may exist within the cell in a non-replicating (latent) state, termed a prophage. As such, phages represent additional genetic elements that may be present within a bacterial cell, forming an important component of several aspects of bacterial genetics at the practical level.

The principal characteristics of experimental bacterial genetics ('crosses') are:

- **the processes are completely distinct from sexual reproduction in eukaryotes;**
- **the processes are *directional*, from a donor cell (exogenote) to a recipient cell (endogenote);**
- **usually, only part of the donor cell's genome is transferred;**
- **in several instances, the recipient cell becomes a merozygote, with more than one copy of a gene, or genes.** The merozygote may be a transient or a stable state, depending on circumstances;
- **recombination (synapsis and 'crossing over') may or may not be involved,** depending on the process and strains involved.

KEY POINT Bacterial crosses are best described as gene *transfer* rather than gene *exchange*, since the latter term suggests reciprocal DNA movement.

Visualising carbon source mutants – chromogenic and fluorogenic enzyme substrates (p. 412) can be used to detect particular phenotypes; for example, *E. coli* mutant strains without a functional *lacZ* gene would give non–coloured colonies on a medium containing X-galactoside (p. 413), in contrast to wild-type strains, which would give blue-green colonies.

Take care when distinguishing between carbon source and nutritional mutants – note that a *Lac⁻* mutant is unable to grow if provided with lactose as the sole carbon source, while a *Trp⁻* mutant is unable to grow unless it is provided with tryptophan as a specific nutrient.

Locating mutants – there are several repositiones of mutant bacterial strains; for example, the Genetic Stock Centre for *E. coli* at http://cgsc.biology.yale.edu.

Working with bacterial mutants

To study bacterial genetics, it is necessary to use mutant strains, which have phenotypic characteristics that allow them to be distinguished from wild-type strains. The principal types of bacterial mutant include the following:

- **Morphological mutants,** with different structural characteristics from the wild-type, e.g. so-called 'rough' mutants of selected bacteria, such as *Streptococcus* and *Klebsiella,* are defective in their synthesis of capsular polysaccharides, giving small, dull colonies on agar-based media. This is in contrast to 'smooth' wild-type strains, where colonies are large and glistening owing to the hydrophilic polysaccharide capsule. At a practical level, it is relatively straightforward to work with such mutants, since wild-type and mutant bacteria will grow on the same medium, the mutants having a feature that visibly distinguishes them from wild-types. However, there are few examples in general use.

- **Resistant mutants,** which grow in the presence of an inhibitory substance such as an antibiotic, a toxic compound or a particular phage. The isolation of an ampicillin-resistant mutant (Amp^r phenotype) is made possible by including ampicillin in the growth medium, since the growth of sensitive wild-type cells (Amp^s phenotype) will be inhibited.

- **Carbon source mutants,** which are unable to use a particular substance as a source of carbon or energy, e.g. *E. coli* mutants unable to use lactose (Lac⁻ phenotype), in contrast to Lac⁺ wild-types. A Lac⁻ mutant would be unable to grow on a minimal medium containing lactose as the principal carbon source. In order to identify and characterise such mutants, a differential medium must be used, e.g. MacConkey agar (a complex medium containing lactose as a *supplementary* carbon source, plus a pH indicator dye – colonies of Lac⁺ and Lac⁻ strains are distinguished on the basis of size and pigmentation, p. 243).

- **Nutritional mutants,** which have an additional requirement for a particular nutrient, compared to the wild-type. For example, a strain auxotrophic for the amino acid tryptophan (Trp⁻) would grow only if the medium contained tryptophan (e.g. in a minimal medium plus tryptophan). Since it is not possible to devise a single medium which will allow an auxotroph to be distinguished from the corresponding prototrophic wild-type, the selection and identification of such mutants require a different approach, involving replica plating from (a) nutrient-rich medium onto (b) minimal medium and (c) minimal medium supplemented with the particular nutrient (Fig. 65.1).

- **Mutants created by transposon mutagenesis,** where a gene has been inactivated by the insertion of a transposon, e.g. *lacZ* inactivated by Tn10 (p. 448).

- **Strains with different forms of a particular gene on the chromosome and also on a plasmid (partial diploids):** the genes carried on the plasmid can be shown as follows: *lacZ*::Tn10/F′*lacZ⁺* with the slash separating the transposon-inactivated chromosomal gene from the F′ plasmid with its functional *lacZ* gene.

- **Conditional lethal mutants,** which have a defect that causes death under a specific set of circumstances (the 'restrictive condition'), e.g. in *E. coli* temperature-sensitive mutants that grow at 30 °C, but not 40 °C, often as a result of the temperature-dependent inactivation of mutant enzymes.

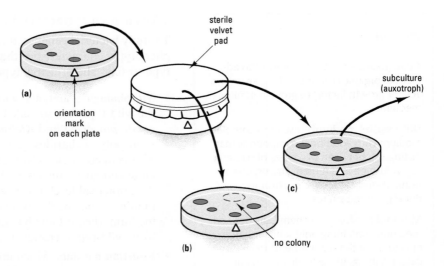

Fig. 65.1 Identification of auxotrophs using replica plating: a mixed suspension of wild-type and mutant cells is first cultured on the surface of a master plate containing nutrient-rich medium (a), then transferred to a sterile pad which is then used to inoculate minimal medium (b) and minimal medium plus a particular nutrient (c). Failure to grow on (b), coupled with growth on (c), implies auxotrophy for that particular nutrient and allows the auxotroph to be further subcultured and studied.

DNA transfer in bacteria

> **KEY POINT** The traditional approach to genetic analysis in prokaryotes involves mapping the position of individual genes using information provided from 'crosses', based on gene transfer. Typically, the recipient cell will be a mutant and the donor DNA will carry wild-type genes, enabling the transfer of wild-type characteristics to be studied in the laboratory.

The principal bacterial gene transfer processes follow.

Natural transformation

In contrast to genetic engineering techniques, where DNA uptake is induced under specific laboratory conditions (p. 478), natural transformation involves the release of DNA to the external medium (e.g. death and lysis of the donor cell) and its subsequent uptake and incorporation into the genome of the recipient cell. Natural transformation is restricted to a limited number of bacterial groups and only occurs if the recipient cells are in the correct physiological state, termed *competence,* often in the early exponential (log) growth phase (p. 231). Transformation is a relatively rare event, occurring at frequencies of ≤ 1 transformant per 10^3 cells. In a competent cell, DNA will be taken up and, if homologous (i.e. from another strain of the same species), may then be incorporated into the genome of the recipient cell by homologous recombination via a double 'crossover' (two recombination events, as shown in Fig. 65.2).

The principal application of transformation has been in mapping the position of genes in those bacteria showing natural competence (e.g. *Bacillus subtilis*). The experiments are often easiest to perform with auxotrophic recipients and prototrophic (wild-type) donor DNA, since transformants can be selectively grown on media lacking one or more individual nutrients.

Fig. 65.2 Homologous recombination between donor DNA (from transformation, transduction or conjugation) and the recipient genome via a double crossover.

The frequency of co-transformation of two genes is a measure of how close together they are likely to be on the donor DNA strand – a high co-transformation frequency implies that they are close together on the chromosome, reaching a recipient cell on the same fragment of DNA.

> **KEY POINT** While co-transformation frequencies are inversely related to map distances, they are not directly equivalent to the recombination frequencies used in mapping eukaryotic genomes (p. 444), because they are also influenced by the size distribution of the fragments of donor DNA and by the likelihood of homologous recombination.

Transformation mapping has several limitations, since it requires a fairly large number of complex replica plating experiments to produce a chromosomal map, and the relative position of genes that are very far apart cannot be determined directly – the 'jigsaw' requires a large number of available pieces, before the underlying structure can be seen. It is also insensitive for small map distances – two genes that are adjacent, or very close together, will give similar high co-transformation values.

Transduction

Here, the DNA exchange is mediated by a phage (p. 228), in one of two processes:

1. **Generalised transduction:** occasionally, a fragment of chromosomal or plasmid DNA within an infected bacterial cell may be packaged within the protein coat of a phage, in place of the phage genome. This fragment might be derived from any part of the host cell genome (exogenote). After release, on lysis of the donor cell, the transducing particle may introduce the DNA fragment into a new recipient cell. The introduced DNA may then be incorporated into the host genome by a double crossover, in a homologous recombination event similar to that shown in Fig. 65.2 for transformation. Generalised transduction can be used to establish gene order and for mapping purposes, using broadly similar principles to transformation, as only closely spaced genes will show co-transduction, with a co-transduction frequency that is inversely related to the distance between the two genes. Since a generalised transducing particle can carry a fairly small amount of DNA, the relative frequency of co-transduction can be used to provide finer detail of gene order over shorter distances than for transformation.

2. **Specialised transduction:** this is mediated only by a temperate phage, e.g. λ phage, which integrates at a specific site on the chromosome by a single crossover event. It involves a restricted number of genes – typically, a pair of genes on either side of the integration site for a particular temperate phage (Fig. 65.3). Specialised transduction will only occur if, on entry into the lytic cycle (phage replication, p. 228), there is an incorrect (abnormal) excision of the prophage and a part of the bacterial genome is incorporated, giving a specialised transducing particle (a modified phage, carrying a bacterial gene). In the case of λ phage (Fig. 65.3), the transducing particle may carry either the biotin gene (λ*bio*), or the galactose gene (λ*gal*, also termed λ*dgal*, since the modified phage is defective owing to the loss of an essential part of the phage genome, and therefore is incapable of replicating without the aid of a co-infecting, non-defective 'helper' phage). On entry into

---

Performing co-transformation tests using auxotrophic recipients – controls must be set up without added donor DNA to determine the number of revertants (spontaneous mutants), giving the actual rate of transformation by subtraction.

Performing generalised transduction in *E. coli* – P1 phage is often used, since it gives a high proportion of transducing particles (≈0.3% of the total), with a large fragment size (≈2.5% of the *E. coli* genome), making it easier to locate individual transductants.

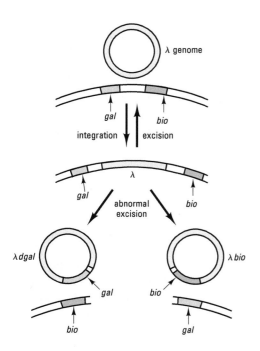

Fig. 65.3 Integration, excision and abnormal excision of λ phage, creating a modified phage genome (either λ*dgal* or λ*bio*), prior to specialised transduction.

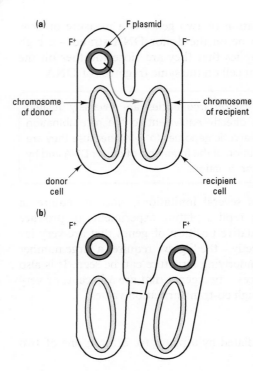

(a)

F plasmid

F⁺   F⁻

chromosome of donor   chromosome of recipient

donor cell   recipient cell

(b)

F⁺   F⁺

Fig. 65.4 Conjugation between F⁺ donor and F⁻ recipient. A single strand of F plasmid DNA is cut and transferred in linear form to the recipient (a), followed by recircularisation and complementary strand synthesis, converting the recipient to F⁺ (b).

the recipient cell, the modified phage may become latent, creating a partial diploid. The practical applications of specialised transduction are limited to those genes flanking the integration sites of temperate phages. The most common investigations are of complementation in merozygotes containing two copies of a particular gene. Where donor and recipients of the same mutant phenotype have mutations in different genes, the creation of a partial diploid can produce a wild-type phenotype.

The integration and excision of phages is likely to be an important means of gene transfer in natural environments, playing a key role in the evolution and pathogenesis of bacteria, as has been demonstrated by recent genome sequencing projects that have demonstrated the role of prophages in toxigenic *Vibrio cholerae* and *E. coli.*

Conjugation

Here, the transfer occurs as a result of cell-to-cell contact, with direct transmission of DNA from donor to recipient cell. In *E. coli,* the donor cell carries specific surface pili (protein microtubules), allowing a donor cell to attach to receptors on the surface of a recipient cell and bringing the paired cells into close contact (Fig. 65.4). In the simplest instance, the donor cell carries an additional plasmid, the F plasmid (originally termed 'F factor'), that encodes the genes responsible for conjugation, including those for the protein subunits of the specialised pili. In conventional notation, the donor is termed F⁺ and the recipient F⁻. During conjugation between F⁺ donor and F⁻ recipient, a single strand of the circular F plasmid is cut and transferred (in linear form) to the recipient cell, which becomes F⁺ once the entire plasmid has been transferred and a complementary strand has been synthesised – this process takes a few minutes.

> **KEY POINT** It is important to understand that the process of bacterial conjugation is completely different from sexual reproduction in eukaryotic organisms, since there are no gametes and no zygote or offspring is formed. Consequently, it is more appropriate to describe the participants as donor and recipient (of genetic information) rather than as male or female.

At a practical level, crosses involving F⁺ donors give little useful information, apart from mapping the position of genes on the F plasmid. However, two other types of donor are more useful:

1. **Hfr strains,** where the F plasmid DNA has become integrated into the chromosome, as a proplasmid, in an analogous manner to a temperate phage such as λ (Fig. 65.3). Such strains show a *high frequency of recombination* (hence Hfr), since chromosomal genes are transferred to the recipient cell at a far greater frequency than in crosses using F⁺ donors. The Hfr × F⁻ cross is illustrated in Fig. 65.5: a part of the F proplasmid is first to be transferred, followed by chromosomal DNA and, finally, the remaining fragment of the F proplasmid. After transfer, donor chromosomal DNA can be integrated into the recipient's genome by homologous recombination (Fig. 65.2). The conjugating pair usually breaks apart before the process is complete and the recipient cell remains F⁻, since only the leading fragment of the F plasmid reaches the recipient cell. A number of different Hfr strains of *E. coli* are available, with the F proplasmid inserted at different chromosomal locations.

Example The *E. coli* strain HfrH has the F proplasmid integrated at a site close to genes for the synthesis of threonine, so these genes would be transferred early during HfrH × F⁻ conjugation.

Example F′ *lac* is an F′ plasmid that incorporates wild-type chromosomal DNA coding for lactose utilisation.

(a)

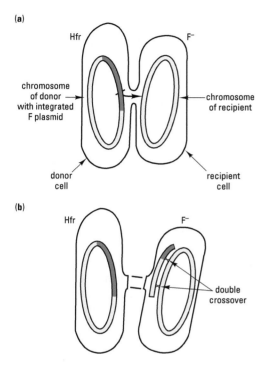

(b)

Fig. 65.5 Conjugation between Hfr donor and F⁻ recipient. The F proplasmid is cut and leads the chromosome (in single-stranded form) from the donor (a), followed by homologous recombination in the recipient (b).

2. **F′ ('F prime') donors,** with a modified F plasmid incorporating one or more chromosomal genes, formed as a result of defective excision of the F proplasmid from the chromosome in a similar manner to λ*bio* (Fig. 65.3). The F′ plasmid will transfer its chromosomal genes at very high frequency to a recipient, which will also become F′ once transfer of the F′ plasmid DNA is complete. This process is sometimes referred to as F-duction (or, less appropriately, as 'sexduction').

Mapping with Hfr donors using interrupted conjugation

Mating an antibiotic-sensitive wild-type Hfr donor with an antibiotic-resistant mutant F⁻ recipient provides a simple means of detecting recombinants, since the donor cells will be unable to grow on a medium containing the antibiotic while unmodified recipient cells will have a different phenotype from the recombinant recipient cells. The method is even simpler for the most common crosses, using nutritional mutants, where a prototrophic Hfr donor is crossed with an auxotrophic F⁻ recipient and only the recombinants are able to grow on minimal medium with added antibiotic; this approach can be extended to multiple genes, using a multiple auxotroph F⁻ recipient. By carrying out a series of experiments where the conjugation process is terminated at different times (e.g. using vigorous mixing in a vortex mixer, p. 156), the time required to transfer a particular characteristic can be determined as the earliest time at which this interruption no longer prevents recombinants from appearing. Typical results from a cross involving several genetic markers are shown in Fig. 65.6(a): by plotting the result graphically and extrapolating the curves to their intersect with the *x* axis, the *time of entry* of a particular marker can be determined accurately.

The different times of entry of each characteristic reflect their relative positions on the chromosome, with genes near to the origin of transfer of the F proplasmid having short times of entry and those further along the chromosome having later times of entry. The times of entry can be used to locate the genes on a chromosome map (Fig. 65.6(b)), representing map distances relative to the origin of transfer. This map differs from those

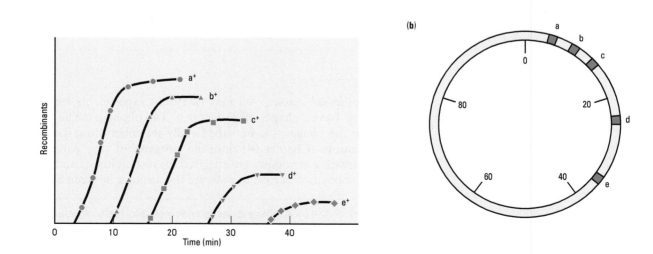

Fig. 65.6 Recombinants obtained from interrupted conjugation for a number of characters (a – e), plotted to show the time of entry of each character (a): the same data have been used to map the position of each character on the bacterial chromosome (b).

Genome mapping using interrupted conjugation – while this technique works well for widely spaced markers, it is less useful than transformation and transduction for closely linked genes.

produced from transformation and transduction analysis, since it is a transfer order map, rather than a linkage map. However, as transfer of the entire chromosome requires ≈ 100 min, it is difficult to map genes far from the origin of transfer, owing to the decreasing number of recombinants with increasing time (Fig. 65.6(a)): this is overcome by combining mapping data from several Hfr strains, each with the F plasmid integrated at a different position. In fact, this approach was originally used to demonstrate the circularity of the *E. coli* chromosome and to determine the location of over 1500 individual genes.

Genetic analysis using F′ plasmids

Since cells containing an F′ plasmid are stable merozygotes, their principal use is in studying the behaviour of genes under diploid conditions; for example:

- to determine whether a particular mutation (on the F′ plasmid) is dominant or recessive over another (on the chromosome);
- to test whether a particular gene complements another: if two mutations are in different genes, then they can complement each other in a stable merozygote, while two mutations in the same gene cannot complement each other in a stable merozygote;
- to examine whether a particular regulatory gene located on an F′ plasmid influences the expression of one or more chromosomal genes, i.e. whether it can operate in the *trans* position. The alternative test is to construct a merozygote with two genes in a chromosomal location (i.e. the *cis* position), for example, using specialised transduction (p. 451). This type of analysis has been used to investigate the molecular basis of gene expression and the function of regulatory regions of the chromosome.

Definitions

Cis (Latin 'here') – on the same DNA molecule.

Trans (Latin 'across') – on separate DNA molecules.

> **KEY POINT**   Note that, while many of the traditional procedures described in this chapter have proved to be useful in the early stages of bacterial genome analysis, the techniques and methods of molecular biology (Chapters 66–68) can be used to provide more detailed genetic information, without the need to obtain certain types of mutant, or carry out specific crosses between particular donors and recipients.

Phage crosses

In practical classes, you may carry out experiments using phages, often using T-even phages of *E. coli* (e.g. T4 coliphage). The methods used to map the phage genome superficially resemble those used for eukaryotic organisms (Chapter 64) since the 'progeny' of two 'parental' phages with contrasting genotypes are analysed for recombinants and the frequency of recombination is used to measure the distance between linked genes.

Performing phage crosses – large numbers of both types of phage must be used to ensure that every bacterial cell is infected with at least one of each type, i.e. there is a high multiplicity of infection.

> **KEY POINT**   A phage cross (sometimes informally described as a 'mating') does not involve meiosis, gamete production or zygote formation, but is a result of a *mixed infection*, where a single bacterial cell is infected with phages of the two original genotypes.

Following a single round of replication, the 'progeny' from this mixed infection are screened for the various possible genotypes, as plaques on a 'lawn' of a susceptible strain of *E. coli,* using conventional phage culture techniques (p. 228).

Phage mutants

The phage genes most widely used in crosses are those affecting:

- **plaque morphology** – for example, mutants may form large plaques as opposed to small plaques, or plaques with a different margin, e.g. a light turbid halo as opposed to a dark halo;
- **conditional lethality,** especially T4 phage mutants of the *rII* locus, unable to grow on *E. coli* K12(λ) in contrast to wild-type T4.
- **host range** – e.g. T-even mutants may infect different strains of *E. coli* from the parental types;

Phage mapping

By crossing a mutant for plaque morphology with a mutant for host range, the relative frequency of recombinants (i.e. progeny with double mutant and wild-type phenotypes) can be expressed in terms of the proportion of the total number of plaques, as a recombination frequency (*R*), where:

$$R = \frac{\text{number of double mutants and wild-type plaques}}{\text{total number of plaques}} \qquad [65.1]$$

This value is sometimes multiplied by 100 and expressed as a percentage, or as map units, equivalent to 'centimorgans' (1 cM = 1% crossing over) in conventional Mendelian crosses (p. 444).

The pioneering work of Seymour Benzer established that the *rII* region of the T4 genome consisted of two distinct genes *(rIIA* and *rIIB),* each of which can give rise to the same mutant phenotype. T4 *rII* mutants are most often used for demonstrating *trans* complementation between these two genes and for demonstrating the principles of fine structure mapping of this region of the genome, since the conditional lethality of the rII phenotype simplifies the search for rare recombinant wild-type phages, as only these recombinants are able to form plaques on *E. coli* K12(λ).

Understanding conditional lethal mutants – an important group of phages are the so-called 'amber' mutants, with polypeptides containing a premature termination codon, UAG (p. 457); such mutant strains are propagated in 'amber suppressor' mutant strains of *E. coli*, which contain a mutant tRNA that recognises UAG as coding for an amino acid.

Learning from Benzer's work – detailed fine structure mapping of the *rII* region of T4 coliphage established that recombination can occur within a single gene, and even between adjacent nucleotides. Before these experiments, a gene was regarded as an indivisible unit – nowadays we appreciate that the smallest unit in genetics is the base pair.

Sources for further study

Anon. *Information about Bacteriophage Lambda.* Available: http://www.asm.org/division/m/fax/LamFax.html Last accessed 01/09/15.
[Provides an overview of phage λ.]

Birge, E.A. (2010) *Bacterial and Bacteriophage Genetics,* 5th edn. Springer, New York.

Casali, N. and Preston, A. (2003) *E. coli Plasmid Vectors: Methods and Applications.* Humana, New York.

Dale, J.W. and Park, S.F. (2003) *Molecular Genetics of Bacteria,* 5th edn. Wiley, Chichester.

Eynard, N. and Teisse, J. (2000) *Electrotransformation of Bacteria.* Springer-Verlag, Berlin.

Snyder, L. and Champness, W. (2013) *Molecular Genetics of Bacteria,* 4th edn. American Society for Microbiology, Washington, DC.

STUDY EXERCISES

65.1 Test your understanding of bacterial and phage genetics. Distinguish between each of the following pairs of terms:

(a) zygote and merozygote;

(b) transformation and transposition;

(c) generalised and specialised transduction;

(d) prototroph and auxotroph;

(e) Hfr and F′ strains.

65.2 Interpret the codes commonly used to describe bacterial mutant strains. Describe the *E. coli* phenotypes represented by the following:

(a) Ampr Lac$^+$ Trp$^-$

(b) His$^-$ Lac$^-$

(c) *glnA*::Tn10

(d) *trpA*::Tn 6/F′*trpA*$^+$

(e) a Gal$^+$ strain lysogenised with λ*dgal*.

65.3 Interpret results from an Hfr mapping experiment. Three different Hfr strains were used to map the genome of a bacterium in a series of interrupted conjugation experiments. The strains transferred genes to the F′ recipient cell in the following orders:

Strain (a): X W C Z A

Strain (b): Y D F X W

Strain (c): Z A E B Y.

Assuming that all of these genes are present only once on a single circular chromosome, what is their order?

65.4 Construct a chromosomal map. The table below represents data for an interrupted conjugation study using an Hfr donor strain containing four 'marker' genes. These were distinguished by plating onto four different media, with the following major components:

Medium A: glucose, histidine (trace amount), tryptophan (trace amount), ampicillin.
Medium B: glucose, histidine (trace amount).
Medium C: glucose, tryptophan (trace amount).
Medium D: lactose, histidine (trace amount), tryptophan (trace amount).

(a) What is tested for by each of these four different media?

(b) Based on the composition of these media and the data below, what is the phenotype of (i) the Hfr donor and (ii) the F$^-$ recipient?

(c) Given a chromosome map size of 60 min, where would each of these genes be located on the bacterial chromosome?

Colony numbers following interrupted mating, with plating on medium A, B, C or D.

Time (min)	Medium A	Medium B	Medium C	Medium D
0	0	0	0	0
10	0	32	0	0
20	0	287	38	0
30	34	339	182	0
40	156	341	226	28
50	179	338	229	89
60	180	340	227	95

Answers to these study exercises are available at www.pearsoned.co.uk/practicalskills.

Nucleic acid bases (abbreviations):

A = adenine
C = cytosine
G = guanine
T = thymine
U = uracil

Definition

Units of nucleic acid size (length) –
Kilobase pair (kbp) = 10^3 base pairs.
Megabase pair (Mbp) = 10^6 base pairs.

Table 66.1 The Genetic Code – combinations of nucleotide bases coding for individual amino acids.

1st Base	2nd Base				3rd Base
	U	C	A	G	
U	F	S	Y	C	U
	F	S	Y	C	C
	L	S	*	*	A
	L	S	*	W	G
C	L	P	H	R	U
	L	P	H	R	C
	L	P	Q	R	A
	L	P	Q	R	G
A	I	T	N	S	U
	I	T	N	S	C
	I	T	K	R	A
	M	T	K	R	G
G	V	A	D	G	U
	V	A	D	G	C
	V	A	E	G	A
	V	A	E	G	G

* = termination codons.
Standard abbreviations for the above amino acids are as given in Table 56.1.
Note that AUG (=M) is the initiation codon. The above codons are given for mRNA – the coding strand of DNA would have T in place of U, while the template strand would have complementary bases to those given above.

Deoxyribonucleic acid (DNA) is the genetic material of all cellular organisms. Structure and assay methods are outlined in Chapter 59.

KEY POINT The sequence of the bases A, G, T and C carries the genetic information of the organism. A section of DNA that encodes the information for a single polypeptide or protein is referred to as a gene, while the entire genetic information of an organism is termed the genome.

The amount of DNA in the genome is usually expressed in terms of base pairs (bp), rather than M_r, and its size depends on the complexity of the organism: for example, the human papillomavirus has a genome of 8×10^3 base pairs (8 kbp), that of *Escherichia coli* is 4×10^6 base pairs (4 Mbp), while the human haploid genome is very large, comprising 3×10^9 base pairs (3000 Mbp). The human genome contains about 23 000 genes (distributed on 23 pairs of chromosomes), which represent only about 1% of the total amount of genomic DNA (i.e. 99% of human DNA is non-coding). Organisms with smaller genomes have smaller amounts of non-coding DNA: some viral genomes have 'overlapping' genes, where the same base sequences carry information for more than one protein.

The size of each gene varies considerably: the largest ones may exceed 10 Mbp. Chromosomes represent the largest organisational units of DNA: in eukaryotes, they are usually linear molecules, complexed with protein and RNA, ranging in length from tens to hundreds of Mbp. The unit used to denote physical distance between genes (base pairs) differs from that used to describe genetic distance (centimorgan, cM), which is based on recombination frequency (p. 444). In humans, 1 cM ≈ 1 Mbp, though this relationship varies widely, depending on recombination frequency within particular regions of a chromosome.

Each DNA template for the synthesis of RNA (transcription) begins at a promoter site upstream of the coding sequence and terminates at a specific site at the end of the gene. The base sequence of this RNA is complementary to the 'template strand' and equivalent to the 'coding strand' of the DNA. In eukaryotic cells, transcription occurs in the nucleus, where the newly synthesised RNA, or primary transcript, is also subject to processing, or 'splicing', in which non-coding regions within the gene (introns) are excised, joining the coding regions (exons) together into a continuous sequence. Further processing results in the addition of a polyadenyl 'tail' at the 3′ end and a 7-methylguanosine 'cap' at the 5′ end of what is now mature eukaryotic messenger RNA (mRNA). The mRNA then migrates from the nucleus to the cytoplasm, where it acts as a template for protein synthesis (translation) at the ribosome: the translated portion of mRNA is read in coding units, termed codons, consisting of three bases.

Each codon corresponds to a specific amino acid, including a codon for the initiation of protein synthesis (Table 66.1). Individual amino acids are brought to the ribosome by specific transfer RNA (tRNA) molecules that recognise particular codons. The amino acids are incorporated into the growing polypeptide chain in the order dictated by the sequence of codons on mRNA until a termination codon is recognised, after which the protein (polypeptide) is released from the ribosome.

The basic techniques described in this and subsequent chapters are used widely in many aspects of molecular biology, including the identification and characterisation of genes, gene cloning and genetic engineering, medical genetics and genetic fingerprinting.

> **KEY POINT** An important characteristic of nucleic acids is their ability to hybridise: two single strands with complementary base pairs will hydrogen bond (anneal) to produce a duplex, as in conventional double-stranded (ds) DNA. This duplex can be converted to single-stranded (ss)DNA (i.e. 'melted') by conditions that disrupt hydrogen bonding, e.g. raising the temperature or addition of salt, and then reannealed by lowering the temperature or by the removal of salt.

DNA can be readily purified and assayed (Chapter 59), and it is relatively stable in its pure form. Most of the problems of working with such a large biomolecule have been overcome by the following steps:

- **Restriction enzymes (see below) can be used to cut DNA precisely and reproducibly:** mammalian DNA may yield millions of fragments, with sizes ranging from a few hundred base pairs to tens of kbp, while small viral genomes may give only a few fragments.

Using restriction digests of viral genomes – the controlled fragmentation pattern obtained when a phage or plasmid is cut with a specific restriction enzyme can be used to create a 'DNA ladder' – a set of DNA fragments of known length that can be used as reference standards in gel electrophoresis (p. 459).

- **Electrophoretic methods for the separation of DNA fragments on the basis of their sizes** (p. 459).

- **Segments of DNA can be detected by 'probes' that specifically hybridise with the DNA of interest** (p. 461).

- **Once separated, specific segments of DNA can be obtained in almost unlimited quantities** by insertion into vectors and multiplication within suitable host cells (DNA cloning, Chapter 68).

- **Methods are now available for rapidly determining the base sequence of DNA segments** (p. 464), together with strategies for combining information from several adjacent segments to give contiguous sequences ('contigs') representing entire genomes in well-studied organisms (Chapter 12).

Example ClustalW software can be used for multiple sequence alignment – it can be downloaded from the European Bioinformatics Institute website at: http://www.ebi.ac.uk/clustalw/

- **Specific target sequences of DNA within a genome can be amplified** by more than a billion-fold by the polymerase chain reaction (PCR, see Chapter 67).

Producing DNA fragments

Fragments of DNA can be produced by mechanical shearing or ultrasonication (p. 268), producing a random array of fragment sizes. Reproducible cleavage can be achieved using type II restriction endonucleases (commonly termed restriction enzymes) which recognise and cleave at a specific palindromic sequence of double-stranded DNA (usually four or six base pairs), known as the restriction site. Each enzyme is given a code name derived from the name of the organism from which it is isolated, e.g. *Hin* dIII was the third restriction enzyme to be obtained from *Haemophilus influenzae* strain Rd (Fig. 66.1). Most restriction enzymes will cut each DNA strand at a slightly different position within the restriction site to produce short, single-stranded regions known as cohesive ends, or 'sticky ends', as shown in Fig. 66.1. A few restriction enzymes cleave DNA to give blunt-ended fragments (e.g. *Hin* dII and *Sma* I).

Using restriction enzymes in genetic engineering – those enzymes that cleave DNA to give single-stranded 'sticky ends' are particularly useful in gene cloning (Chapter 68). Two different molecules of DNA cut with the same restriction enzyme will have complementary single-stranded regions, allowing them to anneal as a result of hydrogen bonding between individual bases within these regions.

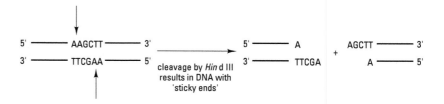

Fig. 66.1 Recognition base sequence and cleavage site for the restriction enzyme *Hin* dIII. This is the conventional representation of double-stranded DNA, showing the individual bases, where A is adenine, C cytosine, G guanine and T thymine. The cleavage site on each strand is shown by an arrow.

Separation of nucleic acids using gel electrophoresis

Separation of DNA by agarose and polyacrylamide gel electrophoresis

Electrophoresis is the term used to describe the movement of ions in an applied electrical field. DNA molecules are negatively charged (p. 398), migrating through an agarose gel towards the anode at a rate that is dependent upon molecular size – smaller, compact DNA molecules can pass through the sieve-like agarose matrix more easily than large, extended fragments. Electrophoresis of plasmid DNA is usually carried out using a submerged agarose gel (Fig. 66.2). The amount of agarose is adjusted, depending on the size of the DNA molecules to be separated, e.g. 0.3% w/v agarose is used for large fragments (>20 000 bp) while 0.8% is used for smaller fragments. Very small fragments are best separated using a polyacrylamide gel (Table 66.2). Note the following practical aspects:

- **Individual samples are added to pre-formed wells using a pipettor.** The volume of sample added to each well is usually less than 25 μL so a steady hand and careful dispensing are needed to pipette each sample.

- **The density of the samples is usually increased by adding a small amount of glycerol or sucrose,** so that each sample is retained within the appropriate well.

- **A water-soluble anionic tracking dye (e.g. bromophenol blue or xylene cyanol) is also added** to each sample, so that migration towards the anode can be followed visually.

Fig. 66.2 Agarose gel electrophoresis of DNA.

Maximising recovery of DNA – these large molecules are easily damaged by mechanical forces, e.g. vigorous shaking or stirring during extraction. In addition, all glassware must be scrupulously cleaned and gloves must be worn, to prevent deoxyribonuclease contamination of solutions.

Working with small volumes in molecular biology – use a pipettor (autopipette) of appropriate size (e.g. P2 or P20 Gilson Pipetman) with a fine tip. For very small volumes, pre-wet the tip before delivering the required volume.

SAFETY NOTE Using an electrophoresis tank – always ensure that the top cover is in position, to prevent evaporation of buffer solution and to reduce the possibility of electric shock.

Table 66.2 Gel concentrations for the separation of DNA of various sizes.

Type of gel	% (w/v)	Range of resolution of DNA (bp)
Polyacrylamide	20.0	5–100
	15.0	20–150
	5.0	75–500
	3.5	100–1000
Agarose	2.0	100–5000
	1.2	200–8000
	0.8	400–20 000
	0.3	1000–70 000

Using DNA fragment size markers – for accurate determination of fragment size (length) your standards must have the same conformation as the DNA in your sample, i.e. linear DNA standards for linear (restriction) fragments and closed circular standards for plasmid DNA.

SAFETY NOTE SYBR Safe and similar fluorescent stains are used in preference to ethidium bromide (Fig. 59.4) due to their reduced toxicity/mutagenicity.

SAFETY NOTE Working with a UV radiation source – always wear suitable UV-filtering plastic safety glasses or goggles to protect your eyes.

Assaying nucleic acids in solution – double-stranded DNA at 50 µg mL^{-1} has an A_{260} of 1, and the same absorbance is obtained for single-stranded DNA at 33 µg mL^{-1} and (single-stranded) RNA at 40 µg mL^{-1}. These values can be used to convert the absorbance of a test solution to a concentration of nucleic acid.

- **DNA fragment size markers ('molecular weight' standards) are added to one or more wells.** After electrophoresis, the relative positions of bands of known molecular weight can be used to prepare a calibration curve (usually by plotting $\log_{10}$ of the size (length) of each band against the distance travelled).

- **The gel should be run until the bromophenol blue tracking dye has migrated across 80% of the gel** (see manufacturer's instructions for appropriate voltages/times).

- **After electrophoresis, the bands of DNA can be visualised by soaking the gel for around 5 min in a suitable fluorescent stain, e.g. SYBR Safe or SafeView.**

- **Under UV light, bands of DNA are visible because of the intense fluorescence of the visualising stain, e.g. SYBR Safe or SafeView fluorescent green in UV and blue light.** The limit of detection using this method is around 1 ng DNA per band. The migration of each band from the well can be measured using a ruler. Alternatively, a photograph can be taken, using a digital camera and adaptor, or a dedicated system, e.g. GelDoc from BioRad.

- **If a particular band is required for further study (e.g. a plasmid), the piece of gel containing that band is cut from the gel using a clean scalpel.** The DNA can be separated from the agarose by solubilising the gel slice and then binding the DNA to an anion exchange resin, followed by elution in water or buffer (e.g. TE buffer, containing 10 mmol L^{-1} TRIS/HC1 at pH 8.0 plus 1 mmol L^{-1} EDTA). Various commercial kits based on this approach are also available, e.g. from Qiagen.

See also Box 52.1 for practical details of agarose gel electrophoresis of DNA.

Electrophoretic separation of RNA

Total cellular RNA or purified mRNA can be separated on the basis of size by electrophoretic separations similar to those used for DNA fragments. However, under the conditions used to separate dsDNA, RNA molecules tend to develop a secondary structure, and this leads to anomalous mobilities. To eliminate RNA secondary structure, samples are pretreated by heating in dilute formamide or glyoxal, and electrophoresis is carried out in 'denaturing gels' which include buffers containing formaldehyde.

Pulsed field gel electrophoresis (PFGE)

If structural information is to be gained about large stretches of genomic DNA, then the order of the relatively short DNA segments (generated by the restriction enzymes described above) needs to be established. This is technically possible by techniques such as 'chromosome walking', but is time-consuming and potentially difficult, especially when dealing with large chromosomes such as those from yeast (a few Mbp) and humans (50–100 Mb), in contrast to the smaller genomes of bacteria and viruses. The technique of PFGE allows separation of DNA fragments of up to ≈12 Mb.

Very large DNA fragments (>100 kbp) can be generated from chromosomal DNA by the use of certain restriction enzymes that recognise base sequences that are present at relatively low frequency, e.g. the enzyme *Not* I, which recognises a sequence of 8 bp rather than 4–6 bp. These enzymes are sometimes called 'rare cutters'. Genomic DNA prepared in the normal way is not suitable for digestion by these enzymes, as shearing during extraction fragments the DNA. Therefore, genomic DNA for analysis by PFGE is prepared as follows:

- **cells are embedded in an agarose block;**
- **the block is incubated in solutions containing detergent, RNase and proteinase K,** lysing the cells and hydrolysing RNA and proteins. The products of RNase and proteinase digestion diffuse away, leaving behind genomic DNA molecules exceeding several thousand kbp;
- **the block is incubated *in situ* in a buffered solution containing an appropriate 'rare cutter':** restriction fragments are produced, of up to ≈ 800 kbp.

PFGE differs from conventional electrophoresis in that it uses two or more alternating electric fields. An explanation for the effectivess of the technique is that large DNA fragments will be distorted by the voltage gradient, tending to elongate in the direction of the electric field and 'snaking' through pores in the gel. If the original electric field (Fig. 66.3 (a)), is removed, and a second is applied at an angle to the first (Fig. 66.3 (b)), the DNA must reorientate before it can migrate in the new direction. Larger (longer) DNA molecules will take more time to reorientate than smaller molecules, resulting in size-dependent separations.

The original configuration of electrodes used in PFGE is shown in Fig. 66.3 (a); this tends to produce 'bent' lanes that make lane-to-lane comparisons difficult. This can be overcome by using one of the many variants of the technique, one of which employs contour-clamped homogeneous electric fields (CHEF). Here, multiple electrodes are arranged in a hexagonal array around the gel (Fig. 66.3 (b)) and these are used to generate homogeneous electric fields with reorientation angles of up to 120°. A further development of CHEF involves programmable, autonomously controlled electrodes (PACE), which allows virtually unlimited variation of field and pulsing configurations, and can fractionate DNA molecules from 100 bp to 6 Mbp.

Identification of specific nucleic acid molecules using blotting and hybridisation techniques

Southern blotting

After separation by conventional agarose gel electrophoresis (p. 459), the fragments of DNA can be denatured and immobilised on a filter membrane using a technique named after its inventor, E.M. Southern.

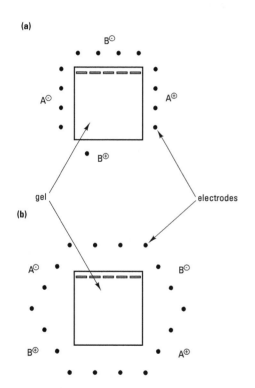

(a)

(b)

gel   electrodes

Fig. 66.3 The configuration of electrodes for conventional PFGE (a) and CHEF/PACE (b).

glass plate

weight

absorbent
material

filter paper
sheets

nitrocellulose
membrane

filter
paper
wick

agarose
gel

support

buffer
solution

glass
plate

cling
film

Fig. 66.4 Components of Southern blotting apparatus.

Biomedical applications of dot blotting – these include the detection of specific pathogenic microbes in samples and the identification of particular genetic markers, e.g. in inherited genetic disorders and forensic science.

Definition

Probe – a labelled DNA or RNA sequence used to detect the presence of a complementary sequence (by molecular hybridisation) in a mixture of DNA or RNA fragments.

Example   For a simple tripeptide containing methionine, aspartate and phenylalanine, the synthetic oligonucleotide probes would include combinations of the following codons:
1st codon (met): ATG
2nd codon (asp): GAT, GAC
3rd codon (phe): TTT, TTC

The main features of the conventional apparatus for Southern blotting are shown in Fig. 66.4. The principal stages in the procedure are as follows:

1. **The gel is first soaked in HCl, leading to random cleavage of the DNA into smaller fragments.**

2. **The gel is subsequently soaked in alkali to denature the dsDNA to ssDNA, then neutralised:** this is necessary to allow hybridisation with probe DNA after blotting.

3. **A nitrocellulose or nylon membrane is then placed directly on the gel, followed by several layers of absorbent paper.** The DNA is 'blotted' onto the filter as the buffer solution soaks into the paper by capillary action.

4. **The filter is baked in a vacuum oven at 80 °C for 3–5 h or exposed to UV light, to 'fix' the DNA.**

5. **Specific DNA fragments are identified by incubation (6–24 h) with complementary labelled probes of ssDNA, which will hybridise with a particular sequence, followed by visualisation,** often using an enzyme-based system (p. 412). If radiolabelled probes are used (see below), the desired fragments are located by autoradiography (p. 301).

Modifications of the method use either a vacuum apparatus, an electric field ('electroblotting') or positive pressure to transfer the DNA fragments from the gel to the membrane. These modifications reduce the time for transfer by around 10-fold.

Northern blotting

This process is similar in principle to Southern blotting, but RNA is the molecule that is separated and probed.

Dot blotting or slot blotting

Here, samples containing denatured DNA or RNA are applied directly to the nitrocellulose membrane (via small individual round or slotlike templated holes) without prior digestion with restriction enzymes or electrophoretic separation. The 'blot' is then probed in a similar manner to that described for Southern blotting. This allows detection of a particular nucleic acid sequence in a sample.

Types of probe

The probes used in blotting and DNA hybridisation can be obtained from a variety of sources, including those listed below:

- **cDNA** (complementary or copy DNA), which is produced from isolated mRNA using reverse transcriptase. This retroviral enzyme catalyses RNA-directed DNA synthesis, rather than the normal transcription of DNA to RNA. After the mRNA has been reverse transcribed, it is degraded by the addition of alkali or ribonuclease, leaving the ssDNA copy. This is then used as a template for a DNA polymerase, which directs the synthesis of the second complementary DNA strand to form dsDNA. This is denatured to ssDNA before use.

- **Oligonucleotide probes** (15–30 nucleotides) can be produced if the amino acid sequence of the gene product is known. Since the genetic code is degenerate, i.e. some amino acids are coded for by more than one codon (see Table 66.1), it may be necessary to synthesise a mixture of oligonucleotides to detect a particular DNA sequence: this mixture of oligonucleotides is termed a 'degenerate probe'.

- **Specific genomic DNA sequences,** where the gene has been characterised.
- **PCR-generated fragments** (Chapter 67).
- **Heterologous probes,** i.e. sequences for the same gene, or its equivalent, in another organism.

Labelling of probes

Detection of very low concentrations of target DNA sequences requires probes that can be detected with high sensitivity. This is achieved by using enzyme-linked methods, or by radiolabelling, e.g. using ^{33}P, which are often available in kit form:

- **Enzyme-linked methods** involve incorporating a modified nucleotide precursor, such as biotinylated dTTP, into the DNA by nick-translation or random priming. When the probe hybridises with the target sequence, it can be detected by addition of an enzyme (e.g. horseradish peroxidase) coupled to streptavidin. The streptavidin binds specifically to the biotin attached to the probe and the addition of a suitable fluorogenic or chromogenic substrate for the enzyme allows the probe to be located.
- **Radiolabelled probes** can be made by several methods, including the nick-translation technique. This uses DNA polymerase I from *E. coli,* which has: (i) an exonuclease activity that 'nicks' dsDNA and removes a nucleotide, and (ii) a polymerase activity that can replace this with a labelled deoxyribonucleotide. After several cycles of nick-translation, the labelled DNA is denatured to ssDNA for use as probes. Newer approaches include random priming of single-stranded template DNA, followed by the synthesis of radiolabelled DNA fragments complementary to the template using a DNA polymerase and a radiolabelled deoxynucleoside triphosphate (dNTP). Probe hybridisation to a target sequence is detected by autoradiography. Oligonucleotide probes cannot be labelled using the above methods and require 'end labelling', where a kinase is used to replace the 5' terminal phosphate group with a radiolabelled group.

Hybridisation of probes

The stability of the duplex formed between the probe and its target is directly proportional to the number and type of complementary base pairs that can be formed between them: stability increases with the amount of G + C, since these bases form three hydrogen bonds per base pair, rather than two (p. 399). Duplex stability is also influenced by temperature, ionic strength and pH of the hybridisation buffer, and these can be varied to suit the stringency of hybridisation required:

- **'Low stringency' hybridisation,** involves duplex formation with less than perfect complementarity is promoted, either by lowering the temperature, or increasing ionic strength.
- **'Stringent' hybridisation conditions (high stringency)** usually involve high temperatures, or decreased ionic strength, and will sustain only perfectly matched duplexes.

Using commercial kits – do not blindly follow the protocol given by the manufacturer without making sure you understand the principles of the method and the reasons for the procedure. This will help you to recognise when things go wrong and what you might be able to do about it.

Using labelled probes – an alternative approach is the chemiluminescent system based on dioxygenin-labelled nucleotides.

Example Low stringency hybridisation can be useful when using a heterologous probe, i.e. from another species.

Example High stringency conditions can be used for detecting a single base change in a mutant gene using a dot-blot hybridisation procedure with a specific oligonucleotide probe.

Fig. 66.5 A 2′,3′-dideoxynucleoside triphosphate.

sequence
from
autoradiograph

ddA ddC ddG ddT

3′	5′
A	T
G	C
T	A
G	C
C	G
G	C
A	T
G	C
5′	3′

sequence of
original
strand

Fig. 66.6 Sanger sequencing gel, showing how the banding pattern is converted into a sequence of nucleotide bases.

DNA sequencing

By fragmenting target DNA with several restriction enzymes and then sequencing the overlapping fragments, it is possible to determine the nucleotide sequences of very large stretches of DNA, including entire genomes. Sequencing methods rely on polyacrylamide gel electrophoresis (p. 352). The Sanger, or chain termination, method is the most widely used in conventional DNA sequencing. This makes use of dideoxynucleoside triphosphates (Fig. 66.5), which have no –OH group at either the C-2 or C-3 of ribose. A dideoxynucleoside triphosphate (ddNTP) can be added to a growing DNA chain, but since it lacks an –OH group at the C-3 position it cannot form a phosphodiester bond with the incoming dNTP of the growing chain. Therefore a dideoxynucleotide acts as a terminator at the site it occupies. Details of the Sanger method are given in Box 66.1 and Fig. 66.6.

Automated DNA sequencing

The development of automated DNA sequencing machines (so-called 'DNA sequenators') has enabled Sanger-based sequencing to be performed several orders of magnitude faster than with manual methods. Base-specificity is achieved by using primers labelled with fluorophores with different fluorescence characteristics in each of the four reaction tubes (Box 66.1). After the reactions are separately completed, the four sets of products can be pooled and fractionated by electrophoresis in a single lane. The fluorophore-labelled fragments are detected as they pass a scanning laser detector and the DNA sequence is determined by using both the specific wavelength emitted by the fragment (indicating the base) and by the migration time (indicating the fragment size, which corresponds to the base location in the DNA sequence). Such, automated DNA sequencing tends to be performed by capillary electrophoresis (p. 361) rather than polyacrylamide gel electrophoresis.

Box 66.1   DNA sequencing using the chain termination (Sanger) method

DNA is first prepared as single-stranded fragments, and sequenced using a commercial kit with the following stages:

1. **Strand synthesis.** Four separate tubes are prepared, each containing (i) target (template) DNA, (ii) a DNA polymerase, (iii) radiolabelled oligonucleotide primer of known sequence, (iv) dGTP, dATP, dCTP, dTTP in excess plus, (v) one type of dideoxynucleotide in limited concentration, e.f. ddTTP to cause chain termination across sites where dTTP would normally be added. The reaction is terminated using a formamide solution, with incubation at 80°C to produce ssDNA.

2. **Fragment separation.** For example, using 4–6% denaturing acrylamide gel electrophoresis, with the products of the four reaction tubes in separate lanes, running at 35–40 W (up to 32 mA, 1.5 kV) for 2.5 h.

3. **Autoradiography.** To locate individual bands. This requires vacuum-drying of the gel, then contact with an X-ray film for 24 h.

4. **Reading the sequence.** Fig. 66.6 shows the process, with the sequence of the template strand being determined from the position of individual bands corresponding to the position of each nucleotide base.

Troubleshooting and other points to note: streaking can be due to damage to wells, air bubbles in wells/gel or contamination; faint/fuzzy gels can be due to insufficient template DNA or dNTPs and to poor annealing of primer and template; bands in the same position in more than one lane can be due to contamination, to multiple primer sites on the template, or to secondary DNA structure, giving 'ghost' banding.

Using PCR (p. 467) for automated sequencing – with the four ddNTPs tagged using different fluorophores, the products can be run in the same lane and detected by laser excitation at the end of the gel, with the sequence of fluorophores giving the base sequence.

DNA microarrays and bioinformatics – interpretation of DNA microarray data is a growing area of bioinformatics, requiring dedicated software, e.g. the TIVU microarray analysis freeware programs available from the Institute of Genomic Research, at: http://www.tm4.org/

Next-generation sequencing (NGS)

The quest for faster sequencing methods has led to a number of recent developments, collectively termed 'next-generation sequencing', making it possible to sequence an entire genome in a fraction of the time required when conventional techniques are used and leading to rapid advances in genomics (Chapter 12). Several specific technologies have been developed, including:

- pyrosequencing
- Roche 454 sequencing
- ion torrent sequencing
- Illumina sequencing.

Irrespective of the underlying chemistry, NGS methods work by simultaneously sequencing separate, clonally amplified DNA fragments (a 'library') in a massive array, with high throughput parallel sequencing operations numbered in the hundred of thousands (for more details of the procedures involved in next-generation sequencing, see Brown, 2015).

DNA microarrays

This area has seen rapid, recent development. DNA microarrays ('gene chips') support the quantitative analysis of many different genes/sequences at the same time. This involves synthesising and fixing single-stranded probes for a large number (>1000) of different nucleotide sequences onto specific sites on a glass, quartz or silicon wafer. When single-stranded labelled nucleic acid is added, it will hybridise only with probes having a complementary sequence and these can be determined by the presence of an attached label, e.g. a fluorescent dye, detected using a scanning laser. One application involves the detection of labelled cDNA produced from the mRNA in an organism by using an 'expression' microchip, which contains probes for all of the expressed genes in an organism, enabling the relative expression of each gene to be determined from the fluorescence signal at each location on the microarray. Other applications include the use of single nucleotide polymorphism (SNP) microarrays to identify mutations and genetic variation, e.g. in screening for genetic diseases and in forensic DNA analysis. A commercial example is the Affymetrix GeneChip system (for more details see Knudsen (2004)).

Text references and sources for further study

Aquino do Muo, A. and Rapley, R. (2010) *Gene Probes: Principles and Protocols.* Humana, New York.

Baldi, P. and Hatfield, G.W. (2002) *DNA Microarrays and Gene Expression: from Experiments to Data Analysis and Modeling.* Cambridge University Press, Cambridge.

Brown, S.M. (2015) *Next-Generation DNA Sequencing Informatics*, 2nd edn. Cold Spring Harbour Lab Press, Cold Spring Harbor.

Darby, I.A. and Hewitson, T.D. (2006) *In Situ Hybridization Protocols,* 3rd edn. Humana, New York.

Knudsen, S. (2004) *Guide to Analysis of DNA Microarray Data,* 2nd edn. Wiley, New York.

Sambrook, J. and Russell, D.W. (2006) *The Condensed Protocols from Molecular Cloning: A Laboratory Manual.* Cold Spring Harbor Lab, Cold Spring Harbor.

STUDY EXERCISES

66.1 Practise the calculations involved in working with nucleic acids. Imagine that you have been supplied with a freeze-dried sample of a single-stranded DNA probe 24 bases in length, which you reconstitute in 1 mL of buffer (reconstituted probe). You take 5 μL, add it to 495 μL of water and determine the A_{260} as 0.14.

(a) What is the mass concentration of the probe in this solution, in μg mL⁻¹?

(b) Assuming the average M_r of a nucleotide is 325, what is the concentration of the probe in the solution, expressed in micromolar terms (i.e. as μmol L⁻¹)?

(c) The concentration of probe required in a blotting procedure is 10 pmoL mL⁻¹ and a total of 10 mL of this solution is required. You have been asked to prepare 5 mL of a stock solution of the DNA probe at 100× the concentration required in the blotting procedure, so that 0.100 mL (100 μL) can be diluted to 10 mL for the final blotting procedure, while the rest will be stored at −20 °C, for subsequent experiments. How would you prepare the 100 × stock solution?
Express all answers to three significant figures.

66.2 Select an electrophoretic technique for separating DNA fragments of different sizes. Which technique would you use to separate DNA fragments in the following ranges: (a) 1–70 kbp; (b) 75–500 bp; (c) 100–6 Mbp.

66.3 Test your understanding of the term 'stringency' in relation to nucleic acid probes. What does the term 'stringency' mean in the context of molecular biology? What factors cause increased stability between a nucleic acid probe and its target sequence? Give examples of when you might use 'low stringency' and 'high stringency' hybridisation.

66.4 Interpret the results of a Sanger sequencing gel. The figure represents a typical gel from a dideoxynucleotide sequencing procedure. Determine the sequence of the original DNA strand.

Answers to these study exercises are available at www.pearsoned.co.uk/practicalskills.

The polymerase chain reaction (PCR) is a rapid, inexpensive and simple means of producing microgram amounts of DNA from minute quantities of template.

> **KEY POINT**   PCR offers an alternative approach to gene cloning (Chapter 68) for the production of many copies of an identical sequence of DNA. The starting material may be genomic DNA (e.g. from a single cell), RNA, DNA from archival specimens, cloned DNA or forensic samples.

The technique uses *in vitro* enzyme-catalysed DNA synthesis to create millions of identical copies of DNA. If the base sequence of the adjacent regions of the DNA to be amplified is known, this enables synthetic oligonucleotide primers to be constructed that are complementary to these so-called 'flanking regions'. Initiation of the PCR occurs when these primers are allowed to hybridise (anneal) to the component single strands of the target DNA, followed by enzymatic extension of the primers (from their 3′ ends) using a thermostable DNA polymerase. A single PCR cycle consists of three distinct steps, carried out at different temperatures (Fig. 67.1), as follows:

1. **Denaturation of dsDNA** by heating to 94–98 °C, separating the individual strands of the target DNA.

2. **Annealing of the primers,** which occurs when the temperature is reduced to 37–65 °C.

3. **Extension of the primers** by a thermostable DNA polymerase (e.g. *Taq* polymerase, isolated from *Thermus aquaticus*) at 72 °C; this step should last long enough to generate the PCR product: approximately 1 min of reaction time is required per kbp of sequence.

In the first cycle, the product from one primer is extended beyond the region of complementarity of the other primer, so each newly synthesised strand can be used as a template for the primers in the second cycle (Fig. 67.1). Successive cycles will thus generate an exponentially increasing number of DNA fragments, the termini of which are bounded by the 5′ ends of the primers (length of each fragment = length of primers + length of target sequence). Since the amount of DNA produced doubles in each cycle, the amount of DNA produced = 2^n, where n is the number of cycles. Up to 1 μg of amplified target DNA can be produced in 25–30 cycles from a single-copy sequence within 50 ng of genomic DNA, assuming close to 100% efficiency during the cycling process. After electrophoresis, the PCR product is normally present in sufficient quantity to be visualised directly, for example with ethidium bromide or SYBR Safe.

The temperature changes in PCR are normally achieved using a thermal cycler, which is simply a purpose-built incubator block that can be programmed to vary temperatures, incubation times and cycle numbers.

> **KEY POINT**   PCR is so sensitive that one of the main problems associated with the technique is contamination and amplification of 'foreign' DNA. Great care is required to avoid sample contamination during *in vitro* amplification (Box 67.1).

Applications of PCR – these include:
- diagnosis and screening of genetic diseases and cancer;
- rapid detection of slowly growing micro-organisms (e.g. mycobacteria) and viruses (e.g. HIV);
- HLA typing in transplantation;
- analysis of DNA in archival material;
- DNA profiling in forensic science;
- preparation of nucleic acid probes;
- clone screening, mapping and subcloning (p. 474).

Understanding the terminology of PCR – cloned (amplified) DNA sequences produced using PCR techniques are often termed amplicons.

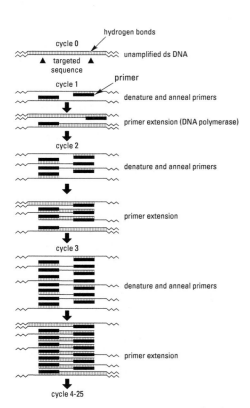

Fig. 67.1 The polymerase chain reaction (PCR).

Designing PCR primers – online tools such as Primer Blast (at: http//www.ncbi.nlm.nih. gov/tools/primer- blast/) can provide primes specific to a particular template sequence.

Storing primers for PCR – primers are best stored in ammonia solution, which remains liquid at –20°C, avoiding the need for repeated freezing and thawing when dispensed. Before use, aliquots of stock solution should be heated in a fume cupboard, to drive off the ammonia.

Using dNTP solutions – make up stock solutions, pool in small volumes (50–100 μL of each dNTP) and store separately at –20 °C.

Establishing the optimum concentration of Mg^{2+} – this can be determined by trial and error, and can be a useful starting point for any new PCR; try a range of different Mg^{2+} concentrations and select the one giving the strongest intensity of the target band following gel electrophoresis.

Understanding codes for primers and probes – oligomeric nucleotides are often referred to by the number of bases, e.g. a 20-mer primer will contain 20 nucleotide bases.

PCR components and conditions

These are readily available in kit form from commercial suppliers; they include a thermostable DNA polymerase, such as *Taq* polymerase, dNTPs, buffer, a detergent (e.g. Triton), KC1, $MgCl_2$ or $MgSO_4$, and primers. Protocols for PCR vary considerably for particular applications – a typical procedure is given in Box 67.1. However, if you are trying to develop your own PCR procedure in a lab project, the following information may be useful:

- **Primers need to be at least 18 nucleotides long** (primers longer than 18 nucleotides should be unique, even in a large eukaryotic genome); both primers should have similar annealing temperatures, T_m, p. 402), with a minimal degree of self-complementarity (to avoid formation of secondary structures), and no complementarity to each other (so that primer dimers are not formed).
- **For most applications, the final concentration of each primer should be 0.1–0.5 μmol L^{-1}**, which gives an excess of primers of about 10^7 with respect to the template, e.g. target genomic DNA at a concentration of 50 ng per 10 μL reaction mixture.
- **Annealing temperatures are based around T_m, the temperature at which 50% of the primers are annealed to their target sequence** (Fig. 59.7). For primers of < 20 bases, T_m can be roughly calculated in °C from the equation:

$$T_m = 4(G + C) + 2(A + T)$$ [67.1]

where G, C, A and T are the number of bases in the primer. Using this as a starting point, the optimum annealing temperature can be determined by trial and error. The T_m values of the two primers should be within 5 °C of each other (and, therefore, be ideally identical in G + C content). The annealing temperature is then set 5 °C below the lowest T_m of the primer pair.
- **dNTPs should be used at equal concentrations of 200 μmol L^{-1}**, which should provide the initial excess required for incorporation into DNA.
- **The last two bases at the 3′ end (where elongation is initiated)** should be either G or C (3 hydrogen bonds, giving strong annealing), rather than either A or T (only 2 hydrogen bonds).
- **One of the key variables in PCR is the Mg^{2+} concentration;** Mg^{2+} is required as a co-factor for the thermostable DNA polymerase (Anon, 2002). Excess Mg^{2+} stabilises dsDNA and may prevent complete denaturation of product at each cycle; it also promotes spurious annealing of primers, leading to the formation of undesired products. However, very low Mg^{2+} concentration impairs polymerisation. The purpose of gelatin and Triton X-100 is to stabilise the DNA polymerase during thermal cycling.
- **The most frequently used thermostable DNA polymerase is *Taq* polymerase,** which extends primers at a rate of 2–4 kbp per min at 72°C. It should be used at a concentration of ≥ 1 nmol L^{-1} (≥0.1 U per 5 μL reaction mixture). A disadvantage of *Taq* polymerase is that it has a relatively high rate of misincorporation of bases (one aberrant nucleotide per 100 000 nucleotides per cycle). Other polymerases are available (e.g. *Pfu* polymerase from *Pyrococcus furiosus,* KOD polymerase from *Thermococcus kodakarensis* or genetically modified forms of *Taq* polymerase), which have lower misincorporation rates.

Box 67.1 How to carry out the polymerase chain reaction (PCR)

The protocol given below is typical for a standard PCR. Note that temperatures, incubation times and the number of cycles will vary with the particular application, as discussed in the text.

1. **Make sure you have the required apparatus and reagents to hand,** including: (i) a thermal cycler; (ii) template thermostable DNA ($\geq$50 ng/µL); (iii) stock solution of all dNTPs (5 mmol L^{-1} for each dNTP); (iv) a DNA polymerase (at 5U/µL); (v) primers at, e.g., 30 µmol L^{-1}; (vi) stock buffer solution, e.g. containing 100 mmol L^{-1} TRIS (pH8.4), 500 mmol L^{-1} KCl, 15 mmol L^{-1} MgCl$_2$, 1% (w/v) gelatin, 1% (v/v) Triton X– 100 (this stock is often termed '10x PCR buffer stock').

2. **Prepare a reaction mixture:** for example, a mixture containing: 1.0 µL target DNA; 2.5 µL stock buffer solution; 1.0 µL primer 1; 1.0 µL primer 2; 1.0 µL of each of the stock solutions of dNTPs, 0.1 µL *Taq* polymerase; 15.4 µL distilled deionised water, to give a total volume of 25.0 µL.

3. **Use appropriate positive and negative controls:** a positive control is a PCR template that is known to work under the conditions used in the laboratory, e.g. a plasmid, with appropriate primers, known to amplify at the annealing temperature to be used. A commonly used negative control is the PCR mixture minus the template DNA, though negative controls can be set up lacking any one of the reaction components.

4. **Cycle in the thermal cycler:** for example, an initial period of 5 min at 94 °C, followed by 30 cycles of 94 °C for 1 min (denaturation), then 50 °C for 1 min (primer annealing; temperature depends upon G + C and A + T content, Eqn [67.1] p. 468), then 72 °C for 1 min (chain extension).

5. **Assess the effectiveness of the PCR:** for example, by gel electrophoresis and SYBR Safe staining.

Troubleshooting

- If no PCR product is detected, repeat the procedure, checking carefully that all components are added to the reaction mixture. If there is still no product, check that the annealing temperature is not too high, or the denaturing temperature is not too low.

- If too many bands are present, this may indicate that: (i) the primers may not be specific, (ii) the annealing temperature is too low, or (iii) there is an excess of Mg^{2+}, dNTPs, primers or enzyme.

- Bands corresponding to primer-dimers indicate that: (i) the 3' ends of the primers show partial complementarity, (ii) the annealing temperature is not high enough, or (iii) the concentration of primers is too high.

Avoiding contamination in diagnostic PCR

The sensitivity of PCR is also the major drawback, since the technique is susceptible to contamination, particularly from DNA from the skin and hair of the operator, from previous PCR products, from airborne microbes and from positive controls. A number of routine precautions can be taken to avoid such contamination:

- **use a laminar-flow cabinet** (p. 225) dedicated to PCR use and located in a separate lab from that used to store PCR products or prepare clones;

- **keep separate supplies of pipettors, tips, microfuge tubes and reagents** – these should be exclusive to the PCR, with separate sets for sample preparation, reagents and product analysis;

- **autoclave all buffers, distilled deionised water, pipette tips and tubes** prior to use;

- **wear disposable gloves at all times and change them frequently:** protective coverings for the face and hair are also advisable;

- **avoid contamination due to carry-over by including dUTP in the PCR mixture instead of dTTP.** Thus copies will contain U rather than T. Before the template denaturing step, treat the mixture with uracil-N-glycosylase (UNG, available commercially as AmpEase): this will destroy any strands containing U, i.e. any strands carried over from a previous reaction, or any contaminating material from another PCR. The target DNA will contain T, rather than U and will not be degraded by UNG. At the first heating step, the UNG will be denatured, so any newly synthesised U-containing copies will remain intact.

- **for very sensitive work, use a strong UVC light (p. 306) inside the PCR workstation for 20–30 minutes before starting your work,** to degrade any contaminating DNA.

SAFETY NOTE Note that you must not expose your skin or eyes to any UVC source due to its mutagenic/carcinogenic properties.

Fig. 67.2 Inverse PCR – basic principles.

> **Example** Random amplification of polymorphic DNA (RAPD-PCR) can be used as a form of 'molecular typing' (p. 252), to identify a particular strain of an organism, e.g. in tracing the route of transmission of a pathogenic microbe.

PCR variations

Nested PCR

This can be used when the target sequence is known, but the number of DNA copies is very small (e.g. a single DNA molecule from a microbial genome), or if the sample is degraded (e.g. a forensic sample). The process involves two consecutive 'rounds' of PCR. The first PCR uses so-called 'external' primers, and the second PCR uses two 'internal' (or 'nested') primers that anneal to sequences within the product of the first PCR. This increases the likelihood of amplification of the target sequences by selecting for it using different primers during each round. Thus, nested PCR also increases the specificity of the reaction, since a single set of primers used in isolation may give a reasonable yield but several bands, while the use of a second set of primers ensures that a unique sequence is amplified, e.g. in microbial diagnostics.

Inverse PCR

This is a useful technique for amplifying a DNA sequence flanking a region of known base sequence (Fig. 67.2), e.g. to provide material for characterising an unknown region of DNA. The DNA is cut with a restriction enzyme so that both the region of known sequence and the flanking regions are included. This restriction fragment is then circularised and cut with a second restriction enzyme with specificity for a region in the known sequence. The now linear DNA will have part of the known sequence at each terminus, and by using primers that anneal to these parts of the known sequence, the unknown region can be amplified by conventional PCR. The product can then be sequenced and characterised (Chapter 66).

Reverse transcriptase-PCR (RT-PCR)

This technique is useful for detecting cell-specific gene expression (as evident by the presence of specific mRNA) when the amount of biological material is limited. Using either an oligo-dT primer to anneal to the 3′ polyadenyl 'tail' of the mRNA, or random hexamer primers, together with reverse transcriptase, cDNA is produced which is then amplified by PCR. RT-PCR is often a useful method of generating a probe, the identity of which can be confirmed by sequencing (p. 464).

Amplification fragment length polymorphism (AFLP)

This term refers to several closely related techniques in which a single oligonucleotide primer of arbitrary sequence is used in a PCR reaction under conditions of low stringency, so that the primer is able to anneal to a large number of different sites within the target DNA. Some of the multiple amplification products will be polymorphic (e.g. the presence or absence of a particular annealing site will result in presence or absence of a particular band on the gel, after PCR and electrophoresis). Such polymorphisms can be used to detect differences between dissimilar DNA sequences.

PCR and DNA profiling techniques

Genetic mapping and sequencing studies have led to the discovery of highly variable regions in the non-coding regions of DNA between different individuals. These hypervariable regions, sometimes termed 'minisatellite

▼ = restriction sites

individual repeat sequence

banding patterns for A and C match, so DNA is likely to be from the same individual

Fig. 67.3 Simplified representation of traditional DNA profiling: (a) tandem repeat sequences within DNA are cut using a restriction enzyme to yield fragments of different size; (b) these fragments are then separated by agarose gel electrophoresis on the basis of their size (M_r), giving a banding pattern that is a characteristic of the DNA used.

DNA', are found at many sites throughout the genome. Each minisatellite contains a defined sequence of nucleotides which is repeated a number of times in a tandem fashion (Fig. 67.3); the greater the number of repeats, the longer the minisatellite. The number of tandem repeats in any particular minisatellite varies from one person to another (i.e. they have a variable number of tandem repeats or VNTRs). This is exploited in identifying individuals on the basis of their DNA profile, by carrying out the following steps:

1. **Extraction of DNA from cells** of the individual (e.g. leukocytes, buccal cells, spermatozoa).

2. **Digestion of the DNA with a restriction enzyme that cuts at sites other than those within the minisatellite,** to produce a series of fragments of different M_r.

3. **Electrophoresis and Southern blotting of the restriction fragments using probes that are specific to the particular minisatellite.**

The size of the fragments identified will depend on the number of minisatellites that each fragment contains, and the pattern obtained in the Southern blot is characteristic of the individual being profiled (Fig. 67.3).

PCR is widely used in DNA profiling in circumstances where there is a limited amount of starting material. By selecting suitable primers, highly variable regions can be amplified from very small amounts of DNA, and the information from several such regions is used to decide, with a very low chance of error, whether any two samples of DNA are from the same individual or not, e.g. in forensic science (Butler, 2009). More recent developments include the use of short tandem repeat (STR) loci, summarized in Fig. 67.4.

Real-time PCR

Conventional PCR techniques rely on end-point detection of amplified product, e.g. by electrophoretic separation and staining (p. 459). However, such methods are time-consuming and are only semi-quantitative, since they are based predominantly on the detection of an amplified fragment (band) in a sample, rather than being designed to give exact information on its abundance (copy number). Quantitative analysis is only feasible during the early stages of PCR, where reagents are in excess and where the amount of amplified product is small, thereby avoiding the problems of product hybridisation, which would compete with primer binding. Real-time PCR enables the simultaneous amplification and quantification of template DNA in a sample by establishing the number of copies present by

Fig. 67.4 Output of multilocus short tandem repeat (STR) analysis in contemporary DNA profiling. Highly polymorphic regions of the human chromosome with repeat sequences of 3–5 bases (STRs) are amplified by PCR and then separated/detected by capillary electrophoresis (p. 361). The profile above shows alleles (peaks) for four loci (shown in the top row of the figure), with two alleles for each locus (one inherited from each parent) - the numeral below each peak shows the number of repeats for the alleles D8, D21 and D18 while the Am (amelogenin) locus shows inheritance of alleles for X and Y chromosomes, which demonstrates that this profile is for a male. Typically, either 13 (US) or 11 (UK) STR loci are used to create an individual human DNA profile.

working in the early exponential phase of amplification. Currently, while several different formats are available for real-time PCR, all rely on the generation of a fluorescent signal, as described below.

Dye fluorescence

This is the simplest and least expensive approach. A fluorescent dye such as SYBR Green is included in the PCR mixture and the level of fluorescence is monitored as the reaction proceeds; since SYBR Green binds strongly to dsDNA, showing an enhancement in fluorescence of over 100-fold, any increase in fluorescence is directly proportional to the amount of dsDNA produced. Calibration is achieved by running a series of standards containing known amounts of dsDNA. This approach works well for optimised single PCR product reactions where non-specific reactions are minimised.

Fluorescent reporter probes

Here a fluorescent reporter dye is covalently bound to the 5′ end of an oligonucleotide probe while a 'quencher' group is attached to the 3′ end (Fig. 67.5). The probe is designed to hybridise to an internal region of the target sequence. During PCR, when the DNA polymerase molecule reaches the hybridised probe, the 5′ nuclease activity of the polymerase will cleave the reporter dye from the rest of the probe, causing an increase in fluorescence with each cycle that is in direct proportion to the amount of PCR product being formed, which is itself directly related to the original number of copies of the target DNA sequence. A commercial example is the TaqMan series of probes – while this approach is more accurate and reliable than dye fluorescence, it is also far more expensive, since a specific reporter probe must be synthesised for each target sequence. Other variants rely on changes in 3D conformation of the probe when it binds to the target sequence, causing an increase in fluorescence, e.g. Molecular Beacons and Scorpion probes.

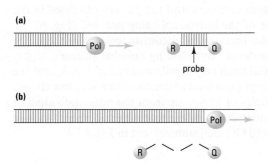

Fig. 67.5 Real-time PCR using a fluorescent reporter probe. In (a) the probe is bound to the target sequence while in (b) the 5′ nuclease activity of the DNA polymerase (Pol) has cleaved the reporter dye (R) and quencher (Q), resulting in fluorescence.

KEY POINT   You should note that the PCR technique is continually being updated, with subtle new variations being produced on a regular basis – novel approaches and novel acronyms are very likely to be reported within the lifetime of this book.

Text references

Anon. (2002) *PCR Amplification.* Available: http://www.promega.com/resources/product-guides-and-selectors/protocols-and-applications-guide/pcr-amplification/ Last accessed 01/09/15. [Details of how to optimise Mg^{2+} and polymerase concentration, etc.]

Butler, J. (2009) *Fundamentals of Forensic DNA Typing.* Academic Press, Oxford.

Sources for further study

Basu, C. (2015) *PCR Primer Design*. Humana Press, New York.

Bustin, S. (2014) *The PCR Revolution: Basic Technologies and Applications*. Cambridge University Press, Cambridge.

Dorak, M.T (2006) *Real-Time PCR*. Taylor & Francis, Abingdon.

Kennedy, S. and Oswald, N. (2006) *PCR Troubleshooting and Optimization: the Essential Guide*. Caister, Wymondham.

McPherson, M.J. and Moller, S.G. (2006) *PCR Basics*. Taylor & Francis, Abingdon.

Van Pelt-Verkuil, E., Van Belkum, A. and Hays, J.P. (2011) *Principles and Technical Aspects of PCR Amplification*. Springer, New York.

STUDY EXERCISES

67.1 Investigate the role of magnesium ions in PCR. Find out why optimising the Mg^{2+} concentration is important for successful amplification of target DNA in PCR.

67.2 Check your knowledge of primer design. List the principal factors that should be taken into account when designing primers for PCR.

67.3 Carry out a T_m calculation for a primer. For the following primer, what would be the temperature at which 50% of the molecules are annealed to their target sequence (i.e. T_m)?

5′–TGCATGGCTGGATTAGCG–3′

67.4 Consider how the risk of contamination can be reduced in PCR. List the most important practical steps that can be taken to reduce the possibility of contamination.

67.5 Find out what types of DNA analysis might be suitable for forensic examination of degraded DNA samples. In 1918 Tsar Nicholas of Russia, his wife Tsarina Alexandra, and their five children were executed and buried in a shallow grave in Yekaterinburg, Russia. Samples from the bones of the family were analysed by the UK Forensic Science Service (FSS). What types of DNA-based tests would you expect to have been carried out to confirm that a family group was present in the grave?

67.6 Select appropriate primers for a specific PCR amplification. The sequence in the box below represents a region of DNA that is to be amplified by PCR.

Which of the following three pairs of primers is most suited to amplify a PCR product from the sequence below?

(a) 5′–TCGCTGAAGGACATGTCG–3′   and
     5′–CGATCTAGGCGACATGTCC–3′

(b) 5′–TCGCTGAAGGACATGTCG–3′   and
     5′–TCCTTAGGTCTCGACCGTA–3′

(c) 5′–TCGCTGAAGGACATGTCG–3′   and
     5′–GTCTCGACCGTAGCTAGCATC–3′

5′–TCGCTGAAGGACATGTCGATGCTAGCTACGGTCGAGACGTAAGGACATGTCGGCTAGATCGC–3′
3′–AGCGACTTCCTGTACAGCTACGATCGATGCCAGCTCTGCATTCCTGTACAGCGGATCTAGCG–5′

Answers to these study exercises are available at www.pearsoned.co.uk/practicalskills.

68 Molecular genetics III – genetic manipulation techniques

Advances in the procedures used to manipulate nucleic acids *in vitro* have increased our understanding of the structure and function of genes at the molecular level. Additionally, these techniques can be used to alter the genome of an organism (genetic manipulation, genetic engineering or gene cloning), e.g. to create a bacterium capable of synthesising a foreign protein such as a potentially useful hormone or vaccine component, or a novel protein. The procedures are often termed 'recombinant DNA technology', since they involve the creation of novel combinations of DNA (i.e. recombinant DNA) under controlled laboratory conditions.

> **KEY POINT** While genetic manipulation must be carried out under strict containment, in accordance with appropriate legislation, the procedures involved in the isolation, amplification, recombination and cloning of DNA are often used at undergraduate level, to illustrate the general features of the techniques.

In the UK, the Genetically Modified Organisms (Contained Use) Regulations (2014) provide the regulatory framework for all research procedures involving the genetic modification of organisms.

Definitions

Bacteriophage (phage) – a bacterial virus. Useful as they possess a means of penetrating the bacterial cell.

Cosmids – hybrid plasmid vectors containing the *cos* (cohesive) sites from phage λ, enabling *in vitro* packaging into phage capsids. Useful for cloning large segments of DNA, typically up to 50 kbp.

Plasmids – circular molecules of DNA capable of autonomous replication within a bacterial cell. Can be isolated, manipulated and reintroduced into bacterial cells.

Subcloning – procedures that isolate and characterise smaller portions of a particular DNA sequence, e.g. during the search for a specific gene.

Transfection – in bacteria: uptake of viral nucleic acid; in eukaryotes: uptake of any naked, foreign DNA.

Transformation – in bacteria: stable incorporation of external DNA, e.g. a plasmid; in eukaryotes: the conversion of a cell culture of finite life to a continuous (immortal) cell line (also occurs in cancer).

Preparing glassware – all glassware for DNA purification must be siliconised, to prevent adsorption of DNA. All glass and plastic items must be sterilised before use.

Basic principles

Gene cloning involves several steps:

1. **Isolation of the DNA sequence (gene) of interest** from the genome of an organism, or from a gene library. This usually involves DNA purification followed by enzymic digestion or mechanical fragmentation, to liberate the target DNA sequence, or PCR amplification (p. 467) of the target DNA sequence.
2. **Creation of an artificial recombinant DNA molecule (rDNA),** by inserting the target sequence into a DNA molecule capable of replicating in a host cell, i.e. a 'cloning vector'. Suitable cloning vectors for bacterial cells include plasmids (p. 476), bacteriophages (p. 477) and cosmids (p. 477).
3. **Introduction of the recombinant DNA molecule into a suitable host,** e.g. *E. coli*. The process is termed transformation when a plasmid is used, or transfection for viral nucleic acid.
4. **Selection and growth of the transformed (or transfected) cell,** using the techniques of cell culture (p. 223). Since a single transformed host cell can be grown to give a clone of genetically identical cells, each carrying the target DNA of interest, the technique is often referred to as 'gene cloning', or molecular cloning.

Extraction and purification of plasmid DNA

Specific details of the steps involved in the isolation of DNA vary, depending upon the source material. However, the following protocol lists the principal stages in the purification of plasmid DNA from bacterial cells:

1. **Cell wall digestion:** incubation of bacteria in a lysozyme solution will hydrolyse the peptidoglycan cell wall. This is often carried out under isotonic conditions, to stop the cells from bursting open and releasing chromosomal DNA. Note that Gram-negative bacteria are relatively insensitive to lysozyme, requiring additional treatment to allow the

Using ultrapure sterile water in molecular biology – to avoid contamination, prepare all solutions using pre-sterilised (autoclaved) deionised purified water (e.g. MilliQ water).

Extracting DNA from organisms other than bacteria – different homogenisation techniques will be required (see Chapter 41). DNA can be extracted from the resulting homogenate by (i) complexation with the detergent cetrylmethylammonium bromide (CTAB), e.g. for plant DNA, or (ii) anion-exchange column chromotography (p. 336). Many laboratories now use small-scale chromatographic columns for routine extraction, e.g. QIA prep Spin Miniprep columns.

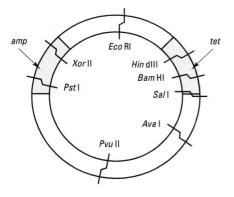

Fig. 68.1 Restriction map of the plasmid pBR322. The position of individual restriction sites is shown together with the genes for ampicillin resistance (*amp*) and tetracycline resistance (*tet*).

Definition

Type II restriction endonucleases – intracellular enzymes, produced by certain strains of bacteria: their function is to restrict the growth of phages within the cell by cutting phage DNA at particular sites (host DNA is protected by methylation at these sites).

enzyme to reach the peptidoglycan layer (p. 266), e.g. osmotic shock, or incubation with a chelating agent, e.g. ethylenediaminetetra-acetate (EDTA). The latter treatment will also inactivate any bacterial deoxyribonucleases (DNases) in the solution, preventing enzymic degradation of plasmid DNA during extraction.

2. **Lysis** using strong alkali (NaOH) and a detergent, e.g. sodium dodecyl sulfate, to solubilise the cellular membranes and partially denature the proteins. Neutralisation of this solution (e.g. using potassium acetate) causes the chromosomal DNA to aggregate as an insoluble mass, leaving the plasmid DNA in solution.

3. **Removal of other macromolecules,** particularly RNA and proteins, by, for example, enzymic digestion using ribonuclease and proteinase. Additional steps give further increases in purity: most laboratories now use small-scale chromatographic columns containing an anion exchange resin for routine extraction of nucleic acids. These are available in kit form, e.g. Wizard miniprep columns. This approach gives more reliable purification than traditional methods, such as phenol/chloroform extraction.

4. **Precipitation of DNA** by adding two volumes of absolute ethanol to one volume of aqueous extract, followed by centrifugation, to recover the DNA as a pellet. Further washing with 70%v/v ethanol will remove any residual salt from the previous stages. After removal of residual ethanol by evaporation, the purified DNA is dissolved in buffer and can be stored for months at –20 °C.

The simplest approach to quantifying the amount of nucleic acid in an aqueous solution is to measure the absorbance of the solution at 260 nm (A_{260}) using a spectrophotometer, as detailed on p. 313. Any contaminating protein would also contribute to A_{260}; protein can be detected by measuring the absorbance of the solution at 280 nm. Purified nucleic acids have a value for A_{260}/A_{280} of 1.8–2.0 and contaminating protein will give a lower ratio. If your solution gives a ratio substantially lower than this, you should carry out further purification steps.

Enzymatic manipulation of DNA

Restriction enzymes

Type II restriction endonucleases can be used to produce linear fragments of DNA with either single-stranded 'sticky' ends or 'blunt' ends (Fig. 66.1): most commercial preparations of restriction enzymes will completely cleave a sample of DNA, producing a 'restriction digest', within 1 h at 37 °C. The position of individual restriction sites can be used to create a diagnostic restriction map for a particular molecule, e.g. a plasmid cloning vector (Fig. 68.1). An important additional feature is that, under appropriate conditions, the sticky ends of any two restriction fragments cut with the same enzyme (Fig. 68.2) can anneal (base pair), because of the formation of hydrogen bonds between individual bases within this region, allowing them to be joined together (ligated), irrespective of the sources of the two restriction fragments. Use online tools to identify useful restriction enzymes for a particular target sequence (eg. see: http://nc2.neb.com/NEBcutter2/).

DNA ligase

To construct a recombinant DNA molecule, a suitable vector must be ligated to the DNA fragment to be cloned: this is performed using another

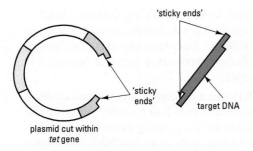

Fig. 68.2 Restriction of plasmid pBR322 and foreign DNA with *Hin* dIII.

Fig. 68.3 Annealing and ligation of plasmid pBR322 and target DNA to give a recombinant plasmid which confers resistance to ampicillin only. As target DNA has been inserted within the *tet* gene (Fig. 68.1), the gene is now discontinuous and inactive.

Understanding the nomenclature for plasmid vectors – pBR322 is named according to standard rules: 'p' indicates the vector is a plasmid, 'BR' identifies the researchers Bolivar and Rodriguez and '322' is the specific code given to this plasmid, to distinguish it from others developed by the same workers.

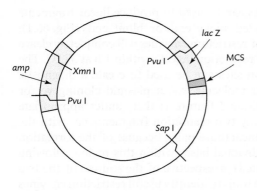

Fig. 68.4 Restriction map of the plasmid pUC19. The position of some individual restriction sites is shown together with the genes for ampicillin resistance (*amp*) and β-galactosidase (*lacZ*). MCS = multiple cloning site for 40 restriction enzymes within the *lacZ* gene.

microbial enzyme, DNA ligase (usually obtained from T4-infected *E. coli*). This ATP-dependent enzyme is capable of forming covalent phosphodiester bonds between annealed DNA molecules, thereby creating recombinant DNA (Fig. 68.3). When the two molecules involved are the cloning vector and the target DNA, the size of the recombinant molecule can be predicted (e.g. a plasmid of 4.5 kbp, plus a target DNA fragment of 2.5 kbp will give a recombinant molecule of 7 kbp), allowing separation and recovery by agarose gel electrophoresis (p. 459). Ligation is usually carried out at lower temperatures (to encourage annealing), over an extended time period of several hours (to allow the enzyme to operate), e.g. an overnight incubation at $\simeq 16°C$. The volume of the ligation mixture is kept as low as possible (typically, <10 µL), with approximately a 2 : 1 ratio of target DNA to vector DNA, to encourage annealing between the two different types of DNA molecule and to reduce the chance of circularisation of vector or target DNA. Treatment of target DNA with alkaline phosphatase reduces the risk of self-ligation.

Choosing a suitable vector

The features to be considered when selecting a suitable vector are:

- **the ease of purification** – e.g. reliable procedures have been developed to allow plasmid DNA to be extracted and purified from bacterial cells;
- **the efficiency of insertion** of the recombinant vector into a new host cell;
- **the presence of single copies of suitable restriction sites;**
- **the presence of selectable markers** (e.g. 'reporter genes', p. 478);
- **the size of the DNA insert to be cloned;**
- **the copy number of the vector** in the host cell.

Plasmids

The simplest bacterial cloning vectors are those based on small plasmids – circular DNA molecules, capable of autonomous replication within a bacterial cell. The general characteristics of plasmids are described in Chapter 65. One of the first cloning vectors to be developed for *E. coli* was pBR322, a genetically engineered plasmid of 4.4 kbp containing an origin of replication, two antibiotic resistance genes and single sites for a range of restriction enzymes (Fig. 68.1). Subsequently, other plasmid vectors have been developed, with additional features: for example, plasmids of the pUC series (Fig. 68.4) are now widely used for transformation of *E. coli,* having the following advantages over earlier types:

- **high copy number:** several hundred identical copies of the plasmid may be present in each cell, giving improved yield of plasmid DNA;
- **single-step selection of recombinants** using the *lacZ'* gene (p. 479);
- **clustering of restriction sites** within a short region of the *lacZ'* gene – restriction using two enzymes (a 'double digest') cuts a small fragment from this 'multiple cloning site' or 'polylinker', producing a cleaved plasmid with two different 'sticky ends' (e.g. *Eco* RI at one end and *Hin* dIII at the other), allowing complementary fragments of target DNA to be ligated in a particular orientation ('directional cloning').

The pET series of plasmid vectors (Novagen; Fig. 68.5) is widely used, since they offer several additional advantages, including:

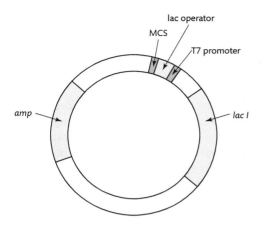

Fig. 68.5 Map of a pET plasmid vector (Novagen) showing the location of the genes for ampicillin resistance (*amp*) and lactose repressor protein (*lacI*) which controls expression of the T7 promoter via the lac operator (binding of the Lac repressor protein is countered by addition of isothiopropylgalactoside, IPTG). Expression of a gene cloned into the multiple cloning site (MCS) can thus be controlled, since DNA transcription only occurs when IPTG is added to the growth medium (inducible expression system).

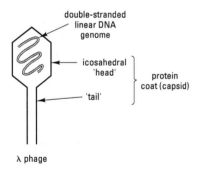

Fig. 68.6 Bacteriophage λ.

Definition

Phagemid – a recombinant entity created by fusion of a plasmid with the origin of replication of a phage: can be propagated and packaged into phage particles with the aid of a 'helper' phage.

- **high levels of protein expression,** as the cloned DNA is under the control of the T7 promoter and the lac operon system;
- **specialised vectors and hosts** for the production of soluble proteins, disulfide bond formation, protein export, etc.;
- **the attachment of a short histidine 'tag'** on the end of an expressed protein, to enable the protein to be purified by immobilised metal affinity chromatography (IMAC), as described on p. 339.

Phages

Bacteriophage vectors can offer several advantages over plasmid vectors:

- some phages have been engineered to carry larger fragments of foreign DNA – plasmid vectors work best with short inserts of 1–2 kbp, while some phage vectors can accept 10 kbp of foreign DNA;
- a recombinant phage has a ready-made mechanism for transferring rDNA to a new host cell: the process is broadly similar to phage transduction, described on p. 451.

A number of vectors have been developed from *E. coli* phage λ (genome = 49 kbp of dsDNA in linear form, Fig. 68.6). One example is λZAPII, an insertion vector capable of accepting up to 10 kbp of additional (foreign) DNA and with the *lacZ'* gene and multiple cloning site for selection. Another is λEMBL4, a replacement vector, where part of the vector DNA is removed and replaced by foreign DNA (up to 23 kbp) during cloning.

Most vectors based on λ are virulent phages, released by lysis of the infected cell at the end of the replication phase: such vectors are cloned using standard techniques for phages, e.g. as plaques of lysis on a 'lawn' of bacteria growing on an agar-based medium (p. 228). Note that suitable strains of *E. coli* must be used for each of the various cloning vectors, e.g. vectors containing *lacZ'* can only be used with those *E. coli* strains that are deficient in this part of their genome (*lacZ'* codes for the α-component of the enzyme β-galactosidase, present in wild-type *E. coli*). In some instances, a second phage (a 'helper phage') must be used along with the vector, to carry out some of the functions removed from the vector genome, e.g. phage packaging and assembly functions. In some instances, more specialised phage vectors may be required, e.g. single-stranded M13mp19, used for sequencing.

Cosmids and other vectors

Cosmid vectors contain the single-stranded cohesive terminal sequences from λ phage (so-called 'cos' sites) inserted into a plasmid cloning vector. They can be used to clone up to 40 kbp of foreign DNA, using the 'cos' sites to allow *in vitro* packaging into the λ phage coat: such large fragments are required for the production of gene libraries, as described below. Other hybrid vectors have been produced by the fusion of a phage and a plasmid vector – these phagemids can be manipulated as plasmids for ease of DNA uptake, or converted to phage-like form for ease of storage.

Creating a gene library

For viruses with small genomes, an individual gene may be identified by hybridisation of a single fragment with a suitable probe, following digestion

Screening clones – a particular gene can be detected using Southern blotting (p. 461), or the expressed gene product (protein) may be detected by a suitable method, e.g. an immunoassay (p. 282). While it is relatively simple to describe the screening process, finding a specific gene (= clone) in a clone library is likely to be a time-consuming process.

Using cloned cDNA – information from the sequence of a particular cDNA clone may be useful in designing a suitable oligonucleotide probe to locate a particular gene in a genomic library.

Fig. 68.7 General structure of a synthetic cationic lipid, e.g. Tfx.

Recognising transformants – after plating bacteria onto medium containing ampicillin, you may notice a few small 'satellite' colonies surrounding a single larger (transformant) colony. These satellite colonies are derived from non-transformed cells which survive to the breakdown of antibiotic in the medium around the transformant colony, and should not be selected for subculture.

with a particular restriction enzyme (Chapter 66). However, for most genomes, such a digest would contain a very complex mixture of fragments and a 'clone library' must be created. The isolation and identification of a particular gene are then carried out by screening individual clones from this library. Two types of clone library can be used.

1. A **genomic library** – prepared from the entire genome of the organism under study. The genome is fragmented to give overlapping fragments, e.g. by partial restriction (incomplete digestion, e.g. at low temperature), or by mechanical shearing. Individual fragments are then incorporated into a suitable vector, e.g. a cosmid, to create a vector library (e.g. a cosmid library). Each recombinant vector is then transferred to a separate host cell, which is cultured, giving a collection of transformants that represents the clone library. This approach is sometimes termed 'shotgun cloning'.

2. A **cDNA library** (cDNA = complementary, or copy DNA) – prepared by converting mRNA into DNA using the retroviral enzyme reverse transcriptase. Such a library consists only of genes expressed in those cells used to create the library, making individual genes easier to locate and identify, since the number of clones represented is usually smaller than for a genomic library. Another feature of cDNA is that it is complementary to the processed transcript (mRNA), so it contains no information on non-coding regions (introns) or transcriptional sequences (promoter regions, etc.).

Transferring rDNA to a suitable host cell

Once a recombinant vector has been produced *in vitro*, it must be introduced into a suitable host cell. Packaged phage vectors have a built-in transfer mechanism, while naked phage DNA and plasmids must be introduced by treatments that cause a temporary increase in membrane permeability resulting in transformation. These treatments include:

- **physicochemical shock treatment** – Box 68.1 gives details of a typical procedure using $CaCl_2$ and heat shock treatment to transform *E. coli* with the plasmid pUC19;

- **electroporation** – cells or protoplasts are subjected to electric shock treatment (typically, >10kV cm^{-1}) for very short periods (<10ms);

- **methods for animal and plant cells** – electroporation of protoplasts, or various micro-injection treatments, either using a microsyringe, DNA-coated microprojectiles ('biolistics') or cationic lipids (Fig. 68.7) which form lipid–DNA complexes that fuse with cell membranes and allow DNA entry. *Agrobacterium* can be used for rDNA transfer in certain plants.

Selection and detection of transformants

As the efficiency of transformation is very low, many of the plasmid vectors used in genetic engineering carry genes coding for antibiotic resistance, e.g. pBR322 carries separate genes for ampicillin resistance, *amp*, and tetracycline resistance, *tet* (Fig. 68.1). These genes act as 'markers' for the vector. One gene (e.g. *amp*) can be used to select for transformants, which would form

Table 68.1 Examples of genes used to detect transformants.

Gene	Product/assay
lacZ/lacZ'	β-galactosidase/chromogenic substrate (e.g. X-GAL) or fluorogenic substrate (e.g. MU-GAL)
uidA	β-glucuronidase/chromogenic substrate (e.g. X-GLUC) or fluorogenic substrate (e.g. MU-GLAC, p. 413)
lux	luciferase/bioluminescence in the presence of luciferin (p. 315)
bla/amp	β-lactamase/resistance to ampicillin
cat	chloramphenicol acetyltransferase/resistance to chloramphenicol
gfp	green fluorescent protein from *Aequorea victoria*, visualised under UV light

colonies on an agar-based medium containing the antibiotic, while non-transformed (ampicillin-sensitive) cells would be killed. The other gene (e.g. *tet*) can be used as a marker for the recombinant plasmid vector, since ligation of the target sequence into this gene causes insertional inactivation (Fig. 68.3). Thus, cells transformed with the recombinant plasmid will be resistant to ampicillin only, while cells transformed with the recircularised ('native') plasmid will be resistant to both antibiotics. These two types of transformant can be distinguished using replica plating (see p. 450). For those genes where the product is an enzyme, the presence or absence of the functional gene can be assessed using a suitable substrate. For example, the insertional inactivation of *lacZ'* can be detected by including a suitable inducer of β-galactosidase (e.g. isopropylthiogalactoside, IPTG) and the chromogenic substrate 5-bromo-4-chloro-3-indolyl-β-D-galactoside (X-GAL) within the agar medium: a transformant colony derived from the native plasmid will be blue-green while a transformant containing the recombinant molecule (inactive *lacZ'*) will grow to produce a white colony (so called 'blue-white screening'). A number of other examples are given in Table 68.1.

Box 68.1 Transformation of *E. coli* and selection of transformants

The following procedure illustrates the principal stages of the process: the efficiency of transformation will depend on the choice of *E. coli* strain and plasmid, and on the handling procedures prior to and during the process – for example, some strains do not require all of the stages listed below (for further details, see Hanahan *et al.*, 1995). Sterile equipment and appropriate technique (Chapter 35) are required at all times.

1. **Grow the cells under appropriate conditions:** the best results are often obtained using actively growing mid-log phase cells, rather than cells that have been grown to stationary phase (p. 231). Actively growing cells can be stored under appropriate conditions, e.g. as a 'frozen stock' – in suspension in concentrated glycerol at –85°C.

2. **Induce cell competence by transfer to a suitable sterile transformation buffer:** this would normally contain divalent and monovalent salts, e.g. a salt solution containing $CaCl_2$ at 50–100 mmol L^{-1}, plus KCl and $MnCl_2$ at 10–20 mmol L^{-1}. Cells are usually kept on ice in this solution for 10–15 min, to encourage the binding of plasmid DNA, added at a later stage.

3. **Add reagents to increase the permeability of cellular membranes:** typically dimethyl sulfoxide (DMSO, at up to 7%v/v) and dithiothreitol (DTT at up to 0.2 mmol L^{-1}), incubated on ice for a few minutes. DMSO is readily oxidised, reducing the effectiveness of the transformation procedure, and should be stored at –80°C when not in use, to minimise oxidation.

4. **Add plasmid DNA:** typically at 10–1000 ng per transformation. Maintain on ice for at least 10 min in a minimal volume of solution, to allow the plasmid to become associated with the cell surface.

5. **Heat shock the cells:** briefly raise the temperature to 42–45°C for 60–120s, then return to ice for a few minutes. It is important that this treatment gives a *rapid* change in temperature – use a small, thin-walled (disposable) sterile plastic tube containing the minimum volume of solution to maximise the rate of temperature change.

6. **Allow the cells to recover from heat shock and to express any new genes** (e.g. to synthesise enzymes conferring antibiotic resistance in transformed cells): add sterile nutrient broth (e.g. 1 mL) and then incubate at 37°C for up to 60 min.

7. **Plate onto an appropriate medium, to allow selection and detection of transformants:** for example, using pUC19, surface spread (p. 226) the suspension onto a medium containing ampicillin (at 25mg L^{-1}), plus IPTG (at 15mg L^{-1}) and X-GAL (at 25 mg L^{-1}), to detect the *lacZ'* gene product, as described in the text.

Text references

Hanahan, D., Jessee, J. and Bloom, F.R. (1995) 'Techniques for transformation of *E. coli*' in Glover, D.M. and Hames, B.D. *DNA Cloning I, Practical Approach Series*. IRL Press, Oxford.

Sources for further study

Brown, T.A. (2015) *Gene Cloning and DNA Analysis: an Introduction,* 7th edn. Wiley, New York.

Dale, J.W., Schantz, M.V. and Plaut, N. (2012) *From Genes to Genomes: Concepts and Applications of DNA Technology,* 3rd edn. Wiley, Chichester.

Hardin, C., Edwards, X., Riell, A., Presutti, D., Miller, W. and Robertson, D. (2001) *Cloning, Gene Expression, and Protein Purification: Experimental Procedures and Process Rationale.* Oxford University Press, Oxford.

[A practical handbook and laboratory manual, with basic experimental protocols together with underlying theoretical principles.]

Howe, C.J. (2007) *Gene Cloning and Manipulation,* 2nd edn. Cambridge University Press, Cambridge.

Lodge, X, Lund, P. and Michin, S. (2007) *Gene Cloning: Principles and Applications.* Taylor & Francis, Abingdon.

Peterson, H. (2013) *Gene Cloning and DNA Analysis: an Introduction.* Koros Press, London.

Russell, P. (2009) *iGenetics: a Molecular Approach,* 3rd edn. Benjamin Cummings, Harlow.

STUDY EXERCISES

68.1 Explain the terminology of restriction enzymes. Why were *Hin* dIII and *Eco* RI given these names?

68.2 Explain the purpose of the various stages involved in the extraction and purification of DNA. Having read through this chapter, briefly describe the main purpose of each of the following reagents within a DNA extraction protocol, and rearrange them into their order of use within the procedure:

 (a) anion exchange resin column chromatography;

 (b) incubation in 70% v/v ethanol: water;

 (c) ribonuclease and proteinase digestion;

 (d) incubation with lysozyme/EDTA;

 (e) sodium dodecyl sulfate incubation;

 (f) NaOH treatment, followed by neutralisation with potassium acetate.

68.3 Construct a restriction map. The figure alongside represents an electrophoretic separation of plasmid pRD, digested with different restriction enzymes, both individually and in combination. All of the enzymes cut the plasmid only once. Use this information to construct a simple restriction map of the plasmid. Key to lanes: 1 = DNA standards (0.5–8 kbp); 2 = *Eco* RI; 3 = *Hin* dIII; 4 = *Pvu* II; 5 = *Eco* RI and *Hin* dIII; 6 = *Eco* RI and *Pvu* II; 7 = *Hin* dIII and *Pvu* II; 8 = *Eco* RI, *Hin* dIII and *Pvu* II; 9 = DNA standards (0.5–8 kbp).

68.4 Calculate transformation efficiency. A volume of 100 µl of *E. coli* cells was transformed using 20 µL of plasmid DNA (prepared at 0.1ng µL^{-1}) and then made up to a final volume of 500 µL in buffer. A 10-fold dilution of this suspension was prepared and 200 µL of this dilution was surface spread (p. 226) onto a suitable medium containing an antibiotic to select for transformants, giving 180 colonies on the spread plate after overnight incubation. What is the transformation efficiency, expressed as the number of transformants (colony-forming units, CFU) per µg DNA? (Give your answer in exponential notation, to three significant figures.)

Answers to these study exercises are available at www.pearsoned.co.uk/practicalskills.

Analysis and presentation of data

The process of discovering the meaning within your results can be fascinating. There are two main elements to this process.

1. **Exploratory data manipulation** – this is used to investigate the nature of your results and suggest possible patterns and relationships within the data set. The aim is to generate hypotheses for further investigation. Exploratory techniques allow you to visualise the form of your data. They are ideal for examining results from pilot 'studies', but should be used throughout your investigations.

2. **Confirmatory analysis** – this is used to test the hypotheses generated during the exploratory phase. The techniques required are generally statistical in nature and are dealt with in Chapter 74.

> **KEY POINT** Spreadsheets (Chapter 13) are invaluable tools for data manipulation and transformation: complex mathematical procedures can be carried out rapidly and the results visualised almost immediately using the inbuilt graphing functions. Spreadsheets also facilitate the statistical analysis of data (see Chapters 73 and 74).

Organising numbers

To organise, manipulate and summarise data, you should:

- **simplify the numbers,** e.g. by rounding or taking means. This avoids the detail becoming overwhelming;
- **rearrange your data** in as many ways as possible for comparison;
- **display in graphical form;** this provides an immediate visual summary that is relatively easy to interpret;
- **look for an overall pattern** in the data – avoid getting lost in the details at this stage;
- **look for any striking exceptions to that pattern (outliers)** they often point to special cases of particular interest or to errors in the data produced through mistakes during the acquisition, recording or copying of data;
- **move from graphical interpretations to well-chosen numerical summaries** and/or verbal descriptions, including, where applicable, an explanatory hypothesis.

After collecting data, the first step is often to count how frequently each value occurs and to produce a frequency table. The frequency is simply the number of times a value occurs in the data set, and is, therefore, a count. The raw data could be acquired using a tally chart system to provide a simple frequency table. To construct a tally chart (e.g. Fig. 69.1):

- **enter only one tally at a time;**
- **if working from a data list, cross out each item on the list as you enter it onto the tally chart,** to prevent double entries;
- **check that all values are crossed out at the end** and that the totals agree.

Summarising your results– original, unsummarised data belong only in your primary record, either in laboratory books or as computer records. You should produce summary tables to organise and condense original data.

Colour	Tally	Total
Green	III	3
Blue	HHI III	8
Red	IIII	4
White	HHI HHI II	12
Black	I	1
Maroon	III	3
Yellow	II	2
		33

Fig. 69.1 An example of a tally chart.

Producing a histogram– a *neatly* constructed tally chart will double as a rough histogram.

Table 69.1 An example of a frequency table.

Size class	Frequency	Relative frequency (%)
0–4.9	7	2.6
5–9.9	23	8.6
10–14.9	56	20.9
15–19.9	98	36.7
20–24.9	50	18.7
25–29.9	30	11.2
30–34.9	3	1.1
Total	267	99.8*

*≠ 100 because of rounding error.

stem	leaves
7	23
7	55
7	6
7	9
8	000
8	233
8	45555
8	77
8	888899
9	0000111111
9	2333333
9	44555555555
9	66677777
9	88888999
10	00

Fig. 69.2 A simple 'stem and leaf' plot of a data set. The 'stem' shows the common component of each number, while the 'leaves' show the individual components, e.g. the top line in this example represents the numbers 72 and 73.

Convert the data to a formal table when complete (e.g. Table 69.1). Because proportions are easier to compare than class totals, the table may contain a column to show the relative frequency of each class. Relative frequency can be expressed in decimal form (as a proportion of 1) or as a percentage (as a proportion of 100).

Graphing data

Graphs are an effective way to investigate trends in data and can reveal features that are difficult to detect from a table, e.g. skewness of a frequency distribution. The construction and use of graphs is described in detail in Chapter 70. When investigating the nature of your data, the main points are as follows:

- **make the values stand out clearly:** attention should focus on the actual data, not the labels, scale markings, etc. (contrast with the requirements for constructing a graph for data presentation; see p. 489);
- **avoid clutter in the graph:** leave out grid lines and try to use the simplest graph possible for your purpose.

> **KEY POINT** Use a computer spreadsheet with graphics options whenever possible: the speed and flexibility of these powerful tools should allow you to explore every aspect of your data rapidly and with relatively little effort (see Chapters 13 and 72).

Displaying distributions

A visual display of a distribution of values is often useful for variables measured on an interval or ratio scale (p. 192). The distribution of a variable can be displayed by a frequency table for each value or, if the possible values are numerous, groups (classes) of values of the variable. Graphically, there are two main ways of viewing such data:

1. **Histograms** (see pp. 489, 492), generally used for large samples.
2. **Stem and leaf plots** (e.g. Fig. 69.2), often used for samples of less than 100: these retain the actual values and are faster to draw by hand. The main drawback is the limitation imposed by the choice of stem values since these class boundaries may obscure some features of the distribution.

These displays allow you to look at the overall shape of a distribution and to observe any significant deviations from the idealised theoretical ones. Where necessary, you can use data transformations to investigate any departures from standard distribution patterns such as the normal distribution or the Poisson distribution (p. 527).

Transforming data

Transformations are mathematical functions applied to data values. They are particularly valuable where your results are related to areas and volumes (e.g. leaf area, body mass).

The most common use of transformations is to prepare data sets so that specific statistical tests may be applied. For instance, if you find that your data distribution is unimodal but not symmetrical (p. 521), it is often useful to apply a transformation that will redistribute the data values to

Using transformations– note that if you wish to conserve the order of your data, you will need to take negative values when using a reciprocal function (i.e. − $1/n^n$). This is essential when using a box plot to compare graphically the effects of transformations on the five-number summary of a data set.

form a symmetrical distribution. The object of this exercise is often to find the function that most closely changes the data into a standard normal distribution, allowing you to apply a wide range of parametric hypothesis-testing statistics (see Chapter 74). A frequently used transformation is to take logarithms of one or more sets of values: if the data then approximate to a normal distribution, the relationship is termed 'log-normal'. Some general points about transformations are:

- **they should be made on the raw data, not on derived data values:** this is simpler, mathematically valid and more easily interpreted;
- **the transformed data can be analysed like any other numbers;**
- **transformed data can be examined for outliers,** which may be more important if they remain after transformation.

Figure 69.3 presents a ladder of transformations that will help you decide which transformations to try (see also Table 74.1). Note that percentage and proportion data are usually arc-sine transformed, which is a more complex procedure; consult Sokal and Rohlf (2011) for details.

Figure 69.4 illustrates the following 'quick-and-easy' way to choose a transformation:

1. **Calculate the 'five-number summary'** for the untransformed data (p. 519).
2. **Present the summary graphically** as a 'box and whisker' plot (p. 519).
3. **Decide whether you need to correct for positive or negative skew (p. 514).**
4. **Apply one of the 'mild' transformations in Fig. 69.3** *on the five-number summary values only.*
5. **Draw a new box and whisker plot** and see whether the skewness has been corrected.
6. **If the skewness has been undercorrected, try again** with a stronger transformation. If it has been overcorrected, try a milder one.
7. **When the distribution appears to be acceptable, transform the full data set** and recalculate the summary statistics. If necessary, use a statistical test to confirm that the transformed data are normally distributed (p. 526).
8. **If no simple transformation works well, you may need to use non-parametric statistics when comparing data sets.**

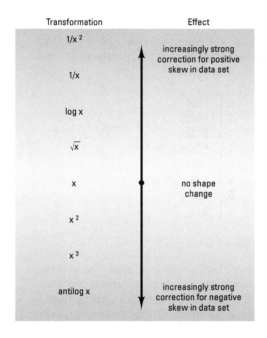

Fig. 69.3 Ladder of transformations (after J.W. Tukey).

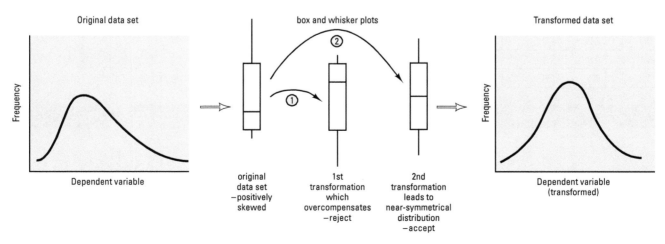

Fig. 69.4 Illustration of the processes of transforming a data set.

Text references and sources for further study

Glass, D.J. (2006) *Experimental Design for Biologists.* Cold Spring Harbor Laboratory Press, Cold Spring Harbor.

Quinn, G.P. and Keough, M.J. (2002) *Experimental Design and Data Analysis for Biologists.* Cambridge University Press, Cambridge.

Sokal, R.R. and Rohlf, F.J. (2011) *Biometry,* 4th edn. W.H. Freeman, San Francisco.

STUDY EXERCISES

The following study exercises can be most easily carried out using a spreadsheet such as Microsoft *Excel*.

69.1 Compare frequency distributions for samples with different sample sizes. Compute relative frequencies for each class in the table below. Graph the data as a frequency polygon.

Frequency distribution data.

Hyphal length (μm)	Treatment B	Treatment A	Hyphal length (μm)	Treatment B	Treatment A
0	0	0	13	6	30
1	1	0	14	3	26
2	2	0	15	1	22
3	4	1	16	1	18
4	8	1	17	0	14
5	13	2	18	0	12
6	22	5	19	0	8
7	36	9	20	0	5
8	48	14	21	0	4
9	46	21	22	0	2
10	32	26	23	0	2
11	18	31	24	0	1
12	11	33	25	0	1

69.2 Use a stem and leaf plot. Work out the mean and standard deviation of the data contained in the following stem and leaf plot:

6	1334
6	56788
7	013344
7	556667899
8	11224
8	569
9	03
9	6

Stem and leaf plot.

69.3 Use transformations to correct skew in a data set. Find a transformation that will make the data of treatment B in Study exercise 69.1 approximately symmetrical about the mean. Demonstrate graphically that you have accomplished this.

Answers to these study exercises are available at www.pearsoned.co.uk/practicalskills.

Fig. 70.1 Effect of antibiotic on yield of two bacterial isolates: ○, sensitive isolate; □, resistant isolate. Vertical bars show standard errors ($n = 6$).

Criteria for preparing graphs
- **Clarity** – show the relationships within the data.
- **Economy** – leave out superfluous information/text.
- **Integrity** – present the data without trying to mislead.

Selecting a title – it is a common fault to use titles that are grammatically incorrect. A widely applicable format is to state the relationship between the dependent and independent variables within the title, e.g. 'The relationship between enzyme activity and external pH'.

Remembering which axis is which – a way of remembering the orientation of the *x* axis is that 'x' is 'a cross', and it runs 'across' the page (horizontal axis) while 'y' is the first letter of 'yacht', with a large vertical mast (vertical axis).

Graphs can be used to show detailed results in an abbreviated form, displaying the maximum amount of information in the minimum space. Graphs and tables present findings in different ways. A graph (figure) gives a visual impression of the content and meaning of your results, while a table provides an accurate numerical record of data values. You must decide whether a graph should be used, e.g. to illustrate a pronounced trend or relationship, or whether a table (Chapter 71) is more appropriate.

A well-constructed graph will combine simplicity, accuracy and clarity. Planning of graphs is needed at the earliest stage in any write-up as your accompanying text will need to be structured so that each graph delivers the appropriate message. Therefore, it is best to decide on the final form for each of your graphs before you write your text. The text, diagrams, graphs and tables in a laboratory write-up or project report should be complementary, each contributing to the overall message. In a formal scientific communication it is rarely necessary to repeat the same data in more than one place (e.g. as a table and as a graph). However, graphical representation of data collected earlier in tabular format may be applicable in laboratory practical reports.

Practical aspects of graph drawing

The following comments apply to graphs drawn for laboratory reports. Figures for publication, or similar formal presentation, are usually prepared according to specific guidelines, provided by the publisher/organiser.

> **KEY POINT**   Graphs should be self-contained – they should include all material necessary to convey the appropriate message without reference to the text. Every graph must have a concise explanatory title to establish the content. If several graphs are used, they should be numbered so they can be quoted in the text.

- **Consider the layout and scale of the axes carefully.** Most graphs are used to illustrate the relationship between two variables (*x* and *y*) and have two axes at right angles (e.g. Fig. 70.1). The horizontal axis is known as the abscissa (*x* axis) and the vertical axis as the ordinate (*y* axis).
- **The axis assigned to each variable must be chosen carefully.** Usually the *x* axis is used for the independent variable (e.g. treatment) while the dependent variable (e.g. biological response) is plotted on the *y* axis (p. 201). When neither variable is determined by the other, or where the variables are interdependent, the axes may be plotted either way round.
- **Each axis must have a descriptive label showing what is represented,** together with the appropriate units of measurement, separated from the descriptive label by a solidus or 'slash' (/), as in Fig. 70.1, or by brackets, as in Fig. 70.2.
- **Each axis must have a scale with reference marks** ('tics') on the axis to show clearly the location of all numbers used.
- **A figure legend should be used to provide explanatory detail,** including a key to the symbols used for each data set.

Fig. 70.2 Frequency distribution (histogram) of masses for a sample of animals (sample size 24085); the size class interval is 2 g.

Example For a data set where the smallest number on the log axis is 12 and the largest number is 9000, three-cycle log-linear paper would be used, covering the range 10–10 000.

Choosing between a histogram and a bar chart – use a histogram for *continuous quantitative* variables and a bar chart for *discrete* variables (see Chapter 30 for details of these types of measurement scale).

Handling very large or very small numbers

To simplify presentation when your experimental data consist of either very large or very small numbers, the plotted values may be the measured numbers multiplied by a power of 10: this multiplying power should be written immediately before the descriptive label on the appropriate axis (as in Fig. 70.2). However, it is often better to modify the primary unit with an appropriate prefix (p. 195) to avoid any confusion regarding negative powers of 10.

Size

Remember that the purpose of your graph is to communicate information. It must not be too small, so use at least half an A4 page and design your axes and labels to fill the available space without overcrowding any adjacent text. If using graph paper, remember that the white space around the grid is usually too small for effective labelling. The shape of a graph is determined by your choice of scale for the x and y axes which, in turn, is governed by your experimental data. It may be inappropriate to start the axes at zero (e.g. Fig. 70.1). In such instances, it is particularly important to show the scale clearly, with scale breaks where necessary, so the graph does not mislead. Note that Fig. 70.1 is drawn with 'floating axes' (i.e. the x and y axes do not meet in the lower left-hand corner), while Fig. 70.2 has clear scale breaks on both x and y axes.

Graph paper

In addition to conventional linear (squared) graph paper, you may need the following:

- **Probability graph paper.** This is useful when one axis is a probability scale (e.g. p. 529).
- **Log-linear graph paper.** This is appropriate when one of the scales shows a logarithmic progression, e.g. the exponential (log) growth of cells in liquid culture (p. 231). Log-linear paper is defined by the number of logarithmic divisions (usually termed 'cycles') covered (e.g. Fig. 70.3), so make sure you use a paper with the appropriate number of cycles for your data. An alternative approach is to plot the log-transformed values on 'normal' graph paper.
- **Log-log graph paper.** This is appropriate when both scales show a logarithmic progression.

Fig. 70.3 Representation of three-cycle log-linear graph paper, marked up to show a y axis (log) scale from 10 to 10 000 and an x axis (linear) scale from 0 to 10.

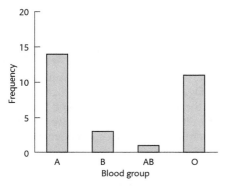

Fig. 70.4 Bar chart, showing the number of students belonging to each ABO blood group (*n* = 29).

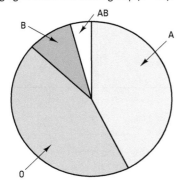

Fig. 70.5 Pie chart: relative abundance of human blood groups.

Choosing graphical symbols – plotted curves are usually drawn using a standard set of symbols: ●, ○, ■, □, ▲, △, ◆, ◇. By convention, paired symbols ('closed' and 'open') are often used to represent 'plus' (treatment) and 'minus' (control) treatments.

Types of graph

Different graphical forms may be used for different purposes.

- **Plotted curves** – used for data where the relationship between two variables can be represented as a continuum (e.g. Fig. 70.1).
- **Scatter diagrams** – used to visualise the relationship between individual data values for two interdependent variables (e.g. Fig. 74.7) often as a preliminary part of a correlation analysis (p. 533).
- **Three-dimensional graphs** show the interrelationships of three variables, often one dependent and two independent (e.g. Fig. 51.1). A contour diagram is an alternative method of representing such data.
- **Histograms** represent frequency distributions of continuous variables (e.g. Fig. 70.2).
- **Frequency polygons** emphasise the form of a frequency distribution by joining the coordinates with straight lines, in contrast to a histogram. This is particularly useful when plotting two or more frequency distributions on the same graph.
- **Bar charts** represent frequency distributions of a discrete qualitative or quantitative variable (e.g. Fig. 70.4). An alternative representation is the line chart (e.g. Fig. 74.3).
- **Pie charts** illustrate portions of a whole (e.g. Fig. 70.5).

The plotted curve

This is the commonest form of graphical representation used in biology. The key features are outlined below and in checklist form in Box 70.1, while Box 70.2 advises on using Microsoft *Excel*.

Data points

Each data point must be shown accurately, so that any reader can determine the exact values of *x* and *y*. The results of each treatment must be readily identifiable. A useful technique is to use a dot for each data point, surrounded by a hollow symbol for each treatment (see Fig. 70.1) or to use symbols only (Fig. 65.6), though the coordinates of each point are defined less accurately. Use the same symbol for the same entity if it occurs in several graphs and provide a key to all symbols.

Box 70.1 **Checklist for the stages in drawing a graph**

The following sequence can be used whenever you need to construct a plotted curve: it will need to be modified for other types of graph.

1. Collect all of the data values and statistical values (in tabular form, where appropriate).
2. Decide on the most suitable form of presentation: this may include transformation to convert data to linear form.
3. Choose a concise descriptive title, together with a reference (figure) number and date, where necessary.
4. Determine which variable is to be plotted on the *x* axis and which on the *y* axis.
5. Select appropriate scales for both axes and make sure that the numbers and their location (scale marks) are clearly shown, together with any scale breaks.
6. Decide on appropriate descriptive labels for both axes, with SI units of measurement, where appropriate.
7. Choose the symbols for each set of data points and decide on the best means of representation for statistical values.
8. Plot the points to show the coordinates of each value with appropriate symbols.
9. Draw a trend line for each set of points. Use a transparent ruler, so you can draw the line to have an equal number of points on either side of it.
10. Write a figure legend, to include a key which identifies all symbols and statistical values and any descriptive footnotes.

Box 70.2 How to create and amend graphs within a spreadsheet (Microsoft *Excel*) for use in coursework reports and dissertations

Microsoft *Excel* can be used to create graphs of reasonable quality, as long as you know how to amend the default settings so that your graph meets the formal standards required for practical and project reports. As with a hand-drawn graph, the basic stages in graph drawing (Box 70.1) still apply. The following instructions explain how to produce an *X–Y* graph (plotted curve, p. 489), bar graph (p. 489), pie graph and histogram using *Excel*, where all types of graphs are termed 'charts'. Earlier versions use broadly similar commands, though not always in the same locations within the software.

Producing an X-Y graph (*Scatter* chart in *Excel*)

1. **Create the appropriate type of graph for your data.** Enter the numeric values for your *X* variable data in the cells of a single column and the equivalent values for the *Y* variable in the adjacent cells of the next column to the right. Then select the whole data array (highlight the appropriate cells by clicking and holding down the left mouse button and dragging the cursor across the cells so that all values are included). Then select the *Insert* tab at the top of the sheet, and select (left-click) the *Scatter* chart from the options provided in the upper ribbon. Note that you should never use the *Line* chart option, as it is based on an *x* axis that is not quantitative, so all *X* values will appear as equally spaced categories, rather than having a true scale. Select the first option from the *Scatter* menu (*Scatter with only Markers*). Once selected, this will produce an embedded scatter chart of the type shown in Fig. 70.6(a). The line is then added later, as described below.

2. **Change the default settings to improve the appearance of your graph.** Consider each element of the image in turn, including the overall size, height and width of the graph (resize by clicking and dragging one of the 'sizing handles' around the edge of the chart). The graph shown in Fig. 70.6(b) was produced by altering the default settings, typically by moving the cursor over the feature and then clicking the right mouse button to reveal an additional menu of editing and formatting options. (Note that the example given below is for illustrative purposes only, and should not necessarily be regarded as prescriptive.)

Example for an X–Y graph (compare Fig. 70.6(a) with Fig. 70.6(b)):

- Unnecessary legend box on right-hand side can be removed using the (right-click) *Delete* option.
- Border to chart can be removed using *Format Chart Area* function (available by right-clicking within chart area).

- Gridlines can either be removed, using the *Delete* function or, if desired (as in Fig. 70.6(b)), changed by clicking on each axis and using the *Add Minor Gridlines* and *Format Gridlines* options to alter the *Colour* and *Style* of the gridlines to make them more like those of conventional graph paper.

- *x* and *y* axes can be reformatted by selecting each in turn, and using the *Format Axis* menu options to select appropriate scales for major and minor units, style of tick marks, etc. Remember that it is better to use a figure legend in *Word* (Chapter 13), rather than the *Chart Title* option within *Excel*.

(a)

(b)

Fig. 70.6 Examples of a plotted curve produced in Microsoft *Excel* using (a) default settings and (b) modified (improved) settings.

(continued)

Box 70.2 (continued)

- *x* and *y* axis labels can be added by selecting the *Layout* tab, then *Axis Titles*, then *Primary Horizontal Axis Title* and *Primary Vertical Axis Title* options, which will produce a text box beside each axis into which can be typed the axis label and any corresponding units. This can then be changed from the default font using the *Home* tab options.

- Data point style can be changed by selecting (right-clicking) any data point and following the *Format Data Series* options to choose appropriate styles, colours and fill of the data markers.

- A straight line of best fit can be added by selecting any data point and using the *Add Trendline* option to choose a *Linear* line type with appropriate colour and style (explore other options within the *Format, Layout* and *Design* tabs at the top of the worksheet).

Producing a bar graph (*Column* chart in *Excel*)

1. **Create the appropriate type of graph for your data.** Enter the category names (for *x* axis) in one column and the numeric values (for *y* axis) in the next column. Select (highlight) the data array, then select the *Insert* tab, and choose the *Column* chart from the options provided. For a standard bar graph, select the first option from the *Column* menu (*Clustered Column*). Once selected, this will produce an embedded bar graph of the type shown in Fig. 70.7(a).

2. **Change the default settings to improve the appearance of your graph.** The bar graph shown in Fig. 70.7(b) was produced by selecting each feature and altering the default settings, as detailed below (illustrative example).

Example for a bar graph (compare Fig. 70.7(a) with Fig. 70.7(b)):

- Unnecessary legend box on right-hand side can be removed using the *Delete* option.

- Border to chart can be removed using *Format Chart Area* function available by right-clicking within chart area.

- Gridlines can either be removed, using the *Delete* function or changed by selecting the gridlines and using the *Format Gridlines* option to alter the *Colour* and *Style*.

- The *y* axis can be reformatted by selecting the axis, then using the *Format Axis* menu options to select appropriate scales, tick marks, line colour, etc. Note that the *x* axis should already contain category labels from the spreadsheet cells (modify original cells to update spreadsheet, if necessary).

- *x* and *y* axis labels can be added by selecting the *Layout* tab, then *Axis Titles*, then *Primary Horizontal Axis Title* and *Primary Vertical Axis Title* options, as detailed for the plotted curve (Fig. 70.6).

- Bar colour can be modified using the *Format Data Series* menu, selecting appropriate *Fill* and *Border* colours, e.g. white and black, respectively, in Fig. 70.7(b).

- Individual *Y* data values can be shown using the *Add Data Labels* option (other options and adjustments can be made using the *Format, Layout* and *Design* tabs at the top of the worksheet).

Note that for all types of graph, it is better not to use the *Chart Title* option within *Excel*, which places the title at the

(a)

(b)

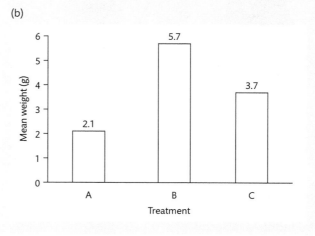

Fig. 70.7 Examples of a bar graph produced in Microsoft *Excel* using (a) default settings and (b) modified (improved) settings.

(*continued*)

Box 70.2 (continued)

top of the chart (as in Fig. 70.7(a)), but to copy and paste your untitled graph into a word-processed document, such as a Microsoft *Word* file (p. 493 gives details of the procedure), and then type a formal figure legend below the graph, as in Fig. 70.6(b) and Fig. 70.7(b). However, once your graph is embedded into a *Word* file, it is generally best not to make further amendments – you should go back to the original *Excel* file, make the required changes and then repeat the copy-paste procedure to reinsert the graph into the *Word* file.

Producing a pie graph (*Pie* chart in *Excel*)

1. **Create the appropriate type of graph for your data.** Enter the category names for each part of the pie chart in one column, and the corresponding numbers (counts, fractions or percentages) in the next column. Select (highlight) the data array, then select (left-click) the *Insert* tab, and choose the *Pie* chart from the options provided. For a standard pie graph, select the first option from the menu (*Pie*). Once selected, this will produce an embedded pie graph.

2. **Change the default settings to improve the appearance of your graph.** For example, you can show the data values (*Data labels*), adjust colours and shading (e.g. switch from multi-colour to shades of grey), remove chart border, etc., as required (an illustrative example is shown as Fig. 70.8).

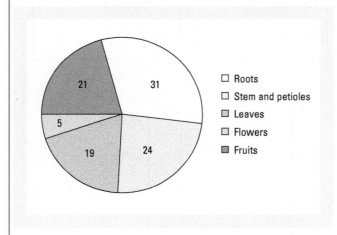

Fig. 70.8 Example of a pie graph produced in Microsoft *Excel* using format modified from the default settings.

Producing a histogram

The histogram function in *Excel* requires a little more effort to master, compared with other chart types. Essentially, a histogram is a graphical display of frequencies (counts) for a continuous quantitative variable (p. 191), where the data values are grouped into classes. It is possible to select the upper limit for each class into which you want the data to be grouped (these are termed 'bin range values' in *Excel*). An alternative approach is to let the software select the class intervals (bin range values) for you: *Excel* selects evenly distributed bins between the minimum and maximum values. However, this is often less effective than selecting your own class intervals.

To use the *Excel* histogram function, you will need to make sure that the *Analysis ToolPak* is loaded (check under the *Data* tab and, if not loaded, select via the *Office Button* > *Excel Options* > *Add-ins* > *Manage Excel Add-ins* > *Analysis ToolPak*. The *Histogram* function should then be available in the *Analysis ToolPak* option on the *Data* tab/ribbon.

The following steps outline the procedure used to create the histograms shown in Fig. 70.9 for the table of data below (length, in mm, of 24 leaf petioles from a single plant).

7.2	6.5	7.1	8.5	6.6	7.2
7.0	7.3	8.6	9.1	7.5	8.3
7.1	5.7	7.3	7.6	6.9	7.1
8.3	7.6	5.4	8.6	7.9	8.0

1. **Enter the raw data values,** e.g. as a single column of numbers, or as an array, as above.

2. **Decide on the class intervals to be used.** Base your choice on the number of data points and the maximum and minimum values (for a small data set such as that shown above, you can do this by visual examination, whereas for a large data set, use the *Excel* functions COUNT, MAX and MIN (find these under the Σ symbol (*More Functions*) on the *Editing* section on the *Home* tab/ribbon, or use the *Descriptive Statistics* > *Summary Statistics* option of the *Data Analysis* component on the *Data* tab/ribbon). A typical histogram would have 4–10 classes, depending on the level of discrimination required. Enter the upper limit for each class (*Bin Range Values*) in ascending order in a separate array of cells, e.g. in a column close to the data values (in the above example, 6, 7, 8 and 9 were chosen – the few data values above the final bin value will be shown on the histogram as a group labelled 'more').

3. **Select the histogram function, then input your data and class interval values.** From the *Data* tab/ribbon, select *Data Analysis* > *Histogram*. A new window will open: input your data into the *Input Range* box (highlight the appropriate cells by clicking on the first data value and dragging to the final data value while holding down the left mouse button). Next, input the *Bin Range* values into

(continued)

Box 70.2 (continued)

the appropriate box (if this is left empty, *Excel* will select default bin range values). Most of the remaining boxes can be left empty, though you must click the last box to get a *Chart Output*, otherwise the software will give the numerical counts for each group, without drawing a histogram. Click *OK* and entries will be created within a new worksheet, showing the upper limits of each group (in a column labelled *Frequency*), plus a poorly constructed chart based on *Excel* default settings, as shown in Fig. 70.9(a) (note that the default output is a bar chart, rather than a histogram, since there are gaps between the groups).

Example for a histogram (compare Fig. 70.9(a) with Fig. 70.9(b)):

- Chart can be resized to increase height, using the 'sizing handles' at the edges of the chart and border line around graph can be removed using *Format Chart Area* > *Border Color* > *No line* (menu available by right-clicking within chart area).
- Title and unnecessary legend box can be removed using the *Delete* option.
- Axis scales can be reformatted using the *Format Axis* menu options (e.g. scales, tick marks, line colour).
- The *x* axis labels (class intervals) can be amended by typing directly into the cells containing the bin range values.
- Axis titles can be changed by typing directly into the axis title box (double-click to access).
- Bar colour can be changed (e.g. to grey, with black outline) using the *Format Data Series* > *Fill* and *Border Colour* options.
- Bar chart converted to correct histogram format (no gaps between bars) using the *Format Data Series* > *Series Options*, setting *Gap Width* to 0%.
- Figure legend can be added below figure in Microsoft *Word*, following copying and pasting of the *Excel* histogram into a *Word* file.

Importing an *Excel* chart into a *Word* document

One problem encountered with current Microsoft *Office* products (but not with earlier versions, e.g. *Office 1997–2003*) is that the standard *Cut* > *Paste* procedure gives a poor quality figure, with grainy appearance and fuzzy lines/text; similar problems occur using the *Insert* tab/ ribbon in *Word*. The simplest approach is to follow the step-wise procedure below:

1. Select your *Excel* chart: right-click outside the chart itself, near to the edge, then choose the *Copy* option from the drop-down menu.

(a)

(b)

Fig. 70.9 Examples of histogram output from Microsoft *Excel* using (a) default settings and (b) modified (improved) settings.

2. Open your *Word* file, go to the *Home* tab/ ribbon and select *Paste Special* > *Microsoft Excel Chart* from the options.

3. This should give a graph with the same crisp axis/ line/ text formatting as the original chart in *Excel*.

The alternative approach is to use *Excel* to print the entire graph (chart) as a single sheet, and then add this to the printout from your word-processed document. However, the disadvantage with this approach is that you cannot produce a professional looking figure legend below your graph.

Note: instructions illustrated here may vary among the different versions of Microsoft *Office* programs. Check the exact functions and syntax using the *Insert function* menu option (*fx* button in some versions).

Statistical measures

If you are plotting average values for several replicates and if you have the necessary statistical knowledge, you can calculate the standard error (p. 520) or the 95% confidence limits (p. 533) for each mean value and show these on your graph as a series of vertical bars (see Fig. 70.1). Make it clear in the legend whether the bars refer to standard errors or 95% confidence limits and quote the value of n (the number of replicates per data point). Another approach is to add a least significant difference bar (p. 530) to the graph.

Interpolation

Once you have plotted each point, you must decide whether to link them by straight lines or a smoothed curve. Each of these techniques conveys a different message to your reader. Joining the points by straight lines may seem the simplest option, but may give the impression that errors are very low or non-existent and that the relationship between the variables is complex. Joining points by straight lines is appropriate in certain graphs involving time sequences (e.g. the number of animals at a particular site each year), or for repeat measurements where measurement error can be assumed to be minimal (e.g. recording a patient's temperature in a hospital, to emphasise any variation from one time point to the next). However, in most plotted curves the best straight line or curved line should be drawn (according to appropriate mathematical or statistical models, or by eye) to highlight the relationship between the variables – after all, your choice of a plotted curve implies that such a relationship exists. Do not worry if some of your points do not lie on the line: this is caused by errors of measurement and by biological variation. Most curves drawn by eye should have an equal number of points lying on either side of the line. You may be guided by 95% confidence limits, in which case your curve should pass within these limits wherever possible.

Curved lines can be drawn using a flexible curve, a set of French curves or freehand. In the latter case, turn your paper so that you can draw the curve in a single, sweeping stroke by a pivoting movement at the elbow (for larger curves) or wrist (for smaller ones). Do not try to force your hand to make complex, unnatural movements as the resulting line will not be smooth.

Extrapolation

Be wary of extrapolation beyond the upper or lower limit of your measured values. This is rarely justifiable and may lead to serious errors. Whenever extrapolation is used, a dotted line ensures that the reader is aware of the uncertainty involved. Any assumptions behind an extrapolated curve should also be stated clearly in your text.

The histogram

While a plotted curve assumes a continuous relationship between the variables by interpolating between individual data points, a histogram involves no such assumptions. Histograms are also used to represent frequency distributions (pp. 488, 514), where the y axis shows the number of times a particular value of x was obtained (e.g. Fig. 70.2). As in a plotted curve, the x axis represents a continuous quantitative variable which can take any value within a given range (e.g. plant height), so the scale must be broken down into discrete classes and the scale marks on the *x* axis should show either the mid-points (mid-values) of each class, or the boundaries between the classes.

The columns are contiguous (adjacent to each other) in a histogram, in contrast to a bar chart, where the columns are separate because the *x* axis of a bar chart represents discrete values.

Interpreting graphs

The process of analysing a graph can be split into five phases:

1. **Consider the context.** Look at the graph in relation to the aims of the study in which it was reported. Why were the observations made? What hypothesis was the experiment set up to test? This information can usually be found in the Introduction or Results section of a report. Also relevant are the general methods used to obtain the results. This might be obvious from the figure title and legend, or from the Materials and Methods section.

2. **Recognise the graph form and examine the axes.** First, what kind of graph is presented (e.g. histogram, plotted curve)? You should be able to recognise the main types summarised on p. 489 and their uses. Next, what do the axes measure? You should check what quantity has been measured in each case and what units are used.

3. **Look closely at the scale of each axis.** What is the starting point and what is the highest value measured? For the *x* axis, this will let you know the scope of the treatments or observations (e.g. whether they lasted for 5 min or 20 years; whether a concentration span was two-fold or 50-fold). For each axis, it is especially important to note whether the values start at zero; if not, then the differences between any treatments shown may be magnified by the scale chosen (see Box 70.3).

4. **Examine the symbols and curves.** Information will be provided in the key or legend to allow you to determine to what these refer. If you have made your own photocopy of the figure, it may be appropriate to note this directly on it. You can now assess what appears to have happened. If, say, two conditions have been observed while a variable is altered, when exactly do they differ from each other; by how much; and for how long?

5. **Evaluate errors and statistics.** It is important to take account of variability in the data. For example, if mean values are presented, the underlying errors may be large, meaning that any difference between two treatments or observations at a given *x*-value could simply have arisen by chance. Thinking about the descriptive statistics used (Chapter 74) will allow you to determine whether apparent differences could be significant in both statistical and biological senses.

Sometimes graphs are used to mislead. This may be unwitting, as in an unconscious favouring of a 'pet' hypothesis of the author. Graphs may be used to 'sell' a product in the field of advertising or to favour a viewpoint as, perhaps, in politics. Experience in drawing and interpreting graphs will help you spot these flawed presentations, and understanding how graphs can be erroneously presented (Box 70.3) will help you avoid the same pitfalls.

Using computers to produce graphs – never allow a computer program to dictate size, shape and other aspects of a graph: find out how to alter scales, labels, axes, etc. and make appropriate selections (see Box 70.2 for Microsoft *Excel*). Draw curves freehand if the program only has the capacity to join the individual points by straight lines.

Interpreting proportions and percentages – you will need to establish how the values have been calculated (see p. 85).

Examining graphs – do not be tempted to look at the data displayed within a graph before you have considered its context, read the legend and the scale of each axis.

Understanding graphs within scientific papers – the legend should be a succinct summary of the key information required to interpret the figure without further reference to the main text. This is a useful approach when 'skimming' a paper for relevant information (p. 21).

Box 70.3 **How graphs can misrepresent and mislead**

1. **The 'volume' or 'area' deception** – this is mainly found in histogram or bar chart presentations where the size of a symbol is used to represent the measured variable. For example, the amount of hazardous waste produced in different years might be represented on a chart by different sizes of a chemical drum, with the *y* axis (height of drum) representing the amount of waste. However, if the symbol retains its *shape* for all heights, as in Fig. 70.10(a), its *volume* will increase as a cubic function of the height, rather than in direct proportion. To the casual observer, a twofold increase may look like an eightfold one, and so on. Strictly, the *height* of the symbol should be the measure used to represent the variable, with no change in symbol width, as in Fig. 70.10(b).

2. **Effects of a non-zero axis** – a non-zero axis acts to emphasise the differences between measures by reducing the range of values covered by the axis. For example, in Fig. 70.11(a), it looks as if there are large differences in mass between males and females; however, if the scale is adjusted to run from zero (Fig. 70.11(b)), then it can be seen that the differences are not large as a proportion of the overall mass. Always scrutinise the scale values carefully when interpreting any graph.

3. **Use of a relative rather than absolute scale** – this is similar to the above, in that data compared using relative scales (e.g. percentage or ratio) can give the wrong impression if the denominator is not the same in all cases. In Fig. 70.12(a), two treatments are shown as equal in *relative* effect, both resulting in 50% relative response compared (say) to the respective controls. However, if treatment A is 50% of a control value of 200 and treatment B is 50% of a control value of 500, then the actual difference in *absolute* response would have been masked, as shown by Fig. 70.12(b).

4. **Effects of a non-linear scale** – when interpreting graphs with non-linear (e.g. logarithmic) scales, you may interpret any changes on an imagined linear scale. For example, the pH scale is logarithmic, and linear changes on this scale mean less in terms of absolute H^+ concentration at high (alkaline) pH than they do at low (acidic) pH. In Fig. 70.13(a), the cell density in two media is compared on a logarithmic scale, while in Fig. 70.13(b), the same data are graphed on a linear scale. Note, also, that the log *y* axis scale in Fig. 70.13(a) cannot be shown to zero, because there is no logarithm for 0.

Fig. 70.10 Increase in pesticide use over a 30-year period.

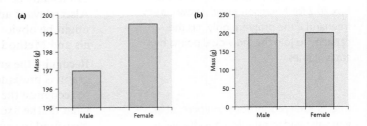

Fig. 70.11 Average mass of males and females in test group.

Fig. 70.12 Responses to treatments A and B.

Fig. 70.13 Effect of different media on cell density.

(continued)

Box 70.3 (continued)

5. **Unwarranted extrapolation** – a graph may be extrapolated to indicate what would happen if a trend continued, as in Fig. 70.14(a). However, this can only be done under certain assumptions (e.g. that certain factors will remain constant or that relationships will hold under new conditions). There may be no guarantee that this will actually be the case. Fig. 70.14(b) illustrates other possible outcomes if the experiment were to be repeated with higher values for the *x* axis.

6. **Failure to account for data point error** – this misrepresentation involves curves that are overly complex in relation to the scatter in the underlying data. When interpreting graphs with complex curves, consider the errors involved in the data values. It is probably unlikely that the curve would pass through all the data points unless the errors were very small. Fig. 70.15(a) illustrates a curve that appears to assume zero error and is thus overly complex, while Fig. 70.15(b) shows a curve that takes possible errors of the points into account.

7. **Failure to reject outlying points** – this is a special case of the previous example. There may be many reasons for outlying data, from genuine mistakes to statistical 'freaks'. If a curve is drawn through such points on a graph, it indicates that the point carries equal weight with the other points, when in fact, it should probably be ignored. To assess this, consider the accuracy of the measurement, the number and position of adjacent points and any special factors that might be involved on a one-off basis. Fig. 70.16(a) shows a curve where an outlier has perhaps been given undue weight when showing the presumed relationship. If there is good reason to think that the point should be ignored, then the curve shown in Fig. 70.16(b) would probably be more valid.

8. **Inappropriate fitted line** – here, the mathematical function chosen to represent a trend in the data might be inappropriate. A straight line might be fitted to the data, when a curve would be more correct, or vice versa. These cases can be difficult to assess. You need to consider the theoretical validity of the model used to generate the curve (this is not always stated clearly). For example, if a straight line is fitted to the points, the implicit underlying model states that one factor varies in direct relation to another, when the true situation may be more complex. In Fig. 70.17(a), the relationship has been shown as a linear relationship, whereas an exponential relationship, as shown in Fig. 70.17(b), could be more correct.

(a) (b)

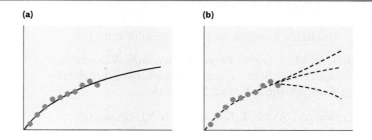

Fig. 70.14 Extrapolation of data under different assumptions.

(a) (b)

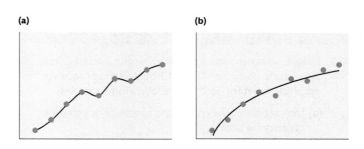

Fig. 70.15 Fitted curves under different assumptions of data error.

(a) (b)

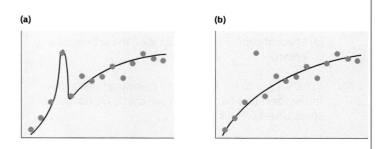

Fig. 70.16 Curves with and without outlier taken into account.

(a) (b)

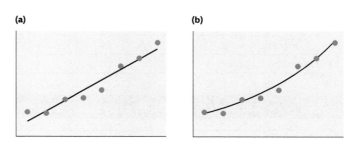

Fig. 70.17 Different mathematical model used to represent trends in data.

Sources for further study

Anon. (2000) *Biological Nomenclature: Recommendations on Terms, Units and Symbols.* Institute of Biology, London. [Includes a section on presentation of data.]

Briscoe, M.H. (1996) *Preparing Scientific Illustrations: A Guide to Better Posters, Presentations and Publications.* Springer-Verlag, New York.

Carter, M., Wiebe, E.N. and Ferzli, M. *Graphing with Excel.* Available: http://www.ncsu.edu/labwrite/res/gt/gt-menu.html

Last accessed 01/09/15.
[Online-tutorial from the US-NSF LabWrite Project.]

Kaufmann, M. and Wagner, D. (2006) *Drawing Graphs: Methods and Models.* Springer, New York.

Robbins, N.B. (2005) *Creating More Effective Graphs.* Wiley, New York.

STUDY EXERCISES

70.1 Select appropriate graphical presentations (see also study exercise 42.3). Choose an appropriate graphical form for each of the following examples.

(a) Interaction between pH and cation concentration on enzyme activity.

(b) Proportion of different eye colours in a survey of undergraduates attending your university.

(c) Relationship between pulse rate and age in humans.

(d) Number of bacteria per field of view for several different samples.

(e) Effect of copper concentration on the activity of an enzyme.

70.2 Create a pie chart. Display the following information in the form of a hand-drawn pie chart. Do *not* use a spreadsheet for this exercise.

Number of bacterial strains isolated from a sample.

Bacterium	Number of strains isolated
Escherichia coli	37
Aeromonas hydrophila	8
Klebsiella spp.	13
Salmonella enterica	2
Enterococcus spp.	25

70.3 Create a frequency distribution histogram. The table below gives data for the haemoglobin levels of 100 people. Plot a histogram showing the frequency distribution of the data. Write a brief description of the important features of the distribution.

Haemoglobin content (g L^{-1}) in blood.

11.1	14.2	13.5	9.8	12.0	13.9	14.1	14.6	11.0	12.3
13.4	12.9	12.9	10.0	13.1	11.8	12.6	10.7	8.1	11.2
13.8	12.4	12.9	11.3	12.7	12.4	14.6	15.1	11.2	9.7
11.3	14.7	10.8	13.3	11.9	11.4	12.5	13.0	11.6	13.1
9.3	13.5	14.6	11.2	11.7	10.9	12.4	12.0	12.1	12.6
10.9	12.1	13.4	9.5	12.5	11.6	12.2	8.8	10.7	11.1
10.2	11.7	10.4	14.0	14.9	11.5	12.0	13.2	12.1	13.3
12.4	9.4	13.2	12.5	10.8	11.7	12.7	14.1	10.4	10.5
13.3	10.6	10.5	13.7	11.8	14.1	10.3	13.6	10.4	13.9
11.7	12.8	10.4	11.9	11.4	10.6	12.7	11.4	12.9	12.1

70.4 Find examples of misleading graphs. Create a portfolio of examples of misleading graphs taken from newspapers. For each graph, state what aspect is misleading (see Box 70.3) and, where possible, attempt to show the data correctly in a new graph.

Answers to these study exercises are available at www.pearsoned.co.uk/practicalskills.

Alternatives to tables for presenting numerical data – if you have only a few numbers, consider simply presenting these within the text; an alternative approach is to show the data values on a bar chart (e.g. Fig. 70.4).

A table is often the most appropriate way to present numerical data in a concise, accurate and structured form. Assignments and project reports should contain tables that have been designed to condense and display results in a meaningful way and to aid numerical comparison. The preparation of tables for recording primary data is discussed on p. 208.

Decide whether you need a table, or whether a graph is more appropriate. Histograms and plotted curves can be used to give a visual impression of the relationships within your data (p. 487). On the other hand, a table gives you the opportunity to make detailed numerical comparisons.

> **KEY POINT** Always remember that the primary purpose of your table is to communicate information and allow appropriate comparison, not simply to put down the results on paper.

Preparation of tables

Title

Constructing titles – take care over titles as it is a common mistake in student practical reports to present tables without titles, or to misconstruct the title.

Every table must have a brief descriptive title. If several tables are used, number them consecutively so they can be quoted in your text. The titles within a report should be compared with one another, making sure they are logical and consistent and that they describe accurately the numerical data contained within them.

Structure

Saving space in tables – you may be able to omit a column of control data if your results can be expressed as percentages of the corresponding control values.

Display the components of each table in a way that will help the reader understand your data and grasp the significance of your results. Organise the columns so that each category of like numbers or attributes is listed vertically, while each horizontal row shows a different experimental treatment, organism, sampling site, etc. (as in Table 71.1). Where appropriate, put control values near the beginning of the table. Columns that need to be compared should be set out alongside each other. Use rulings to subdivide your table appropriately, but avoid cluttering it up with too many lines.

Table 71.1 Characteristics of selected photoautotrophic microbes.

Division	Species	Optimum [NaCl]* (mol m^{-3})	Intracellular carbohydrate Identity	Quantity† (nmol (g dry wt)$^{-1}$)
Chlorophyta	*Scenedesmus quadruplicatum*	340	Sucrose	49.7
	Chlorella emersonii	780	Sucrose	102.3
	Dunaliella salina	4700	Glycerol	910.7
Cyanobacteria	*Microcystis aeruginosa*	<20‡	None	0.0
	Anabaena variabilis	320	Sucrose	64.2
	Rivularia atra	380	Trehalose	ND

*Determined after 28-day incubation in modified Von Stosch medium.
† Individual samples, analysed by gas–liquid chromatography.
‡ Poor growth in all media with added NaCl (minimum NaCl concentration 5 mol m^{-3}).
ND, Sample lost: no quantitative data.

Headings and subheadings

These should identify each set of data and show the units of measurement, where necessary. Make sure that each column is wide enough for the headings and for the longest data value.

Numerical data

Within the table, do not quote values to more significant figures than necessary, as this will imply spurious accuracy (pp. 193, 506). By careful choice of appropriate units for each column you should aim to present numerical data within the range 0 to 1000. As with graphs, it is less ambiguous to use derived SI units, with the appropriate prefixes, in the headings of columns and rows, rather than quoting multiplying factors as powers of 10. Alternatively, include exponents in the main body of the table (see Table 27.1) to avoid any possible confusion regarding the use of negative powers of 10.

Other notations

Avoid using dashes in numerical tables, as their meaning is unclear; enter a zero reading as '0' and use 'NT' not tested or 'ND' if no data value was obtained, with a footnote to explain each abbreviation. Other footnotes, identified by asterisks, superscripts or other symbols in the table, may be used to provide relevant experimental detail (if not given in the text) and an explanation of column headings and individual results, where appropriate. Footnotes should be as condensed as possible. Table 71.1 provides examples.

Statistics

In tables where the dispersion of each data set is shown by an appropriate statistical parameter, you must state whether this is the (sample) standard deviation, the standard error (of the mean) or the 95% confidence limits and you must give the value of n (the number of replicates). Other descriptive statistics should be quoted with similar detail, and hypothesis-testing statistics should be quoted along with the value of P (the probability). Details of any test used should be given in the legend, or in a footnote.

Text

Sometimes a table can be a useful way of presenting textual information in a condensed form (see example on p. 298).

When you have finished compiling your tabulated data, carefully double-check each numerical entry against the original information, to ensure that the final version of your table is free from transcriptional errors. Box 71.1 gives a checklist for the major elements of constructing a table.

Examples If you measured the width of a fungal hypha to the nearest one-tenth of a micrometre, quote the value in the form '52.6 μm'.

Quote the width of a fungal hypha as 52.6 μm, rather than 0.000 052 6 m or 52.6×10^{-6} m.

Saving further space in tables – in some instances a footnote can be used to replace a whole column of repetitive data.

Using spreadsheets and word processing packages – these can be used to prepare high-quality versions of tables for project work (Box 71.2).

Box 71.1 Checklist for preparing a table

Every table should have the following components.
1. **A title**, plus a reference number and date where necessary.
2. **Headings for each column and row**, with appropriate units of measurement.
3. **Data values**, quoted to the nearest significant figure and with statistical parameters, according to your requirements.
4. **Footnotes** to explain abbreviations, modifications and individual details.
5. **Rulings to emphasise groupings** and distinguish items from each other.

Box 71.2 How to use a word processor (Microsoft *Word*) or a spreadsheet (Microsoft *Excel*) to create a table for use in coursework reports and dissertations

Creating tables with Microsoft *Word*

Word-processed tables are suitable for text-intensive or number-intensive tables, although in the second case entering data can be laborious. When working in this way, the natural way to proceed is to create the 'shell' of the table, add the data, and then carry out final formatting on the table.

1. Move the cursor to the desired position in your document – this is where you expect the top left corner of your table to appear. Click the *Insert* tab, then choose *Table*.

2. Select the appropriate number of columns and rows – do not forget to add rows and columns for headings. As default, a full-width table will appear, with single rulings for all cell boundaries, with all columns of equal width and all rows of equal height.

Example of a 4 × 3 table:

3. Customise the columns – by placing the cursor over the vertical rulings then 'dragging', you can adjust their width to suit your heading text entries, which should now be added.

Heading 1	Heading 2	Heading 3	Heading 4

4. Work through the table adding the data – entries can be numbers or text.

Heading 1	Heading 2	Heading 3	Heading 4
XX	XX	XX	XX
XX	XX	XX	

5. Make further adjustments to column and row widths to suit – for example, if text fills several rows within a

cell, consider increasing the column width, and if a column contains only single or double digit numbers, consider shrinking its width. To combine cells, first highlight them, then right-click the mouse button (or equivalent) and click on *Merge Cells*. You may wish to reposition text within a cell by right-clicking the mouse button and choosing an appropriate position in the cell alignment menu.

Heading 1	Heading 2	Heading 3	Heading 4
XX	XX	XX	XX
	XX	XX	

6. Finally, remove selected borders to cells – one way is to highlight the table, then click on the *Design* tab and choose an appropriate style from the template options. If you do not want any shading, click on *Design > Shading* and choose *No Color*. Another way is to highlight the cells in the table, then right-click the mouse button and choose *Borders and Shading*. You can choose the style you wish from this submenu so that your table looks like the examples shown in this chapter.

7. Add a table title – this should be positioned above the table (cf. a figure title and legend, p. 490), legend and footnotes.

Final version of the table:

Table xx. A table of some data

Heading 1	Heading 2[a]	Heading 3	Heading 4
XX	XX	XX	XX
	XX	XX	XX

[a]An example of a footnote.

Creating tables with Microsoft *Excel*

Tables derived from spreadsheets are effective when you have lots of numerical data, especially when these are stored or created using the spreadsheet itself. When working in this way, you can design the table as part of an output or summary section of the spreadsheet, add explanatory headings, format, then possibly export to a word processor when complete.

(continued)

Box 71.2 *(continued)*

1. **Design the output or summary section.** Plan this as if it were a table, including adding text headings within cells.

16				
17				
18	Heading 1	Heading 2	Heading 3	Heading 4
19				
20				
21				
22				

2. **Insert appropriate formulae within cells to produce data.** If necessary, formulae should draw on the other parts of the spreadsheet.

16				
17				
18	Heading 1	Heading 2	Heading 3	Heading 4
19	Aaa	=A1	=C3*5	=SDEV(A1:A12)
20	Bbb	=A2	=F45/G12	=SDEV(B1:B12)
21				
22				

3. **Format the cells.** This is important to control the number of decimal places presented (*Home > Number > Format > Cells*).

4. **Adjust column width to suit.** You can do this via the column headings, by placing the cursor over the rulings between columns then 'dragging'.

16				
17				
18	Heading 1	Heading 2	Heading 3	Heading 4
19	Aaa	=A1	=C3*5	=SDEV(A1:A12)
20	Bbb	=A2	=F45/G12	=SDEV(B1:B12)
21				
22				

5. **Add rulings as appropriate.** One way is to use the *Home > Font > Borders* menu, having selected the relevant cells.

16				
17				
18	Heading 1	Heading 2	Heading 3	Heading 4
19	Aaa	=A1	=C3*5	=SDEV(A1:A12)
20	Bbb	=A2	=F45/G12	=SDEV(B1:B12)
21				

6. **Add 'real' data values to the spreadsheet.** This should result in the summary values within the table being filled. Check that these are presented with the appropriate number of significant figures (pp. 506–508).

7. The table can now be copied and pasted to a *Word* document. For best results, use the *Paste Special... > Microsoft Office Excel Worksheet Object* option.

Note: instructions illustrated here may vary among the different versions of Microsoft *Office* programs. Check the exact functions and syntax using the *Insert function* menu option (*fx* button in some versions).

Sources for further study

Kirkup, L. (1994) *Experimental Methods: An Introduction to the Analysis and Presentation of Data*. Wiley, New York.

Simmonds, D. and Reynolds, L. (1994) *Data Presentation and Visual Literacy in Medicine and Science*. Butterworth-Heinemann, London.

Willis, J. (2004) *Data Analysis and Presentation Skills: An Introduction for the Life and Medical Sciences*. Wiley, Chichester.

71.1 Redesign a table of data. Using the following example, redraft the table to improve layout and correct inconsistencies.

Concentrations of low molecular weight solutes in bacteria.

Bacterium	Constituent	Concentration
Escherichia coli	Proline	21.0 mmol L^{-1}
Escherichia coli	Trehalose	1.547×10^2 kmol m^{-3}
Bacillus subtilis	Proline	39.7 mmol L^{-1}
Bacillus subtilis	Glutamate	0.0521 mmol m^{-3}
*Staphylococcus aureus**	Glutamate	15 260 mmol m^{-3}
Escherichia coli	Glutamate	0.50% w/v
Bacillus subtilis	Trehalose	<0.001% w/v

*Proline and trehalose were not measured.

71.2 Devise a text-based table. After reading through this chapter, and working from memory, draw up a table listing the principal components of a typical table in the first column, and brief comments on the major features of each component in the second column.

71.3 Interpret data from a table. The table below shows representative data for energy, protein and niacin (vitamin B3) as a function of age for human males and females. The typical average mass ('weight') of each class is also shown.

a) In what age class does the mass of females exceed that of males?

b) In what age class is the energy requirement of males greatest in proportion to the energy requirement of females?

c) Over what age range do males and females have the same protein requirement?

d) In what age class do males have their highest niacin requirement, and how does this compare with females?

Selected age-related characteristics of human males and females.

Age (years)	Average mass (kg)		Energy requirement (MJ)		Protein requirement (g)		Niacin requirement (mg)	
	Male	Female	Male	Female	Male	Female	Male	Female
<1	8	8	3.4	3.2	14	14	6	6
1–3	13	13	5.2	4.9	16	16	9	9
4–6	20	20	7.2	6.5	24	24	12	12
7–10	28	28	8.2	7.3	28	28	13	13
11–14	45	46	9.3	7.7	45	46	17	15
15–18	66	55	11.5	8.8	59	44	20	15
19–50	79	63	10.6	8.1	63	50	19	15
>50	77	65	9.5	7.7	63	50	15	13

Answers to these study exercises are available at www.pearsoned.co.uk/practicalskills.

Advice on dealing with numerical procedures – see:

- Box 25.2: preparing solutions (p. 153);
- Box 26.1: molar concentrations (p. 162);
- Box 36.1: cell counting chambers (p. 234);
- Box 36.2: plate (colony) counts (p. 235);
- Box 42.1: calibration curves (p. 276);
- Box 45.1: radioactivity (p. 300).

Biology often requires a numerical or statistical approach. Not only is mathematical modelling an important aid to understanding, but computations are often needed to turn raw data into meaningful information or to compare them with other data sets. Moreover, calculations are part of laboratory routine, perhaps required for making up solutions of known concentration (see p. 153 and below) or for the calibration of a microscope (see p. 180). In research, 'trial' calculations can reveal what input data are required and where errors in their measurement might be amplified in the final result (see p. 193).

> **KEY POINT** If you find numerical work difficult, practice at problem-solving is especially important.

Practising problem-solving:

- **demystifies the procedures involved,** which are normally just the elementary mathematical operations of addition, subtraction, multiplication and division (Table 72.1);
- **allows you to gain confidence** so that you do not become confused when confronted with an unfamiliar or apparently complex form of problem;
- **helps you to recognise the various forms a problem can take** as, for instance, in crossing experiments in classical genetics (Chapter 64).

Table 72.1 Sets of numbers and operations.

Sets of numbers	
Whole numbers:	0, 1, 2, 3, . . .
Natural numbers:	1, 2, 3, . . .
Integers:	. . . −3, −2, −1, 0, 1, 2, 3, . . .
Real numbers:	integers and anything between (e.g. −5, 4.376, 3/16, π, $\sqrt{5}$)
Prime numbers:	subset of natural numbers divisible by 1 and themselves only (i.e. 2, 3, 5, 7, 11, 13, . . .)
Rational numbers:	p/q where p (integer) and q (natural) have no common factor (e.g. 3/4)
Fractions:	p/q where p is an integer and q is natural (e.g. −6/8)
Irrational numbers:	real numbers with no exact value (e.g. π)
Infinity:	(symbol ∞) is larger than any number (technically not a number as it does not obey the laws of algebra)
Operations and symbols	
Basic operators:	$+, -, \times$ and $\div$ will not need explanation; however, / may substitute for $\div$, * may substitute for $\times$ or this operator may be omitted
Powers:	a^n, i.e. 'a to the power n', means a multiplied by itself n times (e.g. $a^2 = a \times a = $ 'a squared', $a^3 = a \times a \times a = $ 'a cubed'), n is said to be the index or exponent. Note $a^0 = 1$ and $a^1 = a$
Logarithms:	the common logarithm (log) of any number $\times$ is the power to which 10 would have to be raised to give x (i.e. the log of 100 is 2; $10^2 = 100$); the antilog of x is 10^x. Note that there is no log for 0, so take this into account when drawing log axes by breaking the axis. Natural or Napierian logarithms (ln) use the base e (= 2.718 28 . . .) instead of 10
Reciprocals:	the reciprocal of a real number a is $1/a$ $(a \neq 0)$
Relational operators:	$a > b$ means 'a is greater (more positive) than b', $<$ means less than, $\leq$ means less-than-or-equal-to and $\geq$ means greater-than-or-equal-to
Proportionality:	$a \propto b$ means 'a is proportional to b' (i.e. $a = kb$, where k is a constant). If $a \propto 1/b$, a is inversely proportional to b $(a = k/b)$
Sums:	Σx_i is shorthand for the sum of all x values from $i = 0$ to $i = n$ (more correctly, the range of the sum is specified under the symbol)
Moduli:	$|x|$ signifies modulus of x, i.e. its absolute value (e.g. $|4| = |-4| = 4$)
Factorials:	$x!$ signifies factorial x, the product of all integers from 1 to x (e.g. 3! = 6). Note 0! = 1! = 1

Steps in tackling a numerical problem

The step-by-step approach outlined below may not be the fastest method of arriving at an answer, but most mistakes occur where steps are missing, combined or not made obvious, so a logical approach is often better. Error tracing is distinctly easier when all stages in a calculation are laid out.

Have the right tools ready

Scientific calculators (p. 132) greatly simplify the numerical part of problem-solving. However, the seeming infallibility of the calculator may lead you to accept an absurd result which could have arisen because of faulty key-pressing or faulty logic. Make sure you know how to use all the features on your calculator, especially how the memory works, how to introduce a constant multiplier or divider, and how to obtain an exponent (note that the 'exp' button on most calculators gives you 10^x, not 1^x or y^x; so 1×10^6 would be entered as $\boxed{1}\,\boxed{\text{exp}}\,\boxed{6}$, *not* $\boxed{10}\,\boxed{\text{exp}}\,\boxed{6}$).

Approach the problem thoughtfully

If the individual steps have been laid out on a worksheet, the 'tactics' will already have been decided. It is more difficult when you have to adopt a strategy on your own, especially if the problem is presented as a story and it is not obvious which equations or rules need to be applied.

- **Read the problem carefully** as the text may give clues as to how it should be tackled. Be certain of what is required as an answer before starting.
- **Analyse what kind of problem it is,** which effectively means deciding which equation(s) or approach will be applicable. If this is not obvious, consider the dimensions/units of the information available and think how they could be fitted to a relevant formula. In examinations, a favourite ploy of examiners is to present a problem such that the familiar form of an equation must be rearranged (see Table 72.2 and Box 72.1). Another is to make you use two or more equations in series (see Box 72.2). If you are unsure whether a recalled formula is correct, a dimensional analysis can help: write in all the units for the variables and make sure that they cancel out to give the expected answer.
- **Check that you have, or can derive, all of the information required** to use your chosen equation(s). It is unusual but not unknown for examiners to supply redundant information. So, if you decide not to use some of the information given, be sure why you do not require it.
- **Decide on what format and in which units the answer should be presented.** This is sometimes suggested to you. If the problem requires many changes in the prefixes to units, it is a good idea to convert all data to base SI units (multiplied by a power of 10) at the outset.
- **If a problem appears complex, break it down into component parts.**

Present your answer clearly

The way you present your answer obviously needs to fit the individual problem. The example shown in Box 72.2 has been chosen to illustrate several important points, but this format would not fit all situations. Guidelines for presenting an answer include the following:

(a) Make your assumptions explicit. Most mathematical models of biological phenomena require that certain criteria are met before they can be

---

Tracing errors in mathematical problems – this is always easier when all the stages in a calculation are laid out clearly.

Using a computer spreadsheet for numerical problems – this may be very useful in repetitive work or for 'what if?' case studies (see Chapter 13) (Adams, 2003).

Table 72.2 Simple algebra – rules for manipulating.

If $a = b + c$, then $b = a - c$ and $c = a - b$

If $a = b \times c$, then $b = a \div c$ and $c = a \div b$

If $a = b^c$, then $b = a^{1/c}$ and $c = \log a \div \log b$

$a^{1/n} = \sqrt[n]{a}$

$a^{-n} = 1 \div a^n$

$a^b \times a^c = a^{(b+c)}$ and $a^b \div a^c = a^{(b-c)}$

$(a^b)^c = a^{(b \times c)}$

$a \times b = \text{antilog}(\log a + \log b)$

Presenting calculations in assessed work – always show the steps in your calculations, as most markers will only penalise a mistake once and part marks will be given if the remaining operations are performed correctly. This can only be done if those operations are visible.

> **Box 72.1** Example of using the algebraic rules of Table 72.2
>
> **Problem:** if $a = (b - c) \div (d + e^n)$, find e.
>
> **1.** Multiply both sides by $(d + e^n)$; formula becomes:
> $$a(d + e^n) = (b - c)$$
>
> **2.** Divide both sides by a; formula becomes:
> $$d + e^n = \frac{b - c}{a}$$
>
> **3.** Subtract d from both sides; formula becomes:
> $$e^n = \frac{b - c}{a} - d$$
>
> **4.** Raise each side to the power $1/n$; formula becomes:
> $$e = \left\{ \frac{b - c}{a} - d \right\}^{1/n}$$

legitimately applied (e.g. 'assuming the tissue is homogeneous . . .'), while some approaches involve approximations that should be clearly stated (e.g. 'to estimate the mouse's skin area, its body was approximated to a cylinder with radius x and height y . . .').

(b) Explain your strategy for answering, perhaps giving the applicable formula or definitions which suit the approach to be taken. Give details of what the symbols mean (and their units) at this point.

(c) Rearrange the formula to the required form with the desired unknown on the left-hand side (see Table 72.2).

(d) Substitute the relevant values into the right-hand side of the formula, using the units and prefixes as given (it may be convenient to convert values to SI beforehand). Convert prefixes to appropriate powers of 10 as soon as possible.

(e) Convert to the desired units step by step, i.e. taking each variable in turn.

(f) When you have the answer in the desired units, rewrite the left-hand side and <u>underline the answer,</u> for emphasis. Make sure that the result is presented to an appropriate number of significant figures (see below).

Check your answer

Having written out your answer, you should check it methodically, answering the following questions:

- **Is the answer realistic?** You should be alerted to an error if a number is absurdly large or small. In repeated calculations, a result standing out from others in the same series should be double-checked.

- **Do the units make sense and match up with the answer required?** For example, do not present a volume in units of m^2.

- **Do you get the same answer if you recalculate in a different way?** If you have time, recalculate the answer using a different 'route', entering the numbers into your calculator in a different form and/or carrying out the operations in a different order.

> **Units** – never write any answer without its unit(s) unless it is truly dimensionless.

> **Rounding off to a specific number of significant figures** – do not round off numbers until you arrive at the final answer or you will introduce 'rounding' errors into the calculation.

Rounding: decimal places and significant figures

In many instances, the answer you produce as a result of a calculation will include more figures than is justified by the accuracy and precision of the original data. Sometimes you will be asked to produce an answer to a specified number of decimal places or significant figures, and at other times you will be expected to decide for yourself what would be appropriate.

Box 72.2 Model answer to a mathematical problem

Problem

Estimate the total length and surface area of the fibrous roots on a maize seedling from measurements of their total fresh weight and mean diameter. Give your answers in m and cm², respectively, to four significant figures.

Measurements

Fresh weight[a] = 5.00 g, mean diameter[b] = 0.5 mm.

Answer

Assumptions: (1) the roots are cylinders with constant radius[c] and the 'ends' have negligible area; (2) the root system has a density of 1000 kg m⁻³ (i.e. that of water[d]).

Strategy: from assumption (1), the applicable equations are those concerned with the volume and surface area of a cylinder, namely:

$$V = \pi r^2 h \qquad [72.1]$$

$$A = 2\pi r h \text{ (ignoring ends)} \qquad [72.2]$$

where V is volume (m³), A is surface area (m²), $\pi \approx 3.141\ 59$, h is height (m) and r is radius (m). The total length of the root system is given by h and its surface area by A. We can find h by rearranging Eqn [72.1] and then substitute its value in Eqn [72.2] to get A.

To calculate total root length: rearranging Eqn [72.1], we have $h = V/\pi r^2$. From measurements[e], $r = 0.25$ mm = 0.25×10^{-3} m.

From density = weight/volume,

V = fresh weight/density

$\quad = 5$ g/1000 kg m⁻³

$\quad = 0.005$ kg/1000 kg m⁻³

$\quad = 5 \times 10^{-6}$ m³

Total root length,

$h = V/\pi r^2$

5×10^{-6} m³/3.141 59 $\times (0.25 \times 10^{-3}$ m)²

Total root length = 25.46 m

To calculate surface area of roots: substituting the value for h obtained above into Eqn [72.2], we have:

Root surface area

$\quad = 2 \times 3.141\ 59 \times 0.25 \times 10^{-3}$ m $\times 25.46$ m

$\quad = 0.04$ m²

$\quad = 0.04 \times 10^4$ cm²

$\quad$ (there being $100 \times 100 = 10^4$ cm² per m²)

Root surface area = 400.0 cm²

Notes

(a) The fresh weight of roots would normally be obtained by washing the roots free of soil, blotting them dry and weighing.

(b) In a real answer you might show the replicate measurements giving rise to the mean diameter.

(c) In reality, the roots will differ considerably in diameter and each root will not have a constant diameter throughout its length.

(d) This will not be wildly inaccurate as about 95% of the fresh weight will be water, but the volume could also be estimated from water displacement measurements.

(e) Note conversion of measurements into base SI units at this stage and on line 3 of the root volume calculation. Forgetting to halve diameter measurements where radii are required is a common error.

KEY POINT Do not simply accept the numerical answer from a calculator or spreadsheet, without considering whether you need to modify this to give an appropriate number of significant figures or decimal places.

Rounding to n *decimal places*

This is relatively easy to do.

1. **Look at the number to the right of the *n*th decimal place.**

2. **If this is less than 5, simply 'cut off all numbers to the right of the *n*th decimal place (i.e. round down).**

3. **If the number is greater than 5, 'cut off all numbers to the right of the *n*th decimal place and add one to the *n*th decimal place (i.e. round up).**

> **Examples**
>
> The number 4.123 correct to two decimal places is 4.12
>
> The number 4.126 correct to two decimal places is 4.13
>
> The number 4.1251 correct to two decimal places is 4.13
>
> The number 4.1250 correct to two decimal places is 4.12
>
> The number 4.1350 correct to two decimal places is 4.14
>
> The number 99.99 correct to one decimal place is 100.0.

> **Examples**
>
> The number of significant figures in 194 is three
>
> The number of significant figures in 2305 is four
>
> The number of significant figures in 0.003 482 is four
>
> The number of significant figures in 210×10^8 is 3 (21×10^9 would be two).

> **Examples**
>
> The number of significant figures in 3051.93 is six
>
> To five significant figures, this number is 3051.9
>
> To four significant figures, this number is 3052
>
> To three significant figures, this number is 3050
>
> To two significant figures, this number is 3100
>
> To one significant figure, this number is 3000
>
> 3051.93 to the nearest 10 is 3050
>
> 3051.93 to the nearest 100 is 3100
>
> Note that in this last case you must include the zeros before the decimal point to indicate the scale of the number (even if the decimal point is not shown). For a number less than 1, the same would apply to the zeros before the decimal point. For example, 0.003 051 93 to three significant figures is 0.003 05. Alternatively, use scientific notation (in this case, 3.05×10^{-3}).

4. **If the number is 5, then look at further numbers to the right to determine whether to round up or not.**

5. **If the number is *exactly* 5 and there are no further numbers to the right, then round the *n*th digit to the nearest even number.** *Note:* When considering a large number of calculations, this procedure will not affect the overall mean value. Some rounding systems do the opposite to this (i.e. round to the nearest odd number), while others always round up where the number is exactly 5 (which *will* affect the mean). Take advice from your tutor and stick to one system throughout a series of calculations.

Whenever you see any numbers quoted, you should assume that the last digit has been rounded. For example, in the number 22.4, the '.4' is assumed to be rounded and the calculated value may have been between 22.35 and 22.45.

Quoting to n *significant figures*

The number of significant figures indicates the degree of approximation in the number. For most cases, it is given by counting all the figures except zeros that occur at the beginning or end of the number. Zeros *within* the number are always counted as significant. The number of significant figures in a number like 200 is ambiguous and could be one, two or three; if you wish to specify clearly, then quote as e.g. 2×10^2 (one significant figure), 2.0×10^2 (two significant figures), etc. to avoid spurious accuracy (p. 193). When quoting a number to a specified number of significant figures, use the same rules as for rounding to a specified number of decimal places, but do not forget to keep zeros before or after the decimal point. The same principle is used if you are asked to quote a number to the 'nearest 10', 'nearest 100', etc.

When deciding for yourself how many significant figures to use, adopt the following rules of thumb:

- **Always round *after* you have completed a calculation.** Use *all* significant figures available in the measured data during a calculation.

- **If adding or subtracting with measured data, then quote the answer to the number of decimal places in the data value with the least number of decimal places** (e.g. $32.1 - 45.67 + 35.6201 = 22.1$, because 32.1 has one decimal place).

- **If multiplying or dividing with measured data, keep as many significant figures as are in the number with the least number of significant figures** (e.g. $34\,901 \div 3445 \times 1.341\,0344 = 13.59$, because 3445 has four significant figures).

- **For the purposes of significant figures, assume 'constants' have an infinite number of significant figures** (e.g. number of millimetres in a metre).

Some reminders of basic mathematics

Errors in calculations sometimes appear because of faults in mathematics rather than computational errors. For reference purposes, Tables 72.1 and 72.2 give some basic mathematical principles that may be useful. Eason *et al.* (1992) or Stephenson (2010) should be consulted for more advanced/specific needs.

> **Examples**
> 1/8 as a percentage is $1 \div 8 \times 100 =$
> $100 \div 8 = 12.5\%$
> 0.602 as a percentage is $0.602 \times 100 = 60.2\%$.

> **Examples**
> 190% as a decimal fraction is
> $190 \div 100 = 1.9$
> 5/2 as a percentage is
> $5 \div 2 \times 100 = 250\%$.

> **Example** A population falls from 4 million
> to 3.85 million. What is the percentage
> change? The decrease in numbers is
> $4 - 3.85 = -0.15$ million. The fractional
> decrease is $-0.15 \div 4 = -0.0375$ and we
> multiply by 100 to get the percentage
> change = minus 3.75%.

> **Example** $2^3 = 2 \times 2 \times 2 = 8$.

> **Example** Avogadro's number,
> $\approx 602\ 352\ 000\ 000\ 000\ 000\ 000\ 000$,
> is more conveniently expressed as
> $6.023\ 52 \times 10^{23}$.

Percentages and proportions

A percentage is just a fraction expressed in terms of hundredths, indicated by putting the percentage sign (%) after the number of hundredths. So 35% simply means 35 hundredths. To convert a fraction to a percentage, just multiply the fraction by 100. When the fraction is in decimal form, multiplying by 100 to obtain a percentage is easily achieved just by moving the decimal point two places to the right.

To convert a percentage to a fraction, just remember that, since a percentage is a fraction multiplied by 100, the fraction is the percentage divided by 100. For example: $42\% = 42/100 = 0.42$. In this example, since we are dealing with a decimal fraction, the division by 100 is just a matter of moving the decimal point two places to the left (42% could be written as 42.0%). Percentages greater than 100% represent fractions greater than 1. Percentages less than 1 may cause confusion. For example, 0.5% means half of one per cent (0.005) and must not be confused with 50% (which is the decimal fraction 0.5).

To find a percentage of a given number, just express the percentage as a decimal fraction and multiply the given number. For example: 35% of 500 is given by $0.35 \times 500 = 175$. To find the percentage change in a quantity, work out the difference (= value 'after' – value 'before'), and divide this difference by the original value to give the fractional change, then multiply by 100.

Exponents

Exponential notation is an alternative way of expressing numbers in the form a^n ('a to the power n'), where a is multiplied by itself n times. The number a is called the base and the number n the exponent (or power or index). The exponent need not be a whole number, and it can be negative if the number being expressed is less than 1. See Table 72.2 for other mathematical relationships involving exponents.

Scientific notation

In scientific notation, also known as 'standard form', the base is 10 and the exponent a whole number. To express numbers that are not whole powers of 10, the form $c \times 10^n$ is used, where the coefficient c is normally between 1 and 10. Scientific notation is valuable when you are using very large numbers and wish to avoid suggesting spurious accuracy. Thus if you write 123 000, this may suggest that you know the number to ± 0.5, whereas 1.23×10^5 might give a truer indication of measurement accuracy (i.e. implied to be ± 500 in this case). Engineering notation is similar, but treats numbers as powers of 10 in groups of 3, i.e. $c \times 10^0$, 10^3, 10^6, 10^9, etc. This corresponds to the SI system of prefixes (p. 195).

A useful property of powers when expressed to the same base is that when multiplying two numbers together, you simply add the powers, while, if dividing, you subtract the powers. Thus, suppose you counted 8 bacteria in a known value of a 10^{-7} dilution, there would be 8×10^7 in the same volume of undiluted solution; if you now dilute this 500-fold (5×10^2), then the number present in the same volume would be $8/5 \times 10^{(7-2)} = 1.6 \times 10^5 = 160\ 000$.

Logarithms

When a number is expressed as a logarithm, this refers to the power n to which the base number a must be raised to give that number. Any

> **Examples** The logarithm to the base 10 ($\log_{10}$) of 1000 is 3, since $10^3 = 1000$. The logarithm to the base e ($\log_e$ or ln) of 1000 is 6.907 755 (to six decimal places).

> **Examples** (use to check the correct use of your own calculator)
> 102 963 as a log (to base 10) = 5.012 681 (to six decimal places) $10^{5.012681} = 1029\,62.96$ (Note loss of accuracy owing to loss of decimal places.) 102 963 as a natural logarithm (ln) = 11.542 125 (to six decimal places) thus $2.718282^{11.542125} = 102\,963$.

> **Example** Using Eqn [72.3], the predicted value for y for a linear function where $a = 2$ and $b = 0.5$, where $x = 8$ is: $y = 2 + (0.5 \times 8) = 6$.

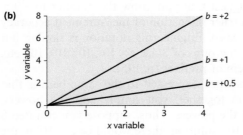

Fig. 72.1 Straight-line relationships ($y = a + bx$), showing the effects of (a) changing the intercept at constant slope, and (b) changing the slope at constant intercept.

base could be used, but the two most common are 10, when the power is referred to as $\log_{10}$ or simply log, and the constant e (2.718 282), used for mathematical convenience in certain situations, when the power is referred to as $\log_e$ or ln (natural logarithm). Note that: (a) logs need not be whole numbers; (b) there is no log value for the number zero; and (c) $\log_{10} = 0$ for the number 1.

To obtain logs, you will need to use the log key on your calculator, or special log tables (now largely redundant). To convert back ('antilog') use

- the $\boxed{10^x}$ key, with $x = $ log value;
- the $\boxed{\text{inverse}}$ then the $\boxed{\text{log}}$ key; or
- the $\boxed{y^x}$ key, with $y = 10$ and $x = $ log value.

If you have used log tables, you will find complementary antilogarithm tables to do this.

There are many uses of logarithms in biology, including pH ($= -\log[H^+]$), where $[H^+]$ is expressed in mol L^{-1} (see p. 169); the exponential growth of micro-organisms, where if log(cell number) is plotted against time, a straight-line relationship is obtained (see p. 231); and allometric studies of growth and development, where, if data are plotted on log axes, a series of straight-line relationships may be found.

Linear functions and straight lines

One of the most straightforward and widely used relationships between two variables x and y is that represented by a straight-line graph, where the corresponding mathematical function is known as the equation of a straight line, where:

$$y = a + bx \qquad [72.3]$$

In this relationship, a represents the intercept of the line on the y-(vertical) axis, i.e. where $x = 0$, and therefore $bx = 0$, while b is equivalent to the slope (gradient) of the line, i.e. the change in y for a change in x of 1. The constants a and b are sometimes given alternative symbols, but the mathematics remains unchanged, e.g. in the equivalent expression for the slope of a straight line, $y = mx + c$. Figure 72.1 shows what happens when these two constants are changed, in terms of the resultant straight lines.

The two main applications of the straight-line relationship are:

1. **Function fitting.** Here, you determine the mathematical form of the function, i.e. you estimate the constants a and b from a data set for x and y, either by drawing a straight line by eye and then working out the slope and y intercept, or by using linear regression (p. 533) to obtain the most probable values for both constants. When putting a straight line of best fit by eye on a hand-drawn graph, note the following:

- **Always use a *transparent* ruler,** so you can see data points on either side of the line.
- **For a data series where the points do not fit a perfect straight line, try to have an equal number of points on either side of the line,** as in Fig. 72.2(a), and try to minimise the average distance of these points from the line.

(a)

(b)

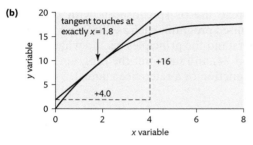

Fig. 72.2 Drawing straight lines. (a) Simple linear relationship, giving a straight line with an intercept of 2.3 and a slope of $-1.6 \div 4.0 = 0.4$. (b) Tangent drawn to a curve at $x = 1.8$, giving a slope of $16 \div 4 = 4$.

Example Using Eqn [72.4], the predicted value for x for a linear function where $a = 1.5$ and $b = 2.5$, where $y = -8.5$ is: $x = (-8.5 - 1.5) \div 2.5 = -4$. Using Eqn [72.4] the predicted x intercept for a linear function where $a = 0.8$ and $b = 3.2$ is: $x = (0 - 0.8) \div 3.2 = -0.25$.

Example A lab schedule states that 5 g of a compound with a relative molecular mass of 220 are dissolved in 400 mL of solvent. For writing up your Materials and Methods, you wish to express this as $mol\ L^{-1}$.

1. If there are 5 g in 400 mL, then there are $5 \div 400$ g in 1 mL.
2. Hence, 1000 mL will contain $5 \div 400 \times 1000$ g $= 12.5$ g.
3. 12.5 g $= 12.5 \div 220$ mol $= 0.0568$ mol, so [solution] $= 56.8$ mmol L^{-1} $(= 56.8\ mol\ m^{-3})$.

- **Once you have drawn the line of best fit use this line, rather than your data values,** in all subsequent procedures (e.g. in a calibration curve, Chapter 42).
- **Tangents drawn to a curve give the slope (gradient) at a particular point,** e.g. in an enzyme reaction progress curve (see Fig. 61.3). These are best drawn by bringing your ruler up to the curve at the exact point where you wish to estimate the slope and then trying to make the two angles immediately on either side of this point approximately the same, by eye (Fig. 72.2(b)).
- Once you have drawn the straight line or tangent, choose two points reasonably far apart at either end of your line and then draw construction lines to represent the change in y and the change in x between these two points: make sure that your construction lines are perpendicular to each other. Determine the slope as the change in y divided by the change in x (Fig. 72.2(b)).

2. **Prediction.** Where a and b are known, or have been estimated, you can use Eqn [72.3] to predict any value of y for a specified value of x, e.g. during exponential growth of a cell culture (p. 231), where $\log_{10}$ cell number (y) increases as a linear function of time (x): note that in this example the dependent variable has been transformed to give a linear relationship. You will need to rearrange Eqn [72.3] in cases where a prediction of x is required for a particular value of y (e.g. in calibration curves, p. 276, or bioassays, p. 236), as follows:

$$x = (y - a) \div b \qquad [72.4]$$

This equation can also be used to determine the intercept on the x-(horizontal) axis, i.e. where $y = 0$.

Hints for some typical problems

Calculations involving proportions or ratios

The 'unitary method' is a useful way of approaching calculations involving proportions or ratios, such as those required when making up solutions from stocks (see also Chapter 26) or as a subsidiary part of longer calculations.

1. **If given a value for a multiple, work out the corresponding value for a single item or unit.**
2. **Use this 'unitary value' to calculate the required new value.**

Calculations involving series

Series (used in, for example, dilutions, see also p. 154) can be of three main forms:

1. **Arithmetic,** where the *difference* between two successive numbers in the series is a constant, e.g. 2, 4, 6, 8, 10, . . .
2. **Geometric,** where the *ratio* between two successive numbers in the series is a constant, e.g. 1, 10, 100, 1000, 10 000, . . .
3. **Harmonic,** where the values are reciprocals of successive whole numbers, e.g. $1, \dfrac{1}{2}, \dfrac{1}{3}, \dfrac{1}{4}, \ldots$

> **Examples** For a geometric dilution series involving tenfold dilution steps, calculation of concentrations is straightforward, e.g. two serial decimal dilutions (= 100-fold dilution) of a solution of NaCl of 250 mmol L^{-1} will produce a dilute solution of $250 \div 100 = 2.5$ mmol L^{-1}. Similarly, for an arithmetic dilution series, divide by the overall dilution to give the final concentration, e.g. a 16-fold dilution of a solution of NaCl of 200 mg mL^{-1} will produce a dilute solution of $200 \div 16 = 12.5$ mg mL^{-1}.

Note that the logs of the numbers in a geometric series will form an arithmetic series (e.g. 0, 1, 2, 3, 4, . . . in the above case). Thus, if a quantity y varies with a quantity x such that the rate of change in y is proportional to the value of y (i.e. it varies in an exponential manner), a semi-log plot of such data will form a straight line. This form of relationship is relevant for exponentially growing cell cultures (p. 231) and radioactive decay (p. 298).

Statistical calculations

The need for long, complex calculations in statistics has largely been removed because of the widespread use of spreadsheets with statistical functions (Chapter 13) and specialised programs such as SPSS and Minitab. It is, however, important to understand the principles behind what you are trying to do (see Chapters 73 and 74) and interpret the program's output correctly, either using the 'help' function or a reference manual.

Problems in Mendelian genetics

These cause difficulties for many students. The key is to recognise the different types of problem and to practise so you are familiar with the techniques for solving them. Chapter 64 deals with the different types of cross you will come across and methods of analysing them, including the use of the χ^2 (Chi^2) test.

Text references

Adams, D.S. (2003) *Lab Math: a Handbook of Measurement, Calculations, and Other Quantitative Skills for Use at the Bench.* Cold Spring Harbor Lab Publishing, Cold Spring Harbor.

Eason, G., Coles, C.W. and Gettinby, G. (1992) *Mathematics and Statistics for the Bio-Sciences.* Ellis Horwood, Chichester.

Stephenson, F.H. (2010) *Calculations for Molecular Biology and Biotechnology,* 2nd edn. Academic Press, London.

Sources for further study

Anon. *S.O.S. Mathematics.*
Available: http://www.sosmath.com
Last accessed 01/09/15.
[A basic web-based guide with very wide coverage.]

Britton, N.F. (2003) *Essential Mathematical Biology.* Springer-Verlag. New York.
[Describes a range of mathematical applications in population dynamics, epidemiology, genetics, biochemistry and medicine.]

Cann, A.J. (2002) *Maths from Scratch for Biologists.* Wiley, Chichester.
[Deals with basic manipulations, formulae, units, molarity, logs and exponents as well as basic statistical procedures.]

Causton, D.R. (1992) *A Biologist's Basic Mathematics.* Cambridge University Press, Cambridge.

Forster, P.C. (1999) *Easy Mathematics for Biologists.* Harwood, Amsterdam.

[Covers basic principles, units, logarithms, ratios and proportions, concentrations, equations, rates and graphs.]

Harris, M., Taylor, G. and Taylor, J. (2005) *Catch Up Maths and Stats for the Life and Medical Sciences.* Scion, Bloxham.
[Covers a range of basic mathematical operations and statistical procedures.]

Koehler, K.R. *College Physics for Students of Biology and Chemistry.* Available: http://www.rwc.uc.edu/koehler/biophys/text.html
Last accessed 01/09/15.
[A 'hypertextbook' written for first-year undergraduates. Assumes that you have a working knowledge of algebra.]

Lawler, G. (2011) *Understanding Maths. Basic Mathematics Explained. Studymates,* 4th edn. Aber Publishing, Abergele.

STUDY EXERCISES

72.1 Rearrange a simple formula. The Beer–Lambert relationship, Eqn [47.3] is written in the form $A = \varepsilon l\, [C]$. Rearrange, in the form:

(a) $[C] =$

(b) $\varepsilon =$

72.2 Rearrange the following formulae:

(a) If $y = ax + b$, find b

(b) If $y = ax + b$, find x

(c) If $x = y^3$, find y

(d) If $x = 3^y$, find y

(e) If $x = (1 - y)(z^p + 3)$, find z

(f) If $x = (y - z)^{1/n}/pq$, find n

72.3 Work with decimal places or significant figures. Give the following numbers to the accuracy indicated:

(a) 214.51 to three significant figures

(b) 107 029 to three significant figures

(c) 0.0450 to one significant figure

(d) 99.817 to two decimal places

(e) 99.897 to two decimal places

(f) 99.997 to two decimal places

(g) 6255 to the nearest 10

(h) 134 903 to the nearest ten thousand

State the following:

(i) the number of significant figures in 3400

(j) the number of significant figures in 3400.3

(k) the number of significant figures in 0.001 67

(l) the number of significant figures in 1.001 67

(m) the number of decimal places in 34.46

(n) the number of decimal places in 0.001 67

72.4 Carry out calculations involving percentages. Answer the following questions, giving your answers to two decimal places:

(a) What is 6/35ths expressed as a percentage?

(b) What is 0.0755 expressed as a percentage?

(c) What is 4.35% of 322?

(d) A rat's weight increased from 55.23 to 75.02 g. What is the % increase in its weight?

72.5 Practise using exponents and logarithms.

(a) Write out: (i) the charge on an electron and (ii) the speed of light *in vacuo* in longhand (i.e. without using powers of 10). See Table 31.4 (p. 197) for values given in scientific notation.

(b) Compute the following values: 10^{3624}; $\log(6.37)$; $e^{-2.32}$; $\ln(1123)$; $6^{3.2}$. A calculator should be used, but round the output to give five significant figures.

72.6 Practise working with linear functions (note also that Chapter 42 includes study exercises based on linear functions and plotting straight lines). Assuming a linear relationship between x and y, calculate the following (give your answers to three significant figures):

(a) x, where $y = 7.0$, $a = 4.5$ and $b = 0.02$;

(b) x, where $y = 15.2$, $a = -2.6$ and $b = -4.46$;

(c) y, where $x = 10.5$, $a = 0.2$ and $b = -0.63$;

(d) y, where $x = 4.5$, $a = -1.8$ and $b = 4.1$.

Answers to these study exercises are available at www.pearsoned.co.uk/practicalskills.

Use of symbols – *Y* is used in Chapters 73 and 74 to signify the dependent variable in statistical calculations (following the example of Sokal and Rohlf, 2011, and Wardlaw, 2000). Note, however, that some authors use *X* or *x* in analogous formulae and many calculators refer to, e.g., $\bar{x}$, Σx^2, etc. for their statistical functions.

Summarising your results – original data belong only in your primary record, either in laboratory books or as computer records. You should produce summary tables to condense and describe original data.

The purpose of most practical work is to observe and measure a particular characteristic of a biomolecule or biological system. However, it would be extremely rare if the same value was obtained every time the characteristic was measured, or with every experimental subject. More commonly, such measurements will show variability, caused by measurement error, sampling variation and/or biological variability (p. 193). Such variability can be displayed as a frequency distribution (e.g. Fig. 70.2), where the *y* axis shows the number of times (frequency, *f*) each particular value of the measured (dependent) variable (*Y*) has been obtained. Descriptive (or summary) statistics quantify aspects of the frequency distribution of a sample. You can use them to condense a large data set, for presentation in figures or tables. An additional application of descriptive statistics is to provide estimates of the true values of the underlying frequency distribution of the population being sampled, allowing the precision and significance of the observations to be assessed (pp. 520, 526).

> **KEY POINT** The appropriate descriptive statistics to choose will depend on both the type of data, i.e. whether quantitative, ranked or qualitative (p. 191), and the nature of the underlying frequency distribution.

In many instances, the normal (Gaussian) distribution (Chapter 74) best describes the observed pattern for a quantitative variable, giving a symmetrical, bell-shaped frequency distribution (p. 528); for example, measurements of a quantitative continuous variable in a number of individuals (e.g. blood plasma pH of several adult males), or replicate measurements of a particular characteristic (e.g. repeated measurements of the blood plasma pH of an individual adult male). If you have no clear theoretical grounds for assuming what the underlying frequency distribution is like, graph one or more sample frequency distributions, ideally with a sample size >100.

Three important features of a frequency distribution that can be summarised by descriptive statistics are:

1. **the sample's location,** i.e. its position along a given dimension representing the dependent (measured) variable (Fig. 73.1);
2. **the dispersion of the data,** i.e. how spread out the values are (Fig. 73.2);
3. **the shape of the distribution,** i.e. whether symmetrical, skewed, U-shaped (Fig. 73.3).

Measuring location

Here, the objective is to pinpoint the 'centre' of the frequency distribution, i.e. the value about which most of the data are grouped. The chief measures of location are the mean, median and mode. Fig. 73.4 shows how to choose among these for a given data set.

Mean

The mean (denoted $\bar{Y}$ and also referred to as the arithmetic mean) is the average value of the data. It is obtained from the sum of all the data values divided by the number of observations (in symbolic terms, $\Sigma Y/n$).

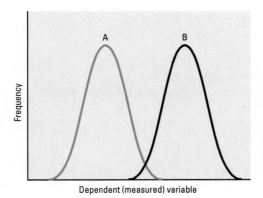

Fig. 73.1 Two distributions with different locations but the same dispersion. The data set labelled B could have been obtained by adding a constant to each datum in the data set labelled A.

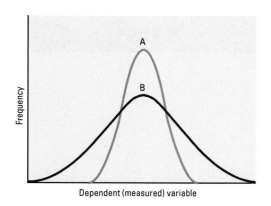

Fig. 73.2 Two distributions with different dispersions but the same location. The data set labelled A covers a relatively narrow range of values of the dependent (measured) variable while that labelled B covers a wider range.

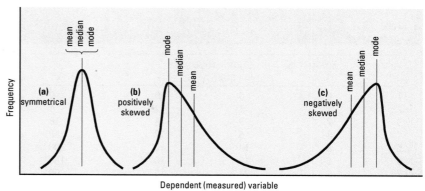

Fig. 73.3 Symmetrical and skewed frequency distributions, showing relative positions of mean, median and mode.

Example   Box 73.1 shows a set of data and the calculated values of the measures of location, dispersion and shape for which methods of calculation are outlined here. Check your understanding by calculating the statistics yourself and confirming that you arrive at the same answers.

Definitions

An outlier – any datum that has a value much smaller or bigger than most of the other data values.

Rank – the position of a data value when all the data are placed in order of ascending magnitude. If ties occur, an average rank of the tied variates is used. Thus, the rank of the datum 6 in the sequence 1,3,5,6,8,8,10 is 4; the rank of each datum with value 8 is 5.5.

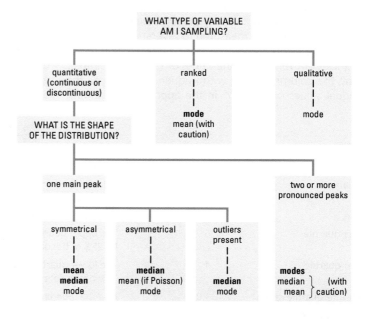

Fig. 73.4 Choosing a statistic for characterising a distribution's location. Statistics written in bold are the preferred option(s).

The mean is a good measure of the centre of symmetrical frequency distributions. It uses all of the numerical values of the sample and therefore incorporates all of the information content of the data. However, the value of a mean is greatly affected by the presence of extreme values (outliers). The arithmetic mean is a widely used statistic in the biosciences, but there are situations when you should be careful about using it (see Box 73.2 for examples).

Median

The median is the mid-point of the observations when ranked in increasing order. For odd-sized samples, the median is the middle observation; for even-sized samples it is the mean of the middle pair of observations. Where data are grouped into classes, the median can only be estimated. This is most simply done from a graph of the cumulative frequency distribution,

Box 73.1 Descriptive statistics for a sample of data – an example

Value (Y)	Frequency (f)	Cumulative frequency	fY	fY²
1	0	0	0	0
2	1	1	2	4
3	2	3	6	18
4	3	6	12	48
5	8	14	40	200
6	5	19	30	180
7	2	21	14	98
8	0	21	0	0
Totals	21 = Σf(=n)		104 = ΣfY	548 = ΣfY²

In this example, for simplicity and ease of calculation, integer values of Y are used. In many practical exercises, where continuous variables are measured to several significant figures and where the number of data values is small, giving frequencies of 1 for most of the values of Y, it may be simpler to omit the column dealing with frequency and list all the individual values of Y and Y^2 in the appropriate columns.

To gauge the underlying frequency distribution of such data sets, you would need to group individual data into broader classes (e.g. all values between 1.0 and 1.9, all values between 2.0 and 2.9, etc.) and then draw a histogram (p. 494). Calculation of certain statistics for data sets that have been grouped in this way (e.g. median, quartiles, extremes) can be tricky, and a statistical text should be consulted.

Statistic	Value*	How calculated
Mean	4.95	$\Sigma fY/n$, i.e. 104/21
Median	5	Value of the $(n + 1)/2$ variate, i.e. the value ranked $(21 + 1)/2 = 11$th (obtained from the cumulative frequency column)
Mode	5	The most common value (Y value with highest frequency)
Upper quartile	6	The upper quartile is between the 16th and 17th values, i.e. the value exceeded by 25% of the data values
Lower quartile	4	The lower quartile is between the 5th and 6th values, i.e. the value exceeded by 75% of the data values
Semi-interquartile range	1.0	Half the difference between the upper and lower quartiles, i.e. $(6 - 4)/2$
Upper extreme	7	Highest Y value in data set
Lower extreme	2	Lowest Y value in data set
Range	5	Difference between upper and lower extremes
Variance (s^2)	1.65	$$s^2 = \frac{\Sigma fY^2 - \left(\Sigma fY\right)^2/n}{n - 1}$$ $$= \frac{548 - (104)^2/21}{20}$$
Standard deviation (s)	1.28	$\sqrt{s^2}$
Standard error (SE)	0.280	$s/\sqrt{n}$
95% confidence limits	4.36–5.54	$\overline{Y} \mp t_{0.05}[20] \times SE$, (where $t_{0.05}[20] = 2.09$, Table 74.2)
Coefficient of variation (cov)	25.9%	$100s/\overline{Y}$

*Rounded to three significant figures (see p. 508), except when it is an exact number.

Box 73.2 **Three examples where simple arithmetic means are inappropriate**

Mean	*n*
6	4
7	7
8	1

1. **If means of samples are themselves meaned, an error can arise if the samples are of different size.** For example, the arithmetic mean of the means in the table shown left is 7, but this does not take account of the different 'reliabilities' of each mean owing to their sample sizes. The correct weighted mean is obtained by multiplying each mean by its sample size (*n*) (a 'weight') and dividing the sum of these values by the total number of observations, i.e. in the case shown, $(24 + 49 + 8)/12 = 6.75$.

2. **When making a mean of ratios (e.g. percentages) for several groups of different sizes, the ratio for the combined total of all the groups is not the mean of the proportions for the individual groups.** For example, if 20 rats from a batch of 50 are male, this implies 40% are male. If 60 rats from a batch of 120 are male, this implies 50% are male. The mean percentage of males $(50 + 40)/2 = 45\%$ is *not* the percentage of males in the two groups combined, because there are $20 + 60 = 80$ males in a total of 170 rats $= 47.1\%$ approx.

pH value	$[H^+]$ (mol L^{-1})
6	1×10^{-6}
7	1×10^{-7}
8	1×10^{-8}
mean	3.7×10^{-7}
$-\log_{10}$ mean	6.43

3. **If the measurement scale is not linear, arithmetic means may give a false value.** For example, if three media had pH values 6, 7 and 8, the appropriate mean pH is not 7 because the pH scale is logarithmic. The definition of pH is $-\log_{10}[H^+]$, where $[H^+]$ is expressed in mol l^{-1} ('molar'); therefore, to obtain the true mean, convert data into $[H^+]$ values (i.e. put them on a linear scale) by calculating $10^{(-pH\,value)}$ as shown. Now calculate the mean of these values and convert the answer back into pH units. Thus, the appropriate answer is pH 6.43 rather than 7. Note that a similar procedure is necessary when calculating statistics of dispersion in such cases, so you will find these almost certainly asymmetric about the mean.

Mean values of log-transformed data are often termed geometric means – they are sometimes used in microbiology and in cell culture studies, where log-transformed values for cell density counts are averaged and plotted (p. 231), rather than using the raw data values. The use of geometric means in such circumstances serves to reduce the effects of outliers on the mean.

but can also be worked out by assuming the data to be evenly spread within the class. The median may represent the location of the main body of data better than the mean when the distribution is asymmetric or when there are outliers in the sample.

Mode

The mode is the most common value in the sample. The mode is easily found from a tabulated frequency distribution as the most frequent value. If data have been grouped into classes then the term modal class is used for the class containing most values. The mode provides a rapidly and easily found estimate of sample location and is unaffected by outliers. However, the mode is affected by chance variation in the shape of a sample's distribution and it may lie distant from the obvious centre of the distribution.

The mean, median and mode have the same units as the variable under discussion. However, whether these statistics of location have the same or

Describing the location of qualitative data – the mode is the only statistic that is suitable for this task. For example, 'the modal (most frequent) eye colour was blue'.

similar values for a given frequency distribution depends on the symmetry and shape of the distribution. If it is near-symmetrical with a single peak, all three will be very similar; if it is skewed or has more than one peak, their values will differ to a greater degree (see Fig. 73.3).

Measuring dispersion

Here, the objective is to quantify the spread of the data about the centre of the distribution. Figure 73.5 indicates how to decide which measure of dispersion to use.

Range

The range is the difference between the largest and smallest data values in the sample (the extremes) and has the same units as the measured variable. The range is easy to determine, but is greatly affected by outliers. Its value may also depend on sample size: in general, the larger this is, the greater will be the range. These features make the range a poor measure of dispersion for many practical purposes.

Semi-interquartile range

The semi-interquartile range is an appropriate measure of dispersion when a median is the appropriate statistic to describe location. For this, you need to determine the first and third quartiles, i.e. the medians for those data values ranked below and above the median of the whole data set (see Fig. 73.6). To calculate a semi-interquartile range for a data set:

1. **rank the observations in ascending order;**
2. **find the values of the first and third quartiles;**
3. **subtract the value of the first quartile from the value of the third;**
4. **halve this number.**

> **Example** In a sample of data with values 3, 7, 15, 8, 5, 10 and 4, the range is 12 (i.e. the difference between the highest value, 15, and the lowest value, 3).

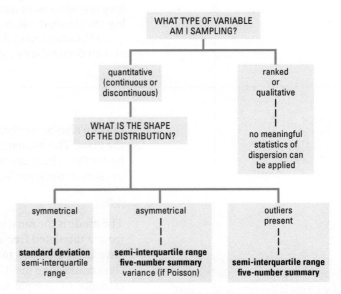

Fig. 73.5 Choosing a statistic for characterising a distribution's dispersion. Statistics written in bold are the preferred option(s). Note that you should match statistics describing dispersion with those you have used to describe location, i.e. standard deviation with mean, semi-interquartile range with median.

Fig. 73.6 Illustration of median, quartiles, range and semi-interquartile range.

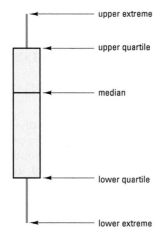

Fig. 73.7 A box and whisker plot, showing the 'five number summary' of a sample as it might be used on a graph.

For data grouped in classes, the semi-interquartile range can only be estimated. Another disadvantage is that it takes no account of the shape of the distribution at its edges. This objection can be countered by using the so-called 'five number summary' of a data set, which consists of the three quartiles and the two extreme values; this can be presented on graphs as a box and whisker plot (see Fig. 73.7) and is particularly useful for summarising skewed frequency distributions. The corresponding 'six number summary' includes the sample's size.

Variance and standard deviation

For symmetrical frequency distributions, an ideal measure of dispersion would take into account each value's deviation from the mean and provide a measure of the average deviation from the mean. Two such statistics are the sample variance, which is the sum of squares ($\Sigma(Y - \overline{Y})^2$) divided by $n - 1$ (where n is the sample size), and the sample standard deviation, which is the positive square root of the sample variance.

The variance (s^2) has units which are the square of the original units, while the standard deviation $(s,$ or SD) is expressed in the original units, one reason s is often preferred as a measure of dispersion. Calculating s or s^2 longhand is a tedious job and is best done with the help of a calculator or computer. If you do not have a calculator that calculates s for you, an alternative formula that simplifies calculations is:

$$s = + \sqrt{\frac{\sum Y^2 - (\sum Y)^2/n}{n - 1}}$$ [73.1]

To calculate s using a calculator:

1. Obtain ΣY, square it, divide by n and store in memory;
2. Square Y values, obtain ΣY^2, subtract the memory value from this;

Using a calculator for statistics – make sure you understand how to enter individual data values and which keys will give the sample mean (usually shown as $\overline{X}$ or x) and sample standard deviation (often shown as σ_{n-1}). In general, you should not use the population standard deviation (usually shown as σ_n).

Example   Consider two methods of bio-assay for a toxin in fresh water. For a given standard, Method A gives a mean result of = 50 'response units' with $s = 8$, while Method B gives a mean result of = 160 'response units' with $s = 18$. Which bioassay gives the more reproducible results? The answer can be found by calculating the CoV values, which are 16 and 11.25%, respectively. Hence, Method B is the more precise (≡ reproducible), even though the absolute value of s is larger.

Fig. 73.8 Frequency distribution of sample means around the population mean (μ). Note that SE is equivalent to the standard deviation of the sample means, for sample size = n.

3. Divide this answer by $n - 1$.

4. Take the positive square root of this value.

Take care to retain significant figures, or errors in the final value of s will result. If continuous data have been grouped into classes, the class mid-values or their squares must be multiplied by the appropriate frequencies before summation (see example in Box 73.1). When data values are large, longhand calculations can be simplified by coding the data, e.g. by subtracting a constant from each datum, and decoding when the simplified calculations are complete (see Sokal and Rohlf, 2011).

Coefficient of variation

The coefficient of variation (CoV) is a dimensionless measure of variability relative to location which expresses the sample standard deviation, usually as a percentage of the sample mean, i.e.

$$\text{CoV} = 100 s / \overline{Y}(\%)$$ [73.2]

This statistic is useful when comparing the relative dispersion of data sets with widely differing means, or where different units have been used for the same or similar quantities.

A useful application of the CoV is to compare different analytical methods or procedures, so that you can decide which involves the least proportional error – create a standard stock solution, then base your comparison on the results from several subsamples analysed by each method. You may find it useful to use the CoV to compare the precision of your own results with those of a manufacturer, e.g. for a pipettor (p. 147). The smaller the CoV, the more precise (repeatable) is the apparatus or technique. *(Note: this does not mean that it is necessarily more accurate, see p. 193.)*

Measuring the precision of the sample mean as an estimate of the true value using the standard error

Most practical exercises are based on a limited number of individual data values (a sample) which are used to make inferences about the population from which they were drawn. For example, the haemoglobin content might be measured in blood samples from 100 adult females and used as an estimate of the adult female haemoglobin content, with the sample mean ($\overline{Y}$) and sample standard deviation (s) providing estimates of the true values of the underlying population mean (μ) and the population standard deviation (σ). The reliability of the sample mean as an estimate of the true (population) mean can be assessed by calculating a statistic termed the standard error of the sample mean (often abbreviated to standard error or SE), from:

$$\text{SE} = s / \sqrt{n}$$ [73.3]

Strictly, the standard error is an estimate of the dispersion of repeated sample means around the true (population) value: if several samples were taken, each with the same number of data values (n), then their means would cluster around the population mean (μ) with a standard deviation equal to SE, as shown in Fig. 73.8. Therefore, the *smaller* the SE, the more reliable the sample mean is likely to be as an estimate of the true value, since the underlying frequency distribution would be more tightly clustered around μ. At a practical level, Eqn [73.3] shows that SE is directly affected by the dispersion of individual data values within the sample, as represented by the sample standard deviation (s). Perhaps more importantly, SE is inversely related to the *square root* of the number of data values (n). Therefore, if you wanted to increase the precision of a sample mean by a factor of 2 (i.e. to reduce SE by half), you would have to increase n by a factor of 2^2 (i.e. fourfold).

Summary descriptive statistics for the sample mean are often quoted as $\overline{Y} \pm \text{SE}$ (n), with the SE being given to one significant figure more than the mean. You can use such information to carry out a *t*-test between two sample means (Box 74.1); the SE is also useful because it allows calculation of confidence limits for the sample mean (p. 533).

Describing the 'shape' of frequency distributions

Frequency distributions may differ in the following characteristics:

- **number of peaks;**
- **skewness or asymmetry;**
- **kurtosis or pointedness.**

The shape of a frequency distribution of a small sample is affected by chance variation and may not be a fair reflection of the underlying population frequency distribution: check this by comparing repeated samples from the same population or by increasing the sample size. If the original shape were caused by random events, it should not appear consistently in repeated samples and should become less obvious as sample size increases.

Genuinely bimodal or polymodal distributions may result from the combination of two or more unimodal distributions, indicating that more than one underlying population is being sampled (Fig. 73.9). An example of a bimodal distribution is the height of adult humans (females and males combined).

A distribution is skewed if it is not symmetrical, a symptom being that the mean, median and mode are not equal (Fig. 73.3). Positive skewness is where the longer 'tail' of the distribution occurs for higher values of the measured variable; negative skewness where the longer tail occurs for lower values. Some biological examples of characteristics distributed in a skewed fashion are volumes of plant protoplasts, insulin levels in human plasma and bacterial colony counts.

Kurtosis is the name given to the 'pointedness' of a frequency distribution. A platykurtic frequency distribution is one with a flattened peak, while a leptokurtic frequency distribution is one with a pointed

---

Example Summary statistics for the sample mean and standard error for the data shown in Box 73.1 would be quoted as 4.95 ± 0.280 ($n = 21$).

Calculating the extent of skew and kurtosis of a data set – use the SKEW and KURT functions in Microsoft *Excel*.

Fig. 73.9 Frequency distributions with different numbers of peaks. A unimodal distribution (a) may be symmetrical or asymmetrical. The dotted lines in (b) indicate how a bimodal distribution could arise from a combination of two underlying unimodal distributions. Note here how the term 'bimodal' is applied to any distribution with two major peaks – their frequencies do not have to be exactly the same.

Fig. 73.10 Examples of the two types of kurtosis.

peak (Fig. 73.10). While descriptive terms can be used, based on visual observation of the shape and direction of skew, the degree of skewness and kurtosis can be quantified and statistical tests exist to test the 'significance' of observed values (see Sokal and Rohlf, 2011), but the calculations required are complex and best done with the aid of a computer.

Using computers to calculate descriptive statistics

There are many specialist statistical packages (e.g. SPSS) that can be used to simplify the process of calculation of statistics. Note that correct interpretation of the output requires an understanding of the terminology used and the underlying process of calculation, and this may best be obtained by working through one or more examples by hand before using these tools. Spreadsheets offer increasingly sophisticated statistical analysis functions, some examples of which are provided in Box 73.3 for Microsoft *Excel*.

Box 73.3 How to use a spreadsheet (Microsoft *Excel*) to calculate descriptive statistics

Method 1: Using spreadsheet functions to generate the required statistics. Suppose you had obtained the following set of data, stored within an array (block of columns and rows) of cells (A2:L6) within a spreadsheet:

	A	B	C	D	E	F	G	H	I	J	K	L
1	My data set											
2	4	4	3	3	5	4	3	7	7	3	5	3
3	6	2	9	7	3	4	5	6	6	9	4	8
4	5	3	2	5	4	5	7	2	8	3	6	3
5	11	3	5	2	4	3	7	8	4	4	4	3
6	3	6	8	5	6	4	3	4	3	6	10	5

Taking Microsoft *Excel* as an example, the following functions could be used to extract descriptive statistics from this data set:

Descriptive statistic	Example of use of function[a,b]	Result for the above data set
Sample size *n*	=COUNT(A2:L6)	60
Mean	=AVERAGE(A2:L6)[c]	4.9
Median	=MEDIAN(A2:L6)	4.0
Mode	=MODE(A2:L6)	3
Upper quartile	=QUARTILE(A2:L6,3)[d]	6.0
Lower quartile	=QUARTILE(A2:L6,1)	3.0
Semi-interquartile range	=QUARTILE(A2: L6,3)–QUARTILE(A2:L6,1)	3.0
Upper extreme	=QUARTILE(A2:L6,4) *or* = MAX(A2:L6)	11
Lower extreme	=QUARTILE(A2:L6,0) *or* = MIN(A2:L6)	2
Range	=MAX(A2:L6)–MIN(A2:L6)[e]	9.0
Variance	=VAR(A2:L6)	4.464
Standard deviation	=STDEV(A2:L6)	2.113
Standard error	=STDEV(A2:L6)/(SQRTCOUNT(A2:L6))[f]	0.273
Coefficient of variation	= 100*STDEV(A2:L6)/AVERAGE(A2:L6)	43.12%

Notes:
[a] Typically for Excel, in an appropriate cell, you would select the *Formulas* tab, then *More Functions > Statistical > COUNT*, then select the input range and press return.

(continued)

Box 73.3 (continued)

[b] Other descriptive statistics can be calculated – these mirror those shown in Box 73.1, but for this specific data set.
[c] There is no 'MEAN' function in Microsoft *Excel*.
[d] The first argument within the brackets relates to the array of data, the second relates to the quartile required (consult the *Help on this function* feature for further information).
[e] There is no direct 'RANGE' function in Microsoft *Excel*.
[f] There is no direct 'STANDARD ERROR' function in Microsoft *Excel*. The SQRT function returns a square root and the COUNT function determines the number of filled data cells in the array.

Method 2: using the *Tools > Data Analysis* option. This can automatically generate a table of descriptive statistics for the data array selected, although the data must be presented as a single row or column. This option might need to be installed for your network or personal computer before it is available to you (consult the *Help* feature for details). Having entered or rearranged your data into a row or column, the steps involved are as follows:

1. Select *Data > Data Analysis*.

2. From the *Data Analysis* box, select *Descriptive Statistics*.

3. Input your data location into the *Input Range* (left click and hold down to highlight the column of data).

4. From the menu options, select *Summary Statistics* and *Confidence Level for Mean: 95%*.

5. When you click *OK*, you should get a new worksheet, with descriptive statistics and confidence limits shown. Alternatively, at step 3, you can select an area of your current worksheet as a data output range (select an area away from any existing content as these cells would otherwise be overwritten by the descriptive statistics output table).

6. Change the format of the cells to show each number to an appropriate number of decimal places. You may also wish to make the columns wider so you can read their content.

7. For the data set shown above, the final output table should look as shown below.

Descriptive statistics for a data set

Column 1[a,b]	
Mean	4.9
Standard error	0.27
Median	4.0
Mode	3
Standard deviation	2.113
Sample variance	4.464
Kurtosis	0.22
Skewness	0.86
Range	9.00
Minimum	2.0
Maximum	11.0
Sum	294
Count	60
Confidence level (95.0%)	0.55

Notes:
[a] These descriptive statistics are specified (and are automatically presented in this order) – any others required can be generated using Method 1.
[b] A more descriptive heading can be added if desired – this is the default.

Note: instructions illustrated here may vary among the different versions of Microsoft *Office* programs. Check the exact functions and syntax using the *Insert function* menu option (*fx* button in some versions).

Text references and sources for further study

Dytham, C. (2011) *Choosing and Using Statistics: A Biologists' Guide,* 3rd edn. Blackwell, Oxford.

Samuels, M.L., Witmer, J.A. and Schaffner, A. (2011) *Statistics for the Life Sciences,* 4th edn. Addison Wesley, Harlow.

Schmuller, J. (2013) *Statistical Analysis with Excel for Dummies,* 3rd edn. Wiley, Hoboken.

Sokal, R.R. and Rohlf, F.J. (2011) *Biometry,* 4th edn. W.H. Freeman and Co., San Francisco.

Wardlaw, A.C. (2000) *Practical Statistics for Experimental Biologists,* 2nd edn. Wiley, New York.

STUDY EXERCISES

73.1 Practise calculating descriptive statistics. Using the data set given in Box 73.3 (p. 522), calculate the following statistics:

(a) range

(b) variance

(c) standard deviation

(d) coefficient of variation

(e) standard error.

    Answers (b) to (e) should be given to three significant figures.

73.2 Calculate and interpret standard errors. Two samples, A and B, gave the following descriptive statistics (measured in the same units): Sample A, mean = 16.2, standard deviation = 12.7, number of data values = 12; Sample B, mean = 13.2, standard deviation = 14.4, number of data values = 20. Which has the lower standard error in absolute terms and in proportion to the sample mean? (Express answers to three significant figures.)

73.3 Compute a mean value correctly. A researcher finds that the mean size of larvae in three replicate samples designated A, B and C is 3.0, 2.5 and 2.0 mm, respectively. He computes the mean larval size as 2.5 mm, but forgets that the sample sizes were 24, 37 and 6, respectively. What is the true mean size of the larvae? (Answer to three significant figures.)

Answers to these study exercises are available at www.pearsoned.co.uk/practicalskills.

Definition

Null hypothesis – a statement of 'no effect' in relation to the comparison under test.

This chapter outlines the philosophy of hypothesis-testing statistics, indicates the steps to be taken when choosing a test, and discusses features and assumptions of some important tests. For details of the mechanics of tests, appropriate texts should be consulted (e.g. Sokal and Rohlf, 2011; Wardlaw, 2000). Most tests are now available in statistical packages for computers (see p. 93) and many in spreadsheets (Chapter 13).

To carry out a statistical test:

1. **decide what it is you wish to test** (create a null hypothesis and its alternative).
2. **determine whether your data fit a standard distribution pattern;**
3. **select a test and apply it to your data.**

Setting up a null hypothesis

Hypothesis-testing statistics are used to compare the properties of samples either with other samples or with some theory about them. For instance, you may be interested in whether two samples can be regarded as having different means, whether the counts of an organism in different locations can be regarded as randomly distributed, or whether characteristic A of an organism is linearly related to characteristic B.

KEY POINT You cannot use statistics to *prove* any hypothesis, but they can be used to assess *how likely* it is to be wrong.

Statistical testing operates in what at first seems a rather perverse manner. Suppose you think a treatment has an effect. The theory you actually test is that it has no effect; the test tells you how improbable your data would be if this theory were true. This 'no effect' theory is the null hypothesis (NH). If your data are very improbable under the NH, then you may suppose it to be wrong, and this would support your original idea (the 'alternative hypothesis'). The concept can be illustrated by an example. Suppose two groups of subjects were treated in different ways, and you observed a difference in the mean value of the measured variable for the two groups. Can this be regarded as a 'true' difference? As Fig. 74.1 shows, it could have arisen in two ways:

1. **Because of the way the subjects were allocated to treatments,** i.e. all the subjects liable to have high values might, by chance, have been assigned to one group and those with low values to the other (Fig. 74.1(a).
2. **Because of a genuine effect of the treatments,** i.e. each group came from a distinct frequency distribution (Fig. 74.1(b).

A statistical test will indicate the probabilities of these options. The NH states that the two groups come from the same population (i.e. the treatment effects are negligible in the context of random variation). To test this, you calculate a test statistic from the data, and compare it with tabulated critical values giving the probability of obtaining the observed or a more extreme result by chance (see Boxes 74.1 and 74.2). This probability is sometimes called the significance of the test.

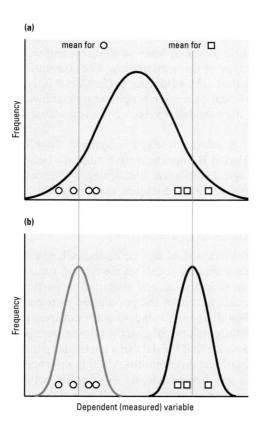

(a)

mean for ○   mean for □

(b)

Dependent (measured) variable

Fig. 74.1 Two explanations for the difference between two means. In case (a) the two samples happen by chance to have come from opposite ends of the same frequency distribution, i.e. there is no true difference between the samples. In case (b) the two samples come from different frequency distributions, i.e. there is a true difference between the samples. In both cases, the means of the two samples are the same.

Note that you must take into account the degrees of freedom (d.f) when looking up critical values of most test statistics. The d.f. is related to the size(s) of the samples studied; formulae for calculating it depend on the test being used. Biologists normally use two-tailed tests, i.e. we have no expectation beforehand that the treatment will have a positive or negative effect compared to the control (in a one-tailed test we expect one particular treatment to be bigger than the other). Be sure to use critical values for the correct type of test.

By convention, the critical probability for rejecting the NH is 5% (i.e. $P = 0.05$). This means we reject the NH if the observed result would have come up by chance a maximum of one time in twenty. If the modulus of the test statistic is less than or equal to the tabulated critical value for $P = 0.05$, then we accept the NH and the result is said to be 'not significant' (NS for short). If the modulus of the test statistic is greater than the tabulated value for $P = 0.05$, then we reject the NH in favour of the alternative hypothesis that the treatments had different effects and the result is 'statistically significant'.

Two types of error are possible when making a conclusion on the basis of a statistical test. The first occurs if you reject the NH when it is true and the second if you accept the NH when it is false. To limit the chance of the first type of error, choose a lower probability, e.g. $P = 0.01$, but note that the critical value of the test statistic increases when you do this and results in the probability of the second type of error increasing. The conventional significance levels given in statistical tables (usually 0.05, 0.01, 0.001) are arbitrary. Increasing use of statistical computer programs now allows the actual probability of obtaining the calculated value of the test statistic to be quoted (e.g. $P = 0.037$).

Note that if the NH is rejected, this does not tell you why many alternative explanations may be possible. Also, it is important to distinguish between statistical significance and biological relevance: identifying a statistically significant difference between two samples does not mean that this will carry any biological importance.

Comparing data with parametric distributions

The distribution pattern of a set of data values may be biologically relevant, but it is also of practical importance because it defines the type of statistical tests that can be used. A parametric test is one that makes particular assumptions about the mathematical nature of the population distribution from which the samples were taken. If these assumptions are not true, then the test is obviously invalid, even though it might give the answer we expect. A non-parametric test does not assume that the data fit a particular pattern, but it may assume characteristics of the distributions. Used in appropriate circumstances, parametric tests are better able to distinguish between true but marginal differences between samples than their non-parametric equivalents (i.e. they have greater 'power').

The properties of the main distribution types found in biology are given below, with both rules-of-thumb and more rigorous tests for deciding whether data fit these distributions.

Binomial distributions

These apply to samples of any size from populations when data values occur independently in only two mutually exclusive classes (e.g. type A or type B). They describe the probability of finding the different

---

Definition

Modulus – the absolute value of a number, e.g. modulus −3.385 = 3.385.

Quoting significance – the convention for quoting significance levels in text, tables and figures is as follows:
$P > 0.05$ = 'not significant' (or NS)
$P \leq 0.05$ = 'significant' (or *)
$P \leq 0.01$ = 'highly significant' (or **)
$P \leq 0.001$ = 'very highly significant' (or ***).

Thus, you might refer to a difference in means as being 'highly significant ($P \leq 0.01$)'. For this reason, the word 'significant' in its everyday meaning of 'important' or 'notable' should be used with care in scientific writing. The asterisk coding shown above is most often used as a shortform in tables and should be explained in the table legend, or in a footnote.

Choosing between parametric and non-parametric tests – always plot your data graphically when determining whether they are suitable for parametric tests as this may save a lot of unnecessary effort later.

Fig. 74.2 Examples of binomial frequency distributions with different probabilities. The distributions show the expected frequency of obtaining n individuals of type A in a sample of 5. Here P is the probability of an individual being type A rather than type B.

possible combinations of the attribute for a specified sample size k (e.g. out of ten specimens, what is the chance of eight being type A). If p is the probability of the attribute being of type A and q the probability of it being type B, then the expected mean sample number of type A is kp and the standard deviation is $\sqrt{kpq}$. Expected frequencies can be calculated using mathematical expressions (see Sokal and Rohlf, 2011). Examples of the shapes of some binomial distributions are shown in Fig. 74.2. Note that they are symmetrical in shape for the special case $p = q = 0.5$ and the greater the disparity between p and q, the more skewed the distribution.

Some biological examples of data likely to be distributed in binomial fashion are: possession of two alleles for seed coat morphology (e.g. smooth and wrinkly); whether an organism is infected with a microbe or not; whether an animal is male or female. Binomial distributions are particularly useful for predicting gene segregation in Mendelian genetics and can be used for testing whether combinations of events have occurred more frequently than predicted (e.g. more siblings being of the same sex than expected). To establish whether a set of data is distributed in binomial fashion, calculate expected frequencies from probability values obtained from theory or observation, then test against observed frequencies using a χ^2-test (p. 445) or a G-test (see Wardlaw, 2000).

Poisson distributions

These apply to discrete characteristics which can assume low whole number values, such as counts of events occurring in area, volume or time. The events should be 'rare' in that the mean number observed should be a small proportion of the total that could possibly be found. Also, finding one count should not influence the probability of finding another. The shape of Poisson distributions is described by only one parameter, the mean number of events observed, and has the special characteristic that the variance is equal to the mean. The shape has a pronounced positive skewness at low mean counts, but becomes more and more symmetrical as the mean number of counts increases (Fig. 74.3).

Some examples of characteristics distributed in a Poisson fashion are: number of microbes per unit volume of medium; number of animals parasitised per unit time; number of radioactive disintegrations per unit time. One of the main uses for the Poisson distribution is to quantify

Quantifying skew – the Microsoft *Excel* SKEW function can be used to assess the extent of skewness in a data set.

Tendency towards the normal distribution – under certain conditions, binomial and Poisson distributions can be treated as normally distributed:
- where samples from a binomial distribution are large (i.e. > 15) and p and q are close to 0.5;
- for Poisson distributions, if the number of counts recorded in each outcome is greater than about 15.

Fig. 74.3 Examples of Poisson frequency distributions differing in mean. The distributions are shown as line charts because the independent variable (events per sample) is discrete.

errors in count data such as estimates of cell densities in dilute suspensions (see p. 234). To decide whether data are Poisson distributed:

- use the rule of thumb that if the coefficient of dispersion ≈ 1, the distribution is likely to be Poisson;

- calculate 'expected' frequencies from the equation for the Poisson distribution and compare with actual values using a χ^2-test or a G-test.

It is sometimes of interest to show that data are *not* distributed in a Poisson fashion, e.g. the distribution of parasite larvae in hosts. If $s^2/\overline{Y} > 1$, the data are 'clumped' and occur together more than would be expected by chance; if $s^2/\overline{Y} < 1$, the data are 'repulsed' and occur together less frequently than would be expected by chance.

Normal distributions (Gaussian distributions)

These occur when random events act to produce variability in a continuous characteristic (quantitative variable). This situation occurs frequently in biology, so normal distributions are very useful and much used. The bell-like shape of normal distributions is specified by the population mean and standard deviation (Fig. 74.4): it is symmetrical and configured

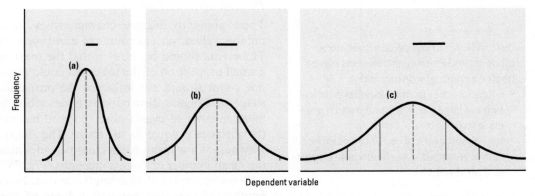

Fig. 74.4 Examples of normal frequency distributions differing in mean and standard deviation. The horizontal bars represent population standard deviations for the curves, increasing from (a) to (c). Vertical dashed lines are population means, while vertical solid lines show positions of values ±1, 2 and 3 standard deviations from the means.

Fig. 74.5 Example of a normal probability plot. The plotted points are from a small data set where the mean $\overline{Y} = 6.93$ and the standard deviation $s = 1.895$. Note that values corresponding to 0% and 100% cumulative frequency cannot be used. The straight line is that predicted for a normal distribution with $\overline{Y} = 6.93$ and $s = 1.895$. This is plotted by calculating the expected positions of points for $\overline{Y} \pm s$. Since 68.3% of the distribution falls within these bounds, the relevant points on the cumulative frequency scale are $50 \pm 34.15\%$; thus this line was drawn using the points (4.495, 15.85) and (8.285, 84.15) as indicated on the plot.

such that 68.27% of the data will lie within ±1 standard deviation of the mean, 95.45% within ±2 standard deviations of the mean, and 99.73% within ±3 standard deviations of the mean.

Some biological examples of data likely to be distributed in a normal fashion are: fresh weight of plants of the same age; linear dimensions of bacterial cells; height of either adult female or male humans. To check whether data come from a normal distribution, you can:

- use the rule of thumb that the distribution should be symmetrical and that nearly all the data should fall within ±3s of the mean and about two-thirds within ±1s of the mean;

- plot the distribution on normal probability graph paper. If the distribution is normal, the data will tend to follow a straight line (see Fig. 74.5). Deviations from linearity reveal skewness and/or kurtosis (see pp. 521–2), the significance of which, be tested statistically (see Sokal and Rohlf, 2011);

- use a suitable statistical computer program to generate predicted normal curves from the $\overline{Y}$ and s values of your sample(s). These can be compared visually with the actual distribution of data and can be used to give 'expected' values for a χ^2-test or a G-test.

The wide availability of tests based on the normal distribution and their relative simplicity means that you may wish to transform your data to make them more like a normal distribution. Table 74.1 provides transformations that might be applied. The transformed data should be tested for normality as described above before proceeding – do not forget that you may need to check that transformed variances are homogeneous for certain tests (see below).

A very important theorem in statistics, the Central Limit Theorem, states that, as sample size increases, the distribution of a series of means from any frequency distribution will become normally distributed. This fact can be used to devise an experimental or sampling strategy that ensures that data are normally distributed, i.e. using means of samples as if they were primary data.

Choosing a suitable statistical test

Comparing location (e.g. means)

If you can assume that your data are normally distributed, the main test for comparing two means from independent samples is Student's t-test

Table 74.1 Suggested transformations altering different types of frequency distribution to the normal type. To use, modify data by the formula shown; then examine effects with the tests described on pp. 526–9.

Type of data; distribution suspected	Suggested transformation(s)
Proportions (including percentages); binomial	arcsine $\sqrt{x}$ (also called the angular transformation)
Scores; Poisson	$\sqrt{x}$ or $\sqrt{(x + 1/2)}$ if zero values present
Measurements; negatively skewed	x^2, x^3, x^4, etc. (in order of increasing strength)
Measurements; positively skewed	$1/\sqrt{x}$, $\sqrt{x}$, ln x, $1/x$ (in order of increasing strength)

Understanding 'degrees of freedom' – this depends on the number of values in the data set analysed, and the method of calculation depends on the statistical test being used. It relates to the number of observations that are free to vary before the remaining quantities for a data set can be determined.

Checking the assumptions of a test – always acquaint yourself with the assumptions of a test. If necessary, test them before using the test.

(see Boxes 74.1–74.3, and Table 74.2). This assumes that the variances of the data sets are homogeneous. Tests based on the t-distribution are also available for comparing means of paired data or for comparing a sample mean with a chosen value.

When comparing means of two or more samples, analysis of variance (ANOVA) is a very useful technique. This method also assumes data are normally distributed and that the variances of the samples are homogeneous. The samples must also be independent (e.g. not subsamples). The test statistic calculated is denoted F and it has two different degrees of freedom related to the number of means tested and the pooled number of replicates per mean. The nested types of ANOVA are useful to determine the relative importance of different sources of variability in your data. Two-way and multi-way ANOVAs are useful for studying interactions between treatments.

For data satisfying the ANOVA requirements, the least significant difference (LSD) is useful for making planned comparisons among several means (see Sokal and Rohlf, 2011). Any two means that differ by more than the LSD will be significantly different. The LSD is useful for showing on graphs.

The chief non-parametric tests for comparing the locations of two samples are the Mann-Whitney U-test and the Kolmogorov–Smirnov test. The former assumes that the frequency distributions of the samples are similar, whereas the latter makes no such assumption. In both cases the sample's size must be ≥ 4 and for the Kolmogorov–Smirnov test the samples must have equal sizes. In the Kolmogorov–Smirnov test, significant differences found with the test could be caused by differences in location or shape of the distribution, or both.

Suitable non-parametric comparisons of location for paired data (sample size ≥ 6) include Wilcoxon's signed rank test, which is used for

Table 74.2 Critical values of Student's t statistic (for two-tailed tests). Reject the null hypothesis at probability P if your calculated t value equals or exceeds the value shown for the appropriate degrees of freedom $= (n_1 - 1) + (n_2 - 1)$.

Degrees of freedom	Critical values for $P = 0.05$	Critical values for $P = 0.01$	Critical values for $P = 0.001$
1	12.71	63.66	636.62
2	4.30	9.92	31.60
3	3.18	5.84	12.94
4	2.78	4.60	8.61
5	2.57	4.03	6.86
6	2.45	3.71	5.96
7	2.36	3.50	5.40
8	2.31	3.36	5.04
9	2.26	3.25	4.78
10	2.23	3.17	4.59
12	2.18	3.06	4.32
14	2.14	2.98	4.14
16	2.12	2.92	4.02
20	2.09	2.85	3.85
25	2.06	2.79	3.72
30	2.04	2.75	3.65
40	2.02	2.70	3.55
60	2.00	2.66	3.46
120	1.98	2.62	3.37
∞	1.96	2.58	3.29

Box 74.1 How to carry out a *t*-test

The *t*-test was devised by a statistician who used the pen-name 'Student', so you may see it referred to as Student's *t*-test. It is used when you wish to decide whether two samples come from the same population or from different ones (Fig. 74.1). The samples might have been obtained by observation, or by applying two different treatments to an originally homogeneous population (Chapter 32).

The null hypothesis (NH) is that the two groups can be represented as samples from the same overlying population (Fig. 74.1(a)). If, as a result of the test, you accept this hypothesis, you can say that there is no significant difference between the group means.

The alternative hypothesis is that the two groups come from different populations (Fig. 74.1(b)). By rejecting the NH as a result of the test, you can accept the alternative hypothesis and say that there is a significant difference between the sample means, or, if an experiment were carried out, that the two treatments affected the samples differently.

How can you decide between these two hypotheses? On the basis of certain assumptions (see below), and some relatively simple calculations, you can work out the probability that the samples came from the same population. If this probability is very low, then you can reasonably reject the NH in favour of the alternative hypothesis, and if it is high, you will accept the NH.

To find out the probability that the observed difference between sample means arose by chance, you must first calculate a '*t* value' for the two samples in question. Some computer programs (e.g. Minitab) provide this probability as part of the output, otherwise you can look up statistical tables (e.g. Table 74.2). These tables show 'critical values' – the borders between probability levels. If your value of *t* equals or exceeds the critical value for probability *P*, you can reject the NH at this probability ('level of significance').

Note that:

- for a given difference in the means of the two samples, the value of *t* will get larger the smaller the scatter within each data set; and

- for a given scatter of the data, the value of *t* will get larger the greater the difference between the means.

So, at what probability should you reject the NH? Normally, the threshold is arbitrarily set at 5% – you quite often see descriptions like 'the sample means were significantly different ($P < 0.05$)'. At this 'significance level' there is still up to a 5% chance of the *t* value arising by chance, so about 1 in 20 times, on average, the conclusion will be wrong. If *P* turns out to be lower, then this kind of error is much less likely.

Tabulated probability levels are generally given for 5%, 1% and 0.1% significance levels (see Table 74.2). Note that this table is designed for 'two-tailed' tests, i.e. where the treatment or sampling strategy could have resulted in either an increase or a decrease in the measured values. These are the most likely situations you will deal with in biology.

Examine Table 74.2 and note the following.

- The larger the size of the samples (i.e. the greater the 'degrees of freedom'), the smaller *t* needs to be to exceed the critical value at a given significance level.

- The lower the probability, the greater *t* needs to be to exceed the critical value.

The mechanics of the test

A calculator that can work out means and standard deviations is helpful.

1. **Work out the sample means $\bar{Y}_1$ and $\bar{Y}_2$ and calculate the difference between them.**

2. **Work out the sample standard deviations S_1 and S_2.** (*Note:* if your calculator offers a choice, choose the '*n*–1' option for calculating *s* – see pp. 519–20).

3. **Work out the sample standard errors $SE_1 = s_1/\sqrt{n_1}$ and $SE_2 = s_2/\sqrt{n_2}$; now square each, add the squares together, then take the positive square root of this** (n_1 and n_2 are the respective sample sizes, which may, or may not, be equal).

4. **Calculate *t* from the formula:**

$$t = \frac{\bar{Y}_1 - \bar{Y}_2}{\sqrt{((SE_1)^2) + ((SE_2)^2)}} \quad \text{[74.1]}$$

The value of *t* can be negative or positive, depending on the values of the means; this does not matter and you should compare the modulus (absolute value) of *t* with the values in tables.

5. **Work out the degrees of freedom $= (n_1 - 1) + (n_2 - 1)$.**

6. **Compare the *t* value with the appropriate critical value (see, e.g., Table 74.2) and decide on the significance of your finding (see p. 526).**

Box 74.2 provides a worked example – use this to check that you understand the above procedures.

Assumptions that must be met before using the test

The most important assumptions are as follows.

- The two samples are independent and randomly drawn (or, if not, drawn in a way that does not create bias). The test assumes that the samples are quite large.

- The underlying distribution of each sample is normal. This can be tested with a special statistical test, but a rule of thumb is that a frequency distribution of the data should be (a) symmetrical about the mean and (b) nearly all of the data should be within three standard deviations of the mean and about two-thirds within one standard deviation of the mean (see p. 529).

- The two samples should have uniform variances. This again can be tested (by an *F*-test), but may be obvious from inspection of the two standard deviations.

Box 74.2 **Worked example of a *t*-test**

Suppose the following data were obtained in an experiment (the units are not relevant):
Control: 6.6, 5.5, 6.8, 5.8, 6.1, 5.9
Treatment: 6.3, 7.2, 6.5, 7.1, 7.5, 7.3
Using the steps outlined in Box 74.1, the following values are obtained (denoting control with subscript 1, treatment with subscript 2):

1. $\bar{Y}_1 = 6.1167$; $\bar{Y}_2 = 6.9833$: difference between means $= \bar{Y}_1 - \bar{Y}_2 = -0.8666$.

2. $s_1 = 0.495\,65$; $s_2 = 0.475\,04$.

3. $SE_1 = 0.495\,65/2.44949 = 0.202\,348$
   $SE_2 = 0.475\,04/2.44949 = 0.193\,934$.

4. $t = \dfrac{-0.8666}{\sqrt{(0.202\,348^2 + 0.193\,934^2)}} = \dfrac{-0.8666}{0.280\,277} = -3.09$

5. $\text{d.f.} = (5 + 5) = 10$.

6. Looking at Table 74.2, we see that the modulus of this *t* value exceeds the tabulated value for $P = 0.05$ at 10 degrees of freedom ($= 2.23$). We therefore reject the NH, and conclude that the means are different at the 5% level of significance. If the modulus of *t* had been ≤ 2.23, we would have accepted the NH. If the modulus of *t* had been > 3.17, we could have concluded that the means are different at the 1% level of significance.

Fig. 74.6 Graphical representation of confidence limits as 'error bars' for (a) a sample mean in a plotted curve, where both upper and lower limits are shown; and (b) a sample mean in a histogram, where, by convention, only the upper value is shown. For data that are assumed to be symmetrically distributed, such representations are often used in preference to the 'box and whisker' plot shown on p. 519. Note that SE is an alternative way of representing sample imprecision/error (e.g. Fig. 70.1). See p. 494 for how to add error bars to Microsoft *Excel* graphs.

Confidence limits for statistics other than the mean – consult an advanced statistical text (e.g. Sokal and Rohlf, 2011) if you wish to indicate the reliability of estimates of, for example, population variances.

quantitative data and assumes that the distributions have similar shape. Dixon and Mood's sign test can be used for paired data scores where one variable is recorded as 'greater than' or 'better than' the other.

Non-parametric comparisons of location for three or more samples include the Kruskal–Wallis *H*-test. Here, the number of samples is without limit and they can be unequal in size, but again the underlying distributions are assumed to be similar. The Friedman *S*-test operates with a maximum of five samples and data must conform to a randomised block design. The underlying distributions of the samples are assumed to be similar.

Comparing dispersions (e.g. variances)

If you wish to compare the variances of two sets of data that are normally distributed, use the .*F*-test. For comparing more than two samples, it may be sufficient to use the F_{max}-test, on the highest and lowest variances. The Scheffe–Box (log-ANOVA) test is recommended for testing the significance of differences between several variances. Non-parametric tests exist but are not widely available: you may need to transform the data and use a test based on the normal distribution.

Determining whether frequency observations fit theoretical expectation

The χ^2-test (Box 64.2) is useful for tests of 'goodness of fit', e.g. comparing expected and observed progeny frequencies in genetic experiments or comparing observed frequency distributions with some theoretical function. One limitation is that simple formulae for calculating χ^2 assume that no expected number is less than 5. The *G*-test (2*I* test) is used in similar circumstances.

Comparing proportion data

When comparing proportions between two small groups (e.g. whether 3/10 is significantly different from 5/10), you can use probability tables such as those of Finney *et al.* (1963) or calculate probabilities from formulae; however, this can be tedious for large sample sizes. Certain proportions can be transformed so that their distribution becomes normal.

(a)

(b)

(c)

(d)

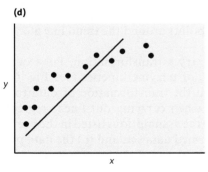

Fig. 74.7 Examples of correlation. The linear regression line is shown. In (a) and (b), the correlation between *x* and *y* is good: for (a) there is a positive correlation and the correlation coefficient, *r*, would be close to 1; for (b) there is a negative correlation and the correlation coefficient would be close to −1. In (c) there is a weak positive correlation and *r* would be close to 0. In (d) the correlation coefficient may be quite large, but the choice of linear regression is clearly inappropriate.

Placing confidence limits on an estimate of a population parameter

On many occasions, a sample statistic is used to provide an estimate of a population parameter, and it is often useful to indicate the reliability of such an estimate. This can be done by putting confidence limits on the sample statistic, i.e. by specifying an interval around the statistic within which you are confident that the true value (the population parameter) is likely to fall, at a specified level of probability. The most common application is to place confidence limits on the mean of a sample taken from a population of normally distributed data values. In practice, you determine a confidence factor for a particular level of probability which is added to and subtracted from the sample mean $(\overline{Y})$ to give the upper confidence limit and lower confidence limit respectively. These are calculated as:

$$\overline{Y} + (t_{P[n-1]} \times \text{SE}) \text{ for the upper limit and}$$
$$\overline{Y} - (t_{P[n-1]} \times \text{SE}) \text{ for the lower limit} \qquad [74.2]$$

where $t_{P[n-1]}$ is the tabulated critical value of Student's *t*-statistic for a two-tailed test with $n-1$ degrees of freedom at a specified probability level (*P*) and SE is the standard error of the sample mean (pp. 519–20). The 95% confidence limits (i.e. $P = 0.05$) tells you that, on average, 95 times out of 100, the interval between the upper and lower limits will contain the true (population) value. Confidence limits are often shown as 'error bars' for individual sample means plotted in graphical form. Figure 74.6 illustrates how this is applied to plotted curves and histograms (note that this can be carried out for data series within a Microsoft *Excel* graph (chart) using the *Layout* tab and *Error Bars* menu).

Correlation and regression

These methods are used when testing the relationship between data values for two variables. Correlation is used to measure the extent to which changes in the two sets of data values occur together in a linear manner. If one variable can be assumed to be dependent upon the other (i.e. a change in *X* causes a particular change in *Y*), then regression techniques can be used to provide a mathematical description of the underlying relationship between the variables, e.g. to find a line of best fit for a data series. If there is no *a priori* reason to assume dependency, then correlation methods alone are appropriate.

A correlation coefficient measures the strength of the linear relationship between two variables, but does not describe the relationship. The coefficient is expressed as a number between −1 and +1: a positive coefficient indicates a direct relationship, where the two variables change in the same direction, while a negative coefficient indicates an inverse relationship, where one variable decreases as the other increases (Fig. 74.7). The nearer the coefficient is to −1 or +1, the stronger the linear relationship between the variables, i.e. the less 'scatter' there would be about a straight line of best fit (note that this does *not* necessarily imply that one variable is dependent upon the other). A coefficient of 0 implies that the two variables show no linear association and therefore the closer the correlation coefficient is to zero, the weaker the linear relationship. The importance of graphing data is shown by the case illustrated in Fig. 74.7(d).

Pearson's product moment correlation coefficient (*r*) is the most commonly used statistic for testing correlations. The test is valid only if both variables are normally distributed. Statistical tests can be used to decide whether the correlation is significant, e.g. using a one-sample t-test to see whether r is significantly different from zero, based on the following equation:

$$t = r \div \sqrt{[(1 - r^2) \div (n - 2)]} \qquad [74.3]$$

at $n - 2$ degrees of freedom, where n is the number of paired observations. If one or both variables are not normally distributed, then you should calculate an alternative non-parametric coefficient, e.g. Spearman's coefficient of rank correlation (r_s) or Kendall's coefficient of rank correlation (τ). These require the two sets of data to be ranked separately, and the calculation can be complex if there are tied (equal) ranks. Spearman's coefficient is said to be better if there is any uncertainty about the reliability of closely ranked data values.

If underlying theory or empirical graphical analysis indicates a linear relationship between a dependent and an independent variable, then linear regression can be used to estimate the mathematical equation that links the two variables. Model I linear regression is the standard approach, and is available within general-purpose software programs such as Microsoft *Excel* (Box 74.3), and on some scientific calculators. It is suitable for experiments where a dependent variable Y varies with an *error-free* independent variable X in accordance with the relationship $Y = a + bX + e_Y$, where e_Y represents the residual (error) variability in the Y variable. For example, this relationship might apply in a laboratory procedure where you have carefully controlled the independent variable and the X values can be assumed to have zero error (e.g. in a calibration curve, see Chapter 42, or in a time course experiment where measurements are made at exact time points). The regression analysis gives estimates for a and b (equivalent to the slope and intercept of the line of best fit, p. 510): computer-based programs usually provide additional features, e.g. residual values for Y (e_Y), estimated errors for a and b, predicted values of Y along with graphical plots of the line of best fit (the trend line) and the residual values. In order for the model to be valid, the residual (error) values should be normally distributed around the trend line and their variance should be uniform (homogeneous), i.e. there should be a similar scatter of data points around the trend line along the x axis (independent variable).

If the relationship is not linear, try a transformation. For example, this is commonly done in analysis of enzyme kinetics (see Fig. 61.5). However, you should be aware that the transformation of data to give a straight line can lead to errors when carrying out linear regression analysis: take care to ensure that (a) the assumptions listed in the previous paragraph are valid for the transformed data set, and (b) the data points are evenly distributed throughout the range of the independent variable. If these criteria cannot be met, non-linear regression may be a better approach, but for this you will require a suitable computer program, e.g. GraphPad Prism.

The strength of the relationship between Y and X in model I linear regression is best estimated by the coefficient of determination (r^2 or R^2), which is equivalent to the square of the Pearson correlation coefficient. The coefficient of determination varies between 0 and +1 and provides a measure of the goodness of fit of the Y data to the regression line: the closer the value is to 1, the better the fit. In effect, r^2 represents the fraction

---

Using more advanced types of regression – these include the following.

- Model II linear regression, which applies to situations where a dependent variable Y varies with an independent variable X, and where both variables may have error terms associated with them.
- Multiple regression, which applies when there is a relationship between a dependent variable and two or more independent variables.
- Non-linear regression, which extends the principles of linear regression to a wide range of functions. Technically, this method is more appropriate than transforming data to allow linear regression.

Advanced statistics books should be consulted for details of these methods, which may also be offered by some statistical computer programs.

Example If a regression analysis gives a value for r^2 of 0.75 (i.e. $r = 0.84$), then 75% of the variance in Y can be explained by the trend line, with $1 - r^2 = 0.25$ (25%) remaining as unexplained (residual) variation.

of the variance in Y that can be accounted for by the regression equation. Conversely, if you subtract this value from 1, you will obtain the residual (error) component, i.e. the fraction of the variance in Y that cannot be explained by the line of best fit. Multiplying the values by 100 allows you to express these fractions in percentage terms.

Using computers to calculate hypothesis-testing statistics

As with the calculation of descriptive statistics (p. 522), specialist statistical packages such as SPSS and Minitab can be used to simplify the calculation of hypothesis-testing statistics. The correct use of the software and interpretation of the output requires an understanding of relevant terminology and of the fundamental principles governing the test, which is probably best obtained by working through one or more examples by hand before using these tools (e.g. Box 64.2; Box 74.2). Spreadsheets offer increasingly sophisticated statistical analysis functions, three examples of which are provided in Box 74.3.

Box 74.3 Using a spreadsheet (Microsoft *Excel*) to calculate hypothesis-testing statistics

Presented below are three examples of the use of a spreadsheet to investigate hypotheses about specific data sets. In each case, there is a brief description of the problem, a table showing the data analysed, an outline of the *Excel* commands used to carry out the analysis and an annotated table of results from the spreadsheet (you can use the same data to check your ability to carry out the procedures).

Example 1: A *t*-test
As part of a project, a student applied a chemical treatment to a series of flasks containing fungal cultures with nutrient solution. An otherwise similar set of control flasks received no chemical treatment. After 3 weeks' growth, the student measured the wet mass of the filtered cultures.

Wet mass of samples (g)

Replicate	1	2	3	4	5	6	7	8	Mean	Variance
Treated with ZH52	2.342	2.256	2.521	2.523	2.943	2.481	2.601	2.449	2.515	0.042
Control	2.658	2.791	2.731	2.402	3.041	2.668	2.823	2.509	2.703	0.038

The student proposed the null hypothesis that there was no difference between the two means and tested this using a *t*-test, and had evidence from other studies that the fungal masses of replicate flasks were normally distributed. The student had also established, by calculation, that the assumption that the populations had homogeneous variances was likely to be valid. She selected *Data* > *Data Analysis* > *t-Test: Two-Sample Assuming Equal Variances*, then enter data ranges as above and

then enter 0 for *Hypothesized Mean Difference* and 0.05 for *Alpha* (= *P*). This returned a table of data as shown below.

t-Test: two-sample assuming equal variances

	Variable 1	Variable 2
Mean	2.515	2.703
Variance	0.042	0.038
Observations	8	8
Pooled variance	0.040	
Hypothesied mean difference	0	
df	14	
t Stat	−1.881	
P (*T≤t*) one-tail	0.040	
t Critical one-tail	1.761	
P (*T≤t*) two-tail	0.081	
t Critical two-tail	2.145	

The value of *t* obtained was −1.881 (row 7 '*t* Stat') and the probability of obtaining this value for a two-tailed test (row 10) was 0.081 (or 8.1%), so the student was able to accept the null hypothesis and conclude that ZH52 had no significant effect on fungal growth in these circumstances.

Example 2: An ANOVA test
A biochemist made six replicate measurements of four different batches (A–D) of alcohol dehydrogenase, obtaining the following data.

(continued)

Box 74.3 (continued)

Alcohol dehydrogenase activity (U L^{-1})

Batch \ replicate	1	2	3	4	5	6	Mean	Variance
A	0.562	0.541	0.576	0.545	0.542	0.551	0.552 833	0.000189
B	0.531	0.557	0.537	0.521	0.559	0.538	0.540 500	0.000221
C	0.572	0.568	0.551	0.549	0.564	0.559	0.560 500	0.000085
D	0.532	0.548	0.541	0.538	0.547	0.536	0.540 333	0.000039

The biochemist wanted to know whether the observed differences were statistically significant, so he carried out an ANOVA test, assuming the samples were normally distributed and the variances in the three populations were homogeneous. He selected *Data > Data Analysis > Anova: Single Factor*, then enter the data ranges, select *Grouped By: Rows* and set *Alpha* to 0.05 (= *P*). This returned a table of data as shown below.

ANOVA: single factor

SUMMARY

Groups	Count	Sum	Average	Variance
A	6	3.317	0.552 833	0.000189
B	6	3.243	0.540 5	0.000 221
C	6	3.363	0.560 5	8.51E-05
D	6	3.242	0.540 333	3.95E-05

ANOVA

Source of variation	SS	df	MS	F	P-value	F crit
Between groups	0.001 761	3	0.000 587	4.397 856	0.015669	3.098391
Within groups	0.002 669	20	0.000133			
Total	0.00 443	23				

The *F* value calculated was 4.397 856. This comfortably exceeds the stated critical value (F_{crit}) of 3.098 391 and the probability of obtaining this result by chance (*P*-value) was calculated as 0.015669 (1.57% to three significant figures), hence the biochemist was able to reject the null hypothesis and conclude that there was a significant difference in average enzyme activity between the four batches, since $P < 0.05$. Such

a finding might lead on to an investigation into why there was batch variation, e.g. had they been stored differently?

Example 3: Testing the significance of a correlation
A researcher wanted to know whether there was any correlation between the levels of tar and nicotine in cigarettes. The researcher made measurements of both consitituents for 10 different brands and obtained the following results.

Tar and nicotine content (mg g^{-1}) of cigarettes

Brand	Tar	Nicotine
1	12.6	0.92
2	8.5	0.58
3	15.7	1.14
4	32.5	2.16
5	14.5	1.03
6	3.1	0.17
7	18.9	1.32
8	7.7	0.98
9	10.7	1.13
10	6.2	0.55

The researcher assumed that both variables were normally distributed and used the Microsoft *Excel* function *PEARSON (array1, array 2)* to obtain a value of +0.950 260 385 for the Pearson's product moment correlation coefficient *r*, specifying the tar content data as array1 and the nicotine content data as array2 (the *Excel CORREL* function can also be used to carry out the same task). The researcher then used a spreadsheet to calculate the *t* statistic (p. 529) for this *r* value, using the formula for Eqn [74.3]. The calculated value of *t* was 8.6296, with 8 degrees of freedom. The critical value from tables (e.g. Table 74.2) at $P = 0.001$ is 5.04, so the researcher concluded that there was a very highly significant positive correlation between the two constituents.

Note: instructions illustrated here may vary among the different versions of Microsoft *Office* programs. Check the exact functions and syntax using the *Insert function* menu option (*fx* button in some versions).

Text references and sources for further study

Dytham, C. (2011) *Choosing and Using Statistics: A Biologist's Guide,* 4th edn. Blackwell, Oxford.

Finney, D.J., Latscha, R., Bennett, B.M. and Hsu, P. (1963) *Tables for Testing Significance in a 2 × 2 Table.* Cambridge University Press, Cambridge.

McKillup, S. (2011) *Statistics Explained: an Introductory Guide for Life Scientists,* 2nd edn. Cambridge University Press, Cambridge.

Maber, J. (1999) *Data Analysis for Biomolecular Sciences.* Longman, Harlow.

Samuels, M.L., Witmer, J.A. and Schaffner, A. (2011) *Statistics for the Life Sciences,* 4th edn. Addison Wesley, Harlow.

Schmuller, J. (2013) *Statistical Analysis with Excel for Dummies,* 3rd edn. Wiley, Hoboken.

Sokal, R.R. and Rohlf, F.J. (2011) *Biometry,* 4th edn. W.H. Freeman and Co., San Francisco.

Wardlaw, A.C. (2000) *Practical Statistics for Experimental Biologists,* 2nd edn. Wiley, New York.

STUDY EXERCISES

74.1 Calculate 95% confidence limits. What are the 95% confidence limits of a sample with a mean = 24.7, standard deviation = 6.8 and number of data values = 16? (Express your answer to three significant figures.)

74.2 Use the Poisson distribution. In a sample of 15 snails, a researcher finds the following number of parasite larvae per snail: 0, 0, 0, 0, 1, 1, 1, 3, 4, 5, 5, 7, 7, 9, 9. Using the rule of thumb on p. 528, decide whether the parasites are 'clumped' or 'repulsed' in distribution on their host. What might this mean in biological terms?

74.3 Practise using a *t*-test. A biology student examined the effect of adding a plant hormone to pea plants. She dissolved an appropriate amount of the compound in ethanol and applied 25 μL of this to the topmost parts of the treated plants. With the controls, she applied the same amount of pure ethanol. After 3 d, she measured the distance between the second and third internodes of the plants and obtained the results shown below. Carry out a *t*-test on the data and draw appropriate conclusions.

Internode distance in cm

Control	7.5	8.1	7.6	6.2	7.5	7.8	8.9
Treatment	5.6	7.5	8.2	6.7	3.5	6.5	5.9

74.4 Interpret the output from Microsoft *Excel* linear regression analysis. The following output represents a regression analysis for an experiment measuring the uptake of an amino acid by a cell suspension (in pmolcell⁻¹) against time (in minutes). Based on this output, what is the form and strength of the underlying linear relationship? (Express the coefficients to three significant figures.)

Output from *Excel* spreadsheet linear regression analysis
Summary output

Regression statistics	
Multiple R	0.985335951
R square	0.970 886937
Adjusted R square	0.963 608672
Standard error	2.133876419
Observations	6

ANOVA

	df	SS	MS	F	Significance F
Regression	1	607.406 285 7	607.4063	133.3954	0.000320975
Residual	4	18.21371429	4.553 429		
Total	5	625.62			

	Coefficients	Standard error	t Stat	P-value
Intercept	1.171428 571	1.544386 367	0.758 507	0.490383
X variable 1	2.945714286	0.255 047 014	11.549 69	0.000321

Answers to these study exercises are available at www.pearsoned.co.uk/practicalskills.

Index

absorbance (A), defined   310
Abstracts, defined   55
accuracy
   defined   379
   in measurement   193
acetic acid   174
active diffusion in solute transport   426–7
activity in solution chemistry   163–4
   equivalent mass   164
   normality   164
   osmolality   164
   osmolarity   164
Adams, D.S.   505
adenine   398, 399
adenosine   399
adenylic acid   399
adjustable pipettor   146, 147
adsorption chromatography   335
   extraction and analysis of lipids   388–9
aerosol formation, minimising risk of   226
aerotolerant anaerobe   244
affinity chromatography   338–9
agar   225
   molten, using   224
   as sterile media   224
agarose gel electrophoresis   351–2
   of DNA   352
agglutination tests
   in antibody production   282
   in identifying microbes   250
air displacement pipettor   146, 147
Airey, P.   298
Alanine   382
algebra, rules   505–6
Alizarin   413
alleles in Mendelian genetics   441, 442
allosteric enzymes
   defined   419
   properties   419–20
alpha decay   297
altruism   136
Ames test on mutagenicity   237
amino acids
   assays of   382
   colorimetric analysis of   380
   detection and quantification of   382
   in Mendelian genetics   441
   side chains of   382
Aminomethyl coumarin   413
ammonium sulfate precipitation
   in purification of proteins   405
   for solubility separation   406–7
ammonium sulfate precipitation in purification of proteins   405
ampicillin   315
amplification fragment length polymorphism (AFLP) in PCR   470

analysis of variance (ANOVA)   530, 535
Anderson, L.W.   24
animal subjects
   cell culture systems   259–60, 262–3
   organs, working with   257–8
   research involving   140
   tissues and cells, working with   257–8, 259–62, 263
      practical procedures   263
      sterile technique   261
Animal Welfare Act (UK, 2006)   140
animalia in taxonomy of organisms   252
Animals Scientific Procedures Acts (UK, 1976, 1986)   238
ANOVA (analysis of variance)   530, 535–6
anthracene   315
antibiotics
   in cell cultures   260
   in microbe isolation   244
antibodies
   defined   281
   production of   281–2
      monoclonal   282
      polyclonal   281
   structure of   281
antigens   281
   typing in identifying microbes   250
apoplasm, defined   258
aqueous solutions, storing   148
Arachidic fatty acid   387
Archaea   252
Arginine   382
Asparagine   382
Aspartate   382
aspirin   315
assertiveness in time management   10
assessment
   of poster display   103
   role of   30–1
atmospheric conditions   158–9
   gas composition   158
   in microbe isolation   243
   pressure   158–9
atomic absorption spectrophotometry   316–17
atomic spectroscopy   316–17
attributes, measurement of   191
authorship
   of information   61–2
   for scientific paper, deciding   123
autoclave tape   224
autonomy   136
autoradiography   301
   in electrophoresis   357
auxotrophs, defined   449
Avogadro's number   509
axes of graphs   487
azide salts in microbe isolation   244
Azure A/eosin B stain   184

bacteria   252
   competence in   450
bacterial genetics   448–53
   bacterial mutants   449–50
   DNA transfer in   450–4
      conjugation   452–3
      F' plasmids, analysis using   454
      interrupted conjugation   453–4
      natural transformation   450–1
      transduction   451–2
   phage crosses   454–5
   phage mutants   455
   principal characteristics   448
bacterial typing, applications of   254
bacterial viruses (phages)   227–9
bacteriocin typing   251
bacteriophages, defined   474
balances   148, 157
ball mills for disruption of cells   269
bar charts   488, 489
barbiturates   315
beakers   148
Beer–Lambert relationship (Beer's Law)   310–11, 312
Behenic fatty acid   387
Belbin, R.M.   13, 14
Benedict's test on monosaccharides   395
beneficence   136
Benzer, S.   455
benzopyrene   315
beta decay   297
bias (systematic error)   193
Bibliography, defined   55
bile salts in microbe isolation   244
bimodal dispersions   521
binomial distribution   526–7
bioassays
   applications   236–8
   defined   236
   legal use of   238
   mutagenicity testing   237
   radioisotopes on   303
   setting up considerations   238
biochemistry, cell and tissue cultures in   259
bioethics   134–41
   animal subjects, research involving   140
   decisions, step-wise approach to   137
   ethical principles   136–7
      autonomy   136
      beneficence   136
      honesty   136
      justice   136
      non-maleficence   136
      precautionary principle   137
      rationality   136
      respect   136
      scientific fraud   136–7
   ethical theories   135–6